D0204143

Engineering Mechanics
DYNAMICS

FOURTH EDITION

Anthony Bedford • Wallace Fowler

University of Texas at Austin

PEARSON

Prentice
Hall

Upper Saddle River, New Jersey 07458

Library of Congress Cataloging-in-Publication Data

Bedford, A.
    Engineering mechanics: dynamics / Anthony Bedford and Wallace Fowler.—4th ed.
      p.   cm.
    Includes index.
    ISBN 0-13-146324-1
    1. Dynamics.   I. Fowler, Wallace.   II. Title.
CIP Data Available.

Vice President and Editorial Director, ECS: *Marcia J. Horton*
Executive Editor: *Eric Svendsen*
Associate Editor: *Dee Bernhard*
Vice President and Director of Production and Manufacturing, ESM: *David W. Riccardi*
Executive Managing Editor: *Vince O'Brien*
Managing Editor: *David A. George*
Senior Marketing Manager: *Holly Stark*
Production Editor: *Craig Little*
Director of Creative Services: *Paul Belfanti*
Assistant Manager, Formatting: *Allyson Graesser*
Electronic Composition: *Clara Bartunek, Judith R. Wilkens*
Creative Director: *Carole Anson*
Art Director: *Kenny Beck*
Interior Designer: *Judith Matz-Coniglio*
Cover Designer: *Susan Anderson-Smith*
Art Editor: *Xiaohong Zhu*
Manufacturing Manager: *Trudy Pisciotti*
Manufacturing Buyer: *Lisa McDowell*

About the Cover: Photograph by Greg Pease / Corbis. Used by permission.

 © 2005 by Pearson Education, Inc.
Pearson Prentice Hall
Pearson Education, Inc.
Upper Saddle River, NJ 07458

All rights reserved. No part of this book may be reproduced, in any format or by any means, without permission in writing from the publisher.

Pearson Prentice Hall® is a trademark of Pearson Education, Inc.

The author and publisher of this book have used their best efforts in preparing this book. These efforts include the development, research, and testing of the theories and programs to determine their effectiveness. The author and publisher make no warranty of any kind, expressed or implied, with regard to these programs or the documentation contained in this book. The author and publisher shall not be liable in any event for incidental or consequential damages in connection with, or arising out of, the furnishing, performance, or use of these programs.

MATLAB is a registered trademark of The MathWorks, Inc., 3 Apple Hill Drive, Natick, MA 01760-2098. Mathcad is a registered trademark of MathSoft Engineering and Education, 101 Main St., Cambridge, MA 02142-1521. All other product names, brand names, and company names are trademarks or registered trademarks of their respective owners.

Printed in the United States of America

10 9 8 7 6 5 4 3 2 1

ISBN 0-13-146324-1

Pearson Education Ltd., *London*
Pearson Education Australia Pty., Limited, *Sydney*
Pearson Education Singapore, Pte. Ltd
Pearson Education North Asia Ltd., *Hong Kong*
Pearson Education Canada, Ltd., *Toronto*
Pearson Educación de Mexico, S.A. de C.V.
Pearson Education—Japan, *Tokyo*
Pearson Education Malaysia, Pte. Ltd.
Pearson Education, *Upper Saddle River, New Jersey*

CONTENTS

APPENDICES

Preface

Our original objective in writing this book was to present the foundations and applications of dynamics as we do in the classroom. We used many sequences of figures, emulating the gradual development of a figure by a teacher explaining a concept. We stressed the importance of visual analysis in gaining understanding, especially through the use of free-body diagrams. Because inspiration is so conducive to learning, we based many of our examples and problems on a variety of modern engineering applications. With encouragement and help from many students and fellow teachers who have used the book, we continue and expand upon these themes in this edition.

Examples That Teach

Each of our examples follows a three-part framework—**Strategy/Solution/Critical Thinking**—that is designed to help students develop engineering problem-solving skills. In the Strategy sections, we demonstrate how to plan the solution to a problem. The Solution presents the detailed steps needed to arrive at the required results. Experienced instructors know that many aspects of a given problem can be explained most effectively after it has been solved. With the Critical Thinking sections, we introduce this important element of classroom teaching. We point out important features of solutions, indicate how the methods used can be extended to other types of problems, and comment on the engineering motivations of particular problems.

Further, you will find each section ends with a section of simple "Study Questions" designed to help students understand their reading.

Engineering Design

In recent years, many teachers have begun to introduce concepts from engineering design in their mechanics courses, and this trend is being further encouraged by ABET requirements. Without compromising our emphasis on fundamental mechanics, we include design considerations in many examples and problems. Optional Design Examples provide more detailed discussions of applications of dynamics in engineering design. Brief sections of problems called Design Experiences are based on the Design Examples and can be assigned at the discretion of the instructor. Many chapters conclude with Design Projects that offer students more extensive participation in design.

Computational Mechanics

Some instructors prefer to teach dynamics without emphasizing the use of a computer. Others use dynamics as an opportunity to introduce students to the use of computers in engineering, having them either write their own programs in a lower level language or use higher level problem-solving software. Our book is suitable for each of these approaches. We provide optional, self-contained Computational Mechanics sections with examples and problems designed for solution by a programmable calculator or computer. In addition, tutorials on using Mathcad and MATLAB in engineering mechanics are available on the web. See the Supplements section for further information.

Consistent Use of Color

To help students recognize and interpret elements of figures, we use consistent identifying colors:

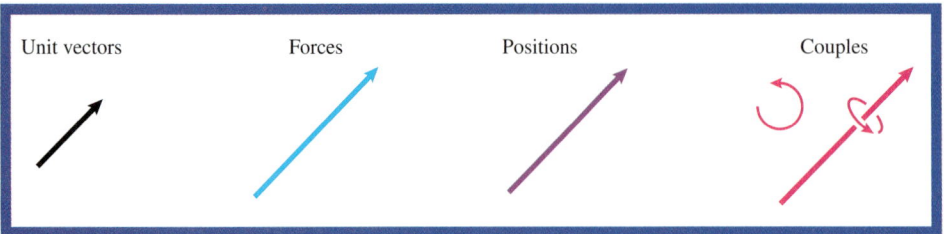

Unit vectors   Forces   Positions   Couples

New to the Fourth Edition

Positive responses from users and reviewers have led us to retain the basic organization, content, and features of previous editions. During our preparation of this edition, we reexamined how we presented each concept, example, figure, summary statement, and problem. Where necessary, we made changes, additions, or deletions to simplify and clarify the presentation. In response to requests, we made the following notable changes:

Critical Thinking Sections–We have revised and expanded the discussion sections featured in our past editions' examples to provide a stronger emphasis on critical thinking. Strategy and Critical Thinking sections are now featured in every example. Each Critical Thinking section was reviewed by Chad M. Landis, Rice University, and Walter Gerstle, University of New Mexico, who provided us with feedback and suggestions for making them more helpful and educational for students.

New Art Program–Nearly every reviewer emphasized that their students needed help visualizing engineering situations. They also noted that students responded more favorably to situations appearing more "real." In response, we have tried to rework the key figures in this text with exceptional "photo-realistic" clarity. We hope this will better help students visualize situations, and provide a stronger connection to the real world.

Sequences of Figures in Examples–In text material and the examples, we continue and expand upon our use of sequences of figures to achieve the clarity of classroom presentations of concepts.

In addition,
- We have included a new section in Chapter 13 on relative motion, and increased our coverage of non-conservative forces in Chapter 15.
- We have added new examples where users indicated more were needed. Many of the new examples continue our emphasis on realistic and motivational applications and engineering design.
- We include 450 new and revised problems. As with the examples, many of the new problems focus on placing dynamics within the context of engineering practice. In this edition, problems that are relatively lengthier or more difficult have been indicated with an asterisk.
- An extensive new supplement program now includes OneKey—a web-based system that helps you better manage your courses, assess student progress, and more. See the Supplements description for complete information.

Commitment to Students and Instructors

In revising the textbook and solution manual, we have taken precautions to ensure accuracy to the best of our ability. We have each solved the new problems in an effort to be sure that their answers are correct and that they are of an appropriate level of difficulty. Karim Nohra of the University of South Florida, Scott Hendricks of VPI, and Kurt Norlin of Laurel Technical Services further verified the text, examples, problems, and solutions manual to help ensure accuracy. Any errors that remain are the responsibility of the authors. We welcome communication from students and instructors concerning errors or areas for improvement. Our mailing address is Department of Aerospace Engineering and Engineering Mechanics, University of Texas at Austin, Austin, Texas 78712. Our electronic mail address is abedford@mail.utexas.edu

Supplements

Student Supplements

The *Dynamics* Study Pack is designed to give students the tools to improve their skills drawing free-body diagrams, and to help them review for tests. (The *Dynamics* Study Pack is available bundled at no additional cost with the fourth edition.) It contains a tutorial on free-body diagrams with 50 practice problems of increasing difficulty with complete solutions. Further strategies and tips help students understand how to use the diagrams in solving the accompanying problems. This supplement and accompanying chapter-by-chapter review material was prepared by Peter Schiavone of the University of Alberta.

The *Dynamics* Study Pack is also available as a stand-alone item. Order stand-alone Study Packs with the ISBN 0-13-150290-5.

MATLAB and Mathcad Tutorials—Each tutorial discusses a basic mechanics concept, and then shows how to solve a specific problem related to this concept using MATLAB and Mathcad. There are 20 tutorials each for MATLAB and Mathcad, and are available to instructors in PDF format for distribution to students. Worksheets were developed by Ronald Larsen and Stephen Hunt of Montana State University–Bozeman.

Instructor Supplements

OneKey–OneKey is an on-line solution perfect for helping manage your class and preparing lectures, quizzes, and tests. Using OneKey, professors can quickly create an online course tailored to their specific needs. OneKey contains complete electronic solutions files of homework and makes it easy for you to post homework solutions at a protected online site for student review. Five hundred additional mechanics problems and solutions—organized by chapter, topic, and level of difficulty—are also available for student study and test prep. Further, OneKey gives you access to a Math Topic review, MATLAB and Mathcad student tutorials, animations, simulations, the Research Navigator, and PHGradeAssist, Prentice Hall's on-line algorithmic homework generator.

To learn more about OneKey, visit www.prenhall.com/onekey, or contact your local PH rep. Prentice Hall will be happy to host a OneKey site for you, or OneKey is available in course management cartridges for schools using this technology. Access and use for professors is free. Student access codes are free with their Bedford/Fowler textbooks, or available for sale as stand-alone items. For further information and ordering ISBNs, contact your local PH rep., or email engineering@prenhall.com

Web Assessment Software—Using **PHGradeAssist**, students solve problems from the text that have randomized parameters, so that each student solves a problem with slightly different numbers. After students have submitted their answers, they receive the correct answers and, if necessary, can continue to work similar problems until they are successful. By using the optional course management system, instructors can have students' results recorded electronically. Contact your PH representative for more information. This supplement is available through the Bedford/Fowler OneKey system.

Instructor's Solutions Manual—This supplement, available to instructors, contains completely worked-out solutions. Each solution comes with the problem statement as well as associated artwork.

Instructors' Resource Center on CD—This CD contains PowerPoint slides and JPEG files of all art from the text. It also contains sets of PowerPoint slides showing each example and electronic files of solutions.

Acknowledgments

The following colleagues provided reviews based on their knowledge and teaching experience that greatly helped us in preparing the fourth edition of *Statics* and *Dynamics*.

Shaaban Abdallah
University of Cincinnati

George G. Adams
Northeastern University

Haim Baruh
Rutgers University

David M. Bayer
University of North Carolina

Glenn Beltz
University of California–Santa Barbara

Mitsunori Denda
Rutgers University

Bogdan I. Epureanu
University of Michigan

Walter Gerstle
University of New Mexico

Paul R. Heyliger
Colorado State University

Robert W. Hinks
Arizona State University

Chad M. Landis
Rice Unversity

John B. Ligon
Michigan Tech University

Mark T. Lusk
Colorado School of Mines

Nels Madsen
Auburn University

James R. Matthews
University of New Mexico

Mohammad Noori
North Carolina State University

Corrado Poli
University of Massachusetts–Amherst

Yitshak Ram
Louisiana State University

Edwin C. Rossow
Northwestern University

Kenneth Sawyers
Lehigh University

Richard A. Scott
University of Michigan

Brian Self
U.S. Air Force Academy

William Semke
University of North Dakota

Dennis VandenBrink
Western Michigan University

Reviewers for the Previous Editions

Students and instructors have made insightful comments and suggested improvements that we have incorporated into this edition. The following academic colleagues have critically reviewed the book and given us many valuable suggestions:

Edward E. Adams
Michigan Technological University

Raid S. Al-Akkad
University of Dayton

Jerry L. Anderson
Memphis State University

James G. Andrews
University of Iowa

Robert J. Asaro
University of California, San Diego

Leonard B. Baldwin
University of Wyoming

Gautam Batra
University of Nebraska

Mary Bergs
Marquette University

Spencer Brinkerhoff
Northern Arizona University

L.M. Brock
University of Kentucky

William (Randy) Burkett
Texas Tech University

Donald Carlson
University of Illinois

Major Robert M. Carpenter
U.S. Military Academy

Douglas Carroll
University of Missouri, Rolla

Paul C. Chan
New Jersey Institute of Technology

Namas Chandra
Florida State University

James Cheney
University of California, Davis

Ravinder Chona
Texas A & M University

Anthony DeLuzio
Merrimack College

Mitsunori Denda
Rutgers University

James F. Devine
University of South Florida

Craig Douglas
University of Massachusetts, Lowell

Marijan Dravinski
University of Southern California

S. Olani Durrant
Brigham Young University

Estelle Eke
California State University, Sacramento

William Ferrante
University of Rhode Island

Robert W. Fitzgerald
Worcester Polytechnic Institute

George T. Flowers
Auburn University

Mark Frisina
Wentworth Institute

Robert W. Fuessle
Bradley University

William Gurley
University of Tennessee, Chattanooga

John Hansberry
University of Massachusetts, Dartmouth

W. C. Hauser
California Polytechnic University, Pomona

Linda Hayes
University of Texas–Austin

R. Craig Henderson
Tennessee Technological University

James Hill
University of Alabama

Allen Hoffman
Worcester Polytechnic Institute

Edward E. Hornsey
University of Missouri, Rolla

Robert A. Howland
University of Notre Dame

Joe Ianelli
University of Tennessee, Knoxville

Ali Iranmanesh
Gadsden State Community College

David B. Johnson
Southern Methodist University

E. O. Jones, Jr.
Auburn University

Serope Kalpakjian
Illinois Institute of Technology

Kathleen A. Keil
California Polytechnic University, San Luis Obispo

Yohannes Ketema
University of Minnesota

Seyyed M. H. Khandani
Diablo Valley College

Charles M. Krousgrill
Purdue University

B. Kent Lall
Portland State University

Kenneth W. Lau
University of Massachusetts, Lowell

Norman Laws
University of Pittsburgh

William M. Lee
U.S. Naval Academy

Donald G. Lemke
University of Illinois, Chicago

Richard J. Leuba
North Carolina State University

Richard Lewis
Louisiana Technological University

Bertram Long
Northeastern University

V. J. Lopardo
U.S. Naval Academy

Frank K. Lu
University of Texas, Arlington

K. Madhaven
Christian Brothers College

Gary H. McDonald
University of Tennessee

James McDonald
Texas Technical University

Jim Meagher
California Polytechnic State University, San Luis Obispo

Lee Minardi
Tufts University

Norman Munroe
Florida International University

Shanti Nair
University of Massachusetts, Amherst

Saeed Niku
California Polytechnic State University, San Luis Obispo

Harinder Singh Oberoi
Western Washington University

James O'Connor
University of Texas, Austin

Samuel P. Owusu-Ofori
North Carolina A& T State University

Venkata Panchakarla
Florida State University

Assimina A Pelegri
Rutgers University
Noel C. Perkins
University of Michigan
David J. Purdy
Rose-Hulman Institute of Technology
Colin E Ratcliffe
U.S. Naval Academy
Daniel Riahi
University of Illinois
Charles Ritz
California Polytechnic State University, Pomona
George Rosborough
University of Colorado, Boulder
Robert Schmidt
University of Detroit
Robert J. Schultz
Oregon State University
Patricia M. Shamamy
Lawrence Technological University
Sorin Siegler
Drexel University
L. N. Tao
Illinois Institute of Technology
Craig Thompson
Western Wyoming Community College

John Tomko
Cleveland State University
Kevin Z. Truman
Washington University
John Valasek
Texas A &M University
Dennis VandenBrink
Western Michigan University
Thomas J. Vasko
University of Hartford
Mark R. Virkler
University of Missouri, Columbia
William H. Walston, Jr.
University of Maryland
Reynolds Watkins
Utah State University
Charles White
Northeastern University
Norman Wittels
Worcester Polytechnic Institute
Julius P. Wong
University of Louisville
T. W. Wu
University of Kentucky
Constance Ziemian
Bucknell University

We have again had the pleasure of working with friendly, professional, and energetic people at Prentice Hall. Our editor, Eric Svendsen, provided philosophical guidance in addition to organizing and managing the large collaborative effort required by a book of this kind. Brian Hoehl provided skillful liaison between Eric and us and kept the work moving smoothly. Dee Bernhard was our interface to the reviewers, and was instrumental in making that process work well. Xiaohong Zhu made important contributions to the new art program and helped with engineer/artist communications. Craig Little devoted large amounts of time and effort to shepherding the book through the production process and to helping us with its many details. Susan Anderson-Smith provided graceful cover designs. Kurt Norlin helped us with checking and made many editorial suggestions. Scott Hendricks and Karim Nohra joined us in checking everything, oversaw the revision of the solutions manual, and prepared the list of answers. We are grateful to the developers of the supplements that are so helpful to students. Peter Schiavone developed the *Dynamics* Study Pack that accompanies the book, and Stephen Hunt and Ronald Larsen wrote the MATLAB/Mathcad tutorials.

We thank the many users of the previous editions who have shared their experiences with us and suggested revisions. And we again thank Nancy and Marsha, who are still waiting for us to stop working weekends.

Anthony Bedford and Wallace Fowler
Austin, Texas

About the Authors

Anthony Bedford is Professor Emeritus with the Department of Aerospace Engineering and Engineering Mechanics at the University of Texas at Austin. He received the B.S. degree from the University of Texas at Austin, the M.S. degree from the California Institute of Technology, and the Ph.D. degree at Rice University in 1967. He has industrial experience at Douglas Aircraft Company, TRW, and Sandia National Laboratories. He has been on the faculty of the University of Texas at Austin since 1968.

Dr. Bedford's main professional activity has been education and research in engineering mechanics. He has written technical papers on mixture theory, wave propagation, and the mechanics of high velocity impacts, and is the author of the books *Hamilton's Principle in Continuum Mechanics, Introduction to Elastic Wave Propagation* (with D. S. Drumheller), and *Mechanics of Materials* (with K. M. Liechti).

Wallace T. Fowler holds the Paul D. and Betty Robertson Meek Professorship in Engineering with the Department of Aerospace Engineering and Engineering Mechanics at the University of Texas at Austin. Dr. Fowler received the B.A., M.S., and Ph.D. degrees from the University of Texas at Austin, and has been on the faculty there since 1965. During the fall of 1976, he was on the staff of the United States Air Force Test Pilot School, Edwards Air Force Base, California, and during 1981–1982, he was a visiting professor at the United States Air Force Academy. He is currently the Director of the Texas Space Grant Consortium.

Dr. Fowler's areas of teaching and research are dynamics, orbital mechanics, and spacecraft mission design. He is author or coauthor of technical papers on trajectory optimization, attitude dynamics, and space mission planning and has also published papers on the theory and practice of engineering teaching. He has received numerous teaching awards including the Chancellor's Council Outstanding Teaching Award, the General Dynamics Teaching Excellence Award, the Halliburton Education Foundation Award of Excellence, the ASEE Fred Merryfield Design Award, and the AIAA-ASEE Distinguished Aerospace Educator Award. He is a member of the Academy of Distinguished Teachers at the University of Texas at Austin. He is a licensed professional engineer, a member of several technical societies, and a Fellow of both the American Institute of Aeronautics and Astronautics and the American Society for Engineering Education. During 2000–2001, he served as president of the American Society for Engineering Education.

Photo Credits

About the Cover: Photograph by Greg Pease / Corbis. Used by permission.

Chapter 12: Opener, David Parker, Photo Researchers, Inc.; Figure 12.05, Courtesy of NASA/JPL/Caltech.

Chapter 13: Opener, Corbis Digital Stock; Figure 13.13, © Harold & Esther Edgerton Foundation, 2002, courtesy of Palm Press, Inc.; Figure 13.38 Sandia National Laboratories; Figures P13.16, P13.27, and P13.68, Corbis Digital Stock.

Chapter 14: Opener, Anthony Saint James/Getty Images Inc./PhotoDisc, Inc.; Figure P14.8, Corbis Digital Stock.

Chapter 15: Opener, Glyn Kirk, Getty Images Inc.; Figure P15.107/15.108, Computer simulation of a volcanic plume on Io courtesy of Victor Austin and David B. Goldstein.

Chapter 16: Opener, U.S. Department of Transportation; Figure 16.5, © Harold & Esther Edgerton Foundation, 2002, courtesy of Palm Press, Inc.; Figure P16.6, Corbis Digital Stock; Figure P16.37, Anthony Saint James/ Getty Images, Inc./PhotoDisc Inc.; Figure P16.70, © Harold & Esther Edgerton Foundation, 2002, courtesy of Palm Press, Inc.

Chapter 17: Opener, John Wilkinson, Ecoscene, CORBIS; Figure 17.46a and b, NASA Headquarters; Figure P17.9, © 2005 General Motors Corp. Used with permission of GM Media Archives.

Chapter 18: Opener and Figure P18.13, Ben Johnson/Photo Researchers, Inc.

Chapter 19: Opener and Figure P19.5, Corbis Digital Stock.

Chapter 20: Opener, Don Farrall, Getty Images, Inc./PhotoDisc, Inc.

Chapter 21: Opener, Leonard Lessin/Peter Arnold, Inc.

Engineering Mechanics

DYNAMICS

Introduction

How do engineers design and construct the devices we use, from simple objects such as chairs and pencil sharpeners to complicated ones such as dams, cars, airplanes, and spacecraft? They must have a deep understanding of the physics underlying the design of these devices, and they must be able to use mathematical models to predict their behavior. Students of engineering begin to learn how to analyze and predict the behaviors of physical systems by studying mechanics.

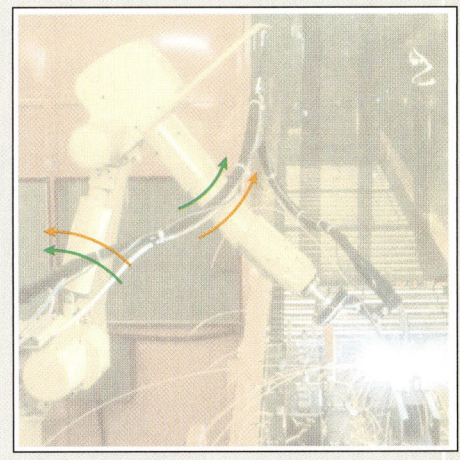

◄ To design and program an industrial robot, engineers must analyze its motion using the principles of dynamics. Dynamics is one of the sciences underlying the design of all machines.

12.1 Engineering and Mechanics

How can engineers design complex systems and predict their characteristics before they are constructed? Engineers have always relied on their knowledge of previous designs, experiments, ingenuity, and creativity to develop new designs. Modern engineers add a powerful technique: They develop mathematical equations based on the physical characteristics of the devices they design. With these mathematical models, engineers predict the behavior of their designs, modify them, and test them prior to their actual construction. Aerospace engineers use mathematical models to predict the paths the space shuttle will follow in flight. Civil engineers use mathematical models to analyze the effects of loads on buildings and foundations.

At its most basic level, mechanics is the study of forces and their effects. Elementary mechanics is divided into *statics*, the study of objects in equilibrium, and *dynamics*, the study of objects in motion. The results obtained in elementary mechanics apply directly to many fields of engineering. Mechanical and civil engineers who design structures use the equilibrium equations derived in statics. Civil engineers who analyze the responses of buildings to earthquakes and aerospace engineers who determine the trajectories of satellites use the equations of motion derived in dynamics.

Mechanics was the first analytical science; consequently fundamental concepts, analytical methods, and analogies from mechanics are found in virtually every field of engineering. Students of chemical and electrical engineering gain a deeper appreciation for basic concepts in their fields, such as equilibrium, energy, and stability, by learning them in their original mechanical contexts. By studying mechanics, they retrace the historical development of these ideas.

12.2 Learning Mechanics

Mechanics consists of broad principles that govern the behavior of objects. In this book we describe these principles and provide examples that demonstrate some of their applications. Although it is essential that you practice working problems similar to these examples, and we include many problems of this kind, our objective is to help you understand the principles well enough to apply them to situations that are new to you. Each generation of engineers confronts new problems.

Problem Solving

In the study of mechanics you learn problem-solving procedures you will use in succeeding courses and throughout your career. Although different types of problems require different approaches, the following steps apply to many of them:

- Identify the information that is given and the information, or answer, you must determine. It's often helpful to restate the problem in your own words. When appropriate, make sure you understand the physical system or model involved.
- Develop a *strategy* for the problem. This means identifying the principles and equations that apply and deciding how you will use them to solve the problem. Whenever possible, draw diagrams to help visualize and solve the problem.
- Whenever you can, try to predict the answer. This will develop your intuition and will often help you recognize an incorrect answer.
- Solve the equations and, whenever possible, interpret your results and compare them with your prediction. This last step is a *reality check*. Is your answer reasonable?

Calculators and Computers

Most of the problems in this book are designed to lead to an algebraic expression with which to calculate the answer in terms of given quantities. A calculator with trigonometric and logarithmic functions is sufficient to determine the numerical value of such answers. The use of a programmable calculator or a computer with problem-solving software such as Mathcad or MATLAB is convenient, but be careful not to become too reliant on tools you will not have during tests.

Sections headed "Computational Mechanics" contain examples and problems that are suitable for solution with a programmable calculator or a computer.

Engineering Applications

Although the problems are designed primarily to help you learn mechanics, many of them illustrate uses of mechanics in engineering. Sections headed "Design Examples" describe how mechanics is applied in various fields of engineering.

We also include problems that emphasize two essential aspects of engineering:

- *Design.* Some problems ask you to choose values of parameters to satisfy stated design criteria.
- *Safety.* Some problems ask you to evaluate the safety of devices and choose values of parameters to satisfy stated safety requirements.

Subsequent Use of This Text

This book contains tables and information you will find useful in subsequent engineering courses and throughout your engineering career. In addition, you will often want to review fundamental engineering subjects, both during the remainder of your formal education and when you are a practicing engineer. The most efficient way to do so is by using the textbooks with which you are familiar. Your engineering textbooks will form the core of your professional library.

12.3 Fundamental Concepts

Some topics in mechanics will be familiar to you from everyday experience or from previous exposure to them in mathematics and physics courses. In this section we briefly review the foundations of elementary mechanics.

Numbers

Engineering measurements, calculations, and results are expressed in numbers. You need to know how we express numbers in the examples and problems and how to express the results of your own calculations.

Significant Digits   This term refers to the number of meaningful (that is, accurate) digits in a number, counting to the right starting with the first nonzero digit. The two numbers 7.630 and 0.007630 are each stated to four significant digits. If only the first four digits in the number 7,630,000 are known to be accurate, this can be indicated by writing the number in scientific notation as 7.630×10^6.

If a number is the result of a measurement, the significant digits it contains are limited by the accuracy of the measurement. If the result of a measurement is stated to be 2.43, this means that the actual value is believed to be closer to 2.43 than to 2.42 or 2.44.

Numbers may be rounded off to a certain number of significant digits. For example, we can express the value of π to three significant digits, 3.14, or we can express it to six significant digits, 3.14159. When you use a calculator or computer, the number of significant digits is limited by the number of digits the machine is designed to carry.

Use of Numbers in This Book You should treat numbers given in problems as exact values and not be concerned about how many significant digits they contain. If a problem states that a quantity equals 32.2, you can assume its value is 32.200 .... We generally express intermediate results and answers in the examples and the answers to the problems to at least three significant digits. If you use a calculator, your results should be that accurate. Be sure to avoid round-off errors that occur if you round off intermediate results when making a series of calculations. Instead, carry through your calculations with as much accuracy as you can by retaining values in your calculator.

Space and Time

Space simply refers to the three-dimensional universe in which we live. Our daily experiences give us an intuitive notion of space and the locations, or positions, of points in space. The distance between two points in space is the length of the straight line joining them.

Measuring the distance between points in space requires a unit of length. We use both the International System of units, or SI units, and U.S. Customary units. In SI units, the unit of length is the meter (m). In U.S. Customary units, the unit of length is the foot (ft).

Time is, of course, familiar—our lives are measured by it. The daily cycles of light and darkness and the hours, minutes, and seconds measured by our clocks and watches give us an intuitive notion of time. Time is measured by the intervals between repeatable events, such as the swings of a clock pendulum or the vibrations of a quartz crystal in a watch. In both SI units and U.S. Customary units, the unit of time is the second (s). The minute (min), hour (h), and day are also frequently used.

If the position of a point in space relative to some reference point changes with time, the rate of change of its position is called its *velocity*, and the rate of change of its velocity is called its *acceleration*. In SI units, the velocity is expressed in meters per second (m/s) and the acceleration is expressed in meters per second per second, or meters per second squared (m/s^2). In U.S. Customary units, the velocity is expressed in feet per second (ft/s) and the acceleration is expressed in feet per second squared (ft/s^2).

Newton's Laws

Elementary mechanics was established on a firm basis with the publication in 1687 of *Philosophiae Naturalis Principia Mathematica*, by Isaac Newton. Although highly original, it built on fundamental concepts developed by many others during a long and difficult struggle toward understanding (Fig. 12.1).

Newton stated three "laws" of motion, which we express in modern terms:

1. *When the sum of the forces acting on a particle is zero, its velocity is constant. In particular, if the particle is initially stationary, it will remain stationary.*

2. *When the sum of the forces acting on a particle is not zero, the sum of the forces is equal to the rate of change of the linear momentum of the particle. If the mass is constant, the sum of the forces is equal to the product of the mass of the particle and its acceleration.*

3. *The forces exerted by two particles on each other are equal in magnitude and opposite in direction.*

Figure 12.1

Chronology of developments in mechanics up to the publication of Newton's
Principia in relation to other events in history.

Notice that we did not define force and mass before stating Newton's laws. The
modern view is that these terms are defined by the second law. To demonstrate,
suppose that we choose an arbitrary object and define it to have unit mass. Then
we define a unit of force to be the force that gives our unit mass an acceleration
of unit magnitude. In principle, we can then determine the mass of any object:
We apply a unit force to it, measure the resulting acceleration, and use the second law to determine the mass. We can also determine the magnitude of any
force: We apply it to our unit mass, measure the resulting acceleration, and use
the second law to determine the force.

Thus Newton's second law gives precise meanings to the terms *mass* and
force. In SI units, the unit of mass is the kilogram (kg). The unit of force is
the newton (N), which is the force required to give a mass of one kilogram an
acceleration of one meter per second squared. In U.S. Customary units, the unit

of force is the pound (lb). The unit of mass is the slug, which is the amount of mass accelerated at one foot per second squared by a force of one pound.

Although the results we discuss in this book are applicable to many of the problems met in engineering practice, there are limits to the validity of Newton's laws. For example, they don't give accurate results if a problem involves velocities that are not small compared to the velocity of light $(3 \times 10^8 \text{ m/s})$. Einstein's special theory of relativity applies to such problems. Elementary mechanics also fails in problems involving dimensions that are not large compared to atomic dimensions. Quantum mechanics must be used to describe phenomena on the atomic scale.

> **Study Questions**
> 1. What is the definition of the significant digits of a number?
> 2. What are the units of length, mass, and force in the SI system?

12.4 Units

The SI system of units has become nearly standard throughout the world. In the United States, U.S. Customary units are also used. In this section we summarize these two systems of units and explain how to convert units from one system to another.

International System of Units

In SI units, length is measured in meters (m) and mass in kilograms (kg). Time is measured in seconds (s), although other familiar measures such as minutes (min), hours (h), and days are also used when convenient. Meters, kilograms, and seconds are called the *base units* of the SI system. Force is measured in newtons (N). Recall that these units are related by Newton's second law: One newton is the force required to give an object of one kilogram mass an acceleration of one meter per second squared:

$$1 \text{ N} = (1 \text{ kg})(1 \text{ m/s}^2) = 1 \text{ kg-m/s}^2.$$

Because the newton can be expressed in terms of the base units, it is called a *derived unit.*

To express quantities by numbers of convenient size, multiples of units are indicated by prefixes. The most common prefixes, their abbreviations, and the multiples they represent are shown in Table 12.1. For example, 1 km is 1 kilometer, which is 1000 m, and 1 Mg is 1 megagram, which is 10^6 g, or 1000 kg. We frequently use kilonewtons (kN).

Table 12.1 The common prefixes used in SI units and the multiples they represent.

Prefix	Abbreviation	Multiple
nano-	n	10^{-9}
micro-	μ	10^{-6}
milli-	m	10^{-3}
kilo-	k	10^{3}
mega-	M	10^{6}
giga-	G	10^{9}

U.S. Customary Units

In U.S. Customary units, length is measured in feet (ft) and force is measured in pounds (lb). Time is measured in seconds (s). These are the base units of the U.S. Customary system. In this system of units, mass is a derived unit. The unit of mass is the slug, which is the mass of material accelerated at one foot per second squared by a force of one pound. Newton's second law states that

$$1 \text{ lb} = (1 \text{ slug})(1 \text{ ft/s}^2).$$

From this expression we obtain

$$1 \text{ slug} = 1 \text{ lb-s}^2/\text{ft}.$$

We use other U.S. Customary units such as the mile (1 mi = 5280 ft) and the inch (1 ft = 12 in). We also use the kilopound (kip), which is 1000 lb.

Angular Units

In both SI and U.S. Customary units, angles are normally expressed in radians (rad). We show the value of an angle θ in radians in Fig. 12.2. It is defined to be the ratio of the part of the circumference subtended by θ to the radius of the circle. Angles are also expressed in degrees. Since there are 360 degrees (360°) in a complete circle, and the complete circumference of the circle is $2\pi R$, 360° equals 2π rad.

Equations containing angles are nearly always derived under the assumption that angles are expressed in radians. Therefore, when you want to substitute the value of an angle expressed in degrees into an equation, you should first convert it into radians. A notable exception to this rule is that many calculators are designed to accept angles expressed in either degrees or radians when you use them to evaluate functions such as sin θ.

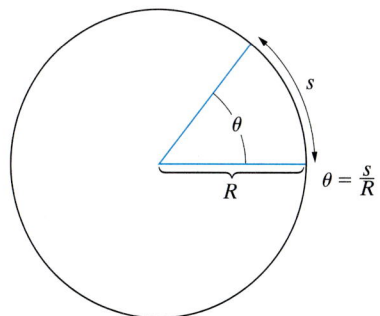

Figure 12.2
Definition of an angle in radians.

Conversion of Units

Many situations arise in engineering practice that require values expressed in one kind of unit to be converted into values in other units. For example, if some of the data to be used in an equation are given in SI units and some are given in U.S. Customary units, they must all be expressed in terms of one system of units before they are substituted into the equation. Converting units is straightforward, although it must be done with care.

Suppose that we want to express 1 mile per hour (mi/h) in terms of feet per second (ft/s). Because 1 mile equals 5280 feet and 1 hour equals 3600 seconds, we can treat the expressions

$$\left(\frac{5280 \text{ ft}}{1 \text{ mi}}\right) \quad \text{and} \quad \left(\frac{1 \text{ h}}{3600 \text{ s}}\right)$$

as ratios whose values are 1. In this way, we obtain

$$1 \text{ mi/h} = (1 \text{ mi/h})\left(\frac{5280 \text{ ft}}{1 \text{ mi}}\right)\left(\frac{1 \text{ h}}{3600 \text{ s}}\right) = 1.47 \text{ ft/s}.$$

Some useful unit conversions are given in Table 12.2.

Table 12.2 Unit conversions.

Time	1 minute	=	60 seconds
	1 hour	=	60 minutes
	1 day	=	24 hours
Length	1 foot	=	12 inches
	1 mile	=	5280 feet
	1 inch	=	25.4 millimeters
	1 foot	=	0.3048 meters
Angle	2π radians	=	360 degrees
Mass	1 slug	=	14.59 kilograms
Force	1 pound	=	4.448 newtons

Study Questions

1. What are the base units of the SI and U.S. Customary systems?
2. What is the definition of an angle in radians?

Example 12.1 Converting Units of Pressure

The pressure exerted at a point of the hull of the deep submersible vehicle in Fig. 12.3 is 3.00×10^6 Pa (pascals). A pascal is 1 newton per square meter. Determine the pressure in pounds per square foot.

Figure 12.3
Deep Submersible Vehicle.

Strategy

From Table 12.2, 1 pound = 4.448 newtons and 1 foot = 0.3048 meters. With these unit conversions we can calculate the pressure in pounds per square foot.

Solution

The pressure (to three significant digits) is

$$3.00 \times 10^6 \text{ N/m}^2 = (3.00 \times 10^6 \text{ N/m}^2)\left(\frac{1 \text{ lb}}{4.448 \text{ N}}\right)\left(\frac{0.3048 \text{ m}}{1 \text{ ft}}\right)^2$$

$$= 62{,}700 \text{ lb/ft}^2.$$

Critical Thinking

How could we have obtained this result in a more direct way? Notice from the table of unit conversions in the inside front cover that 1 Pa = 0.0209 lb/ft^2. Therefore,

$$3.00 \times 10^6 \text{ N/m}^2 = (3.00 \times 10^6 \text{ N/m}^2)\left(\frac{0.0209 \text{ lb/ft}^2}{1 \text{ N/m}^2}\right)$$

$$= 62{,}700 \text{ lb/ft}^2.$$

Example 12.2   Determining Units from an Equation

Suppose that in Einstein's equation

$$E = mc^2,$$

the mass m is in kilograms and the velocity of light c is in meters per second.
(a) What are the SI units of E?
(b) If the value of E in SI units is 20, what is its value in U.S. Customary base units?

Strategy
(a) Since we know the units of the terms m and c, we can deduce the units of E from the given equation.
(b) We can use the unit conversions for mass and length from Table 12.2 to convert E from SI units to U.S. Customary units.

Solution
(a) From the equation for E,

$$E = (m \text{ kg})(c \text{ m/s})^2,$$

the SI units of E are kg-m^2/s^2.

(b) From Table 12.2, 1 slug = 14.59 kg and 1 ft = 0.3048 m. Therefore,

$$1 \text{ kg-m}^2/\text{s}^2 = (1 \text{ kg-m}^2/\text{s}^2)\left(\frac{1 \text{ slug}}{14.59 \text{ kg}}\right)\left(\frac{1 \text{ ft}}{0.3048 \text{ m}}\right)^2$$

$$= 0.738 \text{ slug-ft}^2/\text{s}^2.$$

The value of E in U.S. Customary units is

$$E = (20)(0.738) = 14.8 \text{ slug-ft}^2/\text{s}^2.$$

Critical Thinking
In part (a), how did we know that we could determine the units of E by determining the units of mc^2? The dimensions, or units, of each term in an equation must be the same. For example, in the equation $a + b = c$, the dimensions of each of the terms a, b, and c must be the same. The equation is said to be *dimensionally homogeneous*. This requirement is nicely expressed by the colloquial phrase "Don't compare apples and oranges."

Problems

12.1 Express the fractions $\frac{1}{3}$ and $\frac{2}{3}$ to three significant digits.

12.2 The base of natural logarithms is $e = 2.718281828\ldots$.
(a) Express e to five significant digits.
(b) Determine the value of e^2 to five significant digits.
(c) Use the value of e you obtained in part (a) to determine the value of e^2 to five significant digits.
[Part (c) demonstrates the hazard of using rounded-off values in calculations.]

12.3 A machinist drills a circular hole in a panel with radius $r = 5$ mm. Determine the circumference C and the area A of the hole to four significant digits.

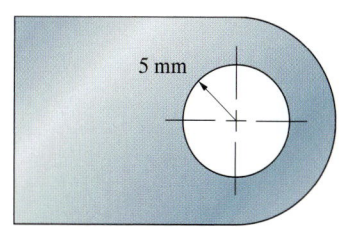

5 mm

Problem 12.3

12.4 The opening in the soccer goal is 24 ft wide and 8 ft high. Use these values to determine its dimensions in meters to three significant digits.

Problem 12.4

12.5 The coordinates (in meters) of point A are $x_A = 3$, $y_A = 7$, and the coordinates of point B are $x_B = 10$, $y_B = 2$. Determine the length of the straight line from A to B to three significant digits.

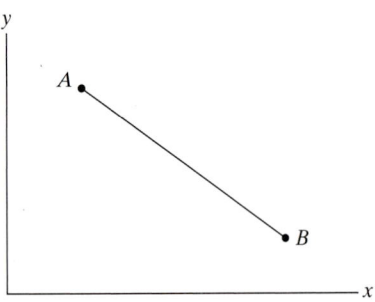

Problem 12.5

12.6 Suppose that you have just purchased a Ferrari F355 coupe and you want to know whether you can use your set of SAE (U.S. Customary unit) wrenches to work on it. You have wrenches with widths $w = 1/4$ in, $1/2$ in, $3/4$ in, and 1 in, and the car has nuts with dimensions $n = 5$ mm, 10 mm, 15 mm, 20 mm, and 25 mm. Defining a wrench to fit if w is no more than 2% larger than n, which of your wrenches can you use?

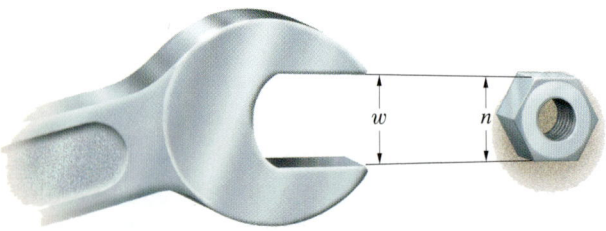

Problem 12.6

12.7 On August 20, 1974, Nolan Ryan threw the first baseball pitch measured at over 100 mi/h. The measured speed was 100.9 mi/h. Determine the speed of the pitch to four significant digits

(a) in ft/s;
(b) in km/h.

12.8 On March 18, 1999, an experimental Maglev (magnetic levitation) train in Japan reached a maximum speed of 552 km/h. What was its velocity in mi/h to three significant digits?

Problem 12.8

12.9 In May 1963, in the last flight of Project Mercury, Astronaut L. Gordon Cooper traveled a distance of 546,167 miles in 1 day, 10 hours, 19 minutes, and 49 seconds. Determine his average speed (the distance traveled divided by the time required) to three significant digits (a) in mi/h; (b) in km/h.

12.10 Engineers who study shock waves sometimes express velocity in millimeters per microsecond (mm/μs). Suppose the velocity of a wavefront is measured and determined to be 5 mm/μs. Determine its velocity (a) in m/s; (b) in mi/s.

12.11 The kinetic energy of a particle of mass m is defined to be $\frac{1}{2}mv^2$, where v is the magnitude of the particle's velocity. If the value of the kinetic energy of a particle at a given time is 200 when m is in kilograms and v is in meters per second, what is the value when m is in slugs and v is in feet per second?

12.12 The acceleration due to gravity at sea level in SI units is $g = 9.81 \text{ m/s}^2$. By converting units, use this value to determine the acceleration due to gravity at sea level in U.S. Customary units.

12.13 A *furlong per fortnight* is a facetious unit of velocity, perhaps made up by a student as a satirical comment on the bewildering variety of units engineers must deal with. A furlong is 660 ft (1/8 mile). A fortnight is 2 weeks (14 nights). If you walk to class at 2 m/s, what is your speed in furlongs per fortnight to three significant digits?

12.14 The cross-sectional area of a beam is 480 in². What is its cross-sectional area in m²?

12.15 The cross-sectional area of the C12×30 American Standard Channel steel beam is $A = 8.81$ in². What is its cross-sectional area in mm²?

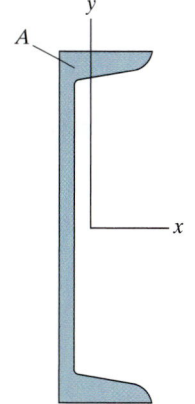

Problem 12.15

12.16 A pressure transducer measures a value of 300 lb/in². Determine the value of the pressure in pascals. A pascal (Pa) is one newton per meter squared.

12.17 A horsepower is 550 ft-lb/s. A watt is 1 N-m/s. Determine the number of watts generated by (a) the Wright brothers' 1903 airplane, which had a 12-horsepower engine, and (b) a modern passenger jet with a power of 100,000 horsepower at cruising speed.

Wright Brothers' Flyer (shown to scale)

Boeing 747

Problem 12.17

12.18 In SI units, the universal gravitational constant $G = 6.67 \times 10^{-11}$ N-m²/kg². Determine the value of G in U.S. Customary base units.

12.19 The moment of inertia of the rectangular area about the x axis is given by the equation
$$I = \tfrac{1}{3}bh^3.$$
The dimensions of the area are $b = 200$ mm and $h = 100$ mm. Determine the value of I to four significant digits in terms of (a) mm⁴, (b) m⁴, and (c) in⁴.

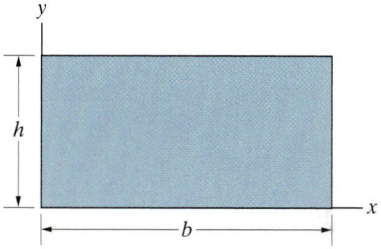

Problem 12.19

12.20 In the equation
$$T = \tfrac{1}{2}I\omega^2,$$
the term I is in kg-m² and ω is in s⁻¹.
(a) What are the SI units of T?
(b) If the value of T is 100 when I is in kg-m² and ω is in s⁻¹, what is the value of T when it is expressed in terms of U.S. Customary base units?

12.21 The equation
$$\sigma = \frac{My}{I}$$
is used in the mechanics of materials to determine normal stresses in beams.
(a) When this equation is expressed in terms of SI base units, M is in newton-meters (N-m), y is in meters (m), and I is in meters to the fourth power (m⁴). What are the SI units of σ?
(b) If $M = 2000$ N-m, $y = 0.1$ m, and $I = 7 \times 10^{-5}$ m⁴, what is the value of σ in U.S. Customary base units?

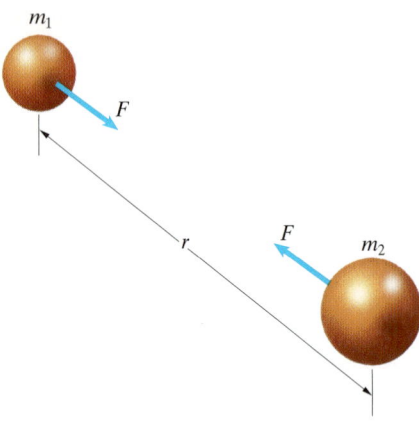

Figure 12.4
The gravitational forces between two particles are equal in magnitude and directed along the line between them.

12.5 Newtonian Gravitation

Newton postulated that the gravitational force between two particles of mass m_1 and m_2 that are separated by a distance r (Fig. 12.4) is

$$F = \frac{Gm_1m_2}{r^2},$$ (12.1)

where G is called the universal gravitational constant. Based on this postulate, he calculated the gravitational force between a particle of mass m_1 and a homogeneous sphere of mass m_2 and found that it is also given by Eq. (12.1), with r denoting the distance from the particle to the center of the sphere. Although the earth is not a homogeneous sphere, we can use this result to approximate the weight of an object of mass m due to the gravitational attraction of the earth. We have

$$W = \frac{Gmm_E}{r^2},$$ (12.2)

where m_E is the mass of the earth and r is the distance from the center of the earth to the object. Notice that the weight of an object depends on its location relative to the center of the earth, whereas the mass of the object is a measure of the amount of matter it contains and doesn't depend on its position.

When an object's weight is the only force acting on it, the resulting acceleration is called the acceleration due to gravity. In this case, Newton's second law states that $W = ma$, and from Eq. (12.2) we see that the acceleration due to gravity is

$$a = \frac{Gm_E}{r^2}.$$ (12.3)

The *acceleration due to gravity at sea level* is denoted by g. Denoting the radius of the earth by R_E, we see from Eq. (12.3) that $Gm_E = gR_E^2$. Substituting this result into Eq. (12.3), we obtain an expression for the acceleration due to gravity at a distance r from the center of the earth in terms of the acceleration due to gravity at sea level:

$$a = g\frac{R_E^2}{r^2}.$$ (12.4)

Since the weight of the object $W = ma$, the weight of an object at a distance r from the center of the earth is

$$W = mg\frac{R_E^2}{r^2}.$$ (12.5)

At sea level ($r = R_E$), the weight of an object is given in terms of its mass by the simple relation

$$W = mg.$$ (12.6)

The value of g varies from location to location on the surface of the earth. The values we use in examples and problems are $g = 9.81$ m/s^2 in SI units and $g = 32.2$ ft/s^2 in U.S. Customary units.

> **Study Questions**
>
> **1.** Does the weight of an object depend on its location?
>
> **2.** If you know an object's mass, how do you determine its weight at sea level?

Example 12.3   Determining an Object's Weight

When the Mars Exploration Rover was fully assembled, its mass was 180 kg. The acceleration due to gravity at the surface of Mars is 3.68 m/s^2 and the radius of Mars is 3390 km.

(a) What was the rover's weight when it was at sea level on Earth?
(b) What is the rover's weight on the surface of Mars?
(c) The entry phase began when the spacecraft reached the Mars atmospheric entry interface point at 3522 km from the center of Mars. What was the rover's weight at that point?

Figure 12.5
Mars Exploration Rover being assembled.

Strategy

The rover's weight at sea level on Earth is given by Eq. (12.6) with $g = 9.81$ m/s^2.

We can determine the weight on the surface of Mars by using Eq. (12.6) with the acceleration due to gravity equal to 3.68 m/s^2.

To determine the rover's weight as it began the entry phase, we can write an equation for Mars equivalent to Eq. (12.5).

Solution

(a) The weight at sea level on Earth is

$$W = mg$$

$$= (180\ \text{kg})(9.81\ \text{m/s}^2)$$

$$= 1770\ \text{N} \ (397\ \text{lb}).$$

(b) Let $g_M = 3.68$ m/s^2 be the acceleration due to gravity at the surface of Mars. Then the weight of the rover on the surface of Mars is

$$W = mg_M$$

$$= (180\ \text{kg})(3.68\ \text{m/s}^2)$$

$$= 662\ \text{N} \ (149\ \text{lb}).$$

(c) Let $R_M = 3390$ km be the radius of Mars. From Eq. (12.5), the rover's weight when it is 3522 km above the center of Mars is

$$W = mg_M \frac{R_M^2}{r^2}$$

$$= (180\ \text{kg})(3.68\ \text{m/s}^2) \frac{(3{,}390{,}000\ \text{m})^2}{(3{,}522{,}000\ \text{m})^2}$$

$$= 614\ \text{N} \ (138\ \text{lb}).$$

Critical Thinking

In part (c), how did we know that we could apply Eq. (12.5) to Mars? Equation (12.5) is applied to Earth based on modeling it as a homogeneous sphere. It can be applied to other celestial objects under the same assumption. The accuracy of the results depends on how aspherical and inhomogeneous the object is.

Problems

12.22 Let W be your weight at sea level in pounds. (a) What is your weight at sea level in newtons? (b) What is your mass in kilograms?

12.23 The acceleration due to gravity is 1.62 m/s² on the surface of the moon and 9.81 m/s² on the surface of the earth. A female astronaut's mass is 57 kg. What is the maximum allowable mass of her spacesuit and equipment if the engineers don't want the total weight on the moon of the woman, her spacesuit, and equipment to exceed 180 N?

12.24 A person has a mass of 50 kg.

(a) The acceleration due to gravity at sea level is $g = 9.81$ m/s². What is the person's weight at sea level?

(b) The acceleration due to gravity on the surface of the moon is $g = 1.62$ m/s². What would the person weigh on the moon?

12.25 The acceleration due to gravity at sea level is $g = 9.81$ m/s². The radius of the earth is 6370 km. The universal gravitational constant $G = 6.67 \times 10^{-11}$ N-m²/kg². Use this information to determine the mass of the earth.

12.26 A person weighs 180 lb at sea level. The radius of the earth is 3960 mi. What force is exerted on the person by the gravitational attraction of the earth if he is in a space station in orbit 200 mi above the surface of the earth?

12.27 The acceleration due to gravity on the surface of the moon is 1.62 m/s². The radius of the moon is $R_M = 1738$ km. Determine the acceleration due to gravity of the moon at a point 1738 km above its surface.

   Strategy: Write an equation equivalent to Eq. (12.4) for the acceleration due to gravity of the moon.

12.28 If an object is near the surface of the earth, the variation of its weight with distance from the center of the earth can often be neglected. The acceleration due to gravity at sea level is $g = 9.81$ m/s². The radius of the earth is 6370 km. The weight of an object at sea level is mg, where m is its mass. At what height above the surface of the earth does the weight of the object decrease to 0.99 mg?

12.29 The centers of two oranges are 1 m apart. The mass of each orange is 0.2 kg. What gravitational force do they exert on each other? (The universal gravitational constant $g = 6.67 \times 10^{-11}$ N-m²/kg².)

12.30 At a point between the earth and the moon, the magnitude of the force exerted on an object by the earth's gravity equals the magnitude of the force exerted on the object by the moon's gravity. What is the distance from the center of the earth to that point to three significant digits? The distance from the center of the earth to the center of the moon is 383,000 km, and the radius of the earth is 6370 km. The radius of the moon is 1738 km, and the acceleration due to gravity at its surface is 1.62 m/s².

Motion of a Point

In this chapter we begin the study of motion. We are not yet concerned with the properties of objects or the causes of their motions—our objective is to describe and analyze the motion of a point in space. After defining the position, velocity, and acceleration of a point, we consider the simplest case, motion along a straight line. We then show how motion of a point along an arbitrary path, or *trajectory*, is expressed and analyzed using various coordinate systems.

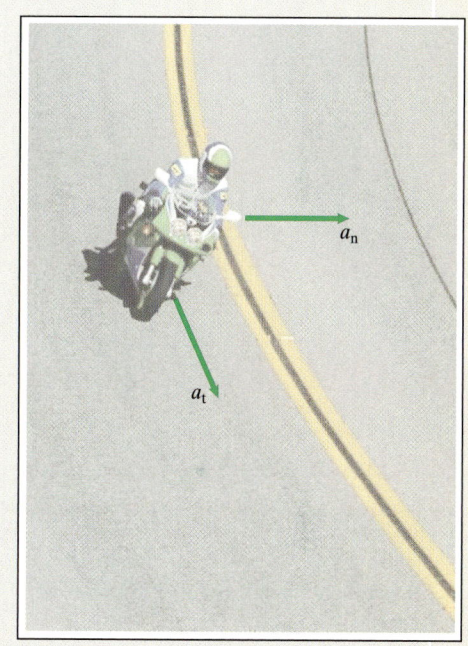

◄ The turning motorcycle has components of acceleration tangential and normal to its path.

13.1 Position, Velocity, and Acceleration

If you observe people in a room, such as a group at a party, you perceive their positions relative to the room. For example, some people may be in the back of the room, some in the middle of the room, and so forth. The colloquial expression is that the room is your "frame of reference." To make this idea precise, we can introduce a cartesian coordinate system with its axes aligned with the walls of the room as in Fig. 13.1a and specify the position of a person (actually, the position of some point of the person, such as his or her center of mass) by specifying the components of the position vector **r** relative to the origin of the coordinate system. This coordinate system is a convenient reference frame for objects in the room. If you are sitting in an airplane, you perceive the positions of objects within the airplane relative to the airplane. In this case, the interior of the airplane is your frame of reference. To precisely specify the position of a person within the airplane, we can introduce a cartesian coordinate system that is fixed relative to the airplane and measure the position of the person's center of mass by specifying the components of the position vector **r** relative to the origin (Fig. 13.1b). A *reference frame* is simply a coordinate system that is suitable for specifying positions of points. You may be familiar only with cartesian coordinates. We discuss other examples in this chapter and continue our discussion of reference frames throughout the book.

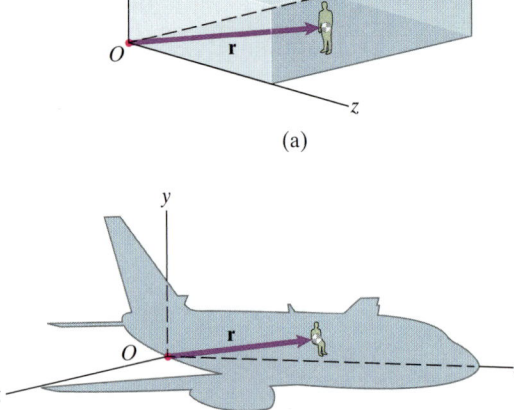

(a)

(b)

Figure 13.1
Convenient reference frames for specifying positions of objects
(a) in a room;
(b) in an airplane.

We can describe the position of a point P relative to a given reference frame with origin O by the *position vector* **r** from O to P (Fig. 13.2a). Suppose that P is in motion relative to the chosen reference frame, so that **r** is a function of time t (Fig. 13.2b). We express this by the notation

$$\mathbf{r} = \mathbf{r}(t).$$

The *velocity* of P relative to the given reference frame at time t is defined by

$$\mathbf{v} = \frac{d\mathbf{r}}{dt} = \lim_{\Delta t \to 0} \frac{\mathbf{r}(t + \Delta t) - \mathbf{r}(t)}{\Delta t}, \tag{13.1}$$

where the vector $\mathbf{r}(t + \Delta t) - \mathbf{r}(t)$ is the change in position, or *displacement*, of P during the interval of time Δt (Fig. 13.2c). Thus, the velocity is the rate of change of the position of P.

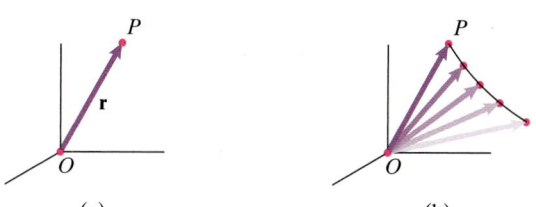

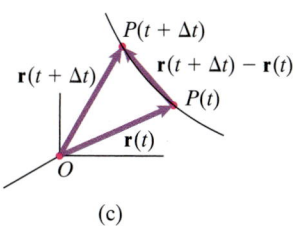

Figure 13.2
(a) The position vector **r** of P relative to O.
(b) Motion of P relative to the reference frame.
(c) Change in position of P from t to t + Δt.

The dimensions of a derivative are determined just as if it is a ratio, so the dimensions of **v** are (distance)/(time). The reference frame being used is often obvious, and we simply call **v** the velocity of P. However, remember that the position and velocity of a point can be specified only relative to some reference frame.

Notice in Eq. (13.1) that the derivative of a vector with respect to time is defined in exactly the same way as is the derivative of a scalar function. As a result, the derivative of a vector shares some of the properties of the derivative of a scalar function. We will use two of these properties: the derivative with respect to time, or time derivative, of the sum of two vector functions **u** and **w** is

$$\frac{d}{dt}(\mathbf{u} + \mathbf{w}) = \frac{d\mathbf{u}}{dt} + \frac{d\mathbf{w}}{dt},$$

and the time derivative of the product of a scalar function f and a vector function **u** is

$$\frac{d(f\mathbf{u})}{dt} = \frac{df}{dt}\mathbf{u} + f\frac{d\mathbf{u}}{dt}.$$

The *acceleration* of P relative to the given reference frame at time t is defined by

$$\mathbf{a} = \frac{d\mathbf{v}}{dt} = \lim_{\Delta t \to 0} \frac{\mathbf{v}(t + \Delta t) - \mathbf{v}(t)}{\Delta t}, \tag{13.2}$$

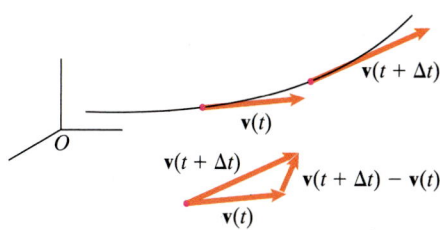

Figure 13.3
Change in the velocity of P from t to t + Δt.

where $\mathbf{v}(t + \Delta t) - \mathbf{v}(t)$ is the change in the velocity of P during the interval of time Δt (Fig. 13.3). The acceleration is the rate of change of the velocity of P at time t (the second time derivative of the displacement), and its dimensions are (distance)/(time)^2.

We have defined the velocity and acceleration of P relative to the origin O of the reference frame. We can show that *a point has the same velocity and acceleration relative to any fixed point in a given reference frame*. Let O′ be an arbitrary fixed point, and let **r**′ be the position vector from O′ to P (Fig. 13.4a). The velocity of P relative to O′ is $\mathbf{v}' = d\mathbf{r}'/dt$. The velocity of P relative to the origin O is $\mathbf{v} = d\mathbf{r}/dt$. We wish to show that $\mathbf{v}' = \mathbf{v}$. Let **R** be the vector from O to O′ (Fig. 13.4b), so that

$$\mathbf{r}' = \mathbf{r} - \mathbf{R}.$$

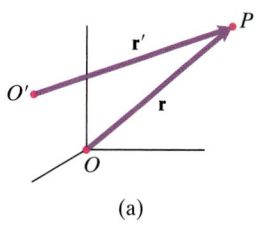

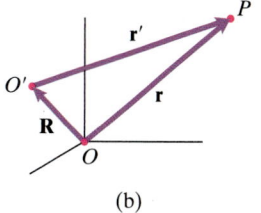

Figure 13.4
(a) Position vectors of P relative to O and O′.
(b) Position vector of O′ relative to O.

Since the vector $\mathbf{R}$ is constant, the velocity of P relative to O' is

$$\mathbf{v}' = \frac{d\mathbf{r}'}{dt} = \frac{d\mathbf{r}}{dt} - \frac{d\mathbf{R}}{dt} = \frac{d\mathbf{r}}{dt} = \mathbf{v}.$$

The acceleration of P relative to O' is $\mathbf{a}' = d\mathbf{v}'/dt$, and the acceleration of P relative to O is $\mathbf{a} = d\mathbf{v}/dt$. Since $\mathbf{v}' = \mathbf{v}$, $\mathbf{a}' = \mathbf{a}$. Thus, the velocity and acceleration of a point P relative to a given reference frame do not depend on the location of the fixed reference point used to specify the position of P.

Study Questions

1. What is a reference frame?

2. What is the definition of the velocity of a point P relative to a given reference frame?

3. Suppose that you know the velocity $\mathbf{v}$ and acceleration $\mathbf{a}$ of a point P relative to the origin O of a given reference frame. What do you know about the velocity and acceleration of P relative to an arbitrary fixed point O' of the reference frame?

13.2 Straight-Line Motion

We discuss this simple type of motion primarily so that you can gain experience and insight before proceeding to the general case of the motion of a point. But engineers must analyze straight-line motions in many practical situations, such as the motion of a vehicle on a straight road or the motion of a piston in an internal combustion engine.

Description of the Motion

Consider a straight line through the origin O of a given reference frame. We assume that the direction of the line relative to the reference frame is fixed. (For example, the x axis of a cartesian coordinate system passes through the origin and has fixed direction relative to the reference frame.) We can specify the position of a point P on such a line relative to O by a coordinate s measured along the line from O to P. In Fig. 13.5a we define s to be positive to the right, so s is positive when P is to the right of O and negative when P is to the left of O. The *displacement* of P during an interval of time from t_0 to t is the change in the position $s(t) - s(t_0)$, where $s(t)$ denotes the position at time t.

By introducing a unit vector $\mathbf{e}$ that is parallel to the line and points in the positive s direction (Fig. 13.5b), we can write the position vector of P relative to O as

$$\mathbf{r} = s\mathbf{e}.$$

Because the magnitude and direction of $\mathbf{e}$ are constant, $d\mathbf{e}/dt = \mathbf{0}$, and so the velocity of P relative to O is

$$\mathbf{v} = \frac{d\mathbf{r}}{dt} = \frac{ds}{dt}\mathbf{e}.$$

We can write the velocity vector as $\mathbf{v} = v\mathbf{e}$, obtaining the scalar equation

$$v = \frac{ds}{dt}.$$

The velocity v of point P along the straight line is the rate of change of the position s. Notice that v is equal to the slope at time t of the line tangent to the graph of s as a function of time (Fig. 13.6).

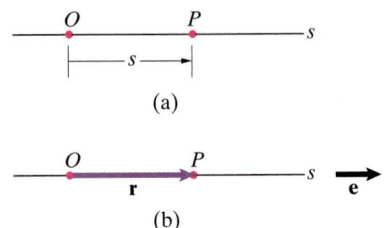

Figure 13.5
(a) The coordinate s from O to P.
(b) The unit vector $\mathbf{e}$ and position vector $\mathbf{r}$.

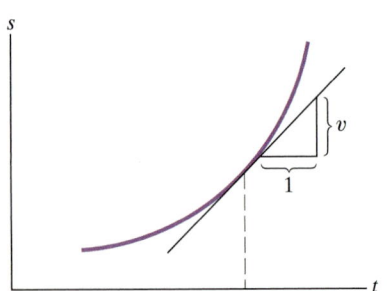

Figure 13.6
The slope of the straight line tangent to the graph of s versus t is the velocity at time t.

The acceleration of *P* relative to *O* is

$$\mathbf{a} = \frac{d\mathbf{v}}{dt} = \frac{d}{dt}(v\mathbf{e}) = \frac{dv}{dt}\mathbf{e}.$$

Writing the acceleration vector as $\mathbf{a} = a\mathbf{e}$, we obtain the scalar equation

$$a = \frac{dv}{dt} = \frac{d^2s}{dt^2}.$$

The acceleration *a* is equal to the slope at time *t* of the line tangent to the graph of *v* as a function of time (Fig. 13.7).

By introducing the unit vector **e**, we have obtained scalar equations describing the motion of *P*. The position is specified by the coordinate *s*, and the velocity and acceleration are governed by the equations

$$v = \frac{ds}{dt} \tag{13.3}$$

and

$$a = \frac{dv}{dt}. \tag{13.4}$$

Applying the *chain rule* of differential calculus, we can write the derivative of the velocity with respect to time as

$$\frac{dv}{dt} = \frac{dv}{ds}\frac{ds}{dt},$$

obtaining an alternative expression for the acceleration that is often useful:

$$a = \frac{dv}{ds}v. \tag{13.5}$$

Analysis of the Motion

In some situations, the position *s* of a point of an object is known as a function of time. Engineers use methods such as radar and laser-Doppler interferometry to measure positions as functions of time. In this case, we can obtain the velocity and acceleration as functions of time from Eqs. (13.3) and (13.4) by differentiation. For example, if the position of the truck in Fig. 13.8 during the interval of time from $t = 2$ s to $t = 4$ s is given by the equation

$$s = 6 + \frac{1}{3}t^3 \text{ m},$$

then the velocity and acceleration of the truck during that interval of time are

$$v = \frac{ds}{dt} = t^2 \text{ m/s}$$

and

$$a = \frac{dv}{dt} = 2t \text{ m/s}^2.$$

However, it is more common to know an object's acceleration than to know its position, because the acceleration of an object can be determined by Newton's second law when the forces acting on it are known. When the acceleration is known, we can determine the velocity and position from Eqs. (13.3)–(13.5) by integration. We discuss a number of important cases in the sections that follow.

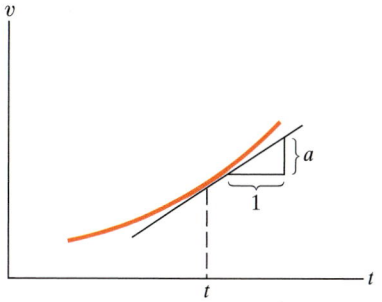

Figure 13.7
The slope of the straight line tangent to the graph of *v* versus *t* is the acceleration at time *t*.

Figure 13.8
The coordinate *s* measures the position of the center of mass of the truck relative to a reference point.

Acceleration Specified as a Function of Time If the acceleration is a known function of time $a(t)$, we can integrate the relation

$$\frac{dv}{dt} = a(t) \tag{13.6}$$

with respect to time to determine the velocity as a function of time. We obtain

$$v = \int a(t)\,dt + A,$$

where A is an integration constant. Then we can integrate the relation

$$\frac{ds}{dt} = v \tag{13.7}$$

to determine the position as a function of time,

$$s = \int v\,dt + B,$$

where B is another integration constant. We would need additional information about the motion, such as the values of v and s at a given time, to determine the constants A and B.

Instead of using indefinite integrals, we can write Eq. (13.6) as

$$dv = a(t)\,dt$$

and integrate in terms of definite integrals:

$$\int_{v_0}^{v} dv = \int_{t_0}^{t} a(t)\,dt. \tag{13.8}$$

The lower limit v_0 is the velocity at time t_0, and the upper limit v is the velocity at an arbitrary time t. Evaluating the integral on the left side of Eq. (13.8), we obtain an expression for the velocity as a function of time:

$$v = v_0 + \int_{t_0}^{t} a(t)\,dt. \tag{13.9}$$

We can then write Eq. (13.7) as

$$ds = v\,dt$$

and integrate in terms of definite integrals to obtain

$$\int_{s_0}^{s} ds = \int_{t_0}^{t} v\,dt,$$

where the lower limit s_0 is the position at time t_0 and the upper limit s is the position at an arbitrary time t. Evaluating the integral on the left side, we obtain the position as a function of time:

$$s = s_0 + \int_{t_0}^{t} v\,dt. \tag{13.10}$$

Although we have shown how to determine the velocity and position when the acceleration is known as a function of time, don't try to remember results such as Eqs. (13.9) and (13.10). As we will demonstrate in the examples, we recommend that straight-line motion problems be solved by using Eqs. (13.3)–(13.5).

We can make some useful observations from Eqs. (13.9) and (13.10):

- The area defined by the graph of the acceleration of P as a function of time from t_0 to t is equal to the change in the velocity from t_0 to t (Fig. 13.9a).
- The area defined by the graph of the velocity of P as a function of time from t_0 to t is equal to the change in position from t_0 to t (Fig. 13.9b).

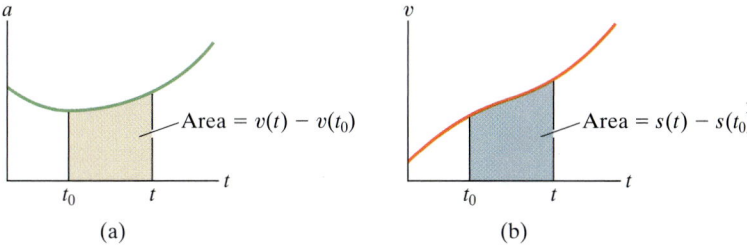

(a) (b)

Figure 13.9
Relations between areas defined by the graphs of the acceleration and velocity of P and changes in its velocity and position.

These relationships can often be used to obtain a qualitative understanding of an object's motion, and in some cases can even be used to determine the object's motion quantitatively.

Constant Acceleration   In some situations, the acceleration of an object is constant or nearly constant. For example, if a dense object such as a golf ball or a rock is dropped and doesn't fall too far, the object's acceleration is approximately equal to the acceleration due to gravity at sea level.

Let the acceleration be a known constant a_0. From Eqs. (13.9) and (13.10), the velocity and position as functions of time are

$$v = v_0 + a_0(t - t_0) \tag{13.11}$$

and

$$s = s_0 + v_0(t - t_0) + \frac{1}{2}a_0(t - t_0)^2, \tag{13.12}$$

where s_0 and v_0 are the position and velocity, respectively, at time t_0. Notice that *if the acceleration is constant, the velocity is a linear function of time.*

From Eq. (13.5), we can write the acceleration as

$$a_0 = \frac{dv}{ds}v.$$

Writing this expression as $v\,dv = a_0\,ds$ and integrating,

$$\int_{v_0}^{v} v\,dv = \int_{s_0}^{s} a_0\,ds,$$

we obtain an equation for the velocity as a function of position:

$$v^2 = v_0^2 + 2a_0(s - s_0). \tag{13.13}$$

Although Eqs. (13.11)–(13.13) can be useful *when the acceleration is constant,* they must not be used otherwise.

Study Questions

1. If you know the position s of a point P in straight-line motion as a function of time, how can you determine the velocity and acceleration of P as functions of time?

2. If you know the acceleration a of a point P in straight-line motion as a function of time, but have no other information, you can't determine the position and velocity of P as functions of time. Why not?

3. Suppose that you know the velocity v of a point P as a function of time. If you calculate the area defined by the graph of v from a time t_0 to a time t, what does that tell you?

Example 13.1 Straight-Line Motion with Constant Acceleration

Engineers testing a vehicle that will be dropped by parachute estimate that the vertical velocity of the vehicle when it reaches the ground will be 6 m/s. If they drop the vehicle from the test rig in Fig. 13.10, from what height h should they drop it to match the impact velocity of the parachute drop?

Figure 13.10

Strategy

If the only significant force acting on an object near the earth's surface is its weight, the acceleration of the object is approximately constant and equal to the acceleration due to gravity at sea level. Therefore, we can assume that the vehicle's acceleration during its short fall is $g = 9.81$ m/s^2. We can integrate Eqs. (13.3) and (13.4) to obtain the vehicle's velocity and position as functions of time and then use them to determine the position of the vehicle when its velocity is 6 m/s.

Solution

Let $t = 0$ be the time at which the vehicle is dropped, and let s be the position of the bottom of the cushioning material beneath the vehicle relative to its position at $t = 0$ (Fig. a). The vehicle's acceleration is $a = 9.81$ m/s^2.

From Eq. (13.4),

$$\frac{dv}{dt} = a = 9.81 \text{ m/s}^2.$$

(a) The coordinate s measures the position of the bottom of the platform relative to its initial position.

Integrating, we obtain

$$v = 9.81t + A,$$

where A is an integration constant. Because the vehicle is at rest when it is released, $v = 0$ at $t = 0$. Therefore $A = 0$, and the vehicle's velocity as a function of time is

$$v = 9.81t \text{ m/s}.$$

We substitute this result into Eq. (13.3) to get

$$\frac{ds}{dt} = v = 9.81t$$

and integrate, obtaining

$$s = 4.91t^2 + B.$$

The position $s = 0$ when $t = 0$, so the integration constant $B = 0$, and the position as a function of time is

$$s = 4.91t^2.$$

From our equation for the velocity as a function of time, the time necessary for the vehicle to reach 6 m/s as it falls is

$$t = \frac{v}{9.81 \text{ m/s}^2} = \frac{6 \text{ m/s}}{9.81 \text{ m/s}^2} = 0.612 \text{ s}.$$

Substituting this time into our equation for the position as a function of time yields the required height h:

$$h = 4.91t^2 = 4.91(0.612)^2 = 1.83 \text{ m}.$$

Critical Thinking

Notice that we could have determined the height h from which the vehicle should be dropped in a simpler way by using Eq. (13.13), which relates the velocity to the position.

$$v^2 = v_0^2 + 2a_0(s - s_0):$$

$$(6 \text{ m/s})^2 = 0 + 2(9.81 \text{ m/s}^2)(h - 0).$$

Solving, we obtain $h = 1.83$ m. But it is essential to remember that Eqs. (13.11)–(13.13) apply *only when the acceleration is constant*, as it is in this example.

Example 13.2   Graphical Solution of Straight-Line Motion

Figure 13.11

The cheetah, *Acinonyx jubatus* (Fig. 13.11), can run as fast as 75 mi/hr. If you assume that the animal's acceleration is constant and that it reaches top speed in 4 s, what distance can the cheetah cover in 10 s?

Strategy

The acceleration has a constant value for the first 4 s and is then zero. We can determine the distance traveled during each of these "phases" of the motion and sum them to obtain the total distance covered. We do so both analytically and graphically.

Solution

The top speed in terms of feet per second is

$$75 \text{ mi/h} = (75 \text{ mi/h})\left(\frac{5280 \text{ ft}}{1 \text{ mi}}\right)\left(\frac{1 \text{ h}}{3600 \text{ s}}\right) = 110 \text{ ft/s}.$$

First Method   Let a_0 be the acceleration during the first 4 s. We integrate Eq. (13.4) to get

$$\int_0^v dv = \int_0^t a_0 \, dt,$$

$$\left[v\right]_0^v = a_0\left[t\right]_0^t,$$

$$v - 0 = a_0(t - 0),$$

obtaining the velocity as a function of time during the first 4 s:

$$v = a_0 t \text{ ft/s}.$$

When $t = 4$ s, $v = 110$ ft/s; so $a_0 = 110/4 = 27.5$ ft/s^2. Therefore, the velocity during the first 4 s is $v = 27.5t$ ft/s. Now we integrate Eq. (13.3),

$$\int_0^s ds = \int_0^t 27.5t \, dt,$$

$$\left[s\right]_0^s = 27.5\left[\frac{t^2}{2}\right]_0^t,$$

$$s - 0 = 27.5\left(\frac{t^2}{2} - 0\right),$$

obtaining the position as a function of time during the first 4 s:

$$s = 13.75t^2 \text{ ft.}$$

At $t = 4$ s, the position is $s = 13.75(4)^2 = 220$ ft.

From $t = 4$ s to $t = 10$ s, the velocity $v = 110$ ft/s. We write Eq. (13.3) as

$$ds = v\, dt = 110\, dt$$

and integrate to determine the distance traveled during the second phase of the motion,

$$\int_0^s ds = \int_4^{10} 110\, dt,$$

$$\left[s \right]_0^s = 110 \left[t \right]_4^{10},$$

$$s - 0 = 110(10 - 4),$$

obtaining $s = 660$ ft. The total distance the cheetah travels is 220 ft $+$ 660 ft $= 880$ ft, or 293 yd, in 10 s.

Second Method   We draw a graph of the cheetah's velocity as a function of time in Fig. a. The acceleration is constant during the first 4 s of motion, so the velocity is a linear function of time from $v = 0$ at $t = 0$ to $v = 110$ ft/s at $t = 4$ s. The velocity is constant during the last 6 s. The total distance covered is the sum of the areas during the two phases of motion:

$$\tfrac{1}{2}(4\text{ s})(110\text{ ft/s}) + (6\text{ s})(110\text{ ft/s}) = 220\text{ ft} + 660\text{ ft} = 880\text{ ft}.$$

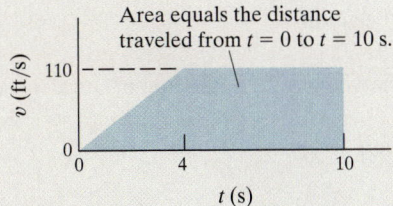

(a) The cheetah's velocity as a function of time.

Critical Thinking
Notice that in the first method we used definite, rather than indefinite, integrals to determine the cheetah's velocity and position as functions of time. You should rework the example using indefinite integrals and compare your results with ours. Whether to use definite or indefinite integrals is primarily a matter of taste, but you need to be familiar with both procedures.

Example 13.3   Acceleration That Is a Function of Time

Figure 13.12

Suppose that the acceleration of the train in Fig. 13.12 during the interval of time from $t = 2$ s to $t = 4$ s is $a = 2t$ m/s^2, and at $t = 2$ s its velocity is $v = 180$ km/h. What is the train's velocity at $t = 4$ s, and what is its displacement (change in position) from $t = 2$ s to $t = 4$ s?

Strategy

We can integrate Eqs. (13.3) and (13.4) to determine the train's velocity and position as functions of time.

Solution

The velocity at $t = 2$ s is

$$(180 \text{ km/h})\left(\frac{1000 \text{ m}}{1 \text{ km}}\right)\left(\frac{1 \text{ h}}{3600 \text{ s}}\right) = 50 \text{ m/s}.$$

We write Eq. (13.4) as

$$dv = a \, dt = 2t \, dt$$

and integrate, introducing the condition $v = 50$ m/s at $t = 2$ s. We have

$$\int_{50}^{v} dv = \int_{2}^{t} 2t \, dt,$$

$$\left[v\right]_{50}^{v} = \left[t^2\right]_{2}^{t},$$

$$v - 50 = t^2 - (2)^2,$$

obtaining

$$v = t^2 + 46 \text{ m/s}.$$

Now that we know the velocity as a function of time, we write Eq. (13.3) as

$$ds = v \, dt = (t^2 + 46) \, dt$$

and integrate, defining the position of the train at $t = 2$ s to be $s = 0$:

$$\int_{0}^{s} ds = \int_{2}^{t} (t^2 + 46) \, dt,$$

$$\left[s\right]_{0}^{s} = \left[\frac{t^3}{3} + 46t\right]_{2}^{t},$$

$$s - 0 = \frac{t^3}{3} + 46t - \frac{(2)^3}{3} - 46(2),$$

yielding the position as a function of time:

$$s = \tfrac{1}{3}t^3 + 46t - 94.7 \text{ m}.$$

Using our equations for the velocity and position, we find that the velocity at $t = 4$ s is

$$v = (4)^2 + 46 = 62 \text{ m/s},$$

and the displacement from $t = 2$ s to $t = 4$ s is

$$\left[\frac{1}{3}(4)^3 + 46(4) - 94.7\right] - 0 = 111 \text{ m}.$$

Critical Thinking

The acceleration in this example is not constant. You must not try to solve such problems by using equations that are valid only when the acceleration is constant. To convince yourself, try applying Eq. (13.11) to this example: Set $a_0 = 2t$ m/s^2, $t_0 = 2$ s, and $v_0 = 50$ m/s, and solve for the velocity at $t = 4$ s.

Problems

The problems that follow involve straight-line motion. The time t is in seconds unless otherwise stated.

13.1 The position of point P relative to point O is given as a function of time by $s = 40 + 2t^3$ ft. Determine the position, velocity, and acceleration of the point at $t = 4$ s.

Strategy: Use Eqs. (13.3) and (13.4) to determine the velocity and acceleration as functions of time.

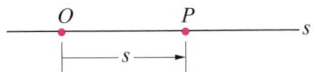

Problem 13.1

13.2 The milling machine is programmed so that the position of its head is given as a function of time by $s = 0.3 - 0.2 \cos(5t)$ m. (When t is in seconds, the argument of the cosine is in radians.)

(a) Determine the velocity of the head as a function of time.
(b) What is the acceleration of the head at $t = 4$ s?

Problem 13.2

13.3 The student drops a ball at time $t = 0$. The ball's position relative to the floor is given as a function of time by $s = -16.1t^2 + 4$ ft.

(a) How long does it take the ball to fall to the floor?
(b) What is the ball's velocity just before it hits the floor?
(c) Determine the ball's acceleration.

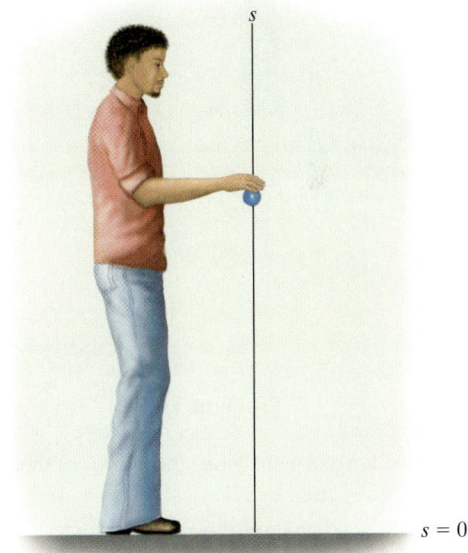

Problem 13.3

13.4 The boat's position during the interval of time from $t = 2$ s to $t = 10$ s is given by $s = 4t + 1.6t^2 - 0.08t^3$ m.

(a) Determine the boat's velocity and acceleration at $t = 4$ s.
(b) What is the boat's maximum velocity during this interval of time, and when does it occur?

Problem 13.4

13.5 The rocket starts from rest at $t = 0$ and travels straight up. Its height above the ground can be approximated by the function $s = bt^2 + ct^4$, where b and c are constants. At $t = 10$ s, the rocket's velocity and acceleration are $v = 229$ m/s and $a = 28.2$ m/s^2, respectively. What are its velocity and acceleration at $t = 5$ s?

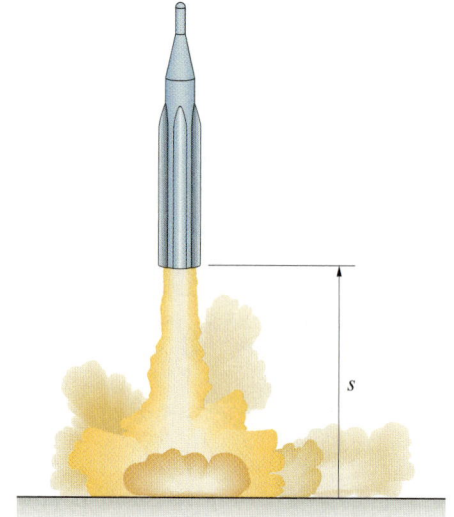

Problem 13.5

13.6 The position of a point during the interval of time from $t = 0$ to $t = 6$ s is given by $s = -\frac{1}{2}t^3 + 6t^2 + 4t$ m.

(a) What is the maximum velocity during this interval of time, and at what time does it occur?
(b) What is the acceleration when the velocity is a maximum?

13.7 The position of a point during the interval of time from $t = 0$ to $t = 3$ s is $s = 12 + 5t^2 - t^3$ ft.

(a) What is the maximum velocity during this interval of time, and at what time does it occur?
(b) What is the acceleration when the velocity is a maximum?

13.8 The rotating crank causes the position of point P as a function of time to be $s = 0.4 \sin(2\pi t)$ m.

(a) Determine the velocity and acceleration of P at $t = 0.375$ s.
(b) What is the maximum magnitude of the velocity of P?
(c) When the magnitude of the velocity of P is a maximum, what is the acceleration of P?

13.9 For the mechanism in Problem 13.8, draw graphs of the position s, velocity v, and acceleration a of point P as functions of time for $0 \le t \le 2$ s. Using your graphs, confirm that the slope of the graph of s is zero at times for which v is zero and that the slope of the graph of v is zero at times for which a is zero.

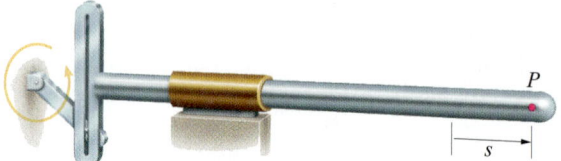

Problems 13.8/13.9

13.10 A seismograph measures the horizontal motion of the ground during an earthquake. An engineer analyzing the data determines that for a 10-s interval of time beginning at $t = 0$, the position is approximated by $s = 100 \cos(2\pi t)$ mm. What are (a) the maximum velocity and (b) the maximum acceleration of the ground during the 10-s interval?

13.11 In an assembly operation, the robot's arm moves along a straight horizontal line. During an interval of time from $t = 0$ to $t = 1$ s, the position of the arm is given by $s = 30t^2 - 20t^3$ mm.

(a) Determine the maximum velocity during this interval of time.
(b) What are the position and acceleration when the velocity is a maximum?

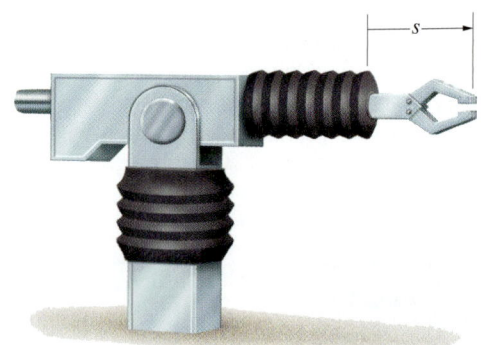

Problem 13.11

13.12 In a test of a prototype car, the driver starts the car from rest at $t = 0$, accelerates, and then applies the brakes. Engineers measuring the position of the car find that from $t = 0$ to $t = 18$ s the position is approximated by $s = 5t^2 + \frac{1}{3}t^3 - \frac{1}{50}t^4$ ft.

(a) What is the maximum velocity, and at what time does it occur?
(b) What is the maximum acceleration, and at what time does it occur?

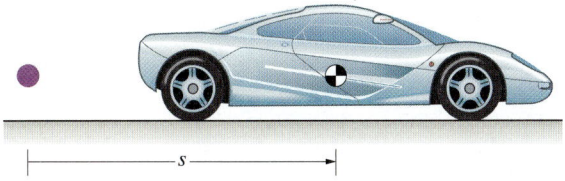

Problem 13.12

13.13 Suppose you want to approximate the position of a vehicle you are testing by the power series $s = A + Bt + Ct^2 + Dt^3$, where A, B, C, and D are constants. The vehicle starts from rest at $t = 0$ and $s = 0$. At $t = 4$ s, $s = 176$ ft, and at $t = 8$ s, $s = 448$ ft.

(a) Determine A, B, C, and D.
(b) What are the approximate velocity and acceleration of the vehicle at $t = 8$ s?

13.14 The acceleration of a point is $a = 20t$ m/s². When $t = 0$, $s = 40$ m and $v = -10$ m/s. What are the position and velocity at $t = 3$ s?

13.15 The acceleration of a point is $a = 60t - 36t^2$ ft/s². When $t = 0$, $s = 0$ and $v = 20$ ft/s. What are the position and velocity as functions of time?

13.16 The snow petrel takes off with constant acceleration. If it requires a distance $s = 4$ m and is moving at 6 m/s when it lifts off, how much time does its takeoff require?

13.17 A bioengineer studying the mechanics of bird flight models the acceleration of the snow petrel by an equation of the form $a = C + Dt$, where C and D are constants. Measurements obtained from videotape indicate that one bird requires 1.42 s and a distance of 4.3 m to take off and is moving at 6.1 m/s when it lifts off. What are the constants C and D?

Problems 13.16/13.17

13.18 Missiles designed for defense against ballistic missiles have attained accelerations in excess of 100 g's, or 100 times the acceleration due to gravity. Suppose that a missile lifts off from the ground and has a constant vertical acceleration of 100 g's.

(a) How long does it take to reach a velocity of 60 mi/h?
(b) How long does it take to reach an altitude of 10,000 ft? How fast is it going when it reaches that altitude?

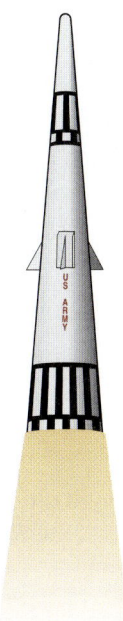

Problem 13.18

13.19 The acceleration due to gravity at sea level on the Earth is 9.81 m/s². The acceleration due to gravity at the surface of the moon is 1.62 m/s².

(a) If an object at sea level on the Earth is given an upward velocity of 10 m/s and aerodynamic drag is negligible, how high does it go?
(b) If the object is on the surface of the moon and is given an upward velocity of 10 m/s, how high does it go?

13.20 The airplane releases its drag parachute at time $t = 0$. Its velocity is given as a function of time by

$$v = \frac{80}{1 + 0.32t} \text{ m/s}.$$

What is the airplane's acceleration at $t = 3$ s?

13.21 How far does the airplane in Problem 13.20 travel during the interval of time from $t = 0$ to $t = 10$ s?

Problems 13.20/13.21

13.22 The velocity of a bobsled is $v = 10t$ ft/s. When $t = 2$ s, the position of the sled is $s = 25$ ft. What is its position when $t = 10$ s?

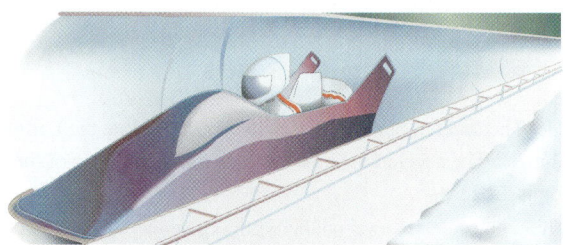

Problem 13.22

13.23 In September 2003, Tony Schumacher started from rest and drove a quarter mile (1320 ft) in 4.498 s in a National Hot Rod Association race. His speed as he crossed the finish line was 328.54 mi/h. Assume that the car's acceleration can be expressed by a linear function of time $a = b + ct$.

(a) Determine the constants b and c.
(b) What was the car's speed 2 s after the start of the race?

13.24 The velocity of an object is $v = 200 - 2t^2$ m/s. When $t = 3$ s, the position of the object is $s = 600$ m. What are the position and acceleration of the object at $t = 6$ s?

13.25 An inertial navigation system measures the acceleration of a vehicle from $t = 0$ to $t = 6$ s and determines it to be $a = 2 + 0.1t$ m/s^2. At $t = 0$, the vehicle's position and velocity are $s = 240$ m and $v = 42$ m/s, respectively. What are the vehicle's position and velocity at $t = 6$ s?

13.26 The missile shown in Problem 13.18 lifts off and accelerates for 3 s at 100 g's. After 3 s, its weight and aerodynamic drag cause it to have a nearly constant deceleration of 4 g's. How long does it take the missile to go from the ground to an altitude of 15 km (approximately 50,000 ft)?

13.27 The graph shows the airplane's acceleration during takeoff. What is the airplane's velocity when it rotates (lifts off) at $t = 30$ s?

13.28 The graph shows an airplane's acceleration during takeoff. What distance has the airplane traveled when it lifts off at $t = 30$ s?

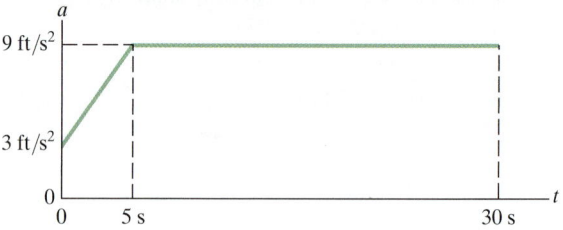

Problems 13.27/13.28

13.29 The car is traveling at 30 mi/h when the traffic light 295 ft ahead turns yellow. The driver takes 1 s to react before he applies the brakes.

(a) After he applies the brakes, what constant rate of deceleration will cause the car to come to a stop just as it reaches the light?
(b) How long does it take the car to travel the 295 ft to the light?

30 mi/h

295 ft

Problem 13.29

13.30 At $t = 0$, a motorist traveling at 100 km/h sees a deer standing in the road 100 m ahead. After a reaction time of 0.3 s, he applies the brakes and decelerates at a constant rate of 4 m/s². If the deer takes 5 s from $t = 0$ to react and leave the road, does the motorist miss it?

13.31 A high-speed rail transportation system has a top speed of 100 m/s. For the comfort of the passengers, the magnitude of the acceleration and deceleration is limited to 2 m/s². Determine the minimum time required for a trip of 100 km.

Strategy: A graphical approach can help you solve this problem. Recall that the change in the position from an initial time t_0 to a time t is equal to the area defined by the graph of the velocity as a function of time from t_0 to t.

Problem 13.31

13.32 The nearest star, Proxima Centauri, is 4.22 light years from the Earth. Ignoring relative motion between the solar system and Proxima Centauri, suppose that a spacecraft accelerates from the vicinity of the Earth at 0.01g (0.01 times the acceleration due to gravity at sea level) until it reaches one-tenth the speed of light, coasts until it is time to decelerate, and then decelerates at 0.01g until it comes to rest in the vicinity of Proxima Centauri. How long does the trip take? (Light travels at 3×10^8 m/s.)

13.33 A race car starts from rest and accelerates at $a = 5 + 2t$ ft/s² for 10 s. The brakes are then applied, and the car has a constant acceleration $a = -30$ ft/s² until it comes to rest. Determine (a) the maximum velocity, (b) the total distance traveled, and (c) the total time of travel.

13.34 When $t = 0$, the position of a point is $s = 6$ m and its velocity is $v = 2$ m/s. From $t = 0$ to $t = 6$ s, the acceleration of the point is $a = 2 + 2t^2$ m/s². From $t = 6$ s until it comes to rest, its acceleration is $a = -4$ m/s².

(a) What is the total time of travel?
(b) What total distance does the point move?

13.35 Zoologists studying the ecology of the Serengeti Plain estimate that the average adult cheetah can run 100 km/h and the average springbok can run 65 km/h. If the animals run along the same straight line, start at the same time, are each assumed to have constant acceleration, and reach top speed in 4 s, how close must a cheetah be when the chase begins to catch a springbok in 15 s?

13.36 Suppose that a person unwisely drives 75 mi/h in a 55 mi/h zone and passes a police car going 55 mi/h in the same direction. If the police officers begin constant acceleration at the instant they are passed and increase their velocity to 80 mi/h in 4 s, how long does it take them to be even with the pursued car?

13.37 If $\theta = 1$ rad and $d\theta/dt = 1$ rad/s, what is the velocity of P relative to O?

Strategy: You can write the position of P relative to O as
$$s = (2 \text{ ft}) \cos \theta + (2 \text{ ft}) \cos \theta$$
and then take the derivative of this expression with respect to time to determine the velocity.

13.38 If $\theta = 1$ rad, $d\theta/dt = -2$ rad/s, and $d^2\theta/dt^2 = 0$, what are the velocity and acceleration of P relative to O?

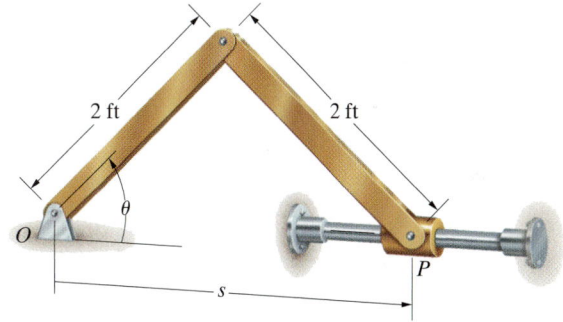

2 ft   2 ft

θ

O

P

s

Problems 13.37/13.38

13.39* If $\theta = 1$ rad and $d\theta/dt = 1$ rad/s, what is the velocity of P relative to O?

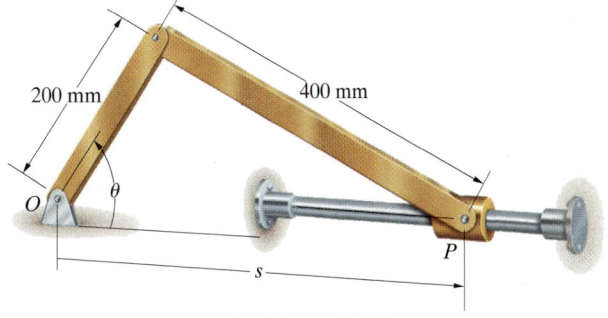

200 mm   400 mm

θ

O

P

s

Problem 13.39

Figure 13.13
Aerodynamic and hydrodynamic forces depend on an object's velocity. As the bullet slows, the aerodynamic drag force resisting its motion decreases.

Acceleration Specified as a Function of Velocity   Aerodynamic and hydrodynamic forces can cause an object's acceleration to depend on its velocity (Fig. 13.13). Suppose that the acceleration is a known function of velocity—that is,

$$\frac{dv}{dt} = a(v). \tag{13.14}$$

We cannot integrate this equation with respect to time to determine the velocity, because $a(v)$ is not known as a function of time. But we can *separate variables*, putting terms involving v on one side of the equation and terms involving t on the other side:

$$\frac{dv}{a(v)} = dt. \tag{13.15}$$

We can now integrate, obtaining

$$\int_{v_0}^{v} \frac{dv}{a(v)} = \int_{t_0}^{t} dt, \tag{13.16}$$

where v_0 is the velocity at time t_0. In principle, we can solve this equation for the velocity as a function of time and then integrate the relation

$$\frac{ds}{dt} = v$$

to determine the position as a function of time.

By using the chain rule, we can also determine the velocity as a function of the position. Writing the acceleration as

$$\frac{dv}{dt} = \frac{dv}{ds}\frac{ds}{dt} = \frac{dv}{ds}v$$

and substituting it into Eq. (13.14), we obtain

$$\frac{dv}{ds}v = a(v).$$

Separating variables yields

$$\frac{v\, dv}{a(v)} = ds.$$

Integrating,

$$\int_{v_0}^{v} \frac{v\, dv}{a(v)} = \int_{s_0}^{s} ds,$$

we can obtain a relation between the velocity and the position. (See Example 13.4.)

Acceleration Specified as a Function of Position   Gravitational forces and forces exerted by springs can cause an object's acceleration to depend on its position. If the acceleration is a known function of position—that is,

$$\frac{dv}{dt} = a(s), \tag{13.17}$$

we cannot integrate with respect to time to determine the velocity, because $a(s)$ is not known as a function of time. Moreover, we cannot separate variables, because the equation contains three variables: v, t, and s. However, by using the chain rule

$$\frac{dv}{dt} = \frac{dv}{ds}\frac{ds}{dt} = \frac{dv}{ds}v,$$

we can write Eq. (13.17) as

$$\frac{dv}{ds}v = a(s).$$

Now we can separate variables,

$$v\, dv = a(s)\, ds, \tag{13.18}$$

and we integrate:

$$\int_{v_0}^{v} v\, dv = \int_{s_0}^{s} a(s)\, ds. \tag{13.19}$$

In principle, we can solve this equation for the velocity as a function of the position:

$$v = \frac{ds}{dt} = v(s). \tag{13.20}$$

Then we can separate variables in this equation and integrate to determine the position as a function of time:

$$\int_{s_0}^{s} \frac{ds}{v(s)} = \int_{t_0}^{t} dt.$$

The procedures we have described when the acceleration is known as a function of velocity or position are summarized in Table 13.1.

TABLE 13.1 Determining the velocity when you know the acceleration is known as a function of velocity or position.

If you know $a = a(v)$	**Separate variables:**
	$\dfrac{dv}{dt} = a(v),$ $\dfrac{dv}{a(v)} = dt.$ Or apply the chain rule $\dfrac{dv}{dt} = \dfrac{dv}{ds}\dfrac{ds}{dt} = \dfrac{dv}{ds}v = a(v),$ and then separate variables: $\dfrac{v\, dv}{a(v)} = ds.$
If you know $a = a(s)$	**Apply the chain rule**
	$\dfrac{dv}{dt} = \dfrac{dv}{ds}\dfrac{ds}{dt} = \dfrac{dv}{ds}v = a(s),$ and then separate variables: $v\, dv = a(s)\, ds.$

Example 13.4 | **Acceleration That Is a Function of Velocity**

After deploying its drag parachute, the airplane shown in Fig. 13.14 has an acceleration $a = -0.004v^2$ m/s^2, where v is the airplane's velocity in m/s.

(a) Determine the time required for the velocity to decrease from 80 m/s to 10 m/s.
(b) What distance does the plane cover during that time?

Figure 13.14

Strategy

(a) The airplane's acceleration is known as a function of its velocity. We can separate variables as shown in Table 13.1 and integrate to determine the velocity as a function of time.
(b) We will first use the chain rule to express the acceleration in terms of a derivative with respect to the position s, then separate variables and integrate to determine the velocity as a function of s.

Solution

(a) The acceleration is

$$a = \frac{dv}{dt} = -0.004v^2.$$

We separate variables,

$$\frac{dv}{v^2} = -0.004 \, dt,$$

and integrate, defining $t = 0$ to be the time at which $v = 80$ m/s:

$$\int_{80}^{v} \frac{dv}{v^2} = \int_{0}^{t} -0.004 \, dt,$$

$$\left[-\frac{1}{v} \right]_{80}^{v} = -0.004 \left[t \right]_{0}^{t},$$

$$-\frac{1}{v} + \frac{1}{80} = -0.004(t - 0).$$

Solving for t, we obtain

$$t = 250 \left(\frac{1}{v} - \frac{1}{80} \right).$$

From this equation, we find that the time required for the plane to slow to $v = 10$ m/s is 21.9 s. The velocity of the airplane as a function of time is shown in Fig. 13.15.

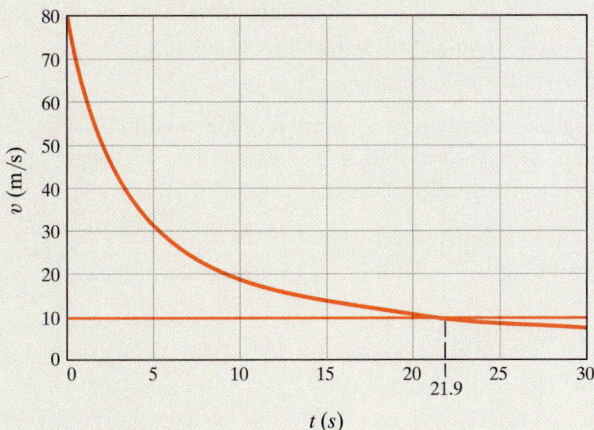

Figure 13.15
Graph of the airplane's velocity as a function of time.

(b) We write the acceleration as

$$a = \frac{dv}{dt} = \frac{dv}{ds}\frac{ds}{dt} = \frac{dv}{ds}v = -0.004v^2.$$

Separating variables yields

$$\frac{dv}{v} = -0.004\ ds.$$

We now integrate, defining $s = 0$ to be the position at which $v = 80$ m/s:

$$\int_{80}^{v}\frac{dv}{v} = \int_{0}^{s}-0.004\ ds,$$

$$\left[\ln v\right]_{80}^{v} = -0.004\left[s\right]_{0}^{s},$$

$$\ln v - \ln 80 = -0.004(s - 0).$$

Solving for s, we obtain

$$s = 250\ln\left(\frac{80}{v}\right)\ \text{m.}$$

The distance required for the plane to slow to $v = 10$ m/s is 520 m.

Critical Thinking

Notice that our results predict that the time elapsed and the distance traveled continue to increase without bound as the airplane's velocity decreases. The reason is that the modeling is incomplete. The equation for the acceleration includes only aerodynamic drag and does not account for other forces, such as friction in the airplane's wheels.

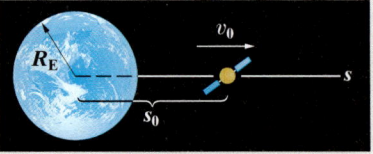

Figure 13.16

Example 13.5   Gravitational (Position-Dependent) Acceleration

In terms of the distance s from the center of the earth, the magnitude of the acceleration due to gravity is gR_E^2/s^2, where R_E is the radius of the earth. (See the discussion of gravity in Section 12.5.) If a spacecraft is a distance s_0 from the center of the earth (Fig. 13.16), what outward velocity v_0 must it be given to reach a specified distance h from the earth's center?

Strategy
The acceleration is known as a function of the position s. We can apply the chain rule and separate variables as shown in Table 13.1, then integrate to determine the velocity as a function of s.

Solution
The acceleration due to gravity is *toward* the center of the earth:

$$a = -\frac{gR_E^2}{s^2}.$$

Applying the chain rule results in

$$a = \frac{dv}{dt} = \frac{dv}{ds}\frac{ds}{dt} = \frac{dv}{ds}v = -\frac{gR_E^2}{s^2}.$$

Separating variables, we obtain

$$v\,dv = -\frac{gR_E^2}{s^2}\,ds.$$

We integrate this equation using the initial condition ($v = v_0$ when $s = s_0$) as the lower limits and the final condition ($v = 0$ when $s = h$) as the upper limits:

$$\int_{v_0}^{0} v\,dv = -\int_{s_0}^{h} \frac{gR_E^2}{s^2}\,ds,$$

$$\left[\frac{v^2}{2}\right]_{v_0}^{0} = gR_E^2\left[\frac{1}{s}\right]_{s_0}^{h},$$

$$0 - \frac{v_0^2}{2} = gR_E^2\left(\frac{1}{h} - \frac{1}{s_0}\right).$$

Solving for v_0, we obtain the initial velocity necessary for the spacecraft to reach a distance h:

$$v_0 = \sqrt{2gR_E^2\left(\frac{1}{s_0} - \frac{1}{h}\right)}.$$

Critical Thinking
We can make an interesting and important observation from the result of this example. Notice that as the distance h increases, the necessary initial velocity v_0 approaches a finite limit. This limit,

$$v_{esc} = \lim_{h \to \infty} v_0 = \sqrt{\frac{2gR_E^2}{s_0}},$$

is called the *escape velocity*. In the absence of other effects, an object with this initial velocity will continue moving outward indefinitely. The existence of an escape velocity makes it feasible to send spacecraft to other planets. Once escape velocity is attained, it isn't necessary to expend additional fuel to keep going.

Problems

13.40 An engineer designing a system to control a router for a machining process models the system so that the router's acceleration during an interval of time is given by $a = -0.4v$ in/s^2, where v is the velocity of the router in in/s. When $t = 0$, the position is $s = 0$ and the velocity is $v = 2$ in/s. What is the velocity at $t = 3$ s?

13.41 What is the position of the router in Problem 13.40 at $t = 3$ s?

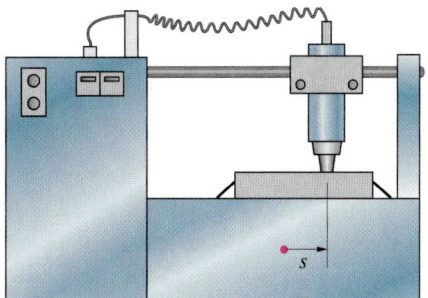

Problems 13.40/13.41

13.42 The boat is moving at 10 m/s when its engine is shut down. Due to hydrodynamic drag, its subsequent acceleration is $a = -0.05v^2$ m/s^2, where v is the velocity of the boat in m/s. What is the boat's velocity 4 s after the engine is shut down?

13.43 In Problem 13.42, what distance does the boat move in the 4 s following the shutdown of its engine?

Problems 13.42/13.43

13.44 A steel ball is released from rest in a container of oil. Its downward acceleration is $a = 2.4 - 0.6v$ in/s^2, where v is the ball's velocity in in/s. What is the ball's downward velocity 2 s after it is released?

13.45 In Problem 13.44, what distance does the ball fall in the first 2 s after its release?

Problems 13.44/13.45

13.46 The greatest ocean depth yet discovered is the Marianas Trench in the western Pacific Ocean. A steel ball released at the surface requires 64 min to reach the bottom. The ball's downward acceleration is $a = 0.9g - cv$, where $g = 9.81$ m/s^2 and the constant $c = 3.02$ s^{-1}. What is the depth of the Marianas Trench in kilometers?

13.47 The acceleration of a regional airliner during its takeoff run is $a = 14 - 0.0003v^2$ ft/s^2, where v is its velocity in ft/s. How long does it take the airliner to reach its takeoff speed of 200 ft/s?

13.48 In Problem 13.47, what distance does the airliner require to take off?

13.49 A sky diver jumps from a helicopter and is falling straight down at 30 m/s when her parachute opens. From then on, her downward acceleration is approximately $a = g - cv^2$, where $g = 9.81$ m/s^2 and c is a constant. After an initial "transient" period, she descends at a nearly constant velocity of 5 m/s.

(a) What is the value of c, and what are its SI units?
(b) What maximum deceleration is the sky diver subjected to?
(c) What is her downward velocity when she has fallen 2 m from the point at which her parachute opens?

Problem 13.49

13.50 The rocket sled starts from rest and accelerates at $a = 30 + 2t$ m/s^2 until its velocity is 400 m/s. It then hits a water brake and its acceleration is $a = -0.003v^2$ m/s^2 until its velocity decreases to 100 m/s. What total distance does the sled travel?

13.51 In Problem 13.50, what is the sled's total time of travel?

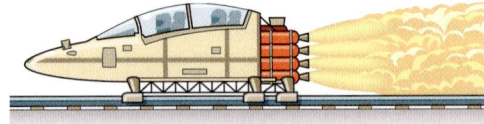

Problems 13.50/13.51

13.52 A car's acceleration is related to its position by $a = 0.01s$ m/s^2. When $s = 100$ m, the car is moving at 12 m/s. How fast is the car moving when $s = 420$ m?

13.53 Engineers analyzing the motion of a linkage determine that the velocity of an attachment point is given by $v = A + 4s^2$ ft/s, where A is a constant. When $s = 2$ ft, the acceleration of the point is measured and determined to be $a = 320$ ft/s^2. What is the velocity of the point when $s = 2$ ft?

13.54 The acceleration of an object is given by the function $a = 2s$ ft/s^2. When $t = 0$, $v = 1$ ft/s. What is the velocity when the object has moved 2 ft from its initial position?

13.55 Gas guns are used to investigate the properties of materials subjected to high-velocity impacts. A projectile is accelerated through the barrel of the gun by gas at high pressure. Assume that the acceleration of the projectile in m/s^2 is given by $a = c/s$, where s is the position of the projectile in the barrel in meters and c is a constant that depends on the initial gas pressure behind the projectile. The projectile starts from rest at $s = 1.5$ m and accelerates until it reaches the end of the barrel at $s = 3$ m. Determine the value of the constant c necessary for the projectile to leave the barrel with a velocity of 200 m/s.

13.56 If the propelling gas in the gas gun described in Problem 13.55 is air, a more accurate modeling of the acceleration of the projectile is obtained by assuming that the acceleration of the projectile is given by $a = c/s^\gamma$, where $\gamma = 1.4$ is the ratio of specific heats for air. (This means that an isentropic expansion process is assumed instead of the isothermal process assumed in Problem 13.55.) Determine the value of the constant c necessary for the projectile to leave the barrel with a velocity of 200 m/s.

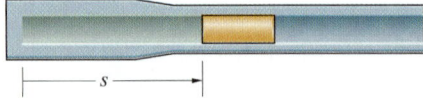

Problems 13.55/13.56

13.57 A spring–mass oscillator consists of a mass and a spring connected as shown. The coordinate s measures the displacement of the mass relative to its position when the spring is unstretched. If the spring is linear, the mass is subjected to a deceleration proportional to s. Suppose that $a = -4s$ m/s^2 and that you give the mass a velocity $v = 1$ m/s in the position $s = 0$.

(a) How far will the mass move to the right before the spring brings it to a stop?

(b) What will be the velocity of the mass when it has returned to the position $s = 0$?

13.58 In Problem 13.57, suppose that at $t = 0$ you release the mass from rest in the position $s = 1$ m. Determine the velocity of the mass as a function of s as it moves from the initial position to $s = 0$.

13.59 In Problem 13.57, suppose that at $t = 0$ you release the mass from rest in the position $s = 1$ m. Determine the position of the mass as a function of time as it moves from its initial position to $s = 0$.

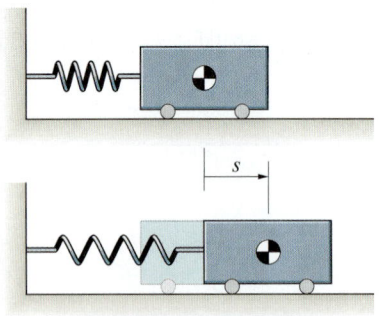

Problems 13.57–13.59

13.60 The mass is released from rest with the springs unstretched. Its downward acceleration is $a = 32.2 - 50s$ ft/s^2, where s is the position of the mass measured from the position in which it is released.

(a) How far does the mass fall?
(b) What is the maximum velocity of the mass as it falls?

13.61 Suppose that the mass in Problem 13.60 is in the position $s = 0$ and is given a downward velocity of 10 ft/s.

(a) How far does the mass fall?
(b) What is the maximum velocity of the mass as it falls?

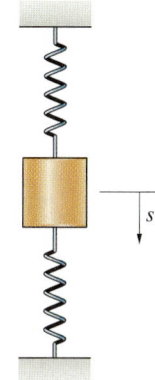

Problems 13.60/13.61

13.62 If a spacecraft is 100 mi above the surface of the earth, what initial velocity v_0 straight away from the earth would be required for the vehicle to reach the moon's orbit, 238,000 mi from the center of the earth? The radius of the earth is 3960 mi. Neglect the effect of the moon's gravity. (See Example 13.5.)

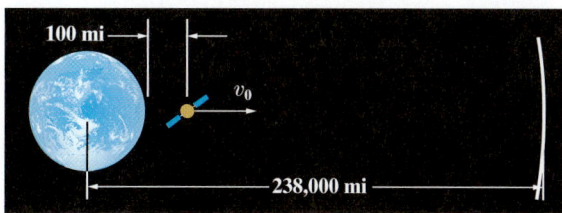

Problem 13.62

13.63 The moon's radius is 1738 km. The magnitude of the acceleration due to gravity of the moon at a distance s from the center of the moon is

$$\frac{4.89 \times 10^{12}}{s^2} \text{ m/s}^2.$$

Suppose that a spacecraft is launched straight up from the moon's surface with a velocity of 2000 m/s.

(a) What will the magnitude of its velocity be when it is 1000 km above the surface of the moon?
(b) What maximum height above the moon's surface will it reach?

13.64* The velocity of an object subjected only to the earth's gravitational field is

$$v = \left[v_0^2 + 2gR_E^2 \left(\frac{1}{s} - \frac{1}{s_0} \right) \right]^{1/2},$$

where s is the object's position relative to the center of the earth, v_0 is the object's velocity at position s_0, and R_E is the earth's radius. Using this equation, show that the object's acceleration is given as a function of s by $a = -gR_E^2/s^2$.

13.65 Suppose that a tunnel could be drilled straight through the earth from the North Pole to the South Pole and the air was evacuated. An object dropped from the surface would fall with acceleration $a = -gs/R_E$, where g is the acceleration of gravity at sea level, R_E is the radius of the earth, and s is the distance of the object from the center of the earth. (The acceleration due to gravity is equal to zero at the center of the earth and increases linearly with distance from the center.) What is the magnitude of the velocity of the dropped object when it reaches the center of the earth?

13.66* Determine the time in seconds required for the object in Problem 13.65 to fall from the surface of the earth to the center. The earth's radius is 6370 km.

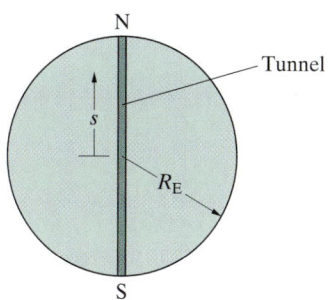

Problems 13.65/13.66

13.3 Curvilinear Motion

The motion of a point along a straight line can be described by the scalars s, v, and a. But if a point describes a *curvilinear* path relative to some reference frame, we must specify its motion in terms of its position, velocity, and acceleration vectors. Although the directions and magnitudes of these vectors do not depend on the coordinate system chosen to express them, we will show that the representations of these vectors are different in different coordinate systems. We can analyze many problems using cartesian coordinates, but some situations, including the motions of satellites and rotating machines, can be analyzed more conveniently using other coordinate systems. In the sections that follow, we show how curvilinear motions of points are analyzed in terms of various coordinate systems.

Cartesian Coordinates

Let $\mathbf{r}$ be the position vector of a point P relative to the origin O of a cartesian reference frame (Fig. 13.17). The components of $\mathbf{r}$ are the x, y, and z coordinates of P:

$$\mathbf{r} = x\mathbf{i} + y\mathbf{j} + z\mathbf{k}.$$

The unit vectors $\mathbf{i}$, $\mathbf{j}$, and $\mathbf{k}$ each have constant magnitude and constant direction relative to the reference frame, so the velocity of P relative to the reference frame is

$$\mathbf{v} = \frac{d\mathbf{r}}{dt} = \frac{dx}{dt}\mathbf{i} + \frac{dy}{dt}\mathbf{j} + \frac{dz}{dt}\mathbf{k}. \tag{13.21}$$

Expressing the velocity in terms of scalar components yields

$$\mathbf{v} = v_x\mathbf{i} + v_y\mathbf{j} + v_z\mathbf{k}, \tag{13.22}$$

from which we obtain scalar equations relating the components of the velocity to the coordinates of P:

$$v_x = \frac{dx}{dt}, \quad v_y = \frac{dy}{dt}, \quad v_z = \frac{dz}{dt}. \tag{13.23}$$

The acceleration of P is

$$\mathbf{a} = \frac{d\mathbf{v}}{dt} = \frac{dv_x}{dt}\mathbf{i} + \frac{dv_y}{dt}\mathbf{j} + \frac{dv_z}{dt}\mathbf{k}.$$

By expressing the acceleration in terms of scalar components as

$$\mathbf{a} = a_x\mathbf{i} + a_y\mathbf{j} + a_z\mathbf{k}, \tag{13.24}$$

we obtain the scalar equations

$$a_x = \frac{dv_x}{dt}, \quad a_y = \frac{dv_y}{dt}, \quad a_z = \frac{dv_z}{dt}. \tag{13.25}$$

Equations (13.23) and (13.25) describe the motion of a point relative to a cartesian coordinate system. Notice that the equations describing the motion in each coordinate direction are identical in form to the equations that describe the motion of a point along a straight line. As a consequence, the motion in each coordinate direction can often be analyzed using the methods we applied to straight-line motion.

The *projectile problem* is the classic example of this kind. If an object is thrown through the air and aerodynamic drag is negligible, the object accelerates

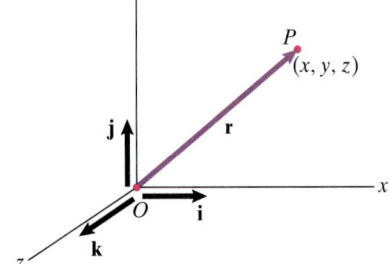

Figure 13.17
A cartesian coordinate system with origin O.

downward with the acceleration due to gravity. In terms of a fixed cartesian co-ordinate system with its y axis upward, the acceleration is given by $a_x = 0$, $a_y = -g$, and $a_z = 0$. Suppose that at $t = 0$ the projectile is located at the origin and has velocity v_0 in the x–y plane at an angle θ_0 above the horizontal (Fig. 13.18). At $t = 0$, $x = 0$ and $v_x = v_0 \cos \theta_0$. The acceleration in the x direction is zero—that is,

$$a_x = \frac{dv_x}{dt} = 0.$$

Therefore v_x is constant and remains equal to its initial value:

$$v_x = \frac{dx}{dt} = v_0 \cos \theta_0. \tag{13.26}$$

(This result may seem unrealistic. The reason is that your intuition, based upon everyday experience, accounts for drag, whereas the analysis presented here does not.) Integrating Eq. (13.26) yields

$$\int_0^x dx = \int_0^t v_0 \cos \theta_0 \, dt,$$

whereupon we obtain the x coordinate of the object as a function of time:

$$x = (v_0 \cos \theta_0)t. \tag{13.27}$$

Thus, we have determined the position and velocity of the projectile in the x direction as functions of time without considering the projectile's motion in the y or z direction.

At $t = 0$, $y = 0$ and $v_y = v_0 \sin \theta_0$. The acceleration in the y direction is

$$a_y = \frac{dv_y}{dt} = -g.$$

Integrating, we obtain

$$\int_{v_0 \sin \theta_0}^{v_y} dv_y = \int_0^t -g \, dt,$$

from which it follows that

$$v_y = \frac{dy}{dt} = v_0 \sin \theta_0 - gt. \tag{13.28}$$

Integrating this equation yields

$$\int_0^y dy = \int_0^t (v_0 \sin \theta_0 - gt) \, dt,$$

and we find that the y coordinate as a function of time is

$$y = (v_0 \sin \theta_0)t - \tfrac{1}{2}gt^2. \tag{13.29}$$

Notice from this analysis that the same vertical velocity and position are obtained by throwing the projectile straight up with initial velocity $v_0 \sin \theta_0$ (Figs. 13.19a, b). The vertical motion is completely independent of the horizontal motion.

By solving Eq. (13.27) for t and substituting the result into Eq. (13.29), we obtain an equation describing the parabolic trajectory of the projectile:

$$y = (\tan \theta_0)x - \frac{g}{2v_0^2 \cos^2 \theta_0}x^2. \tag{13.30}$$

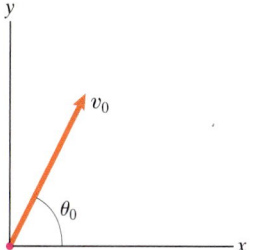

Figure 13.18
Initial conditions for a projectile problem.

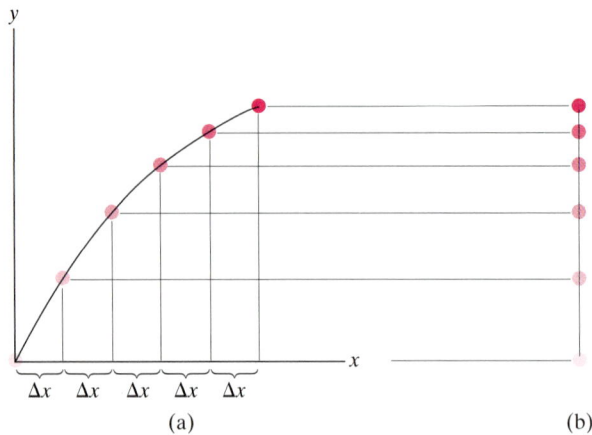

Figure 13.19
(a) Positions of the projectile at equal time
intervals Δt. The distance
$\Delta x = v_0(\cos \theta_0)\, \Delta t$.
(b) Positions at equal time intervals Δt of a
projectile given an initial vertical veloci-
ty equal to $v_0 \sin \theta_0$.

Study Questions

1. What are the components of the position vector $\mathbf{r}$ of a point P in terms of a cartesian reference frame?

2. In Eq. (13.21), we evaluated the time derivative of the position vector to obtain
   the velocity in terms of cartesian components. Why don't the time derivatives of
   the unit vectors appear in the result?

3. In the projectile problem, why is the horizontal component of the velocity constant?

Example 13.6   Analysis of Motion in Terms of Cartesian Components

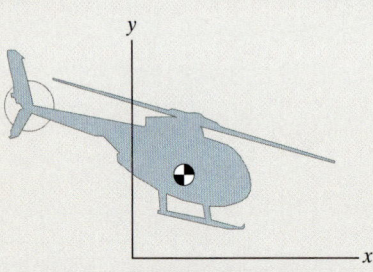

Figure 13.20

During a test flight in which a helicopter starts from rest at $t = 0$ at the origin
of the coordinate system shown in Fig. 13.20, accelerometers mounted on board
the craft indicate that its components of acceleration from $t = 0$ to $t = 10$ s
are closely approximated by

$$a_x = 0.6t \text{ m/s}^2,$$

$$a_y = 1.8 - 0.36t \text{ m/s}^2,$$

$$a_z = 0.$$

Determine the helicopter's velocity and position as functions of time.

Strategy
We can analyze the motion in each coordinate direction independently, integrating the acceleration to determine the velocity and then integrating the velocity to determine the position.

Solution
The velocity is zero at $t = 0$, and we assume that $x = y = z = 0$ at $t = 0$.
The acceleration in the x direction is

$$a_x = \frac{dv_x}{dt} = 0.6t \text{ m/s}^2.$$

Integrating with respect to time yields

$$\int_0^{v_x} dv_x = \int_0^t 0.6t \, dt,$$

and we obtain the velocity component v_x as a function of time:

$$v_x = \frac{dx}{dt} = 0.3t^2 \text{ m/s}.$$

Integrating again results in

$$\int_0^x dx = \int_0^t 0.3t^2 \, dt,$$

and we obtain x as a function of time:

$$x = 0.1t^3 \text{ m}.$$

Now we analyze the motion in the y direction in the same way. The acceleration is

$$a_y = \frac{dv_y}{dt} = 1.8 - 0.36t \text{ m/s}^2.$$

Integrating, we have

$$\int_0^{v_y} dv_y = \int_0^t (1.8 - 0.36t) \, dt,$$

and we obtain the velocity:

$$v_y = \frac{dy}{dt} = 1.8t - 0.18t^2 \text{ m/s}.$$

Integrating again yields

$$\int_0^y dy = \int_0^t (1.8t - 0.18t^2) \, dt,$$

which determines the position:

$$y = 0.9t^2 - 0.06t^3 \text{ m}.$$

You can easily show that the z components of the velocity and position are $v_z = 0$ and $z = 0$. We show the position of the helicopter as a function of time in Fig. a.

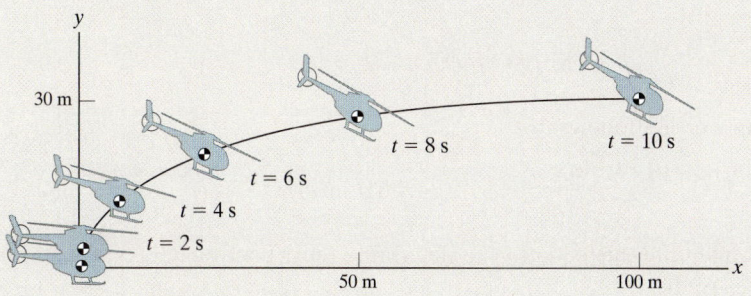

(a) Position of the helicopter at 2-s intervals.

Critical Thinking

This example demonstrates how inertial navigation systems work. They contain accelerometers that measure the x, y, and z components of acceleration. (Gyroscopes maintain the alignments of the accelerometers.) By integrating the acceleration components twice with respect to time, the systems compute changes in the x, y, and z coordinates of the airplane or ship.

Example 13.7 | **A Projectile Problem**

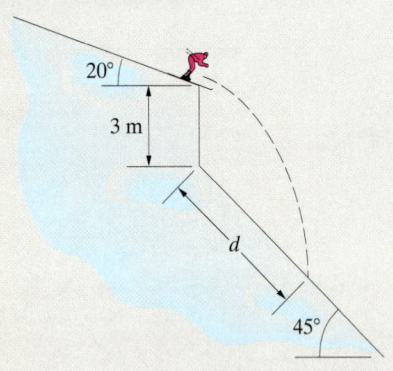

20°

3 m

d

45°

Figure 13.21

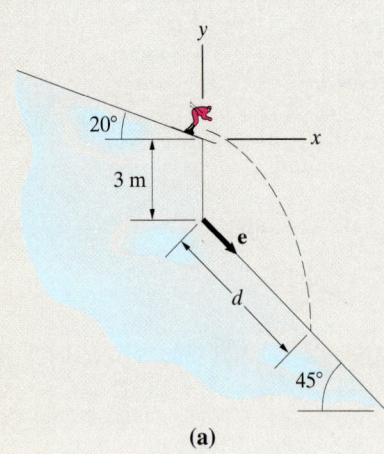

y

20°

3 m

e

d

45°

(a)

The skier in Fig. 13.21 leaves the 20° surface at 10 m/s.
(a) Determine the distance d to the point where he lands.
(b) What are the magnitudes of his components of velocity parallel and perpendicular to the 45° surface just before he lands?

Strategy

(a) By neglecting aerodynamic drag and treating the skier as a projectile, we can determine his velocity and position as functions of time. Using the equation describing the straight surface on which he lands, we can relate his horizontal and vertical coordinates at impact and thereby obtain an equation for the time at which he lands. Knowing the time, we can determine his position and velocity.
(b) We can determine his velocity parallel and perpendicular to the 45° surface by using the result that the component of a vector **U** in the direction of a unit vector **e** is $(\mathbf{e} \cdot \mathbf{U})\mathbf{e}$.

Solution

(a) In Fig. a, we introduce a coordinate system with its origin where the skier leaves the surface. His components of velocity at that instant $(t = 0)$ are

$$v_x = 10 \cos 20° = 9.40 \text{ m/s}$$

and

$$v_y = -10 \sin 20° = -3.42 \text{ m/s}.$$

The x component of acceleration is zero, so v_x is constant and the skier's x coordinate as a function of time is

$$x = 9.40t \text{ m}.$$

The y component of acceleration is

$$a_y = \frac{dv_y}{dt} = -9.81 \text{ m/s}^2.$$

Integrating to determine v_y as a function of time, we obtain

$$\int_{-3.42}^{v_y} dv_y = \int_0^t -9.81 \, dt,$$

from which it follows that

$$v_y = \frac{dy}{dt} = -3.42 - 9.81t \text{ m/s}.$$

We integrate this equation to determine the y coordinate as a function of time. We have

$$\int_0^y dy = \int_0^t (-3.42 - 9.81t) \, dt,$$

yielding

$$y = -3.42t - 4.905t^2 \text{ m}.$$

The slope of the surface on which the skier lands is -1, so the linear equation describing it is $y = (-1)x + A$, where A is a constant. At $x = 0$, the y

coordinate of the surface is -3 m, so $A = -3$ m and the equation describing the 45° surface is

$$y = -x - 3 \text{ m.}$$

Substituting our equations for x and y as functions of time into this equation, we obtain an equation for the time at which the skier lands:

$$-3.42t - 4.905t^2 = -9.40t - 3.$$

Solving for t, we get $t = 1.60$ s. Therefore, his coordinates when he lands are

$$x = 9.40(1.60) = 15.0 \text{ m}$$

and

$$y = -3.42(1.60) - 4.905(1.60)^2 = -18.0 \text{ m,}$$

and the distance d is

$$d = \sqrt{(15.0)^2 + (18.0 - 3)^2} = 21.3 \text{ m.}$$

(b) The components of the skier's velocity just before he lands are

$$v_x = 9.40 \text{ m/s}$$

and

$$v_y = -3.42 - 9.81(1.60) = -19.1 \text{ m/s,}$$

and the magnitude of his velocity is $|\mathbf{v}| = \sqrt{(9.40)^2 + (-19.1)^2} = 21.3$ m/s. Let $\mathbf{e}$ be a unit vector that is parallel to the slope on which he lands (Fig. a):

$$\mathbf{e} = \cos 45°\mathbf{i} - \sin 45°\mathbf{j}.$$

The component of the velocity parallel to the surface is

$$(\mathbf{e} \cdot \mathbf{v})\mathbf{e} = [(\cos 45°\mathbf{i} - \sin 45°\mathbf{j}) \cdot (9.40\mathbf{i} - 19.1\mathbf{j})]\mathbf{e}$$

$$= 20.2\mathbf{e} \text{ (m/s).}$$

The magnitude of the skier's velocity parallel to the surface is 20.2 m/s. Therefore, the magnitude of his velocity perpendicular to the surface is

$$\sqrt{|\mathbf{v}|^2 - (20.2)^2} = 6.88 \text{ m/s.}$$

Critical Thinking

The key to solving this problem was that we knew the skier's acceleration. Knowing the acceleration, we were able to determine the components of his velocity and position as functions of time. Notice how we determined the position at which he landed on the slope. We knew that at the instant he landed, his x and y coordinates specified a point on the straight line defining the surface of the slope. By substituting his x and y coordinates as functions of time into the equation for the straight line defining the slope, we were able to solve for the time at which he landed. Knowing the time, we could determine his velocity and position at that instant.

Problems

13.67 The coordinates (in meters) of a point moving in the x–y plane are given as functions of time by $x = 20t^2 - 160$ and $y = t^3 + 40t$. Determine the magnitudes of the velocity and acceleration of the point at $t = 2$ s.

13.68 In terms of a particular reference frame, the position of the center of mass of the F-14 at the time shown ($t = 0$) is $\mathbf{r} = 10\mathbf{i} + 6\mathbf{j} + 22\mathbf{k}$ (m). The velocity from $t = 0$ to $t = 4$ s is $\mathbf{v} = (52 + 6t)\mathbf{i} + (12 + t^2)\mathbf{j} - (4 + 2t^2)\mathbf{k}$ (m/s). What is the position of the center of mass of the plane at $t = 4$ s?

Problem 13.68

13.69 The acceleration of an object moving in the x–y plane is $\mathbf{a} = (4t - 2)\mathbf{i} + (-2t^2 + 4)\mathbf{j}$ (ft/s^2). At $t = 0$, its position is $\mathbf{r} = 3\mathbf{i} - 2\mathbf{j}$ (ft) and its velocity is $\mathbf{v} = 6\mathbf{i} + 8\mathbf{j}$ (ft/s). What are the position and velocity of the object at $t = 3$ s?

13.70 A projectile is launched from ground level with initial velocity $v_0 = 20$ m/s. Determine its range R if (a) $\theta_0 = 30°$, (b) $\theta_0 = 45°$, and (c) $\theta_0 = 60°$.

13.71 A projectile is launched from ground level with an initial velocity $v_0 = 20$ m/s. What initial angle θ_0 above the horizontal causes the range R to be a maximum, and what is the maximum range?

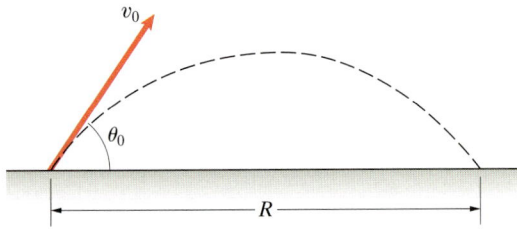

Problems 13.70/13.71

13.72 Suppose that you are designing a mortar to launch a rescue line from coast guard vessels to ships in distress. The light line is attached to a weight fired by the mortar. Neglect aerodynamic drag and the weight of the line for your preliminary analysis. If you want the line to be able to reach a ship 300 ft away when the mortar is fired at 45° above the horizontal, what muzzle velocity is required?

13.73 In Problem 13.72, what maximum height above the point from which it was fired is reached by the weight?

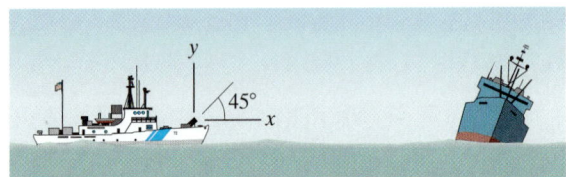

Problems 13.72/13.73

13.74 When the athlete releases the shot, it is 7 ft above the ground. The horizontal distance the shot travels from the point of release to the point where it hits the ground is 60 ft. What was the initial velocity v_0 of the shot?

Problem 13.74

13.75 A pilot wants to drop survey markers at remote locations in the Australian outback. If he flies at at a constant velocity $v_0 = 40$ m/s at altitude $h = 30$ m and the marker is released with zero velocity relative to the plane, at what horizontal distance d from the desired impact point should the marker be released?

Problem 13.75

13.76 If the pitching wedge the golfer is using gives the ball an initial angle $\theta_0 = 50°$, what range of velocities v_0 will cause the ball to land within 3 ft of the hole? (Assume that the hole lies in the plane of the ball's trajectory.)

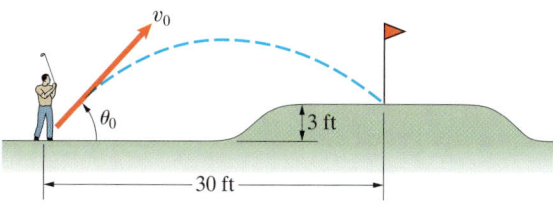

Problem 13.76

13.77 A batter strikes a baseball 3 ft above home plate and pops it up. The second baseman catches it 6 ft above second base 3.68 s after it was hit. What was the ball's initial velocity, and what was the angle between the ball's initial velocity vector and the horizontal?

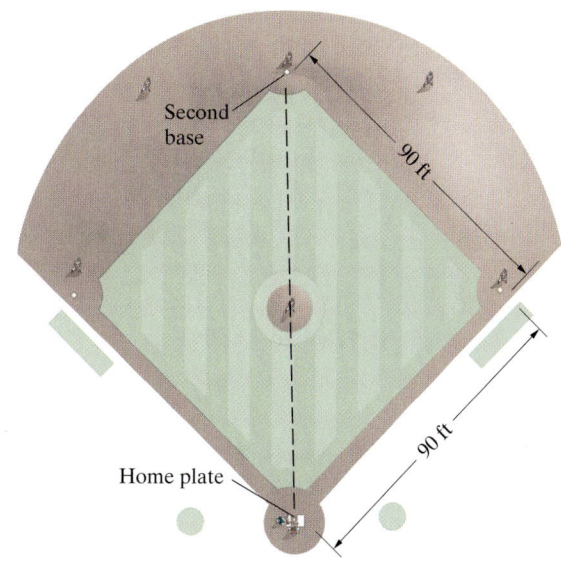

Problem 13.77

13.78 A baseball pitcher releases a fastball with an initial velocity $v_0 = 90$ mi/h. Let θ be the initial angle of the ball's velocity vector above the horizontal. When it is released, the ball is 6 ft above the ground and 58 ft from the batter's plate. The batter's strike zone extends from 1 ft 10 in above the ground to 4 ft 6 in above the ground. Neglecting aerodynamic effects, determine whether the ball will hit the strike zone (a) if $\theta = 1°$ and (b) if $\theta = 2°$.

13.79 In Problem 13.78, assume that the pitcher releases the ball at an angle $\theta = 1°$ above the horizontal, and determine the range of velocities v_0 (in ft/s) within which he must release the ball to hit the strike zone.

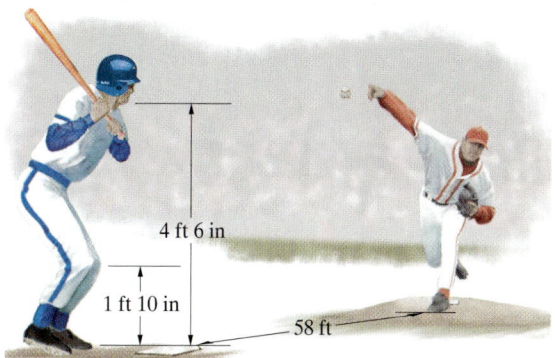

Problems 13.78/13.79

13.80 A zoology graduate student is provided with a bow and an arrow tipped with a syringe of sedative, and is assigned to measure the temperature of a black rhinoceros (*Diceros bicornis*). The range of his bow when it is fully drawn and aimed 45° above the horizontal is 100 m. A truculent rhino suddenly charges straight toward him at 30 km/h. If he fully draws his bow and aims 20° above the horizontal, how far away should the rhino be when the student releases the arrow?

Problem 13.80

13.81 The crossbar of the goalposts in American football is $y_c = 10$ ft above the ground. To kick a field goal, the kicker must make the ball go between the two uprights supporting the crossbar, and the ball must be above the crossbar when it does so. Suppose that the kicker attempts a 40-yd field goal ($x_c = 120$ ft) and kicks the ball with initial velocity $v_0 = 70$ ft/s and angle $\theta_0 = 40°$. By what vertical distance does the ball clear the crossbar?

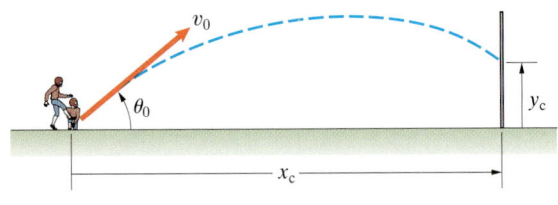

Problem 13.81

13.82* An American football quarterback stands at A. At the instant the quarterback throws the football, the receiver is at B running at 20 ft/s toward C, where he catches the ball. The ball is thrown at an angle of 45° above the horizontal, and it is thrown and caught at the same height above the ground. Determine the magnitude of the ball's initial velocity and the length of time it is in the air.

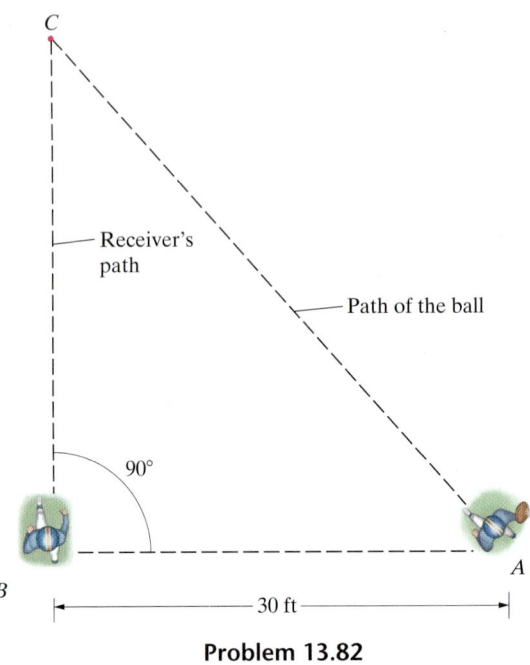

Problem 13.82

13.83 The cliff divers of Acapulco, Mexico, must time their dives so that they enter the water at the crest (high point) of a wave. The crests of the waves are 1 m above the mean water depth $h = 4$ m. The horizontal velocity of the waves is equal to $\sqrt{gh}$. The diver's aiming point is 2 m out from the base of the cliff. Assume that his velocity is horizontal when he begins the dive.

(a) What is the magnitude of the diver's velocity when he enters the water?

(b) How far from his aiming point must a wave crest be when he dives in order for him to enter the water at the crest?

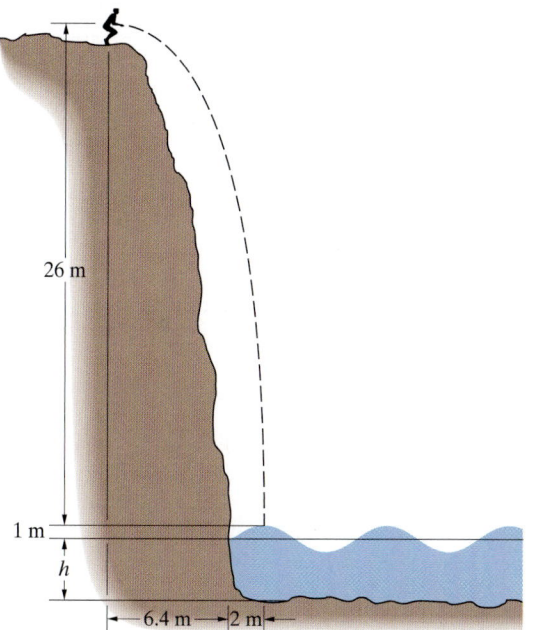

Problem 13.83

13.84 A projectile is launched at 10 m/s from a sloping surface. The angle $\alpha = 80°$. Determine the range R.

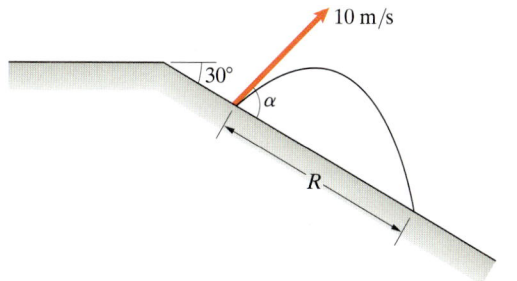

Problem 13.84

13.85 A projectile is launched at 100 ft/s at 60° above the horizontal. The surface on which it lands is described by the equation shown. Determine the x coordinate of the point of impact.

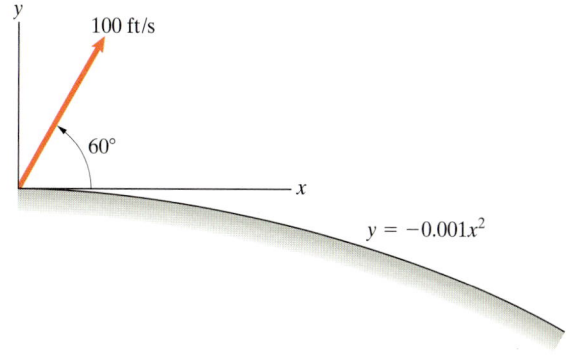

Problem 13.85

13.86 At $t = 0$, a steel ball in a tank of oil is given a horizontal velocity $\mathbf{v} = 2\mathbf{i}$ (m/s). The components of the ball's acceleration, in m/s^2, are $a_x = -1.2v_x$, $a_y = -8 - 1.2v_y$, and $a_z = -1.2v_z$. What is the velocity of the ball at $t = 1$ s?

13.87 In Problem 13.86, what is the position of the ball at $t = 1$ s relative to its position at $t = 0$?

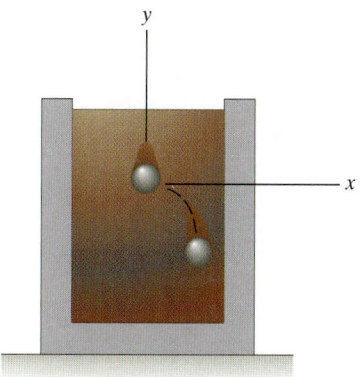

Problems 13.86/13.87

13.88 The point P moves along a circular path with radius R. Show that the magnitude of its velocity is $|\mathbf{v}| = R|d\theta/dt|$.

   Strategy: Use Eqs. (13.23).

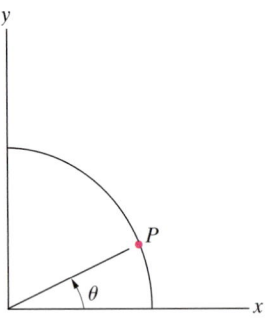

Problem 13.88

13.89 If $y = 150$ mm, $dy/dt = 300$ mm/s, and $d^2y/dt^2 = 0$, what are the magnitudes of the velocity and acceleration of point P?

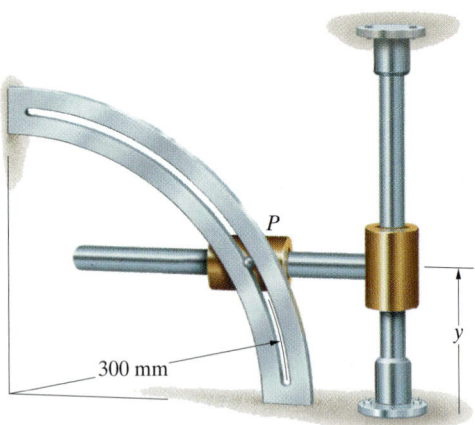

300 mm

Problem 13.89

13.90* A car travels at a constant speed of 100 km/h on a straight road of increasing grade whose vertical profile can be approximated by the equation shown. When the car's horizontal coordinate is $x = 400$ m, what is the car's acceleration?

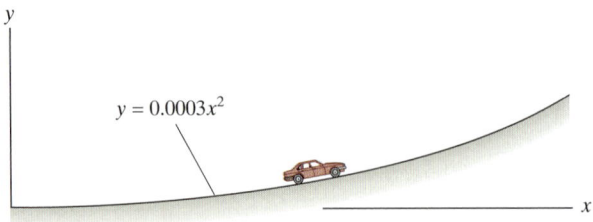

$y = 0.0003x^2$

Problem 13.90

13.91* Suppose that a projectile has the initial conditions shown in Fig. 13.18. Show that in terms of the $x'y'$ coordinate system with its origin at the highest point of the trajectory, the equation describing the trajectory is

$$y' = -\frac{g}{2v_0^2 \cos^2 \theta_0}(x')^2.$$

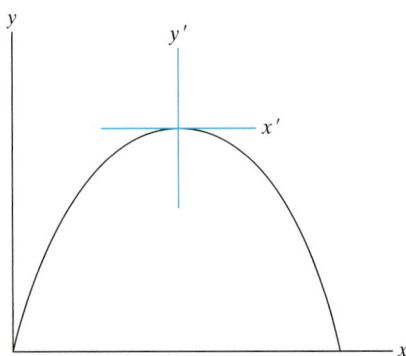

Problem 13.91

13.92* The acceleration components of a point are $a_x = -4\cos 2t$, $a_y = -4\sin 2t$, and $a_z = 0$. At $t = 0$, the position and velocity of the point are $\mathbf{r} = \mathbf{i}$ and $\mathbf{v} = 2\mathbf{j}$. Show that (a) the magnitude of the velocity is constant, (b) the velocity and acceleration vectors are perpendicular, (c) the magnitude of the acceleration is constant and points toward the origin, and (d) the trajectory of the point is a circle with its center at the origin.

Angular Motion

We have seen that in some cases the curvilinear motion of a point can be analyzed by using cartesian coordinates. In the sections that follow, we describe problems that can be analyzed more simply in terms of other coordinate systems. To help you understand our discussion of these alternative coordinate systems, we introduce two preliminary topics in this section: the angular motion of a line in a plane and the time derivative of a unit vector rotating in a plane.

Angular Motion of a Line   We can specify the angular position of a line L in a particular plane relative to a reference line L_0 in the plane by an angle θ (Fig. 13.22). The *angular velocity* of L relative to L_0 is defined by

$$\omega = \frac{d\theta}{dt}, \tag{13.31}$$

and the *angular acceleration* of L relative to L_0 is defined by

$$\alpha = \frac{d\omega}{dt} = \frac{d^2\theta}{dt^2}. \tag{13.32}$$

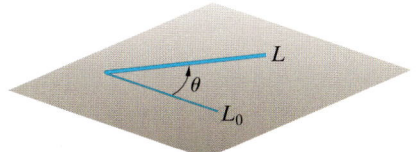

Figure 13.22
A line L and a reference line L_0 in a plane.

The dimensions of the angular position, angular velocity, and angular acceleration are rad, rad/s, and rad/s^2, respectively. Although these quantities are often expressed in terms of degrees or revolutions instead of radians, convert them into radians before using them in calculations.

Notice the analogy between Eqs. (13.31) and (13.32) and the equations relating the position, velocity, and acceleration of a point along a straight line (Table 13.2). In each case, the position is specified by a single scalar coordinate, which can be positive or negative. (In Fig. 13.22, the counterclockwise direction is positive.) Because the equations are identical in form, problems involving angular motion of a line can be analyzed by the same methods we applied to straight-line motion.

TABLE 13.2   The equations governing straight-line motion and the angular motion of a line are identical in form.

Straight-Line Motion	Angular Motion
$v = \dfrac{ds}{dt}$	$\omega = \dfrac{d\theta}{dt}$
$a = \dfrac{dv}{dt} = \dfrac{d^2s}{dt^2}$	$\alpha = \dfrac{d\omega}{dt} = \dfrac{d^2\theta}{dt^2}$

Rotating Unit Vector   The directions of the unit vectors $\mathbf{i}$, $\mathbf{j}$, and $\mathbf{k}$ relative to the cartesian reference frame are constant. However, in other coordinate systems, the unit vectors used to describe the motion of a point rotate as the point moves. To obtain expressions for the velocity and acceleration in such coordinate systems, we must know the time derivative of a rotating unit vector.

We can describe the angular motion of a unit vector $\mathbf{e}$ in a plane just as we described the angular motion of a line. The direction of $\mathbf{e}$ relative to a reference line L_0 is specified by the angle θ in Fig. 13.23a, and the rate of rotation of $\mathbf{e}$ relative to L_0 is specified by the angular velocity

$$\omega = \frac{d\theta}{dt}.$$

The time derivative of $\mathbf{e}$ is defined by

$$\frac{d\mathbf{e}}{dt} = \lim_{\Delta t \to 0} \frac{\mathbf{e}(t + \Delta t) - \mathbf{e}(t)}{\Delta t}.$$

Figure 13.23b shows the vector $\mathbf{e}$ at time t and at time $t + \Delta t$. The change in $\mathbf{e}$ during this interval is $\Delta \mathbf{e} = \mathbf{e}(t + \Delta t) - \mathbf{e}(t)$, and the angle through which $\mathbf{e}$

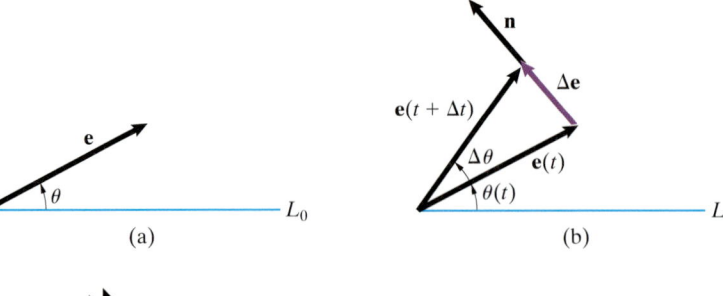

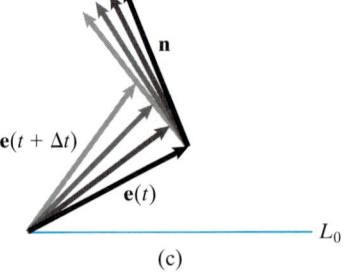

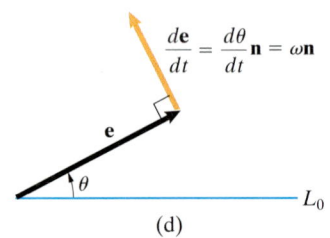

Figure 13.23
(a) A unit vector **e** and reference line L_0.
(b) The change $\Delta\mathbf{e}$ in **e** from t to $t + \Delta t$.
(c) As Δt goes to zero, **n** becomes perpendicular to $\mathbf{e}(t)$.
(d) The time derivative of **e**.

rotates is $\Delta\theta = \theta(t + \Delta t) - \theta(t)$. The triangle in Fig. 13.23b is isosceles, so

$$|\Delta\mathbf{e}| = 2|\mathbf{e}| \sin(\Delta\theta/2) = 2 \sin(\Delta\theta/2).$$

To write the vector $\Delta\mathbf{e}$ in terms of this expression, we introduce a unit vector **n** that points in the direction of $\Delta\mathbf{e}$ (Fig. 13.23b):

$$\Delta\mathbf{e} = |\Delta\mathbf{e}|\mathbf{n} = 2 \sin(\Delta\theta/2)\mathbf{n}.$$

In terms of this expression, the time derivative of **e** is

$$\frac{d\mathbf{e}}{dt} = \lim_{\Delta t\to 0} \frac{\Delta\mathbf{e}}{\Delta t} = \lim_{\Delta t\to 0} \frac{2 \sin(\Delta\theta/2)\mathbf{n}}{\Delta t}.$$

To evaluate the limit, we write it in the form

$$\frac{d\mathbf{e}}{dt} = \lim_{\Delta t\to 0} \frac{\sin(\Delta\theta/2)}{\Delta\theta/2} \frac{\Delta\theta}{\Delta t}\mathbf{n}.$$

In the limit as Δt approaches zero, $\sin(\Delta\theta/2)/(\Delta\theta/2) = 1$, $\Delta\theta/\Delta t = d\theta/dt$, and the unit vector **n** is perpendicular to $\mathbf{e}(t)$ (Fig. 13.23c). Therefore, the time derivative of **e** is

$$\frac{d\mathbf{e}}{dt} = \frac{d\theta}{dt}\mathbf{n} = \omega\mathbf{n}, \tag{13.33}$$

where **n** is a unit vector that is perpendicular to **e** and points in the positive θ direction (Fig. 13.23d). In the sections that follow, we use this result in deriving expressions for the velocity and acceleration of a point in different coordinate systems.

Study Questions

1. What are the definitions of the angular velocity and angular acceleration of a line?
2. Is the time derivative of a rotating unit vector a scalar or a vector?
3. Suppose that a unit vector is rotating in a plane with angular velocity ω. What do you know about the time derivative of the vector?

Example 13.8   Analysis of Angular Motion

The rotor of a jet engine is rotating at 10,000 rpm (revolutions per minute) when the fuel is shut off. The ensuing angular acceleration is $\alpha = -0.02\omega$, where ω is the angular velocity in rad/s.

(a) How long does it take the rotor to slow to 1000 rpm?

(b) How many revolutions does the rotor turn while decelerating to 1000 rpm?

Strategy

To analyze the angular motion of the rotor, we define a line L that is fixed to the rotor and perpendicular to its axis (Fig. 13.24). Then we examine the motion of L relative to the reference line L_0. The angular position, velocity, and acceleration of L describe the angular motion of the rotor.

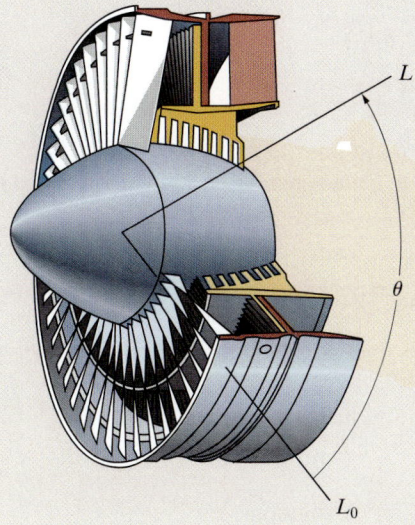

Figure 13.24
Introducing a line L and reference line L_0 to specify the angular position of the rotor.

Solution
The conversion from rpm to rad/s is

$$1 \text{ rpm} = (1 \text{ revolution/min})\left(\frac{2\pi \text{ rad}}{1 \text{ revolution}}\right)\left(\frac{1 \text{ min}}{60 \text{ s}}\right)$$

$$= \frac{\pi}{30} \text{rad/s}.$$

(a) The angular acceleration is

$$\alpha = \frac{d\omega}{dt} = -0.02\omega.$$

We separate variables to get

$$\frac{d\omega}{\omega} = -0.02 \, dt.$$

Then we integrate, defining $t = 0$ to be the time at which the fuel is turned off:

$$\int_{10,000\pi/30}^{1000\pi/30} \frac{d\omega}{\omega} = \int_0^t -0.02 \, dt.$$

Evaluating the integrals and solving for t, we obtain

$$t = \left(\frac{1}{0.02}\right) \ln\left(\frac{10,000\pi/30}{1000\pi/30}\right) = 115 \text{ s}.$$

(b) Writing the angular acceleration as

$$\alpha = \frac{d\omega}{dt} = \frac{d\omega}{d\theta}\frac{d\theta}{dt} = \frac{d\omega}{d\theta}\omega = -0.02\omega,$$

we separate variables to obtain

$$d\omega = -0.02 \, d\theta.$$

Next, we integrate, defining $\theta = 0$ to be the angular position at which the fuel is turned off:

$$\int_{10,000\pi/30}^{1000\pi/30} d\omega = \int_0^\theta -0.02 \, d\theta.$$

Solving for θ yields

$$\theta = \left(\frac{1}{0.02}\right)[(10,000\pi/30) - (1000\pi/30)]$$

$$= 15,000\pi \text{ rad} = 7500 \text{ revolutions}.$$

Critical Thinking
In this example the angular acceleration α of the rotor was specified as a function of its angular velocity ω. Notice that we solved it by applying the techniques described in Table 13.1 for straight-line motion in which the acceleration a is known as a function of the velocity v. Because of the correspondences shown in Table 13.2 between the equations for straight-line motion and those for angular motion, all of the methods you learned in Section 13.2 for solving problems involving straight-line motion can also be applied to angular motion.

Problems

13.93 When an airplane touches down at $t = 0$, a stationary wheel is subjected to a constant angular acceleration $\alpha = 110$ rad/s^2 until $t = 1$ s.

(a) What is the wheel's angular velocity at $t = 1$ s?
(b) At $t = 0$, the angle $\theta = 0$. Determine θ in radians and in revolutions at $t = 1$ s.

Problem 13.93

13.94 Let L be a line from the center of the earth to a fixed point on the equator, and let L_0 be a fixed reference direction. The figure shows the earth seen from above the North Pole.

(a) Is $d\theta/dt$ positive or negative?
(b) What is the magnitude of $d\theta/dt$ in rad/s?

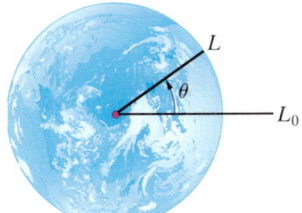

Problem 13.94

13.95 The angular acceleration of the line L relative to the line L_0 is given as a function of time by $\alpha = 2.5 - 1.2t$ rad/s^2. At $t = 0$, $\theta = 0$ and the angular velocity of L relative to L_0 is $\omega = 5$ rad/s. Determine θ and ω at $t = 3$ s.

13.96 The angular acceleration of the line L relative to the line L_0 is given $\alpha = -2\omega^2$ rad/s^2 where ω is the angular velocity in rad/s. When $\theta = 30°$, the angular velocity is 10 rad/s. What is the angular velocity when $\theta = 60°$?

Strategy: Use the chain rule to write the angular acceleration as

$$\alpha = \frac{d\omega}{dt} = \frac{d\omega}{d\theta}\frac{d\theta}{dt} = \frac{d\omega}{d\theta}\omega.$$

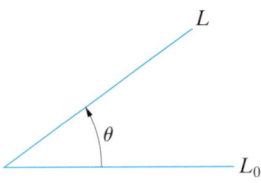

Problems 13.95/13.96

13.97 The stationary astronaut activates hydrogen peroxide jets that give him a constant angular acceleration $\alpha = 0.4$ rad/s^2 for 2 s. He then turns off the jets and rotates with constant angular velocity. From the time he activates the jets, how long does it take him to rotate 180° relative to his original position?

Problem 13.97

13.98 The hydroelectric generator is started from rest. Its angular acceleration is given by $\alpha = 6 - 0.2\omega$ rad/s^2, where ω is the angular velocity in rad/s. What is the angular velocity of the generator 10 s after it is started?

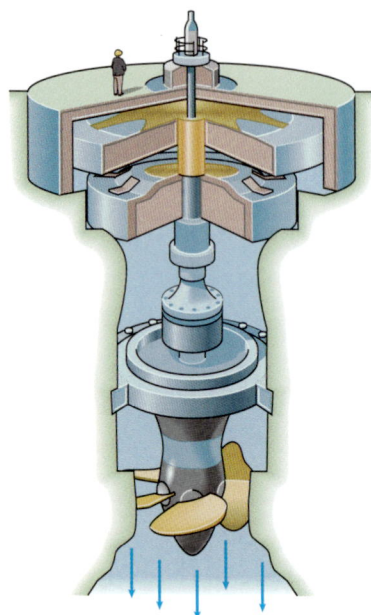

Problem 13.98

13.99 The rotor of an electric generator is rotating at 200 rpm when the motor is turned off. Due to frictional effects, the angular deceleration of the rotor after it is turned off is $\alpha = -0.01\omega$ rad/s^2, where ω is the angular velocity in rad/s. How many revolutions does the rotor turn after the motor is turned off?

13.100 The needle of a measuring instrument is connected to a *torsional spring* that gives it an angular acceleration $\alpha = -4\theta$ rad/s^2, where θ is the needle's angular position in radians relative to a reference direction. The needle is given an angular velocity $\omega = 2$ rad/s in the position $\theta = 0$.

(a) What is the magnitude of the needle's angular velocity when $\theta = 30°$?

(b) What maximum angle θ does the needle reach before it rebounds?

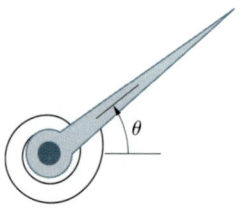

Problem 13.100

13.101 The angle θ measures the direction of the unit vector $\mathbf{e}$ relative to the x axis. The angular velocity of $\mathbf{e}$ is $\omega = d\theta/dt = 2$ rad/s, a constant. Determine the derivative $d\mathbf{e}/dt$ when $\theta = 90°$ in two ways:

(a) Use Eq. (13.33).

(b) Express the vector $\mathbf{e}$ in terms of its x and y components and take the time derivative of $\mathbf{e}$.

13.102 The angle θ measures the direction of the unit vector $\mathbf{e}$ relative to the x axis. The angle θ is given as a function of time by $\theta = 2t^2$ rad. What is the vector $d\mathbf{e}/dt$ at $t = 4$ s?

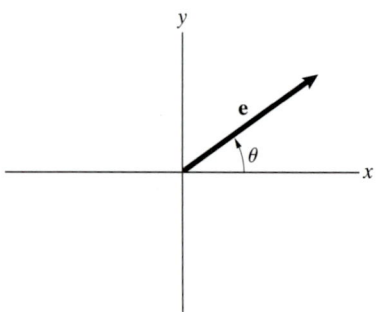

Problems 13.101/13.102

13.103 The line OP is of constant length R. The angle $\theta = \omega_0 t$, where ω_0 is a constant.

(a) Use the relations

$$v_x = \frac{dx}{dt} \quad \text{and} \quad v_y = \frac{dy}{dt}$$

to determine the velocity of point P relative to O.

(b) Use Eq. (13.33) to determine the velocity of point P relative to O, and confirm that your result agrees with the result of part (a).

Strategy: In part (b), write the position vector of P relative to O as $\mathbf{r} = R\mathbf{e}$, where $\mathbf{e}$ is a unit vector that points from O toward P.

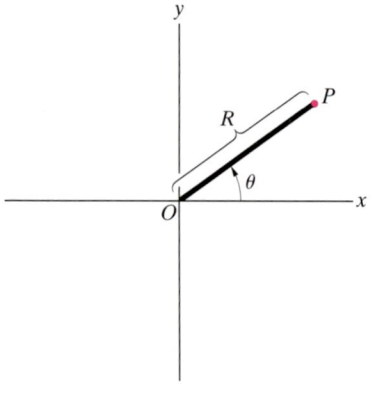

Problem 13.103

Normal and Tangential Components

In this method of describing curvilinear motion, we specify the position of a point by a coordinate measured *along its path* and express the velocity and acceleration in terms of their components tangential and normal (perpendicular) to the path. Normal and tangential components are particularly useful when a point moves along a circular path. Furthermore, they give us unique insight into the character of the velocity and acceleration in curvilinear motion. We first discuss motion in a planar path because of its conceptual simplicity.

Planar Motion   Consider a point P moving along a plane, curvilinear path relative to some reference frame (Fig. 13.25a). The position vector $\mathbf{r}$ specifies the position of P relative to the reference point O, and the coordinate s measures P's position along the path relative to a point O' on the path. The velocity of P relative to O is

$$\mathbf{v} = \frac{d\mathbf{r}}{dt} = \lim_{\Delta t \to 0} \frac{\mathbf{r}(t + \Delta t) - \mathbf{r}(t)}{\Delta t} = \lim_{\Delta t \to 0} \frac{\Delta \mathbf{r}}{\Delta t}, \tag{13.34}$$

where $\Delta \mathbf{r} = \mathbf{r}(t + \Delta t) - \mathbf{r}(t)$ (Fig. 13.25b). We denote the distance traveled along the path from t to $t + \Delta t$ by Δs. By introducing a unit vector $\mathbf{e}$ defined to point in the direction of $\Delta \mathbf{r}$, we can write Eq. (13.34) as

$$\mathbf{v} = \lim_{\Delta t \to 0} \frac{\Delta s}{\Delta t} \mathbf{e}.$$

As Δt approaches zero, $\Delta s / \Delta t$ becomes ds/dt and $\mathbf{e}$ becomes a unit vector tangent to the path at the position of P at time t, which we denote by $\mathbf{e}_t$ (Fig. 13.25c):

$$\mathbf{v} = v\mathbf{e}_t = \frac{ds}{dt} \mathbf{e}_t. \tag{13.35}$$

The velocity of a point in curvilinear motion is a vector whose magnitude equals the rate of change of distance traveled along the path and whose direction is tangent to the path.

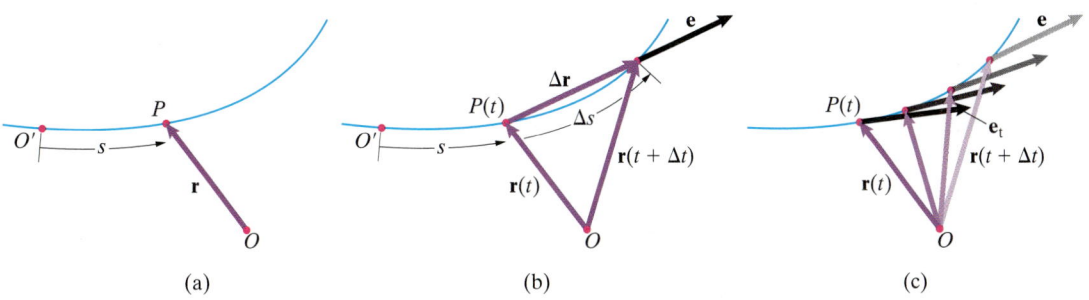

(a) (b) (c)

Figure 13.25
(a) The position of P along its path is specified by the coordinate s.
(b) Position of P at time t and at time $t + \Delta t$.
(c) The limit of $\mathbf{e}$ as $\Delta t \to 0$ is a unit vector tangent to the path.

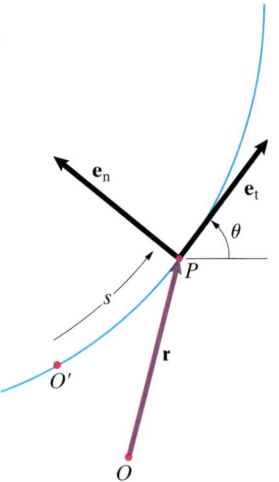

Figure 13.26
The path angle θ.

To determine the acceleration of P, we take the time derivative of Eq. (13.35):

$$\mathbf{a} = \frac{d\mathbf{v}}{dt} = \frac{dv}{dt}\mathbf{e}_t + v\frac{d\mathbf{e}_t}{dt}. \tag{13.36}$$

If the path is not a straight line, the unit vector $\mathbf{e}_t$ rotates as P moves. As a consequence, the time derivative of $\mathbf{e}_t$ is not zero. In the previous section, we derived an expression for the time derivative of a rotating unit vector in terms of the unit vector's angular velocity [Eq. (13.33)]. To use that result, we define the *path angle* θ specifying the direction of $\mathbf{e}_t$ relative to a reference line (Fig. 13.26). Then, from Eq. (13.33), the time derivative of $\mathbf{e}_t$ is

$$\frac{d\mathbf{e}_t}{dt} = \frac{d\theta}{dt}\mathbf{e}_n,$$

where $\mathbf{e}_n$ is a unit vector that is normal to $\mathbf{e}_t$ and points in the positive θ direction if $d\theta/dt$ is positive. Substituting this expression into Eq. (13.36), we obtain the acceleration of P:

$$\mathbf{a} = \frac{dv}{dt}\mathbf{e}_t + v\frac{d\theta}{dt}\mathbf{e}_n. \tag{13.37}$$

We can derive this result in another way that is less rigorous, but that gives additional insight into the meanings of the tangential and normal components of the acceleration. Figure 13.27a shows the velocity of P at times t and $t + \Delta t$. In Fig. 13.27b, you can see that the change in the velocity, $\mathbf{v}(t + \Delta t) - \mathbf{v}(t)$, consists of two components. The component Δv, which is tangent to the path at time t, is due to the change in the magnitude of the velocity. The component $v\,\Delta\theta$, which is perpendicular to the path at time t, is due to the change in the direction of the velocity vector. Thus, the change in the velocity is (approximately)

$$\mathbf{v}(t + \Delta t) - \mathbf{v}(t) = \Delta v\,\mathbf{e}_t + v\Delta\theta\,\mathbf{e}_n.$$

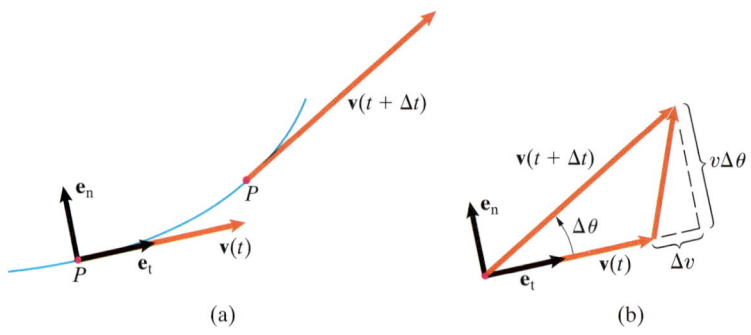

(a)   (b)

Figure 13.27
(a) Velocity of P at t and at $t + \Delta t$.
(b) The tangential and normal components of the change in the velocity.

To obtain the acceleration, we divide this expression by Δt and take the limit as $\Delta t \to 0$:

$$\mathbf{a} = \lim_{\Delta t \to 0} \frac{\Delta \mathbf{v}}{\Delta t} = \lim_{\Delta t \to 0} \left(\frac{\Delta v}{\Delta t} \mathbf{e}_t + v \frac{\Delta \theta}{\Delta t} \mathbf{e}_n \right)$$

$$= \frac{dv}{dt} \mathbf{e}_t + v \frac{d\theta}{dt} \mathbf{e}_n.$$

Thus, we again obtain Eq. (13.37). However, this derivation clearly points out that the tangential component of the acceleration arises from the rate of change of the magnitude of the velocity, whereas the normal component arises from the rate of change in the direction of the velocity vector. Notice that if the path is a straight line at time t, the normal component of the acceleration equals zero, because in that case $d\theta/dt$ is zero.

We can express the acceleration in another form that often is more convenient to use. Figure 13.28 shows the positions on the path reached by P at times t and $t + dt$. If the path is curved, straight lines extended from these points perpendicular to the path will intersect as shown. The distance ρ from the path to the point where these two lines intersect is called the *instantaneous radius of curvature* of the path. (If the path is circular, ρ is simply the radius of the path.) The angle $d\theta$ is the change in the path angle, and ds is the distance traveled from t to $t + \Delta t$. You can see from the figure that ρ is related to ds by

$$ds = \rho \, d\theta.$$

Dividing by dt, we obtain

$$\frac{ds}{dt} = v = \rho \frac{d\theta}{dt}.$$

Using this relation, we can write Eq. (13.37) as

$$\mathbf{a} = \frac{dv}{dt} \mathbf{e}_t + \frac{v^2}{\rho} \mathbf{e}_n.$$

For a given value of v, the normal component of the acceleration depends on the instantaneous radius of curvature. The greater the curvature of the path, the greater is the normal component of acceleration. When the acceleration is expressed in this way, the unit vector $\mathbf{e}_n$ must be defined to point toward the *concave* side of the path (Fig. 13.29).

Thus, the velocity and acceleration in terms of normal and tangential components are (Fig. 13.30)

$$\mathbf{v} = v\mathbf{e}_t = \frac{ds}{dt} \mathbf{e}_t \tag{13.38}$$

and

$$\mathbf{a} = a_t \mathbf{e}_t + a_n \mathbf{e}_n, \tag{13.39}$$

where

$$a_t = \frac{dv}{dt} \quad \text{and} \quad a_n = v \frac{d\theta}{dt} = \frac{v^2}{\rho}. \tag{13.40}$$

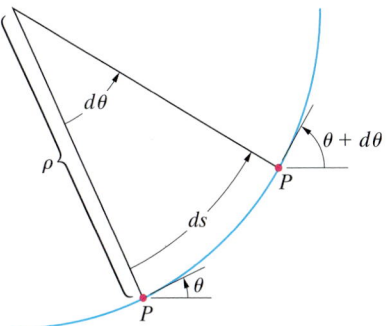

Figure 13.28
The instantaneous radius of curvature, ρ.

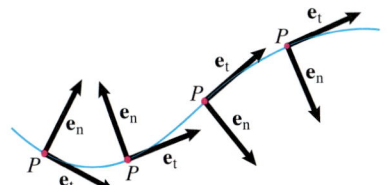

Figure 13.29
The unit vector normal to the path points toward the concave side.

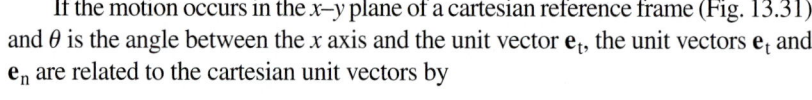

(a) (b)

Figure 13.30
Normal and tangential components of the velocity (a) and acceleration (b).

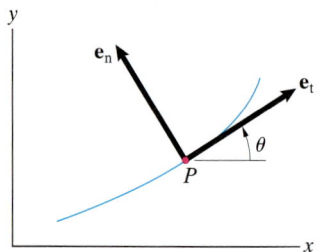

Figure 13.31
A point P moving in the x–y plane.

If the motion occurs in the x–y plane of a cartesian reference frame (Fig. 13.31) and θ is the angle between the x axis and the unit vector $\mathbf{e}_t$, the unit vectors $\mathbf{e}_t$ and $\mathbf{e}_n$ are related to the cartesian unit vectors by

$$\mathbf{e}_t = \cos\theta\,\mathbf{i} + \sin\theta\,\mathbf{j}$$

and

$$\mathbf{e}_n = -\sin\theta\,\mathbf{i} + \cos\theta\,\mathbf{j}. \tag{13.41}$$

If the path in the x–y plane is described by a function $y = y(x)$, it can be shown that the instantaneous radius of curvature is given by

$$\rho = \frac{\left[1 + \left(\dfrac{dy}{dx}\right)^2\right]^{3/2}}{\left|\dfrac{d^2y}{dx^2}\right|}. \tag{13.42}$$

Circular Motion If a point P moves in a plane circular path of radius R (Fig. 13.32), the distance s is related to the angle θ by

$$s = R\theta \qquad \text{(circular path).}$$

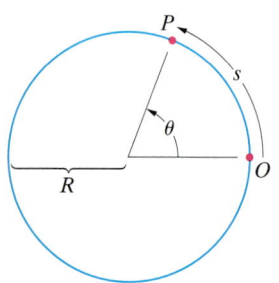

Figure 13.32
A point moving in a circular path.

Using this relation, then, we can specify the position of P along the circular path by either s or θ. Taking the time derivative of the equation, we obtain a relation between $v = ds/dt$ and the angular velocity of the line from the center of the path to P:

$$v = R\frac{d\theta}{dt} = R\omega \qquad \text{(circular path)} \tag{13.43}$$

Taking another time derivative, we obtain a relation between the tangential component of the acceleration $a_t = dv/dt$ and the angular acceleration:

$$a_t = R\frac{d\omega}{dt} = R\alpha \qquad \text{(circular path)} \tag{13.44}$$

For this circular path, the instantaneous radius of curvature $\rho = R$, so the normal component of the acceleration is

$$a_n = \frac{v^2}{R} = R\omega^2 \qquad \text{(circular path)} \tag{13.45}$$

Because problems involving circular motion of a point are so common, these relations are worth remembering. But you must be careful to use them *only* when the path is circular.

Three-Dimensional Motion Although most applications of normal and tangential components involve the motion of a point in a plane, we briefly discuss three-dimensional motion for the insight it provides into the nature of the velocity and acceleration. If we consider the motion of a point along a three-dimensional path relative to some reference frame, the steps leading to Eq. (13.38) are unaltered. The velocity is

$$\mathbf{v} = v\mathbf{e}_t = \frac{ds}{dt}\mathbf{e}_t, \tag{13.46}$$

where $v = ds/dt$ is the rate of change of distance along the path and the unit vector $\mathbf{e}_t$ is tangent to the path and points in the direction of motion. We take the time derivative of this equation to obtain the acceleration:

$$\mathbf{a} = \frac{d\mathbf{v}}{dt} = \frac{dv}{dt}\mathbf{e}_t + v\frac{d\mathbf{e}_t}{dt}.$$

As the point moves along its three-dimensional path, the direction of the unit vector $\mathbf{e}_t$ changes. In the case of motion of a point in a plane, this unit vector rotates in the plane, but in three-dimensional motion, the picture is more complicated. Figure 13.33a shows the path seen from a viewpoint perpendicular to the plane containing the vector $\mathbf{e}_t$ at times t and $t + dt$. This plane is called the *osculating plane*. It can be thought of as the instantaneous plane of rotation of the unit vector $\mathbf{e}_t$, and its orientation will generally change as P moves along its path. Since $\mathbf{e}_t$ is rotating in the osculating plane at time t, its time derivative is

$$\frac{d\mathbf{e}_t}{dt} = \frac{d\theta}{dt}\mathbf{e}_n, \tag{13.47}$$

where $d\theta/dt$ is the angular velocity of $\mathbf{e}_t$ in the osculating plane and the unit vector $\mathbf{e}_n$ is defined as shown in Fig. 13.33b. The vector $\mathbf{e}_n$ is perpendicular to $\mathbf{e}_t$, parallel to the osculating plane, and directed toward the concave side of the path. Therefore the acceleration is

$$\mathbf{a} = \frac{dv}{dt}\mathbf{e}_t + v\frac{d\theta}{dt}\mathbf{e}_n. \tag{13.48}$$

In the same way as in the case of motion in a plane, we can also express the acceleration in terms of the instantaneous radius of curvature of the path (Fig. 13.33c):

$$\mathbf{a} = \frac{dv}{dt}\mathbf{e}_t + \frac{v^2}{\rho}\mathbf{e}_n. \tag{13.49}$$

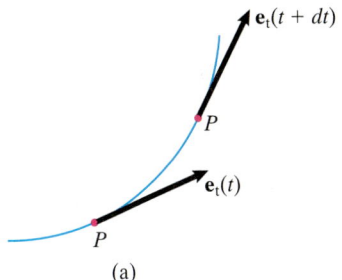

(a)

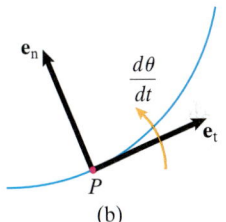

(b)

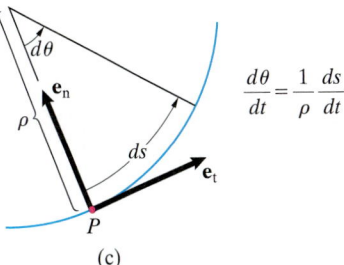

(c)

Figure 13.33
(a) Defining the osculating plane.
(b) Definition of the unit vector $\mathbf{e}_n$.
(c) The instantaneous radius of curvature.

We see that the expressions for the velocity and acceleration in normal and tangential components for three-dimensional motion are identical in form to the expressions for planar motion. The velocity is a vector whose magnitude equals the rate of change of distance traveled along the path and whose direction is tangent to the path. The acceleration has a component tangential to the path equal to the rate of change of the magnitude of the velocity and a component perpendicular to the path that depends on the magnitude of the velocity and the instantaneous radius of curvature of the path. In planar motion, the unit vector $\mathbf{e}_n$ is parallel to the plane of the motion. In three-dimensional motion, $\mathbf{e}_n$ is parallel to the osculating plane, whose orientation depends on the nature of the path. Notice from Eq. (13.47) that $\mathbf{e}_n$ can be expressed in terms of $\mathbf{e}_t$ by

$$\mathbf{e}_n = \frac{\dfrac{d\mathbf{e}_t}{dt}}{\left|\dfrac{d\mathbf{e}_t}{dt}\right|}.$$

As the final step necessary to establish a three-dimensional coordinate system, we introduce a third unit vector that is perpendicular to both $\mathbf{e}_t$ and $\mathbf{e}_n$ by the definition

$$\mathbf{e}_p = \mathbf{e}_t \times \mathbf{e}_n.$$

The unit vector $\mathbf{e}_p$ is perpendicular to the osculating plane (Fig. 13.34).

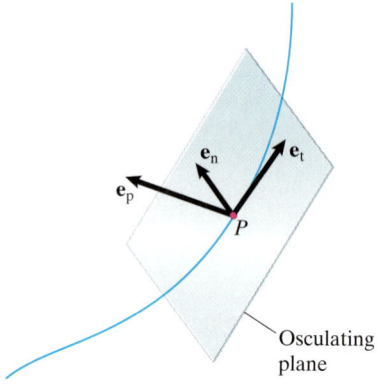

Figure 13.34
Defining the third unit vector $\mathbf{e}_p$.

Study Questions

1. What is the velocity vector of a point in terms of normal and tangential components?
2. If the magnitude of the velocity of a point moving in a curved path is constant, is the acceleration of the point zero?
3. What is the instantaneous radius of curvature?
4. If a point P moves in a circular path of radius R, what is the relation between the magnitude of the velocity of the point and the angular velocity of the line from the center of the path to P?

Example 13.9   Motion in Terms of Normal and Tangential Components

Figure 13.35

The motorcycle in Fig. 13.35 starts from rest at $t = 0$ on a circular track of 400-m radius. The tangential component of the cycle's acceleration is $a_t = 2 + 0.2t \ \text{m/s}^2$. At $t = 10$ s, determine (a) the distance the motorcycle has moved along the track and (b) the magnitude of its acceleration.

Strategy

Let s be the distance along the track from the initial position O of the motorcycle to its position at time t (Fig. a). Knowing the tangential acceleration as a function of time, we can integrate to determine v and s as functions of time.

Solution

(a) The tangential acceleration is

$$a_t = \frac{dv}{dt} = 2 + 0.2t \ \text{m/s}^2.$$

Integrating,

$$\int_0^v dv = \int_0^t (2 + 0.2t)\, dt,$$

we obtain v as a function of time:

$$v = \frac{ds}{dt} = 2t + 0.1t^2 \ \text{m/s}.$$

Integrating again,

$$\int_0^s ds = \int_0^t (2t + 0.1t^2)\, dt,$$

we find that the coordinate s as a function of time is

$$s = t^2 + \frac{0.1}{3}t^3 \ \text{m}.$$

At $t = 10$ s, the distance moved along the track is

$$s = (10)^2 + \frac{0.1}{3}(10)^3 = 133 \ \text{m}.$$

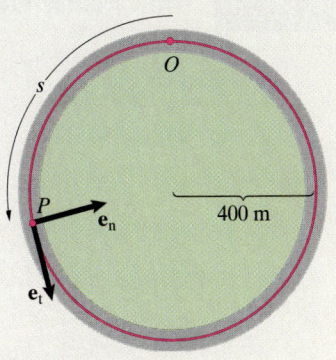

(a) The coordinate s measures the distance along the track.

(b) At $t = 10$ s, the tangential component of the acceleration is

$$a_t = 2 + 0.2(10) = 4 \ \text{m/s}^2.$$

We must also determine the normal component of acceleration. The instantaneous radius of curvature of the path is the radius of the circular track, $\rho = 400$ m. The magnitude of the velocity at $t = 10$ s is

$$v = 2(10) + 0.1(10)^2 = 30 \ \text{m/s}.$$

Therefore, the normal acceleration is

$$a_n = \frac{v^2}{\rho} = \frac{(30)^2}{400} = 2.25 \ \text{m/s}^2.$$

The magnitude of the acceleration at $t = 10$ s is

$$|\mathbf{a}| = \sqrt{a_t^2 + a_n^2} = \sqrt{(4)^2 + (2.25)^2} = 4.59 \ \text{m/s}^2.$$

Critical Thinking

In curvilinear motion, the magnitude of the velocity is related to the position s measured along the path by

$$v = \frac{ds}{dt}$$

and the tangential component of the acceleration is related to the magnitude of the velocity by

$$a_t = \frac{dv}{dt}.$$

These equations are identical in form to Eqs. (13.3) and (13.4) for straight-line motion. This means that you can apply the same methods developed for straight-line motion in Section 13.2 to curvilinear motion. In this example the motorcycle's tangential acceleration was known as a function of time. Just as we would in straight-line motion, we integrated the tangential acceleration to determine the velocity as a function of time and then integrated the velocity to determine the position as a function of time.

Example 13.10 The Circular-Orbit Problem

A satellite is in a circular orbit of radius R around the earth. What is its velocity?

Strategy

The acceleration due to gravity at a distance R from the center of the earth is $g R_E^2 / R^2$, where R_E is the radius of the earth. (See Eq. 12.4.) By using this expression together with the equation for the acceleration in terms of normal and tangential components, we can obtain an equation for the satellite's velocity.

Solution

In terms of normal and tangential components (Fig. a), the acceleration of the satellite is

$$\mathbf{a} = \frac{dv}{dt}\mathbf{e}_t + \frac{v^2}{R}\mathbf{e}_n.$$

This expression must equal the acceleration due to gravity toward the center of the earth:

$$\frac{dv}{dt}\mathbf{e}_t + \frac{v^2}{R}\mathbf{e}_n = \frac{g R_E^2}{R^2}\mathbf{e}_n. \tag{1}$$

Because there is no $\mathbf{e}_t$ component on the right side of Eq. (1), we conclude that the magnitude of the satellite's velocity is constant:

$$\frac{dv}{dt} = 0.$$

Equating the $\mathbf{e}_n$ components in Eq. (1) and solving for v, we obtain

$$v = \sqrt{\frac{g R_E^2}{R}}.$$

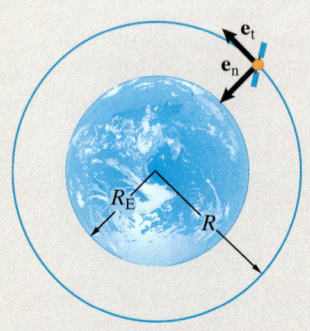

(a) Describing the satellite's motion in terms of normal and tangential components.

Critical Thinking

Why are different kinds of vector components and reference frames used in dynamics? This example is a good illustration. By expressing the acceleration of the satellite as it moves in its circular path in terms of normal and tangential components, we were able to determine the relationship between the magnitude of the velocity and the radius of the orbit very easily.

In Example 13.5, we determined the escape velocity of an object traveling straight away from the earth in terms of its initial distance from the center of the earth. The escape velocity for an object at a distance R from the center of the earth, $v_{esc} = \sqrt{2g R_E^2 / R}$, is only $\sqrt{2}$ times the velocity of an object in a circular orbit of radius R. This explains why it was possible to begin launching probes to other planets not long after the first satellites were placed in earth orbit.

Example 13.11 Relating Cartesian Components to Normal and Tangential Components

During a flight in which a helicopter starts from rest at $t = 0$, the cartesian components of its acceleration are

$$a_x = 0.6t \text{ m/s}^2$$

and

$$a_y = 1.8 - 0.36t \text{ m/s}^2.$$

What are the normal and tangential components of the helicopter's acceleration and the instantaneous radius of curvature of its path at $t = 4$ s?

Strategy

We can integrate the cartesian components of acceleration to determine the cartesian components of the velocity at $t = 4$ s, and then we can determine the components of the tangential unit vector $\mathbf{e_t}$ by dividing the velocity vector by its magnitude: $\mathbf{e_t} = \mathbf{v}/|\mathbf{v}|$. Next, we can determine the tangential component of the acceleration by evaluating the dot product of the acceleration vector with $\mathbf{e_t}$. Knowing the tangential component of the acceleration, we can then evaluate the normal component and determine the radius of curvature of the path from the relation $a_n = v^2/\rho$.

Solution

Integrating the components of acceleration with respect to time (see Example 13.6), we find that the cartesian components of the velocity are

$$v_x = 0.3t^2 \text{ m/s}$$

and

$$v_y = 1.8t - 0.18t^2 \text{ m/s}.$$

At $t = 4$ s, $v_x = 4.80$ m/s and $v_y = 4.32$ m/s. The tangential unit vector $\mathbf{e_t}$ at $t = 4$ s is (Fig. a)

$$\mathbf{e_t} = \frac{\mathbf{v}}{|\mathbf{v}|} = \frac{4.80\mathbf{i} + 4.32\mathbf{j}}{\sqrt{(4.80)^2 + (4.32)^2}} = 0.743\mathbf{i} + 0.669\mathbf{j}.$$

The components of the acceleration at $t = 4$ s are

$$a_x = 0.6(4) = 2.4 \text{ m/s}^2$$

and

$$a_y = 1.8 - 0.36(4) = 0.36 \text{ m/s}^2,$$

so the tangential component of the acceleration at $t = 4$ s is

$$a_t = \mathbf{e_t} \cdot \mathbf{a}$$

$$= (0.743\mathbf{i} + 0.669\mathbf{j}) \cdot (2.4\mathbf{i} + 0.36\mathbf{j})$$

$$= 2.02 \text{ m/s}^2.$$

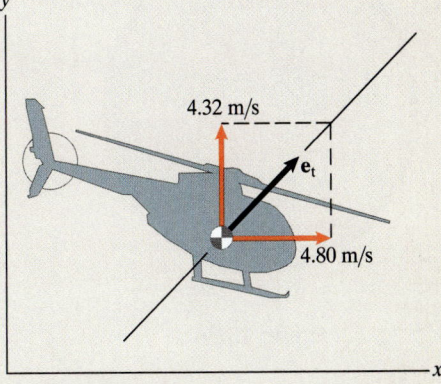

(a) Cartesian components of the velocity and the vector $\mathbf{e_t}$.

The magnitude of the acceleration is $\sqrt{(2.4)^2 + (0.36)^2} = 2.43 \text{ m/s}^2$, so the magnitude of the normal component of the acceleration is

$$a_n = \sqrt{|\mathbf{a}|^2 - a_t^2} = \sqrt{(2.43)^2 - (2.02)^2} = 1.34 \text{ m/s}^2.$$

The radius of curvature of the path is thus

$$\rho = \frac{|\mathbf{v}|^2}{a_n} = \frac{(4.80)^2 + (4.32)^2}{1.34} = 31.2 \text{ m}.$$

Critical Thinking

The cartesian components of a vector are parallel to the axes of the cartesian coordinate system, whereas the normal and tangential components are normal and tangential to the *path*. In this example the cartesian components of the acceleration of the helicopter were given as functions of time. How could we determine the normal and tangential components of the acceleration at $t = 4$ s without knowing the path? Notice that we used the fact that *the velocity vector is tangent to the path*. We integrated the cartesian components of the acceleration to determine the cartesian components of the velocity at $t = 4$ s. That told us the direction of the path. By dividing the velocity vector by its magnitude, we obtained a unit vector tangent to the path that pointed in the direction of the motion, which is the vector $\mathbf{e}_t$.

Design Example 13.12 Centrifuge Design

The distance from the center of the medical centrifuge shown in Fig. 13.36 to its samples is 300 mm. When the centrifuge is turned on, its motor and control system give it an angular acceleration $\alpha = A - B\omega^2$. Choose the constants A and B so that the samples will be subjected to a maximum horizontal acceleration of 12,000 g's and the centrifuge will reach 90% of its maximum operating speed in 2 min.

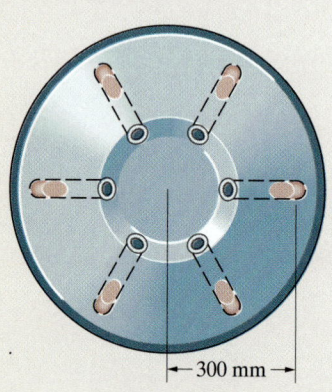

Figure 13.36

— 300 mm —

Strategy

Since we know both the radius of the circular path in which the samples move and the horizontal acceleration to which they are to be subjected, we can solve for the operating angular velocity of the centrifuge. We will use the given angular acceleration to determine the centrifuge's angular velocity as a function of time in terms of the constants A and B. We can then use the operating angular velocity and the condition that the centrifuge reach 90% of the operating angular velocity in 2 min to determine the constants A and B.

Solution

From Eq. (13.45), the samples are subjected to a normal acceleration

$$a_n = R\omega^2.$$

Setting $a_n = (12,000)(9.81)$ m/s^2 and $R = 0.3$ m and solving for the angular velocity, we find that the desired maximum operating speed is $\omega_{max} = 626$ rad/s.

The angular acceleration is

$$\alpha = \frac{d\omega}{dt} = A - B\omega^2.$$

We separate variables to get

$$\frac{d\omega}{A - B\omega^2} = dt.$$

Then we integrate to determine ω as a function of time, assuming that the centrifuge starts from rest at $t = 0$:

$$\int_0^\omega \frac{d\omega}{A - B\omega^2} = \int_0^t dt.$$

Evaluating the integrals, we obtain

$$\frac{1}{2\sqrt{AB}} \ln\left(\frac{A + \sqrt{AB}\omega}{A - \sqrt{AB}\omega}\right) = t.$$

The solution of this equation for ω is

$$\omega = \sqrt{\frac{A}{B}}\left(\frac{e^{2\sqrt{AB}t} - 1}{e^{2\sqrt{AB}t} + 1}\right).$$

As t becomes large, ω approaches $\sqrt{A/B}$, so we have the condition that

$$\sqrt{\frac{A}{B}} = \omega_{max} = 626 \text{ rad/s}, \tag{1}$$

and we can write the equation for ω as

$$\omega = \omega_{max}\left(\frac{e^{2\sqrt{AB}t} - 1}{e^{2\sqrt{AB}t} + 1}\right). \tag{2}$$

We also have the condition that $\omega = 0.9\omega_{max}$ after 2 min. Setting $\omega = 0.9\omega_{max}$ and $t = 120$ s in Eq. (2) and solving for $\sqrt{AB}$, we obtain

$$\sqrt{AB} = \frac{\ln(19)}{240}.$$

We solve this equation together with Eq. (1), obtaining $A = 7.69$ rad/s^2 and $B = 1.96 \times 10^{-5}$ rad^{-1}. Figure 13.37 shows the angular velocity of the centrifuge as a function of time.

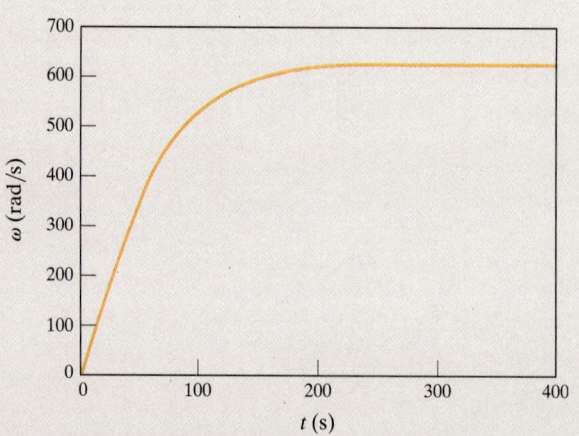

Figure 13.37
Graph of the angular velocity as a function of time.

Design Issues

A centrifuge is a device designed to rotate and thereby subject objects to large normal accelerations. The common spin dryer used to dry clothing is an example. Centrifuges are used extensively in biological and medical testing and research and also in the chemical processing industry to separate constituents of different densities. By using a centrifuge to separate the solid matter in blood (primarily red cells) from the liquid plasma, diagnostic tests can be carried out on the plasma, and the patient's *hematocrit*—the percentage of solid matter—can be determined. During the early days of manned space flight, there was concern that the large accelerations to which the astronauts would be subjected during their boost to orbit might injure them or affect their ability to carry out essential tasks. Large centrifuges consisting of rotating beams with "gondolas" at the end capable of carrying a person were built and used to study the effects of acceleration on human subjects. Centrifuges have also been designed to investigate the effects of large accelerations on mechanisms and to facilitate the modeling of geophysical phenomena such as slope stability and cratering. One such centrifuge is shown in Fig. 13.38. Consisting of a large beam that rotates samples in a horizontal circular path with a radius of 8 m, it can subject samples to 150 g's of acceleration.

Centrifuges intended for commercial use, such as medical testing, must be designed so that samples can be installed easily, and they must have an electric motor and control system that permits specified accelerations to be achieved quickly and accurately. Their bearings must be designed to sustain the very high angular velocities and have long lifetimes. (The *ultracentrifuge*, used in biological research, reaches angular velocities of millions of revolutions per minute, and is usually designed with an air bearing so that there is no solid contact between the spinning centrifuge and its support.) The structure of a centrifuge must be designed to support the large forces resulting from its rapid rotational motion, and the objects being accelerated must be adequately supported.

An essential consideration in the design of a centrifuge is the possibility of structural failure, which could result in projectiles moving outward at high velocities. A centrifuge must be designed with a surrounding structure strong enough that such projectiles would be safely contained and not cause additional damage or injury. This design requirement was achieved in the case of the large centrifuge shown in Fig. 13.38 by locating it underground.

Figure 13.38
A centrifuge designed to subject test articles to large accelerations.
(Photograph courtesy of Sandia National Laboratories.)

Problems

13.104 The armature of an electric motor rotates with a constant angular velocity of 400 rpm (revolutions per minute).
(a) What is the magnitude of the velocity of point P relative to point O?
(b) What are the normal and tangential components of the acceleration of P relative to O?

13.105 The armature starts from rest at $t = 0$. Its angular acceleration is given as a function of time by $\alpha = 2t$ rad/s^2. Determine the velocity and acceleration of point P relative to point O in terms of normal and tangential components at $t = 10$ s.

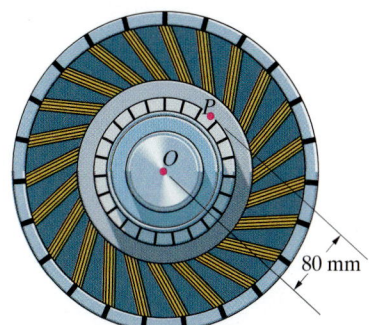

80 mm

Problems 13.104/13.105

13.106 Suppose that you want to design a medical centrifuge to subject samples to normal accelerations of 1000 g's.
(a) If the distance from the center of the centrifuge to the sample is 300 mm, what speed of rotation in rpm is necessary?
(b) If you want the centrifuge to reach its design rpm in 1 min, what constant angular acceleration is necessary?

13.107 The medical centrifuge starts from rest at $t = 0$ and is subjected to a constant angular acceleration $\alpha = 3$ rad/s^2. What is the magnitude of the total acceleration to which the samples are subjected at $t = 1$ s?

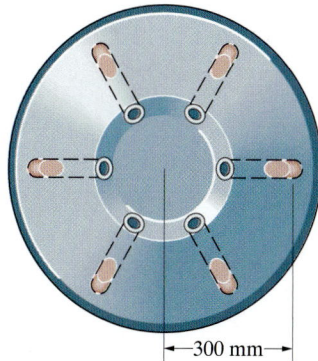

—300 mm—

Problems 13.106/13.107

13.108 The radial distance from the axis of the centrifuge shown in Fig. 13.38 to the sample container is 8 m. Suppose that you want to subject a sample to a normal acceleration of 100 g's.
(a) What speed of rotation in rpm is necessary?
(b) If you want the centrifuge to reach the rpm determined in part (a) in 2 min, what constant angular acceleration is necessary?

13.109 A powerboat being tested for maneuverability is started from rest at $t = 0$ and driven in a circular path 12 m in radius. The tangential component of the boat's acceleration as a function of time is $a_t = 0.4t$ m/s^2.
(a) What are the boat's velocity and acceleration in terms of normal and tangential components at $t = 4$ s?
(b) What distance does the boat move along its circular path from $t = 0$ to $t = 4$ s?

Problem 13.109

13.110 The angle $\theta = 2t^2$ rad.
(a) What are the velocity and acceleration of point P in terms of normal and tangential components at $t = 1$ s?
(b) What distance along the circular path does point P move from $t = 0$ to $t = 1$ s?

13.111 The angle $\theta = 2t^2$ rad. What are the velocity and acceleration of point P in terms of normal and tangential components when P has gone one revolution around the circular path starting at $t = 0$?

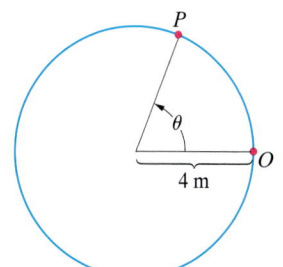

Problems 13.110/13.111

13.112 At the instant shown, the crank AB is rotating with a counterclockwise angular velocity of 5000 rpm. Determine the velocity of point B (a) in terms of normal and tangential components and (b) in terms of cartesian components.

13.113 The crank AB is rotating with a constant counterclockwise angular velocity of 5000 rpm. Determine the acceleration of point B (a) in terms of normal and tangential components and (b) in terms of cartesian components.

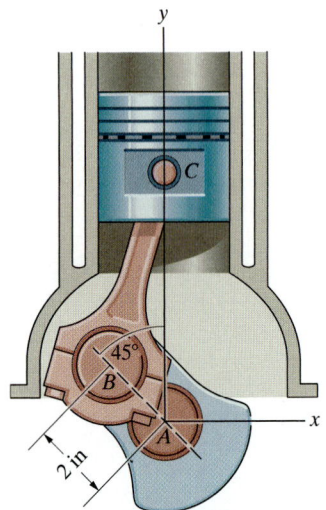

Problems 13.112/13.113

13.114 Suppose that you are standing at point P at 30° north latitude (that is, a point 30° north of the equator). The radius of the earth is $R_E = 6370$ km. What are the magnitudes of your velocity and acceleration relative to a nonrotating reference frame with its origin at the center of the earth?

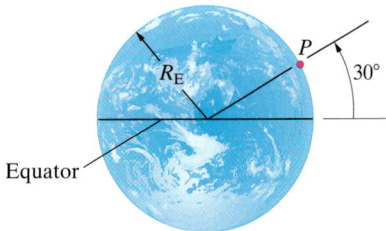

Problem 13.114

13.115 At the instant shown, the magnitude of the airplane's velocity is 130 m/s, its tangential component of acceleration is $a_t = -4$ m/s², and the rate of change of its path angle is $d\theta/dt = 5°/s$.

(a) What are the airplane's velocity and acceleration in terms of normal and tangential components?
(b) What is the instantaneous radius of curvature of the airplane's path?

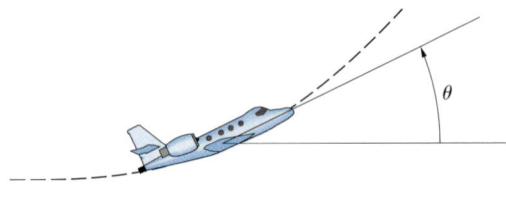

Problem 13.115

13.116 In the preliminary design of a sun-powered car, a group of engineering students estimates that the car's acceleration will be 0.6 m/s². Suppose that the car starts from rest at A and the tangential component of its acceleration is $a_t = 0.6$ m/s². What are the car's velocity and acceleration in terms of normal and tangential components when it reaches B?

13.117 After subjecting a car design to wind tunnel testing, the students estimate that the tangential component of the car's acceleration will be $a_t = 0.6 - 0.002v^2$ m/s², where v is the car's velocity in m/s. If the car starts from rest at A, what are its velocity and acceleration in terms of normal and tangential components when it reaches B?

13.118 Suppose that the tangential component of acceleration of a car is given in terms of the car's position by $a_t = 0.4 - 0.001s$ m/s², where s is the distance the car travels along the track from point A. What are the car's velocity and acceleration in terms of normal and tangential components at point B?

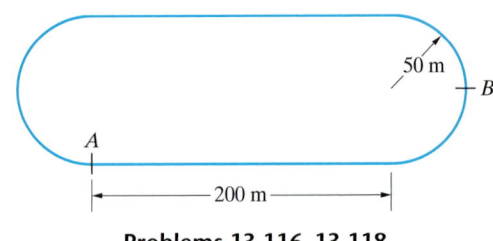

Problems 13.116–13.118

13.119 The car increases its speed at a constant rate from 40 mi/h at A to 60 mi/h at B. What is the magnitude of its acceleration 2 s after it passes point A?

13.120 The car increases its speed at a constant rate from 40 mi/h at A to 60 mi/h at B. Determine the magnitude of its acceleration when it has traveled along the road a distance (a) 120 ft from A and (b) 160 ft from A.

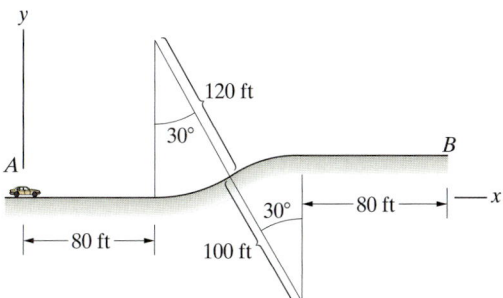

Problems 13.119/13.120

13.121 Astronaut candidates are to be tested in a centrifuge with 10-m radius that rotates in the horizontal plane. Test engineers want to subject the candidates to an acceleration of 5 g's, or five times the acceleration due to gravity. Earth's gravity effectively exerts an acceleration of 1 g in the vertical direction. Determine the angular velocity of the centrifuge in revolutions per second so that the magnitude of the total acceleration is 5 g's.

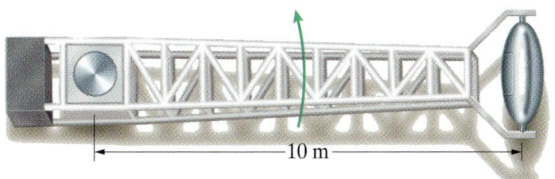

Problem 13.121

13.122 After first-stage separation and before the second-stage engines have fired, a rocket is moving at $v = 3000$ m/s and the angle between its velocity vector and the vertical is 60°. Because aerodynamic forces are negligible, the rocket's acceleration is that due to gravity, which is 9.50 m/s² at the rocket's altitude. Determine (a) the normal and tangential components of the rocket's acceleration and (b) the instantaneous radius of curvature of the rocket's path.

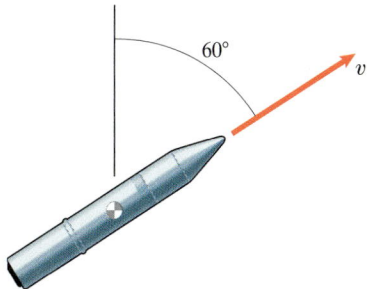

Problem 13.122

13.123 The athlete releases the shot with velocity $v = 16$ m/s.

(a) What are the velocity and acceleration of the shot in terms of normal and tangential components when it is at the highest point of its trajectory?

(b) What is the instantaneous radius of curvature of the shot's path when it is at the highest point of its trajectory?

13.124 At $t = 0$, the athlete releases the shot with velocity $v = 16$ m/s.

(a) What are the velocity and acceleration of the shot in terms of normal and tangential components at $t = 0.3$ s?

(b) Use the relation $a_n = v^2/\rho$ to determine the instantaneous radius of curvature of the shot's path at $t = 0.3$ s.

13.125 At $t = 0$, the athlete releases the shot with velocity $v = 16$ m/s. Use Eq. (13.42) to determine the instantaneous radius of curvature of the shot's path at $t = 0.3$ s.

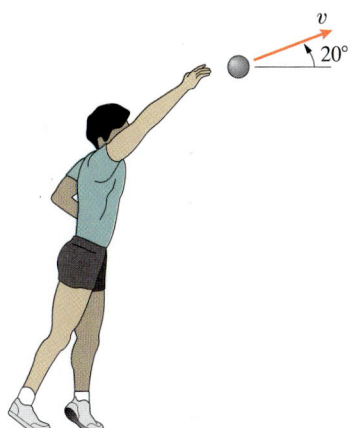

Problems 13.123–13.125

13.126 The cartesian coordinates of a point moving in the x–y plane are

$$x = 20 + 4t^2 \text{ m} \quad \text{and} \quad y = 10 - t^3 \text{ m}.$$

What is the instantaneous radius of curvature of the path of the point at $t = 3$ s?

13.127 The helicopter starts from rest at $t = 0$. The cartesian components of its acceleration are $a_x = 0.6t$ m/s^2 and $a_y = 1.8 - 0.36t$ m/s^2. Determine the tangential and normal components of the acceleration at $t = 6$ s.

13.128 Use Eq. (13.42) to determine the instantaneous radius of curvature of the path of the helicopter in Problem 13.127 at $t = 6$ s.

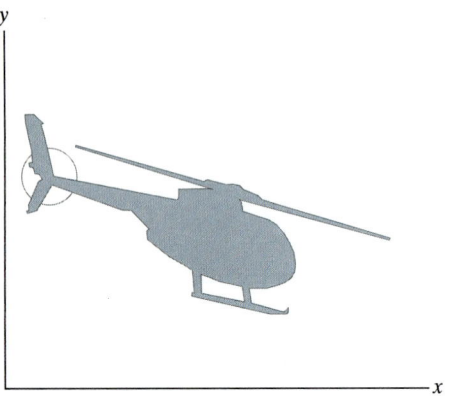

Problems 13.127/13.128

13.129* For astronaut training, the airplane shown is to achieve "weightlessness" for a short period of time by flying along a path such that its acceleration is $a_x = 0$ and $a_y = -g$. If the velocity of the plane at O at time $t = 0$ is $\mathbf{v} = v_0\mathbf{i}$, show that the autopilot must fly the airplane so that its tangential component of acceleration as a function of time is

$$a_t = g \frac{(gt/v_0)}{\sqrt{1 + (gt/v_0)^2}}.$$

13.130* In Problem 13.129, what is the airplane's normal component of acceleration as a function of time?

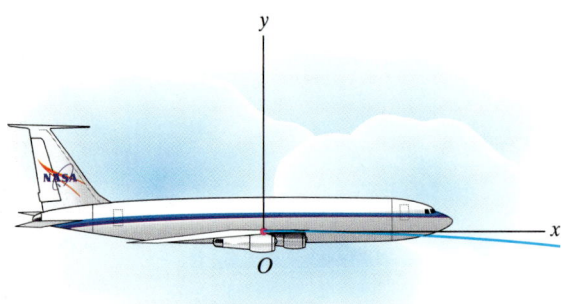

Problems 13.129/13.130

13.131 If $y = 100$ mm, $dy/dt = 200$ mm/s, and $d^2y/dt^2 = 0$, what are the velocity and acceleration of P in terms of normal and tangential components?

13.132* Suppose that the point P moves upward in the slot with velocity $\mathbf{v} = 300\mathbf{e}_t$ (mm/s). When $y = 150$ mm, what are dy/dt and d^2y/dt^2?

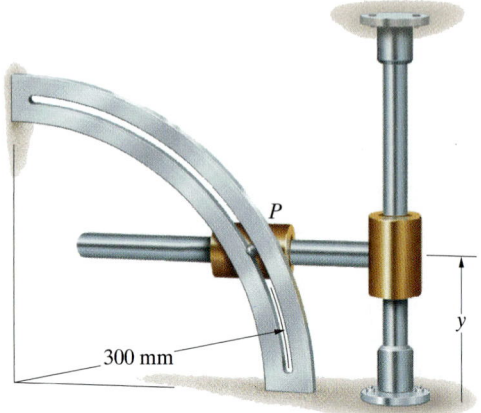

Problems 13.131/13.132

13.133* A car travels at 100 km/h on a straight road of increasing grade whose vertical profile can be approximated by the equation shown. When the car's horizontal coordinate is $x = 400$ m, what are the tangential and normal components of the car's acceleration?

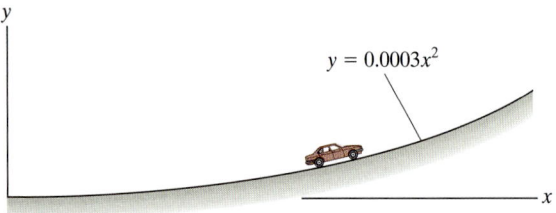

Problem 13.133

13.134 A boy rides a skateboard on the concrete surface of an empty drainage canal described by the equation shown. He starts at $y = 20$ ft, and the magnitude of his velocity is approximated by $v = \sqrt{2(32.2)(20 - y)}$ ft/s.

(a) Use Eq. (13.42) to determine the instantaneous radius of curvature of the boy's path when he reaches the bottom.

(b) What is the normal component of his acceleration when he reaches the bottom?

13.135 In Problem 13.134, what is the normal component of the boy's acceleration when he has passed the bottom and reached $y = 10$ ft?

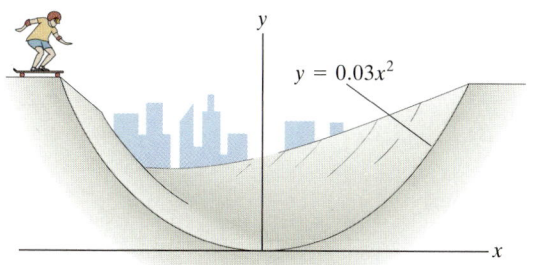

Problems 13.134/13.135

13.136* By using Eqs. (13.41): (a) show that the relations between the cartesian unit vectors and the unit vectors $\mathbf{e_t}$ and $\mathbf{e_n}$ are

$$\mathbf{i} = \cos\theta\,\mathbf{e_t} - \sin\theta\,\mathbf{e_n}$$

$$\mathbf{j} = \sin\theta\,\mathbf{e_t} + \cos\theta\,\mathbf{e_n}.$$

(b) Show that

$$\frac{d\mathbf{e_t}}{dt} = \frac{d\theta}{dt}\mathbf{e_n} \quad \text{and} \quad \frac{d\mathbf{e_n}}{dt} = -\frac{d\theta}{dt}\mathbf{e_t}.$$

Polar and Cylindrical Coordinates

Polar coordinates are often used to describe the curvilinear motion of a point. Circular motion, certain orbit problems, and, more generally, *central-force* problems, in which the acceleration of a point is directed toward a given point, can be expressed conveniently in polar coordinates.

Consider a point P in the x–y plane of a cartesian coordinate system. We can specify the position of P relative to the origin O either by its cartesian coordinates x, y or by its polar coordinates r, θ (Fig. 13.39a). To express vectors in terms of polar coordinates, we define a unit vector $\mathbf{e_r}$ that points in the direction of the radial line from the origin to P and a unit vector $\mathbf{e_\theta}$ that is perpendicular to $\mathbf{e_r}$ and points in the direction of increasing θ (Fig. 13.39b). In terms of these vectors, the position vector $\mathbf{r}$ from O to P is

$$\mathbf{r} = r\mathbf{e_r}. \tag{13.52}$$

(Notice that $\mathbf{r}$ has no component in the direction of $\mathbf{e_\theta}$.)

We can determine the velocity of P in terms of polar coordinates by taking the time derivative of Eq. (13.52):

$$\mathbf{v} = \frac{d\mathbf{r}}{dt} = \frac{dr}{dt}\mathbf{e_r} + r\frac{d\mathbf{e_r}}{dt}. \tag{13.53}$$

As P moves along a curvilinear path, the unit vector $\mathbf{e_r}$ rotates with angular velocity $\omega = d\theta/dt$. Therefore, from Eq. (13.33), we can express the time derivative of $\mathbf{e_r}$ in terms of $\mathbf{e_\theta}$ as

$$\frac{d\mathbf{e_r}}{dt} = \frac{d\theta}{dt}\mathbf{e_\theta}. \tag{13.54}$$

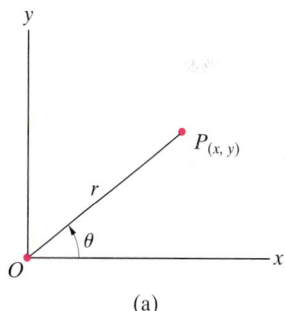

(a)

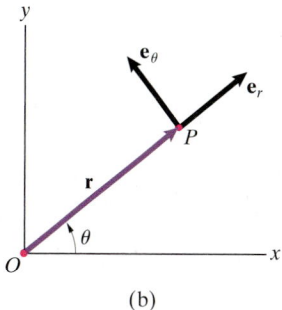

(b)

Figure 13.39
(a) The polar coordinates of P.
(b) The unit vectors $\mathbf{e_r}$ and $\mathbf{e_\theta}$ and the position vector $\mathbf{r}$.

Substituting this result into Eq. (13.53), we obtain the velocity of P:

$$\mathbf{v} = \frac{dr}{dt}\mathbf{e}_r + r\frac{d\theta}{dt}\mathbf{e}_\theta = \frac{dr}{dt}\mathbf{e}_r + r\omega\mathbf{e}_\theta. \tag{13.55}$$

We can get this result in another way that is less rigorous, but more direct and intuitive. Figure 13.40 shows the position vector of P at times t and $t + \Delta t$. The change in the position vector, $\mathbf{r}(t + \Delta t) - \mathbf{r}(t)$, consists of two components. The component Δr is due to the change in the radial position r and is in the $\mathbf{e}_r$ direction. The component $r\Delta\theta$ is due to the change in θ and is in the $\mathbf{e}_\theta$ direction. Thus, the change in the position of P is (approximately)

$$\mathbf{r}(t + \Delta t) - \mathbf{r}(t) = \Delta r\mathbf{e}_r + r\Delta\theta\mathbf{e}_\theta.$$

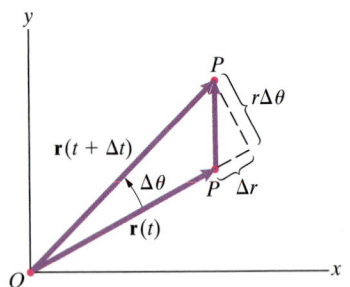

Figure 13.40
The position vector of P at t and $t + \Delta t$.

Dividing this expression by Δt and taking the limit as $\Delta t \to 0$, we obtain the velocity of P:

$$\mathbf{v} = \lim_{\Delta t \to 0}\left(\frac{\Delta r}{\Delta t}\mathbf{e}_r + r\frac{\Delta\theta}{\Delta t}\mathbf{e}_\theta\right)$$
$$= \frac{dr}{dt}\mathbf{e}_r + r\omega\mathbf{e}_\theta.$$

One component of the velocity is in the radial direction and is equal to the rate of change of the radial position r. The other component is normal, or *transverse*, to the radial direction and is proportional to the radial distance and to the rate of change of θ.

We obtain the acceleration of P by taking the time derivative of Eq. (13.55):

$$\mathbf{a} = \frac{d\mathbf{v}}{dt} = \frac{d^2r}{dt^2}\mathbf{e}_r + \frac{dr}{dt}\frac{d\mathbf{e}_r}{dt} + \frac{dr}{dt}\frac{d\theta}{dt}\mathbf{e}_\theta + r\frac{d^2\theta}{dt^2}\mathbf{e}_\theta + r\frac{d\theta}{dt}\frac{d\mathbf{e}_\theta}{dt}. \tag{13.56}$$

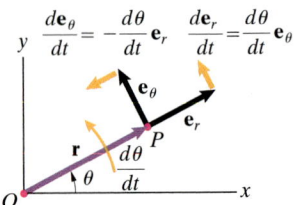

Figure 13.41
Time derivatives of $\mathbf{e}_r$ and $\mathbf{e}_\theta$.

The time derivative of the unit vector $\mathbf{e}_r$ due to the rate of change of θ is given by Eq. (13.54). As P moves, $\mathbf{e}_\theta$ also rotates with angular velocity $d\theta/dt$ (Fig. 13.41). You can see from this figure that the time derivative of $\mathbf{e}_\theta$ is in the $-\mathbf{e}_r$ direction if $d\theta/dt$ is positive:

$$\frac{d\mathbf{e}_\theta}{dt} = -\frac{d\theta}{dt}\mathbf{e}_r.$$

Substituting this expression and Eq. (13.54) into Eq. (13.56), we obtain the acceleration of P:

$$\mathbf{a} = \left[\frac{d^2r}{dt^2} - r\left(\frac{d\theta}{dt}\right)^2\right]\mathbf{e}_r + \left[r\frac{d^2\theta}{dt^2} + 2\frac{dr}{dt}\frac{d\theta}{dt}\right]\mathbf{e}_\theta.$$

Thus, the velocity and acceleration are respectively (Fig. 13.42)

$$\mathbf{v} = v_r\mathbf{e}_r + v_\theta\mathbf{e}_\theta = \frac{dr}{dt}\mathbf{e}_r + r\omega\mathbf{e}_\theta \tag{13.57}$$

and

$$\mathbf{a} = a_r\mathbf{e}_r + a_\theta\mathbf{e}_\theta, \tag{13.58}$$

where

$$a_r = \frac{d^2r}{dt^2} - r\left(\frac{d\theta}{dt}\right)^2 = \frac{d^2r}{dt^2} - r\omega^2$$

(13.59)

$$a_\theta = r\frac{d^2\theta}{dt^2} + 2\frac{dr}{dt}\frac{d\theta}{dt} = r\alpha + 2\frac{dr}{dt}\omega.$$

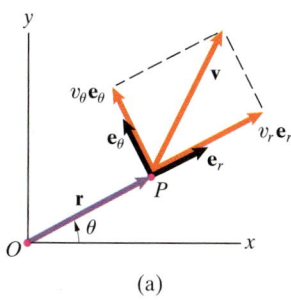 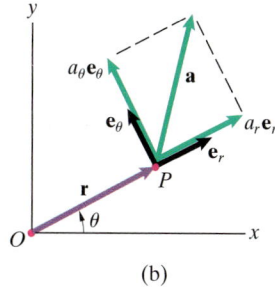

(a)   (b)

Figure 13.42
Radial and transverse components of the velocity (a) and acceleration (b).

The term $-r\omega^2$ in the radial component of the acceleration is called the *centripetal acceleration*, and the term $2(dr/dt)\omega$ in the transverse component is called the *Coriolis acceleration*.

The unit vectors $\mathbf{e}_r$ and $\mathbf{e}_\theta$ are related to the cartesian unit vectors by

$$\mathbf{e}_r = \cos\theta\mathbf{i} + \sin\theta\mathbf{j}$$

and

(13.60)

$$\mathbf{e}_\theta = -\sin\theta\mathbf{i} + \cos\theta\mathbf{j}.$$

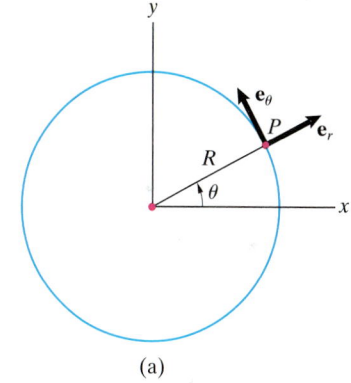

(a)

Circular Motion   Circular motion can be conveniently described using either radial and transverse or normal and tangential components. Let us compare these two methods of expressing the velocity and acceleration of a point P moving in a circular path of radius R (Fig. 13.43). Because the polar coordinate $r = R$ is constant, Eq. (13.57) for the velocity reduces to

$$\mathbf{v} = R\omega\mathbf{e}_\theta.$$

In terms of normal and tangential components, the velocity is

$$\mathbf{v} = v\mathbf{e}_t.$$

Notice in Fig. 13.43 that $\mathbf{e}_\theta = \mathbf{e}_t$. Comparing these two expressions for the velocity, we obtain the relation between the velocity and the angular velocity in circular motion:

$$v = R\omega.$$

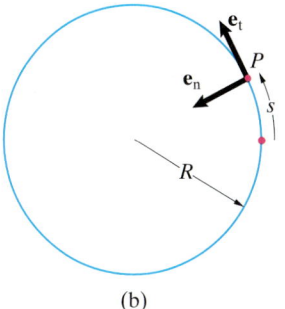

(b)

Figure 13.43
A point P moving in a circular path.
(a) Polar coordinates.
(b) Normal and tangential components.

From Eqs. (13.58) and (13.59), the acceleration for a circular path of radius R in terms of polar coordinates is

$$\mathbf{a} = -R\omega^2\mathbf{e}_r + R\alpha\mathbf{e}_\theta,$$

and the acceleration in terms of normal and tangential components is

$$\mathbf{a} = \frac{dv}{dt}\mathbf{e}_t + \frac{v^2}{R}\mathbf{e}_n.$$

The unit vector $\mathbf{e}_r = -\mathbf{e}_n$. Because of the relation $v = R\omega$, the normal components of acceleration are equal: $v^2/R = R\omega^2$. Equating the transverse and tangential components, we obtain the relation

$$\frac{dv}{dt} = a_t = R\alpha.$$

Cylindrical Coordinates Polar coordinates describe the motion of a point P in the x–y plane. We can describe three-dimensional motion by using *cylindrical coordinates* r, θ, and z (Fig. 13.44). The cylindrical coordinates r and θ are the polar coordinates of P, measured in the plane parallel to the x–y plane, and the definitions of the unit vectors $\mathbf{e}_r$ and $\mathbf{e}_\theta$ are unchanged. The position of P perpendicular to the x–y plane is measured by the coordinate z, and the unit vector $\mathbf{e}_z$ points in the positive z axis direction.

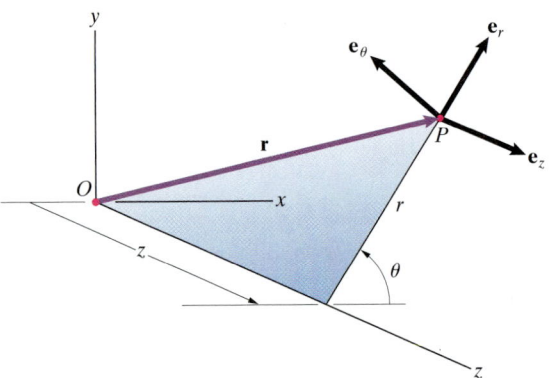

Figure 13.44
Cylindrical coordinates r, θ, and z of point P and the unit vectors $\mathbf{e}_r$, $\mathbf{e}_\theta$, and $\mathbf{e}_z$.

In terms of cylindrical coordinates, the position vector $\mathbf{r}$ is the sum of the expression for the position vector in polar coordinates and the z component:

$$\mathbf{r} = r\mathbf{e}_r + z\mathbf{e}_z. \tag{13.61}$$

(The polar coordinate r is not the magnitude of $\mathbf{r}$, except when P lies in the x–y plane.) By taking time derivatives, we obtain the velocity

$$\mathbf{v} = \frac{d\mathbf{r}}{dt} = v_r\mathbf{e}_r + v_\theta\mathbf{e}_\theta + v_z\mathbf{e}_z$$

$$= \frac{dr}{dt}\mathbf{e}_r + r\omega\,\mathbf{e}_\theta + \frac{dz}{dt}\mathbf{e}_z \tag{13.62}$$

and acceleration

$$\mathbf{a} = \frac{d\mathbf{v}}{dt} = a_r\mathbf{e}_r + a_\theta\mathbf{e}_\theta + a_z\mathbf{e}_z, \tag{13.63}$$

where

$$a_r = \frac{d^2r}{dt^2} - r\omega^2, \quad a_\theta = r\alpha + 2\frac{dr}{dt}\omega, \quad \text{and} \quad a_z = \frac{d^2z}{dt^2}. \tag{13.64}$$

Notice that Eqs. (13.62) and (13.63) reduce to the polar coordinate expressions for the velocity and acceleration, Eqs. (13.57) and (13.58), when P moves along a path in the x–y plane.

> ## Study Questions
>
> 1. What are the definitions of the unit vectors $\mathbf{e}_r$ and $\mathbf{e}_\theta$?
> 2. What is the position vector of a point in terms of polar coordinates?
> 3. If a point moves in a circular path about the origin of a coordinate system, what is its Coriolis acceleration? What is its centripetal acceleration?

Example 13.13 Expressing Motion in Terms of Polar Coordinates

Suppose that you are standing on a large disk (say, a merry-go-round) rotating with constant angular velocity ω_0 and you start walking at constant speed v_0 along a straight radial line painted on the disk (Fig. 13.45). What are your velocity and acceleration when you are a distance r from the center of the disk?

Strategy

We can describe your motion in terms of polar coordinates (Fig. a). By using the information given about your motion and the motion of the disk, we can evaluate the terms in the expressions for the velocity and acceleration in terms of polar coordinates.

Solution

The speed with which you walk along the radial line is the rate of change of r, $dr/dt = v_0$, and the angular velocity of the disk is the rate of change of θ, $\omega = \omega_0$. Your velocity is

$$\mathbf{v} = \frac{dr}{dt}\mathbf{e}_r + r\omega\mathbf{e}_\theta = v_0\mathbf{e}_r + r\omega_0\mathbf{e}_\theta.$$

Your velocity consists of two components: a radial component due to the speed at which you are walking and a transverse component due to the disk's rate of rotation. The transverse component increases as your distance from the center of the disk increases.

Your walking speed $v_0 = dr/dt$ is constant, so $d^2r/dt^2 = 0$. Also, the disk's angular velocity $\omega_0 = d\theta/dt$ is constant, so $d^2\theta/dt^2 = 0$. The radial component of your acceleration is

$$a_r = \frac{d^2r}{dt^2} - r\omega^2 = -r\omega_0^2,$$

and the transverse component is

$$a_\theta = r\alpha + 2\frac{dr}{dt}\omega = 2v_0\omega_0.$$

Critical Thinking

Why didn't we use normal and tangential components to determine your velocity and acceleration? The reason they would not be convenient in this example is that the path is not known, and normal and tangential components are defined in terms of the path.

If you have ever tried walking on a merry-go-round, you know that it is a difficult proposition. This example indicates why. Subjectively, you are walking along a straight line with constant velocity, but you are actually experiencing the centripetal acceleration a_r and the Coriolis acceleration a_θ due to the disk's rotation.

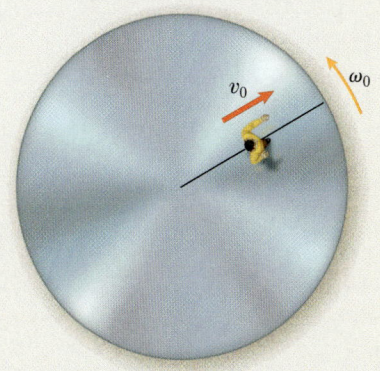

Figure 13.45

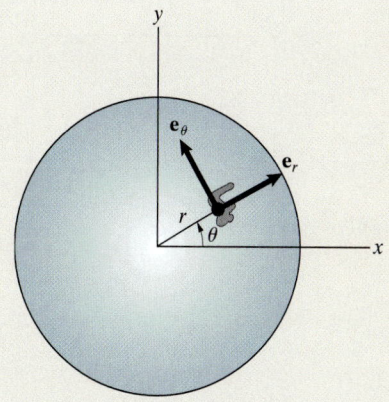

(a) Your position in terms of polar coordinates.

Example 13.14 Relating Polar Components to Cartesian Components

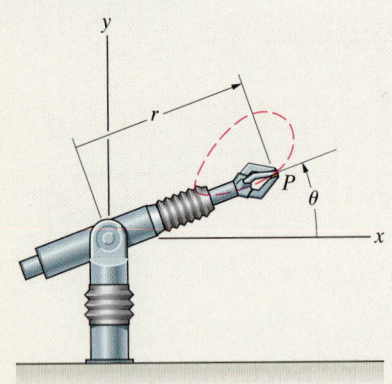

Figure 13.46

The robot arm in Fig. 13.46 is programmed so that point P traverses the path described by

$$r = 1 - 0.5 \cos 2\pi t \text{ m}$$

and

$$\theta = 0.5 - 0.2 \sin 2\pi t \text{ rad.}$$

At $t = 0.8$ s, determine (a) the velocity of P in terms of polar coordinates and (b) the cartesian components of the velocity of P.

Strategy

(a) Since we are given r and θ as functions of time, we can calculate the derivatives in the expression for the velocity in terms of polar coordinates and obtain the velocity as a function of time.

(b) By determining the value of θ at $t = 0.8$ s, we can use trigonometry to determine the cartesian components in terms of the polar components.

Solution

(a) From Eq. (13.57), the velocity is

$$\mathbf{v} = \frac{dr}{dt}\mathbf{e}_r + r\frac{d\theta}{dt}\mathbf{e}_\theta$$

$$= (\pi \sin 2\pi t)\mathbf{e}_r + (1 - 0.5 \cos 2\pi t)(-0.4\pi \cos 2\pi t)\mathbf{e}_\theta.$$

At $t = 0.8$ s,

$$\mathbf{v} = -2.99\mathbf{e}_r - 0.328\mathbf{e}_\theta \text{ (m/s)}.$$

(b) At $t = 0.8$ s, $\theta = 0.690$ rad $= 39.5°$ (Fig. a). The x component of the velocity of P is

$$v_x = v_r \cos 39.5° - v_\theta \sin 39.5°$$

$$= (-2.99) \cos 39.5° - (-0.328) \sin 39.5° = -2.09 \text{ m/s},$$

and the y component is

$$v_y = v_r \sin 39.5° + v_\theta \cos 39.5°$$

$$= (-2.99) \sin 39.5° + (-0.328) \cos 39.5° = -2.16 \text{ m/s}.$$

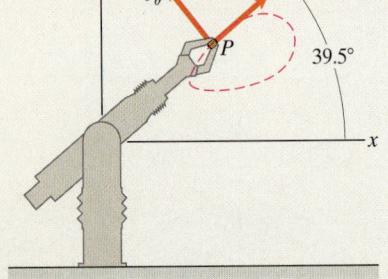

(a) Position at $t = 0.8$ s.

Critical Thinking

When you determine components of a vector in terms of different coordinate systems, you should check the components to confirm that they give the same magnitude. In this example,

$$|\mathbf{v}| = \sqrt{(-2.99)^2 + (-0.328)^2} = \sqrt{(-2.09)^2 + (-2.16)^2} = 3.01 \text{ m/s.}$$

Remember that although the components of the velocity are different in the two coordinate systems, those components describe the same velocity vector.

Example 13.15   Velocity in Terms of Polar and Cartesian Components

In the cam–follower mechanism shown in Fig. 13.47, the slotted bar rotates with constant angular velocity $\omega = 4$ rad/s and the radial position of the follower is determined by the elliptic profile of the stationary cam. The path of the follower is described by the polar equation

$$r = \frac{0.15}{1 + 0.5 \cos \theta} \text{ m.}$$

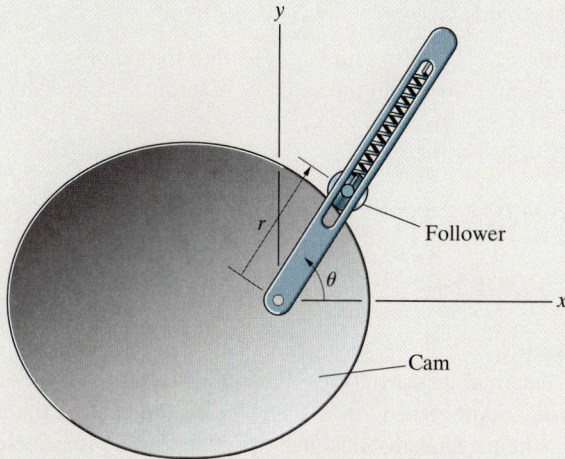

Figure 13.47

Determine the velocity of the follower when $\theta = 45°$ in terms of (a) polar coordinates and (b) cartesian coordinates.

Strategy

By taking the time derivative of the polar equation for the profile of the cam, we can obtain a relation between the known angular velocity and the radial component of velocity that permits us to evaluate the velocity in terms of polar coordinates. Then, by using Eqs. (13.60), we can obtain the velocity in terms of cartesian coordinates.

Solution

(a) The polar equation for the cam profile is of the form $r = r(\theta)$. Taking its derivative with respect to time, we obtain

$$\frac{dr}{dt} = \frac{dr(\theta)}{d\theta} \frac{d\theta}{dt}$$

$$= \frac{d}{d\theta} \left(\frac{0.15}{1 + 0.5 \cos \theta} \right) \frac{d\theta}{dt}$$

$$= \left[\frac{0.075 \sin \theta}{(1 + 0.5 \cos \theta)^2} \right] \frac{d\theta}{dt}.$$

The velocity of the follower in polar coordinates is therefore

$$\mathbf{v} = \frac{dr}{dt}\mathbf{e}_r + r\frac{d\theta}{dt}\mathbf{e}_\theta$$

$$= \left[\frac{0.075\sin\theta}{(1 + 0.5\cos\theta)^2}\right]\frac{d\theta}{dt}\mathbf{e}_r + \left(\frac{0.15}{1 + 0.5\cos\theta}\right)\frac{d\theta}{dt}\mathbf{e}_\theta.$$

The angular velocity $\omega = d\theta/dt = 4$ rad/s, so we can evaluate the polar components of the velocity when $\theta = 45°$, obtaining

$$\mathbf{v} = 0.116\mathbf{e}_r + 0.443\mathbf{e}_\theta \ (\text{m/s}).$$

(b) Substituting Eqs. (13.60) with $\theta = 45°$ into the polar coordinate expression for the velocity, we obtain the velocity in terms of cartesian coordinates:

$$\mathbf{v} = 0.116\mathbf{e}_r + 0.443\mathbf{e}_\theta$$

$$= 0.116(\cos 45°\mathbf{i} + \sin 45°\mathbf{j}) + 0.443(-\sin 45°\mathbf{i} + \cos 45°\mathbf{j})$$

$$= -0.232\mathbf{i} + 0.395\mathbf{j} \ (\text{m/s}).$$

Critical Thinking

Notice that in determining the velocity of the follower, we made the tacit assumption that it stays in contact with the surface of the cam as the bar rotates. Designers of cam mechanisms must insure that the spring is sufficiently strong so that the follower does not lose contact with the surface. In Chapter 14 we introduce the concepts needed to analyze such problems.

Problems

13.137 The polar coordinates of the collar A as functions of time are $r = 1 + 0.2t^2$ m and $\theta = 2t$ rad. Determine the velocity of the collar in terms of polar coordinates at $t = 1$ s.

13.138 In Problem 13.137, what is the acceleration of the collar in terms of polar coordinates at $t = 1$ s?

13.139 The polar coordinates of point A of the crane are given as functions of time by $r = 12 + 0.4t^2$ ft and $\theta = 0.02t^3$ rad. Determine the velocity of A in terms of polar coordinates at $t = 2$ s.

13.140 In Problem 13.139, determine the acceleration of A of the crane in terms of polar coordinates at $t = 2$ s.

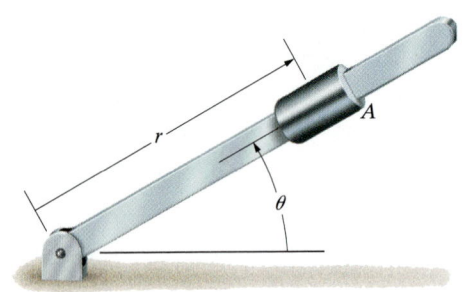

Problems 13.137/13.138

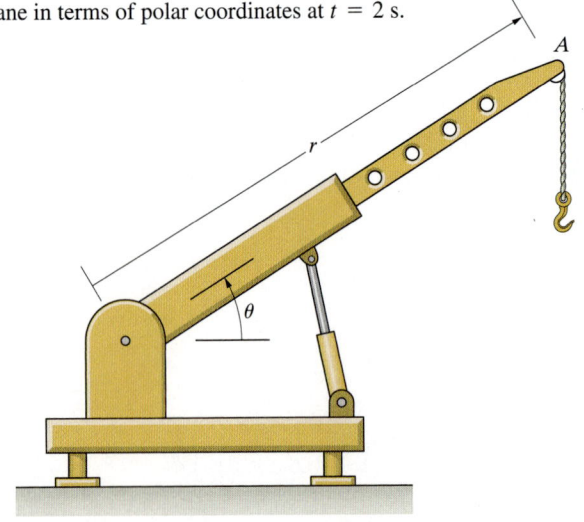

Problems 13.139/13.140

13.141 The radial line rotates with a constant angular velocity of 2 rad/s. Point P moves along the line at a constant speed of 4 m/s. Determine the magnitudes of the velocity and acceleration of P when $r = 2$ m.

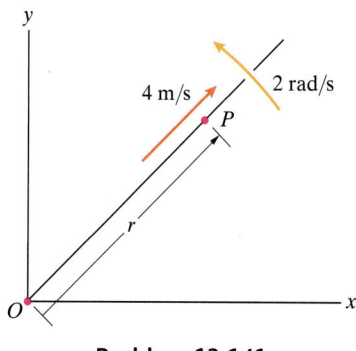

Problem 13.141

13.142 The collar A slides on the vertical bar. At the instant shown, its coordinates are $x = 0.9$ m, $y = 0.6$ m, and its velocity and acceleration are $\mathbf{v} = 4\mathbf{j}$ (m/s) and $\mathbf{a} = -9.81\mathbf{j}$ (m/s^2). Determine the velocity of A in terms of polar coordinates.

13.143 Determine the acceleration of the collar A in Problem 13.142 in terms of polar coordinates.

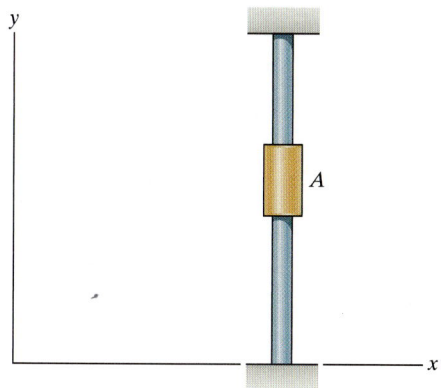

Problems 13.142/13.143

13.144* A boat searching for underwater archaeological sites in the Aegean Sea moves at 4 knots and follows the path $r = 10\theta$ m, where θ is in radians. (A knot is one nautical mile, or 1852 meters, per hour.) When $\theta = 2\pi$ rad, determine the boat's velocity (a) in terms of polar coordinates and (b) in terms of cartesian coordinates.

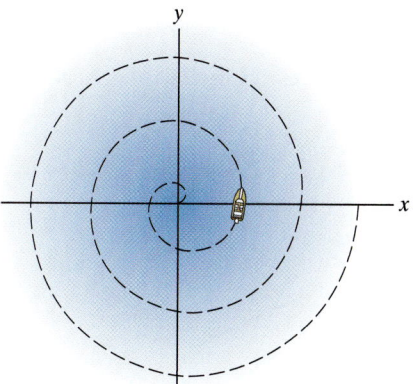

Problem 13.144

13.145 The collar A slides on the circular bar. The radial position of A (in meters) is given as a function of θ by $r = 2 \cos \theta$. At the instant shown, $\theta = 25°$ and $d\theta/dt = 4$ rad/s. Determine the velocity of A in terms of polar coordinates.

13.146 In Problem 13.145, $d^2\theta/dt^2 = 0$ at the instant shown. Determine the acceleration of A in terms of polar coordinates.

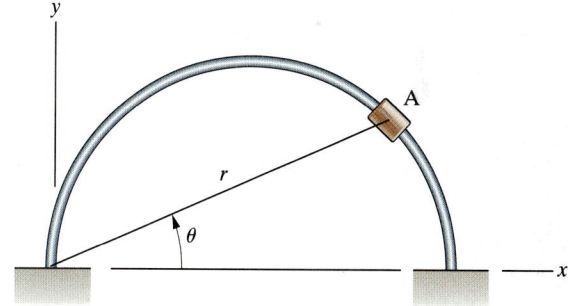

Problems 13.145/13.146

13.147 The radial coordinate of the earth satellite is related to its angular position θ by

$$r = \frac{1.91 \times 10^7}{1 + 0.5 \cos \theta} \text{ m}.$$

The product of the radial position and the transverse component of the velocity is

$$r v_\theta = 8.72 \times 10^{10} \text{ m}^2/s.$$

What is the satellite's velocity in terms of polar coordinates when $\theta = 90°$?

13.148* In Problem 13.147, what is the satellite's acceleration in terms of polar coordinates when $\theta = 90°$?

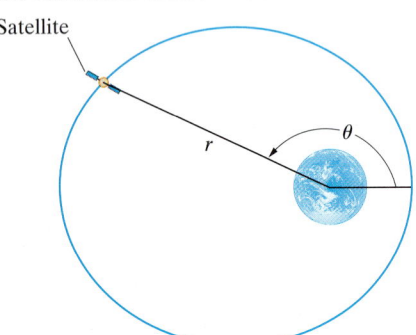

Problems 13.147/13.148

13.149 A bead slides along a wire that rotates in the x–y plane with constant angular velocity ω_0. The radial component of the bead's acceleration is zero. The radial component of its velocity is v_0 when $r = r_0$. Determine the polar components of the bead's velocity as a function of r.

Strategy: The radial component of the bead's velocity is

$$v_r = \frac{dr}{dt},$$

and the radial component of its acceleration is

$$a_r = \frac{d^2 r}{dt^2} - r\left(\frac{d\theta}{dt}\right)^2 = \frac{dv_r}{dt} - r\omega_0^2.$$

By using the chain rule,

$$\frac{dv_r}{dt} = \frac{dv_r}{dr}\frac{dr}{dt} = \frac{dv_r}{dr}v_r,$$

you can express the radial component of the acceleration in the form

$$a_r = \frac{dv_r}{dr}v_r - r\omega_0^2.$$

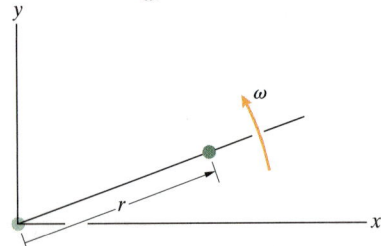

Problem 13.149

13.150 If the motion of a point in the x–y plane is such that its transverse component of acceleration a_θ is zero, show that the product of its radial position and its transverse velocity is constant: $r v_\theta = $ constant.

13.151* From astronomical data, Kepler deduced that the line from the sun to a planet traces out equal areas in equal times (Fig. a). Show that this result follows from the fact that the transverse component a_θ of the planet's acceleration is zero. [When r changes by an amount dr and θ changes by an amount $d\theta$ (Fig. b), the resulting differential element of area is $dA = \frac{1}{2}r(r\,d\theta)$.]

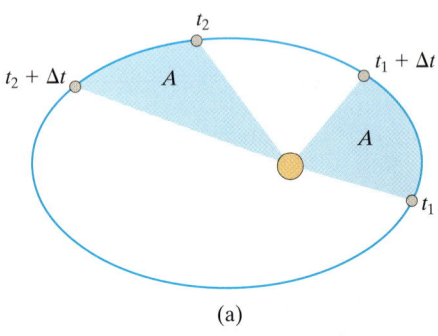

(a)

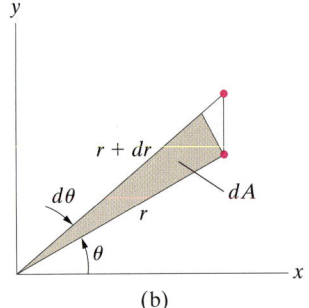

(b)

Problem 13.151

13.152 The bar rotates in the x–y plane with constant angular velocity ω_0. The radial component of acceleration of the collar C is $a_r = -Kr$, where K is a constant. When $r = r_0$, the radial component of velocity of C is v_0. Determine the polar components of the velocity of C as functions of r.

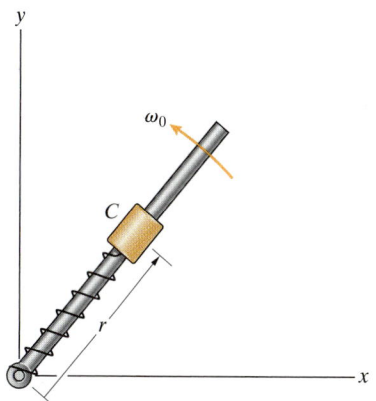

Problem 13.152

13.153 The hydraulic actuator moves the pin P upward with velocity $\mathbf{v} = 2\mathbf{j}$ (m/s). Determine the velocity of the pin in terms of polar coordinates and the angular velocity of the slotted bar when $\theta = 35°$.

13.154 The hydraulic actuator moves the pin P upward with constant velocity $\mathbf{v} = 2\mathbf{j}$ (m/s). Determine the acceleration of the pin in terms of polar coordinates and the angular acceleration of the slotted bar when $\theta = 35°$.

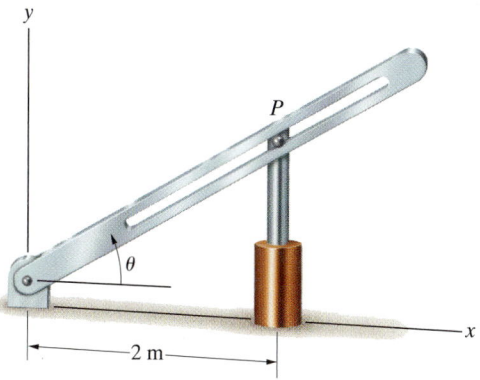

Problems 13.153/13.154

13.155 In Example 13.15, determine the velocity of the cam follower when $\theta = 135°$ (a) in terms of polar coordinates and (b) in terms of cartesian coordinates.

13.156* In Example 13.15, determine the acceleration of the cam follower when $\theta = 135°$ (a) in terms of polar coordinates and (b) in terms of cartesian coordinates.

13.157 In the cam–follower mechanism, the slotted bar rotates with constant angular velocity $\omega = 10$ rad/s and the radial position of the follower A is determined by the profile of the stationary cam. The path of the follower is described by the polar equation

$$r = 1 + 0.5 \cos 2\theta \text{ ft.}$$

Determine the velocity of the cam follower when $\theta = 30°$ (a) in terms of polar coordinates and (b) in terms of cartesian coordinates.

13.158* In Problem 13.157, determine the acceleration of the cam follower when $\theta = 30°$ (a) in terms of polar coordinates and (b) in terms of cartesian coordinates.

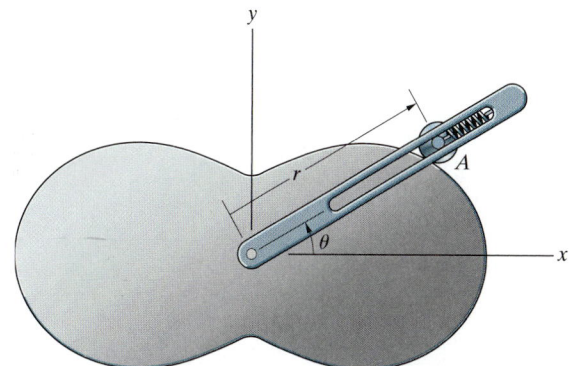

Problems 13.157/13.158

13.159* The cartesian coordinates of a point P in the x–y plane are related to the polar coordinates of the point by the equations $x = r \cos \theta$ and $y = r \sin \theta$.

(a) Show that the unit vectors $\mathbf{i}$ and $\mathbf{j}$ are related to the unit vectors $\mathbf{e}_r$ and $\mathbf{e}_\theta$ by

$$\mathbf{i} = \cos \theta \, \mathbf{e}_r - \sin \theta \, \mathbf{e}_\theta$$

and

$$\mathbf{j} = \sin \theta \, \mathbf{e}_r + \cos \theta \, \mathbf{e}_\theta.$$

(b) Beginning with the expression for the position vector of P in terms of cartesian coordinates, $\mathbf{r} = x\mathbf{i} + y\mathbf{j}$, derive Eq. (13.52) for the position vector in terms of polar coordinates.

(c) By taking the time derivative of the position vector of point P expressed in terms of cartesian coordinates, derive Eq. (13.55) for the velocity in terms of polar coordinates.

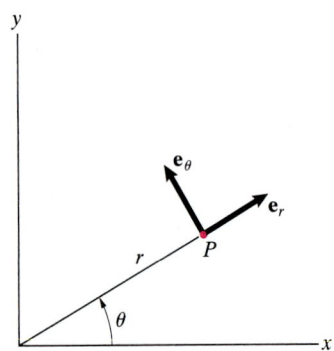

Problem 13.159

13.160 The airplane flies in a straight line at 400 mi/h. The radius of its propeller is 5 ft, and the propeller turns at 2000 rpm in the counterclockwise direction when seen from the front of the airplane. Determine the velocity and acceleration of a point on the tip of the propeller in terms of cylindrical coordinates. (Let the z axis be oriented as shown in the figure.)

Problem 13.160

13.161 A charged particle P in a magnetic field moves along the spiral path described by $r = 1$ m, $\theta = 2z$ rad, where z is in meters. The particle moves along the path in the direction shown with constant speed $|\mathbf{v}| = 1$ km/s. What is the velocity of the particle in terms of cylindrical coordinates?

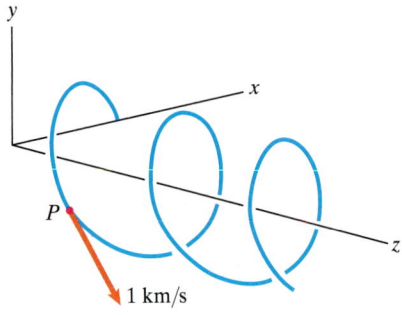

Problem 13.161

13.4 Relative Motion

We have discussed the curvilinear motion of a point relative to a given reference frame. In many applications, it is necessary to analyze the motions of two or more points relative to a reference frame and also their motions relative to each other. As a simple example, consider a passenger on a moving bus. If he walks down the aisle, the position and velocity that are important to him are his position in the bus and how fast he is moving down the aisle. His subjective motion is relative to the bus. But he also has a position and velocity relative to the earth. It would be convenient to have a framework for analyzing the bus's motion relative to the earth, the passenger's motion relative to the bus, and his motion relative to the earth. We develop such a framework in this section, introducing concepts and terminology that will be used in many contexts throughout the book.

Let A and B be two points whose motions we want to describe relative to a reference frame with origin O. We denote the positions of A and B relative to O by $\mathbf{r}_A$ and $\mathbf{r}_B$ (Fig.13.48). We also want to describe the motion of point A relative to point B, and denote the position of A relative to B by $\mathbf{r}_{A/B}$. These vectors are related by

$$\mathbf{r}_A = \mathbf{r}_B + \mathbf{r}_{A/B}. \tag{13.65}$$

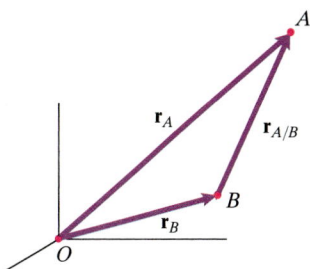

Figure 13.48

Stated in words, *the position of A is equal to the position of B plus the position of A relative to B*. Notice that when we simply say the "position of A" or "position of B," we mean their positions relative to O. The derivative of Eq. (13.65) with respect to time is

$$\frac{d\mathbf{r}_A}{dt} = \frac{d\mathbf{r}_B}{dt} + \frac{d\mathbf{r}_{A/B}}{dt}.$$

We write this equation as

$$\mathbf{v}_A = \mathbf{v}_B + \mathbf{v}_{A/B}, \tag{13.66}$$

where $\mathbf{v}_A$ is the velocity of A relative to O, $\mathbf{v}_B$ is the velocity of B relative to O, and $\mathbf{v}_{A/B} = d\mathbf{r}_{A/B}/dt$ is the velocity of A relative to B. *The velocity of A is equal to the velocity of B plus the velocity of A relative to B*. We now take the derivative of Eq. (13.66) with respect to time,

$$\frac{d\mathbf{v}_A}{dt} = \frac{d\mathbf{v}_B}{dt} + \frac{d\mathbf{v}_{A/B}}{dt},$$

and write this equation as

$$\mathbf{a}_A = \mathbf{a}_B + \mathbf{a}_{A/B}. \tag{13.67}$$

The term $\mathbf{a}_A$ is the acceleration of A relative to O, $\mathbf{a}_B$ is the acceleration of B relative to O, and $\mathbf{a}_{A/B} = d\mathbf{v}_{A/B}/dt$ is the acceleration of A relative to B. *The acceleration of A is equal to the acceleration of B plus the acceleration of A relative to B*.

Although they are simple in form, Eqs. (13.65)–(13.67) and the underlying concepts are extremely useful, and we apply them in a variety of contexts throughout the book.

Example 13.16 Motion of a Ship in a Current

A ship moving at 5 m/s relative to the water is in a uniform current flowing east at 2 m/s. If the helmsman wants to travel northwest relative to the earth, what direction must he point the ship? What is the resulting magnitude of the ship's velocity relative to the earth?

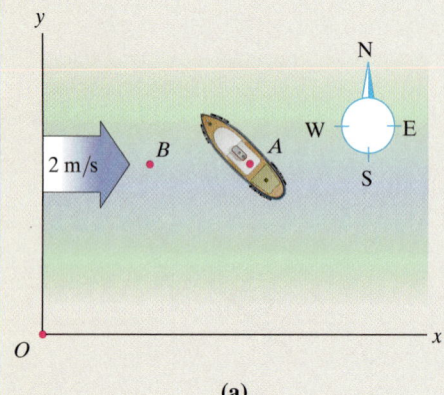

(a)

Strategy

Let the reference frame in Fig. a be stationary with respect to the earth. We denote the ship by A and define B to be a point that is stationary *relative to the water*. That is, point B moves toward the east at 2 m/s. Therefore, we know $\mathbf{v}_B$, the magnitude of $\mathbf{v}_{A/B}$ (the ship's speed relative to the water), and the desired direction of $\mathbf{v}_A$. We can use Eq. (13.66) to determine the direction of $\mathbf{v}_{A/B}$ and the magnitude of $\mathbf{v}_A$.

Solution

The ship's velocity relative to the earth is equal to the sum of the velocity of the water and the ship's velocity relative to the water:

$$\mathbf{v}_A = \mathbf{v}_B + \mathbf{v}_{A/B}. \tag{1}$$

In Fig. b we show these velocities together with the information that is known about them. Point B, which is assumed to be fixed with respect to the water, moves toward the east at 2 m/s. The magnitude of the velocity of the ship relative to the water is 5 m/s, and the direction of the ship's velocity relative to the earth is northwest.

In terms of the reference frame shown, $\mathbf{v}_B = 2\mathbf{i}$ (m/s). Let v_A be the unknown magnitude of $\mathbf{v}_A$. We can write $\mathbf{v}_A$ in terms of components as

$$\mathbf{v}_A = -v_A \cos 45°\mathbf{i} + v_A \sin 45°\mathbf{j}.$$

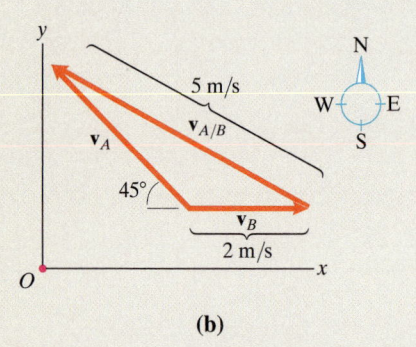

(b)

From Eq. (1), the velocity of the ship relative to the water is

$$\mathbf{v}_{A/B} = \mathbf{v}_A - \mathbf{v}_B$$

$$= -(v_A \cos 45° + 2 \text{ m/s})\mathbf{i} + v_A \sin 45°\mathbf{j}. \tag{2}$$

The magnitude of this vector is known to be 5 m/s:

$$|\mathbf{v}_{A/B}| = \sqrt{(v_A \cos 45° + 2 \text{ m/s})^2 + (v_A \sin 45°)^2} = 5 \text{ m/s}.$$

Solving this equation, we find that the magnitude of the ship's velocity relative to the earth is $v_A = 3.38$ m/s. Substituting this result into Eq. (2), the ship's velocity relative to the water is

$$\mathbf{v}_{A/B} = -4.39\mathbf{i} + 2.39\mathbf{j} \text{ (m/s)}.$$

The helmsman must point the ship at $\arctan(4.39/2.39) = 61.4°$ west of north to travel northwest relative to the earth.

Critical Thinking

When you apply Eqs. (13.65)–(13.67), the points A and B will usually be specific points whose motions you want to analyze. But in this example, notice how we analyzed the ship's motion relative to the water and relative to the earth by defining a fictitious point B that moved with the water.

Problems

13.162 At $t = 0$, two projectiles A and B are simultaneously launched from O with the initial velocities and elevation angles shown. Determine the velocity of projectile A relative to projectile B (a) at $t = 0.5$ s and (b) at $t = 1$ s.

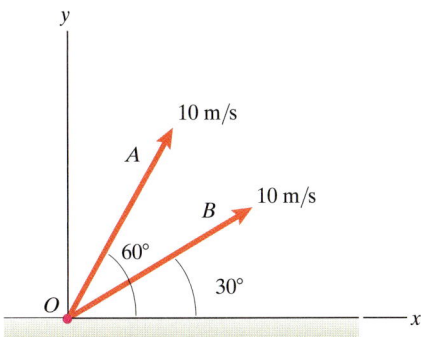

Problem 13.162

13.163 Relative to the earth-fixed coordinate system, the disk rotates about the fixed point O at 10 rad/s. What is the velocity of point A relative to point B at the instant shown?

13.164 Relative to the earth-fixed coordinate system, the disk rotates about the fixed point O with a constant angular velocity of 10 rad/s. What is the acceleration of point A relative to point B at the instant shown?

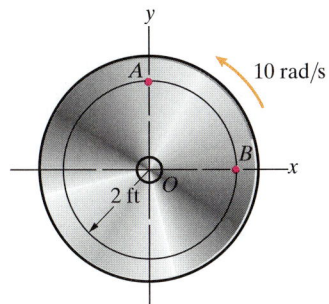

Problems 13.163/13.164

13.165 The train on the circular track is traveling at 50 ft/s. The train on the straight track is traveling at 20 ft/s. In terms of the earth-fixed coordinate system shown, what is the velocity of passenger A relative to passenger B?

13.166 The train on the circular track is traveling at a constant speed of 50 ft/s. The train on the straight track is traveling at 20 ft/s and is increasing its speed at 2 ft/s². In terms of the earth-fixed coordinate system shown, what is the acceleration of passenger A relative to passenger B?

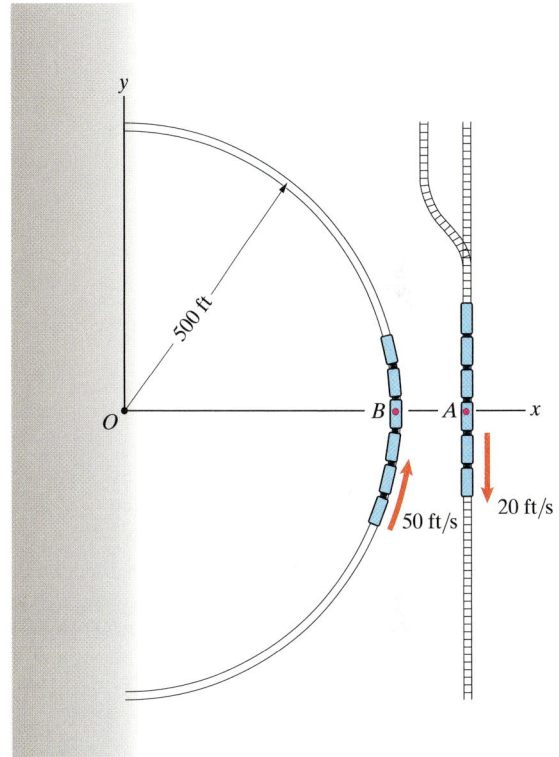

Problems 13.165/13.166

13.167 Each bar is 0.4-m long and rotates in the x–y plane. They are connected by a pin at A. Relative to the reference frame shown, bar OA has a counterclockwise angular velocity of 5 rad/s and bar AB has a counterclockwise angular velocity of 10 rad/s. What is the velocity of point B relative to the reference frame at the instant shown?

Strategy: Point A moves in a circular path about O. *Relative to point A*, point B moves in a circular path. (Imagine yourself sitting on point A with point B rotating around you.) Determine the x and y components of the velocity of A relative to O and the velocity of B relative to A. Then you can determine the velocity of B relative to O.

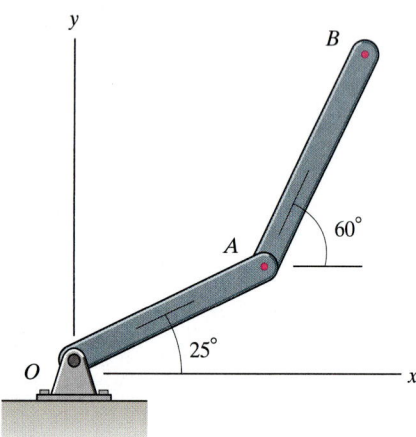

Problem 13.167

13.168 A private pilot wishes to fly from a city P to a city Q that is 200 km directly north of city P. The airplane will fly with an airspeed of 290 km/h. At the altitude at which the airplane will be flying, there is an east wind (that is, the wind's direction is west) with a speed of 50 km/h. What direction should the pilot point the airplane to fly directly from city P to city Q? How long will the trip take?

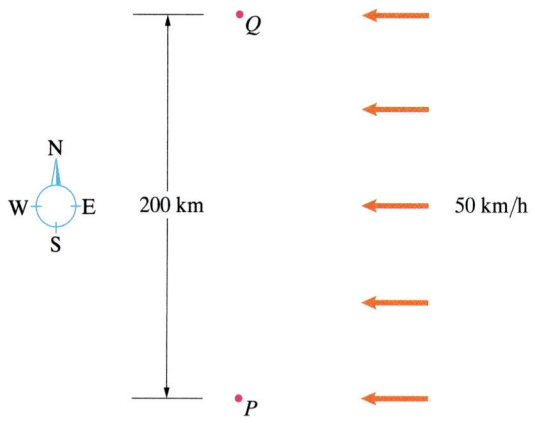

Problem 13.168

13.169 The river flows north at 3 m/s. (Assume that the current is uniform.) If you want to travel in a straight line from point C to point D in a boat that moves at a constant speed of 10 m/s relative to the water, in what direction should you point the boat? How long does it take to make the crossing?

13.170 The river flows north at 3 m/s. (Assume that the current is uniform.) What minimum speed must a boat have relative to the water in order to travel in a straight line from point C to point D? How long does it take to make the crossing?

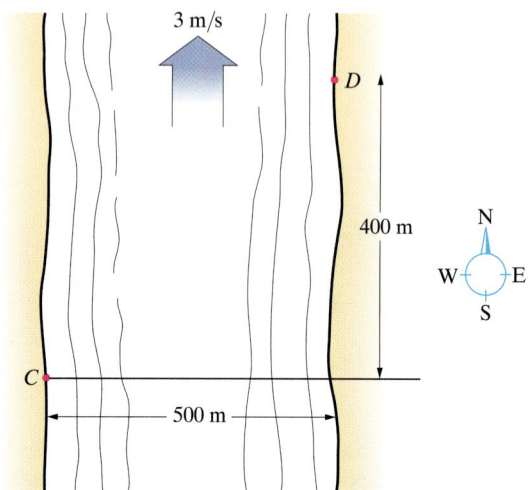

Problems 13.169/13.170

13.171* Relative to the earth, the sailboat sails north with speed $v_0 = 6$ knots (nautical miles per hour) and then sails east at the same speed. The telltale indicates the direction of the wind *relative to the boat*. Determine the direction and magnitude of the wind's velocity (in knots) relative to the earth.

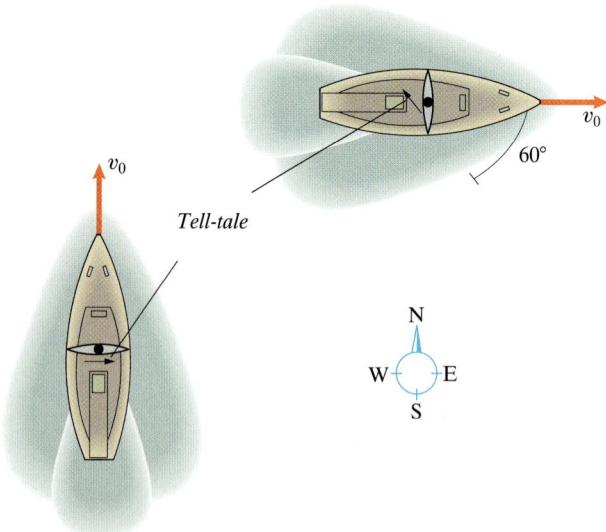

Problem 13.171

COMPUTATIONAL MECHANICS

The following examples and problems are designed to be worked with the use of a programmable calculator or computer.

Computational Example 13.17

With buoyancy accounted for, the downward acceleration of a steel ball falling in the container of liquid in Fig. 13.49 is $a = 0.9g - cv$, where c is a constant that is proportional to the viscosity of the liquid. To determine the viscosity, a rheologist releases the ball from rest at the surface of the liquid. If the ball requires 2 s to fall the 2 m to the bottom, what is the value of c?

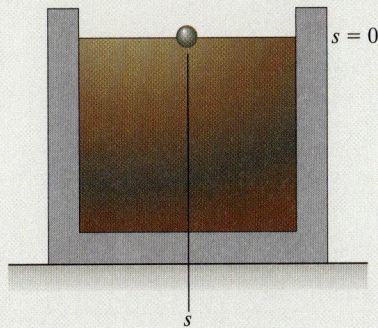

$s = 0$

s

Figure 13.49

Strategy
We can obtain an equation for c by determining the distance the ball falls as a function of time.

Solution
We measure the ball's position s downward from the point of release and let $t = 0$ be the time of release. The acceleration is

$$a = \frac{dv}{dt} = 0.9g - cv.$$

Separating variables and integrating gives

$$\int_0^v \frac{dv}{0.9g - cv} = \int_0^t dt,$$

from which it follows that

$$\left[\frac{1}{-c} \ln(0.9g - cv) \right]_0^v = \left[t \right]_0^t,$$

or

$$\frac{1}{-c}[\ln(0.9g - cv) - \ln(0.9g)] = \frac{1}{-c}\ln\left(\frac{0.9g - cv}{0.9g} \right) = t.$$

Solving for v, we obtain

$$v = \frac{ds}{dt} = \frac{0.9g}{c}(1 - e^{-ct}).$$

Integrating this equation yields

$$\int_0^s ds = \int_0^t \frac{0.9g}{c}(1 - e^{-ct})\, dt,$$

or

$$\left[s \right]_0^s = \frac{0.9g}{c}\left[t - \frac{e^{-ct}}{(-c)} \right]_0^t,$$

from which we obtain the distance the ball has fallen as a function of the time from its release:

$$s = \frac{0.9g}{c^2}(ct - 1 + e^{-ct}).$$

We know that $s = 2$ m when $t = 2$ s, so determining c requires solving the equation

$$f(c) = \frac{(0.9)(9.81)}{c^2}(2c - 1 + e^{-2c}) - 2 = 0.$$

We can't solve this transcendental equation in closed form to determine c. Many computer programs are designed to obtain roots of such equations. Another approach is to compute the value of $f(c)$ for a range of values of c and plot the results, as we have done in Fig. 13.50. From the graph, we estimate that $c = 8.3$ s^{-1}.

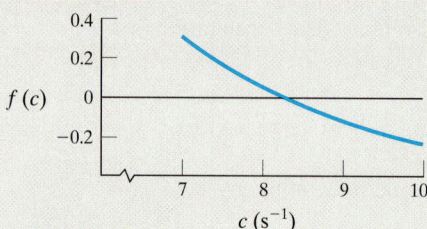

Figure 13.50
Graph of the function $f(c)$.

Critical Thinking

The information required in a problem can dictate whether an analytical or numerical solution is appropriate. Notice that if c was known in this example, our analytical solution determines the distance s the ball has fallen as a function of time. But if c is unknown and must be determined by knowing the distance s the ball falls in a given time, a nonlinear equation must be solved numerically.

Computational Example 13.18

In an industrial process, the sprayer in Fig. 13.51 projects a stream of liquid onto the horizontal surface. An engineer wants to place the sprayer at a horizontal position x_e that maximizes its coverage of the horizontal surface. That is, she wants to maximize the distance $x_s - x_e$ reached by the spray. The angle θ_0 can be varied, but the spray velocity $v_0 = 3$ m/s is fixed. The dimensions $H = 0.2$ m, $L = 0.12$ m, and $h = 0.06$ m. If the droplets of the spray are treated as projectiles, what is the desired position x_e?

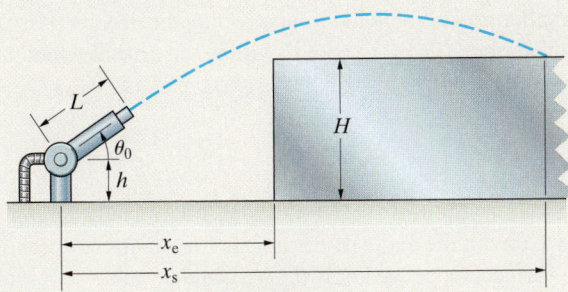

Figure 13.51

Strategy

We first observe that the trajectory that maximizes the coverage of the horizontal surface will be one that just clears the edge of the surface. For if a trajectory clears the edge by some horizontal distance, as in Fig. 13.51, the sprayer can be moved that distance to the right, increasing its coverage by that amount. Our procedure will be to choose a distance x_e, determine the angle θ_0 that causes the trajectory to just clear the edge, and calculate $x_s - x_e$. By doing this for a range of values of x_e, we can determine the horizontal placement that maximizes $x_s - x_e$.

Solution

Let $t = 0$ be the time at which a droplet leaves the nozzle. In terms of the coordinate system shown in Fig. a, the x and y coordinates of the droplet are

$$x = L \cos \theta_0 + v_0 \cos \theta_0 t$$

and

$$y = h + L \sin \theta_0 + v_0 \sin \theta_0 t - \frac{1}{2} g t^2.$$

By setting $y = H$, we can solve for the times at which the droplet is at the height of the horizontal surface. The two resulting solutions for t are

$$t_1, t_2 = \frac{v_0 \sin \theta_0 \pm \sqrt{v_0^2 \sin^2 \theta_0 - 2g(H - h - L \sin \theta_0)}}{g}.$$

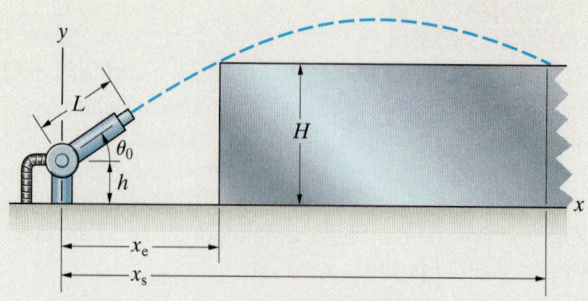

(a)

We wish the smaller root t_1 to be the time at which the droplet passes the edge of the horizontal surface. Then the root t_2 will be the time at which the droplet lands on the surface. Setting $x = x_e$ when $t = t_1$, we obtain

$$x_e = L \cos \theta_0 + v_0 \cos \theta_0 t_1,$$

which is an equation we can use to determine θ_0 numerically for a given value of x_e. That is, we obtain the angle of the nozzle for which the spray just clears the edge of the surface. Once we know θ_0, we can determine x_s by substituting t_2 into the equation for x:

$$x_s = L \cos \theta_0 + v_0 \cos \theta_0 t_2.$$

Figure 13.52 shows the values of $x_s - x_e$ we obtained for a range of values of x_e. From the graph we estimate that a maximum coverage of $x_s - x_e = 0.82$ m is obtained at $x_e = 0.12$ m. The trajectory that gives the maximum coverage, obtained with a nozzle angle $\theta_0 = 50°$, is shown in Fig. 13.53.

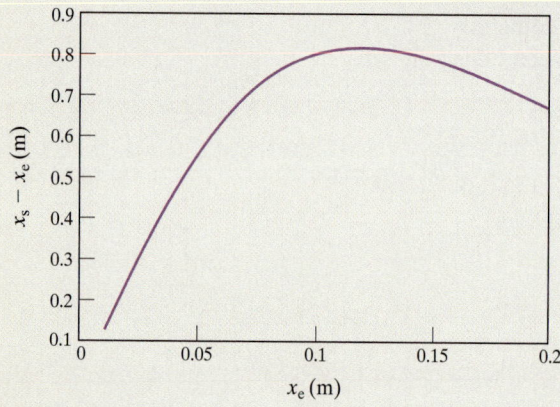

Figure 13.52
Graph of the extent of coverage as a function of the placement of the nozzle.

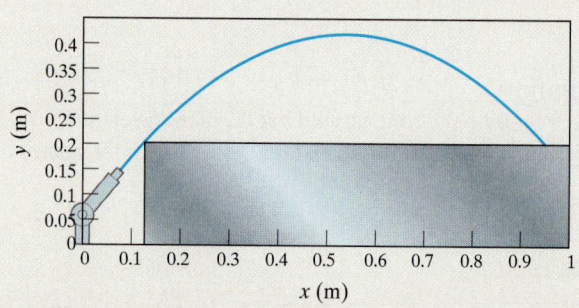

Figure 13.53
Placement of the nozzle and the trajectory that maximizes coverage of the surface.

Critical Thinking

This example demonstrates how the solution to a problem in dynamics and computational results can be combined to analyze an industrial process. In particular, notice how the graphical presentation of computed results can give insight into the sensitivity of the coverage to the placement of the nozzle.

Computational Problems

13.172 An engineer analyzing a large-scale machining process determines that a tool which moves in a straight line starts from rest at time $t = 0$ and position $s = 0$ and moves with acceleration

$$a = 2 + t^{1/2} - t^{3/2} \text{ m/s}^2$$

from $t = 0$ to $t = 4$ s.

(a) Draw a graph of the tool's position from $t = 0$ to $t = 4$ s.

(b) What is the maximum velocity of the tool during this time interval, and at what time does it occur?

13.173 A projectile is launched at 10 m/s from a sloping surface.

(a) Determine the values of the angle α for which the range $R = 15$m.

(b) Determine the angle α for which the range R is a maximum. What is the maximum range?

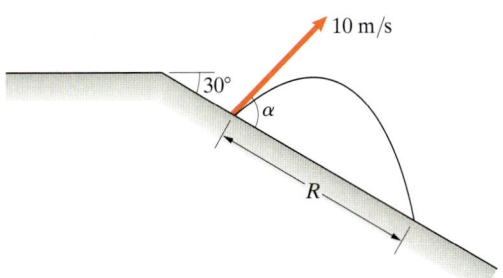

Problem 13.173

13.174 A catapult designed to throw a line to ships in distress hurls a projectile with initial velocity $v_0(1 - 0.4 \sin \theta_0)$, where θ_0 is the angle above the horizontal. Determine the value of θ_0 for which the distance the projectile is thrown is a maximum, and show that the maximum distance is $0.559 v_0^2/g$.

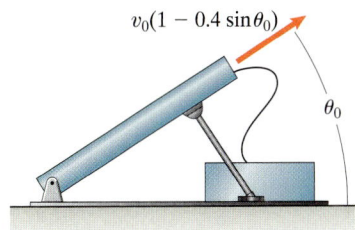

Problem 13.174

13.175 At $t = 0$, a projectile is located at the origin and has a velocity of 20 m/s at $40°$ above the horizontal. The profile of the surface the projectile strikes can be approximated by the equation $y = 0.4x - 0.006x^2$, where x and y are in meters. Determine the approximate coordinates of the point where the projectile hits the ground.

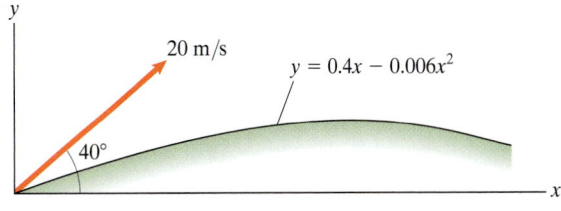

Problem 13.175

13.176 A carpenter working on a house asks his apprentice to throw him an apple. The apple is thrown at 32 ft/s. What two values of θ_0 will cause the apple to land in the carpenter's hand, 12 ft horizontally and 12 ft vertically from the point where it is thrown?

Problem 13.176

13.177 A motorcycle starts from rest at $t = 0$ and moves along a circular track with a 400-m radius. The tangential component of its acceleration is $a_t = 2 + 0.2t$ m/s². When the magnitude of its total acceleration reaches 6 m/s², friction can no longer keep the motorcycle on the circular track, and it spins out. How long after it starts does it spin out, and how fast is it going?

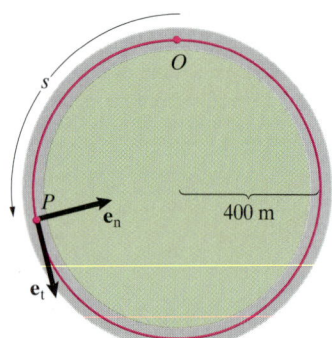

Problem 13.177

13.178 At $t = 0$, a steel ball in a tank of oil is given a horizontal velocity $\mathbf{v} = 2\mathbf{i}$ m/s. The components of the ball's acceleration are $a_x = -cv_x$, $a_y = -0.8g - cv_y$, and $a_z = -cv_z$, where c is a constant. When the ball hits the bottom of the tank, its position relative to its position at $t = 0$ is $\mathbf{r} = 0.8\mathbf{i} - \mathbf{j}$ (m). What is the value of c?

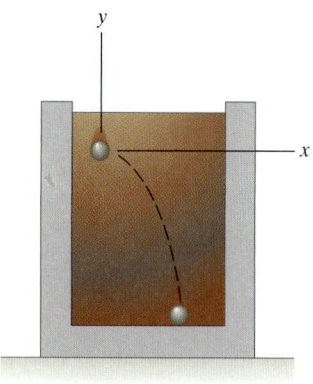

Problem 13.178

13.179 The polar coordinates of a point P moving in the x–y plane are $r = t^3 - 4t$ m, $\theta = t^2 - t$ rad.

(a) Draw a graph of the magnitude of the velocity of P from $t = 0$ to $t = 2$ s.

(b) Estimate the minimum magnitude of the velocity and the time at which it occurs.

13.180 (a) Draw a graph of the magnitude of the acceleration of the point P in Problem 13.179 from $t = 0$ to $t = 2$ s.

(b) Estimate the minimum magnitude of the acceleration and the time at which it occurs.

13.181 The robot is programmed so that point P describes the path
$$r = 1 - 0.5 \cos 2\pi t \text{ m}$$
$$\theta = 0.5 - 0.2 \sin[2\pi(t - 0.1)] \text{ rad.}$$
Determine the values of r and θ at which the magnitude of the velocity of P attains its maximum value.

13.182 In Problem 13.181, determine the values of r and θ at which the magnitude of the acceleration of P attains its maximum value.

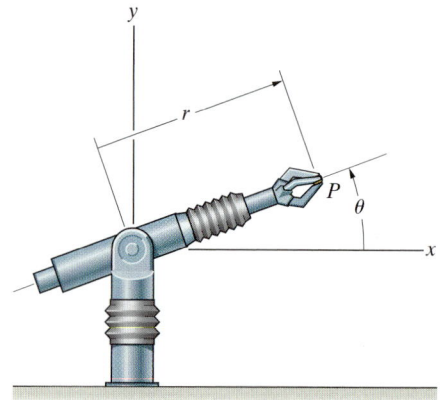

Problems 13.181/13.182

13.183 In the cam–follower mechanism, the slotted bar rotates with constant angular velocity $\omega = 10$ rad/s, and the radial position of the follower A is determined by the profile of the stationary cam. The path of the follower is described by the polar equation
$$r = 1 + 0.5 \cos 2\theta \text{ ft.}$$

(a) Draw a graph of the magnitude of the follower's acceleration as a function of θ for $0 < \theta < 360°$.

(b) Use your graph to estimate the maximum magnitude of the follower's acceleration and the angle(s) at which it occurs.

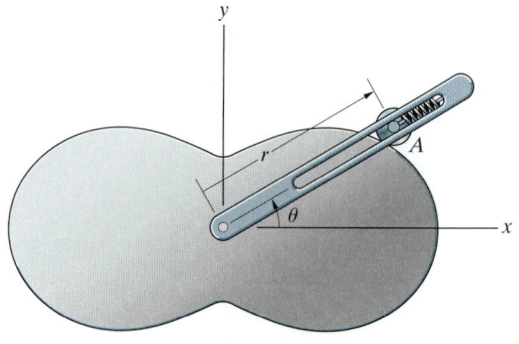

Problem 13.183

CHAPTER SUMMARY

In this chapter we were concerned with the motion of a point relative to a given reference frame. We defined the position, velocity, and acceleration of the point and showed how to express them in terms of cartesian coordinates, normal and tangential components, and polar coordinates. We demonstrated how to determine the velocity and acceleration by differentiation when the position is known and how to determine the velocity and position by integration when the acceleration is known. In Chapter 14, we will use Newton's second law to determine the acceleration of an object when the forces acting on it are known. Once the acceleration is known, the methods we developed in the current chapter will be applied to obtain information about the velocity and position of the object.

The position of a point P relative to a given reference frame with origin O can be specified by the *position vector* $\mathbf{r}$ from O to P. The *velocity* of P relative to the reference frame is

$$\mathbf{v} = \frac{d\mathbf{r}}{dt}, \tag{13.1}$$

and the *acceleration* of P relative to the reference frame is

$$\mathbf{a} = \frac{d\mathbf{v}}{dt}. \tag{13.2}$$

Straight-Line Motion

Consider a point P and a reference point O on a straight line. The position of P relative to O can be specified by a coordinate s measured along the line from O to P. The velocity of P relative to O is

$$v = \frac{ds}{dt}, \tag{13.3}$$

and the acceleration of P relative to O is

$$a = \frac{dv}{dt}. \tag{13.4}$$

Applying the chain rule to Eq. (13.4) yields an expression for the acceleration that is often useful:

$$a = \frac{dv}{ds}v. \tag{13.5}$$

If the acceleration is known as a function of time, Eq. (13.4) can be integrated to determine the velocity as a function of time, and then Eq. (13.3) can be integrated to determine the position as a function of time.

If the acceleration is known as a function of velocity $dv/dt = a(v)$, the velocity can be determined as a function of time by separating variables and integrating:

$$\int_{v_0}^{v} \frac{dv}{a(v)} = \int_{t_0}^{t} dt. \tag{13.16}$$

If the acceleration is specified as a function of position, $a = a(s)$, Eq. (13.5) can be integrated as a function of position:

$$\int_{v_0}^{v} v \, dv = \int_{s_0}^{s} a(s) \, ds. \tag{13.19}$$

Cartesian Coordinates

The position, velocity, and acceleration relative to the cartesian coordinate system in Fig. a are [Eqs. (13.21)–(13.25)]

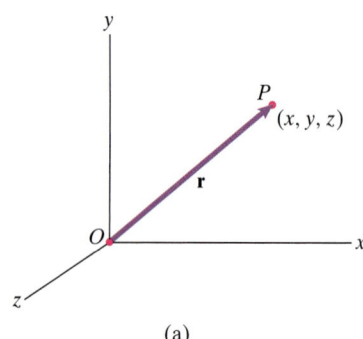

(a)

$$\mathbf{r} = x\mathbf{i} + y\mathbf{j} + z\mathbf{k},$$

$$\mathbf{v} = v_x\mathbf{i} + v_y\mathbf{j} + v_z\mathbf{k} = \frac{dx}{dt}\mathbf{i} + \frac{dy}{dt}\mathbf{j} + \frac{dz}{dt}\mathbf{k},$$

$$\mathbf{a} = a_x\mathbf{i} + a_y\mathbf{j} + a_z\mathbf{k} = \frac{dv_x}{dt}\mathbf{i} + \frac{dv_y}{dt}\mathbf{j} + \frac{dv_z}{dt}\mathbf{k}.$$

The equations describing the motion in each coordinate direction are identical in form to the equations that describe the motion of a point along a straight line.

Angular Motion

The angular velocity ω and angular acceleration α of L relative to L_0 are (Fig. b)

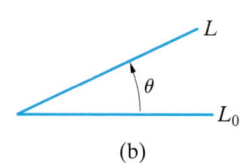

(b)

$$\omega = \frac{d\theta}{dt}, \tag{13.31}$$

$$\alpha = \frac{d\omega}{dt} = \frac{d^2\theta}{dt^2}. \tag{13.32}$$

Normal and Tangential Components

The velocity and acceleration are (Fig. c),

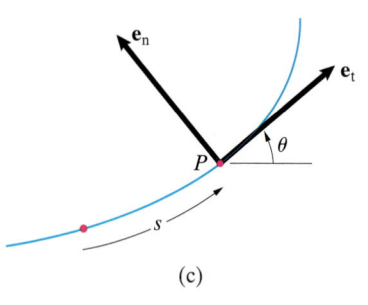

(c)

$$\mathbf{v} = v\mathbf{e}_t = \frac{ds}{dt}\mathbf{e}_t \tag{13.38}$$

and

$$\mathbf{a} = a_t\mathbf{e}_t + a_n\mathbf{e}_n, \tag{13.39}$$

where

$$a_t = \frac{dv}{dt} \quad \text{and} \quad a_n = v\frac{d\theta}{dt} = \frac{v^2}{\rho}. \tag{13.40}$$

The unit vector $\mathbf{e}_n$ points toward the concave side of the path. The term ρ is the instantaneous radius of curvature of the path.

Polar Coordinates

The position, velocity, and acceleration are (Fig. d),

$$\mathbf{r} = r\mathbf{e}_r, \tag{13.52}$$

$$\mathbf{v} = v_r\mathbf{e}_r + v_\theta\mathbf{e}_\theta = \frac{dr}{dt}\mathbf{e}_r + r\omega\mathbf{e}_\theta, \tag{13.57}$$

and

$$\mathbf{a} = a_r\mathbf{e}_r + a_\theta\mathbf{e}_\theta, \tag{13.58}$$

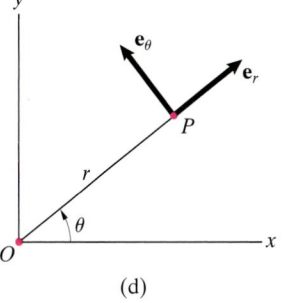

(d)

where

$$a_r = \frac{d^2r}{dt^2} - r\left(\frac{d\theta}{dt}\right)^2 = \frac{d^2r}{dt^2} - r\omega^2$$

and

$$a_\theta = r\frac{d^2\theta}{dt^2} + 2\frac{dr}{dt}\frac{d\theta}{dt} = r\alpha + 2\frac{dr}{dt}\omega. \tag{13.59}$$

Relative Motion

Let $\mathbf{r}_A$ and $\mathbf{r}_B$ be the positions of points A and B relative to a reference frame with origin O, and let $\mathbf{r}_{A/B}$ be the position of A relative to B (Fig. e). The position of A is equal to the position of B plus the position of A relative to B:

$$\mathbf{r}_A = \mathbf{r}_B + \mathbf{r}_{A/B}. \tag{13.65}$$

Differentiating this equation with respect to time yields the result

$$\mathbf{v}_A = \mathbf{v}_B + \mathbf{v}_{A/B}. \tag{13.66}$$

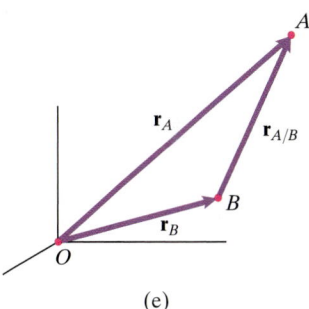

(e)

The velocity of A is equal to the velocity of B plus the velocity of A relative to B. Differentiating Eq. (13.66) with respect to time yields

$$\mathbf{a}_A = \mathbf{a}_B + \mathbf{a}_{A/B}. \tag{13.67}$$

The acceleration of A is equal to the acceleration of B plus the acceleration of A relative to B.

Review Problems

13.184 Suppose that you throw a ball straight up at 10 m/s and release it at 2 m above the ground.

(a) What maximum height above the ground does the ball reach?
(b) How long after you release it does the ball hit the ground?
(c) What is the magnitude of its velocity just before it hits the ground?

13.185 Suppose that you must determine the duration of the yellow light at a highway intersection. Assume that cars will be approaching the intersection traveling as fast as 65 mi/h, that drivers' reaction times are as long as 0.5 s, and that cars can safely achieve a deceleration of at least 0.4 g.

(a) How long must the light remain yellow to allow drivers to come to a stop safely before the light turns red?
(b) What is the minimum distance cars must be from the intersection when the light turns yellow to come to a stop safely at the intersection?

13.186 The acceleration of a point moving along a straight line is $a = 4t + 2$ m/s^2. When $t = 2$ s, the position of the point is $s = 36$ m, and when $t = 4$ s, its position is $s = 90$ m. What is the velocity of the point when $t = 4$ s?

13.187 A model rocket takes off straight up. Its acceleration during the 2 s its motor burns is 25 m/s^2. Neglect aerodynamic drag, and determine

(a) the maximum velocity of the rocket during the flight and
(b) the maximum altitude the rocket reaches.

13.188 In Problem 13.187, if the rocket's parachute fails to open, what is the total time of flight from takeoff until the rocket hits the ground?

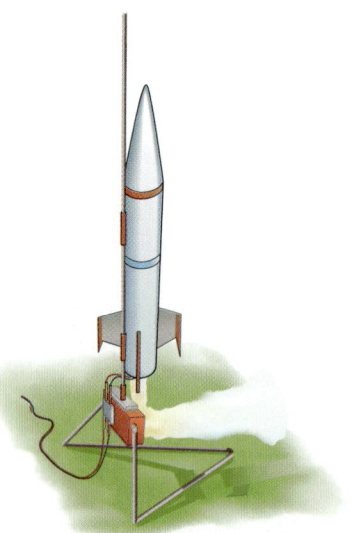

Problems 13.187/13.188

13.189 The acceleration of a point moving along a straight line is $a = -cv^3$, where c is a constant. If the velocity of the point is v_0, what distance does the point move before its velocity decreases to $v_0/2$?

13.190 Water leaves the nozzle at $20°$ above the horizontal and strikes the wall at the point indicated. What is the velocity of the water as it leaves the nozzle?

Strategy: Determine the motion of the water by treating each particle of water as a projectile.

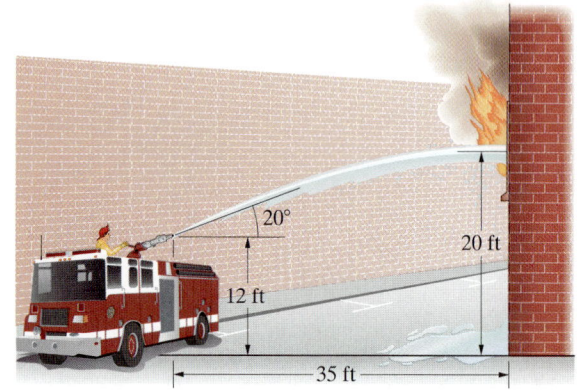

Problem 13.190

13.191 In practice, the quarterback throws the football with velocity v_0 at $45°$ above the horizontal. At the same instant, the receiver standing 20 ft in front of him starts running straight downfield at 10 ft/s and catches the ball. Assume that the ball is thrown and caught at the same height above the ground. What is the velocity v_0?

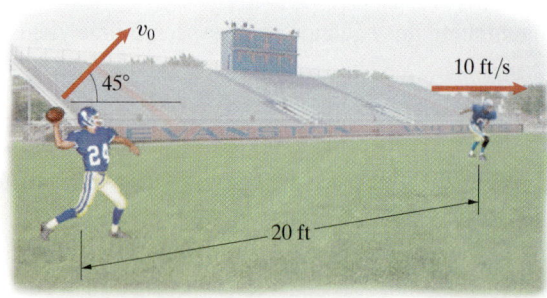

Problem 13.191

13.192 The constant velocity $v = 2$ m/s. What are the magnitudes of the velocity and acceleration of point P when $x = 0.25$ m?

13.193 The constant velocity $v = 2$ m/s. What is the acceleration of point P in terms of normal and tangential components when $x = 0.25$ m?

13.194 The constant velocity $v = 2$ m/s. What is the acceleration of point P in terms of polar coordinates when $x = 0.25$ m?

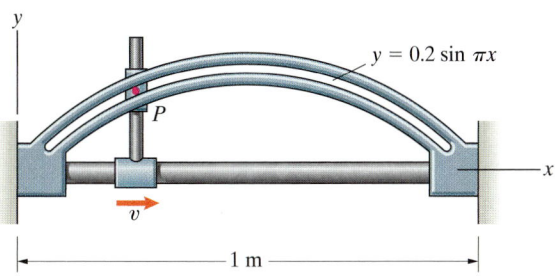

$y = 0.2 \sin \pi x$

Problems 13.192–13.194

13.195 A point P moves along the spiral path $r = (0.1)\theta$ ft, where θ is in radians. The angular position $\theta = 2t$ rad, where t is in seconds, and $r = 0$ at $t = 0$. Determine the magnitudes of the velocity and acceleration of P at $t = 1$ s.

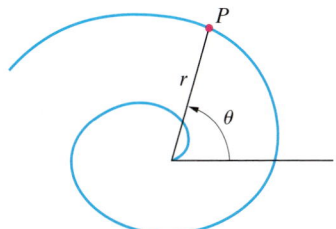

Problem 13.195

13.196 In the cam–follower mechanism, the slotted bar rotates with constant angular velocity $\omega = 12$ rad/s, and the radial position of the follower A is determined by the profile of the stationary cam. The slotted bar is pinned a distance $h = 0.2$ m to the left of the center of the circular cam. The follower moves in a circular path 0.42 m in radius. Determine the velocity of the follower when $\theta = 40°$ (a) in terms of polar coordinates and (b) in terms of cartesian coordinates.

13.197* In Problem 13.196, determine the acceleration of the follower when $\theta = 40°$ (a) in terms of polar coordinates and (b) in terms of cartesian coordinates.

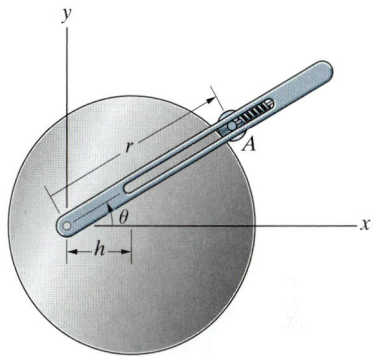

Problems 13.196/13.197

Design Project

Design and carry out experiments to measure the acceleration due to gravity. Galileo (1564–1642) did so by measuring the motions of falling objects. Use his method, but also try to devise other approaches that may result in improved accuracy. Investigate the repeatability of your measurements. Write a brief report describing your experiments, discussing possible sources of error, and presenting your results.

Force, Mass, and Acceleration

Until now, we have analyzed motions of objects without considering the forces that cause them. In this chapter we relate cause
and effect: By drawing the free-body diagram of an object to
identify the forces acting on it, we can use Newton's second law
to determine the acceleration of the object. Alternatively, when
we know an object's acceleration, we can use Newton's second
law to obtain information about the forces acting on it.

◄ When the forces acting on the soccer ball are known, Newton's second law
can be used to determine the acceleration of the ball's center of mass.

14.1 Newton's Second Law

Newton stated that the total force on a particle is equal to the rate of change of its *linear momentum*, which is the product of its mass and velocity:

$$\mathbf{f} = \frac{d}{dt}(m\mathbf{v}).$$

If the particle's mass is constant, the total force equals the product of its mass and acceleration:

$$\mathbf{f} = m\frac{d\mathbf{v}}{dt} = m\mathbf{a}. \tag{14.1}$$

We pointed out in Chapter 12 that the second law gives precise meanings to the terms *force* and *mass*. Once a unit of mass is chosen, a unit of force is defined to be the force necessary to give one unit of mass an acceleration of unit magnitude. For example, the unit of force in SI units, the newton, is the force necessary to give a mass of one kilogram an acceleration of one meter per second squared. In principle, the second law then gives the value of any force and the mass of any object. By subjecting a one-kilogram mass to an arbitrary force and measuring the acceleration of the mass, we can solve the second law for the direction of the force and its magnitude in newtons. By subjecting an arbitrary mass to a one-newton force and again measuring the acceleration, we can solve the second law for the value of the mass in kilograms.

    If the mass of a particle and the total force acting on it are known, Newton's second law determines its acceleration. In Chapter 13, we described how to determine the velocity, position, and trajectory of a point whose acceleration is known. Therefore, with the second law, a particle's motion can be determined when the total force acting on it is known, or the total force can be determined when the motion is known.

14.2 Equation of Motion for the Center of Mass

Newton's second law is postulated for a particle, or small element of matter, but an equation of precisely the same form describes the motion of the center of mass of an arbitrary object. We can show that the total external force on an arbitrary object is equal to the product of its mass and the acceleration of its center of mass.

    To do so, we consider an arbitrary system of N particles. Let m_i be the mass of the ith particle, and let $\mathbf{r}_i$ be its position vector (Fig. 14.1a). Let m be the total mass of the particles; that is,

$$m = \sum_i m_i,$$

where the summation sign with subscript i means "the sum over i from 1 to N." The position of the center of mass of the system is

$$\mathbf{r} = \frac{\sum_i m_i \mathbf{r}_i}{m}.$$

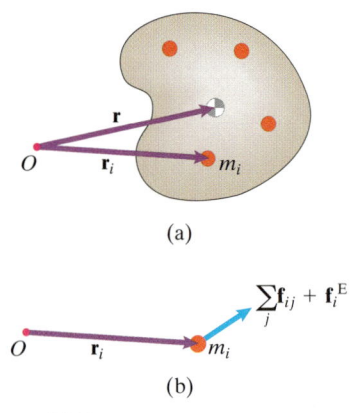

(a)

(b)

Figure 14.1
(a) Dividing an object into particles. The vector $\mathbf{r}_i$ is the position vector of the ith particle, and $\mathbf{r}$ is the position vector of the object's center of mass.
(b) Forces on the ith particle.

By taking two time derivatives of this expression, we obtain

$$\sum_i m_i \frac{d^2 \mathbf{r}_i}{dt^2} = m \frac{d^2 \mathbf{r}}{dt^2} = m\mathbf{a}, \tag{14.2}$$

where $\mathbf{a}$ is the acceleration of the center of mass.

The ith particle of the system may be acted upon by forces exerted by the other particles. Let $\mathbf{f}_{ij}$ be the force exerted on the ith particle by the jth particle. Then Newton's third law states that the ith particle exerts a force on the jth particle of equal magnitude and opposite direction: $\mathbf{f}_{ji} = -\mathbf{f}_{ij}$. If the external force on the ith particle (i.e., the total force exerted on the ith particle by objects other than the object we are considering) is denoted by $\mathbf{f}_i^{\mathrm{E}}$, Newton's second law for the ith particle is (Fig. 14.1b)

$$\sum_j \mathbf{f}_{ij} + \mathbf{f}_i^{\mathrm{E}} = m_i \frac{d^2 \mathbf{r}_i}{dt^2}.$$

We can write this equation for each particle of the system. Summing the resulting equations from $i = 1$ to N, we obtain

$$\sum_i \sum_j \mathbf{f}_{ij} + \sum_i \mathbf{f}_i^{\mathrm{E}} = \sum_i m_i \frac{d^2 \mathbf{r}_i}{dt^2}. \tag{14.3}$$

The first term on the left side, the sum of the internal forces on the system, is zero due to Newton's third law:

$$\sum_i \sum_j \mathbf{f}_{ij} = \mathbf{f}_{12} + \mathbf{f}_{21} + \mathbf{f}_{13} + \mathbf{f}_{31} + \cdots = \mathbf{0}.$$

The second term on the left side of Eq. (14.3) is the sum of the external forces on the system. Denoting this sum by $\Sigma \mathbf{F}$ and using Eq. (14.2), we conclude that *the sum of the external forces equals the product of the total mass and the acceleration of the center of mass*:

$$\Sigma \mathbf{F} = m\mathbf{a}. \tag{14.4}$$

Because this equation is identical in form to Newton's postulate for a single particle, for convenience we also refer to it as Newton's second law.

Notice that we made no assumptions restricting the nature of the system of particles or its state of motion in obtaining Eq. (14.4). The sum of the external forces on any object or collection of objects, solid, liquid, or gas, equals the product of the total mass and the acceleration of the center of mass.

For example, suppose that the space shuttle is in orbit and has fuel remaining in its tanks. If its engines are turned on, the fuel sloshes in a complicated manner, affecting the shuttle's motion due to internal forces between the fuel and the shuttle. Nevertheless, we can use Eq. (14.4) to determine the exact acceleration of the center of mass of the shuttle, including the fuel it contains, and thereby determine the velocity, position, and trajectory of the center of mass.

14.3 Inertial Reference Frames

When we discussed the motion of a point in Chapter 13, we specified the position, velocity, and acceleration of the point relative to an arbitrary reference frame. But Newton's second law cannot be expressed in terms of just any reference frame. Suppose that no force acts on a particle and that we measure the particle's motion relative to a particular reference frame and determine that its acceleration is zero. In terms of this reference frame, Newton's second law agrees with our observation. But if we then measure the particle's motion relative to a second reference frame that is accelerating or rotating with respect to the first one, we would find that the particle's acceleration is *not* zero. In terms of the second reference frame, Newton's second law, at least in the form given by Eq. (14.4), does not predict the correct result.

A well-known example is a person riding in an elevator. Suppose that you conduct an experiment in which you ride in an elevator while standing on a set of scales that measure your weight (Fig. 14.2a). The forces acting on you are your weight W and the force N exerted on you by the scales (Fig. 14.2b). You exert an equal and opposite force N on the scales, which is the force they measure. If the elevator is stationary, you observe that the scales read your weight, $N = W$. The sum of the forces on you is zero, and Newton's second law correctly states that your acceleration relative to the elevator is zero. If the elevator has an upward acceleration a (Fig. 14.2c), you know you will feel heavier, and indeed, you observe that the scales read a force greater than your weight, $N > W$. In terms of an earth-fixed reference frame, Newton's second law correctly relates the forces acting on you to your acceleration: $\Sigma F = N - W = ma$. But suppose that you use the elevator as your frame of reference. Then the sum of the forces acting on you is not zero, so Newton's second law states that you are accelerating relative to the elevator. But you are stationary relative to the elevator. Thus, expressed in terms of this accelerating reference frame, *Newton's second law gives an erroneous result.*

Newton stated that the second law should be expressed in terms of a reference frame at rest with respect to the "fixed stars." Even if the stars were fixed that would not be practical advice, because virtually every convenient reference frame accelerates, rotates, or both. Newton's second law *can* be applied rigorously using reference frames that accelerate and rotate by properly accounting for the acceleration and rotation. We explain how to do this in Chapter 17, but for now, we need to give some guidance on when Newton's second law can be applied.

Fortunately, in nearly all "down-to-earth" situations, applying Eq. (14.4) in terms of a reference frame that is fixed relative to the earth results in sufficiently accurate answers. For example, if a piece of chalk is thrown across a room, a reference frame that is fixed relative to the room can be used to predict the chalk's motion. While the chalk is in motion, the earth rotates, and therefore the reference frame rotates. But *because the chalk's flight is brief*, the effect on the prediction is very small. (The earth rotates slowly—its angular velocity is one-half that of a clock's hour hand.) Equation (14.4) can usually be applied using a reference frame that translates (moves without rotating) at constant velocity relative to the earth. For example, if two people play tennis on the deck of a cruise ship moving with constant velocity relative to the earth, Eq. (14.4) can be expressed in terms of a reference frame fixed relative to the ship to analyze the ball's motion. But such a "ship-fixed" reference frame cannot be used if the ship is turning or changing its speed.

A reference frame in which Eq. (14.4) can be applied is said to be *Newtonian*, or *inertial*. We discuss inertial reference frames in greater detail in Chapter 17. For now, it should be assumed examples and problems are expressed in terms of inertial reference frames.

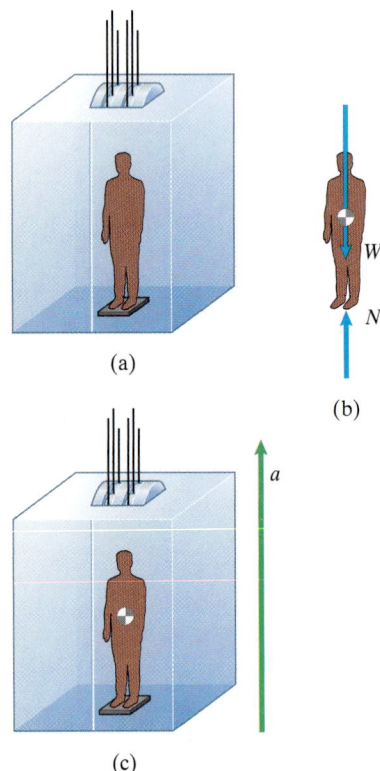

(a)

(b)

(c)

Figure 14.2
(a) Riding in an elevator while standing on scales.
(b) Your free-body diagram.
(c) Upward acceleration of the elevator.

14.4 Applications

In statics we described different types of forces, including the weights of objects, the normal and friction forces exerted by contacting surfaces, and forces exerted by linear springs. By showing these forces acting on free-body diagrams, information was obtained about the systems of forces acting on objects in equilibrium. In this section, we show that the same types of forces are dealt with in dynamics. Furthermore, the techniques used to draw free-body diagrams in statics also apply to objects that are not in equilibrium.

By drawing the free-body diagram of an object, the external forces acting on it can be identified and Newton's second law used to determine the object's acceleration. Conversely, if the motion of an object is known, Newton's second law can be used to determine the total external force on the object. In particular, if an object's acceleration in a particular direction is known to be zero, the sum of the external forces in that direction must equal zero.

To apply Newton's second law in a particular situation, a coordinate system must be chosen. Often, the forces acting on an object can be resolved into components most conveniently in terms of a particular coordinate system, or the choice may be determined by the object's path. In the sections that follow, we show how to use different types of coordinate systems to analyze the motions of objects and the forces acting on them.

Cartesian Coordinates and Straight-Line Motion

If we express the sum of the forces acting on an object of mass m and the acceleration of its center of mass in terms of their components in a cartesian reference frame (Fig. 14.3), Newton's second law states that

$$\Sigma \mathbf{F} = m\mathbf{a},$$

or

$$(\Sigma F_x \mathbf{i} + \Sigma F_y \mathbf{j} + \Sigma F_z \mathbf{k}) = m(a_x \mathbf{i} + a_y \mathbf{j} + a_z \mathbf{k}).$$

Equating x, y, and z components, we obtain three scalar equations of motion:

$$\Sigma F_x = ma_x, \qquad \Sigma F_y = ma_y, \qquad \Sigma F_z = ma_z. \qquad (14.5)$$

The total force in each coordinate direction equals the product of the mass and the component of the acceleration in that direction.

An important example is the projectile problem, in which an object is launched through the air and aerodynamic forces are neglected, so that the only force on the object is its weight. If we describe the motion of the object by using

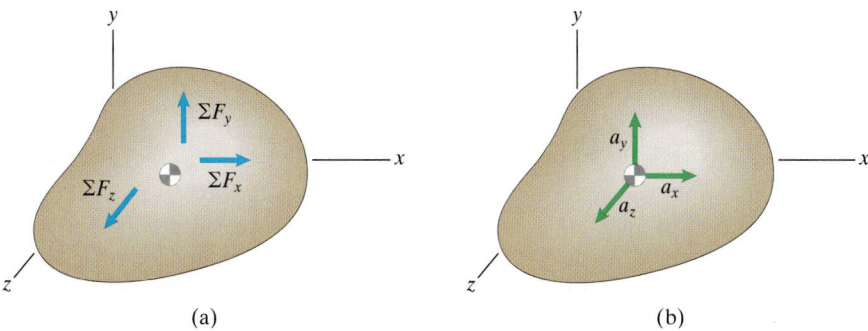

(a) (b)

Figure 14.3
(a) Cartesian components of the sum of the forces on an object.
(b) Components of the acceleration of the center of mass of the object.

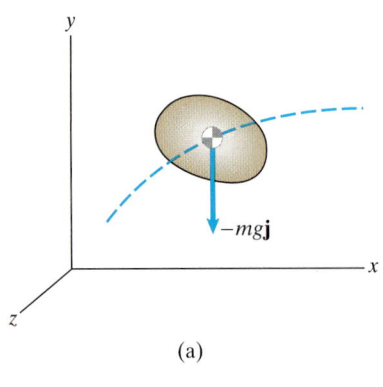

(a)

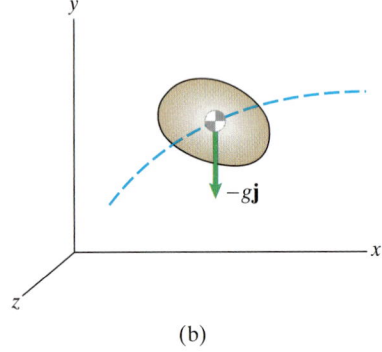

(b)

Figure 14.4
(a) Free-body diagram of a projectile.
(b) The resulting acceleration of the center of mass of the projectile.

an earth-fixed coordinate system with the y axis upward (Fig. 14.4a), the sum of the forces is $\Sigma \mathbf{F} = -mg\mathbf{j}$. Therefore, $\Sigma F_x = 0$, $\Sigma F_y = -mg$, and $\Sigma F_z = 0$, and from Eqs. (14.5) we obtain $a_x = 0$, $a_y = -g$, and $a_z = 0$. This was the basis for our assumption in Chapter 13 that a projectile accelerates downward with the acceleration due to gravity and has no horizontal acceleration (Fig. 14.4b).

If an object's motion is confined to the x–y plane, $a_z = 0$, so the sum of the forces in the z direction is zero. Thus, when the motion is confined to a fixed plane, the component of the total force normal to that plane equals zero. For straight-line motion along the x axis (Fig. 14.5a), Eqs. (14.5) are

$$\Sigma F_x = ma_x, \qquad \Sigma F_y = 0, \quad \text{and} \quad \Sigma F_z = 0.$$

We see that in straight-line motion, the components of the total force perpendicular to the line equal zero, and the component of the total force tangent to the line equals the product of the mass and the acceleration along the line (Fig. 14.5b).

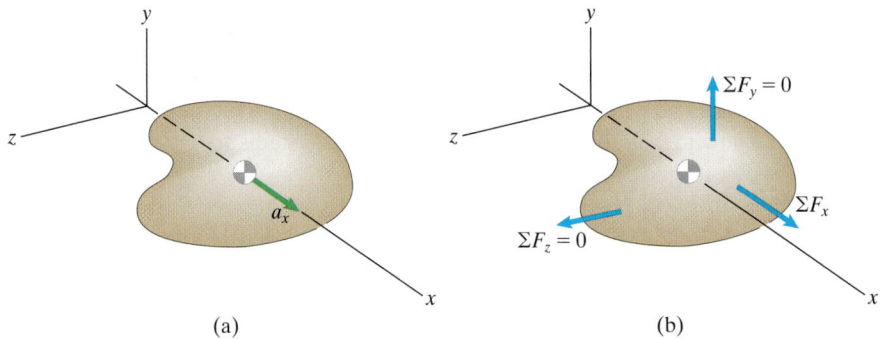

(a)   (b)

Figure 14.5
(a) Acceleration of an object in straight-line motion along the x axis.
(b) The y and z components of the total force acting on the object equal zero.

Study Questions

1. Explain how Newton's second law gives precise meanings to the terms *force* and *mass*.

2. Newton's second law states that the total external force on an object equals the product of the object's mass and the acceleration of what point?

3. What is an inertial reference frame?

4. If you know that the sum of the external forces on an object is zero in a particular direction, what does that tell you?

Example 14.1   Application to Straight-Line Motion

The 100-lb crate in Fig. 14.6 is released from rest on the smooth inclined surface at time $t = 0$. (We say that a surface is "smooth" when it is assumed to exert negligible friction force.) Determine how fast the crate is moving and how far it has moved at $t = 1$ s.

Strategy
We must draw the free-body diagram of the crate to identify the external forces acting on it. By applying Newton's second law, we can determine the crate's acceleration. Once the acceleration is known, we can determine the velocity and position of the crate as functions of time.

20°

Figure 14.6

Solution

We draw the free-body diagram of the crate and introduce a coordinate system in Fig. a. The external forces on the crate are its weight and the normal force exerted by the smooth surface.

The crate's mass is

$$m = \frac{W}{g} = \frac{100 \text{ lb}}{32.2 \text{ ft/s}^2} = 3.11 \text{ slug}.$$

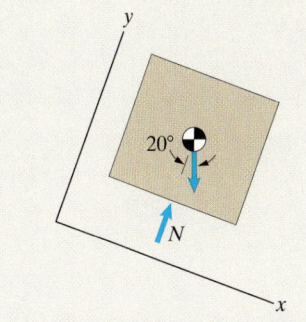

(a) Free-body diagram of the crate.

When the crate is released, it will accelerate down the inclined surface. Its acceleration parallel to the surface is shown in Fig. b. Applying Newton's second law in the direction parallel to the surface,

$$\Sigma F_x = ma_x:$$

$$(100 \text{ lb}) \sin 20° = (3.11 \text{ slug})a_x,$$

we find that the crate's acceleration is $a_x = 11.0 \text{ ft/s}^2$.

Because the acceleration is constant, we can use Eqs. (13.11) and (13.12) to determine how fast the crate is moving and how far it has moved at $t = 1$ s:

$$v = (11.0 \text{ ft/s}^2)(1 \text{ s}) = 11.0 \text{ ft/s},$$

$$s = \frac{1}{2}(11.0 \text{ ft/s}^2)(1 \text{ s})^2 = 5.51 \text{ ft}.$$

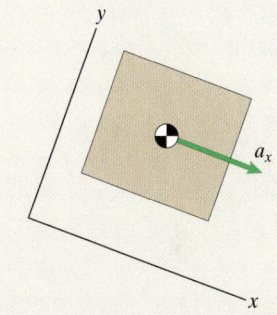

(b) The crate's acceleration.

Critical Thinking

This simple example demonstrates the typical steps you must carry out to apply Newton's second law to an object. You must first draw a free-body diagram to identify the external forces acting on the object and choose a coordinate system. (Notice that we aligned the x axis of the coordinate system with the direction of the crate's motion.) Then by applying Newton's second law, you can determine an object's acceleration. Knowing the acceleration, you can use the methods described in Chapter 13 to analyze the object's motion.

Example 14.2   Connected Objects in Straight-Line Motion

The two crates in Fig. 14.7 are released from rest. Their masses are $m_A = 40$ kg and $m_B = 30$ kg, and the coefficients of friction between crate A and the inclined surface are $\mu_s = 0.2$ and $\mu_k = 0.15$. What is the acceleration of the crates?

Strategy

We must first determine whether A slips. We will assume that the crates remain stationary and see whether the force of friction necessary for equilibrium exceeds the maximum friction force. If slip occurs, we can determine the resulting acceleration by drawing free-body diagrams of the crates and applying Newton's second law to them individually.

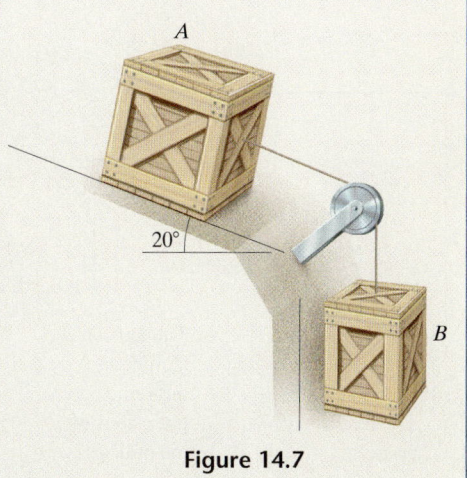

Figure 14.7

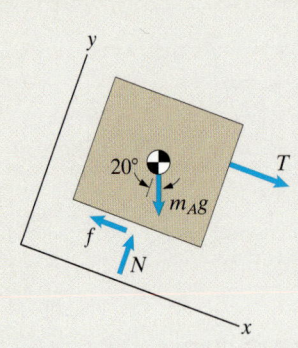

(a) Free-body diagram of crate A.

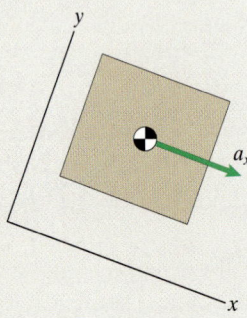

(b) The crate's acceleration.

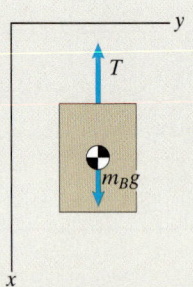

(c) Free-body diagram of crate B.

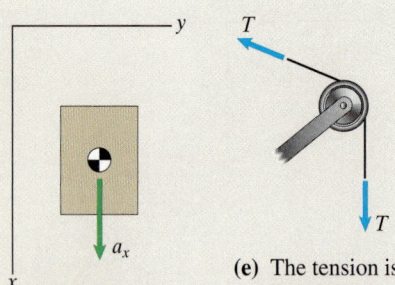

(d) Vertical acceleration of crate B.

(e) The tension is assumed to be the same on both sides of the pulley.

Solution

We draw the free-body diagram of crate A and introduce a coordinate system in Fig. a. If we assume that the crate does not slip, the following equilibrium equations apply:

$$\Sigma F_x = T + m_A g \sin 20° - f = 0;$$

$$\Sigma F_y = N - m_A g \cos 20° = 0.$$

In the first equation, the tension T equals the weight of crate B; therefore, the friction force necessary for equilibrium is

$$f = m_B g + m_A g \sin 20°$$

$$= (30 \text{ kg})(9.81 \text{ m/s}^2) + (40 \text{ kg})(9.81 \text{ m/s}^2) \sin 20°$$

$$= 429 \text{ N}.$$

The normal force $N = m_A g \cos 20°$, so the maximum friction force the surface will support is

$$f_{max} = \mu_s N$$

$$= (0.2)[(40 \text{ kg})(9.81 \text{ m/s}^2) \cos 20°]$$

$$= 73.7 \text{ N}.$$

Crate A will therefore slip, and the friction force is $f = \mu_k N$. We show the crate's acceleration down the plane in Fig. b. Its acceleration perpendicular to the plane is zero (i.e., $a_y = 0$). Applying Newton's second law yields

$$\Sigma F_x = T + m_A g \sin 20° - \mu_k N = m_A a_x$$

$$\Sigma F_y = N - m_A g \cos 20° = 0.$$

In this case, *we do not know* the tension T, because crate B is not in equilibrium. We show the free-body diagram of crate B and its vertical acceleration in Figs. c and d. The equation of motion is

$$\Sigma F_x = m_B g - T = m_B a_x.$$

(In terms of the two coordinate systems we use, the two crates have the same acceleration a_x.) Thus, by applying Newton's second law to both crates, we have obtained three equations in terms of the unknowns T, N, and a_x. Solving for a_x, we obtain $a_x = 5.33 \text{ m/s}^2$.

Critical Thinking

Notice that we assumed the tension in the cable to be the same on each side of the pulley (Fig. e). In fact, however, the tensions must be different, because a moment is necessary to cause angular acceleration of the pulley. For now, our only recourse is to assume that the pulley is light enough that the moment necessary to accelerate it is negligible. In Chapter 18 we include the analysis of the angular motion of the pulley in problems of this type and obtain more realistic solutions.

Example 14.3   Application to Straight-Line Motion

The airplane in Fig. 14.8 touches down on the aircraft carrier with a horizontal velocity of 50 m/s relative to the carrier. The arresting gear exerts a horizontal force of magnitude $T_x = 10{,}000v$ newtons (N), where v is the plane's velocity in meters per second. The plane's mass is 6500 kg.
(a) What maximum horizontal force does the arresting gear exert on the plane?
(b) If other horizontal forces can be neglected, what distance does the plane travel before coming to rest?

Figure 14.8

Strategy

(a) Since the plane begins to decelerate when it contacts the arresting gear, the maximum force occurs at first contact when $v = 50$ m/s.
(b) The horizontal force exerted by the arresting gear equals the product of the plane's mass and its acceleration. Once we know the acceleration, we can integrate to determine the distance required for the plane to come to rest.

Solution

(a) We draw the free-body diagram of the airplane and introduce a coordinate system in Fig. a. The forces T_x and T_y are the horizontal and vertical components of force exerted by the arresting gear, and N is the vertical force on the landing gear. The horizontal force on the plane is $\Sigma F_x = -T_x = -10{,}000v$ N. The magnitude of the maximum force is

$$10{,}000v = (10{,}000)(50) = 500{,}000 \text{ N},$$

or 112,400 lb.

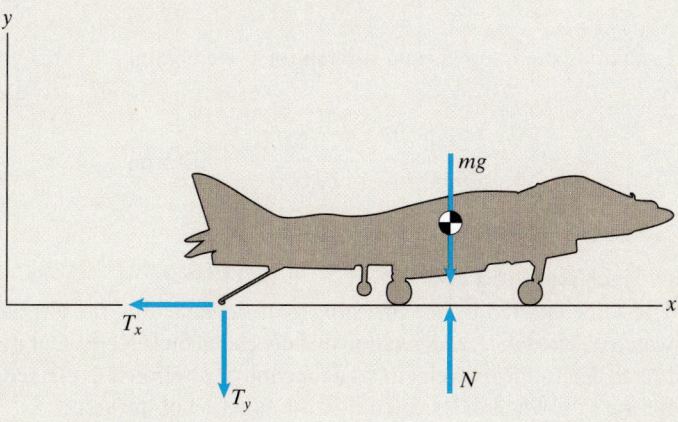

(a) Introducing a coordinate system with the x axis parallel to the horizontal force.

(b) In terms of the plane's horizontal component of acceleration (Fig. b), we obtain the equation of motion

$$\Sigma F_x = ma_x:$$

$$-10{,}000v_x = ma_x.$$

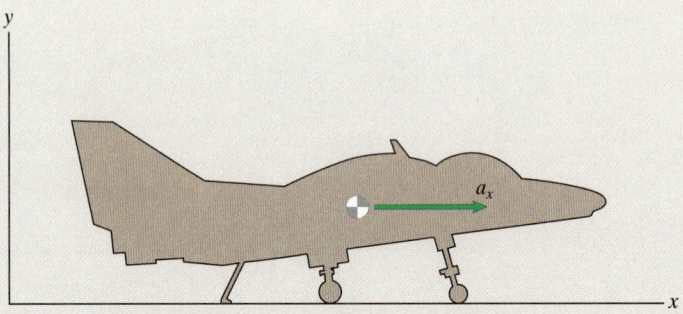

(**b**) The airplane's horizontal acceleration.

The airplane's acceleration' is a function of its velocity. We use the chain rule to express the acceleration in terms of a derivative with respect to x:

$$ma_x = m\frac{dv_x}{dt} = m\frac{dv_x}{dx}\frac{dx}{dt} = m\frac{dv_x}{dx}v_x = -10{,}000v_x.$$

Now we separate variables and integrate, defining $x = 0$ to be the position at which the plane contacts the arresting gear:

$$\int_{50}^{0} m\,dv_x = -\int_{0}^{x} 10{,}000\,dx.$$

Evaluating the integrals and solving for x, we obtain

$$x = \frac{50m}{10{,}000} = \frac{(50)(6500)}{10{,}000} = 32.5 \text{ m}.$$

Critical Thinking

The force exerted by the arresting gear depended on the airplane's velocity, which resulted in an acceleration that depended on velocity. Our use of the chain rule to determine the velocity as a function of position is explained in Table 13.1 for the case when the acceleration is a function of the velocity.

Problems

14.1 The 100-lb crate is released from rest on the smooth inclined surface. How long does it take the crate to slide 4 ft down the surface from its initial position? What is the magnitude of its velocity at that time?

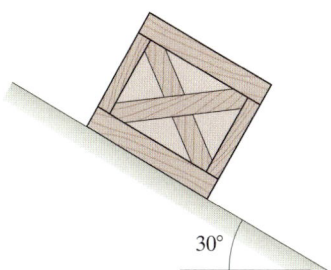

30°

Problem 14.1

14.2 The Sikorsky UH-60A helicopter weighs 20,500 lb. It takes off vertically with its rotor exerting a constant thrust of 24,000 lb. How high does it rise in the first 6 s of flight?

14.3 The 20,500-lb helicopter shown in Problem 14.2 is at rest on the ground at time $t = 0$. Its pilot advances the throttle so that the upward thrust exerted by the rotor as a function of time is $T = 18,000 + 1000t$ lb. How high has the helicopter risen at $t = 6$ s?

Problems 14.2/14.3

14.4 The 8-kg object is initially stationary on the smooth horizontal surface. At $t = 0$, it is subjected to a constant force $F = 10$ N.

(a) How fast is it moving at $t = 3$ s?

(b) What distance has it moved at $t = 3$ s?

14.5 Solve Problem 14.4 if the coefficient of kinetic friction between the object and the surface is $\mu_k = 0.1$.

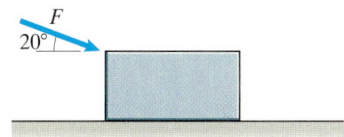

F
20°

Problems 14.4/14.5

14.6 The 8-kg object starts from rest on the smooth inclined surface at $t = 0$ and is subjected to a constant horizontal force $F = 40$ N.

(a) How fast is it moving at $t = 3$ s?

(b) What distance has it moved at $t = 3$ s?

14.7 Solve Problem 14.6 if the coefficient of kinetic friction between the object and the surface is $\mu_k = 0.1$.

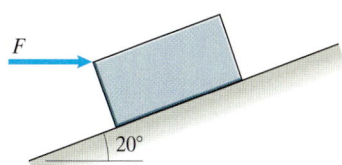

F

20°

Problems 14.6/14.7

14.8 The schussing skier is on a 25° slope. His mass is 80 kg. The kinetic coefficient of friction between his skis and the snow is $\mu_s = 0.08$. His velocity is 9 m/s.

(a) What is his acceleration in the direction parallel to the slope?

(b) What is his velocity when he has gone 20 m down the slope?

14.9 Aerodynamic drag exerts a force on the skier in Problem 14.8 of magnitude $0.6v^2$, where v is the magnitude of his velocity. Determine the skier's velocity when he has gone 20 m down the slope, and compare your answer with the answer to part (b) of Problem 14.8.

Strategy: After using Newton's second law to determine the skier's acceleration in the direction parallel to the slope, use the chain rule to express it as the product of the velocity and the derivative of the velocity with respect to position.

Problems 14.8/14.9

14.10 The total external force on the 10-kg object is constant and equal to $\Sigma\mathbf{F} = 90\mathbf{i} - 60\mathbf{j} + 20\mathbf{k}$ (N). At time $t = 0$, its velocity is $\mathbf{v} = -14\mathbf{i} + 26\mathbf{j} + 32\mathbf{k}$ (m/s). What is its velocity at $t = 4$ s?

14.11 The total external force on the 10-kg object shown in Problem 14.10 is given as a function of time by $\Sigma\mathbf{F} = (-20t + 90)\mathbf{i} - 60\mathbf{j} + (10t + 40)\mathbf{k}$ (N). At time $t = 0$, its position is $\mathbf{r} = 40\mathbf{i} + 30\mathbf{j} - 360\mathbf{k}$ (m) and its velocity is $\mathbf{v} = -14\mathbf{i} + 26\mathbf{j} + 32\mathbf{k}$ (m/s). What is its position at $t = 4$ s?

14.12 The position of the 10-kg object is given as a function of time by $\mathbf{r} = (20t^3 - 300)\mathbf{i} + 60t^2\mathbf{j} + (6t^4 - 40t^2)\mathbf{k}$ (m). What is the total external force on the object at $t = 2$ s?

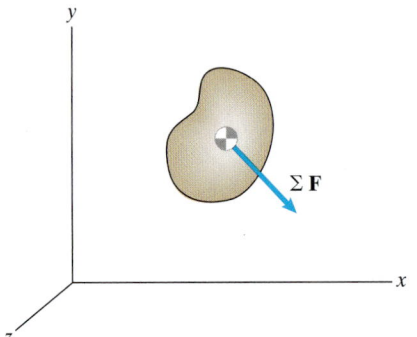

Problems 14.10–14.12

14.13 The velocity of an 800-kg missile is measured by radar from $t = 0$ to $t = 6$ s and is determined to be $\mathbf{v} = (15t - 0.5t^2)\mathbf{i} + (50t - 2t^2)\mathbf{j}$ (m/s).

(a) What is the change in position of the missile (its displacement) from $t = 0$ to $t = 6$ s?
(b) Determine the total force acting on the missile at $t = 3$ s.

Problem 14.13

14.14 At the instant shown, the horizontal component of acceleration of the 26,000-lb airplane due to the sum of the external forces acting on it is 14 ft/s². If the pilot suddenly increases the magnitude of the thrust T by 4000 lb, what is the horizontal component of the plane's acceleration immediately afterward?

Problem 14.14

14.15 The rocket is traveling straight up at low altitude. Its mass at the instant shown is 90,000 kg and the thrust of its engine is 1200 kN. An onboard accelerometer indicates that its upward acceleration is 3 m/s².

(a) Draw the free-body diagram of the rocket showing the forces acting on it, including the aerodynamic drag force.
(b) What is the magnitude of the aerodynamic drag force?

Problem 14.15

14.16 A 2-kg cart containing 8 kg of water is initially stationary (Fig. P14.16a). The center of mass of the "object" consisting of the cart and water is at $x = 0$. The cart is subjected to the time-dependent force shown in Fig. P14.16b, where $F_0 = 5$ N and $t_0 = 2$ s. Assume that no water spills out of the cart and that the horizontal forces exerted on the wheels by the floor are negligible.

(a) Do you know the acceleration of the cart during the period $0 < t < t_0$?
(b) Do you know the acceleration of the center of mass of the "object" consisting of the cart and water during the period $0 < t < t_0$?
(c) What is the x coordinate of the center of mass of the "object" when $t > 2t_0$?

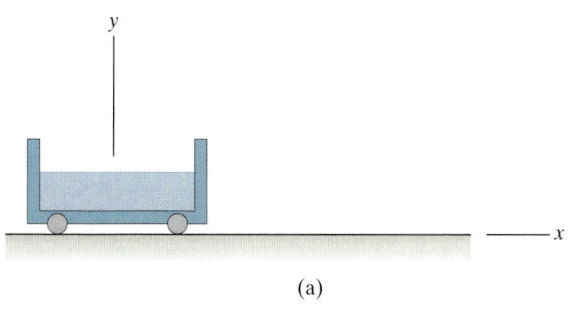

(a)

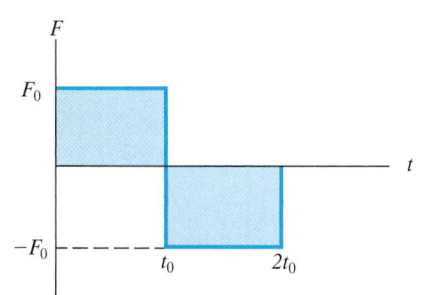

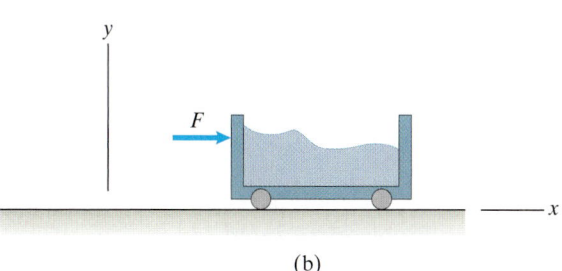

(b)

Problem 14.16

14.17 The combined weight of the motorcycle and rider is 360 lb. The coefficient of kinetic friction between the tires and the road is $\mu_k = 0.8$. The rider starts from rest, spinning the rear wheel. Neglect the horizontal force exerted on the front wheel by the road. In two seconds, the motorcycle moves 35 ft. What was the normal force between the rear wheel and the road?

Problem 14.17

14.18 The mass of the bucket B is 180 kg. From $t = 0$ to $t = 2$ s, the x and y coordinates of the center of mass of the bucket are

$$x = -0.2t^3 + 0.05t^2 + 10 \text{ m},$$
$$y = 0.1t^2 + 0.4t + 6 \text{ m}.$$

Determine the x and y components of the force exerted on the bucket by its supports at $t = 1$ s.

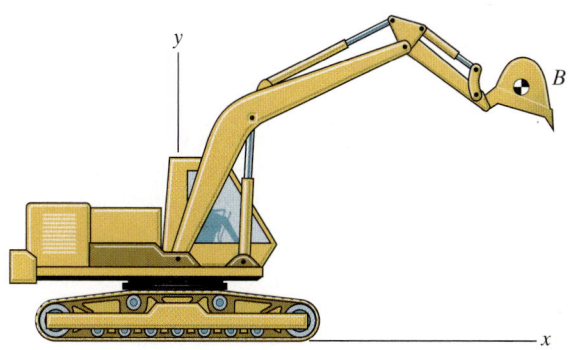

Problem 14.18

14.19 During a test flight in which a 9000-kg helicopter starts from rest at $t = 0$, the acceleration of its center of mass from $t = 0$ to $t = 10$ s is

$$\mathbf{a} = 0.6t\mathbf{i} + (1.8 - 0.36t)\mathbf{j}\ (\text{m/s}^2).$$

What is the magnitude of the total external force on the helicopter (including its weight) at $t = 6$ s?

14.20 The engineers conducting the test described in Problem 14.19 want to express the total force on the helicopter at $t = 6$ s in terms of three forces: the weight W, a component T tangent to the path, and a component L normal to the path. What are the values of W, T, and L?

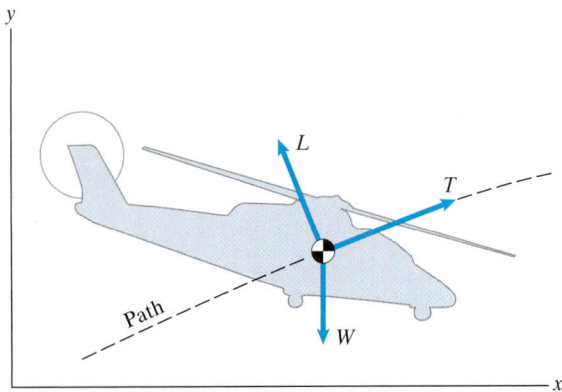

Problems 14.19/14.20

14.21 At the instant shown, the 11,000-kg airplane's velocity is $\mathbf{v} = 270\mathbf{i}$ (m/s). The forces acting on the plane are its weight, the thrust $T = 110$ kN, the lift $L = 260$ kN, and the drag $D = 34$ kN. (The x axis is parallel to the airplane's path.) Determine the magnitude of the airplane's acceleration.

14.22 At the instant shown, the 11,000-kg airplane's velocity is $\mathbf{v} = 300\mathbf{i}$ (m/s). The rate of change of the magnitude of the velocity is $dv/dt = 5$ m/s². The radius of curvature of the airplane's path is 4500 m, and the y axis points toward the concave side of the path. The thrust is $T = 120,000$ N. Determine the lift L and drag D.

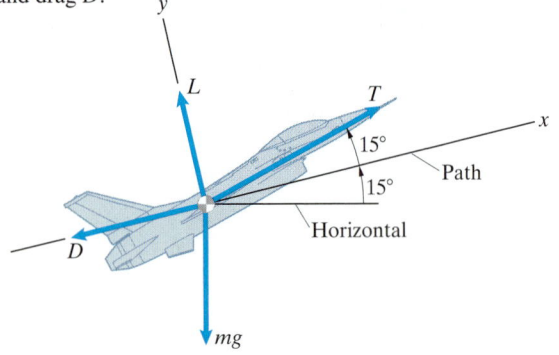

Problems 14.21/14.22

14.23 The coordinates in meters of the 360-kg sport plane's center of mass relative to an earth-fixed reference frame during an interval of time are

$$x = 20t - 1.63t^2,$$

$$y = 35t - 0.15t^3,$$

and

$$z = -20t - 1.38t^2,$$

where t is the time in seconds. The y axis points upward. The forces exerted on the plane are its weight, the thrust vector $\mathbf{T}$ exerted by its engine, the lift force vector $\mathbf{L}$, and the drag force vector $\mathbf{D}$. At $t = 4$ s, determine $\mathbf{T} + \mathbf{L} + \mathbf{D}$.

14.24 The force in newtons exerted on the 360-kg sport plane in Problem 14.23 by its engine, the lift force, and the drag force during an interval of time is

$$\mathbf{T} + \mathbf{L} + \mathbf{D} = (-1000 + 280t)\mathbf{i} + (4000 - 430t)\mathbf{j}$$

$$+ (720 + 200t)\mathbf{k},$$

where t is the time in seconds. If the coordinates of the plane's center of mass are $(0, 0, 0)$ and its velocity is $20\mathbf{i} + 35\mathbf{j} - 20\mathbf{k}$ (m/s) at $t = 0$, what are the coordinates of the center of mass at $t = 4$ s?

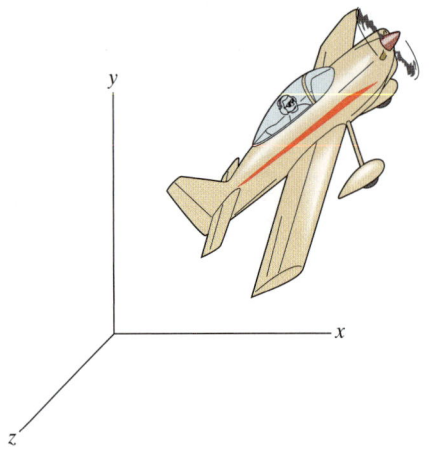

Problems 14.23/14.24

14.25 The robot manipulator is programmed so that $x = 40 + 24t^2$ mm, $y = 4t^3$ mm, and $z = 0$ during the interval of time from $t = 0$ to $t = 4$ s. The y axis points upward. What are the x and y components of the total force exerted by the jaws of the manipulator on the 2-kg widget A at $t = 3$ s?

14.26 The robot manipulator is programmed so that it is stationary at $t = 0$ and the components of the acceleration of A are $a_x = 400 - 0.8v_x$ mm/s^2 and $a_y = 200 - 0.4v_y$ mm/s^2 from $t = 0$ to $t = 2$ s, where v_x and v_y are the components of the velocity in mm/s. The y axis points upward. What are the x and y components of the total force exerted by the jaws of the manipulator on the 2-kg widget A at $t = 1$ s?

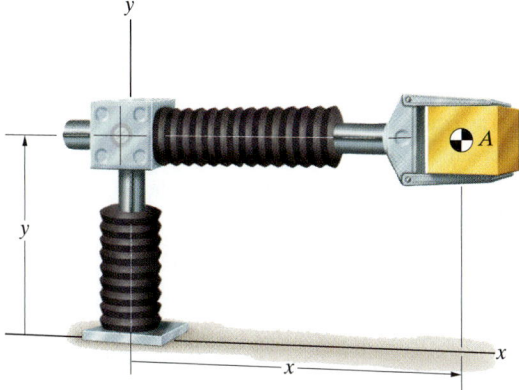

Problems 14.25/14.26

14.27 In the sport of curling, the object is to slide a "stone" weighing 44 lb onto the center of a target located 31 yards from the point of release. If $\mu_k = 0.01$ and the stone is thrown directly toward the target, what initial velocity would result in a perfect shot?

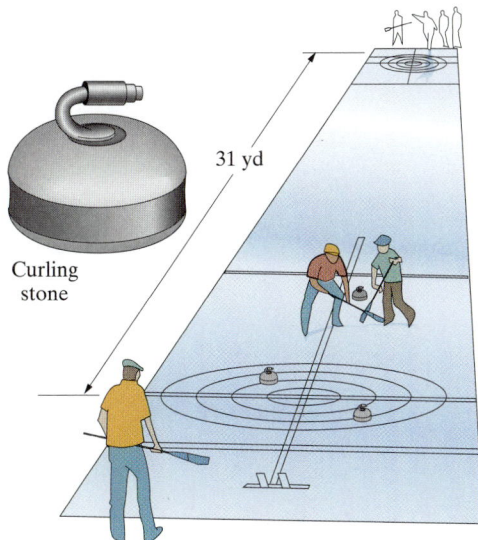

Problem 14.27

14.28 The two masses are released from rest. How fast are they moving when the 5-kg mass has fallen 0.2 m?

Strategy: Draw individual free-body diagrams of the masses. (See the discussion at the end of Example 14.2.)

Problem 14.28

14.29 The two weights are released from rest. The coefficient of kinetic friction between the horizontal surface and the 5-lb weight is $\mu_k = 0.2$. How far does the 10-lb weight fall in 0.5 s?

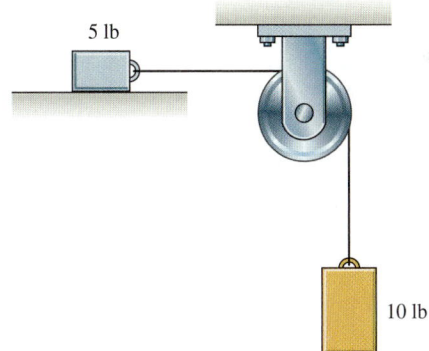

Problem 14.29

14.30 The mass of each box is 4 kg. Friction is negligible. The boxes start from rest at $t = 0$. Determine the magnitude of their velocity and the distance they have moved from their initial positions at $t = 1$ s. (See the discussion at the end of Example 14.2.)

14.31 In Problem 14.30, determine the magnitude of the velocity of the boxes and the distance they have moved from their initial positions at $t = 1$ s if the coefficient of kinetic friction between the boxes and the surface is $\mu_k = 0.15$.

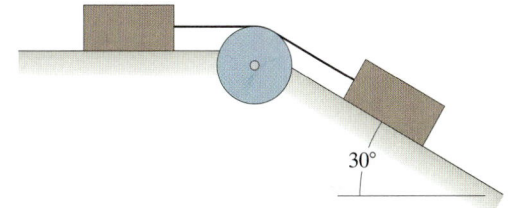

Problems 14.30/14.31

14.32 The masses $m_A = 15$ kg and $m_B = 30$ kg, and the coefficients of friction between all of the surfaces are $\mu_s = 0.4$ and $\mu_k = 0.35$. What is the largest force F that can be applied without causing A to slip relative to B? What is the resulting acceleration?

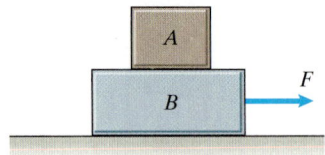

Problem 14.32

14.33 The crane's trolley at A moves to the right with constant acceleration, and the 800-kg load moves without swinging.
(a) What is the acceleration of the trolley and load?
(b) What is the sum of the tensions in the parallel cables supporting the load?

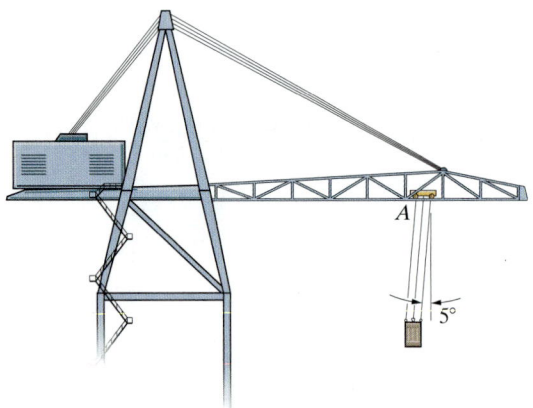

Problem 14.33

14.34 The mass of A is 30 kg and the mass of B is 5 kg. The force $F = 150$ N. The mass A slides on the smooth surface and the angle θ is constant. What is θ?

14.35 In Problem 14.34, determine the angle θ if the coefficient of kinetic friction between mass A and the horizontal surface is $\mu_k = 0.24$.

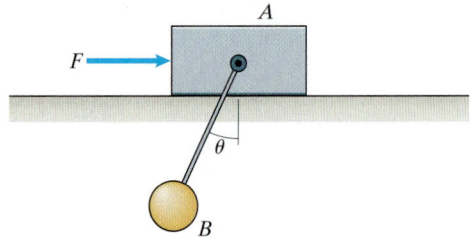

Problems 14.34/14.35

14.36 The 100-lb crate is initially stationary. The coefficients of friction between the crate and the inclined surface are $\mu_s = 0.2$ and $\mu_k = 0.16$. Determine how far the crate moves from its initial position in 2 s if the horizontal force $F = 90$ lb.

14.37 In Problem 14.36, determine how far the crate moves from its initial position in 2 s if the horizontal force $F = 30$ lb.

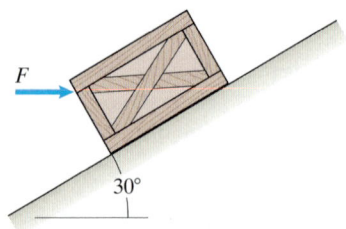

Problems 14.36/14.37

14.38 The crate has a mass of 120 kg, and the coefficients of friction between it and the sloping dock are $\mu_s = 0.6$ and $\mu_k = 0.5$.
(a) What tension must the winch exert on the cable to start the stationary crate sliding up the dock?
(b) If the tension is maintained at the value determined in part (a), what is the magnitude of the crate's velocity when it has moved 2 m up the dock?

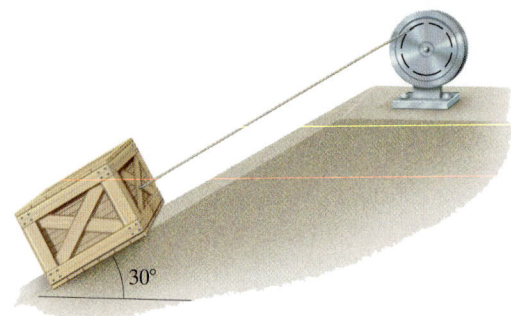

Problem 14.38

14.39 The utility vehicle is moving forward at 10 ft/s. The coefficients of friction between its load A and the bed of the vehicle are $\mu_s = 0.5$ and $\mu_k = 0.45$. If $\theta = 0$, determine the shortest distance in which the vehicle can be brought to a stop without causing the load to slide on the bed.

14.40 In Problem 14.39, determine the shortest distance if the angle θ is (a) 15° and (b) −15°.

Problems 14.39/14.40

14.41 The package starts from rest and slides down the smooth ramp. The hydraulic device B exerts a constant 2000-N force and brings the package to rest in a distance of 100 mm from the point at which it makes contact. What is the mass of the package?

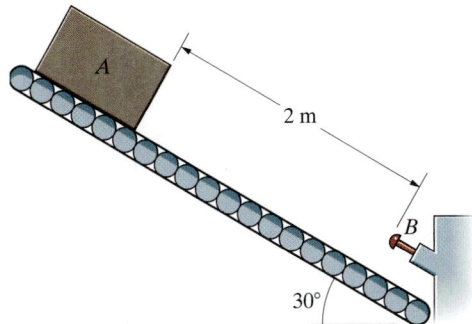

Problem 14.41

14.42 The force exerted on the 10-kg mass by the linear spring is $F = -ks$, where k is the spring constant and s is the displacement of the mass from its position when the spring is unstretched. The value of k is 50 N/m. The mass is released from rest in the position $s = 1$ m.

(a) What is the acceleration of the mass at the instant it is released?
(b) What is the velocity of the mass when it reaches the position $s = 0$?

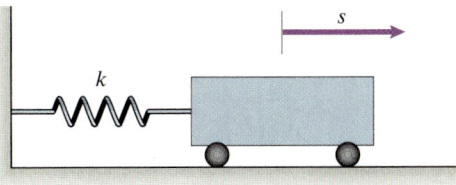

Problem 14.42

14.43 The 450-kg boat is moving at 10 m/s when its engine is shut down. The magnitude of the hydrodynamic drag force (in newtons) is $20v^2$, where v is the velocity of the boat in m/s. What is the boat's velocity 4 s after the engine is shut down?

Problem 14.43

14.44 A sky diver and his parachute weigh 200 lb. He is falling vertically at 100 ft/s when his parachute opens. With the parachute open, the magnitude of the drag force (in pounds) is $0.5v^2$.

(a) What is the magnitude of the sky diver's acceleration at the instant the parachute opens?
(b) What is the magnitude of his velocity when he has descended 20 ft from the point where his parachute opens?

Problem 14.44

14.45 The Panavia Tornado, with a mass of 18,000 kg, lands at a speed of 213 km/h. The decelerating force (in newtons) exerted on it by its thrust reversers and aerodynamic drag is $80,000 + 2.5v^2$, where v is the velocity in m/s. What is the length of the plane's landing roll?

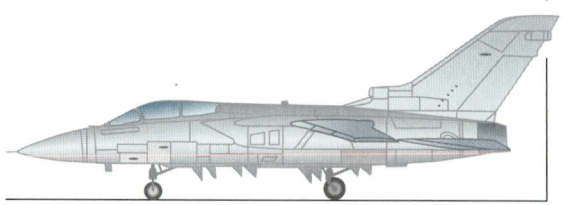

Problem 14.45

14.46 A 200-lb bungee jumper jumps from a bridge 130 ft above a river. The bungee cord has an unstretched length of 60 ft and has a spring constant $k = 14$ lb/ft.
(a) How far above the river is the jumper when the cord brings him to a stop?
(b) What maximum force does the cord exert on him?

Problem 14.46

14.47 A helicopter weighs 20,500 lb. It takes off vertically from sea level, and its upward velocity in ft/s is given as a function of its altitude h in feet by $v = 66 - 0.01h$.
(a) How long does it take the helicopter to climb to an altitude of 4000 ft?
(b) What is the sum of the vertical forces on the helicopter when its altitude is 2000 ft?

14.48 In a cathode-ray tube, an electron (mass $= 9.11\times10^{-31}$ kg) is projected at O with velocity $\mathbf{v} = (2.2 \times 10^7)\mathbf{i}$ (m/s). While the electron is between the charged plates, the electric field generated by the plates subjects it to a force $\mathbf{F} = -eE\mathbf{j}$, where the charge of the electron $e = 1.6 \times 10^{-19}$ C (coulombs) and the electric field strength $E = 15$ kN/C. External forces on the electron are negligible when it is not between the plates. Where does the electron strike the screen?

14.49 In Problem 14.48, determine where the electron strikes the screen if the electric field strength is $E = 15 \sin(\omega t)$ kN/C, where the frequency $\omega = 2 \times 10^9$ s^{-1}.

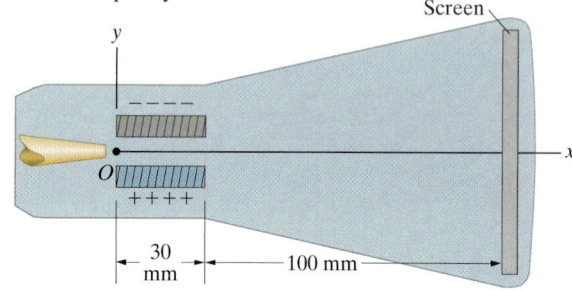

Problems 14.48/14.49

14.50 An astronaut wants to travel from a space station to a satellite S that needs repair. She departs the space station at O. A spring-loaded launching device gives her maneuvering unit an initial velocity of 1 m/s (relative to the space station) in the y direction. At that instant, the position of the satellite is $x = 70$ m, $y = 50$ m, $z = 0$, and it is drifting at 2 m/s (relative to the station) in the x direction. The astronaut intercepts the satellite by applying a constant thrust parallel to the x axis. The total mass of the astronaut and her maneuvering unit is 300 kg.
(a) How long does it take the astronaut to reach the satellite?
(b) What is the magnitude of the thrust she must apply to make the intercept?
(c) What is the astronaut's velocity *relative to the satellite* when she reaches it?

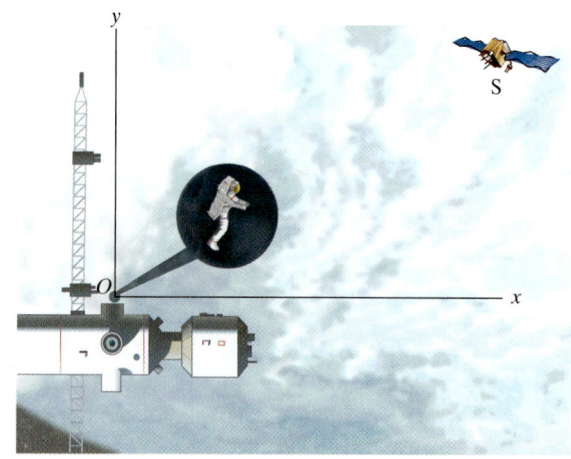

Problem 14.50

14.51 What is the acceleration of the 8-kg collar A relative to the smooth bar?

14.52 Determine the acceleration of the 8-kg collar A relative to the bar if the coefficient of kinetic friction between the collar and the bar is $\mu_k = 0.1$.

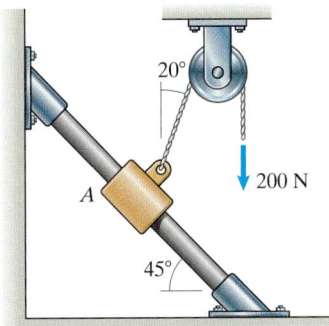

Problems 14.51/14.52

14.53 The force $F = 50$ lb. What is the magnitude of the acceleration of the 20-lb collar A along the smooth bar at the instant shown?

14.54* In Problem 14.53, determine the magnitude of the acceleration of the 20-lb collar A along the bar at the instant shown if the coefficient of static friction between the collar and the bar is $\mu_k = 0.2$.

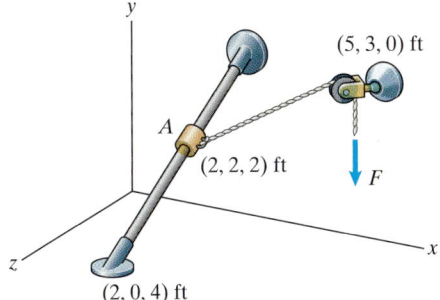

Problems 14.53/14.54

14.55 The 6-kg collar starts from rest at position A, where the coordinates of its center of mass are (400, 200, 200) mm, and slides up the smooth bar to position B, where the coordinates of its center of mass are (500, 400, 0) mm, under the action of a constant force $\mathbf{F} = -40\mathbf{i} + 70\mathbf{j} - 40\mathbf{k}$ (N). How long does the collar take to go from A to B?

14.56* In Problem 14.55, how long does the collar take to go from A to B if the coefficient of kinetic friction between the collar and the bar is $\mu_k = 0.2$?

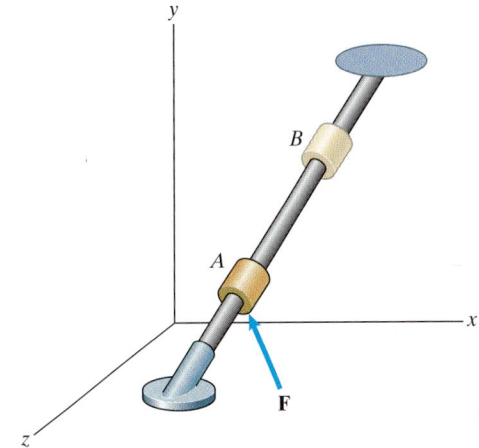

Problems 14.55/14.56

14.57 The crate is drawn across the floor by a winch that retracts the cable at a constant rate of 0.2 m/s. The crate's mass is 120 kg, and the coefficient of kinetic friction between the crate and the floor is $\mu_k = 0.24$.

(a) At the instant shown, what is the tension in the cable?
(b) Obtain a "quasi-static" solution for the tension in the cable by ignoring the crate's acceleration. Compare this solution with your result in part (a).

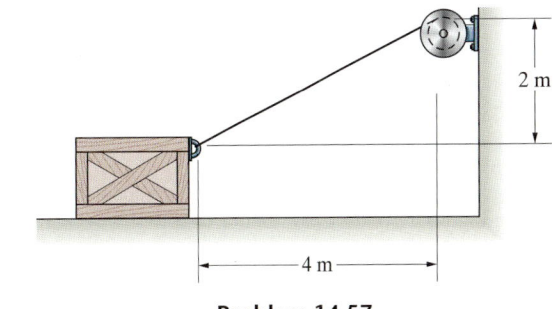

Problem 14.57

14.58 If $y = 100$ mm, $dy/dt = 600$ mm/s, and $d^2y/dt^2 = -200$ mm/s^2, what horizontal force is exerted on the 0.4-kg slider A by the smooth circular slot?

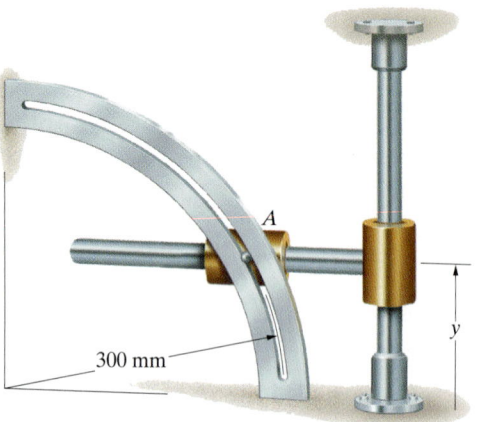

300 mm

Problem 14.58

14.59 The 1-kg collar P slides on the vertical bar and has a pin that slides in the curved slot. The vertical bar moves with constant velocity $v = 2$ m/s. The y axis points upward. What are the x and y components of the total force exerted on the collar by the vertical bar and the slotted bar when $x = 0.25$ m?

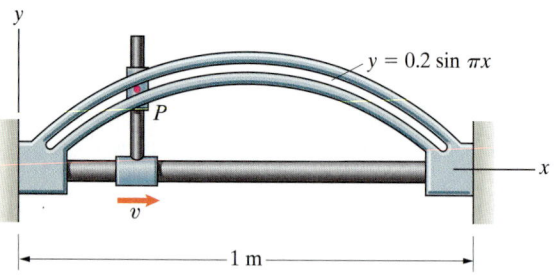

$y = 0.2 \sin \pi x$

1 m

Problem 14.59

14.60* The 1360-kg car travels along a straight road of increasing grade whose vertical profile is given by the equation shown. The magnitude of the car's velocity is a constant 100 km/h. When $x = 200$ m, what are the x and y components of the total force acting on the car (including its weight)?

Strategy: You know that the tangential component of the car's acceleration is zero. You can use this condition together with the equation for the profile of the road to determine the x and y components of the car's acceleration.

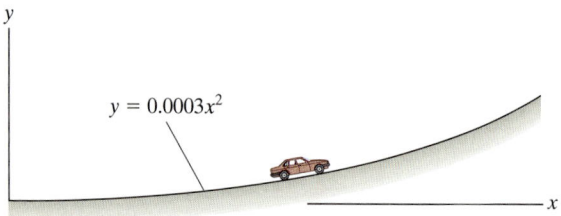

$y = 0.0003x^2$

Problem 14.60

14.61* The two 100-lb blocks are released from rest. Determine the magnitudes of their accelerations if friction at all the contacting surfaces is negligible.

Strategy: Use the fact that the components of the accelerations of the blocks perpendicular to their mutual interface must be equal.

14.62* The two 100-lb blocks are released from rest. The coefficient of kinetic friction between all contacting surfaces is $\mu_k = 0.1$. How long does it take block A to fall 1 ft?

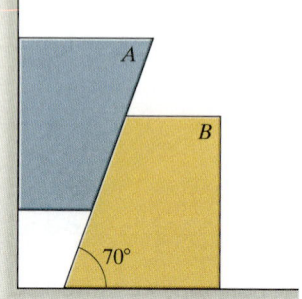

70°

Problems 14.61/14.62

14.63 The 3000-lb vehicle has left the ground after driving over a rise. At the instant shown, it is moving horizontally at 30 mi/h and the bottoms of its tires are 24 in above the (approximately) level ground. The earth-fixed coordinate system is placed with its origin 30 in above the ground, at the height of the vehicle's center of mass when the tires first contact the ground. (Assume that the vehicle remains horizontal.) When that occurs, the vehicle's center of mass initially continues moving downward and then rebounds upward due to the flexure of the suspension system. While the tires are in contact with the ground, the force exerted on them by the ground is $-2400\mathbf{i} - 18000y\mathbf{j}$ (lb), where y is the vertical position of the center of mass in feet. When the vehicle rebounds, what is the vertical component of the velocity of the center of mass at the instant the wheels leave the ground? (The wheels leave the ground when the center of mass is at $y = 0$.)

24 in

30 in

30 in

24 in

Problem 14.63

14.64* A steel sphere in a tank of oil is given an initial velocity $\mathbf{v} = 2\mathbf{i}$ (m/s) at the origin of the coordinate system shown. The radius of the sphere is 15 mm. The density of the steel is 8000 kg/m^3 and the density of the oil is 980 kg/m^3. If V is the sphere's volume, the (upward) buoyancy force on the sphere is equal to the weight of a volume V of oil. The magnitude of the hydrodynamic drag force $\mathbf{D}$ on the sphere as it falls is $|\mathbf{D}| = 1.6|\mathbf{v}|$ N, where $|\mathbf{v}|$ is the magnitude of the sphere's velocity in m/s. What are the x and y components of the sphere's velocity at $t = 0.1$ s?

14.65* In Problem 14.64, what are the x and y coordinates of the sphere at $t = 0.1$ s?

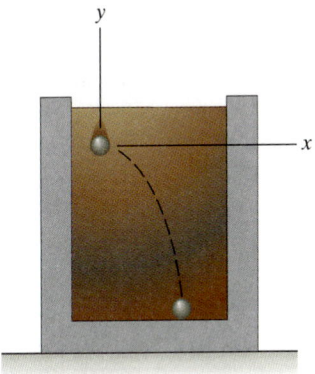

Problems 14.64/14.65

Normal and Tangential Components

When an object moves in a curved path, we can resolve the sum of the forces acting on it into normal and tangential components (Fig. 14.9a). We can also express the object's acceleration in terms of normal and tangential components (Fig. 14.9b) and write Newton's second law, $\Sigma\mathbf{F} = m\mathbf{a}$, in the form

$$\Sigma F_t \mathbf{e}_t + \Sigma F_n \mathbf{e}_n = m(a_t \mathbf{e}_t + a_n \mathbf{e}_n), \tag{14.6}$$

where

$$a_t = \frac{dv}{dt} \quad \text{and} \quad a_n = \frac{v^2}{\rho}.$$

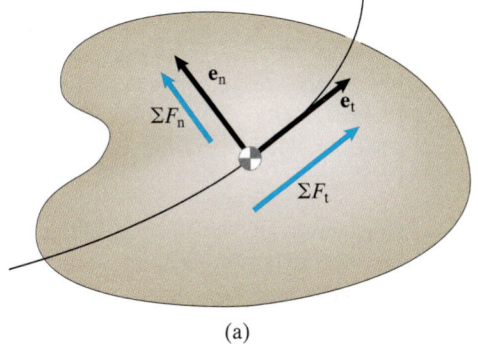

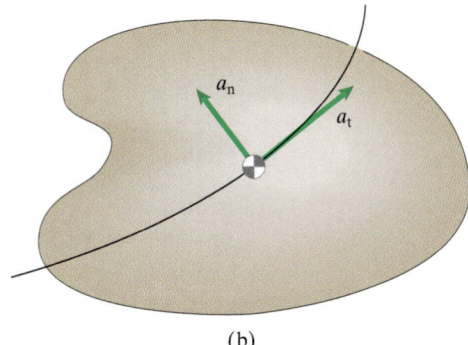

(a)   (b)

Equating the normal and tangential components in Eq. (14.6), we obtain two scalar equations of motion:

$$\Sigma F_t = ma_t = m\frac{dv}{dt}, \qquad \Sigma F_n = ma_n = m\frac{v^2}{\rho}. \tag{14.7}$$

Figure 14.9
(a) Normal and tangential components of the sum of the forces on an object.
(b) Normal and tangential components of the acceleration of the center of mass of the object.

The sum of the forces in the tangential direction equals the product of the mass and the rate of change of the magnitude of the velocity, and the sum of the forces in the normal direction equals the product of the mass and the normal component of acceleration. If the path of the object's center of mass lies in a plane, the acceleration of the center of mass perpendicular to the plane is zero, so the sum of the forces perpendicular to the plane is zero.

When an object moves in a circular path, normal and tangential components are usually the simplest choice for analyzing the motion of the object.

Example 14.4 Newton's Second Law in Normal and Tangential Components

Future space stations may be designed to rotate in order to provide simulated gravity for their inhabitants (Fig. 14.10). If the distance from the axis of rotation of the station to the occupied outer ring is $R = 100$ m, what rotation rate is necessary to simulate one-half of earth's gravity?

Figure 14.10

Strategy

By drawing the free-body diagram of a person and expressing Newton's second law in terms of normal and tangential components, we can relate the force exerted on the person by the floor to the angular velocity of the station. The person exerts an equal and opposite force on the floor, which is his effective weight.

Solution

We draw the free-body diagram of a person standing in the outer ring in Fig. a, where N is the force exerted on him by the floor. Relative to a nonrotating reference frame with its origin at the center of the station, the person moves in a circular path of radius R. His normal and tangential components of acceleration are shown in Fig. b. Applying Eqs. (14.7), we obtain

$$\Sigma F_t = 0 = m\frac{dv}{dt}$$

and

$$\Sigma F_n = N = m\frac{v^2}{R}.$$

The first equation simply indicates that the magnitude of the person's velocity is constant. The second equation tells us the force N. The magnitude of his velocity is $v = R\omega$, where ω is the angular velocity of the station. If one-half of earth's gravity is simulated, $N = \frac{1}{2}mg$. Therefore

$$N = \frac{1}{2}mg = m\frac{(R\omega)^2}{R}.$$

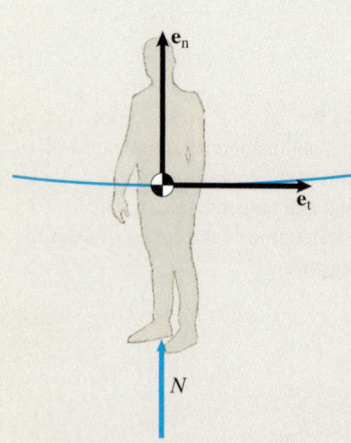

(a) Free-body diagram of a person standing in the occupied ring.

Solving for ω, we obtain the necessary angular velocity of the station:

$$\omega = \sqrt{\frac{g}{2R}} = \sqrt{\frac{9.81 \text{ m/s}^2}{2(100 \text{ m})}} = 0.221 \text{ rad/s}.$$

This is one revolution every 28.4 s.

Critical Thinking

When you are standing in a room, the floor pushes upward on you with a force N equal to your weight. The effect of gravity on your body is indistinguishable from the effect of a force of magnitude N pushing on your feet and accelerating you upward with acceleration g in the absence of gravity. (This observation was one of Einstein's starting points in developing his general theory of relativity.) This is the basis of simulating gravity by using rotation, and it explains why we set $N = mg/2$ in this example to simulate one-half of earth's gravity.

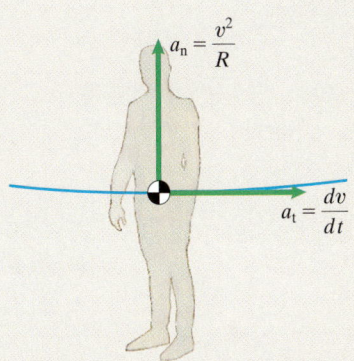

(b) The person's normal and tangential components of acceleration.

Example 14.5   Forces on an Object in Circular Motion

The experimental magnetically levitated train in Fig. 14.11 is supported by magnetic repulsion forces exerted in a direction normal to the tracks. Motion of the train transverse to the tracks is prevented by lateral supports. The 20,000-kg train is traveling at 30 m/s on a circular segment of track of radius $R = 150$ m, and the bank angle of the track is 40°. What force must the magnetic levitation system exert to support the train, and what total force is exerted by the lateral supports?

Figure 14.11

Strategy

We know the train's velocity and the radius of its circular path, so we can determine its normal component of acceleration. By expressing Newton's second law in terms of normal and tangential components, we can determine the components of force normal and transverse to the track.

Solution

Figure a shows the train viewed from above. The unit vector $\mathbf{e}_n$ is horizontal and points toward the center of the train's circular path, and $\mathbf{e}_t$ is tangential to the path. In Fig. b we draw the free-body diagram of the train seen from the front, where M is the magnetic force normal to the tracks and S is the transverse force. In Fig. c we show the train's acceleration, which is perpendicular to the circular path of the train and toward the center of the path. The sum of the forces in the vertical direction (perpendicular to the train's circular path) must equal zero:

$$M \cos 40° + S \sin 40° - mg = 0. \tag{1}$$

The sum of the forces in the $\mathbf{e}_n$ direction equals the product of the mass and the normal component of the acceleration; that is,

$$\Sigma F_n = m \frac{v^2}{\rho},$$

or

$$M \sin 40° - S \cos 40° = m \frac{v^2}{R}. \tag{2}$$

Solving Eqs. (1) and (2) for M and S, we obtain $M = 227.4$ kN and $S = 34.2$ kN.

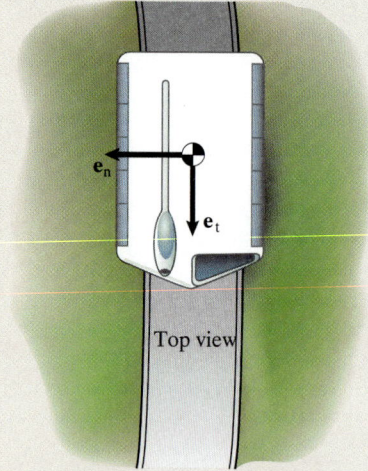

(a) The train's circular path viewed from above.

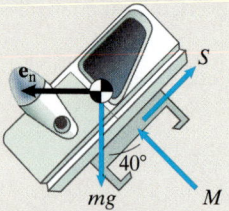

Front view

(b) Free-body diagram of the train.

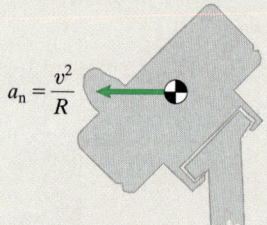

Front view

(c) The train's acceleration.

Critical Thinking

Suppose that we ask the following question: For what speed v of the train would the lateral force S be zero? Setting $S = 0$ in Eqs. (1) and (2) and solving for v, we obtain

$$v = \sqrt{Rg \tan 40°} = \sqrt{(150 \text{ m})(9.81 \text{ m/s}^2) \tan 40°} = 35.1 \text{ m/s}.$$

This is the optimum speed for the train to travel on the banked curve when it is carrying passengers. If you were a passenger, you would not need to exert any lateral force to remain in place in your seat.

Design Example 14.6 — Motor Vehicle Dynamics

A civil engineer's preliminary design for a freeway off-ramp is circular with radius $R = 60$ m (Fig. 14.12). If she assumes that the coefficient of static friction between tires and road is at least $\mu_s = 0.4$, what is the maximum speed at which vehicles can enter the ramp without losing traction?

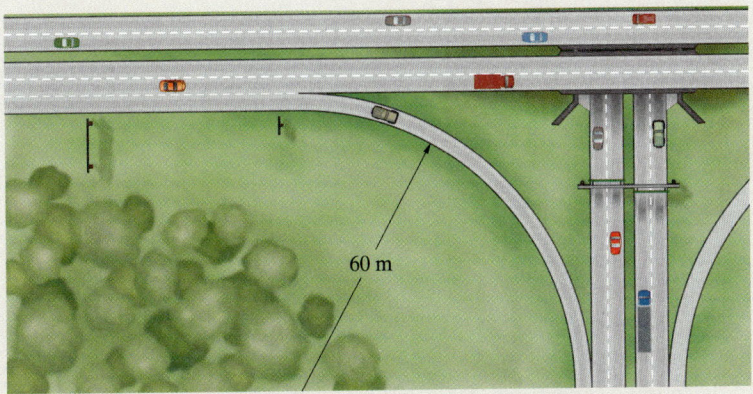

60 m

Figure 14.12

Strategy

Since a vehicle on the off-ramp moves in a circular path, it has a normal component of acceleration that depends on its velocity. The necessary normal component of force is exerted by friction between the tires and the road, and the friction force cannot be greater than the product of μ_s and the normal force. By assuming that the friction force is equal to this value, we can determine the maximum velocity for which slipping will not occur.

Solution

We view the free-body diagram of a car on the off-ramp from above the car in Fig. a and from the front of the car in Fig. b. In Fig. c, we show the car's acceleration, which is perpendicular to the circular path of the car and toward

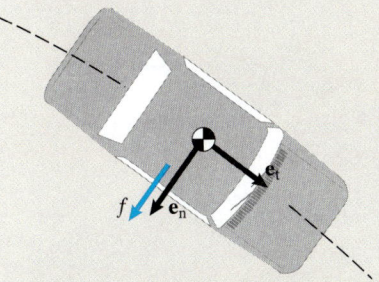

(a) Top view of the free-body diagram.

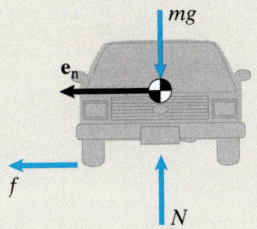

(b) Front view of the free-body diagram.

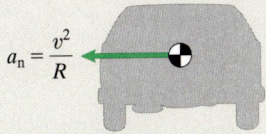

$$a_n = \frac{v^2}{R}$$

(c) The acceleration seen in the front view.

the center of the path. The sum of the forces in the $\mathbf{e}_n$ direction equals the product of the mass and the normal component of the acceleration; that is,

$$\Sigma F_n = ma_n = m\frac{v^2}{R},$$

or

$$f = m\frac{v^2}{R}.$$

The required friction force increases as v increases. The maximum friction force the surfaces will support is $f_{\max} = \mu_s N = \mu_s mg$. Therefore, the maximum velocity for which slipping does not occur is

$$v = \sqrt{\mu_s g R} = \sqrt{0.4(9.81 \text{ m/s}^2)(60 \text{ m})} = 15.3 \text{ m/s},$$

or 55.2 km/h (34.3 mi/h).

Design Issues

Automotive engineers, civil engineers who design highways, and engineers who study traffic accidents and their prevention must analyze and measure the motions of vehicles under different conditions. By using the methods discussed in this chapter, they can relate the forces acting on vehicles to their motions and study, for example, the factors influencing the distance necessary for a car to be brought to a stop in an emergency or the effects of banking and curvature on the velocity at which a car can safely be driven on a curved road (Fig. 14.13).

In this example, the analysis indicates that vehicles will lose traction if they enter the freeway off-ramp at speeds greater than 34.3 mi/h. This result can be used as an indication of the speed limit that must be posted in order for vehicles to enter the ramp safely, or the off-ramp could be designed for a greater speed by increasing the radius of curvature of the road.

Figure 14.13
Tests of the capabilities of vehicles to negotiate curves influence the design of both vehicles and highways.

Problems

14.66 The boat and its passengers weigh 1200 lb. The boat is moving in a circular path with radius $R = 40$ ft at a constant speed of 20 ft/s. What are the tangential and normal components of the total force acting on the boat?

14.67 The boat and its passengers weigh 1200 lb. The boat is moving in a circular path with radius $R = 40$ ft at a speed of 20 ft/s. The driver suddenly increases the throttle so that the tangential component of the total force acting on the boat increases to 100 lb and remains constant. He continues following the same circular path. Two seconds after he increases the throttle, what are (a) the magnitude of the boat's velocity and (b) the tangential and normal components of the boat's acceleration?

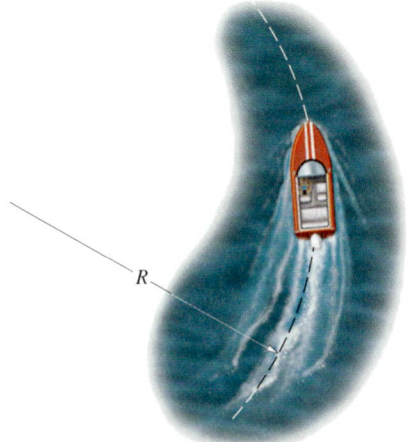

Problems 14.66/14.67

14.68 In preliminary design studies for a sun-powered car, it is estimated that the mass of the car and driver will be 100 kg and the torque produced by the engine will result in a 60-N tangential force on the car. Suppose that the car starts from rest at A and it is subjected to a constant 60-N tangential force. What are the tangential and normal components of the car's acceleration when it reaches B?

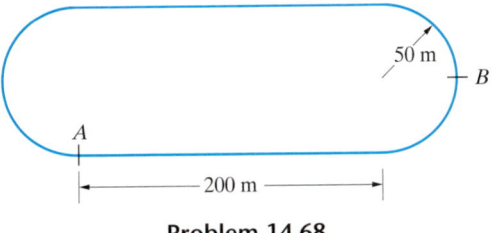

Problem 14.68

14.69 An astronaut candidate with a mass of 72 kg is tested in a centrifuge with a radius of 10 m. The centrifuge rotates in the horizontal plane. It starts from rest at time $t = 0$ and has constant angular acceleration $\alpha = 0.2$ rad/s^2.

(a) What is the magnitude of his horizontal acceleration at $t = 10$ s?
(b) What are the tangential and normal components of the force exerted on him by the centrifuge at $t = 10$ s?

Strategy: If a point moves in a circular path of radius R, the magnitude of its velocity v is related to its angular velocity ω by $v = R\omega$, so the normal component of its acceleration can be written $a_n = v^2/R = \omega^2 R$. (See Eq. 13.45.) You can use this expression to determine the normal component of force acting on the candidate.

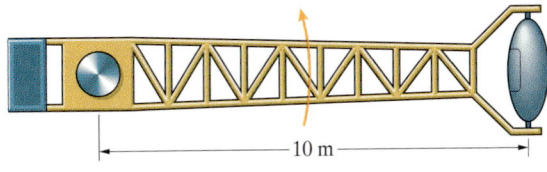

Problem 14.69

14.70 The circular disk lies *in the horizontal plane*. At the instant shown, the disk rotates with a counterclockwise angular velocity of 4 rad/s and a counterclockwise angular acceleration of 2 rad/s^2. The 0.5-kg slider A is supported horizontally by the smooth slot and the string attached at B. Determine the tension in the string and the magnitude of the horizontal force exerted on the slider by the slot.

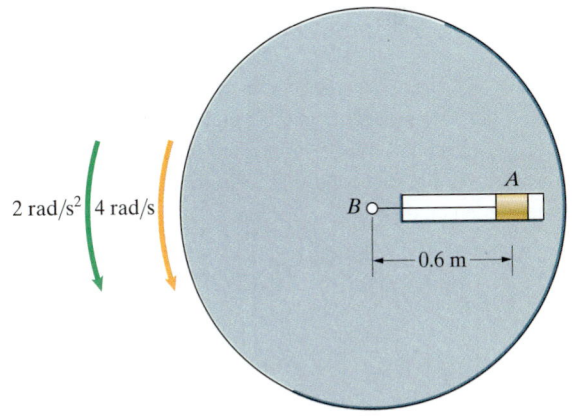

Problem 14.70

14.71 The circular disk lies *in the horizontal plane* and rotates with a constant counterclockwise angular velocity of 4 rad/s. The 0.5-kg slider A is supported horizontally by the smooth slot and the string attached at B. Determine the tension in the string and the magnitude of the horizontal force exerted on the slider by the slot.

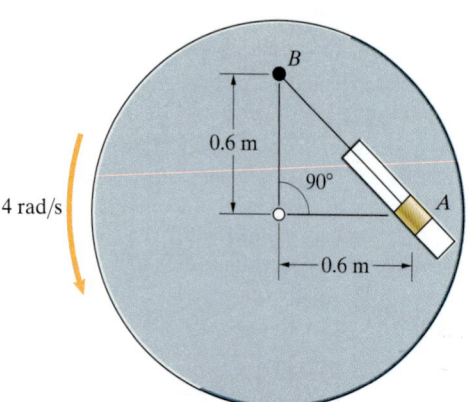

Problem 14.71

14.72 The 32,000-lb airplane is flying in the vertical plane at 420 ft/s. At the instant shown the angle $\theta = 30°$ and the cartesian components of the plane's acceleration are $a_x = -6$ ft/s², $a_y = 30$ ft/s².
(a) What are the tangential and normal components of the total force acting on the airplane (including its weight)?
(b) What is $d\theta/dt$ in degrees per second?

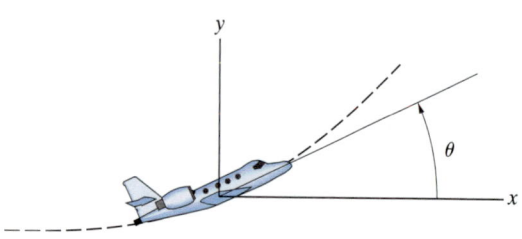

Problem 14.72

14.73 The 2-kg slider A starts from rest and slides *in the horizontal plane* along the smooth circular bar under the action of a tangential force $F_t = 4t$ N. At $t = 4$ s, determine (a) the magnitude of the velocity of the slider and (b) the magnitude of the horizontal force exerted on the slider by the bar.

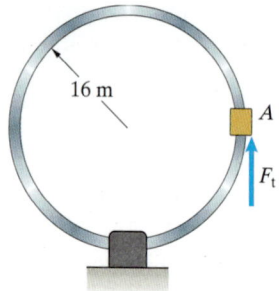

Top view

Problem 14.73

14.74 Small parts on a conveyer belt moving with constant velocity v are allowed to drop into a bin. Show that the angle θ at which the parts start sliding on the belt satisfies the equation

$$\cos \theta - \frac{1}{\mu_s} \sin \theta = \frac{v^2}{gR},$$

where μ_s is the coefficient of static friction between the parts and the belt.

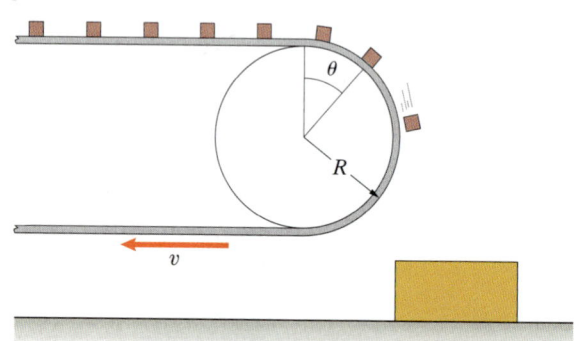

Problem 14.74

14.75 The 1-slug mass m rotates around the vertical pole in a horizontal circular path. The angle $\theta = 30°$ and the length of the string is $L = 4$ ft. What is the magnitude of the velocity of the mass?

Strategy: Notice that the vertical acceleration of the mass is zero. Draw the free-body diagram of the mass and write Newton's second law in terms of tangential and normal components.

14.76 The 1-slug mass m rotates around the vertical pole in a horizontal circular path. The length of the string is $L = 4$ ft. Determine the magnitude of the velocity of the mass and the angle θ if the tension in the string is 50 lb.

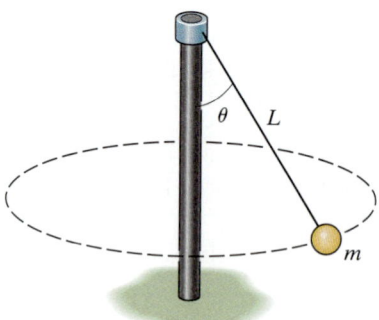

Problems 14.75/14.76

14.77 The 10-kg mass *m* rotates around the vertical pole in a horizontal circular path of radius $R = 1$ m. If the magnitude of the velocity of the mass is $v = 3$ m/s, what are the tensions in the strings *A* and *B*?

14.78 The 10-kg mass *m* rotates around the vertical pole in a horizontal circular path of radius $R = 1$ m. For what range of values of the velocity *v* of the mass will the mass remain in the circular path described?

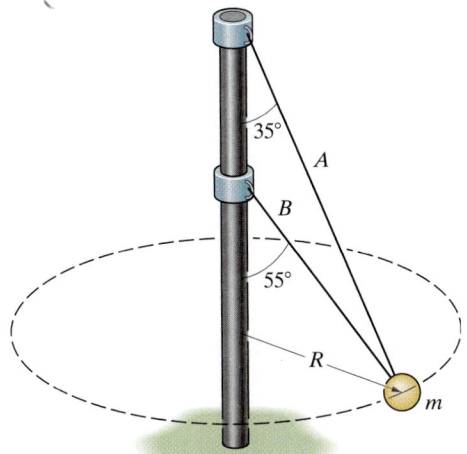

Problems 14.77/14.78

14.79 Suppose you are designing a monorail transportation system that will travel at 50 m/s and you decide that the angle θ that the cars swing out from the vertical when they go through a turn must not be larger than 20°. If the turns in the track consist of circular arcs of constant radius *R*, what is the minimum allowable value of *R*?

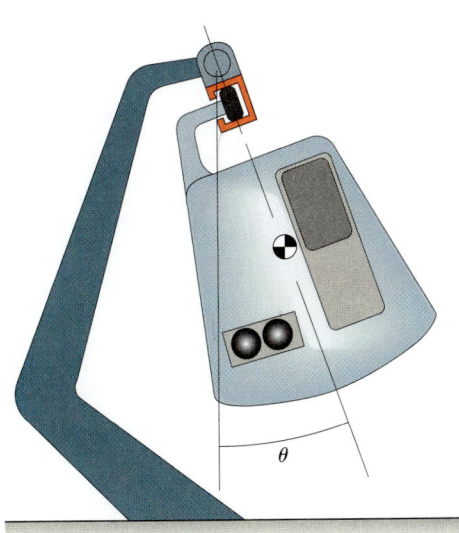

Problem 14.79

14.80 An airplane of weight $W = 200,000$ lb makes a turn at constant altitude and at constant velocity $v = 600$ ft/s. The bank angle is 15°.
(a) Determine the lift force *L*.
(b) What is the radius of curvature of the plane's path?

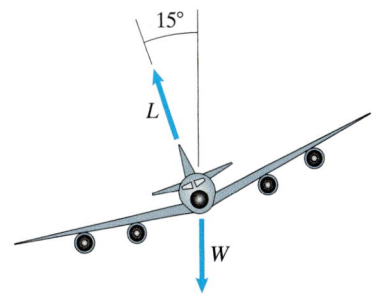

Problem 14.80

14.81 The suspended 2-kg mass *m* is stationary.
(a) What are the tensions in the strings *A* and *B*?
(b) If string *A* is cut, what is the tension in string *B* immediately afterward?

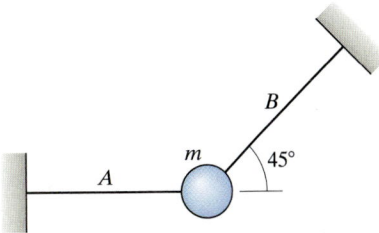

Problem 14.81

14.82 The airplane flies with constant velocity *v* along a circular path in the vertical plane. The radius of the airplane's circular path is 2000 m. The mass of the pilot is 72 kg.
(a) The pilot will experience "weightlessness" at the top of the circular path if the airplane exerts no net force on him at that point. Draw a free-body diagram of the pilot and use Newton's second law to determine the velocity *v* necessary to achieve this condition.
(b) Suppose that you don't want the force exerted on the pilot by the airplane to exceed four times his weight. If he performs this maneuver at $v = 200$ m/s, what is the minimum acceptable radius of the circular path?

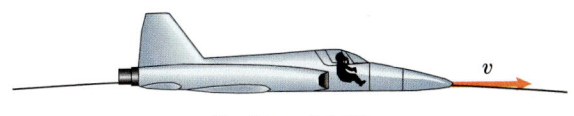

Problem 14.82

14.83 The smooth circular bar rotates with constant angular velocity ω_0 about the vertical axis AB. The radius $R = 0.5$ m. The mass m remains stationary relative to the circular bar at $\beta = 40°$. Determine ω_0.

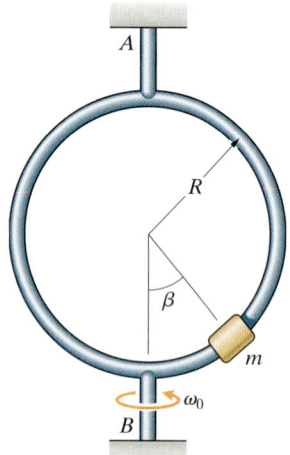

Problem 14.83

14.84 The force exerted on a charged particle by a magnetic field is

$$\mathbf{F} = q\mathbf{v} \times \mathbf{B},$$

where q and $\mathbf{v}$ are the charge and velocity vector of the particle and $\mathbf{B}$ is the magnetic field vector. A particle of mass m and positive charge q is projected at O with velocity $\mathbf{v} = v_0\mathbf{i}$ into a uniform magnetic field $\mathbf{B} = B_0\mathbf{k}$. Using normal and tangential components, show that (a) the magnitude of the particle's velocity is constant and (b) the particle's path is a circle with radius mv_0/qB_0.

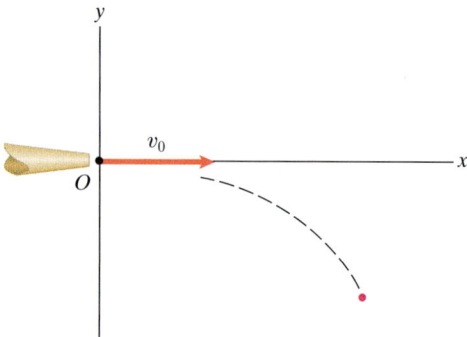

Problem 14.84

14.85 The mass m is attached to a string that is wrapped around the fixed post of radius R. At $t = 0$, the mass is given a velocity v_0 as shown. Neglect external forces on m other than the force exerted by the string. Determine the tension in the string as a function of the angle θ.

Strategy: The velocity vector of the mass is perpendicular to the string. Express Newton's second law in terms of normal and tangential components.

14.86 The mass m is attached to a string that is wrapped around the fixed post of radius R. At $t = 0$, the mass is given a velocity v_0 as shown. Neglect external forces on m other than the force exerted by the string. Determine the angle θ as a function of time.

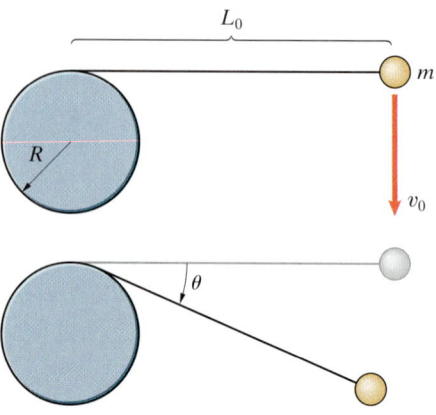

Problems 14.85/14.86

14.87 The sum of the forces in newtons exerted on the 360-kg sport plane (including its weight) during an interval of time is

$$(-1000 + 280t)\mathbf{i} + (480 - 430t)\mathbf{j} + (720 + 200t)\mathbf{k},$$

where t is the time in seconds. At $t = 0$, the velocity of the plane's center of mass relative to the earth-fixed reference frame is $20\mathbf{i} + 35\mathbf{j} - 20\mathbf{k}$ (m/s). If you resolve the sum of the forces on the plane into components tangent and normal to the plane's path at $t = 2$ s, what are the values of ΣF_t and ΣF_n?

14.88 In Problem 14.87, what is the instantaneous radius of curvature of the plane's path at $t = 2$ s? The vector components of the sum of the forces in the directions tangential and normal to the path lie in the osculating plane. Determine the components of a unit vector perpendicular to the osculating plane at $t = 2$ s.

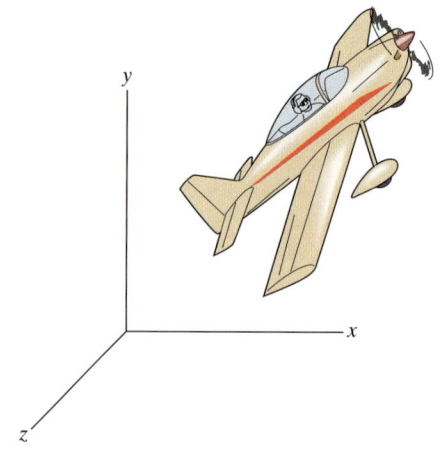

Problems 14.87/14.88

Design Experience

Problems 14.89–14.93 are related to Design Example 14.6.

14.89 The freeway off-ramp is circular with 60-m radius (Fig. a). The off-ramp has a slope $\beta = 15°$ (Fig. b). If the coefficient of static friction between the tires of a car and the road is $\mu_s = 0.4$, what is the maximum speed at which it can enter the ramp without losing traction?

14.90* The freeway off-ramp is circular with 60-m radius (Fig. a). The off-ramp has a slope β (Fig. b). If the coefficient of static friction between the tires of a car and the road is $\mu_s = 0.4$ what minimum slope β is needed so that the car could (in theory) enter the off-ramp at any speed without losing traction?

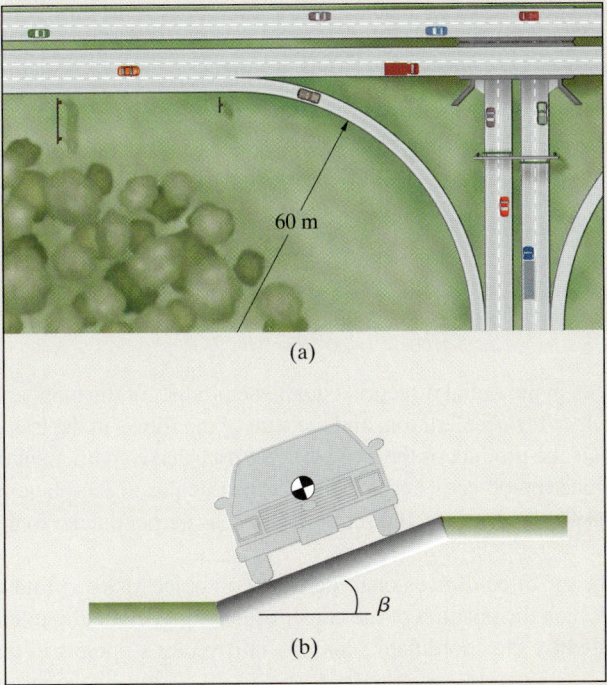

(a)

(b)

Problems 14.89/14.90

14.91 A car traveling at 30 m/s is at the top of a hill. The coefficient of kinetic friction between the tires and the road is $\mu_k = 0.8$. The instantaneous radius of curvature of the car's path is 200 m. If the driver applies the brakes and the car's wheels lock, what is the resulting deceleration of the car in the direction tangent to its path?

Problem 14.91

14.92 A car traveling at 30 m/s is at the bottom of a depression. The coefficient of kinetic friction between the tires and the road is $\mu_k = 0.8$. The instantaneous radius of curvature of the car's path is 200 m. If the driver applies the brakes and the car's wheel's lock, what is the resulting deceleration of the car in the direction tangential to its path? Compare your answer to that of Problem 14.91.

Problem 14.92

14.93 The combined mass of the motorcycle and rider is 160 kg. The motorcycle starts from rest at $t = 0$ and moves along a circular track with a 400-m radius. The tangential component of acceleration of the motorcycle as a function of time is $a_t = 2 + 0.2t$ m/s². The coefficient of static friction between the tires and the track is $\mu_s = 0.8$. How long after it starts does the motorcycle reach the limit of adhesion, which means that its tires are on the verge of slipping? How fast is the motorcycle moving when that occurs?

Strategy: Draw a free-body diagram showing the tangential and normal components of force acting on the motorcycle.

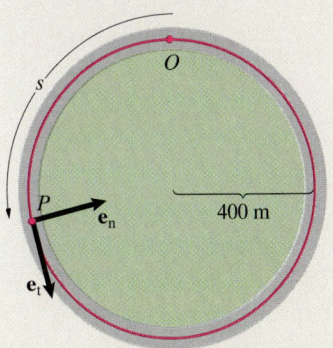

Problem 14.93

Polar and Cylindrical Coordinates

When an object moves in a planar curved path, we can describe the motion of the center of mass of the object in terms of polar coordinates. Resolving the sum of the forces parallel to the plane into polar components (Fig. 14.14a) and expressing the acceleration of the center of mass in terms of polar components (Fig. 14.14b), we can write Newton's second law, $\Sigma \mathbf{F} = m\mathbf{a}$, in the form

$$\Sigma F_r \mathbf{e}_r + \Sigma F_\theta \mathbf{e}_\theta = m(a_r \mathbf{e}_r + a_\theta \mathbf{e}_\theta), \tag{14.8}$$

where

$$a_r = \frac{d^2 r}{dt^2} - r\left(\frac{d\theta}{dt}\right)^2 = \frac{d^2 r}{dt^2} - r\omega^2$$

and

$$a_\theta = r\frac{d^2\theta}{dt^2} + 2\frac{dr}{dt}\frac{d\theta}{dt} = r\alpha + 2\frac{dr}{dt}\omega.$$

Equating the $\mathbf{e}_r$ and $\mathbf{e}_\theta$ components in Eq. (14.8), we obtain the scalar equations

$$\Sigma F_r = ma_r = m\left(\frac{d^2 r}{dt^2} - r\omega^2\right) \tag{14.9}$$

and

$$\Sigma F_\theta = ma_\theta = m\left(r\alpha + 2\frac{dr}{dt}\omega\right). \tag{14.10}$$

The sum of the forces in the radial direction equals the product of the mass and the radial component of the acceleration, and the sum of the forces in the transverse direction equals the product of the mass and the transverse component of the acceleration. Since the object's acceleration perpendicular to the plane in which the motion takes place is zero, the sum of the forces perpendicular to the plane is zero.

We can describe the three-dimensional motion of an object using cylindrical coordinates, in which the position of the center of mass perpendicular to the x–y plane is measured by the coordinate z and the unit vector $\mathbf{e}_z$ points in the

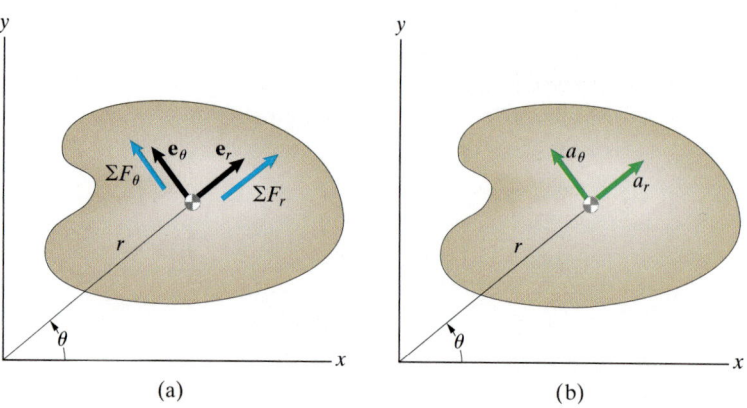

Figure 14.14
Polar components of (a) the sum of the forces and (b) the acceleration of the center of mass

(a) (b)

positive z direction. We resolve the sum of the forces into radial, transverse, and z components (Fig. 14.15a) and express the acceleration of the center of mass in terms of radial, transverse, and z components (Fig. 14.15b). The three scalar equations of motion are the polar equations (14.9) and (14.10) and the equation of motion in the z direction,

$$\Sigma F_z = ma_z = m\frac{dv_z}{dt} = m\frac{d^2z}{dt^2}. \qquad (14.11)$$

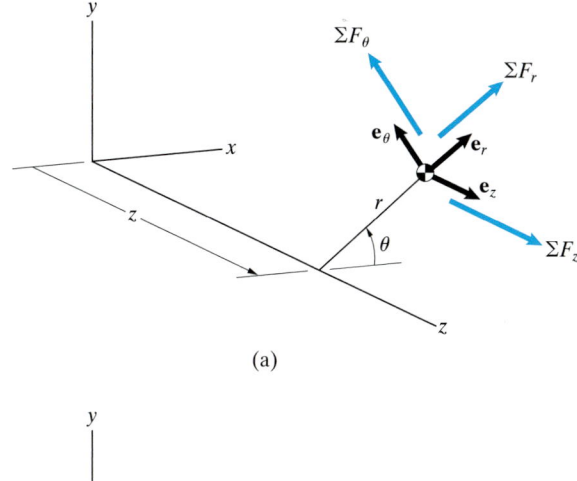

(a)

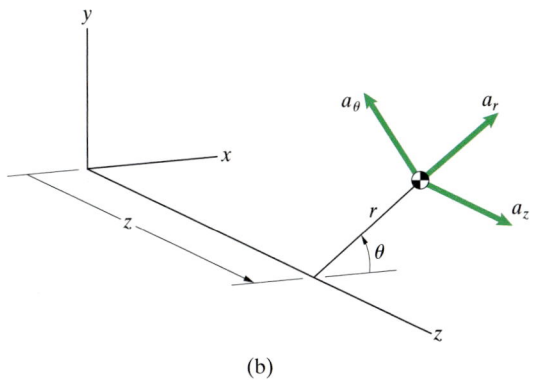

(b)

Figure 14.15
(a) Components of the sum of the forces on an object in cylindrical coordinates.
(b) Components of the acceleration of the center of mass.

Example 14.7	Newton's Second Law in Polar Coordinates

The smooth bar in Fig. 14.16 rotates *in the horizontal plane* with constant angular velocity ω_0. The unstretched length of the linear spring is r_0. The collar A has mass m and is released at $r = r_0$ with no radial velocity.

(a) Determine the radial velocity of the collar as a function of r.
(b) Determine the horizontal force exerted on the collar by the bar as a function of r.

Strategy

(a) The only force on the collar in the radial direction is the spring force, which we can express in polar coordinates in terms of r. By integrating Eq. (14.9), we can determine the radial velocity v_r as a function of r.
(b) Once $v_r = dr/dt$ is known in terms of r, we can use Eq. (14.10) to determine the transverse force exerted on the collar by the bar.

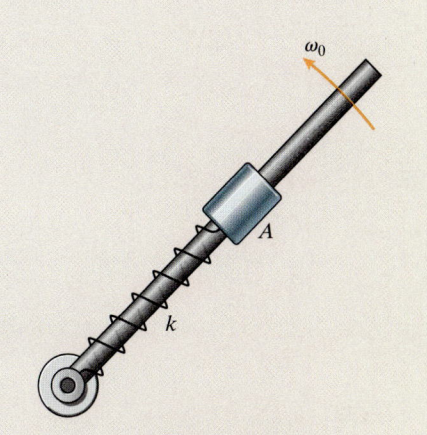

Figure 14.16

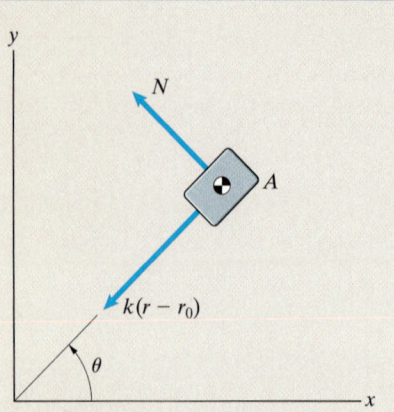

(a) Radial and transverse forces on A.

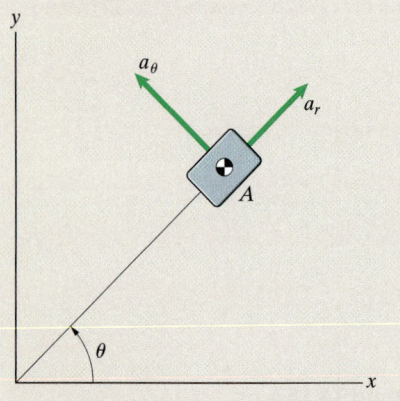

(b) Radial and transverse components of the acceleration of the center of mass.

Solution

(a) The spring exerts a radial force $k(r - r_0)$ in the negative r direction (Fig. a). Since the bar is smooth, it exerts no radial force on A, but may exert a transverse force N. Figure b shows the radial and transverse components of the collar's acceleration. Newton's second law in the radial direction is

$$\Sigma F_r = ma_r,$$

or

$$-k(r - r_0) = m\left(\frac{d^2r}{dt^2} - r\omega^2\right) = m\left(\frac{dv_r}{dt} - r\omega_0^2\right).$$

We solve this equation for the time derivative of v_r:

$$\frac{dv_r}{dt} = r\omega_0^2 - \frac{k}{m}(r - r_0).$$

Applying the chain rule,

$$\frac{dv_r}{dt} = \frac{dv_r}{dr}\frac{dr}{dt} = \frac{dv_r}{dr}v_r,$$

we obtain

$$v_r\,dv_r = \left[\left(\omega_0^2 - \frac{k}{m}\right)r + \frac{k}{m}r_0\right]dr.$$

Finally, integrating

$$\int_0^{v_r} v_r\,dv_r = \int_{r_0}^{r}\left[\left(\omega_0^2 - \frac{k}{m}\right)r + \frac{k}{m}r_0\right]dr$$

yields the radial velocity as a function of r:

$$v_r = \sqrt{\left(\omega_0^2 - \frac{k}{m}\right)(r^2 - r_0^2) + \frac{2k}{m}r_0(r - r_0)}.$$

(b) To determine the force N, we use Newton's second law in the transverse direction,

$$\Sigma F_\theta = ma_\theta,$$

or

$$N = m\left(r\alpha + 2\frac{dr}{dt}\omega\right) = 2m\omega_0 v_r.$$

Substituting our expression for v_r as a function of r, we obtain the transverse force exerted by the bar as a function of r:

$$N = 2m\omega_0\sqrt{\left(\omega_0^2 - \frac{k}{m}\right)(r^2 - r_0^2) + \frac{2k}{m}r_0(r - r_0)}.$$

Critical Thinking

Why did we solve this example using polar coordinates instead of normal and tangential components? Normal and tangential components are oriented with the path of the object, and we did not know the path the collar A would follow. In addition, the forces acting on the collar are aligned in the radial and transverse directions (Fig a).

Problems

14.94 The center of mass of the 12-kg object moves in the x–y plane. Its polar coordinates are given as functions of time by $r = 12 - 0.4t^2$ m, $\theta = 0.02t^3$ rad. Determine the polar components of the total force acting on the object at $t = 2$ s.

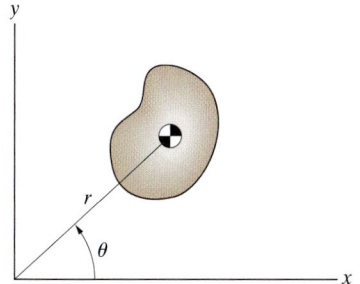

Problem 14.94

14.95 A 100-lb person walks on a large disk that rotates with constant angular velocity $\omega_0 = 0.3$ rad/s. He walks at a constant speed $v_0 = 5$ ft/s along a straight radial line painted on the disk. Determine the polar components of the horizontal force exerted on him when he is 6 ft from the center of the disk. (How are these forces exerted on him?)

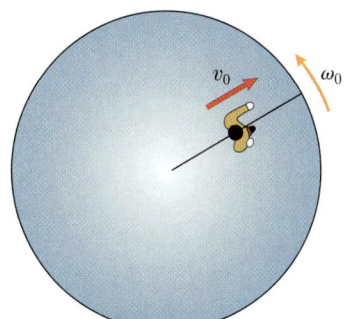

Problem 14.95

14.96 The robot is programmed so that the 0.4-kg part A describes the path

$$r = 1 - 0.5 \cos 2\pi t \text{ m,}$$
$$\theta = 0.5 - 0.2 \sin 2\pi t \text{ rad.}$$

Determine the polar components of force exerted on A by the robot's jaws at $t = 2$ s.

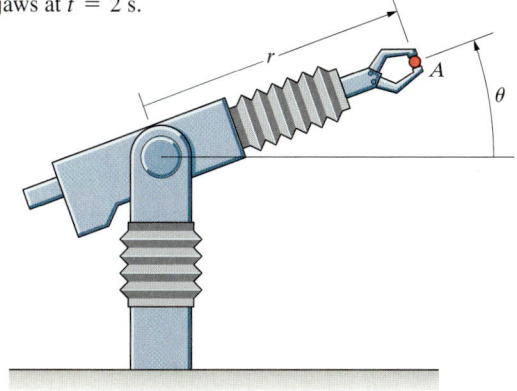

Problem 14.96

14.97 A 50-lb object P moves along the spiral path $r = (0.1)\theta$ ft, where θ is in radians. Its angular position is given as a function of time by $\theta = 2t$ rad, and $r = 0$ at $t = 0$. Determine the polar components of the total force acting on the object at $t = 4$ s.

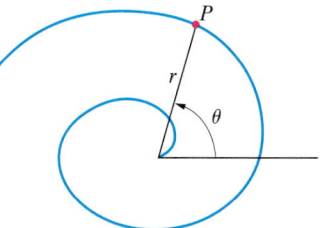

Problem 14.97

14.98 The smooth bar rotates *in the horizontal plane* with constant angular velocity $\omega_0 = 60$ rpm. If the 1-kg collar A is released at $r = 1$ m with no radial velocity, what is its radial velocity when it reaches the end of the bar?

Strategy: Notice that no horizontal force is exerted on the collar in the direction parallel to the bar, so $a_r = 0$. Write the term d^2r/dt^2 as dv_r/dt and apply the chain rule:

$$\frac{dv_r}{dt} = \frac{dv_r}{dr}\frac{dr}{dt} = \frac{dv_r}{dr}v_r.$$

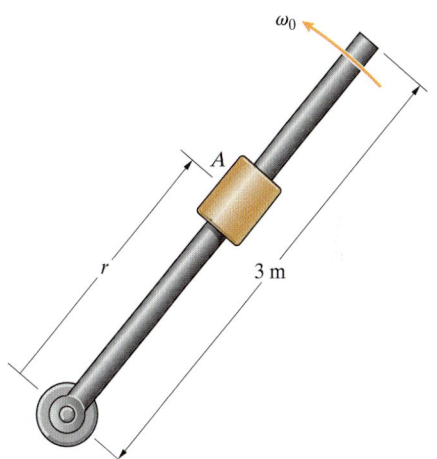

Problem 14.98

14.99 The smooth bar rotates *in the horizontal plane* with constant angular velocity $\omega_0 = 60$ rpm. The spring constant is $k = 20$ N/m, and the unstretched length of the spring is 3 m. If the 1-kg collar A is released at $r = 1$ m with no radial velocity, what is its radial velocity when $r = 2$ m?

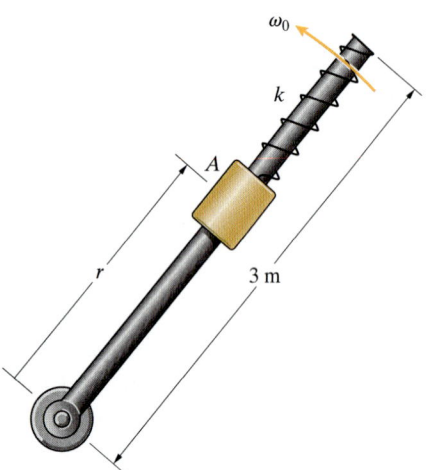

Problem 14.99

14.100 The 2-kg mass m is released from rest with the string horizontal. The length of the string is $L = 0.6$ m. By using Newton's second law in terms of polar coordinates, determine the magnitude of the velocity of the mass and the tension in the string when $\theta = 45°$.

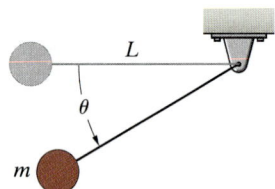

Problem 14.100

14.101 The 1-lb block A is given an initial velocity $v_0 = 14$ ft/s to the right when it is in the position $\theta = 0$, causing it to slide up the smooth circular surface. By using Newton's second law in terms of polar coordinates, determine the magnitude of the velocity of the block when $\theta = 60°$.

14.102 The 1-lb block is given an initial velocity $v_0 = 14$ ft/s to the right when it is in the position $\theta = 0$, causing it to slide up the smooth circular surface. Determine the normal force exerted on the block by the surface when $\theta = 60°$.

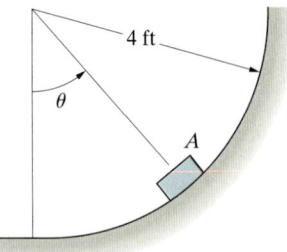

Problems 14.101/14.102

14.103 The skier passes point A going 17 m/s. From A to B, the radius of his circular path is 6 m. By using Newton's second law in terms of polar coordinates, determine the magnitude of the skier's velocity as he leaves the jump at B. Neglect tangential forces other than the tangential component of his weight.

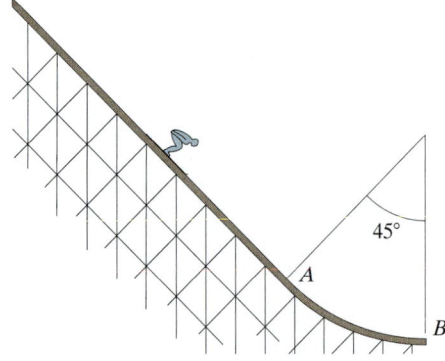

Problem 14.103

14.104* A 2-kg mass rests on a flat horizontal bar. The bar begins rotating *in the vertical plane* about O with a constant angular acceleration of 1 rad/s^2. The mass is observed to slip relative to the bar when the bar is 30° above the horizontal. What is the static coefficient of friction between the mass and the bar? Does the mass slip toward or away from O?

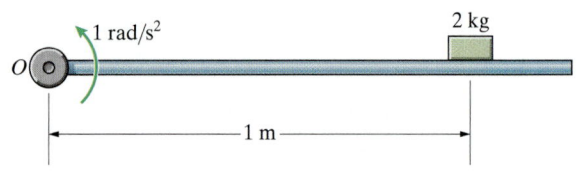

Problem 14.104

14.105* The 1/4-lb slider A is pushed along the circular bar by the slotted bar. The circular bar lies *in the horizontal plane*. The angular position of the slotted bar is $\theta = 10t^2$ rad. Determine the polar components of the total external force exerted on the slider at $t = 0.2$ s.

14.106* The 1/4-lb slider A is pushed along the circular bar by the slotted bar. The circular bar lies *in the vertical plane*. The angular position of the slotted bar is $\theta = 10t^2$ rad. Determine the polar components of the total force exerted on the slider by the circular and slotted bars at $t = 0.25$ s.

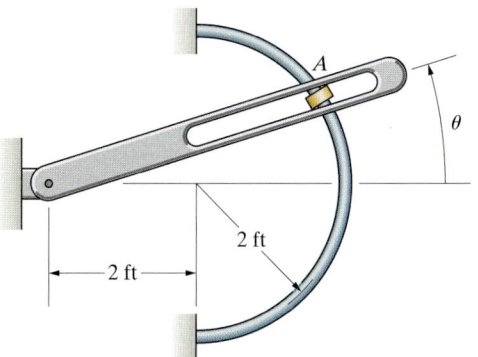

Problems **14.105/14.106**

14.107* The slotted bar rotates *in the horizontal plane* with constant angular velocity ω_0. The mass m has a pin that fits in the slot of the bar. A spring holds the pin against the surface of the fixed cam. The surface of the cam is described by $r = r_0(2 - \cos \theta)$. Determine the polar components of the total external force exerted on the pin as functions of θ.

14.108* In Problem 14.107, suppose that the unstretched length of the spring is r_0. Determine the smallest value of the spring constant k for which the pin will remain on the surface of the cam.

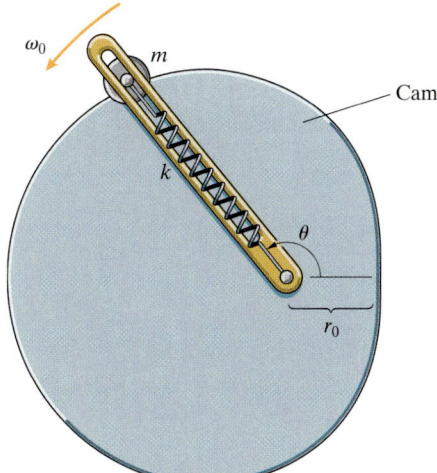

Problems **14.107/14.108**

14.109 A charged particle P in a magnetic field moves along the spiral path described by $r = 1$ m, $\theta = 2z$ rad, where z is in meters. The particle moves along the path in the direction shown with constant speed $|\mathbf{v}| = 1$ km/s. The mass of the particle is 1.67×10^{-27} kg. Determine the sum of the forces on the particle in terms of cylindrical coordinates.

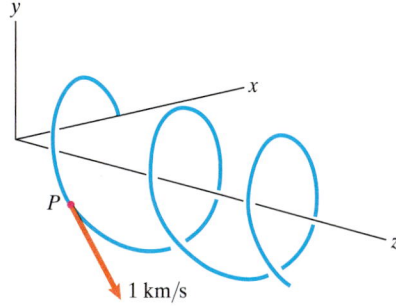

Problem **14.109**

14.110 At the instant shown, the cylindrical coordinates of the 4-kg part A held by the robotic manipulator are $r = 0.6$ m, $\theta = 25°$, and $z = 0.8$ m. (The coordinate system is fixed with respect to the earth, and the y axis points upward.) A's radial position is increasing at $dr/dt = 0.2$ m/s, and $d^2r/dt^2 = -0.4$ m/s². The angle θ is increasing when $d\theta/dt = 1.2$ rad/s, and $d^2\theta/dt^2 = 2.8$ rad/s². The base of the manipulator arm is accelerating in the z direction at $d^2z/dt^2 = 2.5$ m/s². Determine the force vector exerted on A by the manipulator in terms of cylindrical coordinates.

14.111 Suppose that the robotic manipulator is used in a space station to investigate zero-g manufacturing techniques. During an interval of time, the manipulator is programmed so that the cylindrical coordinates of the 4-kg part A are $\theta = 0.15t^2$ rad, $r = 0.5(1 + \sin \theta)$ m, and $z = 0.8(1 + \theta)$ m. Determine the force vector exerted on A by the manipulator at $t = 2$ s in terms of cylindrical coordinates.

14.112* In Problem 14.111, draw a graph of the magnitude of the force exerted on part A by the manipulator as a function of time from $t = 0$ to $t = 5$ s, and use the graph to estimate the maximum force during that interval of time.

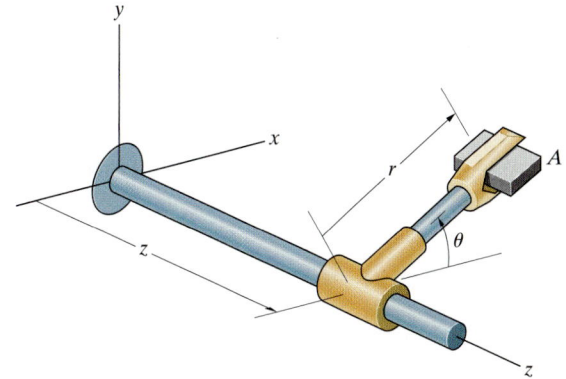

Problems **14.110–14.112**

14.5 Orbital Mechanics

It is appropriate to include a discussion of orbital mechanics in our chapter on applications of Newton's second law. Newton's analytical determination of the elliptical orbits of the planets, which had been deduced from observational data by Johannes Kepler, was a triumph for Newtonian mechanics and confirmation of the inverse-square relation for gravitational acceleration.

We can use Newton's second law expressed in polar coordinates to determine the orbit of an earth satellite or a planet. Suppose that at $t = 0$ a satellite has an initial velocity v_0 at a distance r_0 from the center of the earth (Fig. 14.17a). We assume that the initial velocity is perpendicular to the line from the center of the earth to the satellite. The satellite's position during its subsequent motion is specified by its polar coordinates (r, θ), where θ is measured from the satellite's position at $t = 0$ (Fig. 14.17b). Our objective is to determine r as a function of θ.

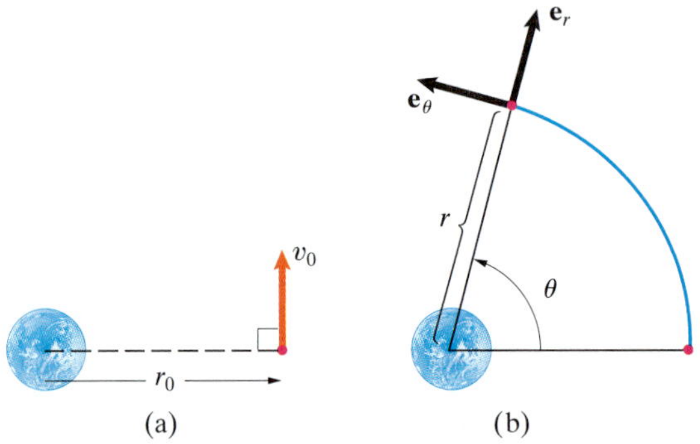

Figure 14.17
(a) Initial position and velocity of an earth satellite.
(b) Specifying the subsequent path in terms of polar coordinates.

(a)

(b)

Determination of the Orbit

If we model the earth as a homogeneous sphere, the force exerted on the satellite by gravity at a distance r from the center of the earth is mgR_E^2/r^2, where R_E is the earth's radius. (See Eq. 12.5.) From Eq. (14.9), the equation of motion in the radial direction is

$$\Sigma F_r = ma_r:$$

$$-\frac{mgR_E^2}{r^2} = m\left[\frac{d^2r}{dt^2} - r\left(\frac{d\theta}{dt}\right)^2\right].$$

From Eq. (14.10), the equation of motion in the transverse direction is

$$\Sigma F_\theta = ma_\theta:$$

$$0 = m\left(r\frac{d^2\theta}{dt^2} + 2\frac{dr}{dt}\frac{d\theta}{dt}\right).$$

We therefore obtain the two equations

$$\frac{d^2r}{dt^2} - r\left(\frac{d\theta}{dt}\right)^2 = -\frac{gR_E^2}{r^2} \tag{14.12}$$

and

$$r\frac{d^2\theta}{dt^2} + 2\frac{dr}{dt}\frac{d\theta}{dt} = 0. \tag{14.13}$$

We can write Eq. (14.13) in the form

$$\frac{1}{r}\frac{d}{dt}\left(r^2\frac{d\theta}{dt}\right) = 0,$$

which indicates that

$$r^2\frac{d\theta}{dt} = rv_\theta = \text{constant}. \tag{14.14}$$

At $t = 0$, the components of the velocity are $v_r = 0$ and $v_\theta = v_0$, and the radial position is $r = r_0$. We can therefore write the constant in Eq. (14.14) in terms of the initial conditions:

$$r^2\frac{d\theta}{dt} = rv_\theta = r_0v_0. \tag{14.15}$$

Using this equation to eliminate $d\theta/dt$ from Eq. (14.12), we obtain

$$\frac{d^2r}{dt^2} - \frac{r_0^2v_0^2}{r^3} = -\frac{gR_E^2}{r^2}. \tag{14.16}$$

We can solve this differential equation by introducing the change of variable

$$u = \frac{1}{r}. \tag{14.17}$$

In doing so, we will also change the independent variable from t to θ, because we want to determine r as a function of the angle θ instead of time. To express Eq. (14.16) in terms of u, we must determine d^2r/dt^2 in terms of u. Using the chain rule, we write the derivative of r with respect to time as

$$\frac{dr}{dt} = \frac{d}{dt}\left(\frac{1}{u}\right) = -\frac{1}{u^2}\frac{du}{dt} = -\frac{1}{u^2}\frac{du}{d\theta}\frac{d\theta}{dt}. \tag{14.18}$$

Notice from Eq. (14.15) that

$$\frac{d\theta}{dt} = \frac{r_0v_0}{r^2} = r_0v_0u^2. \tag{14.19}$$

Substituting this expression into Eq. (14.18), we obtain

$$\frac{dr}{dt} = -r_0v_0\frac{du}{d\theta}. \tag{14.20}$$

We differentiate Eq. (14.20) with respect to time and apply the chain rule again:

$$\frac{d^2r}{dt^2} = \frac{d}{dt}\left(-r_0v_0\frac{du}{d\theta}\right) = -r_0v_0\frac{d\theta}{dt}\frac{d}{d\theta}\left(\frac{du}{d\theta}\right) = -r_0v_0\frac{d\theta}{dt}\frac{d^2u}{d\theta^2}.$$

Using Eq. (14.19) to eliminate $d\theta/dt$ from this expression, we obtain the second time derivative of r in terms of u:

$$\frac{d^2r}{dt^2} = -r_0^2v_0^2u^2\frac{d^2u}{d\theta^2}.$$

Substituting this result into Eq. (14.16) yields a linear differential equation for u as a function of θ:

$$\frac{d^2u}{d\theta^2} + u = \frac{gR_E^2}{r_0^2v_0^2}.$$

The general solution of this equation is

$$u = A \sin \theta + B \cos \theta + \frac{gR_E^2}{r_0^2 v_0^2}, \tag{14.21}$$

where A and B are constants. We can use the initial conditions to determine A and B. When $\theta = 0$, $u = 1/r_0$. Also, when $\theta = 0$ the radial component of velocity $v_r = dr/dt = 0$, so from Eq. (14.20) we see that $du/d\theta = 0$. From these two conditions, we obtain

$$A = 0 \quad \text{and} \quad B = \frac{1}{r_0} - \frac{gR_E^2}{r_0^2 v_0^2}.$$

Substituting these results into Eq. (14.21), we can write the resulting solution for $r = 1/u$ as

$$\frac{r}{r_0} = \frac{1 + \varepsilon}{1 + \varepsilon \cos \theta}, \tag{14.22}$$

where

$$\varepsilon = \frac{r_0 v_0^2}{gR_E^2} - 1. \tag{14.23}$$

Types of Orbits

The curve called a *conic section* (Fig. 14.18) has the property that the ratio of r to the perpendicular distance d to a straight line called the *directrix* is constant. This ratio, $r/d = r_0/d_0$, is called the *eccentricity* of the curve. From the figure, we see that

$$r \cos \theta + d = r_0 + d_0,$$

which we can write as

$$\frac{r}{r_0} = \frac{1 + (r_0/d_0)}{1 + (r_0/d_0) \cos \theta}.$$

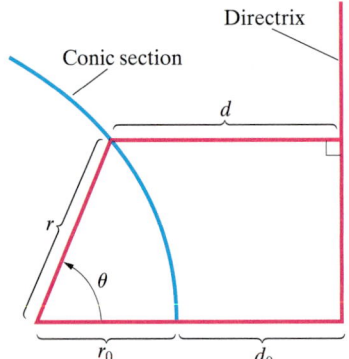

Figure 14.18
If the ratio r/d is constant, the curve describes a conic section.

Comparing this expression with Eq. (14.22), we see that *the satellite's orbit describes a conic section with eccentricity ε*. The value of the eccentricity determines the character of the orbit.

Circular Orbit If the initial velocity v_0 is chosen so that $\varepsilon = 0$, Eq. (14.22) reduces to $r = r_0$ and the orbit is circular (Fig. 14.19). Setting $\varepsilon = 0$ in Eq. (14.23) and solving for v_0, we obtain

$$v_0 = \sqrt{\frac{gR_E^2}{r_0}}, \tag{14.24}$$

which agrees with the velocity for a circular orbit we obtained by a different method in Example 13.10.

Elliptic Orbit If $0 < \varepsilon < 1$, the orbit is an ellipse (Fig. 14.19). The maximum radius of the ellipse occurs when $\theta = 180°$. Setting θ equal to $180°$ in Eq. (14.22), we obtain an expression for the maximum radius of the ellipse in terms of the initial radius and ε:

$$r_{\max} = r_0 \left(\frac{1 + \varepsilon}{1 - \varepsilon} \right). \tag{14.25}$$

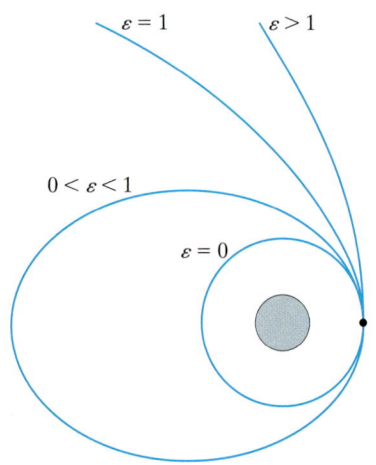

Figure 14.19
Orbits for different eccentricities.

Parabolic Orbit   Notice from Eq. (14.25) that the maximum radius of the elliptic orbit increases without limit as $\varepsilon \rightarrow 1$. When $\varepsilon = 1$, the orbit is a parabola (Fig. 14.19). The corresponding velocity v_0 is the minimum initial velocity for which the radius r increases without limit, which is the escape velocity. Setting $\varepsilon = 1$ in Eq. (14.23) and solving for v_0, we obtain

$$v_0 = \sqrt{\frac{2gR_E^2}{r_0}}.$$

This is the same value for the escape velocity we obtained in Example 13.5 for the case of an object moving in a straight path directly away from the center of the earth.

Hyperbolic Orbit   If $\varepsilon > 1$, the orbit is a hyperbola (Fig. 14.19).

The solution we have presented, based on the assumption that the earth is a homogeneous sphere, approximates the orbit of an earth satellite. Determining the orbit accurately requires taking into account the variations in the earth's gravitational field due to its actual mass distribution. Similarly, depending on the accuracy required, determining the orbit of a planet around the sun may require accounting for perturbations due to the gravitational attractions of the other planets.

Example 14.8   Orbit of an Earth Satellite

An earth satellite is in an elliptic orbit with a minimum radius of 6600 km and a maximum radius of 16,000 km. The earth's radius is 6370 km.
(a) Determine the satellite's velocity when it is at perigee (its minimum radius) and when it is at apogee (its maximum radius).
(b) Draw a graph of the orbit.

Strategy
We can regard the radius and velocity of the satellite at perigee as the initial conditions r_0 and v_0 used in obtaining Eq. (14.22). Since the maximum radius of the orbit is given, we can solve Eq. (14.25) for the eccentricity of the orbit and then use Eq. (14.23) to determine v_0. From Eq. (14.14), the product of r and the transverse component of the velocity is constant. From this condition, we can determine the velocity at apogee.

Solution
(a) The ratio of the radius at apogee to the radius at perigee is

$$\frac{r_{max}}{r_0} = \frac{1.60 \times 10^7 \text{ m}}{6.60 \times 10^6 \text{ m}} = 2.42.$$

Solving Eq. (14.25) for ε, we find that the eccentricity is

$$\varepsilon = \frac{\dfrac{r_{max}}{r_0} - 1}{\dfrac{r_{max}}{r_0} + 1} = \frac{2.42 - 1}{2.42 + 1} = 0.416.$$

From Eq. (14.23), the velocity at perigee is

$$v_0 = \sqrt{\frac{(\varepsilon + 1)gR_E^2}{r_0}}$$

$$= \sqrt{\frac{(0.416 + 1)(9.81 \text{ m/s}^2)(6.37 \times 10^6 \text{ m})^2}{6.60 \times 10^6 \text{ m}}}$$

$$= 9240 \text{ m/s}.$$

At both perigee and apogee, the velocity has only a transverse component. From Eq. (14.14), the velocity at apogee, v_a, is related to the velocity v_0 by

$$r_0 v_0 = r_{\max} v_a.$$

Therefore, the velocity at apogee is

$$v_a = \frac{v_0}{r_{\max}/r_0} = \frac{9240 \text{ m/s}}{2.42} = 3810 \text{ m/s}.$$

(b) By plotting Eq. (14.22) with $\varepsilon = 0.416$, we obtain the graph of the orbit (Fig. 14.20).

Figure 14.20
Orbit of an earth satellite with a perigee of 6600 km and an apogee of 16,000 km.

Critical Thinking

In this example the radii of the perigee and apogee of the elliptic orbit were given and the task was to determine the corresponding velocities. When a satellite is launched, it is given an initial velocity at an initial radial distance. If the initial velocity is in the transverse direction, as shown in Fig. 14.17a, Eqs. (14.22) and (14.23) determine the resulting orbit. By extending the analysis we used to obtain Eq. (14.22), the orbit can be determined when the initial velocity is in an arbitrary direction (see Problem 14.119).

Our analysis of the orbit problem was based on modeling the earth as a homogeneous sphere. To obtain accurate predictions of the orbits of objects near the earth, the effects of the actual mass distribution of the earth must be taken into account.

Problems

Use the values $R_E = 6370$ km $= 3960$ mi **for the radius of the earth.**

14.113 The International Space Station is in a circular orbit 225 miles above the earth's surface.

(a) What is the magnitude of the velocity of the space station?

(b) How long does it take to complete one revolution?

14.114 The moon is approximately 383,000 km from the earth. Assume that the moon's orbit around the earth is circular with velocity given by Eq. (14.24).

(a) What is the magnitude of the moon's velocity?

(b) How long does it take to complete one revolution around the earth?

14.115 A satellite is given an initial velocity $v_0 = 6700$ m/s at a distance $r_0 = 2R_E$ from the center of the earth as shown in Fig. 14.17a.

(a) What is the maximum radius of the resulting elliptic orbit?

(b) What is the magnitude of the satellite's velocity when the satellite is at its maximum radius?

14.116 A satellite is given an initial velocity $v_0 = 6700$ m/s at a distance $r_0 = 2R_E$ from the center of the earth as shown in Fig. 14.17a. Draw a graph of the resulting orbit.

14.117 A satellite is given an initial velocity v_0 at a distance $r_0 = 6800$ km from the center of the earth, as shown in Fig. 14.17a. The resulting elliptic orbit has a maximum radius of 20,000 km. What is v_0?

14.118* You can send a spacecraft from the earth to the moon in the following way: First, launch the spacecraft into a circular "parking" orbit of radius r_0 around the earth (Fig. a). Then, increase its velocity in the direction tangent to the circular orbit to a value v_0 such that it will follow an elliptic orbit whose maximum radius is equal to the radius r_M of the moon's orbit around the earth (Fig. b). The radius $r_M = 238,000$ mi. Let $r_0 = 4160$ mi. What velocity v_0 is necessary to send a spacecraft to the moon? (This description is simplified in that it disregards the effect of the moon's gravity.)

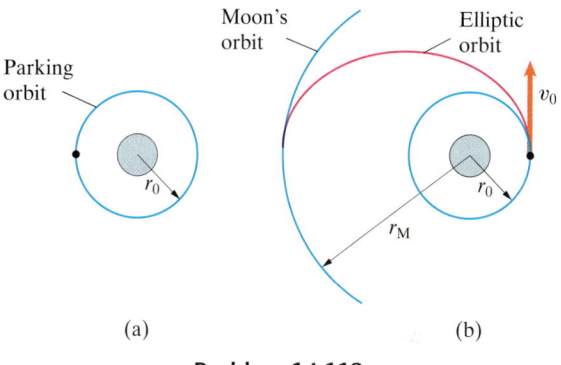

(a)   (b)

Problem 14.118

14.119* At $t = 0$, an earth satellite is a distance r_0 from the center of the earth and has an initial velocity v_0 in the direction shown. Show that the polar equation for the resulting orbit is

$$\frac{r}{r_0} = \frac{(\varepsilon + 1)\cos^2\beta}{[(\varepsilon+1)\cos^2\beta - 1]\cos\theta - (\varepsilon+1)\sin\beta\cos\beta\sin\theta + 1},$$

where $\varepsilon = (r_0 v_0^2 / g R_E^2) - 1$.

14.120 Draw graphs of the orbits given by the polar equation obtained in Problem 14.119 for $\varepsilon = 0$ and $\beta = 0$, 30°, and 60°.

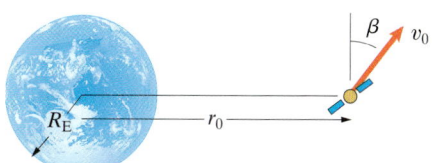

Problems 14.119/14.120

14.6 Numerical Solutions

In this chapter we have described many situations in which we were able to determine the motion of an object by a simple procedure: After using Newton's second law to determine the acceleration, we integrated to obtain analytical, or *closed-form*, expressions for the object's velocity and position. Such examples are very valuable—they demonstrate how to use free-body diagrams and formulate problems using different coordinate systems, and they develop intuitive understanding of forces and motions. But it would be misleading if we presented examples of this kind only, because most problems that must be dealt with in engineering cannot be solved in that way. The functions describing the forces, and therefore the accelerations, are often too complicated to integrate and obtain closed-form solutions. In other situations, the forces are not known in terms of functions, but instead are specified in terms of data, either as a continuous recording of force as a function of time (analog data) or as values of force measured at discrete times (digital data).

We can obtain approximate solutions to problems for which closed-form solutions are not possible by using numerical integration. Consider an object of mass m in straight-line motion along the x axis (Fig. 14.21), and assume that the x component of the total force may depend on the time as well as the position and velocity of the object:

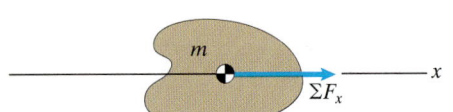

Figure 14.21
An object moving along the x axis.

$$\Sigma F_x = \Sigma F_x(t, x, v_x). \tag{14.26}$$

Suppose that at a particular time t_0 we know the position $x(t_0)$ and velocity $v_x(t_0)$. The acceleration at t_0 is

$$\frac{dv_x}{dt}(t_0) = \frac{1}{m}\Sigma F_x(t_0, x(t_0), v_x(t_0)). \tag{14.27}$$

To determine the velocity at time $t_0 + \Delta t$, we express it as a Taylor series:

$$v_x(t_0 + \Delta t) = v_x(t_0) + \frac{dv_x}{dt}(t_0)\,\Delta t + \frac{1}{2}\frac{d^2v_x}{dt^2}(t_0)(\Delta t)^2 + \cdots.$$

By choosing a sufficiently small value of Δt, we can neglect terms in this equation that are of second and higher order in Δt and substitute Eq. (14.27) to obtain an approximation for the velocity at time $t_0 + \Delta t$:

$$v_x(t_0 + \Delta t) = v_x(t_0) + \frac{1}{m}\Sigma F_x(t_0, x(t_0), v_x(t_0))\,\Delta t. \tag{14.28}$$

We can approximate the position at $t_0 + \Delta t$ in the same way. Expressing it as a Taylor series, we have

$$x(t_0 + \Delta t) = x(t_0) + \frac{dx}{dt}(t_0)\,\Delta t + \frac{1}{2}\frac{d^2x}{dt^2}(t_0)(\Delta t)^2 + \cdots,$$

and neglecting higher order terms in Δt, we obtain

$$x(t_0 + \Delta t) = x(t_0) + v_x(t_0)\,\Delta t. \tag{14.29}$$

Thus, if we know the position and velocity at a time t_0, we can approximate their values at $t_0 + \Delta t$ by using Eqs. (14.28) and (14.29). We can then repeat the procedure, using $x(t_0 + \Delta t)$ and $v_x(t_0 + \Delta t)$ as initial conditions to determine the approximate position and velocity at time $t_0 + 2\Delta t$. By continuing in this way, we obtain approximate solutions for the position and velocity in terms of time. This procedure is easy to carry out using a calculator or a computer. It is called a *finite-difference method*, because it determines changes in the dependent variables over finite intervals of time. The particular finite-difference method we have described, due to Leonhard Euler (1707–1783), is called *forward differencing*: The value of the derivative of a function at t_0 is approximated by using its value at t_0 and its value forward in time at $t_0 + \Delta t$.

More elaborate finite-difference methods, based on retaining more terms in the Taylor series, produce smaller errors in each time step than Euler's method. For example, in the fourth-order Runge–Kutta method, terms through the fourth order in Δt are retained. However, Euler's method is adequate for introducing numerical solutions of problems in dynamics.

Notice that Eq. (14.26) does not need to be a functional expression to carry out the process we have described. The values of the total force must be known at times $t_0, t_0 + \Delta t, \ldots$, and can be determined either from a function or from analog or digital data.

We can use the same approach to determine the velocity and position of an object in curvilinear motion. Suppose that the object moves in the x–y plane and that the components of force may depend on the time as well as the position and velocity of the object; that is,

$$\Sigma F_x = \Sigma F_x(t, x, y, v_x, v_y) \quad \text{and} \quad \Sigma F_y = \Sigma F_y(t, x, y, v_x, v_y).$$

If the position and velocity are known at a time t_0, we can use the same steps leading to Eqs. (14.28) and (14.29) to obtain approximate expressions for the components of position and velocity at $t_0 + \Delta t$, yielding

$$x(t_0 + \Delta t) = x(t_0) + v_x(t_0)\,\Delta t,$$

$$y(t_0 + \Delta t) = y(t_0) + v_y(t_0)\,\Delta t,$$

$$\tag{14.30}$$

$$v_x(t_0 + \Delta t) = v_x(t_0) + \frac{1}{m}\Sigma F_x(t_0, x(t_0), y(t_0), v_x(t_0), v_y(t_0))\,\Delta t,$$

$$v_y(t_0 + \Delta t) = v_y(t_0) + \frac{1}{m}\Sigma F_y(t_0, x(t_0), y(t_0), v_x(t_0), v_y(t_0))\,\Delta t.$$

COMPUTATIONAL MECHANICS

The following example and problems are designed to be worked with the use of a programmable calculator or computer.

Computational Example 14.9

A 50-kg projectile is launched from $x = 0, y = 0$ with initial velocity $v_x = 100$ m/s, $v_y = 100$ m/s. (The y axis is positive upward.) The aerodynamic drag force on the projectile is of magnitude $C|\mathbf{v}|^2$, where C is a constant. Determine the trajectory of the projectile for $C = 0.005, 0.01,$ and 0.02.

Strategy

We will determine the x and y components of the forces exerted on the projectile by its weight and aerodynamic drag. Then we can use Eqs. (14.30) to estimate the projectile's position and velocity as functions of time.

Solution

To apply Eqs. (14.30), we must determine the x and y components of the total force on the projectile. Let $\mathbf{D}$ be the drag force (Fig. 14.22). Because $\mathbf{v}/|\mathbf{v}|$ is a unit vector in the direction of $\mathbf{v}$, we can write the drag force as

$$\mathbf{D} = -C|\mathbf{v}|^2 \frac{\mathbf{v}}{|\mathbf{v}|} = -C|\mathbf{v}|\mathbf{v}.$$

The external forces on the projectile are its weight and the drag, so we have

$$\Sigma \mathbf{F} = -mg\mathbf{j} - C|\mathbf{v}|\mathbf{v},$$

and the components of the total force are

$$\Sigma F_x = -C\sqrt{v_x^2 + v_y^2}\, v_x, \qquad \Sigma F_y = -mg - C\sqrt{v_x^2 + v_y^2}\, v_y.$$

Consider the case in which $C = 0.005$, and let $\Delta t = 0.2$ s. At the initial time $t_0 = 0$, the position coordinates and the components of the velocity are $x(t_0) = 0, y(t_0) = 0, v_x(t_0) = 100$ m/s, and $v_y(t_0) = 100$ m/s. The x coordinate after the first time step is

$$x(t_0 + \Delta t) = x(t_0) + v_x(t_0)\,\Delta t,$$

or

$$x(0.2) = x(0) + v_x(0)\,\Delta t$$

$$= 0 + (100)(0.2)$$

$$= 20 \text{ m}.$$

The y coordinate after the first time step is

$$y(t_0 + \Delta t) = y(t_0) + v_y(t_0)\,\Delta t,$$

or

$$y(0.2) = y(0) + v_y(0)\,\Delta t$$

$$= 0 + (100)(0.2)$$

$$= 20 \text{ m}.$$

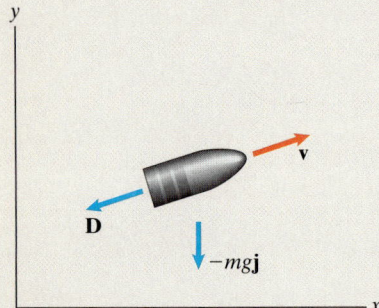

Figure 14.22
The forces on the projectile are its weight and the drag force $\mathbf{D}$.

The x component of the velocity after the first time step is

$$v_x(t_0 + \Delta t) = v_x(t_0) + \frac{1}{m}\Sigma F_x(t_0, x(t_0), y(t_0), v_x(t_0), v_y(t_0))\,\Delta t,$$

or

$$v_x(0.2) = v_x(0) + \frac{1}{m}\left\{-C\sqrt{[v_x(0)]^2 + [v_y(0)]^2}\,v_x(0)\right\}\Delta t$$

$$= 100 + \frac{1}{50}\left[-0.005\sqrt{(100)^2 + (100)^2}(100)\right](0.2)$$

$$= 99.72 \text{ m/s},$$

and the y component of the velocity after the first time step is

$$v_y(t_0 + \Delta t) = v_y(t_0) + \frac{1}{m}\Sigma F_y(t_0, x(t_0), y(t_0), v_x(t_0), v_y(t_0))\,\Delta t,$$

or

$$v_y(0.2) = v_y(0) + \frac{1}{m}\left\{-mg - C\sqrt{[v_x(0)]^2 + [v_y(0)]^2}\,v_y(0)\right\}\Delta t$$

$$= 100 + \frac{1}{50}\left[-(50)(9.81) - 0.005\sqrt{(100)^2 + (100)^2}(100)\right](0.2)$$

$$= 97.76 \text{ m/s}.$$

Continuing in this way, we obtain the following results for the first five time steps:

Time (s)	x (m)	y (m)	v_x (m/s)	v_y (m/s)
0.0	0.00	0.00	100.00	100.00
0.2	20.00	20.00	99.72	97.76
0.4	39.94	39.55	99.44	95.52
0.6	59.83	58.66	99.16	93.29
0.8	79.66	77.31	98.89	91.08
1.0	99.44	95.53	98.63	88.87

When there is no drag ($C = 0$), we can obtain the closed-form solution for the trajectory. Figure 14.23 compares the closed-form solution with numerical solutions obtained using $\Delta t = 2.0$ s, 1.0 s, and 0.2 s. The numerical solutions approach the exact solution as Δt decreases.

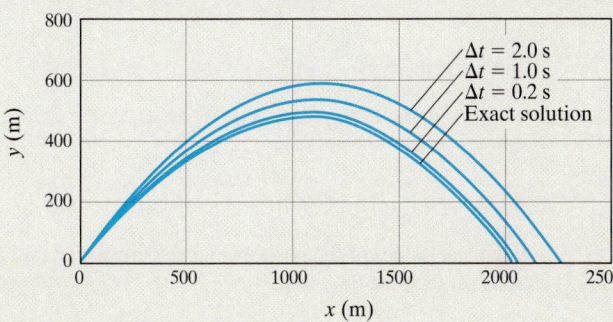

Figure 14.23
The closed-form solution for the trajectory when $C = 0$, compared with numerical solutions.

Figure 14.24 shows numerical solutions (obtained using $\Delta t = 0.01$ s) for the various values of C. As expected, the range of the projectile decreases as C increases. Also, notice that drag changes the shape of the trajectory: The angle at which the projectile descends is steeper than the angle at which it was launched.

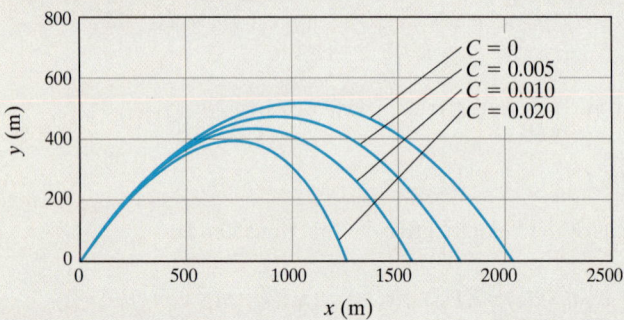

Figure 14.24
Trajectories for various values of C.

The development of the first completely electronic digital computer, the ENIAC (Electronic Numerical Integrator and Computer), at the University of Pennsylvania between 1943 and 1945 was motivated in part by the need to calculate trajectories of projectiles. A room-sized machine with 18,000 vacuum tubes, the ENIAC had 20 bytes of RAM and 450 bytes of ROM.

Critical Thinking

Two important limitations apply to the analysis we presented in this example. We have pointed out that Euler's method is primitive, and much more accurate and stable finite-difference methods are available. We used Euler's method because it is simple, intuitive, and provides insight into how such methods work. The other limitation is that our physical modeling of the problem is incomplete. The aerodynamic force on the projectile depends on the velocity in a more complex way than we assumed, and it also depends on the projectile's orientation as well as the density and temperature of the air. But these factors can be accounted for, and engineers can obtain very accurate predictions of the motions of objects by using iterative numerical calculations.

Computational Problems

14.121 A 1-kg object moves along the x axis under the action of the force $F_x = 6t$ N. At $t = 0$, the position and velocity of the object are $x = 0$ and $v_x = 10$ m/s. Using numerical integration with $\Delta t = 0.1$ s, determine the position and velocity of the object for the first five time steps.

Strategy: At the initial time $t_0 = 0$, $x(t_0) = 0$ and $v_x(t_0) = 10$ m/s. You can use Eqs. (14.28) and (14.29) to determine the velocity and position at time $t_0 + \Delta t = 0.1$ s. The position is

$$x(t_0 + \Delta t) = x(t_0) + v_x(t_0)\,\Delta t,$$

or

$$x(0.1) = x(0) + v_x(0)\,\Delta t$$
$$= 0 + (10)(0.1) = 1 \text{ m},$$

and the velocity is

$$v_x(t_0 + \Delta t) = v_x(t_0) + \frac{1}{m}F_x(t_0)\,\Delta t,$$

or

$$v_x(0.1) = 10 + \frac{1}{(1)}6(0)(0.1) = 10 \text{ m/s}.$$

Use these values of the position and velocity as the initial conditions for the next time step.

14.122 For the 1-kg object described in Problem 14.121, draw a graph comparing the closed-form solution from $t = 0$ to $t = 10$ s with the solutions obtained using numerical integration with $\Delta t = 2$ s, $\Delta t = 0.5$ s, and $\Delta t = 0.1$ s.

14.123 At $t = 0$, an object released from rest falls with constant acceleration $g = 9.81$ m/s^2.

(a) Using the closed-form solution, determine the velocity of the object and the distance it has fallen at $t = 2$ s.

(b) Approximate the answers to (a) by using numerical integration with $\Delta t = 0.2$ s.

14.124 In Problem 14.123, draw a graph of the distance the object falls as a function of time from $t = 0$ to $t = 4$ s, comparing the closed-form solution, the numerical solution using $\Delta t = 0.5$ s, and the numerical solution using $\Delta t = 0.05$ s.

14.125 The 9000-kg rocket starts from rest and travels straight up. The thrust of its engine is 4400 kN. The magnitude of the aerodynamic drag force **D** is $|\mathbf{D}| = 0.24v^2$ N, where v is the magnitude of the rocket's velocity in m/s. Due to the rate at which the rocket burns fuel, its mass as a function of time is $m = 9000 - 600t$ kg. Using numerical integration with $\Delta t = 0.1$ s, determine the rocket's height and velocity for the first five time steps.

Problem 14.125

14.126 The force exerted on the 50-kg mass by the linear spring is $F = -kx$, where x is the displacement of the mass from its position when the spring is unstretched. The spring constant k is 50 N/m. The mass is released from rest in the position $x = 1$ m. Use numerical integration with $\Delta t = 0.01$ s to determine the position and velocity of the mass for the first five time steps.

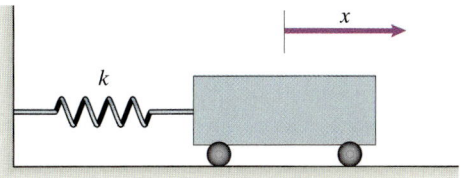

Problem 14.126

14.127 In Problem 14.126, use numerical integration with $\Delta t = 0.01$ s to determine the position and velocity of the mass in terms of time from $t = 0$ to $t = 10$ s. Draw graphs of your results.

14.128 At $t = 0$, the velocity of a 50-slug machine element that moves along the x axis is $v_x = 22$ ft/s. Measurements of the total force ΣF_x acting on the element at 0.1-s intervals from $t = 0$ to $t = 0.9$ s give the following values:

Time (s)	Force (lb)	Time (s)	Force (lb)
0.0	50.0	0.5	58.8
0.1	51.1	0.6	57.6
0.2	56.0	0.7	55.4
0.3	57.2	0.8	52.1
0.4	58.5	0.9	49.9

Determine approximately how far the element moves from $t = 0$ to $t = 1$ s, and determine the approximate velocity of the element at $t = 1$ s.

14.129 The lateral supports of a 100-kg structural element exert the horizontal force components

$$F_x = -2000x \quad \text{and} \quad F_y = -2000y,$$

where x and y are the coordinates of the center of mass of the element in meters. At $t = 0$, the coordinates and components of velocity of the center of mass are $x = 0.1$ m, $y = 0$, $v_x = 0$, and $v_y = 1$ m/s. Using $\Delta t = 0.1$ s, determine the approximate position and velocity of the center of mass for the first five time steps.

14.130 In Problem 14.129, use numerical integration with $\Delta t = 0.001$ s to determine the elliptical path described by the center of mass of the element, and draw a graph of the path.

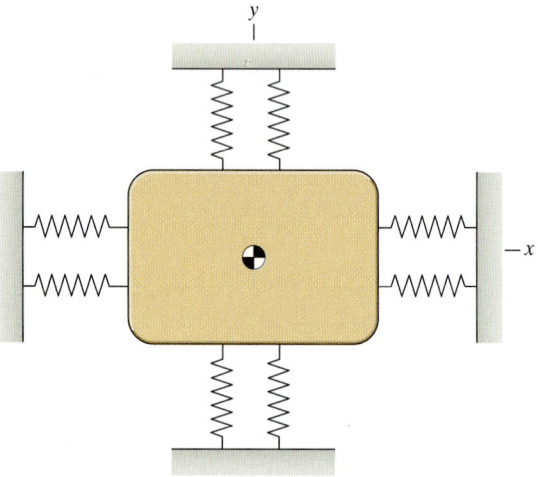

Problems 14.129/14.130

14.131 A car starts from rest at $t = 0$. Its acceleration is

$$a = 10 + 2t - 0.0185t^3 \text{ ft/s}^2.$$

(a) Using the closed-form solution of this equation, determine the distance the car has traveled and the car's velocity at $t = 6$ s.

(b) Use numerical integration with $\Delta t = 0.1$ s to approximate the answers obtained in (a).

(c) Use numerical integration with $\Delta t = 0.01$ s to approximate the answers obtained in (a).

14.132 A 20-kg projectile is launched from the ground with velocity components $v_x = 100$ m/s and $v_y = 49$ m/s. The magnitude of the aerodynamic drag force is $C|\mathbf{v}|^2$, where C is a constant. If the range of the projectile is 600 m, what is the constant C? (Use numerical integration with $\Delta t = 0.01$ s to compute the trajectory.)

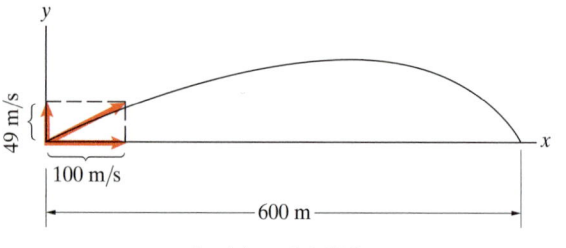

Problem 14.132

14.133 The 3000-lb vehicle has left the ground after driving over a rise. At the instant shown, it is moving horizontally at 30 mi/h and the bottoms of its tires are 24 in above the (approximately) level ground. The earth-fixed coordinate system is placed with its origin 30 in above the ground, at the height of the vehicle's center of mass when the tires first contact the ground. (Assume that the vehicle remains horizontal.) When that occurs, the vehicle's center of mass initially continues moving downward and then rebounds upward due to the flexure of the suspension system. While the tires are in contact with the ground, the force exerted on them by the ground is

$$-2400\mathbf{i} - (18{,}000y + 6000y^3 + 600v_y)\mathbf{j} \text{ (lb)},$$

where y is the vertical position of the center of mass in feet. Letting $t = 0$ be the instant the tires contact the ground, use numerical integration with $\Delta t = 0.001$ to determine the vertical position y and velocity v_y for the first five time steps.

14.134 In Problem 14.133, use numerical integration with $\Delta t = 0.001$ to estimate the magnitude of the maximum acceleration to which the vehicle is subjected during its impact with the ground.

14.135 In Problem 14.133, use numerical integration with $\Delta t = 0.001$ to estimate the components of velocity of the vehicle's center of mass at the instant the wheels leave the ground. (The wheels leave the ground when the center of mass is at $y = 0$.)

Problems 14.133–14.135

CHAPTER SUMMARY

We have used Newton's second law both to determine the acceleration of an object when the sum of the forces acting on it is known and to determine the sum of the forces when the acceleration is known. Once the acceleration of an object was known, we used the methods developed in Chapter 13 to obtain information about the object's velocity and position. In applying Newton's second law, we expressed it in terms of different coordinate systems. The choice of coordinate system was sometimes dictated by the nature of the forces acting on the object. When an object's path is known—especially when the object is constrained to move in a circle—normal and tangential components are often advantageous. In Chapter 15, we will use Newton's second law to derive a technique called the method of work and energy, which can greatly simplify the solution of particular types of problems in dynamics.

The total external force on an object is equal to the product of its mass and the acceleration of its center of mass relative to an inertial reference frame:

$$\Sigma \mathbf{F} = m\mathbf{a}. \tag{14.4}$$

A reference frame is said to be inertial if it is one in which the second law can be applied in this form. A reference frame translating at constant velocity relative to an inertial reference frame is also inertial.

Expressing Newton's second law in terms of a coordinate system yields the following scalar equations of motion:

Cartesian Coordinates

$$\Sigma F_x = ma_x, \quad \Sigma F_y = ma_y, \quad \Sigma F_z = ma_z. \tag{14.5}$$

Normal and Tangential Components

$$\Sigma F_{\mathrm{t}} = m\frac{dv}{dt}, \quad \Sigma F_{\mathrm{n}} = m\frac{v^2}{\rho}. \tag{14.7}$$

Polar Coordinates

$$\Sigma F_r = m\left(\frac{d^2 r}{dt^2} - r\omega^2\right), \tag{14.9}$$

$$\Sigma F_\theta = m\left(r\alpha + 2\frac{dr}{dt}\omega\right). \tag{14.10}$$

If the motion of an object is confined to a fixed plane, the component of the total force normal to the plane equals zero. In straight-line motion, the components of the total force perpendicular to the line equal zero and the component of the total force tangent to the line equals the product of the mass and the acceleration of the object along the line.

Review Problems

14.136 The Acura NSX can brake from 60 mi/h to a stop in a distance of 112 ft. The car weighs 3250 lb. (a) If you assume that the vehicle's deceleration is constant, what are its deceleration and the magnitude of the horizontal force its tires exert on the road? (b) If the car's tires are at the limit of adhesion (i.e., slip is impending), and the normal force exerted on the car by the road equals the car's weight, what is the coefficient of friction μ_s? (This analysis neglects the effects of horizontal and vertical aerodynamic forces.)

14.137 Using the coefficient of friction obtained in Problem 14.136, determine the highest constant speed at which the NSX could drive on a flat, circular track of 600-ft radius without skidding.

14.138 A "cog" engine hauls three cars of sightseers to a mountaintop in Bavaria. The mass of each car, including its passengers, is 10,000 kg, and the friction forces exerted by the wheels of the cars are negligible. Determine the forces in the couplings 1, 2, and 3 if (a) the engine is moving at constant velocity and (b) the engine is accelerating up the mountain at 1.2 m/s^2.

Problem 14.138

14.139 In a future mission, a spacecraft approaches the surface of an asteroid passing near the earth. Just before it touches down, the spacecraft is moving downward at a constant velocity relative to the surface of the asteroid and its downward thrust is 0.01 N. The computer decreases the downward thrust to 0.005 N, and an onboard laser interferometer determines that the acceleration of the spacecraft relative to the surface becomes 5×10^{-6} m/s^2 downward. What is the gravitational acceleration of the asteroid near its surface?

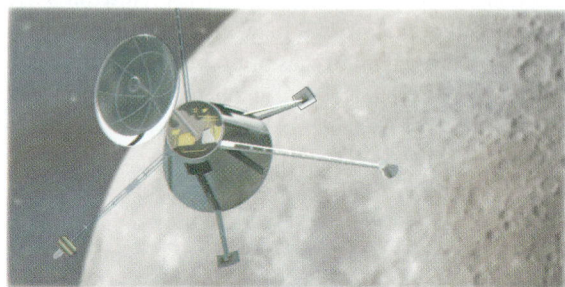

Problem 14.139

14.140 A car with a mass of 1470 kg, including its driver, is driven at 130 km/h over a slight rise in the road. At the top of the rise, the driver applies the brakes. The coefficient of static friction between the tires and the road is $\mu_s = 0.9$, and the radius of curvature of the rise is 160 m. Determine the car's deceleration at the instant the brakes are applied, and compare it with the deceleration on a level road.

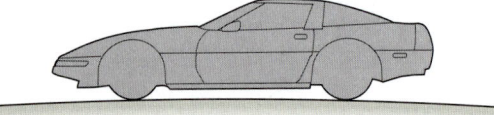

Problem 14.140

14.141 The car drives at constant velocity up the straight segment of road on the left. If the car's tires continue to exert the same tangential force on the road after the car has gone over the crest of the hill and is on the straight segment of road on the right, what will be the car's acceleration?

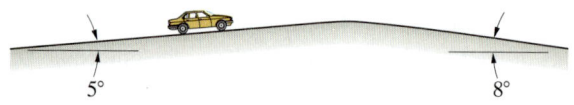

Problem 14.141

14.142 The aircraft carrier *Nimitz* weighs 91,000 tons. (A ton is 2000 lb.) Suppose that it is traveling at its top speed of approximately 30 knots (a knot is 6076 ft/h) when its engines are shut down. If the water exerts a drag force of magnitude $20{,}000v$ lb, where v is the carrier's velocity in feet per second, what distance does the carrier move before coming to rest?

14.143 If $m_A = 10$ kg, $m_B = 40$ kg, and the coefficient of kinetic friction between all surfaces is $\mu_k = 0.11$, what is the acceleration of B down the inclined surface?

14.144 If A weighs 20 lb, B weighs 100 lb, and the coefficient of kinetic friction between all surfaces is $\mu_k = 0.15$, what is the tension in the cord as B slides down the inclined surface?

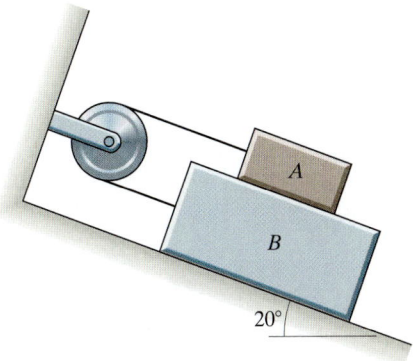

Problems 14.143/14.144

14.145 A gas gun is used to accelerate projectiles to high velocities for research on material properties. The projectile is held in place while gas is pumped into the tube to a high pressure p_0 on the left and the tube is evacuated on the right. The projectile is then released and is accelerated by the expanding gas. Assume that the pressure p of the gas is related to the volume V it occupies by $pV^\gamma = $ constant, where γ is a constant. If friction can be neglected, show that the velocity of the projectile at the position x is

$$v = \sqrt{\frac{2p_0 A x_0^\gamma}{m(\gamma - 1)}\left(\frac{1}{x_0^{\gamma-1}} - \frac{1}{x^{\gamma-1}}\right)},$$

where m is the mass of the projectile and A is the cross-sectional area of the tube.

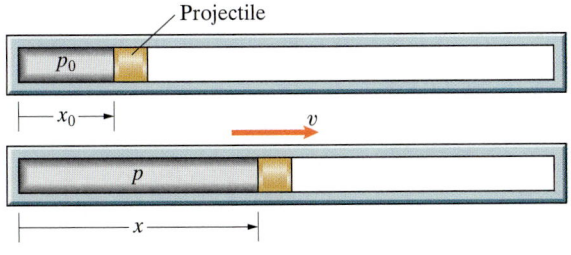

Problem 14.145

14.146 The weights of the blocks are $W_A = 120$ lb and $W_B = 20$ lb, and the surfaces are smooth. Determine the acceleration of block A and the tension in the cord.

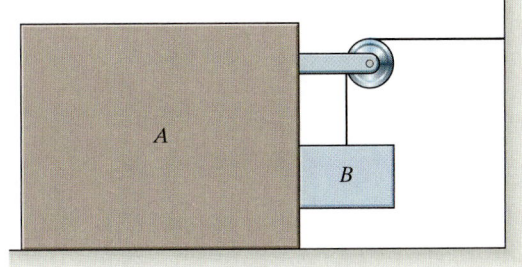

Problem 14.146

14.147 The 100-Mg space shuttle is in orbit when its engines are turned on, exerting a thrust force $\mathbf{T} = 10\mathbf{i} - 20\mathbf{j} + 10\mathbf{k}$ (kN) for 2 s. Neglect the resulting change in mass of the shuttle. At the end of the 2-s burn, fuel is still sloshing back and forth in the shuttle's tanks. What is the change in the velocity of the center of mass of the shuttle (including the fuel it contains) due to the 2-s burn?

14.148 The water skier contacts the ramp with a velocity of 25 mi/h parallel to the surface of the ramp. Neglecting friction and assuming that the tow rope exerts no force on him once he touches the ramp, estimate the horizontal length of the skier's jump from the end of the ramp.

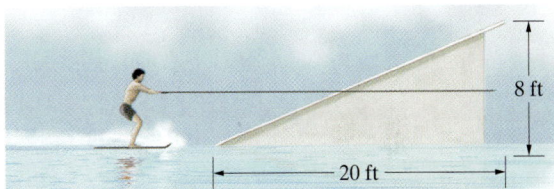

Problem 14.148

14.149 Suppose you are designing a roller-coaster track that will take the cars through a vertical loop of 40-ft radius. If you decide that, for safety, the downward force exerted on a passenger by his or her seat at the top of the loop should be at least one-half the passenger's weight, what is the minimum safe velocity of the cars at the top of the loop?

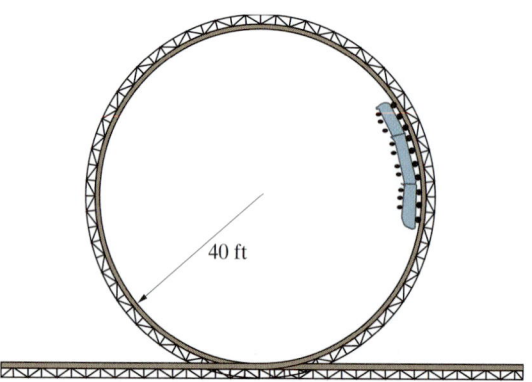

40 ft

Problem 14.149

14.150 As the smooth bar rotates *in the horizontal plane*, the string winds up on the fixed cylinder and draws the 1-kg collar A inward. The bar starts from rest at $t = 0$ in the position shown and rotates with constant angular acceleration. What is the tension in the string at $t = 1$ s?

14.151 In Problem 14.150, suppose that the coefficient of kinetic friction between the collar and the bar is $\mu_k = 0.2$. What is the tension in the string at $t = 1$ s?

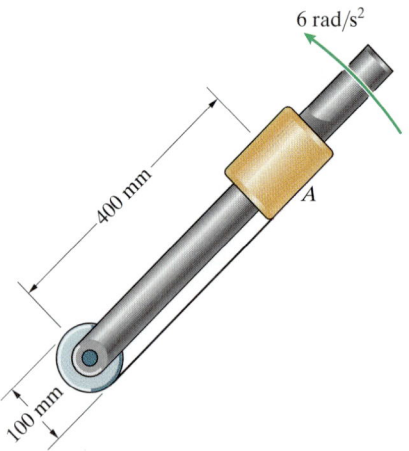

6 rad/s²

400 mm

A

100 mm

Problems 14.150/14.151

14.152 If you want to design the cars of a train to tilt as the train goes around curves in order to achieve maximum passenger comfort, what is the relationship between the desired tilt angle θ, the velocity v of the train, and the instantaneous radius of curvature, ρ, of the track?

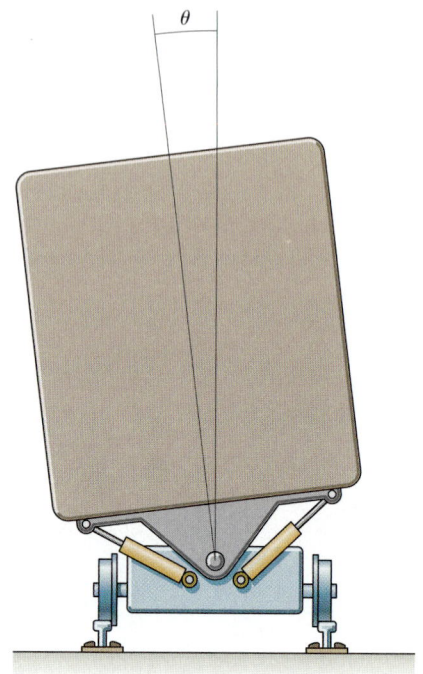

θ

Problem 14.152

14.153 To determine the coefficient of static friction between two materials, an engineer at the U.S. National Institute of Standards and Technology places a small sample of one material on a horizontal disk whose surface is made of the other material and then rotates the disk from rest with a constant angular acceleration of 0.4 rad/s^2. If she determines that the small sample slips on the disk after 9.903 s, what is the coefficient of friction?

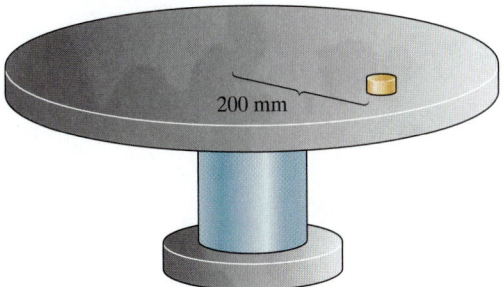

Problem 14.153

14.154* The 1-kg slider A is pushed along the curved bar by the slotted bar. The curved bar lies *in the horizontal plane*, and its profile is described by $r = 2(\theta/2\pi + 1)$ m, where θ is in radians. The angular position of the slotted bar is $\theta = 2t$ rad. Determine the polar components of the total external force exerted on the slider when $\theta = 120°$.

14.155* In Problem 14.154, suppose that the curved bar lies *in the vertical plane*. Determine the polar components of the total force exerted on A by the curved and slotted bars at $t = 0.5$ s.

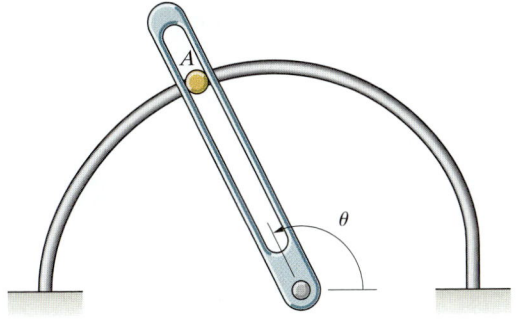

Problems 14.154/14.155

Design Project

The proposed design for an energy-absorbing bumper for a car exerts a decelerating force of magnitude $bs + cv$ on the car when it collides with a rigid obstacle, where s is the distance the car travels from the point where it contacts the obstacle and v is the car's velocity. Thus the force exerted on the car by the bumper is a function of the car's position and velocity.

(a) Suppose that at $t = 0$ the car contacts the obstacle with initial velocity v_0. Prove that the car's position is given as a function of time by

$$s = \frac{v_0}{2h}\left[e^{-(d-h)t} - e^{-(d+h)t}\right],$$

where $d = c/2m$, $h = \sqrt{d^2 - b/m}$, and m is the mass of the car. To do this, first show that this equation satisfies Newton's second law. Then confirm that it satisfies the initial conditions $s = 0$ and $v = v_0$ at $t = 0$.

(b) Investigate the effects of the car's mass, the initial velocity, and the constants b and c on the motion of the car when it strikes the obstacle. (Assume that $d^2 > b/m$.) Pay particular attention to how your choices for the constants b and c affect the maximum deceleration to which the occupants of the car would be subjected. Write a brief report presenting the results of your analysis and giving your conclusions concerning the design of energy-absorbing bumpers.

CHAPTER
15

Energy Methods

The concepts of energy and conservation of energy originated in large part from the study of classical mechanics. A simple transformation of Newton's second law results in an equation that motivates the definitions of work, kinetic energy (energy due to an object's motion), and potential energy (energy due to an object's position). This equation can greatly simplify the solution of problems involving certain forces that depend on an object's position, including gravitational forces and forces exerted by springs.

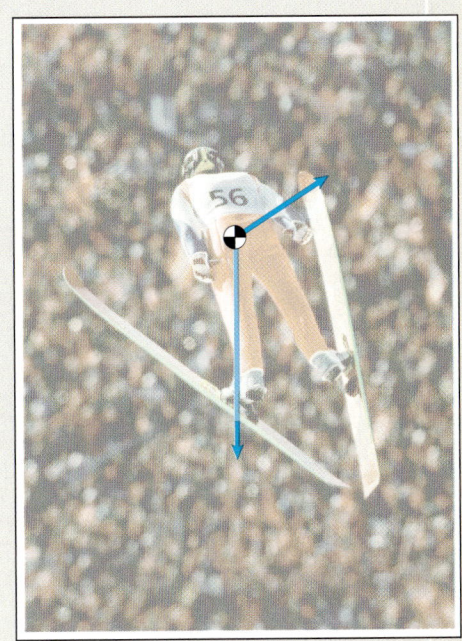

◄ The ski jumper's kinetic energy is determined by the change in his gravitational potential energy and the work done on him by aerodynamic forces. In this chapter, we use energy methods to analyze motions of objects.

WORK AND KINETIC ENERGY

15.1 Principle of Work and Energy

We have used Newton's second law to relate the acceleration of an object's center of mass to external forces acting on it. We will now show how Newton's second law, which is a vector equation, can be transformed into a scalar equation that is extremely useful in particular circumstances. We begin with Newton's second law in the form

$$\Sigma \mathbf{F} = m\frac{d\mathbf{v}}{dt}, \tag{15.1}$$

and take the dot product of both sides with the velocity:

$$\Sigma \mathbf{F} \cdot \mathbf{v} = m\frac{d\mathbf{v}}{dt} \cdot \mathbf{v}. \tag{15.2}$$

We write the left side of this equation as

$$\Sigma \mathbf{F} \cdot \mathbf{v} = \Sigma \mathbf{F} \cdot \frac{d\mathbf{r}}{dt}$$

and write the right side as

$$m\frac{d\mathbf{v}}{dt} \cdot \mathbf{v} = \tfrac{1}{2}m\frac{d}{dt}(\mathbf{v} \cdot \mathbf{v}),$$

obtaining

$$\Sigma \mathbf{F} \cdot d\mathbf{r} = \tfrac{1}{2}m\,d(v^2), \tag{15.3}$$

where $v^2 = \mathbf{v} \cdot \mathbf{v}$ is the square of the magnitude of the velocity. The term on the left side of Eq. (15.3) is the *work* expressed in terms of the total external force on the object and an infinitesimal displacement $d\mathbf{r}$ of its center of mass. Integrating Eq. (15.3) yields

$$\int_{\mathbf{r}_1}^{\mathbf{r}_2} \Sigma \mathbf{F} \cdot d\mathbf{r} = \tfrac{1}{2}mv_2^2 - \tfrac{1}{2}mv_1^2, \tag{15.4}$$

where v_1 and v_2 are the magnitudes of the velocity of the center of mass of the object when it is at positions $\mathbf{r}_1$ and $\mathbf{r}_2$, respectively. The term $\tfrac{1}{2}mv^2$ is called the *kinetic energy* associated with the motion of the center of mass. Denoting the work done as the center of mass moves from $\mathbf{r}_1$ to $\mathbf{r}_2$ by

$$U_{12} = \int_{\mathbf{r}_1}^{\mathbf{r}_2} \Sigma \mathbf{F} \cdot d\mathbf{r}, \tag{15.5}$$

we obtain the principle of work and energy:

> *The work done on an object as it moves between two positions equals the change in its kinetic energy.*

$$U_{12} = \tfrac{1}{2}mv_2^2 - \tfrac{1}{2}mv_1^2. \tag{15.6}$$

The dimensions of work, and therefore the dimensions of kinetic energy, are (force) × (length). In SI units, work is usually expressed in N-m, or joules (J). In U.S. Customary units, work is usually expressed in ft-lb.

If the work done on an object as it moves between two positions can be evaluated, the principle of work and energy permits us to determine the change in the magnitude of the object's velocity. We can also apply this principle to a system of objects, equating the total work done by external forces to the change in the total kinetic energy of the system. But the principle must be applied with caution, because, as we demonstrate in Example 15.3, net work can be done on a system by internal forces.

Although the principle of work and energy relates a change in the position of an object to the change in its velocity, it is not convenient for obtaining other information about the motion of the object, such as the time it takes the object to move from one position to another. Furthermore, since the work is an integral with respect to position, we can usually evaluate it only when the forces doing work are known as functions of position. Despite these limitations, the principle is extremely useful for certain problems because the work can be determined very easily.

15.2 Work and Power

In this section, we discuss how to determine the work done on an object. We also define the power transferred to or from an object by the forces acting on it and show how the power is calculated.

Evaluating the Work

Let us consider an object in curvilinear motion relative to an inertial reference frame (Fig. 15.1a) and specify its position by the coordinate s measured along its path from a reference point O. In terms of the tangential unit vector $\mathbf{e}_t$, the object's velocity is

$$\mathbf{v} = \frac{ds}{dt}\mathbf{e}_t.$$

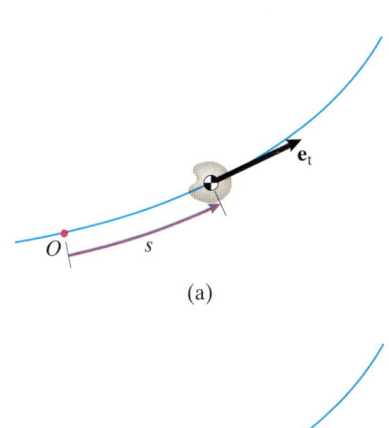

(a)

Because $\mathbf{v} = d\mathbf{r}/dt$, we can multiply the velocity by dt to obtain an expression for the vector $d\mathbf{r}$ describing an infinitesimal displacement along the path (Fig. 15.1b):

$$d\mathbf{r} = \mathbf{v}\,dt = ds\,\mathbf{e}_t.$$

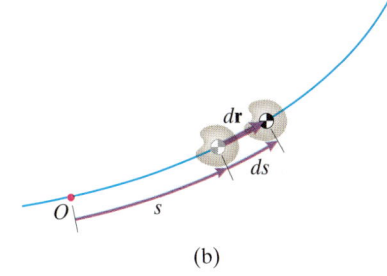

(b)

The work done by the external forces acting on the object as a result of the displacement $d\mathbf{r}$ is

$$\Sigma\mathbf{F}\cdot d\mathbf{r} = (\Sigma\mathbf{F}\cdot\mathbf{e}_t)\,ds = \Sigma F_t\,ds,$$

where ΣF_t is the tangential component of the total force. Therefore, as the object moves from a position s_1 to a position s_2 (Fig. 15.1c), the work is

$$U_{12} = \int_{s_1}^{s_2} \Sigma F_t\,ds. \tag{15.7}$$

The work is equal to the integral of the tangential component of the total force with respect to distance along the path. Thus, the work done is equal to the area

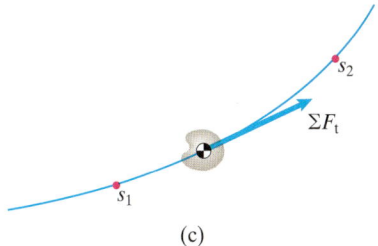

(c)

Figure 15.1
(a) The coordinate s and tangential unit vector.
(b) An infinitesimal displacement $d\mathbf{r}$.
(c) The work done from s_1 to s_2 is determined by the tangential component of the external forces.

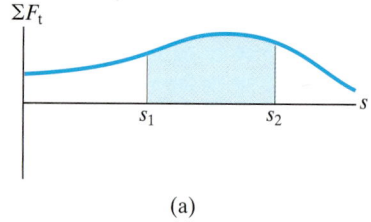

(a)

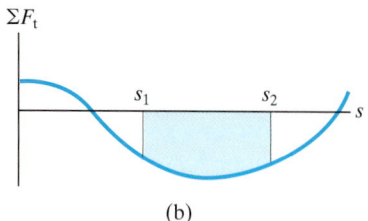

(b)

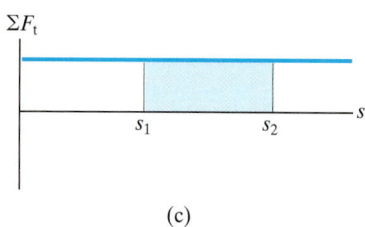

(c)

Figure 15.2
(a) The work equals the area defined by the graph of the tangential force as a function of the distance along the path.
(b) Negative work is done if the tangential force is opposite to the direction of the motion.
(c) The work done by a constant tangential force equals the product of the force and the distance.

defined by the graph of the tangential force from s_1 to s_2 (Fig. 15.2a). *Components of force perpendicular to the path do no work.* Notice that if ΣF_t is opposite to the direction of motion over some part of the path, which means that the object is decelerating, the work is negative (Fig. 15.2b). If ΣF_t is constant between s_1 and s_2, the work is simply the product of the total tangential force and the displacement (Fig. 15.2c):

$$U_{12} = \Sigma F_t(s_2 - s_1). \qquad \text{Constant tangential force} \qquad (15.8)$$

Power

Power is the rate at which work is done. The work done by the external forces acting on an object during an infinitesimal displacement $d\mathbf{r}$ is

$$\Sigma \mathbf{F} \cdot d\mathbf{r}.$$

We obtain the power P by dividing this expression by the interval of time dt during which the displacement takes place:

$$P = \Sigma \mathbf{F} \cdot \mathbf{v}. \qquad (15.9)$$

This is the power transferred to or from the object, depending on whether P is positive or negative. In SI units, power is expressed in newton-meters per second, which is joules per second (J/s) or watts (W). In U.S. Customary units, power is expressed in foot-pounds per second or in the anachronistic horsepower (hp), which is 746 W, or approximately 550 ft-lb/s.

Notice from Eq. (15.3) that the power equals the rate of change of the kinetic energy of the object:

$$P = \frac{d}{dt}\left(\tfrac{1}{2}mv^2\right).$$

Transferring power to and from an object causes its kinetic energy to increase and decrease, respectively. Using the preceding relation, we can write the average with respect to time of the power during an interval of time from t_1 to t_2 as

$$P_{av} = \frac{1}{t_2 - t_1}\int_{t_1}^{t_2} P\, dt = \frac{1}{t_2 - t_1}\int_{v_1{}^2}^{v_2{}^2} \tfrac{1}{2}m\, d(v^2).$$

Performing the integration, we find that the average power transferred to or from an object during an interval of time is equal to the change in its kinetic energy, or the work done, divided by the interval of time:

$$P_{av} = \frac{\tfrac{1}{2}mv_2^2 - \tfrac{1}{2}mv_1^2}{t_2 - t_1} = \frac{U_{12}}{t_2 - t_1}. \qquad (15.10)$$

Study Questions

1. What is the definition of the kinetic energy associated with the motion of the center of mass of an object?

2. What is the principle of work and energy?

3. If the tangential component of the total force on an object is constant, what do you know about the work done as the object moves a given distance along its path?

4. If an object is subjected only to forces that are perpendicular to its path, what do you know about its kinetic energy?

Example 15.1 Work and Energy in Straight-Line Motion

The 180-kg container A in Fig. 15.3 starts from rest at position $s = 0$. The hydraulic cylinder exerts a horizontal force on the container that is given as a function of position by $F = 700 - 150s$ N. The coefficient of kinetic friction between the container and the floor is $\mu_k = 0.26$. What is the velocity of the container when it has reached the position $s = 1$ m?

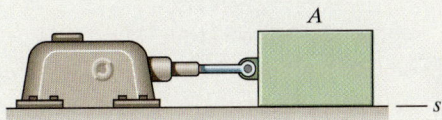

Figure 15.3

Strategy
We are asked to determine the change in the velocity of the container, given a change in its position, so we can apply the method of work and energy. We will determine the forces on the container in the direction tangent to its path and use Eq. (15.7) to evaluate the work.

Solution
Identify the Forces That Do Work We draw the free-body diagram of the container in Fig. a. The forces tangent to the path of the container are the force exerted by the hydraulic cylinder and the friction force. The container's acceleration in the vertical direction is zero, so

$$N = (180 \text{ kg})(9.81 \text{ m/s}^2) = 1770 \text{ N}.$$

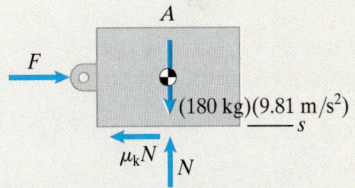

(a) Free-body diagram of the container.

Apply Work and Energy Let v be the container's velocity (Fig. b). At the initial position $s_1 = 0$, the velocity is $v_1 = 0$. We can apply the principle of work and energy to determine the container's kinetic energy at the position $s_2 = 1$ m, using Eq. (15.7) to evaluate the work.

$$\int_{s_1}^{s_2} \Sigma F_t \, ds = \tfrac{1}{2}mv_2^2 - \tfrac{1}{2}mv_1^2:$$

$$\int_0^1 (F - \mu_k N) \, ds = \tfrac{1}{2}mv_2^2 - \tfrac{1}{2}mv_1^2,$$

$$\int_0^1 [(700 - 150s) - (0.26)(1770)] \, ds = \tfrac{1}{2}(180)v_2^2 - 0.$$

Evaluating the integral and solving for the velocity, we obtain $v_2 = 1.36$ m/s.

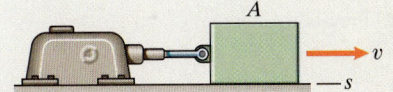

(b) Magnitude v of the container's velocity.

Critical Thinking
For comparison, you should determine the velocity of the container when it has moved 1 m by using Newton's second law. For problems involving straight-line motion, such as this example, the effort required to use Newton's second law is comparable to using work and energy. In the next section, we discuss problems involving curvilinear motion for which using work and energy is far simpler.

Example 15.2 | **Applying Work and Energy to a System**

The two crates in Fig. 15.4 are released from rest. Their masses are $m_A = 40$ kg and $m_B = 30$ kg, and the kinetic coefficient of friction between crate A and the inclined surface is $\mu_k = 0.15$. What is the magnitude of the velocity of the crates when they have moved 400 mm?

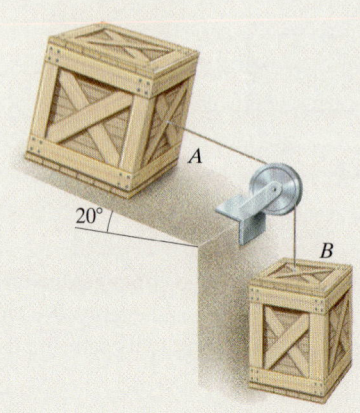

Figure 15.4

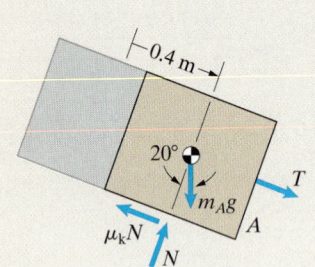

(a) Free-body diagram of A.

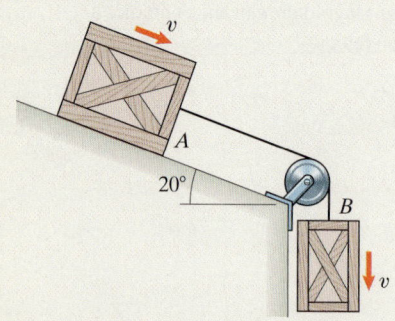

(b) The magnitude of the velocity of each crate is the same.

Strategy

We will determine the velocity in two ways.

First Method By drawing free-body diagrams of each of the crates and applying the principle of work and energy to them individually, we can obtain two equations in terms of the magnitude of the velocity and the tension in the cable.

Second Method We can draw a single free-body diagram of the two crates, the cable, and the pulley, and apply the principle of work and energy to the entire system.

Solution

First Method We draw the free-body diagram of crate A in Fig. a. The forces that do work as the crate moves down the plane are the forces tangential to its path: the tension T; the tangential component of the weight, $m_A g \sin 20°$; and the friction force $\mu_k N$. Because the acceleration of the crate normal to the surface is zero, $N = m_A g \cos 20°$. The magnitude v of the velocity at which A moves parallel to the surface equals the magnitude of the velocity at which B falls (Fig. b). Using Eq. (15.7) to determine the work, we equate the work done on A as it moves from $s_1 = 0$ to $s_2 = 0.4$ m to the change in the kinetic energy of A.

$$\int_{s_1}^{s_2} \Sigma F_t \, ds = \tfrac{1}{2} m v_2^2 - \tfrac{1}{2} m v_1^2:$$

$$\int_0^{0.4} \left[T + m_A g \sin 20° - \mu_k (m_A g \cos 20°) \right] ds = \tfrac{1}{2} m_A v_2^2 - 0. \quad (1)$$

The forces that do work on crate B are its weight $m_B g$ and the tension T (Fig. c). The magnitude of B's velocity is the same as that of crate A. The work done on B equals the change in its kinetic energy.

$$\int_{s_1}^{s_2} \Sigma F_t \, ds = \tfrac{1}{2} m v_2^2 - \tfrac{1}{2} m v_1^2:$$

$$\int_0^{0.4} (m_B g - T) \, ds = \tfrac{1}{2} m_B v_2^2 - 0. \tag{2}$$

By summing Eqs. (1) and (2), we eliminate T, obtaining

$$\int_0^{0.4} (m_A g \sin 20° - \mu_k m_A g \cos 20° + m_B g) \, ds = \tfrac{1}{2}(m_A + m_B) v_2^2:$$

$$[40 \sin 20° - (0.15)(40) \cos 20° + 30](9.81)(0.4) = \tfrac{1}{2}(40 + 30) v_2^2.$$

Solving for the velocity, we get $v_2 = 2.07 \text{ m/s}$.

Second Method We draw the free-body diagram of the system consisting of the crates, cable, and pulley in Fig. d. Notice that the cable tension does not appear in this diagram. The reactions at the pin support of the pulley do no work, because the support does not move. The total work done by external forces on the system as the boxes move 400 mm is equal to the change in the total kinetic energy of the system.

$$\int_0^{0.4} [m_A g \sin 20° - \mu_k (m_A g \cos 20°)] \, ds + \int_0^{0.4} m_B g \, ds$$

$$= \tfrac{1}{2} m_A v_2^2 + \tfrac{1}{2} m_B v_2^2 - 0:$$

$$[40 \sin 20° - (0.15)(40) \cos 20° + 30](9.81)(0.4) = \tfrac{1}{2}(40 + 30) v_2^2.$$

This equation is identical to the one we obtained by applying the principle of work and energy to the individual crates.

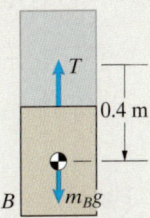

(c) Free-body diagram of B.

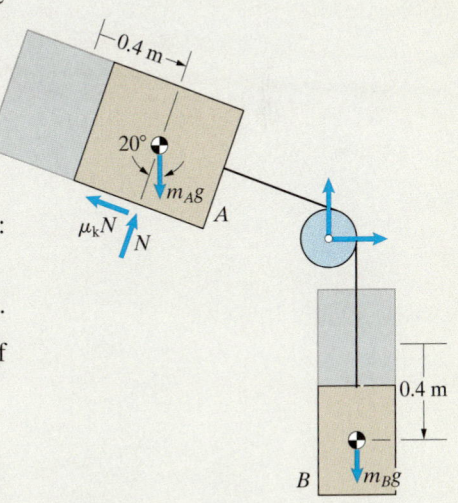

(d) Free-body diagram of the system.

Critical Thinking

You will often find it simpler to apply the principle of work and energy to an entire system instead of its separate parts. However, as we demonstrate in the next example, you need to be aware that internal forces in a system can do net work.

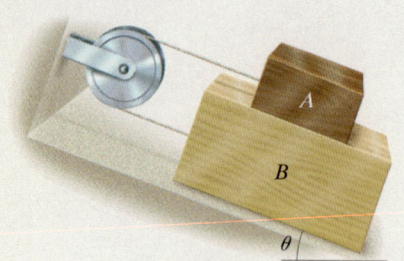

Example 15.3	Net Work by Internal Forces

Crates A and B in Fig. 15.5 are released from rest. The coefficient of kinetic friction between A and B is μ_k, and friction between B and the inclined surface can be neglected. What is the velocity of the crates when they have moved a distance b?

Figure 15.5

Strategy
By applying the principle of work and energy to each crate, we can obtain two equations in terms of the tension in the cable and the velocity.

Solution
We draw the free-body diagrams of the crates in Figs. a and b. The acceleration of A normal to the inclined surface is zero, so $N = m_A g \cos \theta$. The magnitudes of the velocities of A and B are equal (Fig. c). The work done on A equals the change in its kinetic energy.

$$U_{12} = \tfrac{1}{2}m_A v_2^2 - \tfrac{1}{2}m_A v_1^2:$$

$$\int_0^b (T - m_A g \sin \theta - \mu_k m_A g \cos \theta)\, ds = \tfrac{1}{2}m_A v_2^2. \qquad (1)$$

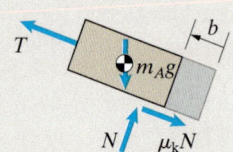

(a) Free-body diagram of A.

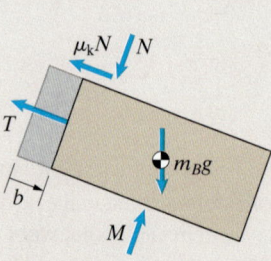

(b) Free-body diagram of B.

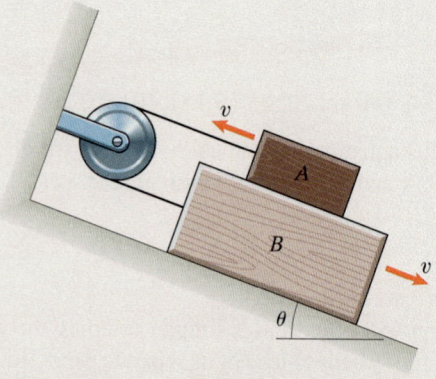

(c) The magnitude of the velocity of each crate is the same.

The work done on B equals the change in *its* kinetic energy.

$$U_{12} = \tfrac{1}{2}m_B v_2^2 - \tfrac{1}{2}m_B v_1^2:$$

$$\int_0^b (-T + m_B g \sin\theta - \mu_k m_A g \cos\theta)\, ds = \tfrac{1}{2}m_B v_2^2. \qquad (2)$$

Summing these equations to eliminate T and solving for v_2, we obtain

$$v_2 = \sqrt{2gb[(m_B - m_A)\sin\theta - 2\mu_k m_A \cos\theta]/(m_A + m_B)}.$$

Critical Thinking

If we attempt to solve this example by applying the principle of work and energy to the system consisting of the crates, the cable, and the pulley (Fig. d), we obtain an incorrect result. Equating the work done by external forces to the change in the total kinetic energy of the system, we obtain

$$\int_0^b m_B g \sin\theta\, ds - \int_0^b m_A g \sin\theta\, ds = \tfrac{1}{2}m_A v_2^2 + \tfrac{1}{2}m_B v_2^2:$$

$$(m_B g \sin\theta)b - (m_A g \sin\theta)b = \tfrac{1}{2}m_A v_2^2 + \tfrac{1}{2}m_B v_2^2.$$

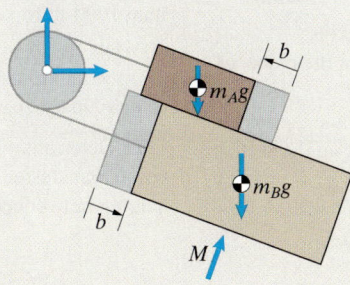

(d) Free-body diagram of the system.

But if we sum our work and energy equations for the individual crates—Eqs. (1) and (2)—we obtain the correct equation:

$$\underbrace{[(m_B g \sin\theta)b - (m_A g \sin\theta)b]}_{\substack{\text{Work done by}\\ \text{external forces}}} + \underbrace{[-(2\mu_k m_A g \cos\theta)b]}_{\substack{\text{Work done by}\\ \text{internal forces}}} = \tfrac{1}{2}m_A v_2^2 + \tfrac{1}{2}m_B v_2^2.$$

The internal frictional forces the crates exert on each other do net work on the system. We did not account for this work in applying the principle of work and energy to the free-body diagram of the entire system.

Problems

15.1 The airplane weighs 120,000 lb. During its takeoff roll it accelerates to a velocity of 255 ft/s. How much work is done on the airplane by the forces acting on it during takeoff?

Problem 15.1

15.2 The mass of the helicopter is 9300 kg. It takes off vertically with its rotor exerting a constant thrust of 107,000 N. Use the principle of work and energy to determine how fast the helicopter is moving when it has risen 10 m.

Problem 15.2

15.3 The 100-lb object starts from rest on the smooth inclined surface and is subjected to a constant force $F = 65$ lb.

(a) How much work is done by the forces acting on the object as it slides 3 ft up the surface from its initial position?
(b) Use the principle of work and energy to determine the magnitude of the object's velocity when it has moved 3 ft up the surface.

15.4 Solve Problem 15.3 if the coefficient of kinetic friction between the object and the inclined surface is $\mu_k = 0.2$.

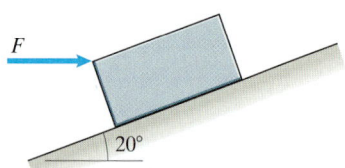

Problems 15.3/15.4

15.5 The 0.45-kg soccer ball is 1 m above the ground when it is kicked upward at 12 m/s.

(a) How much work is done on the ball by its weight as it rises from 1 m above the ground to 4 m above the ground?
(b) Use the principle of work and energy to determine the magnitude of the ball's velocity when it is 4 m above the ground.
(c) Use the principle of work and energy to determine the maximum height above the ground the ball will reach.

15.6 Assume that the soccer ball is stationary the instant before it is kicked upward at 12 m/s. The duration of the kick is 0.02 s. What average power is transferred to the ball during the kick?

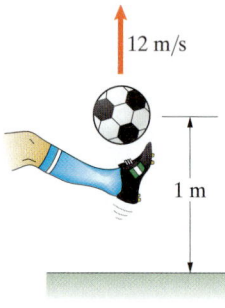

Problems 15.5/15.6

15.7 The 2000-lb drag racer starts from rest and travels a quarter-mile course. Friction between the tires and the road exerts a constant 1690-lb force on it.

(a) How much work is done on the car as it travels the course?
(b) Use the principle of work and energy to determine the car's velocity in mi/h as it crosses the finish line.

15.8 Determine (a) the maximum power and (b) the average power transferred to the car in Problem 15.7 as it travels the quarter-mile course.

Problems 15.7 /15.8

15.9 As a 10,000-kg airplane takes off, the tangential component of force exerted on it by its engines is constant and equal to 60 kN. Neglecting other forces acting on the airplane, use the principle of work and energy to determine its velocity when it has rolled 500 m.

15.10 Determine (a) the maximum power and (b) the average power transferred to the airplane in Problem 15.9 during its 500-m roll.

15.11 When aerodynamic drag is accounted for, the tangential component of force exerted on the airplane in Problem 15.9 is given as a function of the distance s (in meters) traveled by the airplane by

$$\Sigma F_t = 60e^{-0.0024s} \text{ kN.}$$

Use the principle of work and energy to determine its velocity when it has rolled 500 m.

Problems 15.9–15.11

15.12 The spring ($k = 20 \text{ N/m}$) is unstretched when $s = 0$. The 5-kg cart is given a velocity of 4 m/s down the sloped surface at position $s = 0$.
(a) How much work is done on the cart by its weight and the spring as it moves from $s = 0$ to $s = 2$ m?
(b) What is the magnitude of the cart's velocity at $s = 2$ m?

15.13 What maximum distance down the sloped surface does the cart in Problem 15.12 travel relative to its initial position?

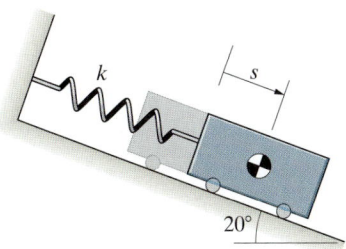

Problems 15.12/15.13

15.14 The force exerted on a car by a prototype crash barrier as the barrier crushes is $F = -(4400 + 50,000s)$ N, where s is the distance in meters from the initial contact. Suppose that you want to design the barrier so that it can stop a 2400-kg car traveling at 100 km/h. What is the necessary effective length of the barrier? That is, what is the distance required for the barrier to bring the car to a stop?

15.15 The duration of the impact of the car with the barrier described in Problem 15.14 is 0.33 s. What average power is transferred from the car during the impact?

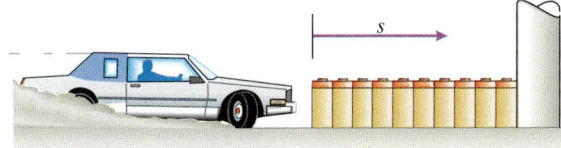

Problems 15.14/15.15

15.16 A group of engineering students constructs a sun-powered car and tests it on a circular track with 1000-ft radius. The car, with a weight of 460 lb including its occupant, starts from rest. The total tangential component of force on the car is

$$\Sigma F_t = 30 - 0.2s \text{ lb,}$$

where s is the distance (in ft) the car travels along the track from the position where it starts.
(a) Determine the work done on the car when it has gone a distance $s = 120$ ft.
(b) Determine the magnitude of the *total* horizontal force exerted on the car's tires by the road when it is at the position $s = 120$ ft.

15.17 At the instant shown, the 160-lb vaulter's center of mass is 8.5 ft above the ground, and the vertical component of his velocity is 4 ft/s. As his pole straightens, it exerts a vertical force on the vaulter of magnitude $180 + 2.8y^2$ lb, where y is the vertical position of his center of mass *relative to its position at the instant shown*. This force is exerted on him from $y = 0$ to $y = 4$ ft, when he releases the pole. What is the maximum height above the ground reached by the vaulter's center of mass?

Problem 15.17

15.18 The springs are unstretched when $s = 0$. The 4-kg mass is released from rest with $s = 0$, and it falls 0.2 m before rebounding.
(a) Use the principle of work and energy to determine the spring constant k.
(b) What maximum velocity does the mass attain as it falls, and at what position does it occur?

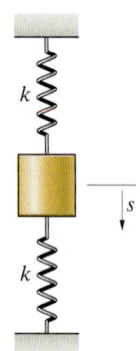

Problem 15.18

15.19 The coefficients of friction between the 160-kg crate and the ramp are $\mu_s = 0.3$ and $\mu_k = 0.28$.
(a) What tension T_0 must the winch exert to start the crate moving up the ramp?
(b) If the tension remains at the value T_0 after the crate starts sliding, what total work is done on the crate as it slides a distance $s = 3$ m up the ramp, and what is the resulting velocity of the crate?

15.20 In Problem 15.19, if the winch exerts a tension $T = T_0(1 + 0.1s)$ after the crate starts sliding, what total work is done on the crate as it slides a distance $s = 3$ m up the ramp, and what is the resulting velocity of the crate?

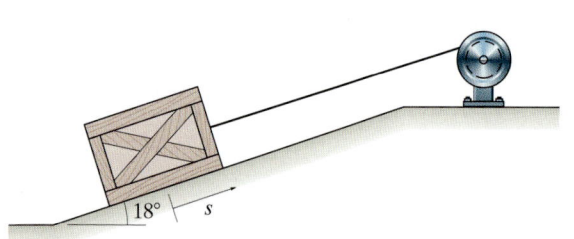

Problems 15.19/15.20

15.21 The 200-mm-diameter gas gun is evacuated on the right of the 8-kg projectile. On the left of the projectile, the tube contains gas with pressure $p_0 = 1 \times 10^5$ Pa (N/m^2). The force F is slowly increased, moving the projectile 0.5 m to the left from the position shown. The force is then removed and the projectile accelerates to the right. If you neglect friction and assume that the pressure of the gas is related to its volume by $pV = $ constant, what is the velocity of the projectile when it has returned to its original position?

15.22 In Problem 15.21, if you assume that the pressure of the gas is related to its volume by $pV = $ constant while the gas is compressed (an isothermal process) and by $pV^{1.4} = $ constant while it is expanding (an isentropic process), what is the velocity of the projectile when it has returned to its original position?

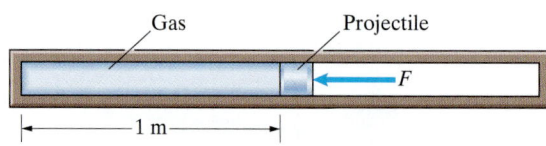

Problems 15.21/15.22

15.23 The system is released from rest. By applying the principle of work and energy to each mass, determine the magnitude of the velocity of the masses when the right mass has fallen 1 m.

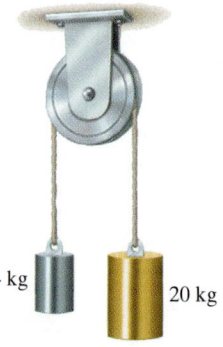

Problem 15.23

15.24 The system is released from rest. The 4-kg mass slides on the smooth horizontal surface. By using the principle of work and energy, determine the magnitude of the velocity of the masses when the 20-kg mass has fallen 1 m.

15.25 Solve Problem 15.24 if the coefficient of kinetic friction between the 4-kg mass and the horizontal surface is $\mu_k = 0.4$.

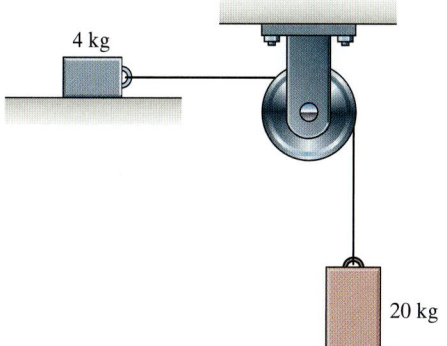

4 kg

20 kg

Problems 15.24/15.25

15.26 Each box weighs 50 lb and the inclined surfaces are smooth. The system is released from rest. Determine the magnitude of the velocities of the boxes when they have moved 1 ft.

15.27 Solve Problem 15.26 if the coefficient of kinetic friction between the boxes and the inclined surfaces is $\mu_k = 0.05$.

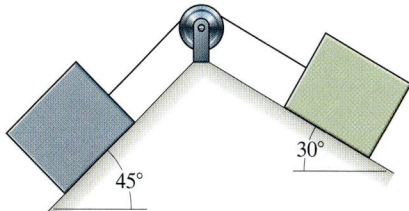

30°

45°

Problems 15.26/15.27

15.28 The masses of the three blocks are $m_A = 40$ kg, $m_B = 16$ kg, and $m_C = 12$ kg. Neglect the mass of the bar holding C in place. Friction is negligible. By applying the principle of work and energy to A and B individually, determine the magnitude of their velocity when they have moved 500 mm.

15.29 Solve Problem 15.28 by applying the principle of work and energy to the system consisting of A, B, the cable connecting them, and the pulley.

15.30 In Problem 15.28, determine the magnitude of the velocity of A and B when they have moved 500 mm if the coefficient of kinetic friction between all surfaces is $\mu_k = 0.1$.

 Strategy: The simplest approach is to apply the principle of work and energy to A and B individually. If you treat them as a single system, you must account for the work done by internal friction forces. (See Example 15.3.)

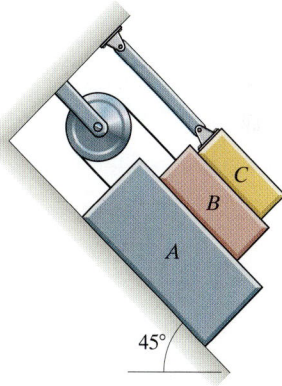

C

B

A

45°

Problems 15.28–15.30

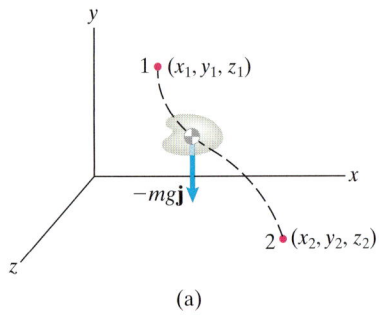

(a)

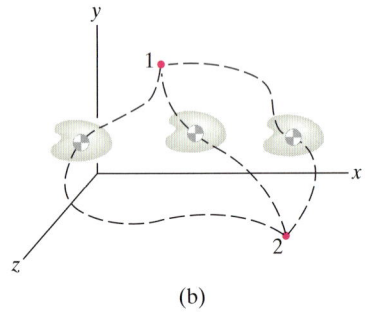

(b)

Figure 15.6
(a) An object moving between two positions.
(b) The work done by the weight is the same for any path.

15.3 Work Done by Particular Forces

We have seen that if the tangential component of the total external force on an object is known as a function of distance along the object's path, the principle of work and energy can be used to relate a change in the position of the object to the change in its velocity. For certain types of forces, however, not only can we determine the work without knowing the tangential component of the force as a function of distance along the path, but we don't even need to know the path. Two important examples are weight and the force exerted by a spring.

Weight

To evaluate the work done by an object's weight, we orient a cartesian coordinate system with the y axis upward and suppose that the object moves from position 1 with coordinates (x_1, y_1, z_1) to position 2 with coordinates (x_2, y_2, z_2) (Fig. 15.6a). The force exerted by the object's weight is $\mathbf{F} = -mg\mathbf{j}$. (Other forces may act on the object, but we are concerned only with the work done by its weight.) Because $\mathbf{v} = d\mathbf{r}/dt$, we can multiply the velocity, expressed in cartesian coordinates, by dt to obtain an expression for the vector $d\mathbf{r}$:

$$d\mathbf{r} = \left(\frac{dx}{dt}\mathbf{i} + \frac{dy}{dt}\mathbf{j} + \frac{dz}{dt}\mathbf{k} \right) dt = dx\,\mathbf{i} + dy\,\mathbf{j} + dz\,\mathbf{k}.$$

Taking the dot product of $\mathbf{F}$ and $d\mathbf{r}$ yields

$$\mathbf{F} \cdot d\mathbf{r} = (-mg\mathbf{j}) \cdot (dx\,\mathbf{i} + dy\,\mathbf{j} + dz\,\mathbf{k}) = -mg\,dy.$$

The work done as the object moves from position 1 to position 2 reduces to an integral with respect to y:

$$U_{12} = \int_{\mathbf{r}_1}^{\mathbf{r}_2} \mathbf{F} \cdot d\mathbf{r} = \int_{y_1}^{y_2} -mg\,dy.$$

Evaluating the integral, we obtain the work done by the weight of an object as it moves between two positions:

$$U_{12} = -mg(y_2 - y_1). \tag{15.11}$$

The work is simply the product of the weight and the change in the object's height. The work done is negative if the height increases and positive if it decreases. Notice that *the work done is independent of the path the object follows from position 1 to position 2* (Fig. 15.6b). Thus, we don't need to know the path to determine the work done by an object's weight—we only need to know the relative heights of the initial and final positions.

What work is done by an object's weight if we account for its variation with distance from the center of the earth? In terms of polar coordinates, we can write the weight of an object at a distance r from the center of the earth as (Fig. 15.7)

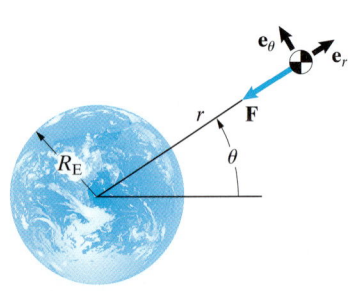

Figure 15.7
Expressing an object's weight in polar coordinates.

$$\mathbf{F} = -\frac{mgR_E^2}{r^2}\mathbf{e}_r.$$

Using the expression for the velocity in polar coordinates, we obtain, for the vector $d\mathbf{r} = \mathbf{v}\,dt$,

$$d\mathbf{r} = \left(\frac{dr}{dt}\mathbf{e}_r + r\frac{d\theta}{dt}\mathbf{e}_\theta\right)dt = dr\,\mathbf{e}_r + r\,d\theta\,\mathbf{e}_\theta. \qquad (15.12)$$

The dot product of $\mathbf{F}$ and $d\mathbf{r}$ is

$$\mathbf{F}\cdot d\mathbf{r} = \left(-\frac{mgR_E^2}{r^2}\mathbf{e}_r\right)\cdot(dr\,\mathbf{e}_r + r\,d\theta\,\mathbf{e}_\theta) = -\frac{mgR_E^2}{r^2}dr,$$

so the work reduces to an integral with respect to r:

$$U_{12} = \int_{\mathbf{r}_1}^{\mathbf{r}_2}\mathbf{F}\cdot d\mathbf{r} = \int_{r_1}^{r_2} -\frac{mgR_E^2}{r^2}dr.$$

Evaluating the integral, we obtain the work done by an object's weight, accounting for the variation of the weight with height:

$$U_{12} = mgR_E^2\left(\frac{1}{r_2} - \frac{1}{r_1}\right). \qquad (15.13)$$

Again, the work is independent of the path from position 1 to position 2. To evaluate it, we only need to know the object's radial distance from the center of the earth at the two positions.

Springs

Suppose that a linear spring connects an object to a fixed support. In terms of polar coordinates (Fig. 15.8), the force exerted on the object is

$$\mathbf{F} = -k(r - r_0)\mathbf{e}_r,$$

where k is the spring constant and r_0 is the unstretched length of the spring. Using Eq. (15.12), we get the dot product of $\mathbf{F}$ and $d\mathbf{r}$:

$$\mathbf{F}\cdot d\mathbf{r} = [-k(r - r_0)\mathbf{e}_r]\cdot(dr\,\mathbf{e}_r + r\,d\theta\,\mathbf{e}_\theta) = -k(r - r_0)\,dr.$$

It is convenient to express the work done by a spring in terms of its *stretch*, defined by $S = r - r_0$. (Although the word *stretch* usually means an increase in length, we use the term more generally to denote the change in length of the spring. A negative stretch is a decrease in length.) In terms of this variable, $\mathbf{F}\cdot d\mathbf{r} = -kS\,dS$, and the work is

$$U_{12} = \int_{\mathbf{r}_1}^{\mathbf{r}_2}\mathbf{F}\cdot d\mathbf{r} = \int_{S_1}^{S_2} -kS\,dS.$$

The work done on an object by a spring attached to a fixed support is

$$U_{12} = -\tfrac{1}{2}k(S_2^2 - S_1^2), \qquad (15.14)$$

where S_1 and S_2 are the values of the stretch at the initial and final positions. We don't need to know the object's path to determine the work done by the spring. Remember, however, that Eq. (15.14) applies only to a linear spring. In Fig. 15.9, we determine the work done in stretching a linear spring by calculating the area defined by the graph of the force as a function of S.

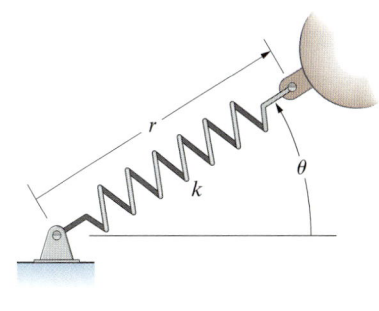
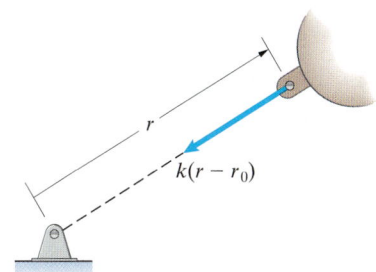

Figure 15.8
Expressing the force exerted by a linear spring in polar coordinates.

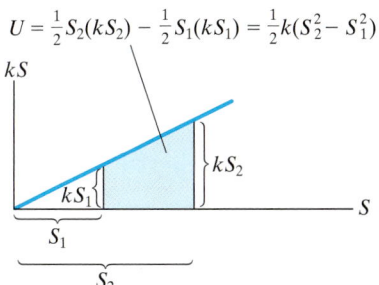

Figure 15.9
Work done in stretching a linear spring from S_1 to S_2. (If $S_2 > S_1$, the work done *on* the spring is positive, so the work done *by* the spring is negative.)

Study Questions

1. If an object moves from a position 1 to a position 2, what do you need to know to calculate the work done by the object's weight?

2. If the height of an object increases, is the work done by its weight positive or negative?

3. What is the definition of the *stretch* of a spring?

4. If the stretch of a spring connecting an object to a fixed support changes from S_1 to S_2, how much work is done on the object by the spring?

Example 15.4 | Work Done by Weight

At position 1, the skier in Fig. 15.10 is approaching his jump at 15 m/s. When he reaches the horizontal end of the ramp at position 2, 20 m below position 1, he jumps upward, achieving a vertical component of velocity of 3 m/s. (Disregard the small change in the vertical position of his center of mass due to his jumping motion.) Neglect aerodynamic drag and the frictional forces on his skis. (a) What is the magnitude of the skier's velocity as he leaves the ramp at position 2? (b) At the highest point of his jump, position 3, what are the magnitude of his velocity and the height of his center of mass above position 2?

Figure 15.10

Strategy

(a) If we neglect aerodynamic and frictional forces, the only force doing work from position 1 to position 2 is the skier's weight. The normal force exerted on his skis by the ramp does no work because it is perpendicular to his path. We need to know only the change in the skier's height from position 1 to position 2 to determine the work done by his weight, so we can apply the principle of work and energy to determine his velocity at position 2 before he jumps.

(b) From the time he leaves the ramp at position 2 until he reaches position 3, the only force acting on the skier is his weight, so the horizontal component of his velocity is constant. This means that we know the magnitude of his velocity at position 3, because he is moving horizontally at that point. Therefore, we can apply the principle of work and energy to his motion from position 2 to position 3 to determine his height above position 2.

Solution

(a) We will use Eq. (15.11) to evaluate the work done by the skier's weight, measuring the height of his center of mass relative to position 2 (Fig. a). The principle of work and energy from position 1 to position 2 is

$$U_{12} = -mg(y_2 - y_1) = \tfrac{1}{2}mv_2^2 - \tfrac{1}{2}mv_1^2:$$

$$-m(9.81)(0 - 20) = \tfrac{1}{2}mv_2^2 - \tfrac{1}{2}m(15)^2.$$

Solving for v_2, we find that the skier's horizontal velocity at position 2 before he jumps upward is 24.8 m/s. After he jumps upward, the magnitude of his velocity at position 2 is $v_2' = \sqrt{(24.8)^2 + (3)^2} = 25.0$ m/s.

(b) The magnitude of the skier's velocity at position 3 is equal to the horizontal component of his velocity at position 2: $v_3 = v_2 = 24.8$ m/s. Applying work and energy to his motion from position 2 to position 3, we obtain

$$U_{23} = -mg(y_3 - y_2) = \tfrac{1}{2}mv_3^2 - \tfrac{1}{2}m(v_2')^2:$$

$$-m(9.81)(y_3 - 0) = \tfrac{1}{2}m(24.8)^2 - \tfrac{1}{2}m(25.0)^2,$$

from which it follows that $y_3 = 0.459$ m.

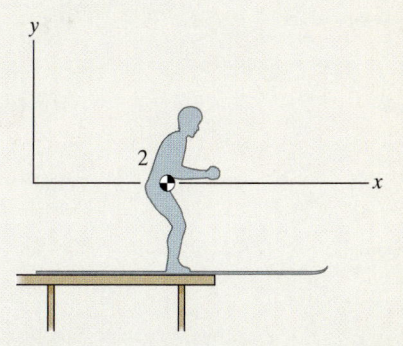

(a) The height of the skier's center of mass is measured relative to position 2.

Critical Thinking

Why didn't we need to include the effect of the normal force exerted on the skier by the ramp? The reason is that *it is perpendicular to his path and so does no work*. To obtain an accurate prediction of the skier's motion, we would need to account for the friction force exerted by the ramp and aerodynamic forces. Nevertheless, our approximate analysis in this example provides useful insight, showing how the work done by gravity as he descends increases his kinetic energy. Notice that the work done by gravity is determined by his change in height, not the length of his path.

Example 15.5   Work Done by Weight and Springs

In the forging device shown in Fig. 15.11, the 40-kg hammer is lifted to position 1 and released from rest. It falls and strikes a workpiece when it is in position 2. The spring constant $k = 1500$ N/m, and the tension in each spring is 150 N when the hammer is in position 2. Neglect friction.

(a) What is the velocity of the hammer just before it strikes the workpiece?
(b) Assuming that all the hammer's kinetic energy is transferred to the workpiece, what average power is transferred if the duration of the impact is 0.02 s?

Strategy

Work is done on the hammer by its weight and by the forces exerted by the springs (Fig. a). We can apply the principle of work and energy to the motion of the hammer from position 1 to position 2 to determine its velocity at position 2.

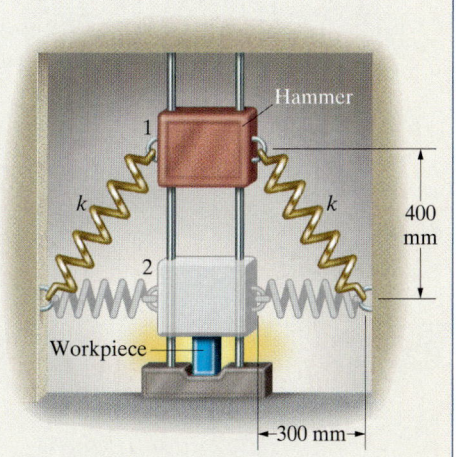

Figure 15.11

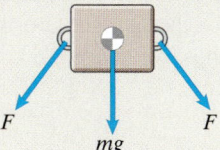

(a) Free-body diagram of the hammer.

Solution

(a) Let r_0 be the unstretched length of one of the springs. In position 2, the tension in the spring is 150 N and its length is 0.3 m. From the relation between the tension in a linear spring and its stretch, we obtain

$$150 \text{ N} = k(0.3 \text{ m} - r_0)$$

$$= (1500 \text{ N/m})(0.3 \text{ m} - r_0),$$

from which it follows that $r_0 = 0.2$ m. The values of the stretch of each spring in positions 1 and 2 are $S_1 = \sqrt{(0.4)^2 + (0.3)^2} - 0.2 = 0.3$ m and $S_2 = 0.3 - 0.2 = 0.1$ m. From Eq. (15.14), the total work done on the hammer by the two springs from position 1 to position 2 is

$$U_{\text{springs}} = 2\left[-\tfrac{1}{2}k(S_2^2 - S_1^2)\right]$$

$$= -(1500 \text{ N/m})\left[(0.1 \text{ m})^2 - (0.3 \text{ m})^2\right]$$

$$= 120 \text{ N-m.}$$

The work done by the weight from position 1 to position 2 is positive and equal to the product of the weight and the change in height:

$$U_{\text{weight}} = mg(0.4 \text{ m})$$

$$= (40 \text{ kg})(9.81 \text{ m/s}^2)(0.4 \text{ m})$$

$$= 157 \text{ N-m.}$$

From the principle of work and energy, we have

$$U_{\text{springs}} + U_{\text{weight}} = \tfrac{1}{2}mv_2^2 - \tfrac{1}{2}mv_1^2:$$

$$120 \text{ N-m} + 157 \text{ N-m} = \tfrac{1}{2}(40 \text{ kg})v_2^2 - 0,$$

so that $v_2 = 3.72$ m/s.

(b) All the hammer's kinetic energy is transferred to the workpiece, so Eq. (15.10) indicates that the average power equals the kinetic energy of the hammer divided by the duration of the impact:

$$P_{\text{avg}} = \frac{(1/2)(40 \text{ kg})(3.72 \text{ m/s})^2}{0.02 \text{ s}} = 13.8 \text{ kW (kilowatts).}$$

Critical Thinking

You will gain appreciation for the method of work and energy if you attempt to use Newton's second law to determine the velocity of the hammer when it reaches the workpiece. Notice that the directions and magnitudes of the forces exerted on the hammer by the springs depend on the hammer's position as it descends. To determine the work done by the springs, we needed to know only their initial and final lengths.

Design Example 15.6   Automated Machining

The device in Fig. 15.12a machines a sample of material with the cutting tool at A. The hydraulic actuator attached at B pushes the toolholder to the right on its horizontal guide rail. The spring attached at C returns the toolholder and hydraulic actuator to their initial positions when the cut is completed. The position of the sample of material is changed, and the cycle is repeated. In the free-body diagram of the assembly consisting of the cutting tool and toolholder (Fig. 15.12b), A_x and A_y are the forces exerted on the cutting tool by the machined sample, B is the force exerted by the actuator, F is the force exerted by the spring, and N is the normal force exerted by the guide rail. The mass of the assembly is $m = 22$ kg, the unstretched length of the spring is 200 mm, and the spring constant is $k = 1800$ N/m.
(a) Suppose the assembly starts from rest in the position shown, the horizontal force on the cutting tool is $A_x = 1330$ N, and the force exerted by the actuator is $B = 1700$ N. Determine the velocity of the assembly when it has moved 0.05 m to the right.
(b) Determine the maximum power transferred from the actuator as the assembly moves from $x = 0$ to $x = 0.15$ m.

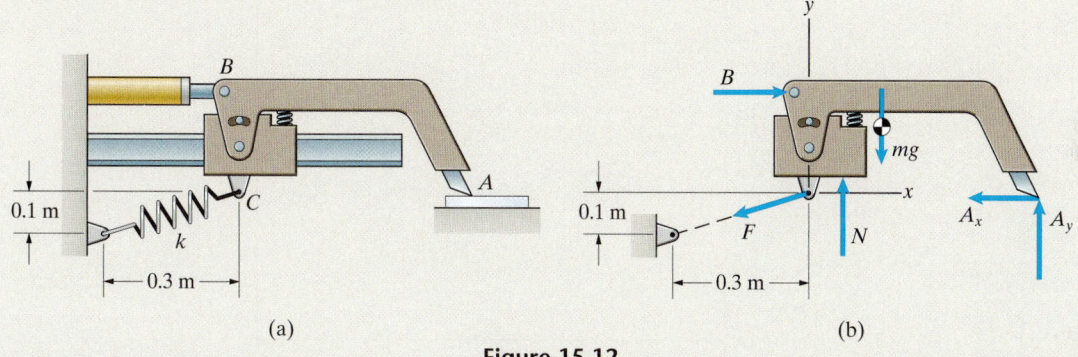

(a)   (b)

Figure 15.12

Strategy

(a) By determining the total work done by the spring and the horizontal forces A_x and B as a function of x, we can use work and energy to determine the velocity of the assembly as a function of x.
(b) The power transferred equals the product of the force exerted by the actuator and the velocity. We will draw a graph of the velocity as a function of x to determine the maximum velocity.

Solution

(a) Let the initial position of the assembly be position 1, and let its position when it has moved a distance x to the right be position 2. The stretch of the spring in position 1 is

$$S_1 = \sqrt{(0.1)^2 + (0.3)^2} - 0.2 \text{ m.}$$

and the stretch in position 2 is

$$S_2 = \sqrt{(0.1)^2 + (0.3 + x)^2} - 0.2 \text{ m.}$$

The work done from position 1 to position 2 by the horizontal forces A_x and B and the spring is

$$U_{12} = (B - A_x)x - \tfrac{1}{2}k(S_2^2 - S_1^2).$$

We apply the principle of work and energy:

$$U_{12} = \tfrac{1}{2}mv_2^2 - \tfrac{1}{2}mv_1^2;$$

$$(B - A_x)x - \tfrac{1}{2}k(S_2^2 - S_1^2) = \tfrac{1}{2}mv_2^2 - 0.$$

Setting $x = 0.05$ m, $A_x = 1330$ N, $B = 1700$ N, and $m = 22$ kg and solving for the velocity, we obtain $v_2 = 0.766$ m/s.

(b) In Fig. 15.13, we draw a graph of the velocity v_2 as a function of x for $0 \leq x \leq 0.15$ m. From the graph, we estimate that the maximum velocity occurs at $x = 0.1$ m. Solving for the velocity at this position, we obtain $v_2 = 0.884$ m/s. The power transferred from the actuator at the maximum velocity is

$$P = Bv_2 = (1700\ \text{N})(0.884\ \text{m/s}) = 1500\ \text{N-m/s (watts)}.$$

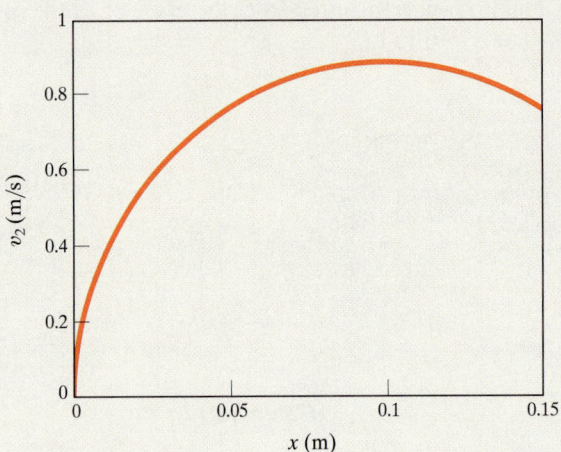

Figure 15.13
Velocity of the assembly as a function of x.

Design Issues

In developing a machine, the engineers must ensure that it will perform the tasks for which it is designed and do so reliably through a design lifetime that may be many years and millions of cycles of use. The machine should be designed for ease of use and maintenance and must be safe for its operators and other persons. It must also be economical. If it is an industrial machine, its original cost and the cost of its use and maintenance must be acceptable in comparison to the income derived through its use. The choices implied by these requirements are often contradictory, and the design engineers must make successful compromises.

The use of dynamics to study the motions of machines is essential to designers who analyze the behavior of the machines, predict their performance, and determine whether their structures will support the dynamic loads to which they will be subjected. Employing a simple context, we demonstrated in this example that the principle of work and energy can be useful in analyzing a machine's motion. It can also provide information on energy requirements of machines. The work done by the hydraulic actuator in this example provided an estimate of the energy that must be supplied (in the form of electrical energy to power the hydraulic pump) for each cycle of operation of the machine tool.

Example 15.7   Work Done by the Earth's Gravity

A spacecraft at a distance $r_1 = 2R_E$ from the center of the earth has a velocity of magnitude $v_1 = \sqrt{2gR_E/3}$ relative to a nonrotating reference frame with its origin at the center of the earth (Fig. 15.14). Determine the magnitude of the spacecraft's velocity when it is at a distance $r_2 = 4R_E$ from the center of the earth.

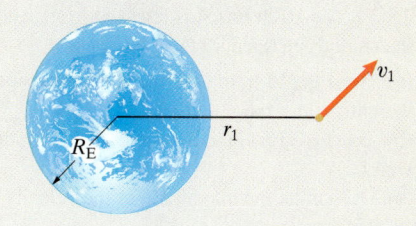

Figure 15.14

Strategy

By applying Eq. (15.13) to determine the work done by the gravitational force on the spacecraft, we can use the principle of work and energy to determine the magnitude of the spacecraft's velocity.

Solution

From Eq. (15.13), the work done by gravity as the spacecraft moves from a distance r_1 from the center of the earth to a distance r_2 is

$$U_{12} = mgR_E^2\left(\frac{1}{r_2} - \frac{1}{r_1}\right).$$

Let v_2 be the magnitude of the velocity of the spacecraft when it is at a distance r_2 from the center of the earth. Applying the principle of work and energy yields

$$U_{12} = mgR_E^2\left(\frac{1}{r_2} - \frac{1}{r_1}\right) = \tfrac{1}{2}mv_2^2 - \tfrac{1}{2}mv_1^2.$$

We solve for v_2, obtaining

$$v_2 = \sqrt{v_1^2 + 2gR_E^2\left(\frac{1}{r_2} - \frac{1}{r_1}\right)}$$

$$= \sqrt{\left(\frac{2gR_E}{3}\right) + 2gR_E^2\left(\frac{1}{4R_E} - \frac{1}{2R_E}\right)}$$

$$= \sqrt{\frac{gR_E}{6}}.$$

The velocity $v_2 = v_1/2$.

Critical Thinking

Notice that we did not need to specify the direction of the spacecraft's initial velocity to determine the magnitude of its velocity at a different distance from the center of the earth. This illustrates the power of the principle of work and energy, as well as one of its limitations. Even if we know the direction of the initial velocity, the principle of work and energy tells us only the *magnitude* of the velocity at a different distance.

Problems

15.31 The 10-lb box A is released from rest and slides 2 ft down the smooth inclined surface.

(a) Use Eq. (15.8) to determine the work done on the box by its weight.

(b) Use Eq. (15.11) to determine the work done on the box by its weight.

(c) Determine the magnitude of the velocity the box attains.

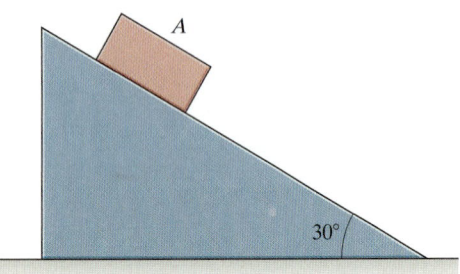

Problem 15.31

15.32 Suppose that you stand at the edge of a 200-ft cliff and throw rocks at 30 ft/s in the three directions shown. Neglecting aerodynamic drag, use the principle of work and energy to determine the magnitude of the velocity of the rock just before it hits the ground in each case.

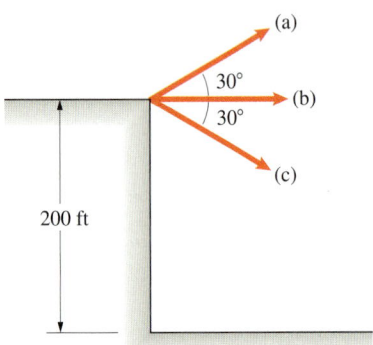

Problem 15.32

15.33 The 30-kg box is sliding down the smooth surface at 1 m/s when it is in position 1. Determine the magnitude of the box's velocity at position 2 in each case.

15.34 Solve Problem 15.33 if the coefficient of kinetic friction between the box and the inclined surface is $\mu_k = 0.2$.

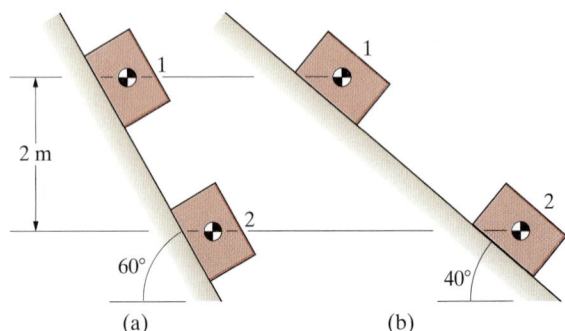

Problems 15.33/15.34

15.35 In case (a), a 5-lb ball is released from rest at position 1 and falls to position 2. In case (b), the ball is released from rest at position 1 and swings to position 2. For each case, use the principle of work and energy to determine the magnitude of the ball's velocity at position 2. [In case (b), notice that the force exerted on the ball by the string is perpendicular to the ball's path.]

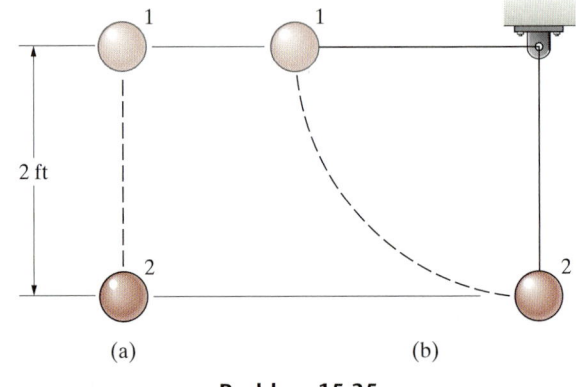

Problem 15.35

15.36 The 2-kg ball is released from rest in position 1 with the string horizontal. The length of the string is $L = 1$ m. What is the magnitude of the ball's velocity when it is in position 2?

15.37 The 2-kg ball is released from rest in position 1 with the string horizontal. The length of the string is $L = 1$ m. What is the tension in the string when the ball is in position 2?

Strategy: Draw the free-body diagram of the ball when it is in position 2 and write Newton's second law in terms of normal and tangential components.

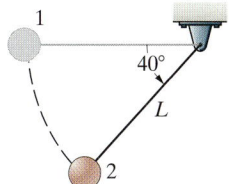

Problems 15.36/15.37

15.38 The 400-lb wrecker's ball swings at the end of a 25-ft cable. If the magnitude of the ball's velocity at position 1 is 4 ft/s, what is the magnitude of its velocity just before it hits the wall at position 2?

15.39 The 400-lb wrecker's ball swings at the end of a 25-ft cable. If the magnitude of the ball's velocity at position 1 is 4 ft/s, what is the maximum tension in the cable as the ball swings from position 1 to position 2?

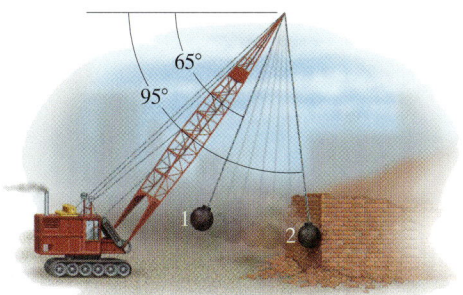

Problems 15.38/15.39

15.40 A stunt driver wants to drive a car through the circular loop of radius $R = 5$ m. Determine the minimum velocity v_0 at which the car can enter the loop and coast through without losing contact with the track. What is the car's velocity at the top of the loop?

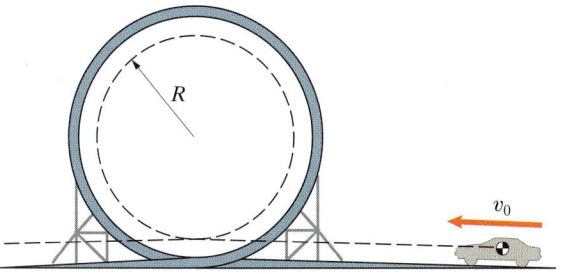

Problem 15.40

15.41 The 2-kg collar starts from rest at position 1 and slides down the smooth rigid wire. The y axis points upward. What is the magnitude of the velocity of the collar when it reaches position 2?

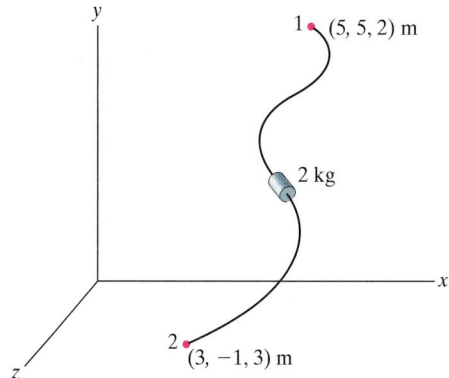

Problem 15.41

15.42 The 4-lb collar slides down the smooth rigid wire from position 1 to position 2. When it reaches position 2, the magnitude of its velocity is 24 ft/s. What was the magnitude of its velocity at position 1?

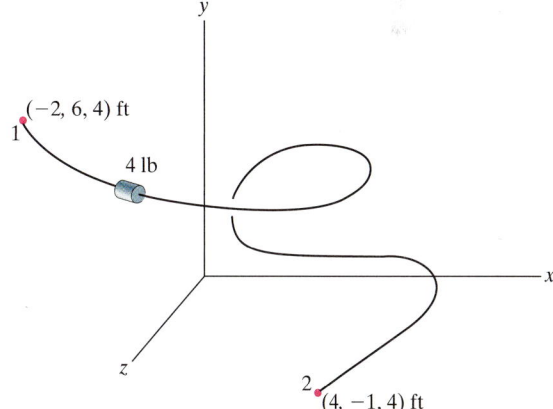

Problem 15.42

15.43 The forces acting on the 28,000-lb airplane are the thrust T and drag D, which are parallel to the airplane's path, the lift L, which is perpendicular to the path, and the weight, W. The airplane climbs from an altitude of 3000 ft to an altitude of 10,000 ft. During the climb, the magnitude of its velocity decreases from 800 ft/s to 600 ft/s.

(a) What work is done on the airplane by its lift during the climb?
(b) What work is done by the thrust and drag combined?

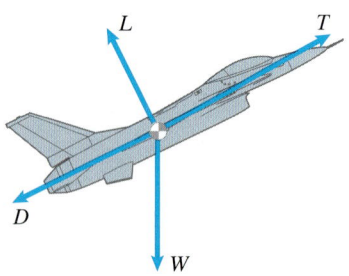

Problem 15.43

15.44 The 2400-lb car is traveling 40 mi/h at position 1. If the combined effect of the aerodynamic drag on the car and the tangential force exerted on its wheels by the road is that they exert no net tangential force on the car, what is the magnitude of the car's velocity at position 2?

15.45 The 2400-lb car is traveling 40 mi/h at position 1. If the combined effect of the aerodynamic drag on the car and the tangential force exerted on its wheels by the road is that they exert a constant 400-lb tangential force on the car in the direction of its motion, what is the magnitude of the car's velocity at position 2?

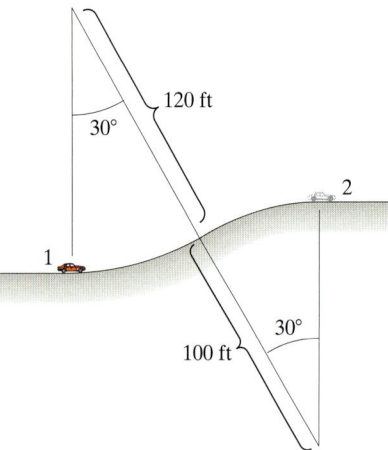

Problems 15.44/15.45

15.46 The mass of the rocket is 250 kg. Its engine has a constant thrust of 45 kN. The length of the launching ramp is 10 m. If the magnitude of the rocket's velocity when it reaches the end of the ramp is 52 m/s, how much work is done on the rocket by friction and aerodynamic drag?

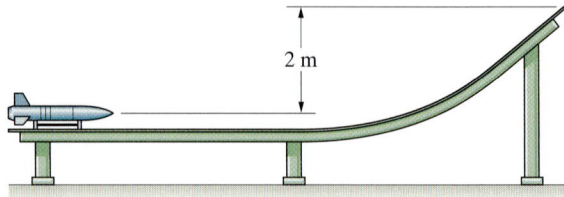

Problem 15.46

15.47 A bioengineer interested in the energy requirements of sports determines from videotape that when the athlete begins his motion to throw the 7.25-kg shot (Fig. P15.47a), the shot is stationary and 1.50 m above the ground. At the instant the athlete releases it (Fig. P15.47b), the shot is 2.10 m above the ground. The shot reaches a maximum height of 4.60 m above the ground and travels a horizontal distance of 18.66 m from the point where it was released. How much work does the athlete do on the shot from the beginning of his motion to the instant he releases it?

(a) (b)

Problem 15.47

15.48 A small pellet of mass $m = 0.2$ kg starts from rest at position 1 and slides down the smooth surface of the cylinder to position 2, where $\theta = 30°$.
(a) What work is done on the pellet as it slides from position 1 to position 2?
(b) What is the magnitude of the pellet's velocity at position 2?

15.49 A small pellet of mass $m = 0.2$ kg starts from rest at position 1 and slides down the smooth surface of the cylinder. What is the value of the angle θ at which the pellet loses contact with the surface of the cylinder?
    Strategy: Draw the free-body diagram of the pellet in position 2 and write Newton's second law in terms of normal and tangential components.

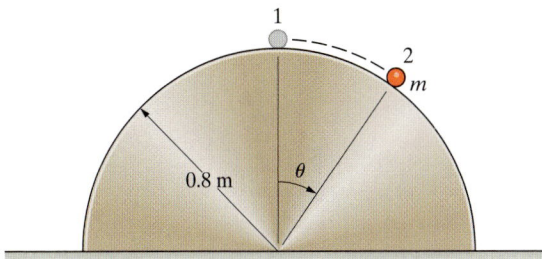

Problems 15.48/15.49

15.50 Suppose that you want to design a bumper that will bring a 50-lb package moving at 10 ft/s to rest 6 in from the point of contact with the bumper. If friction is negligible, what is the necessary spring constant k?

15.51 In Problem 15.50, what spring constant is necessary if the coefficient of kinetic friction between the package and the floor is $\mu_k = 0.3$ and the package contacts the bumper moving at 10 ft/s?

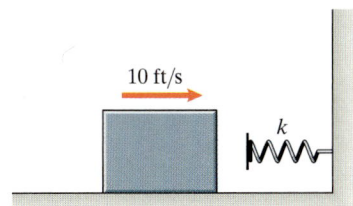

Problems 15.50/15.51

15.52 An astronaut in an excursion module approaches a space station docking collar. The designer of the collar incorporated a spring to attenuate the shock due to docking. The spring constant is $k = 4800$ N/m. The combined mass of the astronaut and module is 780 kg. If the module contacts the docking collar moving at 0.1 m/s relative to the collar, what distance is required for the spring to decrease its relative velocity to zero? What is the module's maximum relative deceleration?

15.53 In Problem 15.52, suppose that you choose the spring constant so that a 10,000-kg vehicle moving at 0.2 m/s relative to the collar will be brought to rest in a distance of 0.15 m. If the module contacts the docking collar moving at 0.1 m/s, what is the module's maximum relative deceleration?

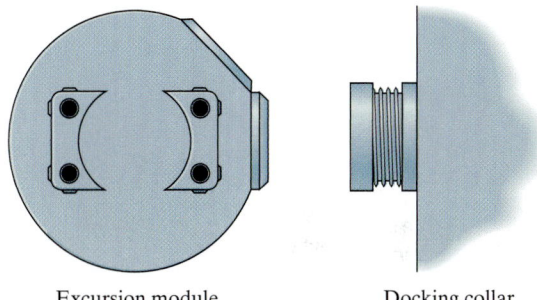

Excursion module   Docking collar

Problems 15.52/15.53

15.54 The system is released from rest with the spring unstretched. The spring constant is $k = 200$ N/m. Determine the magnitude of the velocity of the masses when the right mass has fallen 1 m.

15.55 The system is released from rest with the spring unstretched. The spring constant is $k = 200$ N/m. What maximum downward velocity does the right mass attain as it falls?

Problems 15.54/15.55

15.56 The system is released from rest. The 4-kg mass slides on the smooth horizontal surface. The spring constant is $k = 100$ N/m, and the tension in the spring when the system is released is 50 N. By using the principle of work and energy, determine the magnitude of the velocity of the masses when the 20-kg mass has fallen 1 m.

15.57 Solve Problem 15.56 if the coefficient of kinetic friction between the 4-kg mass and the horizontal surface is $\mu_k = 0.4$.

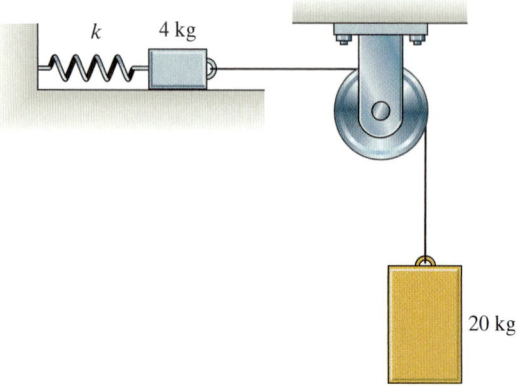

k 4 kg

20 kg

Problems 15.56/15.57

15.58 The 40-lb crate is released from rest on the smooth inclined surface with the spring unstretched. The spring constant is $k = 8$ lb/ft.
(a) How far down the inclined surface does the crate slide before it stops?
(b) What maximum velocity does the crate attain on its way down?

15.59 Solve Problem 15.58 if the coefficient of kinetic friction between the 4-kg mass and the horizontal surface is $\mu_k = 0.2$.

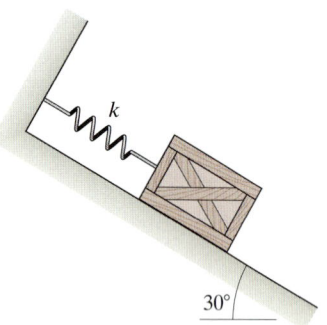

k

30°

Problems 15.58/15.59

15.60 The 4-kg collar starts from rest in position 1 on the smooth bar with the spring unstretched. The spring constant is $k = 100$ N/m. How far does the collar fall relative to position 1?

15.61 In position 1 on the smooth bar, the 4-kg collar has a downward velocity of 1 m/s and the spring is unstretched. The spring constant is $k = 100$ N/m. What maximum downward velocity does the collar attain as it falls?

15.62 The 4-kg collar starts from rest in position 1 on the smooth bar. The tension in the spring in position 1 is 20 N. The spring constant is $k = 100$ N/m. How far does the collar fall relative to position 1?

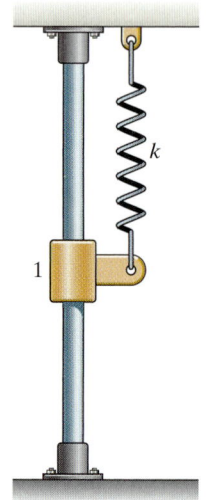

k

1

Problems 15.60–15.62

15.63 The 4-kg collar is released from rest in position 1 on the smooth bar. If the spring constant is $k = 6$ kN/m and the spring is unstretched in position 2, what is the velocity of the collar when it has fallen to position 2?

15.64 The 4-kg collar is released from rest in position 1 on the smooth bar. The spring constant is $k = 4$ kN/m. The tension in the spring in position 2 is 500 N. What is the velocity of the collar when it has fallen to position 2?

15.65 The 4-kg collar starts from rest in position 1 on the smooth bar. Its velocity when it has fallen to position 2 is 4 m/s. The spring is unstretched when the collar is in position 2. What is the spring constant k?

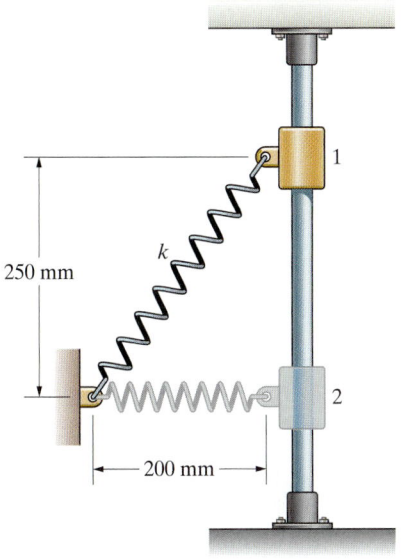

Problems 15.63–15.65

15.66 The 10-kg collar starts from rest at position 1 and slides along the smooth bar. The y axis points upward. The spring constant is $k = 100$ N/m and the unstretched length of the spring is 2 m. What is the velocity of the collar when it reaches position 2?

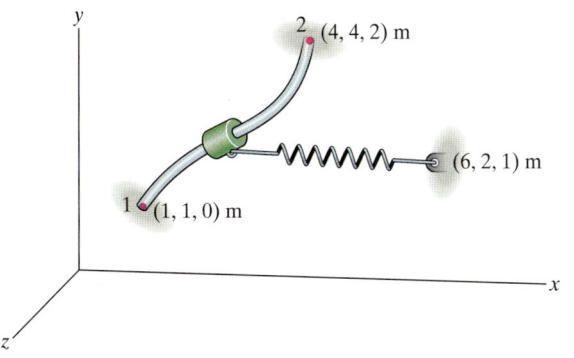

Problem 15.66

15.67 A spring-powered mortar is used to launch 10-lb packages of fireworks into the air. The package starts from rest with the spring compressed to a length of 6 in. The unstretched length of the spring is 30 in. If the spring constant is $k = 1300$ lb/ft, what is the magnitude of the velocity of the package as it leaves the mortar?

15.68 Suppose you want to design the mortar in Problem 15.67 to throw the package to a height of 150 ft above its initial position. Neglecting friction and drag, determine the necessary spring constant.

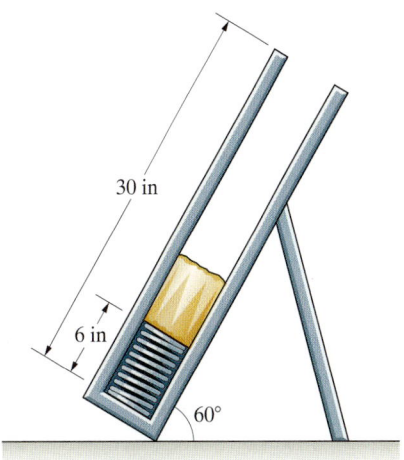

Problems 15.67/15.68

15.69 Suppose an object has a string or cable with *constant* tension T attached as shown. The force exerted on the object can be expressed in terms of polar coordinates as $\mathbf{F} = -T\mathbf{e}_r$. Show that the work done on the object as it moves along an *arbitrary* plane path from a radial position r_1 to a radial position r_2 is $U_{12} = -T(r_2 - r_1)$.

Problem 15.69

15.70 The 2-kg collar is initially at rest at position 1. A constant 100-N force is applied to the string, causing the collar to slide up the smooth vertical bar. What is the velocity of the collar when it reaches position 2? (See Problem 15.69.)

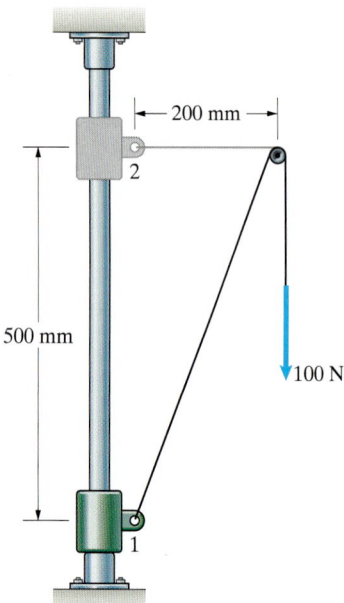

200 mm

2

500 mm

100 N

1

Problem 15.70

15.71 The 10-kg collar starts from rest at position 1. The tension in the string is 200 N, and the y axis points upward. If friction is negligible, what is the magnitude of the velocity of the collar when it reaches position 2? (See Problem 15.69.)

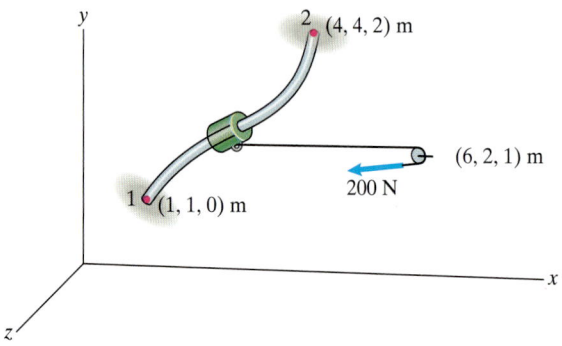

2 (4, 4, 2) m

y

(6, 2, 1) m

200 N

1 (1, 1, 0) m

x

z

Problem 15.71

15.72 As the F/A-18 lands at 210 ft/s, the cable from A to B engages the airplane's arresting hook at C. The arresting mechanism maintains the tension in the cable at a constant value, bringing the 26,000-lb airplane to rest in a distance of 72 ft. What is the tension in the cable? (See Problem 15.69.)

15.73 If the airplane in Problem 15.72 lands at 240 ft/s, what distance does it roll before the arresting system brings it to rest?

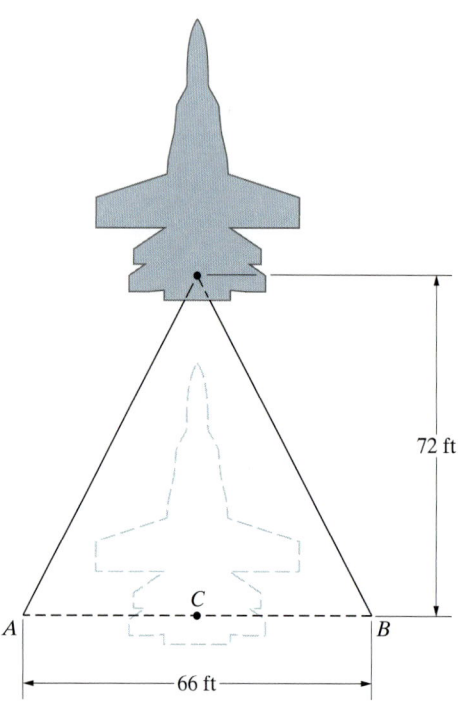

72 ft

C

A

B

66 ft

Problems 15.72/15.73

15.74 A spacecraft 320 km above the surface of the earth is moving at escape velocity $v_{esc} = 10,900$ m/s. What is the magnitude of the spacecraft's velocity when it reaches the moon's orbit 383,000 km from the center of the earth? Assume that the spacecraft is affected only by the earth's gravity. The radius of the earth is 6370 km.

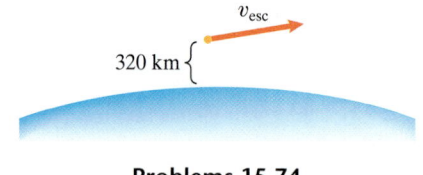

v_{esc}

320 km

Problems 15.74

15.75 A piece of ejecta is thrown up by the impact of a meteor on the moon. When it is 1000 km above the moon's surface, the magnitude of its velocity (relative to a nonrotating reference frame with its origin at the center of the moon) is 200 m/s. What is the magnitude of its velocity just before it strikes the moon's surface? The acceleration due to gravity at the surface of the moon is 1.62 m/s^2. The moon's radius is 1738 km.

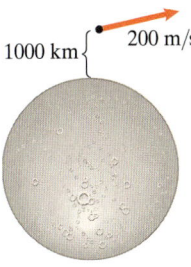

1000 km {   200 m/s

Problem 15.75

15.76 A satellite in a circular orbit of radius r around the earth has velocity $v = \sqrt{gR_E^2/r}$, where $R_E = 6370$ km is the radius of the earth. Suppose you are designing a rocket to transfer a 900-kg communication satellite from a circular parking orbit with 6700-km radius to a circular geosynchronous orbit with 42,222-km radius. How much work must the rocket do on the satellite?

15.77 The force exerted on a charged particle by a magnetic field is

$$\mathbf{F} = q\mathbf{v} \times \mathbf{B},$$

where q and $\mathbf{v}$ are the charge and velocity of the particle and $\mathbf{B}$ is the magnetic field vector. Suppose that other forces on the particle are negligible. Use the principle of work and energy to show that the magnitude of the particle's velocity is constant.

POTENTIAL ENERGY

15.4 Work and Potential Energy

The work done on an object by some forces can be expressed as the change of a function of the object's position called the potential energy. When all the forces that do work on a system have this property, we can state the principle of work and energy as a conservation law: The sum of the kinetic and potential energies is constant.

When we derived the principle of work and energy in Section 15.1 by integrating Newton's second law, we were able to evaluate the integral on one side of the equation, obtaining the change in the kinetic energy:

$$U_{12} = \int_{\mathbf{r}_1}^{\mathbf{r}_2} \Sigma \mathbf{F} \cdot d\mathbf{r} = \tfrac{1}{2}mv_2^2 - \tfrac{1}{2}mv_1^2. \tag{15.15}$$

Suppose we could determine a scalar function of position V such that

$$dV = -\Sigma \mathbf{F} \cdot d\mathbf{r}. \tag{15.16}$$

Then we could also evaluate the integral defining the work:

$$U_{12} = \int_{\mathbf{r}_1}^{\mathbf{r}_2} \Sigma \mathbf{F} \cdot d\mathbf{r} = \int_{V_1}^{V_2} -dV = -(V_2 - V_1), \tag{15.17}$$

where V_1 and V_2 are the values of V at the positions $\mathbf{r}_1$ and $\mathbf{r}_2$, respectively. Substituting this expression into Eq. (15.15), we obtain the principle of work and energy in the form

$$\tfrac{1}{2}mv_1^2 + V_1 = \tfrac{1}{2}mv_2^2 + V_2. \tag{15.18}$$

If the kinetic energy increases as the object moves from position 1 to position 2, the function V must decrease, and vice versa, as if V represents a reservoir of "potential" kinetic energy. For this reason, V is called the *potential energy*.

Equation (15.18) states that the sum of the kinetic and potential energies of an object has the same value at any two points. Energy is *conserved*. However, *there is an important restriction on the use of this result*. We arrived at Eq. (15.18) by assuming that a function V, the potential energy, exists that satisfies Eq. (15.16). This is true only for a limited class of forces, which are said to be *conservative*. We discuss conservative forces in the next section. If *all* of the forces that do work on an object are conservative, Eq. (15.18) can be applied, where V is the sum of the potential energies of the forces that do work on the object. Otherwise, Eq. (15.18) cannot be used. A system is said to be conservative if all of the forces that do work on the system are conservative. The sum of the kinetic and potential energies of a conservative system is conserved.

An object may be subjected to both conservative and nonconservative forces. When that is the case, it is often convenient to introduce the potential energies of the forces that are conservative into the statement of the principle of work and energy. To allow for this option, we write Eq. (15.15) as

$$\tfrac{1}{2}mv_1^2 + V_1 + U_{12} = \tfrac{1}{2}mv_2^2 + V_2.$$

$$(15.19)$$

When the principle of work and energy is written in this form, the term U_{12} includes the work done by all nonconservative forces acting on the object. If a conservative force does work on the object, there is a choice. The work can be calculated and included in U_{12}, *or* the force's potential energy can be included in V. This procedure can also be applied to a system that is subjected to both conservative and nonconservative forces. In words, the sum of the kinetic and potential energies of a system in position 1, plus the work done as the system moves from position 1 to position 2, is equal to the total sum of the kinetic and potential energies in position 2.

15.5 Conservative Forces

We can apply conservation of energy only if the forces doing work on an object or system are conservative and we know (or can determine) their potential energies. In this section, we determine the potential energies of some conservative forces and use the results to demonstrate applications of conservation of energy. But before discussing forces that are conservative, we demonstrate with a simple example that *frictional forces are not conservative*.

The work done by a conservative force as an object moves from a position 1 to a position 2 is independent of the object's path. This result follows from Eq. (15.17), which states that the work depends only on the values of the potential energy at positions 1 and 2. Equation (15.17) also implies that if the object moves along a closed path, returning to position 1, the work done by a conservative force is zero. Suppose that a book of mass m rests on a table and you push it horizontally so that it slides along a path of length L. The magnitude of the force of friction is $\mu_k mg$, and the direction of the force is opposite to that of the book's motion (Fig. 15.15). The work done is

$$U_{12} = \int_0^L -\mu_k mg \, ds = -\mu_k mgL.$$

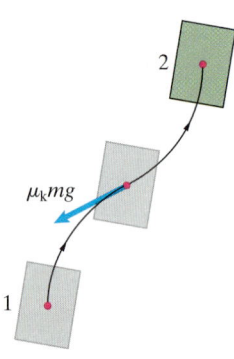

Figure 15.15
The book's path from position 1 to position 2. The force of friction points opposite to the direction of the motion.

The work is proportional to the length of the object's path and therefore is not independent of the path. As this simple example demonstrates, friction forces are not conservative.

Potential Energies of Particular Forces

The weight of an object and the force exerted by a spring attached to a fixed support are conservative forces. Using them as examples, we demonstrate how you can determine the potential energies of other conservative forces. We also use the potential energies of these forces in examples of the use of conservation of energy to analyze the motions of conservative systems.

Weight To determine the potential energy associated with an object's weight, we use a cartesian coordinate system with its y axis pointing upward (Fig. 15.16). The weight is $\mathbf{F} = -mg\mathbf{j}$, and its dot product with the vector $d\mathbf{r}$ is

$$\mathbf{F} \cdot d\mathbf{r} = (-mg\mathbf{j}) \cdot (dx\,\mathbf{i} + dy\,\mathbf{j} + dz\,\mathbf{k}) = -mg\,dy.$$

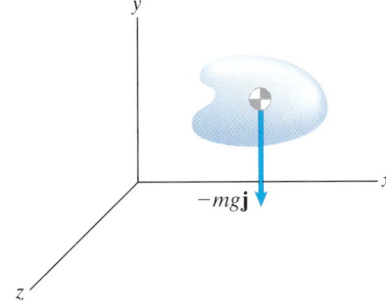

Figure 15.16
Weight of an object expressed in terms of a coordinate system with the y axis pointing upward.

From Eq. (15.16), the potential energy V must satisfy the relation

$$dV = -\mathbf{F} \cdot d\mathbf{r} = mg\,dy, \tag{15.20}$$

which we can write as

$$\frac{dV}{dy} = mg.$$

Integrating this equation, we obtain

$$V = mgy + C,$$

where C, the constant of integration, is arbitrary. This expression satisfies Eq. (15.20) for any value of C. Another way of understanding why C is arbitrary is to notice in Eq. (15.18) that it is the difference in the potential energy between two positions that determines the change in the kinetic energy. We will let $C = 0$ and write the potential energy of the weight of an object as

$$V = mgy. \tag{15.21}$$

The potential energy is the product of the object's weight and height. The height can be measured from any convenient reference level, or *datum*. Since it is the difference in potential energy that determines the change in the kinetic energy, it is the difference in height that matters, not the level from which the height is measured.

The roller coaster (Fig. 15.17a) is a classic example of conservation of energy. If aerodynamic and frictional forces are neglected, the weight is the only force doing work, and the system is conservative. The potential energy of the roller coaster is proportional to the height of the track relative to a datum. In Fig. 15.17b, we assume that the roller coaster started from rest at the datum level. The sum of the kinetic and potential energies is constant, so the kinetic energy "mirrors" the potential energy. At points of the track that have equal heights, the magnitudes of the velocities are equal.

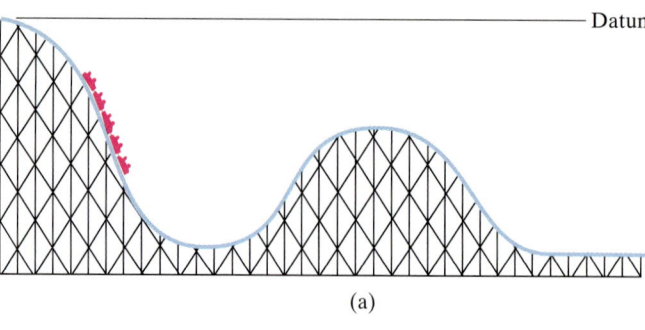

(a)

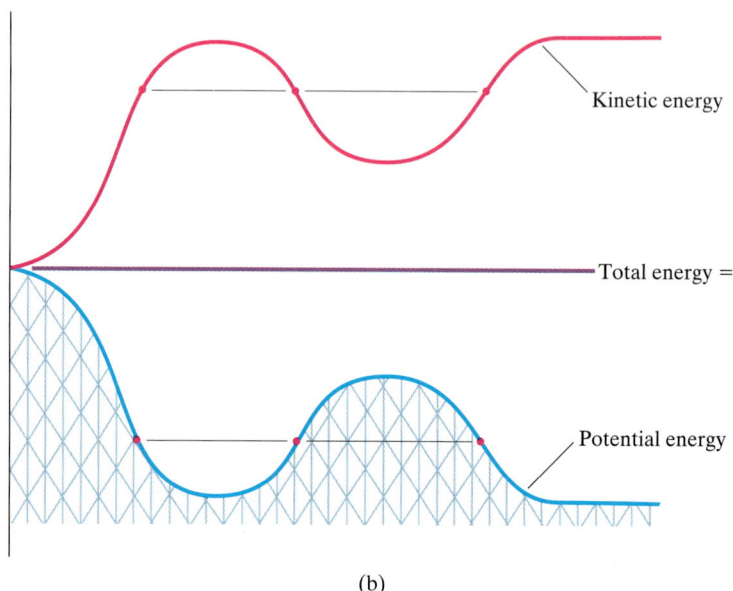

Kinetic energy

Total energy = 0

Potential energy

(b)

Figure 15.17
(a) Roller coaster and a reference level, or datum.
(b) The sum of the potential and kinetic energies is constant.

To account for the variation of weight with distance from the center of the earth, we can express the weight in polar coordinates as

$$\mathbf{F} = -\frac{mgR_\mathrm{E}^2}{r^2}\mathbf{e}_r,$$

where r is the distance from the center of the earth (Fig. 15.18). From Eq. (15.12), the vector $d\mathbf{r}$ in terms of polar coordinates is

$$d\mathbf{r} = dr\,\mathbf{e}_r + r\,d\theta\,\mathbf{e}_\theta. \tag{15.22}$$

The potential energy must satisfy

$$dV = -\mathbf{F}\cdot d\mathbf{r} = \frac{mgR_\mathrm{E}^2}{r^2}dr,$$

or

$$\frac{dV}{dr} = \frac{mgR_\mathrm{E}^2}{r^2}.$$

We integrate this equation and let the constant of integration be zero, obtaining the potential energy

$$V = -\frac{mgR_\mathrm{E}^2}{r}. \tag{15.23}$$

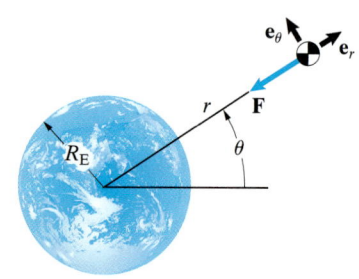

Figure 15.18
Expressing weight in terms of polar coordinates.

Compare this expression with the gravitational potential energy given by Eq. (15.21), in which the variation of the gravitational force with height is neglected. (See Problem 15.109.)

Springs In terms of polar coordinates, the force exerted on an object by a linear spring is

$$\mathbf{F} = -k(r - r_0)\mathbf{e}_r,$$

where r_0 is the unstretched length of the spring (Fig. 15.19). Using Eq. (15.22), we see that the potential energy must satisfy

$$dV = -\mathbf{F} \cdot d\mathbf{r} = k(r - r_0)\, dr.$$

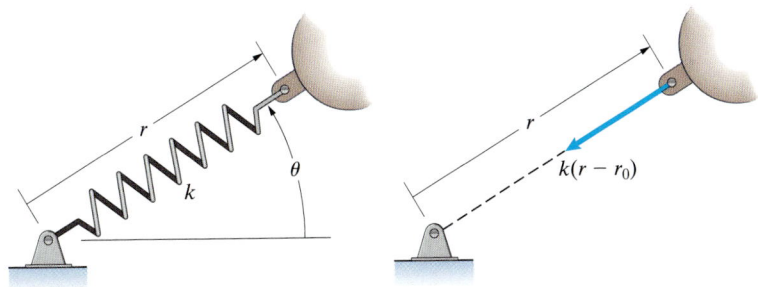

Figure 15.19
Expressing the force exerted by a linear spring in polar coordinates.

Expressed in terms of the stretch of the spring $S = r - r_0$, this equation is $dV = kS\, dS$, or

$$\frac{dV}{dS} = kS.$$

Integrating, we obtain the potential energy of a linear spring:

$$V = \tfrac{1}{2}kS^2. \tag{15.24}$$

Using conservation of energy to relate changes in the positions of conservative systems to changes in their kinetic energies typically involves three steps:

1. *Determine whether the system is conservative.* Draw a free-body diagram to identify the forces that do work, and confirm that they are conservative.
2. *Determine the potential energy.* Evaluate the potential energies of the forces in terms of the position of the system.
3. *Apply conservation of energy.* Equate the sum of the kinetic and potential energies of the system at two positions to obtain an expression for the change in the kinetic energy.

Study Questions

1. What is the definition of a conservative force?
2. What condition is necessary for the total energy of a system to be conserved?
3. If you know the change in the total potential energy as an object subjected to conservative forces moves from a position 1 to a position 2, what can you infer about the work done on the object as it moves between the two positions? What can you infer about the change in the total energy?
4. If an object moves upward, does the potential energy associated with its weight increase or decrease? (If you don't know the answer, consider what happens when you throw an object upward.)

| Example 15.8 | Potential Energy of Weight and Springs |

In Example 15.5, the 40-kg hammer is lifted into position 1 and released from rest. Its weight and the two springs ($k = 1500$ N/m) accelerate the hammer downward to position 2, where it strikes a workpiece. Use conservation of energy to determine the hammer's velocity when it reaches position 2.

Strategy
We must draw a free-body diagram of the hammer and confirm that all of the forces that do work on it are conservative. If they are, we can apply conservation of energy to determine the velocity of the hammer in position 2. The potential energy associated with the weight of the hammer is given by Eq. (15.21), and the potential energy associated with each spring is given by Eq. (15.24).

Solution
Determine whether the System Is Conservative From the free-body diagram of the hammer (Fig. a), we see that work is done only by its weight and the forces exerted by the springs. Therefore, the system is conservative.

Determine the Potential Energy The potential energy of each spring is $\frac{1}{2}kS^2$, where S is the stretch, so the total potential energy of the two springs is

$$V_{\text{springs}} = 2\left(\tfrac{1}{2}kS^2\right).$$

In Example 15.5, the stretches in positions 1 and 2 were determined to be $S_1 = 0.3$ m and $S_2 = 0.1$ m, respectively. The potential energy associated with the weight is

$$V_{\text{weight}} = mgy,$$

where y is the height relative to a convenient datum (Fig. b).

Apply Conservation of Energy The sums of the potential and kinetic energies at positions 1 and 2 must be equal.

$$2\left(\tfrac{1}{2}kS_1^2\right) + mgy_1 + \tfrac{1}{2}mv_1^2 = 2\left(\tfrac{1}{2}kS_2^2\right) + mgy_2 + \tfrac{1}{2}mv_2^2:$$

$$(1500 \text{ N/m})(0.3 \text{ m})^2 + (40 \text{ kg})(9.81 \text{ m/s}^2)(0.4 \text{ m}) + 0$$

$$= (1500 \text{ N/m})(0.1 \text{ m})^2 + 0 + \tfrac{1}{2}(40 \text{ kg})v_2^2.$$

Solving this equation, we obtain $v_2 = 3.72$ m/s.

Critical Thinking
From the graphs of the total potential energy associated with the springs and the weight and the kinetic energy of the hammer as functions of y (Fig. 15.20), you can see the transformation of the potential energy into kinetic energy as the hammer falls. Notice that the total energy of the conservative system remains constant.

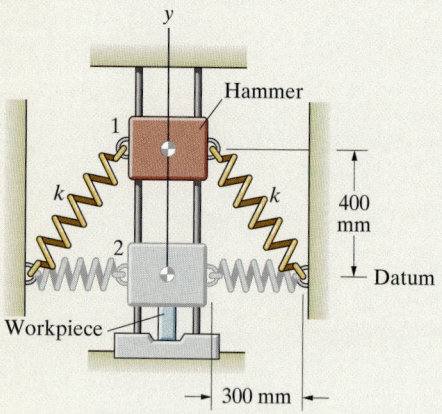

(a) Free-body diagram of the hammer.

(b) Measuring the height of the hammer relative to position 2.

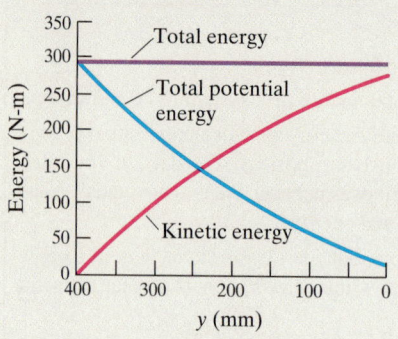

Figure 15.20
The potential and kinetic energies as functions of the y coordinate of the hammer.

Example 15.9   Conservation of Energy of a System

The spring in Fig. 15.21 ($k = 300$ N/m) is connected to the floor and to the 90-kg collar A. Collar A is at rest, supported by the spring, when the 135-kg box B is released from rest in the position shown. What are the velocities of A and B when B has fallen 1 m?

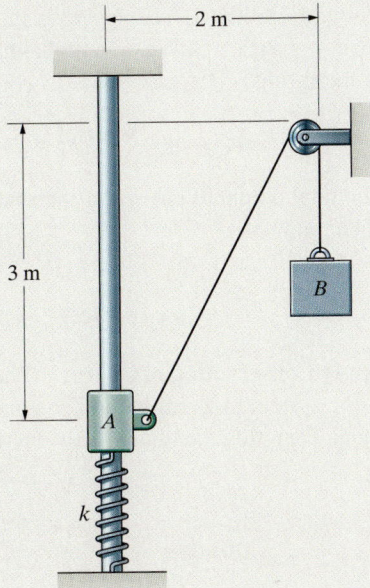

Figure 15.21

Strategy

If all of the forces that do work on the system are conservative, we can apply conservation of energy to obtain one equation in terms of the velocities of A and B when B has fallen 1 m. To complete the solution, we must also use kinematics to determine the relationship between the velocities of A and B.

Solution

Determine whether the System Is Conservative   We consider the collar A, box B, and pulley as a single system. From the free-body diagram of the system in Fig. a, we see that work is done only by the weights of the collar and box and the spring force F. The system is therefore conservative.

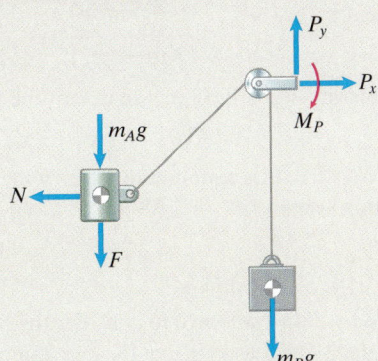

(a) Free-body diagram of the system.

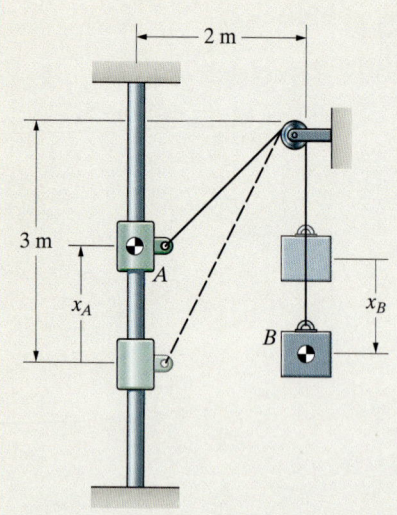

(b) Displacements of the collar and box.

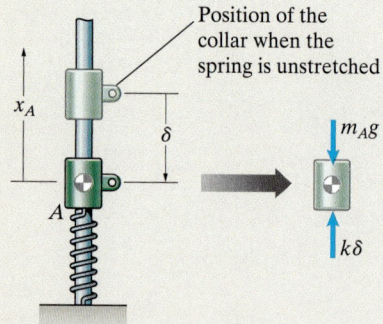

(c) Determining the initial compression of the spring.

Determine the Potential Energy Using the initial position of collar A as its datum, the potential energy associated with the weight of A when it has risen a distance x_A (Fig. b) is $V_A = m_A g x_A$. Using the initial position of box B as its datum, the potential energy associated with its weight when it has fallen a distance x_B is $V_B = -m_B g x_B$. (The minus sign is necessary because x_B is positive downward.)

To determine the potential energy associated with the spring force, we must account for the fact that in the initial position the spring is compressed by the weight of collar A. The spring is initially compressed a distance δ such that $m_A g = k\delta$ (Fig. c). When the collar has moved upward a distance x_A, the stretch of the spring is $S = x_A - \delta = x_A - m_A g/k$, so its potential energy is

$$V_S = \tfrac{1}{2}kS^2 = \tfrac{1}{2}k\left(x_A - \frac{m_A g}{k}\right)^2.$$

The total potential energy of the system in terms of the displacements of the collar and box is

$$V = V_A + V_B + V_S$$
$$= m_A g x_A - m_B g x_B + \tfrac{1}{2}k\left(x_A - \frac{m_A g}{k}\right)^2.$$

Apply Conservation of Energy The sum of the kinetic and potential energies of the system in its initial position and in the position shown in Fig. b must be equal. Denoting the total kinetic energy by T, we have

$$T_1 + V_1 = T_2 + V_2:$$

$$0 + \tfrac{1}{2}k\left(-\frac{m_A g}{k}\right)^2 = \tfrac{1}{2}m_A v_A^2 + \tfrac{1}{2}m_B v_B^2$$
$$+ \; m_A g x_A - m_B g x_B + \tfrac{1}{2}k\left(x_A - \frac{m_A g}{k}\right)^2. \tag{1}$$

We want to determine v_A and v_B when $x_B = 1$ m, but we have only one equation in terms of x_A, x_B, v_A, and v_B. To complete the solution, we must relate the displacement and velocity of the collar A to the displacement and velocity of the box B.

From Fig. b, the decrease in the length of the rope from A to the pulley as the collar rises must equal the distance the box falls:

$$\sqrt{(3\text{ m})^2 + (2\text{ m})^2} - \sqrt{(3\text{ m} - x_A)^2 + (2\text{ m})^2} = x_B.$$

Solving this equation for the value of x_A when $x_B = 1$ m, we obtain $x_A = 1.33$ m. By taking the derivative of this equation with respect to time, we also obtain a relation between v_A and v_B:

$$\left[\frac{3\text{ m} - x_A}{\sqrt{(3\text{ m} - x_A)^2 + (2\text{ m})^2}}\right]v_A = v_B.$$

Setting $x_A = 1.33$ m, we determine from the preceding equation that

$$0.641 v_A = v_B.$$

We solve this equation together with Eq. (1) for the velocities of the collar and box when $x_A = 1.33$ m and $x_B = 1$ m, obtaining $v_A = 3.82$ m/s and $v_B = 2.45$ m/s.

Critical Thinking

Why didn't we have to consider the forces exerted on the collar and box by the rope? The reason is that they are internal forces when the collar, box, and pulley are regarded as a single system. This example clearly demonstrates the advantage of applying conservation of energy to an entire system whenever possible.

Example 15.10 Conservation of Energy of a Spacecraft

A spacecraft at a distance $r_0 = 2R_E$ from the center of the earth is moving outward with initial velocity $v_0 = \sqrt{2gR_E/3}$ (Fig. 15.22). Determine the velocity of the craft as a function of its distance from the center of the earth.

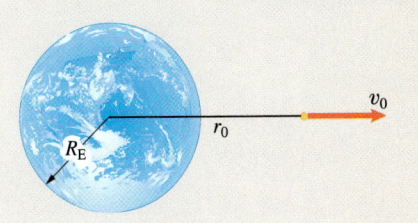

Figure 15.22

Strategy
The potential energy associated with the earth's gravity is given by Eq. (15.23). The initial radial position and velocity of the spacecraft are given, so we can use conservation of energy to determine its velocity as a function of its radial position.

Solution
Determine whether the System Is Conservative If work is done on the spacecraft by gravity alone, the system is conservative.

Determine the Potential Energy The potential energy associated with the weight of the spacecraft is given in terms of its distance r from the center of the earth by Eq. (15.23):

$$V = -\frac{mgR_E^2}{r}.$$

Apply Conservation of Energy Let v be the magnitude of the spacecraft's velocity at an arbitrary distance r. The sums of the potential and kinetic energies at r_0 and at r must be equal.

$$-\frac{mgR_E^2}{r_0} + \tfrac{1}{2}mv_0^2 = -\frac{mgR_E^2}{r} + \tfrac{1}{2}mv^2:$$

$$-\frac{mgR_E^2}{2R_E} + \tfrac{1}{2}m\left(\tfrac{2}{3}gR_E\right) = -\frac{mgR_E^2}{r} + \tfrac{1}{2}mv^2.$$

Solving for v, we find that the spacecraft's velocity as a function of r is

$$v = \sqrt{gR_E\left(\frac{2R_E}{r} - \frac{1}{3}\right)}.$$

Critical Thinking
We show graphs of the kinetic energy, potential energy, and total energy as functions of r/R_E in Fig. 15.23. The kinetic energy decreases and the potential energy increases as the spacecraft moves outward until its velocity decreases to zero at $r = 6R_E$.

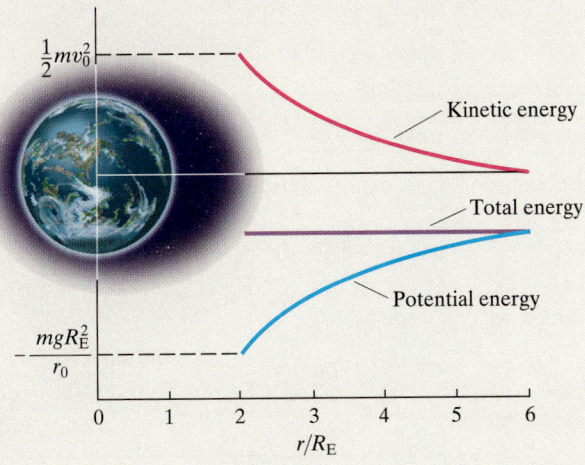

Figure 15.23
Energies as functions of the radial coordinate.

| Example 15.11 | Potential Energy of a Compressed Gas |

The bore of the gas gun shown in Fig. 15.24 has cross-sectional area A. The bore is evacuated on the right of the projectile of mass m, and on the left it contains gas at pressure p. Let the value of the pressure when $s = s_0$ be p_0, and assume that the pressure of the gas is related to its volume V by $pV = $ constant.

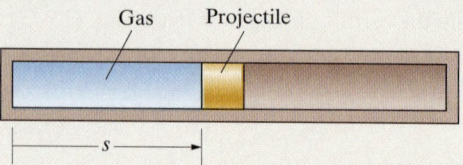

Gas   Projectile

Figure 15.24

s

(a) Determine the potential energy associated with the force exerted on the projectile in terms of s.

(b) If the projectile starts from rest at $s = s_0$ and friction is negligible, what is the velocity the projectile as a function of s?

Strategy

(a) We can determine the force exerted on the piston by the pressure of the gas as a function of s and then use Eq. (15.16) to determine the potential energy.

(b) Knowing the potential energy, we can use conservation of energy to determine the velocity of the projectile as a function of s.

Solution

(a) The volume of the gas to the left of the projectile is $V = sA$. We know that pV is constant, so psA is constant. The value of the pressure when $s = s_0$ is p_0, so $psA = p_0 s_0 A$. The force F exerted on the projectile is the product of the pressure p and the cross-sectional area A of the projectile. Therefore,

$$F = pA = \frac{p_0 s_0 A}{s}.$$

To apply Eq. (15.16), we need to express the total force on the piston and an infinitesimal displacement $d\mathbf{r}$ of the piston as vectors. Let $\mathbf{e}_t$ be a unit vector that points to the right. Neglecting friction, we observe that the total force on the piston is

$$\Sigma \mathbf{F} = F\mathbf{e}_t = \frac{p_0 s_0 A}{s} \mathbf{e}_t,$$

and we can express $d\mathbf{r}$ as

$$d\mathbf{r} = ds\,\mathbf{e}_t.$$

Substituting these expressions into Eq. (15.16), we obtain

$$dV = -\Sigma \mathbf{F} \cdot d\mathbf{r} = -\left(\frac{p_0 s_0 A}{s}\mathbf{e}_t\right) \cdot (ds\,\mathbf{e}_t) = -\frac{p_0 s_0 A}{s} ds,$$

which we can write as

$$\frac{dV}{ds} = -\frac{p_0 s_0 A}{s}.$$

We integrate this equation, setting the constant of integration equal to zero, to obtain the potential energy:

$$V = -p_0 s_0 A \ln s.$$

(b) If the projectile starts from rest at $s = s_0$, we can use conservation of energy to determine its velocity v at an arbitrary position s.

$$\tfrac{1}{2}mv_1^2 + V_1 = \tfrac{1}{2}mv_2^2 + V_2:$$

$$0 - p_0 s_0 A \ln s_0 = \tfrac{1}{2}mv^2 - p_0 s_0 A \ln s.$$

Solving for the velocity, we obtain

$$v = \sqrt{\frac{2 p_0 s_0 A}{m} \ln\!\left(\frac{s}{s_0}\right)}.$$

Critical Thinking

We designed this example to demonstrate how you can determine the potential energy of a given force by using Eq. (15.16). (See Problems 15.99 and 15.101.)

Problems

15.78 The 10-lb box is released from rest at position 1 and slides down the smooth inclined surface to position 2.

(a) If the datum is placed at the level of the floor as shown, what is the sum of the kinetic and potential energies of the box when it is in position 1?

(b) What is the sum of the kinetic and potential energies of the box when it is in position 2?

(c) Use conservation of energy to determine the magnitude of the box's velocity when it is in position 2.

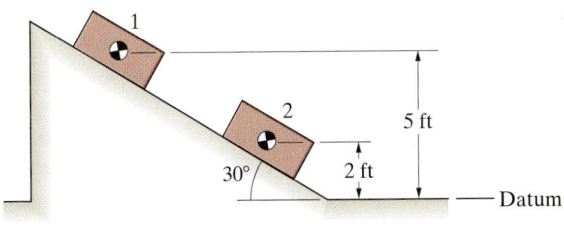

Problem 15.78

15.79 The 0.45-kg soccer ball is 1 m above the ground when it is kicked upward at 12 m/s. Use conservation of energy to determine the magnitude of the ball's velocity when it is 4 m above the ground. Obtain the answer by placing the datum (a) at the level of the ball's initial position and (b) at ground level.

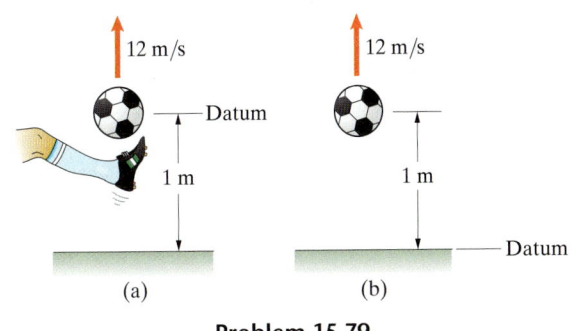

Problem 15.79

15.80 The Lunar Module used in the Apollo moon landings could make a safe landing if the magnitude of its vertical velocity at impact was no greater than 5 m/s. Use conservation of energy to determine the maximum height h at which the pilot could shut off the engine if the vertical velocity of the lander is (a) 2 m/s downward and (b) 2 m/s upward. The acceleration due to gravity at the moon's surface is 1.62 m/s².

Problem 15.80

15.81 The 0.4-kg collar starts from rest at position 1 and slides down the smooth rigid wire. The y axis points upward. Use conservation of energy to determine the magnitude of the velocity of the collar when it reaches point 2.

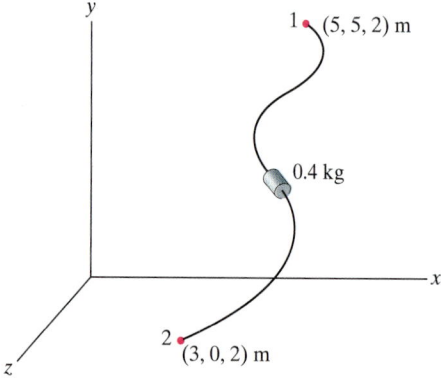

Problem 15.81

15.82 At the instant shown, the 20-kg mass is moving downward at 1.6 m/s. Let d be the downward displacement of the mass relative to its present position. Use conservation of energy to determine the magnitude of the velocity of the 20-kg mass when $d = 1$ m.

Problem 15.82

15.83 The mass of the ball is $m = 2$ kg and the string's length is $L = 1$ m. The ball is released from rest in position 1 and swings to position 2, where $\theta = 40°$.

(a) Use conservation of energy to determine the magnitude of the ball's velocity at position 2.

(b) Draw graphs of the kinetic energy, the potential energy, and the total energy for values of θ from zero to 180°.

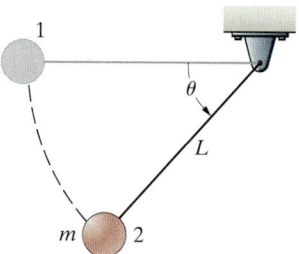

Problem 15.83

15.84 The mass of the ball is $m = 2$ kg and the string's length is $L = 1$ m. The ball is released from rest in position 1. When the string is vertical, it hits the fixed peg shown.

(a) Use conservation of energy to determine the minimum angle θ necessary for the ball to swing to position 2.

(b) If the ball is released at the minimum angle θ determined in part (a), what is the tension in the string just before and just after it hits the peg?

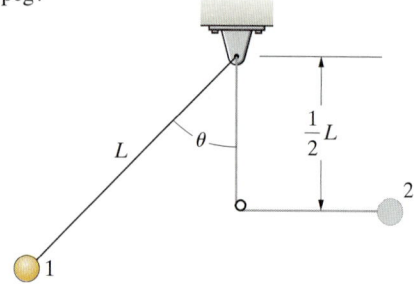

Problem 15.84

15.85 A small pellet of mass $m = 0.2$ kg starts from rest at position 1 and slides down the smooth surface of the cylinder to position 2. The radius $R = 0.8$ m. Use conservation of energy to determine the magnitude of the pellet's velocity at position 2 if $\theta = 45°$.

15.86 In Problem 15.85, what is the value of the angle θ at which the pellet loses contact with the surface of the cylinder?

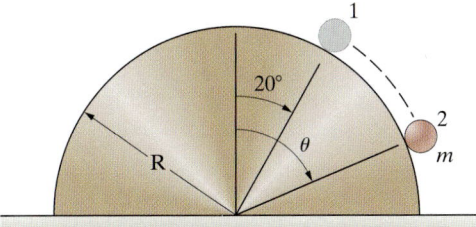

Problems 15.85/15.86

15.87 The bar is smooth. Use conservation of energy to determine the minimum velocity the 10-kg slider must have at A (a) to reach C; (b) to reach D.

15.88 In Problem 15.87, what normal force does the bar exert on the slider at B in cases (a) and (b)?

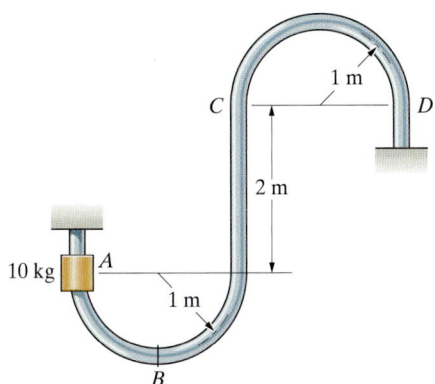

Problems 15.87/15.88

15.89 The 10-kg collar starts from rest at position 1 and slides along the bar. The y axis points upward. The spring constant is $k = 100$ N/m, and the unstretched length of the spring is 2 m. Use conservation of energy to determine the magnitude of the velocity of the collar when it reaches position 2.

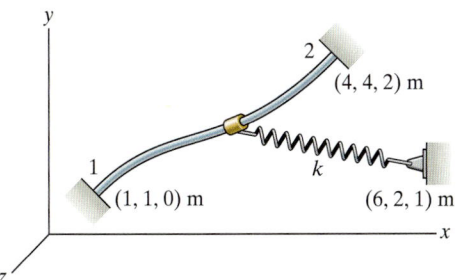

Problem 15.89

15.90 A rock climber of weight W has a rope attached a distance h below him for protection. Suppose that he falls, and assume that the rope behaves like a linear spring with unstretched length h and spring constant $k = C/h$, where C is a constant. Use conservation of energy to determine the maximum force exerted on the climber by the rope. (Notice that the maximum force is independent of h, which is a reassuring result for climbers: The maximum force resulting from a long fall is the same as that resulting from a short one.)

Problem 15.90

15.91 The spring constant $k = 700$ N/m, $m_A = 14$ kg, and $m_B = 18$ kg. The collar A slides on the smooth horizontal bar. The system is released from rest with the spring unstretched. Use conservation of energy to determine the velocity of the collar A when it has moved 0.2 m to the right.

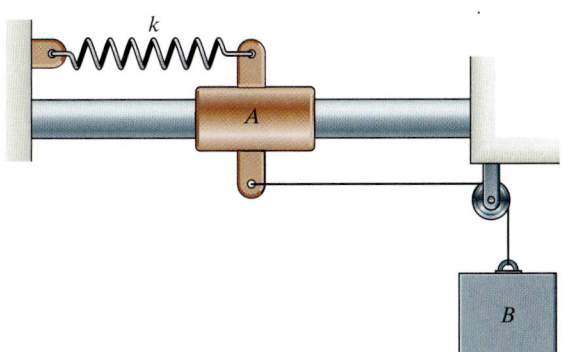

Problem 15.91

15.92 The spring constant $k = 700$ N/m, $m_A = 14$ kg, and $m_B = 18$ kg. The collar A slides on the smooth horizontal bar. The system is released from rest with the spring unstretched. Use conservation of energy to determine the velocity of the collar A when it has moved 0.2 m to the right.

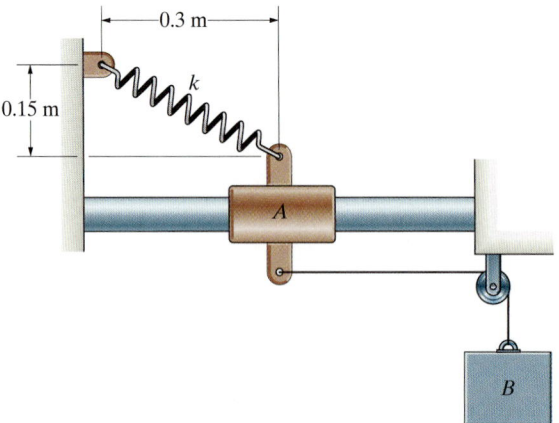

Problem 15.92

15.93 The 5-lb collar starts from rest at A and slides along the semicircular bar. The spring constant is $k = 100$ lb/ft, and the unstretched length of the spring is 1 ft. Use conservation of energy to determine the velocity of the collar at B.

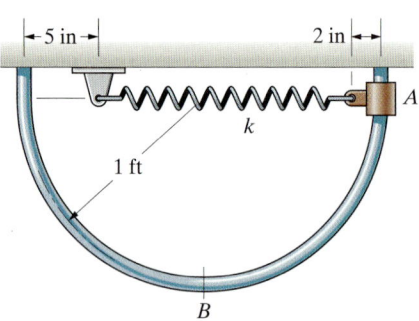

Problem 15.93

15.94 The mass $m = 1$ kg, the spring constant $k = 200$ N/m, and the unstretched length of the spring is 0.1 m. When the system is released from rest in the position shown, the spring contracts, pulling the mass to the right. Use conservation of energy to determine the magnitude of the velocity of the mass when the string and the spring are parallel.

15.95 In Problem 15.94, what is the tension in the string when the string and spring are parallel?

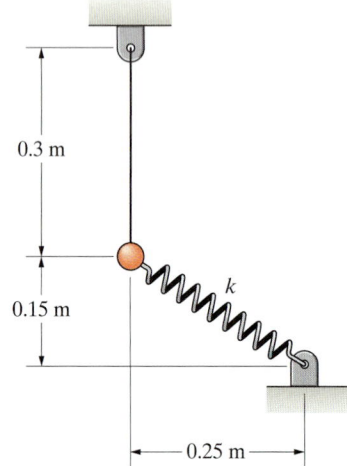

Problems 15.94/15.95

15.96 The force exerted on an object by a *nonlinear* spring is

$$\mathbf{F} = -[k(r - r_0) + q(r - r_0)^3]\mathbf{e}_r,$$

where k and q are constants and r_0 is the unstretched length of the spring. Determine the potential energy of the spring in terms of its stretch $S = r - r_0$.

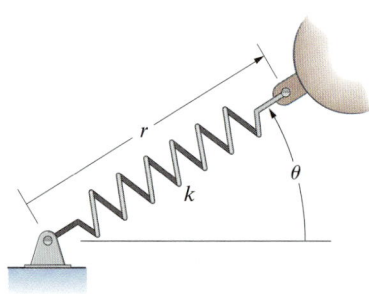

Problem 15.96

15.97 The 20-kg cylinder is released at the position shown and falls onto the linear spring ($k = 3000$ N/m). Use conservation of energy to determine how far down the cylinder moves after contacting the spring.

15.98 The 20-kg cylinder is released at the position shown and falls onto the *nonlinear* spring. In terms of the stretch S of the spring, its potential energy is $V = \frac{1}{2}kS^2 + \frac{1}{4}qS^4$, where $k = 3000$ N/m and $q = 4000$ N/m^3. What is the velocity of the cylinder when the spring has been compressed 0.5 m?

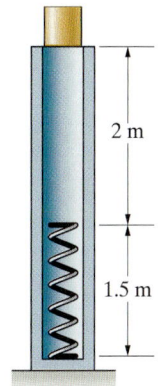

Problems 15.97/15.98

15.99 The string exerts a force of constant magnitude T on the object. Use polar coordinates to show that the potential energy associated with this force is $V = Tr$.

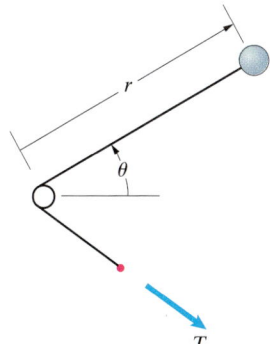

Problem 15.99

15.100 The system is at rest in the position shown, with the 12-lb collar A resting on the spring ($k = 20$ lb/ft), when a constant 30-lb force is applied to the cable. What is the velocity of the collar when it has risen 1 ft? (See Problem 15.99.)

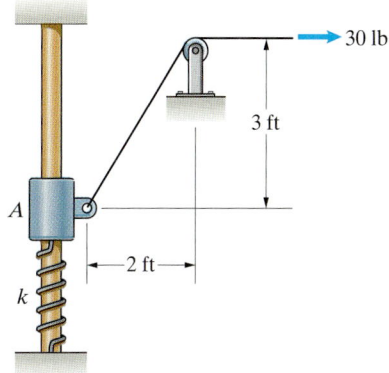

Problem 15.100

15.101 As the 26,000-lb F/A-18 lands at 210 ft/s, the cable from A to B engages the airplane's arresting hook at C. The arresting mechanism maintains the tension in the cable at a constant value T. The dimension $h = 33$ ft.

(a) Determine the potential energy V associated with the force exerted on the airplane by the cable in terms of the tension T and the distance s. (See Problem 15.99.)

(b) The arresting system brings the airplane to rest in a distance $s = 72$ ft. Use conservation of energy to determine the tension T.

15.102 Suppose that the airplane in Problem 15.101 lands at 240 ft/s. Use conservation of energy to determine the distance it rolls before the arresting system brings it to rest.

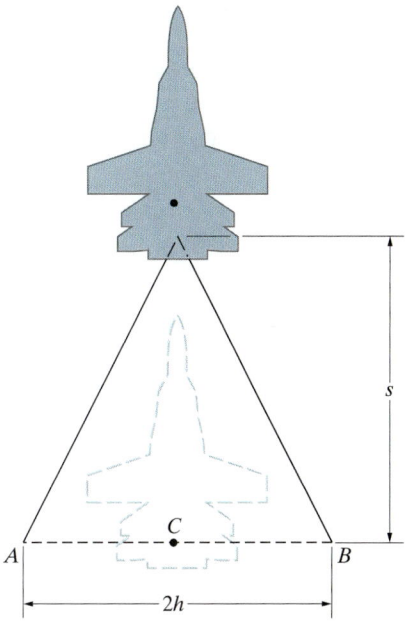

Problems 15.101/15.102

15.103 A satellite at a distance $r_0 = 17,000$ km from the center of the earth has a velocity of magnitude $v_0 = 7000$ m/s. Use conservation of energy to determine the magnitude of the velocity v of the satellite when it is at a distance $r = 36,000$ km from the center of the earth. The radius of the earth is 6370 km.

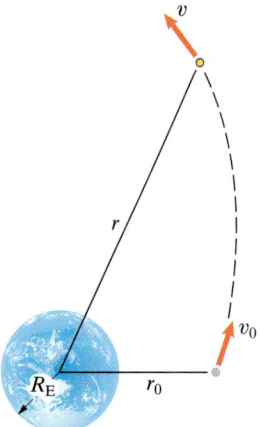

Problem 15.103

15.104 Astronomers detect an asteroid 100,000 km from the earth moving at 2 km/s relative to the center of the earth. Suppose the asteroid strikes the earth. Use conservation of energy to determine the magnitude of its velocity as it enters the atmosphere. (You can neglect the thickness of the atmosphere in comparison to the earth's 6370-km radius.)

15.105 A satellite is in the elliptic earth orbit shown. Its velocity at the perigee A is 8640 m/s. Use conservation of energy to determine the magnitude of the satellite's velocity at B. The radius of the earth is 6370 km.

15.106 Use conservation of energy to determine the magnitude of the velocity of the satellite in Problem 15.105 at the apogee C. Using your result, confirm numerically that the velocities at perigee and apogee satisfy the relation $r_A v_A = r_C v_C$.

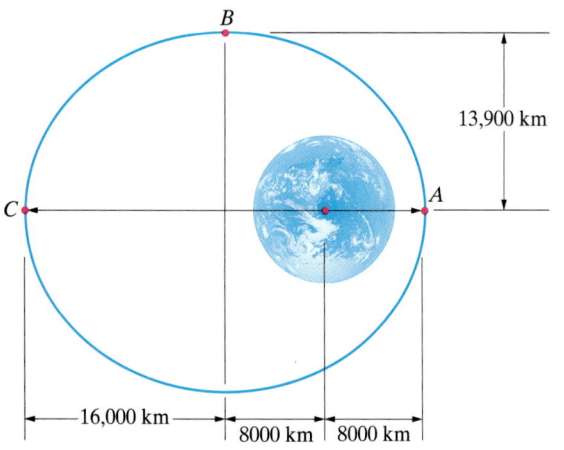

Problems 15.105/15.106

15.107 The *Voyager* and *Galileo* spacecraft observed volcanic plumes, believed to consist of condensed sulfur or sulfur dioxide gas, above the surface of the Jovian satellite Io. The plume observed above a volcano named Prometheus was estimated to extend 50 km above the surface. The acceleration due to gravity at the surface is 1.80 m/s^2. Using conservation of energy and neglecting the variation of gravity with height, determine the velocity at which a solid particle would have to be ejected to reach 50 km above Io's surface.

15.108 Solve Problem 15.107 using conservation of energy and accounting for the variation of gravity with height. The radius of Io is 1815 km.

Problems 15.107/15.108

15.109* What is the relationship between Eq. (15.21), which is the gravitational potential energy neglecting the variation of the gravitational force with height, and Eq. (15.23), which accounts for the variation? Express the distance from the center of the earth as $r = R_E + y$, where R_E is the earth's radius and y is the height above the surface, so that Eq. (15.23) can be written as

$$V = -\frac{mgR_E}{1 + \dfrac{y}{R_E}}.$$

By expanding this equation as a Taylor series in terms of y/R_E and assuming that $y/R_E \ll 1$, show that you obtain a potential energy equivalent to Eq. (15.21).

15.6 Relationships between Force and Potential Energy

Here we consider two questions: (1) Given a potential energy, how can we determine the corresponding force? (2) Given a force, how can we determine whether it is conservative? That is, how can we tell whether an associated potential energy exists?

The potential energy V of a force $\mathbf{F}$ is a function of position that satisfies the relation

$$dV = -\mathbf{F} \cdot d\mathbf{r}. \tag{15.25}$$

Let us express V in terms of a cartesian coordinate system:

$$V = V(x, y, z).$$

The differential of V is

$$dV = \frac{\partial V}{\partial x}dx + \frac{\partial V}{\partial y}dy + \frac{\partial V}{\partial z}dz. \tag{15.26}$$

Expressing $\mathbf{F}$ and $d\mathbf{r}$ in terms of cartesian components and taking their dot product yields

$$\mathbf{F} \cdot d\mathbf{r} = (F_x\mathbf{i} + F_y\mathbf{j} + F_z\mathbf{k}) \cdot (dx\mathbf{i} + dy\mathbf{j} + dz\mathbf{k})$$

$$= F_x\,dx + F_y\,dy + F_z\,dz.$$

Substituting this expression and Eq. (15.26) into Eq. (15.25), we obtain

$$\frac{\partial V}{\partial x}dx + \frac{\partial V}{\partial y}dy + \frac{\partial V}{\partial z}dz = -(F_x\,dx + F_y\,dy + F_z\,dz),$$

which implies that

$$F_x = -\frac{\partial V}{\partial x}, \quad F_y = -\frac{\partial V}{\partial y}, \quad \text{and} \quad F_z = -\frac{\partial V}{\partial z}. \tag{15.27}$$

Given a potential energy V expressed in cartesian coordinates, we can use Eqs. (15.27) to determine the corresponding force. The force

$$\mathbf{F} = -\left(\frac{\partial V}{\partial x}\mathbf{i} + \frac{\partial V}{\partial y}\mathbf{j} + \frac{\partial V}{\partial z}\mathbf{k}\right) = -\nabla V, \tag{15.28}$$

where ∇V is the *gradient* of V. By using expressions for the gradient in terms of other coordinate systems, we can determine the force $\mathbf{F}$ when we know the potential energy in terms of those coordinate systems. For example, in terms of cylindrical coordinates,

$$\mathbf{F} = -\left(\frac{\partial V}{\partial r}\mathbf{e}_r + \frac{1}{r}\frac{\partial V}{\partial \theta}\mathbf{e}_\theta + \frac{\partial V}{\partial z}\mathbf{e}_z\right). \tag{15.29}$$

If a force $\mathbf{F}$ is conservative, its *curl* $\nabla \times \mathbf{F}$ is zero. The expression for the curl of $\mathbf{F}$ in cartesian coordinates is

$$\nabla \times \mathbf{F} = \begin{vmatrix} \mathbf{i} & \mathbf{j} & \mathbf{k} \\ \dfrac{\partial}{\partial x} & \dfrac{\partial}{\partial y} & \dfrac{\partial}{\partial z} \\ F_x & F_y & F_z \end{vmatrix}. \tag{15.30}$$

Substituting Eqs. (15.27) into this expression confirms that $\nabla \times \mathbf{F} = \mathbf{0}$ when $\mathbf{F}$ is conservative. The converse is also true. A force $\mathbf{F}$ is conservative if its curl is zero. We can use this condition to determine whether a given force is conservative. In terms of cylindrical coordinates, the curl of $\mathbf{F}$ is

$$
\nabla \times \mathbf{F} = \frac{1}{r}
\begin{vmatrix}
\mathbf{e}_r & r\mathbf{e}_\theta & \mathbf{e}_z \\
\dfrac{\partial}{\partial r} & \dfrac{\partial}{\partial \theta} & \dfrac{\partial}{\partial z} \\
F_r & rF_\theta & F_z
\end{vmatrix}.
\tag{15.31}
$$

Example 15.12 Determining the Force from a Potential Energy

From Eq. (15.23), the potential energy associated with the weight of an object of mass m at a distance r from the center of the earth is (in polar coordinates)

$$
V = - \frac{mgR_E^2}{r},
$$

where R_E is the radius of the earth. Use this expression to determine the force exerted on the object by its weight.

Strategy
The force $\mathbf{F} = -\nabla V$. The potential energy is expressed in terms of polar coordinates, so we can use Eq. (15.29) to determine the force.

Solution
The partial derivatives of V with respect to r, θ, and z are

$$
\frac{\partial V}{\partial r} = \frac{mgR_E^2}{r^2}, \qquad \frac{\partial V}{\partial \theta} = 0, \quad \text{and} \quad \frac{\partial V}{\partial z} = 0.
$$

From Eq. (15.29), the force is

$$
\mathbf{F} = -\nabla V = - \frac{mgR_E^2}{r^2}\mathbf{e}_r.
$$

Critical Thinking
We already know that the force is conservative, because we know its potential energy, but we can use Eq. (15.31) to confirm that its curl is zero:

$$
\nabla \times \mathbf{F} = \frac{1}{r}
\begin{vmatrix}
\mathbf{e}_r & r\mathbf{e}_\theta & \mathbf{e}_z \\
\dfrac{\partial}{\partial r} & \dfrac{\partial}{\partial \theta} & \dfrac{\partial}{\partial z} \\
-\dfrac{mgR_E^2}{r^2} & 0 & 0
\end{vmatrix} = \mathbf{0}.
$$

Although we used cylindrical coordinates in determining $\mathbf{F}$ and in evaluating the cross product, the expression for V and our resulting expression for $\mathbf{F}$ are valid only if the object remains in the plane $z = 0$.

Problems

15.110 The potential energy associated with a force **F** acting on an object is $V = 2x^2 - y$ N-m, where x and y are in meters.

(a) Determine **F**.

(b) Suppose the object moves from position 1 to position 2 along the paths A and B. Determine the work done by **F** along each path.

15.111 An object is subjected to the force $\mathbf{F} = y\mathbf{i} - x\mathbf{j}$ (N), where x and y are in meters.

(a) Show that **F** is *not* conservative.

(b) Suppose the object moves from point 1 to point 2 along the paths A and B shown in Problem 15.110. Determine the work done by **F** along each path.

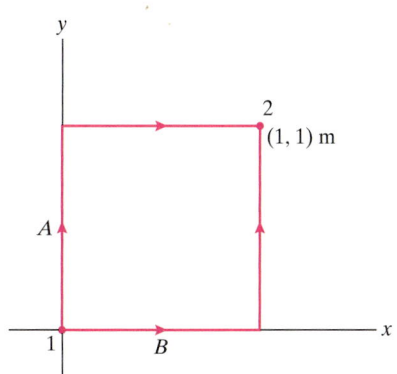

Problems 15.110/15.111

15.112 In terms of polar coordinates, the potential energy associated with the force **F** exerted on an object by a *nonlinear* spring is

$$V = \frac{1}{2}k(r - r_0)^2 + \frac{1}{4}q(r - r_0)^4,$$

where k and q are constants and r_0 is the unstretched length of the spring. Determine **F** in terms of polar coordinates.

15.113 In terms of polar coordinates, the force exerted on an object by a *nonlinear* spring is

$$\mathbf{F} = -[k(r - r_0) + q(r - r_0)^3]\mathbf{e}_r,$$

where k and q are constants and r_0 is the unstretched length of the spring. Use Eq. (15.31) to show that **F** is conservative.

15.114 The potential energy associated with a force **F** acting on an object is $V = -r \sin \theta + r^2 \cos^2 \theta$ ft-lb, where r is in feet.

(a) Determine **F**.

(b) If the object moves from point 1 to point 2 along the circular path, how much work is done by **F**?

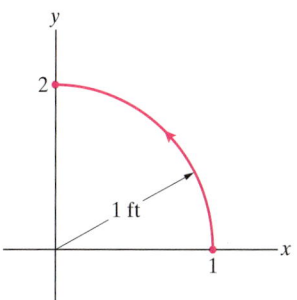

Problem 15.114

15.115 In terms of polar coordinates, the force exerted on an object of mass m by the gravity of a hypothetical two-dimensional planet is $\mathbf{F} = -(mg_T R_T/r)\mathbf{e}_r$, where g_T is the acceleration due to gravity at the surface, R_T is the radius of the planet, and r is the distance of the object from the center of the planet.

(a) Determine the potential energy associated with this gravitational force.

(b) If the object is given a velocity v_0 at a distance r_0, what is its velocity v as a function of r?

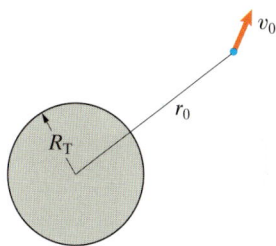

Problem 15.115

15.116 By substituting Eqs. (15.27) into Eq. (15.30), confirm that $\nabla \times \mathbf{F} = \mathbf{0}$ if **F** is conservative.

15.117 Determine which of the following forces are conservative:

(a) $\mathbf{F} = (3x^2 - 2xy)\mathbf{i} - x^2\mathbf{j}$;

(b) $\mathbf{F} = (x - xy^2)\mathbf{i} + x^2y\mathbf{j}$;

(c) $\mathbf{F} = (2xy^2 + y^3)\mathbf{i} + (2x^2y - 3xy^2)\mathbf{j}$.

15.118 Determine which of the following forces are conservative:

(a) $\mathbf{F} = 3r^2 \sin^2 \theta \, \mathbf{e}_r + 2r^2 \sin \theta \cos \theta \, \mathbf{e}_\theta$;

(b) $\mathbf{F} = (2r \sin \theta - \cos \theta)\mathbf{e}_r + (r \cos \theta - \sin \theta)\mathbf{e}_\theta$;

(c) $\mathbf{F} = (\sin \theta + r \cos^2\theta)\mathbf{e}_r + (\cos \theta - r \sin \theta \cos \theta)\mathbf{e}_\theta$.

COMPUTATIONAL MECHANICS

The following example and problems are designed to be worked with the use of a programmable calculator or computer.

Computational Example 15.13

In the mechanical delay switch shown in Fig. 15.25, an electromagnet releases the 1-kg slider at position 1. Under the actions of gravity and the linear spring, the slider moves along the smooth bar from position 1 to position 2, closing the switch. The constant of the spring is $k = 40$ N/m, and the unstretched length of the spring is $r_0 = 50$ mm. The dimensions are $R = 200$ mm and $h = 100$ mm. What is the magnitude of the slider's maximum velocity, and where does it occur?

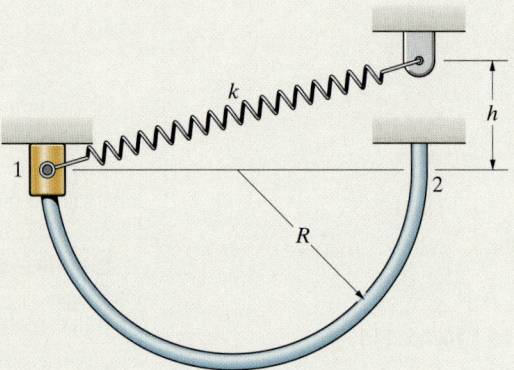

Figure 15.25

Strategy

We can use conservation of energy to obtain an equation relating the magnitude of the velocity of the slider to its position. By drawing a graph of the velocity as a function of the position, we can estimate the maximum velocity and the position where it occurs.

Solution

We can specify the slider's position by the angle θ through which it has moved relative to position 1 (Fig. a). In position 1, the stretch of the spring equals its length in position 1 minus its unstretched length:

$$S_1 = \sqrt{(2R)^2 + h^2} - r_0.$$

When the slider has moved through the angle θ, the stretch of the spring is

$$S = \sqrt{(R + R\cos\theta)^2 + (h + R\sin\theta)^2} - r_0.$$

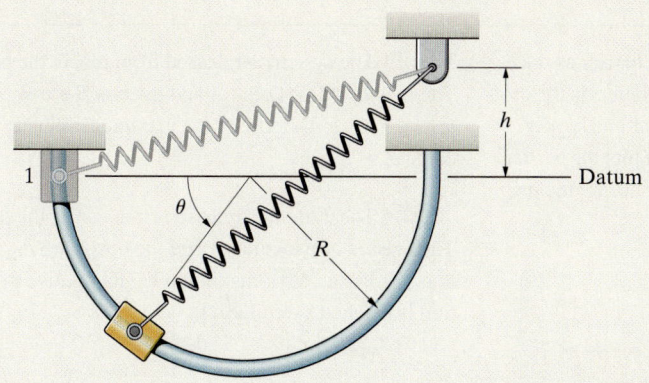

(a) The angle θ specifies the slider's position.

We express the potential energy of the slider's weight using the datum shown in Fig. a. The sum of the potential and kinetic energies at position 1 must equal the sum of the potential and kinetic energies when the slider has moved through the angle θ.

$$\tfrac{1}{2}kS_1^2 + mgy_1 + \tfrac{1}{2}mv_1^2 = \tfrac{1}{2}kS^2 + mgy + \tfrac{1}{2}mv^2:$$

$$\tfrac{1}{2}k\left[\sqrt{(2R)^2 + h^2} - r_0\right]^2 + 0 + 0$$

$$= \tfrac{1}{2}k\left[\sqrt{(R + R\cos\theta)^2 + (h + R\sin\theta)^2} - r_0\right]^2$$

$$- mgR\sin\theta + \tfrac{1}{2}mv^2.$$

Solving for v yields

$$v = \{(k/m)\left[\sqrt{(2R)^2 + h^2} - r_0\right]^2$$

$$- (k/m)\left[\sqrt{(R + R\cos\theta)^2 + (h + R\sin\theta)^2} - r_0\right]^2 + 2gR\sin\theta\}^{1/2}.$$

Computing the values of this expression as a function of θ, we obtain the graph shown in Fig. 15.26. The graph indicates that the velocity is a maximum at approximately $\theta = 135°$, so we examine the computed results near 135°:

θ	v, m/s
132°	2.5393
133°	2.5397
134°	2.5399
135°	2.5398
136°	2.5394
137°	2.5389
138°	2.5380

We estimate that a maximum velocity of 2.54 m/s occurs at $\theta = 134°$.

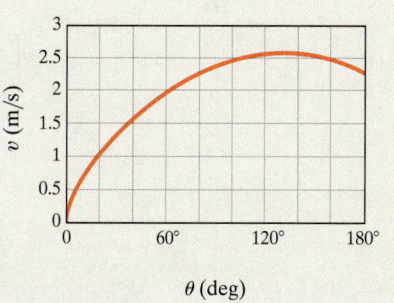

Figure 15.26
Magnitude of the velocity as a function of θ.

Critical Thinking

The placement of the spring and the geometry of the circular bar causes the dependence of the velocity on the angle θ to be very nonlinear. Notice that it is straightforward to determine the velocity for a given value of θ. But determining the maximum velocity, or determining the value of θ corresponding to a particular value of the velocity, requires a numerical solution.

Computational Problems

15.119 The force exerted on the car by the crash barrier as the barrier crushes is $F = -(4400 + 50{,}000s)$ N, where s is the distance in meters from the initial contact between car and barrier. Suppose that a 2400-kg car traveling at 100 km/h hits the barrier. What maximum power is transferred from the car during the impact, and at what position s does it occur?

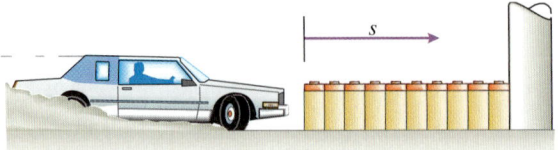

Problem 15.119

15.120 The 6-kg collar is released from rest in the position shown on the smooth bar. The spring constant is $k = 4$ kN/m, and the unstretched length of the spring is 150 mm. How far does the mass fall relative to its initial position before rebounding?

15.121 The 6-kg collar is released from rest in the position shown on the smooth bar. The spring constant is $k = 4$ kN/m, and the unstretched length of the spring is 150 mm. How far below its initial position does the mass reach its maximum velocity, and what is the maximum velocity?

15.122 The 6-kg collar is released from rest in the position shown on the smooth bar. The spring constant is $k = 4$ kN/m, and the unstretched length of the spring is 150 mm. How far below its initial position does the power being transferred to the collar reach its maximum, and what is the maximum power?

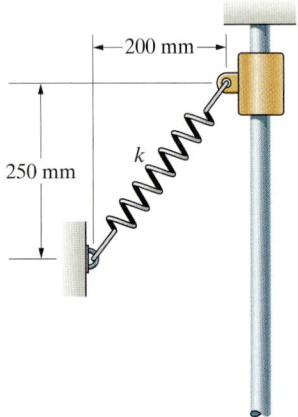

Problems 15.120–122

15.123 The system is released from rest in the position shown. The masses of the collar A and the box B are $m_A = 90$ kg and $m_B = 136$ kg. Determine the maximum velocity attained by the collar A as it rises.

15.124 The system is released from rest in the position shown. The masses of the collar A and the box B are $m_A = 90$ kg and $m_B = 136$ kg. What maximum height relative to its initial position is reached by collar A?

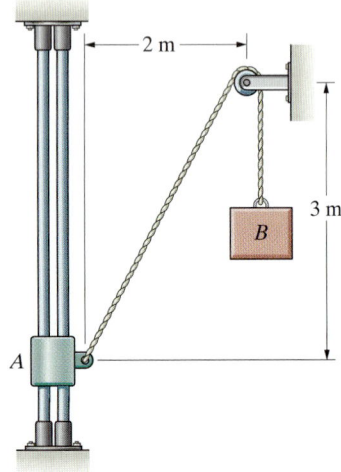

Problems 15.123/15.124

15.125 The spring constant is $k = 2000$ N/m, and the masses $m_A = 14$ kg and $m_B = 18$ kg. The collar A slides on the smooth horizontal bar. The system is released from rest in the position shown with the spring unstretched. (a) Use conservation of energy to determine the velocity of the collar A when it has moved a distance x to the right. Draw a graph of the velocity for $0 \leq x \leq 0.15$ m. (b) Use your graph to estimate the maximum velocity of the collar.

15.126 The spring constant is $k = 2000$ N/m, and the masses $m_A = 14$ kg and $m_B = 18$ kg. The collar A slides on the smooth horizontal bar. The system is released from rest in the position shown with the spring unstretched. Estimate the maximum distance the collar A slides to the right.

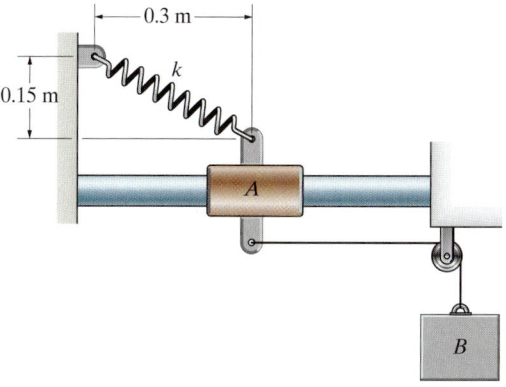

Problems 15.125/15.126

15.127 The 16-kg cylinder is released at the position shown and falls onto the *nonlinear* spring. In terms of the stretch S of the spring, its potential energy is $V = \frac{1}{2}kS^2 + \frac{1}{4}qS^4$, where $k = 2400$ N/m and $q = 3000$ N/m^3. Determine how far down the cylinder moves after contacting the spring.

15.128 The 16-kg cylinder is released at the position shown and falls onto the *nonlinear* spring. In terms of the stretch S of the spring, its potential energy is $V = \frac{1}{2}kS^2 + \frac{1}{4}qS^4$, where $k = 2400$ N/m and $q = 3000$ N/m^3. What is the maximum downward velocity attained by the cylinder?

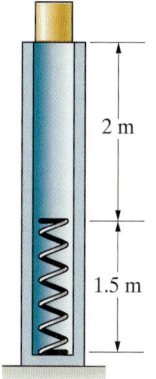

Problems 15.127/15.128

15.129 The system shown in Fig. P15.129 a is released from rest. The mass $m = 1$ kg, the spring constant is $k = 200$ N/m, and the unstretched length of the spring is 0.1 m. (a) Use conservation of energy to determine the magnitude of the velocity of the mass as a function of the angle θ between the string and the vertical (Fig. P15.129b), and draw a graph of the velocity for $0 \leq \theta \leq 40°$. (b) Use your graph to estimate the maximum velocity of the mass.

15.130 The system shown in Fig. a is released from rest. The mass $m = 1$ kg, the spring constant is $k = 200$ N/m, and the unstretched length of the spring is 0.1 m. Estimate the maximum value of θ (Fig. b) reached by the mass.

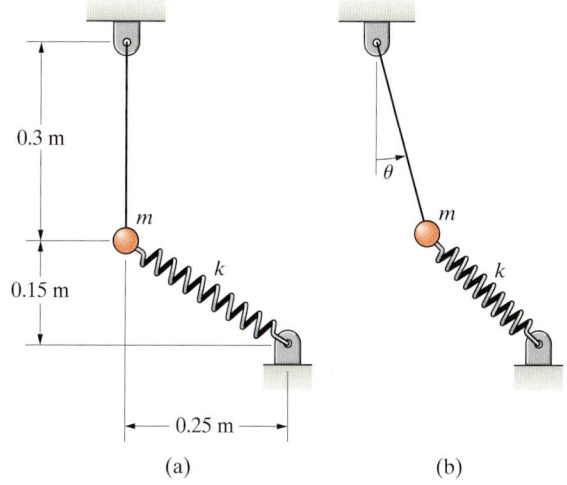

(a)   (b)

Problems 15.129/15.130

15.131 A student runs at 15 ft/s, grabs a rope, and swings out over a lake. Determine the angle θ at which he should release the rope to maximize the horizontal distance b. What is the resulting value of b?

Problem 15.131

CHAPTER SUMMARY

In Chapter 14 we used Newton's second law to determine an object's acceleration when the forces acting on it were known. Once the acceleration was known, it could be integrated to obtain information about the object's velocity and position. In this chapter, we have used Newton's second law to derive a new technique, the principle of work and energy, that relates the work done during a change in an object's position to the change in its kinetic energy. This principle is usually applicable only when the forces that do work are known as functions of the object's position, and the only information obtained is the change in the magnitude of the object's velocity. Nevertheless, the principle can be highly useful because the work done by certain forces, including gravitational forces and forces exerted by springs, is easy to determine and is independent of an object's path. This property motivated the definitions of the potential energy and conservative forces and resulted in the concept of conservation of energy. In Chapter 16 we will use Newton's second law to derive momentum principles, which are especially useful for applications involving collisions between objects.

Principle of Work and Energy

Defining the *work* done on an object as its center of mass moves from a position $\mathbf{r}_1$ to a position $\mathbf{r}_2$ by

$$U_{12} = \int_{\mathbf{r}_1}^{\mathbf{r}_2} \Sigma \mathbf{F} \cdot d\mathbf{r}, \tag{15.5}$$

where $\Sigma \mathbf{F}$ is the sum of the forces acting on the object, *the principle of work and energy* states that the work equals the change in the kinetic energy:

$$U_{12} = \tfrac{1}{2} m v_2^2 - \tfrac{1}{2} m v_1^2. \tag{15.6}$$

The total work done by external forces on a system of objects equals the change in the total kinetic energy of the system if no net work is done by internal forces.

Power

The *power* is the rate at which work is done. The power transferred to an object by the external forces acting on it is

$$P = \Sigma \mathbf{F} \cdot \mathbf{v}. \tag{15.9}$$

The power equals the rate of change of the object's kinetic energy. The average with respect to time of the power during an interval of time from t_1 to t_2 is equal to the change in kinetic energy of the object, or the work done, divided by the interval of time:

$$P_{\text{avg}} = \frac{\tfrac{1}{2} m v_2^2 - \tfrac{1}{2} m v_1^2}{t_2 - t_1} = \frac{U_{12}}{t_2 - t_1}. \tag{15.10}$$

Evaluating the Work

Let s be the position of an object's center of mass along its path. The work done on the object from a position s_1 to a position s_2 is

$$U_{12} = \int_{s_1}^{s_2} \Sigma F_t \, ds, \qquad (15.7)$$

where ΣF_t is the tangential component of the total external force on the object. Components of force perpendicular to the path do no work.

Weight   In terms of a coordinate system with the positive y axis upward, the work done by an object's weight as the center of mass of the object moves from position 1 to position 2 is

$$U_{12} = -mg(y_2 - y_1). \qquad (15.11)$$

The work is the product of the weight and the change in the height of the center of mass of the object. The work is negative if the height increases and positive if it decreases.

When the variation of an object's weight with distance r from the center of the earth is accounted for, the work done by the weight of the object is

$$U_{12} = mgR_E^2\left(\frac{1}{r_2} - \frac{1}{r_1}\right), \qquad (15.13)$$

where R_E is the radius of the earth.

Springs   The work done on an object by a spring attached to a fixed support is

$$U_{12} = -\tfrac{1}{2}k(S_2^2 - S_1^2), \qquad (15.14)$$

where S_1 and S_2 are the values of the stretch at the initial and final positions of the object.

Potential Energy

For a given force $\mathbf{F}$ acting on an object, if a function V of the object's position exists such that

$$dV = -\mathbf{F} \cdot d\mathbf{r}, \qquad (15.16)$$

then $\mathbf{F}$ is said to be *conservative* and V is called the *potential energy* associated with $\mathbf{F}$. The work done by $\mathbf{F}$ from a position 1 to a position 2 is

$$U_{12} = -(V_2 - V_1). \qquad (15.17)$$

If all the forces that do work on an object are conservative, the total energy—the sum of the kinetic energy of the object and the potential energies of the forces that do work on it—is conserved:

$$\tfrac{1}{2}mv_1^2 + V_1 = \tfrac{1}{2}mv_2^2 + V_2. \qquad (15.18)$$

If an object is subjected to both conservative and nonconservative forces, the principle of work and energy can be written

$$\tfrac{1}{2}mv_1^2 + V_1 + U_{12} = \tfrac{1}{2}mv_2^2 + V_2. \tag{15.19}$$

The term U_{12} includes the work done by all nonconservative forces acting on the object. The work done by a conservative force can be calculated and included in U_{12}, *or* the force's potential energy can be included in V.

Weight In terms of a cartesian coordinate system with its y axis pointing upward, the potential energy of the weight of an object is

$$V = mgy. \tag{15.21}$$

The potential energy is the product of the object's weight and the height of its center of mass, measured from any convenient reference level, or *datum*. When the variation of an object's weight with distance r from the center of the earth is accounted for, the potential energy of the weight is

$$V = -\frac{mgR_E^2}{r}, \tag{15.23}$$

where R_E is the radius of the earth.

Springs The potential energy of the force exerted on an object by a linear spring is

$$V = \tfrac{1}{2}kS^2, \tag{15.24}$$

where S is the stretch of the spring.

Relationships between Force and Potential Energy A force $\mathbf{F}$ is related to its associated potential energy by

$$\mathbf{F} = -\left(\frac{\partial V}{\partial x}\mathbf{i} + \frac{\partial V}{\partial y}\mathbf{j} + \frac{\partial V}{\partial z}\mathbf{k}\right) = -\nabla V. \tag{15.28}$$

A force $\mathbf{F}$ is conservative if and only if its *curl* is zero:

$$\nabla \times \mathbf{F} = \begin{vmatrix} \mathbf{i} & \mathbf{j} & \mathbf{k} \\ \dfrac{\partial}{\partial x} & \dfrac{\partial}{\partial y} & \dfrac{\partial}{\partial z} \\ F_x & F_y & F_z \end{vmatrix} = \mathbf{0}.$$

Review Problems

15.132 The driver of a 3000-lb car moving at 40 mi/h applies an increasing force on the brake pedal. The magnitude of the resulting frictional force exerted on the car by the road is $f = 250 + 6s$ lb, where s is the car's horizontal position (in feet) relative to its position when the brakes were applied. Assuming that the car's tires do not slip; determine the distance required for the car to stop (a) by using Newton's second law and (b) by using the principle of work and energy.

15.133 Suppose that the car in Problem 15.132 is on wet pavement and the coefficients of friction between the tires and the road are $\mu_s = 0.4$ and $\mu_k = 0.35$. Determine the distance required for the car to stop.

15.134 An astronaut in a small rocket vehicle (combined mass = 450 kg) is hovering 100 m above the surface of the moon when he discovers that he is nearly out of fuel and can exert the thrust necessary to cause the vehicle to hover for only 5 more seconds. He quickly considers two strategies for getting to the surface: (a) Fall 20 m, turn on the thrust for 5 s, and then fall the rest of the way; (b) fall 40 m, turn on the thrust for 5 s, and then fall the rest of the way. Which strategy gives him the best chance of surviving? How much work is done by the engine's thrust in each case? ($g_{\text{moon}} = 1.62 \text{ m/s}^2$.)

15.135 The coefficients of friction between the 20-kg crate and the inclined surface are $\mu_s = 0.24$ and $\mu_k = 0.22$. If the crate starts from rest and the horizontal force $F = 200$ N, what is the magnitude of the velocity of the crate when it has moved 2 m?

15.136 The coefficients of friction between the 20-kg crate and the inclined surface are $\mu_s = 0.24$ and $\mu_k = 0.22$. If the crate starts from rest and the horizontal force $F = 40$ N, what is the magnitude of the velocity of the create when it has moved 2 m?

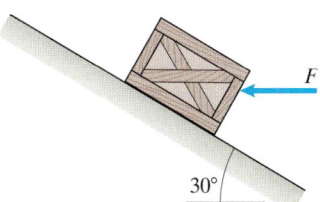

Problems 15.135/15.136

15.137 The Union Pacific Big Boy locomotive weighs 1.19 million lb, and the tractive effort (tangential force) of its drive wheels is 135,000 lb. If you neglect other tangential forces, what distance is required for the train to accelerate from zero to 60 mi/h?

15.138 In Problem 15.137, suppose that the acceleration of the locomotive as it accelerates from zero to 60 mi/h is $(F_0/m)(1 - v/88)$, where $F_0 = 135,000$ lb, m is the mass of the locomotive, and v is its velocity in feet per second.

(a) How much work is done in accelerating the train to 60 mi/h?
(b) Determine the locomotive's velocity as a function of time.

Problems 15.137/15.138

15.139 A car traveling 65 mi/h hits the crash barrier described in Problem 15.14. Determine the maximum deceleration to which the passengers are subjected if the car weighs (a) 2500 lb and (b) 5000 lb.

15.140 In a preliminary design for a mail-sorting machine, parcels moving at 2 ft/s slide down a smooth ramp and are brought to rest by a linear spring. What should the spring constant be if you don't want a 10-lb parcel to be subjected to a maximum deceleration greater than 10 g's?

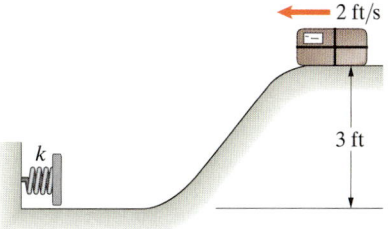

Problem 15.140

15.141 When the 1-kg collar is in position 1, the tension in the spring is 50 N, and the unstretched length of the spring is 260 mm. If the collar is pulled to position 2 and released from rest, what is its velocity when it returns to position 1?

15.142 When the 1-kg collar is in position 1, the tension in the spring is 100 N, and when the collar is in position 2, the tension in the spring is 400 N.

(a) What is the spring constant k?

(b) If the collar is given a velocity of 15 m/s at position 1, what is the magnitude of its velocity just before it reaches position 2?

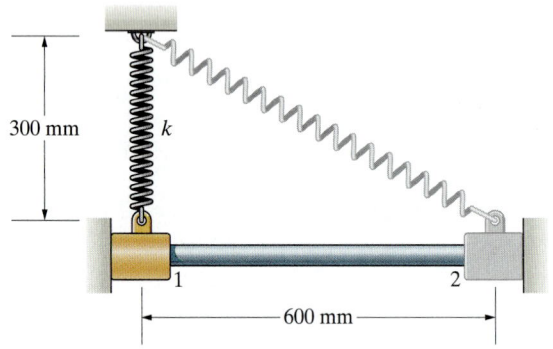

Problems 15.141/15.142

15.143 The 30-lb weight is released from rest with the two springs ($k_A = 30$ lb/ft, $k_B = 15$ lb/ft) unstretched.

(a) How far does the weight fall before rebounding?

(b) What maximum velocity does it attain?

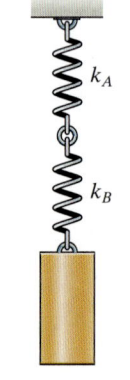

Problem 15.143

15.144 The piston and the load it supports are accelerated upward by the gas in the cylinder. The total weight of the piston and load is 1000 lb. The cylinder wall exerts a constant 50-lb frictional force on the piston as it rises. The net force exerted on the piston by pressure is $(p_2 - p_{atm})A$, where p is the pressure of the gas, $p_{atm} = 2117$ lb/ft^2 is atmospheric pressure, and $A = 1$ ft^2 is the cross-sectional area of the piston. Assume that the product of p and the volume of the cylinder is constant. When $s = 1$ ft, the piston is stationary and $p = 5000$ lb/ft^2. What is the velocity of the piston when $s = 2$ ft?

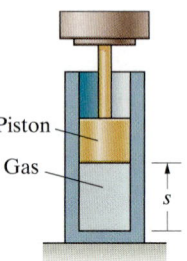

Problem 15.144

15.145 When a 22,000-kg rocket's engine burns out at an altitude of 2 km, the velocity of the rocket is 3 km/s and it is traveling at an angle of 60° relative to the horizontal. Neglect the variation in the gravitational force with altitude.

(a) If you neglect aerodynamic forces, what is the magnitude of the velocity of the rocket when it reaches an altitude of 6 km?

(b) If the actual velocity of the rocket when it reaches an altitude of 6 km is 2.8 km/s, how much work is done by aerodynamic forces as the rocket moves from 2 km to 6 km altitude?

15.146 The 12-kg collar A is at rest in the position shown at $t = 0$ and is subjected to the tangential force $F = 24 - 12t^2$ N for 1.5 s. Neglecting friction, what maximum height h does the collar reach?

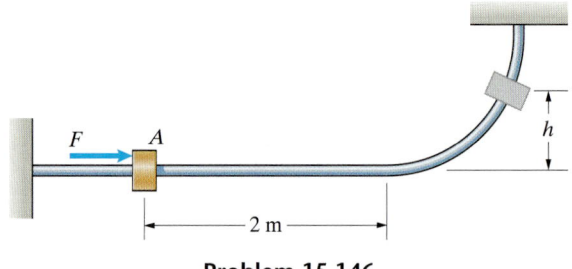

Problem 15.146

15.147 Suppose that, in designing a loop for a roller coaster's track, you establish as a safety criterion that at the top of the loop the normal force exerted on a passenger by the roller coaster should equal 10 percent of the passenger's weight. (That is, the passenger's "effective weight" pressing him down into his seat is 10 percent of his actual weight.) The roller coaster is moving at 62 ft/s when it enters the loop. What is the necessary instantaneous radius of curvature ρ of the track at the top of the loop?

Problem 15.147

15.148 A 180-lb student runs at 15 ft/s, grabs a rope, and swings out over a lake. He releases the rope when his velocity is zero.

(a) What is the angle θ when he releases the rope?
(b) What is the tension in the rope just before he releases it?
(c) What is the maximum tension in the rope?

15.149 If the student in Problem 15.148 releases the rope when $\theta = 25°$, what maximum height does he reach relative to his position when he grabs the rope?

Problems 15.148/15.149

15.150 A boy takes a running start and jumps on his sled at position 1. He leaves the ground at position 2 and lands in deep snow at a distance $b = 25$ ft. How fast was he going at position 1?

15.151 In Problem 15.150, if the boy starts at position 1 going 15 ft/s, what distance b does he travel through the air?

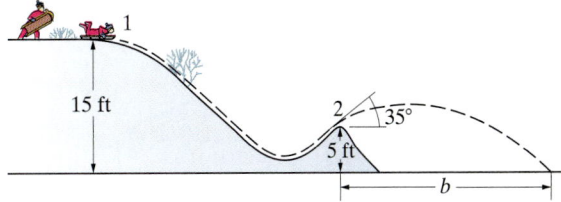

Problems 15.150/15.151

15.152 The 1-kg collar A is attached to the linear spring ($k = 500$ N/m) by a string. The collar starts from rest in the position shown, and the initial tension in the string is 100 N. What distance does the collar slide up the smooth bar?

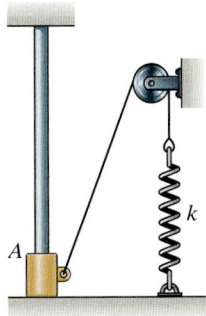

Problem 15.152

15.153 The masses $m_A = 40$ kg and $m_B = 60$ kg. The collar A slides on the smooth horizontal bar. The system is released from rest. Use conservation of energy to determine the velocity of the collar A when it has moved 0.5 m to the right.

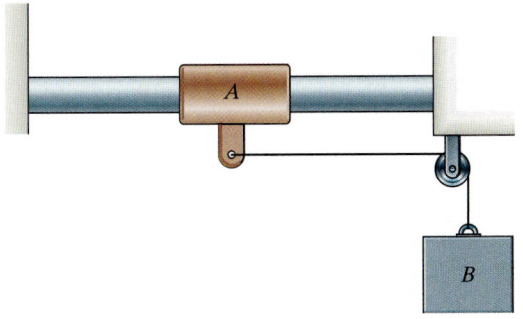

Problem 15.153

15.154 The spring constant is $k = 850$ N/m, $m_A = 40$ kg, and $m_B = 60$ kg. The collar A slides on the smooth horizontal bar. The system is released from rest in the position shown with the spring unstretched. Use conservation of energy to determine the velocity of the collar A when it has moved 0.5 m to the right.

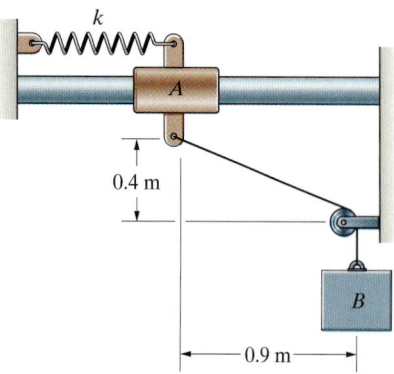

Problem 15.154

15.155 The y axis is vertical and the curved bar is smooth. If the magnitude of the velocity of the 4-lb slider is 6 ft/s at position 1, what is the magnitude of its velocity when it reaches position 2?

15.156 In Problem 15.155, determine the magnitude of the velocity of the slider when it reaches position 2 if it is subjected to the additional force $\mathbf{F} = 3x\mathbf{i} - 2\mathbf{j}$ (lb) during its motion.

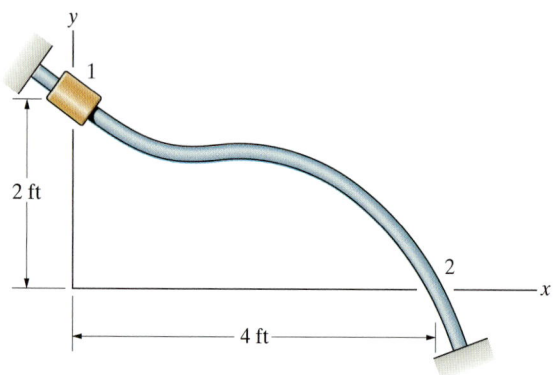

Problems 15.155/15.156

15.157 Suppose that an object of mass m is beneath the surface of the earth. In terms of a polar coordinate system with its origin at the earth's center, the gravitational force on the object is $-(mgr/R_E)\mathbf{e}_r$, where R_E is the radius of the earth. Show that the potential energy associated with the gravitational force is $V = mgr^2/2R_E$.

15.158 It has been pointed out that if tunnels could be drilled straight through the earth between points on the surface, trains could travel between those points using gravitational force for acceleration and deceleration. (The effects of friction and aerodynamic drag could be minimized by evacuating the tunnels and using magnetically levitated trains.) Suppose that such a train travels from the North Pole to a point on the equator. Determine the magnitude of the velocity of the train (a) when it arrives at the equator and (b) when it is halfway from the North Pole to the equator. The radius of the earth is $R_E = 3960$ mi. (See Problem 15.157.)

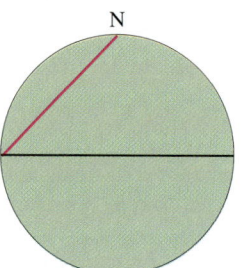

Problem 15.158

15.159 In Problem 15.137, what is the maximum power transferred to the locomotive during its acceleration?

15.160 Just before it lifts off, a 10,500-kg airplane is traveling at 60 ft/s. The total horizontal force exerted by the plane's engines is 189 kN, and the plane is accelerating at 15 m/s^2.

(a) How much power is being transferred to the plane by its engines?
(b) What is the total power being transferred to the plane?

Problem 15.160

15.161 The "Paris Gun" used by Germany in World War I had a range of 120 km, a 37.5-m barrel, and a muzzle velocity of 1550 m/s and fired a 120-kg shell.

(a) If you assume the shell's acceleration to be constant, what maximum power was transferred to the shell as it traveled along the barrel?
(b) What average power was transferred to the shell?

Problem 15.161

Design Project

Determine the specifications (unstretched length and spring constant k) for the elastic cord to be used at a bungee-jumping facility. Participants are to jump from a platform 150 ft above the ground. When they rebound, they must avoid an obstacle that extends 15 ft below the point at which they jumped. In determining the specifications for the cord, establish reasonable safety limits for the minimum distances by which participants must avoid the ground and obstacle. Account for the fact that participants will have different weights. If necessary, specify a maximum allowable weight for participants. Write a brief report presenting your analyses and making a design recommendation for the specifications of the cord.

Momentum Methods

Integrating Newton's second law with respect to time yields a relation between the time integral of the forces acting on an object and the change in the object's linear momentum. With this result, called the principle of impulse and momentum, we can determine the change in an object's velocity when the external forces are known as functions of time, analyze impacts between objects, and evaluate forces exerted by continuous flows of mass.

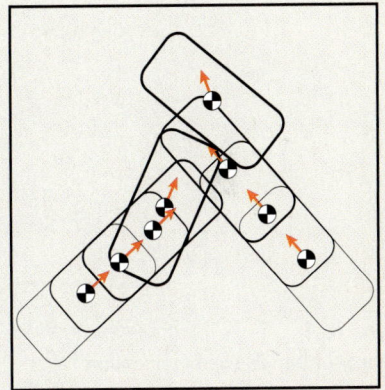

◀ The total linear momentum of the test vehicles is approximately the same before and after their collision. In this chapter, we use methods based on linear and angular momentum to analyze motions of objects.

16.1 Principle of Impulse and Momentum

The principle of work and energy is a very useful tool in mechanics. We can derive another useful tool for the analysis of motion by integrating Newton's second law with respect to time. We express Newton's second law in the form

$$\Sigma \mathbf{F} = m\frac{d\mathbf{v}}{dt}.$$

Then we integrate with respect to time to obtain

$$\int_{t_1}^{t_2} \Sigma \mathbf{F}\, dt = m\mathbf{v}_2 - m\mathbf{v}_1, \tag{16.1}$$

where $\mathbf{v}_1$ and $\mathbf{v}_2$ are the velocities of the center of mass of the object at the times t_1 and t_2. The term on the left is called the *linear impulse*, and $m\mathbf{v}$ is the *linear momentum*. Equation (16.1) is called the *principle of impulse and momentum*: The impulse applied to an object during an interval of time is equal to the change in the object's linear momentum (Fig. 16.1). The dimensions of the linear impulse and linear momentum are (mass) $\times$ (length)/(time).

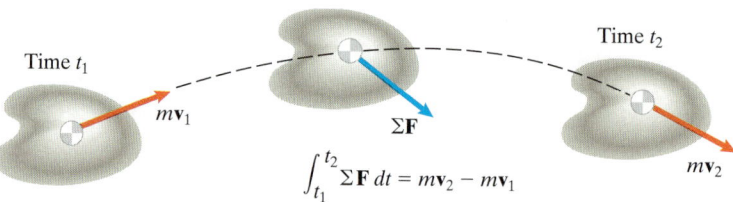

Figure 16.1
Principle of impulse and momentum.

The average with respect to time of the total force acting on an object from t_1 to t_2 is

$$\Sigma \mathbf{F}_{av} = \frac{1}{t_2 - t_1}\int_{t_1}^{t_2} \Sigma \mathbf{F}\, dt,$$

so we can write Eq. (16.1) as

$$(t_2 - t_1)\Sigma \mathbf{F}_{av} = m\mathbf{v}_2 - m\mathbf{v}_1. \tag{16.2}$$

With this equation, we can determine the average value of the total force acting on an object during a given interval of time if we know the change in the object's velocity.

A force that acts over a small interval of time but exerts a significant linear impulse is called an *impulsive force*. An impulsive force and its average with respect to time are shown in Fig. 16.2. Determining the time history of such a force is often impractical, but with Eq. (16.2) its average value can sometimes be determined. For example, a golf ball struck by a club is subjected to an impulsive force. By making high-speed motion pictures, the duration of the impact and the ball's velocity after the impact can be measured. Knowing the duration of the impact and the ball's change in linear momentum, we can determine the average force exerted by the club. (See Example 16.3.)

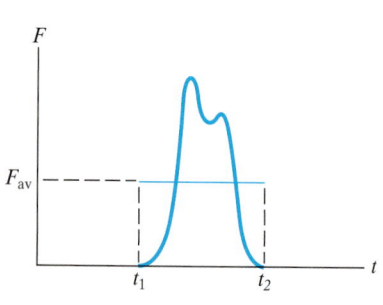

Figure 16.2
An impulsive force and its average value.

We can express Eqs. (16.1) and (16.2) in scalar forms that are often useful. The sum of the forces in the direction tangent to an object's path equals the product of the object's mass and the rate of change of its velocity along the path (see Eq. 14.7):

$$\Sigma F_t = ma_t = m\frac{dv}{dt}.$$

Integrating this equation with respect to time, we obtain

$$\int_{t_1}^{t_2} \Sigma F_t \, dt = mv_2 - mv_1, \tag{16.3}$$

where v_1 and v_2 are the velocities along the path at the times t_1 and t_2. The impulse applied to an object by the sum of the forces tangent to its path during an interval of time is equal to the change in the object's linear momentum along the path. In terms of the average with respect to time of the sum of the forces tangent to the path, or

$$\Sigma F_{t\,\text{av}} = \frac{1}{t_2 - t_1}\int_{t_1}^{t_2} \Sigma F_t \, dt,$$

we can write Eq. (16.3) as

$$(t_2 - t_1)\Sigma F_{t\,\text{av}} = mv_2 - mv_1. \tag{16.4}$$

This equation relates the average of the sum of the forces tangent to the path during an interval of time to the change in the velocity along the path.

Notice that Eq. (16.1) and the principle of work and energy, Eq. (15.6), are quite similar. They both relate an integral of the external forces to the change in an object's velocity. Equation (16.1) is a vector equation that determines the change in both the magnitude and direction of the velocity, whereas the principle of work and energy, a scalar equation, gives only the change in the magnitude of the velocity. But there is a greater difference between the two methods: In the case of impulse and momentum, there is no class of forces equivalent to the conservative forces that make the principle of work and energy so easy to apply.

When the external forces acting on an object are known as functions of time, the principle of impulse and momentum can be applied to determine the change in velocity of the object during an interval of time. Although this is an important result, it is not new. In Chapter 14, when we used Newton's second law to determine an object's acceleration and then integrated the acceleration with respect to time to determine the object's velocity, we were effectively applying the principle of impulse and momentum. However, in the rest of this chapter we show that this principle can be extended to new and interesting applications.

Study Questions

1. What is the definition of the linear momentum?

2. What is the principle of impulse and momentum?

3. If you know the change in the velocity of an object during an interval of time, how can you determine the average of the total force exerted on the object during that interval?

4. How does the principle of impulse and momentum differ from the principle of work and energy?

Example 16.1 Applying Impulse and Momentum

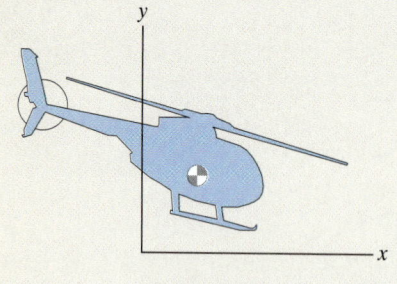

Figure 16.3

A 1200-kg helicopter starts from rest at $t = 0$ (Fig. 16.3).
(a) The components of the total force on the helicopter from $t = 0$ to $t = 10$ s are given by

$$\Sigma F_x = 720t \text{ N},$$

$$\Sigma F_y = 2160 - 360t \text{ N},$$

$$\Sigma F_z = 0.$$

Determine the helicopter's velocity at $t = 10$ s.
(b) At $t = 20$ s, the helicopter's velocity is $36\mathbf{i} + 8\mathbf{j}$ (m/s). What is the average of the total force acting on the craft from $t = 10$ s to $t = 20$ s?

Strategy

(a) Because we know the helicopter's velocity at $t = 0$ and the components of the total force acting on the helicopter as functions of time, we can use the principle of impulse and momentum, Eq. (16.1), to determine the velocity at $t = 10$ s. (b) Knowing the velocity at $t = 10$ s and at $t = 20$ s, we can determine the average of the total force from Eq. (16.2).

Solution

(a) Applying the principle of impulse and momentum from $t = 0$ to $t = 10$ s, we obtain

$$\int_{t_1}^{t_2} \Sigma \mathbf{F} \, dt = m\mathbf{v}_2 - m\mathbf{v}_1:$$

$$\int_0^{10} [720t\mathbf{i} + (2160 - 360t)\mathbf{j}] \, dt = (1200)\mathbf{v}_2 - (1200)(0),$$

$$36{,}000\mathbf{i} + 3600\mathbf{j} = 1200\mathbf{v}_2.$$

We find that the velocity at $t = 10$ s is $30\mathbf{i} + 3\mathbf{j}$ (m/s).
(b) To determine the average total force from $t = 10$ s to $t = 20$ s, we apply Eq. (16.2) to that interval of time:

$$(t_2 - t_1)\Sigma \mathbf{F}_{av} = m\mathbf{v}_2 - m\mathbf{v}_1:$$

$$(20 - 10)\Sigma \mathbf{F}_{av} = 1200(36\mathbf{i} + 8\mathbf{j}) - 1200(30\mathbf{i} + 3\mathbf{j}),$$

$$\Sigma \mathbf{F}_{av} = 720\mathbf{i} + 600\mathbf{j} \text{ (N)}.$$

Critical Thinking

Notice that even though we did not know the forces acting on the helicopter during the interval of time from $t = 10$ s to $t = 20$ s, we were able to determine the average force because we knew the mass of the helicopter and its velocity at the beginning and end of the interval of time.

Example 16.2 Impulse and Momentum Tangent to the Path

The motorcycle in Fig. 16.4 starts from rest at $t = 0$ and travels along a circular track with a 400-m radius. The tangential component of the total force on the motorcycle from $t = 0$ to $t = 30$ s is $\Sigma F_t = 200 - 6t$ N. The combined mass of the motorcycle and rider is 150 kg.
(a) What is the magnitude of the velocity at $t = 30$ s?
(b) What is the average of the tangential component of the total force from $t = 0$ to $t = 30$ s?

Strategy
(a) Because the tangential component of the total force is given as a function of time, we can use Eq. (16.3) to determine the magnitude of the velocity as a function of time.
(b) We can simply calculate the average of the tangential component of the total force, or we can use Eq. (16.4).

Figure 16.4

Solution
(a) We use Eq. (16.3) to determine the magnitude v of the velocity as a function of time.

$$\int_{t_1}^{t_2} \Sigma F_t \, dt = mv_2 - mv_1:$$

$$\int_0^t (200 - 6t) \, dt = 150v - 0.$$

Evaluating the integral, we obtain

$$v = \frac{1}{150}(200t - 3t^2) \text{ m/s.}$$

At $t = 30$ s, the magnitude of the velocity is $v = 22$ m/s.
(b) We use Eq. (16.4) to determine the average of the tangential component of the total force from $t = 0$ to $t = 30$ s.

$$(t_2 - t_1)\Sigma F_{t\,av} = mv_2 - mv_1:$$

$$(30 \text{ s} - 0)\Sigma F_{t\,av} = (150 \text{ kg})(22 \text{ m/s}) - 0.$$

The result is $\Sigma F_{t\,av} = 110$ N.

Critical Thinking
By knowing the tangential component of the force on the motorcycle as a function of time, we were able to determine the magnitude of its velocity as a function of time. Because the motorcycle is on a circular track, you know that it has a component of acceleration normal to its path, which means it is subjected to a component of force normal to its path. But as this example emphasizes, a force normal to the path of an object changes the direction of its motion, but has no effect on the magnitude of its velocity.

| **Example 16.3** | **Determining an Impulsive Force** |

A golf ball in flight is photographed at intervals of 0.001 s (Fig. 16.5). The 1.62-oz ball is 1.68 in in diameter. If the club was in contact with the ball for 0.0006 s, estimate the average value of the impulsive force exerted by the club.

Figure 16.5

Strategy

By measuring the distance traveled by the ball in one of the 0.001-s intervals, we can estimate its velocity after being struck and then use Eq. (16.2) to determine the average total force on the ball.

Solution

By comparing the distance moved during one of the 0.001-s intervals with the known diameter of the ball, we estimate that the ball traveled 1.9 in and that its direction is 21° above the horizontal (Fig. a). The magnitude of the ball's velocity is

$$\frac{(1.9/12)\text{ ft}}{0.001\text{ s}} = 158\text{ ft/s}.$$

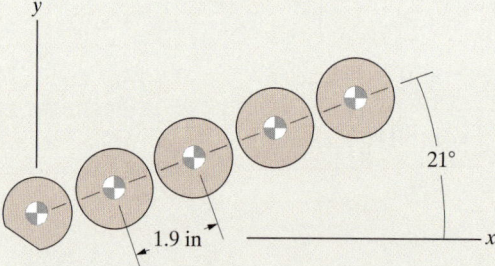

(a) Estimating the distance traveled during one 0.001-s interval.

The weight of the ball is $1.62/16 = 0.101$ lb, so its mass is $0.101/32.2 = 3.14 \times 10^{-3}$ slug. From Eq. (16.2), we obtain

$$(t_2 - t_1)\Sigma\mathbf{F}_{\text{av}} = m\mathbf{v}_2 - m\mathbf{v}_1:$$

$$(0.0006\text{ s})\Sigma\mathbf{F}_{\text{av}} =$$
$$(3.14 \times 10^{-3}\text{ slug})(158\text{ ft/s})(\cos 21°\mathbf{i} + \sin 21°\mathbf{j}) - 0,$$

which yields

$$\Sigma \mathbf{F}_{av} = 775\mathbf{i} + 297\mathbf{j} \text{ (lb)}.$$

Critical Thinking

The average force during the time the club is in contact with the ball includes both the impulsive force exerted by the club and the ball's weight. In comparison with the large average impulsive force exerted by the club, the weight $(-0.101\mathbf{j}$ lb) is negligible.

Determining the time history of the force exerted on the ball by the club would require a complicated analysis accounting for the ball's deformation during the impact. In contrast, we were able to determine the *average* force exerted on the ball by a straightforward application of impulse and momentum.

Problems

16.1 The 100-lb crate is released from rest on the smooth inclined surface. Use the principle of impulse and momentum to determine the magnitude of its velocity 1 s after it is released.

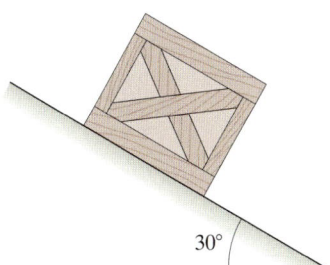

30°

Problem 16.1

16.2 The Sikorsky UH-60A helicopter weighs 20,500 lb. It takes off vertically with its rotor exerting a constant thrust of 24,000 lb. Use the principle of impulse and momentum to determine the magnitude of its velocity 6 s after it takes off.

Problem 16.2

16.3 The 8-kg object is initially stationary on the horizontal surface. The coefficient of kinetic friction between the object and the surface is $\mu_k = 0.1$. At $t = 0$, it is subjected to a constant force $F = 10$ N. Use the principle of impulse and momentum to determine the magnitude of its velocity at $t = 3$ s.

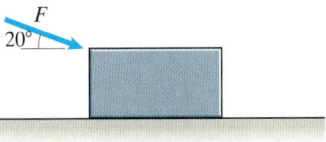

F
20°

Problem 16.3

16.4 The 21,900-kg Gloster Saro Protector is designed for rapid response to airport emergencies. The vehicle starts from rest at $t = 0$ and the total horizontal force (in newtons) exerted on it is given as a function of time by $F = 12,000 + 1000t$.

(a) What impulse is applied to the vehicle during the first 4 s of its motion?

(b) Use the principle of impulse and momentum to determine the magnitude of its velocity at $t = 4$ s.

Problem 16.4

16.5 The combined mass of the motorcycle and rider is 136 kg. The coefficient of kinetic friction between the motorcycle's tires and the road is $\mu_k = 0.6$. The rider starts from rest and spins the rear (drive) wheel. The normal force between the rear wheel and the road is 790 N.

(a) What impulse does the friction force on the rear wheel exert in 2 s?

(b) If you neglect other horizontal forces, what velocity is attained by the motorcycle in 2 s?

Problem 16.5

16.6 The snow petrel begins its takeoff at $t = 0$. The total horizontal force (in newtons) on the 0.15-kg bird as it takes off is given as a function of time by $F = 0.65[1 - \cos(24t)]$. Use the principle of impulse and momentum to determine the bird's velocity when it lifts off at $t = 1.42$ s.

Problem 16.6

16.7 An astronaut drifts toward a space station at 8 m/s. He carries a maneuvering unit (a small hydrogen peroxide rocket) that can exert a total impulse of 720 N-s. The total mass of the astronaut, his suit, and the maneuvering unit is 120 kg. If he uses all of the impulse to slow himself down, what will be the velocity of the astronaut relative to the station?

Problem 16.7

16.8 The total external force on the 10-kg object is given as a function of time by

$$\Sigma \mathbf{F} = (-20t + 90)\mathbf{i} - 60\mathbf{j} + (10t + 40)\mathbf{k} \ (\text{N}).$$

At time $t = 0$, its velocity is $\mathbf{v} = -14\mathbf{i} + 26\mathbf{j} + 32\mathbf{k}$ (m/s). Use the principle of impulse and momentum to determine its velocity at $t = 4$ s.

16.9 What is the average total force on the object in Problem 16.8 during the interval of time from $t = 0$ to $t = 4$ s?

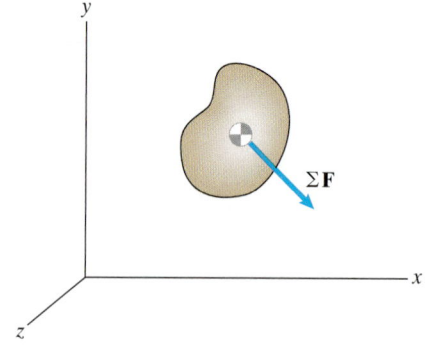

Problems 16.8/16.9

16.10 The 1-lb collar A is initially at rest in the position shown on the smooth horizontal bar. At $t = 0$, a force $\mathbf{F} = \frac{1}{20}t^2\mathbf{i} + \frac{1}{10}t\mathbf{j} - \frac{1}{30}t^3\mathbf{k}$ (lb) is applied to the collar, causing it to slide along the bar. What is the velocity of the collar at $t = 2$ s?

16.11 (a) In Problem 16.10, use the principle of impulse and momentum to determine the collar's velocity as a function of time. (b) Use the result of part (a) to determine the time at which the collar reaches the right end of the bar.

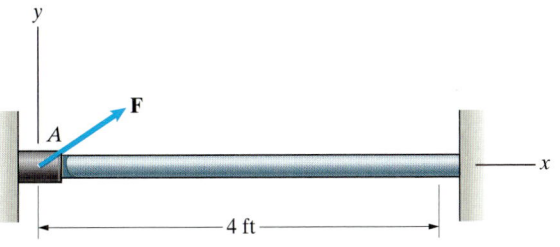

Problems 16.10/16.11

16.12 During the first 5 s of the 14,200-kg airplane's takeoff roll, the pilot increases the engine's thrust at a constant rate from 22 kN to its full thrust of 112 kN.

(a) What impulse does the thrust exert on the airplane during the 5 s?
(b) If you neglect other forces, what total time is required for the airplane to reach its takeoff speed of 46 m/s?

Problem 16.12

16.13 The 10-kg box starts from rest and is subjected to the horizontal force described by the graph. If friction is negligible, what is the magnitude of the box's velocity at $t = 8$ s?

16.14 Suppose that the coefficients of static and kinetic friction between the box in Problem 16.13 and the floor are $\mu_s = \mu_k = 0.2$. What maximum velocity does the box attain?

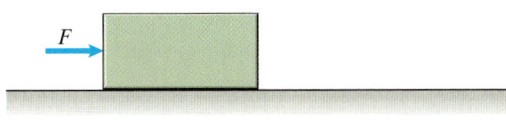

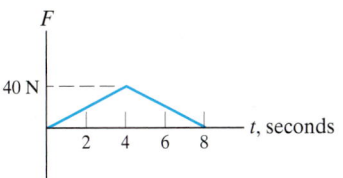

Problems 16.13/16.14

16.15 The crate has a mass of 120 kg, and the coefficients of friction between it and the sloping dock are $\mu_s = 0.6$ and $\mu_k = 0.5$. The crate starts from rest, and the winch exerts a tension $T = 1220$ N.
(a) What impulse is applied to the crate during the first second of motion?
(b) What is the crate's velocity after 1 s?

16.16 Solve Problem 16.15 if the crate starts from rest at $t = 0$ and the winch exerts a tension $T = 1220 + 200t$ N.

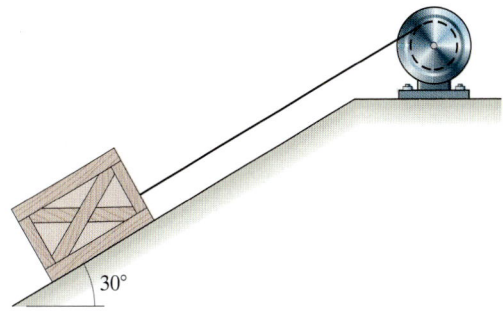

Problems 16.15/16.16

16.17 In an assembly-line process, the 20-kg package A starts from rest and slides down the smooth ramp. Suppose that you want to design the hydraulic device B to exert a constant force of magnitude F on the package and bring it to rest in 0.15 s. What is the required force F?

16.18 The 20-kg package A starts from rest and slides down the smooth ramp. If the hydraulic device B exerts a force of magnitude $F = 540(1 + 0.4t^2)$ N on the package, where t is in seconds measured from the time of first contact, what time is required to bring the package to rest?

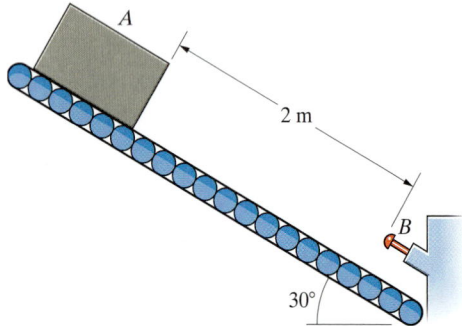

Problems 16.17/16.18

16.19 In a cathode-ray tube, an electron (mass $= 9.11 \times 10^{-31}$ kg) is projected at O with velocity $\mathbf{v} = (2.2 \times 10^7)\mathbf{i}$ (m/s). While the electron is between the charged plates, the electric field generated by the plates subjects it to a force $\mathbf{F} = -eE\mathbf{j}$. The charge on the electron is $e = 1.6 \times 10^{-19}$ C (coulombs), and the electric field strength is $E = 15 \sin(\omega t)$ kN/C, where the frequency $\omega = 2 \times 10^9$ s^{-1}.

(a) What impulse does the electric field exert on the electron while it is between the plates?
(b) What is the velocity of the electron as it leaves the region between the plates?

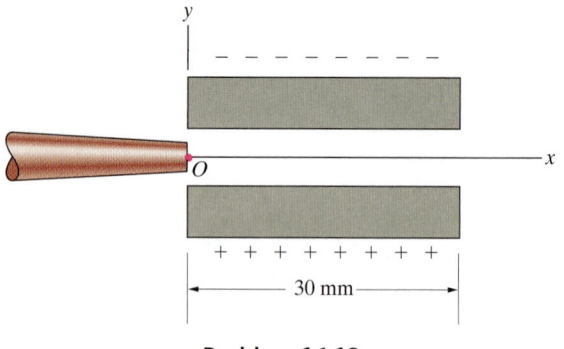

Problem 16.19

16.20 The two weights are released from rest at time $t = 0$. The coefficient of kinetic friction between the horizontal surface and the 5-lb weight is $\mu_k = 0.4$. Use the principle of impulse and momentum to determine the magnitude of the velocity of the 10-lb weight at $t = 1$ s.

Strategy: Apply the principle to each weight individually.

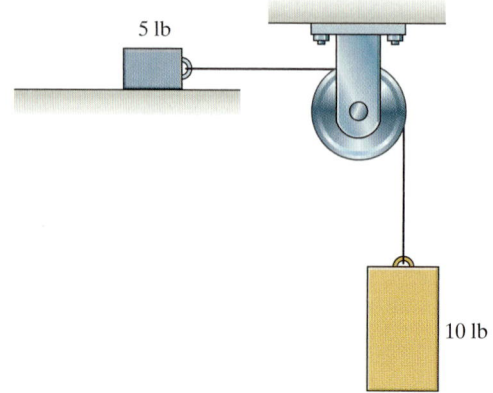

Problem 16.20

16.21 The two crates are released from rest. Their masses are $m_A = 40$ kg and $m_B = 30$ kg, and the coefficient of kinetic friction between crate A and the inclined surface is $\mu_k = 0.15$. What is the magnitude of the velocity of the crates after 1 s?

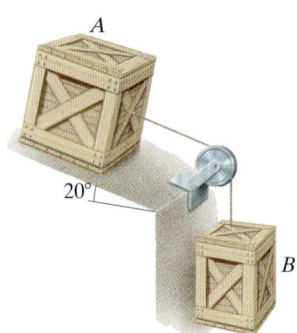

Problem 16.21

16.22 The two crates are released from rest. Their masses are $m_A = 20$ kg and $m_B = 80$ kg, and the surfaces are smooth. The angle $\theta = 20°$. What is the magnitude of the velocity of crate A after 1 s?

Strategy: Apply the principle of impulse and momentum to each crate individually.

16.23 The two crates are released from rest. Their masses are $m_A = 20$ kg and $m_B = 80$ kg. The coefficient of kinetic friction between the contacting surfaces is $\mu_k = 0.1$. The angle $\theta = 20°$. What is the magnitude of the velocity of crate A after 1 s?

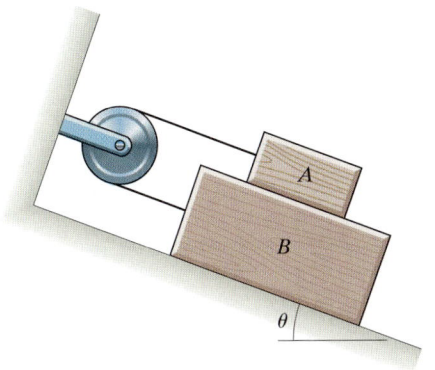

Problems 16.22/16.23

16.24 At $t = 0$, a 20-kg projectile is given an initial velocity $v_0 = 20$ m/s at $\theta_0 = 60°$ above the horizontal.

(a) By using Newton's second law to determine the acceleration of the projectile, determine its velocity at $t = 3$ s.
(b) What impulse is applied to the projectile by its weight from $t = 0$ to $t = 3$ s?
(c) Use the principle of impulse and momentum to determine the projectile's velocity at $t = 3$ s.

16.25 The 20-kg projectile is accelerated from rest to its initial velocity $v_0 = 20$ m/s by an impulsive force with a duration of 0.0006 s. What was the average magnitude of the impulsive force?

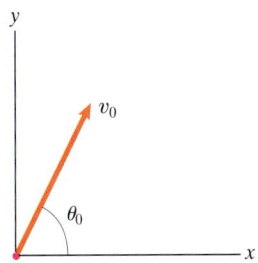

Problems 16.24/16.25

16.26 An object of mass $m = 2$ kg slides with constant velocity $v_0 = 4$ m/s on a horizontal table (seen from above in the figure). The object is connected by a string of length $L = 1$ m to the fixed point O and is in the position shown, with the string parallel to the x axis, at $t = 0$.

(a) Determine the x and y components of the force exerted on the mass by the string as functions of time.
(b) Use your results from part (a) and the principle of impulse and momentum to determine the velocity vector of the mass at $t = 1$ s.

Strategy: To do part (a), write Newton's second law for the mass in terms of polar coordinates.

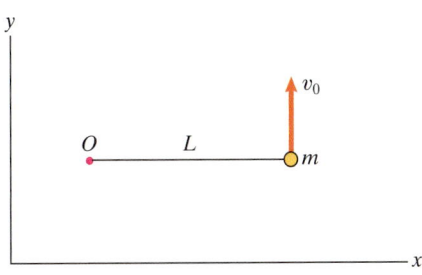

Problem 16.26

16.27 A rail gun, which uses an electromagnetic field to accelerate an object, accelerates a 30-g projectile from zero to 5 km/s in 0.0004 s. What is the magnitude of the average force exerted on the projectile?

16.28 The powerboat is traveling at 14 m/s when its motor is turned off. In 5 s, its speed decreases to 3 m/s. The mass of the boat and its passenger is 420 kg. What is the magnitude of the average horizontal force exerted on the boat by hydrodynamic and aerodynamic drag during the 5 s?

Problem 16.28

16.29 The motorcycle starts from rest at $t = 0$ and travels along a circular track with 300-m radius. From $t = 0$ to $t = 10$ s, the component of the total force on the motorcycle tangential to its path is $\Sigma F_t = 600$ N. The combined mass of the motorcycle and rider is 150 kg. Use the principle of impulse and momentum to determine the magnitude of the motorcycle's velocity at $t = 10$ s.

16.30 The motorcycle starts from rest at $t = 0$ and travels along a circular track with 300-m radius. From $t = 0$ to $t = 10$ s, the component of the total tangential force on the motorcycle is given as a function of time by $\Sigma F_t = 460 + 3t^2$ N. The combined mass of the motorcycle and rider is 150 kg. Use the principle of impulse and momentum to determine the magnitude of the motorcycle's velocity at $t = 10$ s.

Problems 16.29/16.30

16.31 The flight-path angle θ is the angle between the horizontal and the airplane's path. Suppose that the airplane climbs at a constant angle $\theta = 30°$ from $t = 0$ to $t = 6$ s. During this period, the airplane's thrust and aerodynamic drag are balanced, so that the only force exerted on the airplane in the direction tangent to its path is due to its weight. The magnitude of the airplane's velocity at $t = 0$ is 120 m/s. Use the principle of impulse and momentum to determine the magnitude of the airplane's velocity at $t = 6$ s.

16.32 The angle θ between the horizontal and the airplane's path varies from $\theta = 0$ to $\theta = 30°$ at a constant rate of 5 degrees per second. During this maneuver, the airplane's thrust and aerodynamic drag are balanced, so that the only force exerted on the airplane in the direction tangent to its path is due to its weight. The magnitude of the airplane's velocity when $\theta = 0$ is 120 m/s. Use the principle of impulse and momentum to determine the magnitude of the velocity when $\theta = 30°$.

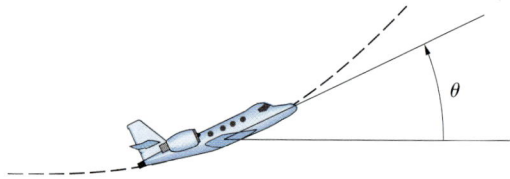

Problems 16.31/16.32

16.33 The 77-kg skier is traveling at 10 m/s at position 1, from which he goes to position 2 in 0.7 s.

(a) Neglecting friction and aerodynamic drag, what is the time average of the tangential component of force exerted on the skier as he moves from position 1 to position 2?

(b) If his actual velocity is measured at position 2 and determined to be 13.1 m/s, what is the time average of the tangential component of force exerted on the skier as he moves from position 1 to position 2?

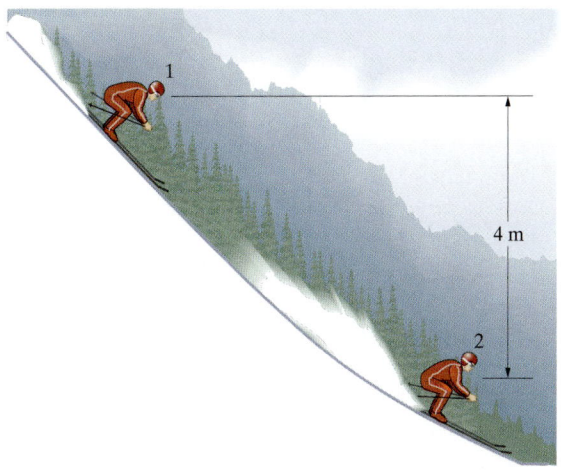

Problem 16.33

16.34 In a test of an energy-absorbing bumper, a 1270-kg car is driven into a barrier at 8 km/h. The duration of the impact is 0.4 s, and the car bounces back from the barrier at 1.6 km/h.

(a) What is the magnitude of the average horizontal force exerted on the car during the impact?

(b) What is the average deceleration of the car during the impact?

Problem 16.34

16.35 A bioengineer, using an instrumented dummy to test a protective mask for a hockey goalie, launches the 170-g puck so that it strikes the mask moving horizontally at 40 m/s. From photographs of the impact, she estimates its duration to be 0.02 s and observes that the puck rebounds at 5 m/s.

(a) What linear impulse does the puck exert?

(b) What is the average value of the impulsive force exerted on the mask by the puck?

Problem 16.35

16.36 A fragile object dropped onto a hard surface breaks because it is subjected to a large impulsive force. If you drop a 2-oz watch from 4 ft above the floor, the duration of the impact is 0.001 s, and the watch bounces 2 in above the floor, what is the average value of the impulsive force?

16.37 The 0.45-kg soccer ball is given a kick with a 0.12-s duration that accelerates it from rest to a velocity of 12 m/s at 60° above the horizontal.

(a) What is the magnitude of the average total force exerted on the ball during the kick?

(b) What is the magnitude of the average force exerted on the ball by the player's foot during the kick?

   Strategy: Use Eq. (16.2) to determine the average total force on the ball. To determine the average force exerted by the player's foot, you must subtract the ball's weight from the average total force.

Problem 16.37

16.38 An entomologist measures the motion of a 3-g locust during its jump and determines that the insect accelerates from rest to 3.4 m/s in 25 ms (milliseconds). The angle of takeoff is 55° above the horizontal. What are the horizontal and vertical components of the average impulsive force exerted by the locust's hind legs during the jump?

16.39 A 5-oz baseball is 3 ft above the ground when it is struck by a bat. The horizontal distance to the point where the ball strikes the ground is 180 ft. Photographic studies indicate that the ball was moving approximately horizontally at 100 ft/s before it was struck, the duration of the impact was 0.015 s, and the ball was traveling at 30° above the horizontal after it was struck. What was the magnitude of the average impulsive force exerted on the ball by the bat?

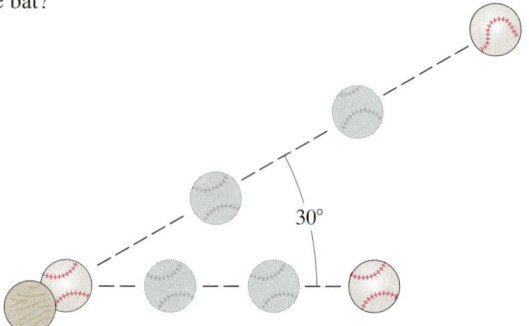

Problem 16.39

16.40 A 1-kg ball is given a horizontal velocity of 1.2 m/s at *A*. Photographic measurements indicate that $b = 1.2$ m, $h = 1.3$ m, and the duration of the bounce at *B* was 0.1 s. What are the components of the average impulsive force exerted on the ball by the floor at *B*?

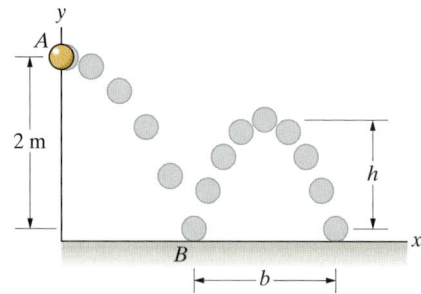

Problem 16.40

16.41 At time $t = 0$, the two masses are released from rest on the smooth surface with the spring stretched. Show that at any later time t, the velocities of the masses are related by
$m_A v_A + m_B v_B = 0$.

   Strategy: Write the principle of impulse and momentum for each mass.

16.42 The masses $m_A = 40$ kg and $m_B = 30$ kg, and $k = 400$ N/m. The two masses are released from rest on the smooth surface with the spring stretched 1 m. What are the magnitudes of the velocities of the masses when the spring is unstretched?

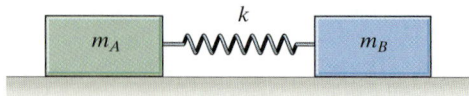

Problems 16.41/16.42

16.2 Conservation of Linear Momentum

In this section, we consider the motions of several objects and show that if the effects of external forces can be neglected, the total linear momentum of the objects is conserved. (By *external forces*, we mean forces that are not exerted by the objects under consideration.) This result provides a powerful tool for analyzing interactions between objects, such as collisions, and also permits us to determine forces exerted on objects as a result of gaining or losing mass.

Consider the objects A and B in Fig. 16.6. $\mathbf{F}_{AB}$ is the force exerted on A by B, and $\mathbf{F}_{BA}$ is the force exerted on B by A. These forces could result from the two objects being in contact, for example, or could be exerted by a spring connecting them. As a consequence of Newton's third law, the two forces are equal and opposite, so that

$$\mathbf{F}_{AB} + \mathbf{F}_{BA} = \mathbf{0}. \tag{16.5}$$

Suppose that no external forces act on A and B or that the external forces are negligible in comparison with the forces that A and B exert on each other. Then we can apply the principle of impulse and momentum to each object for arbitrary times t_1 and t_2:

$$\int_{t_1}^{t_2} \mathbf{F}_{AB} \, dt = m_A \mathbf{v}_{A2} - m_A \mathbf{v}_{A1},$$

$$\int_{t_1}^{t_2} \mathbf{F}_{BA} \, dt = m_B \mathbf{v}_{B2} - m_B \mathbf{v}_{B1}.$$

If we sum these equations, the terms on the left cancel, and we obtain

$$m_A \mathbf{v}_{A1} + m_B \mathbf{v}_{B1} = m_A \mathbf{v}_{A2} + m_B \mathbf{v}_{B2},$$

which means that the *total linear momentum of A and B is conserved*:

$$m_A \mathbf{v}_A + m_B \mathbf{v}_B = \text{constant}. \tag{16.6}$$

We can show that the velocity of the combined center of mass of the objects A and B (that is, the center of mass of A and B regarded as a single object) is also constant. Let $\mathbf{r}_A$ and $\mathbf{r}_B$ be the position vectors of their individual centers of mass (Fig. 16.7). The position of the combined center of mass is

$$\mathbf{r} = \frac{m_A \mathbf{r}_A + m_B \mathbf{r}_B}{m_A + m_B}.$$

By taking the derivative of this equation with respect to time and using Eq. (16.6), we obtain

$$\mathbf{v} = \frac{m_A \mathbf{v}_A + m_B \mathbf{v}_B}{m_A + m_B} = \text{constant}, \tag{16.7}$$

where $\mathbf{v} = d\mathbf{r}/dt$ is the velocity of the combined center of mass. Although the goal will usually be to determine the individual motions of the objects, knowing that the velocity of the combined center of mass is constant can contribute to our understanding of a problem, and in some instances the motion of the combined center of mass may be the only information that can be obtained.

Even when significant external forces act on A and B, if the external forces are negligible in a particular direction, Eqs. (16.6) and (16.7) apply in that direction. These equations also apply to an arbitrary number of objects: If the external forces acting on any collection of objects are negligible, the total linear momentum of the objects is conserved, and the velocity of their center of mass is constant.

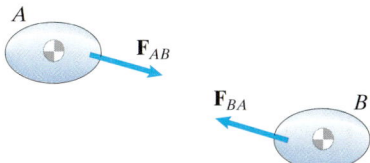

Figure 16.6
Two objects and the forces they exert on each other.

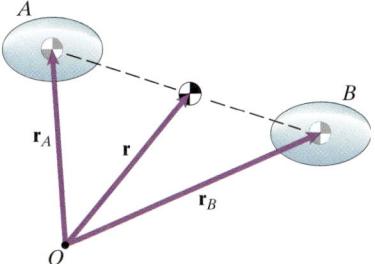

Figure 16.7
Position vector $\mathbf{r}$ of the common center of mass of A and B.

Study Questions

1. What condition is necessary for the total linear momentum of two objects to be conserved?

2. If the only forces exerted on two objects are the forces they exert on each other, what can be inferred about the motion of their common center of mass?

Example 16.4	Conservation of Linear Momentum

A person of mass m_P stands at the center of a stationary barge of mass m_B (Fig. 16.8). Neglect horizontal forces exerted on the barge by the water.

(a) If the person starts running to the right with velocity v_P relative to the water, what is the resulting velocity of the barge relative to the water?

(b) If the person stops when he reaches the right-hand end of the barge, what are his position and the barge's position relative to their original positions?

Figure 16.8

Strategy

(a) The only horizontal forces exerted on the person and the barge are the forces they exert on each other. Therefore, their total linear momentum in the horizontal direction is conserved, and we can use Eq. (16.6) to determine the barge's velocity while the person is running.

(b) The combined center of mass of the person and the barge is initially stationary, so it must remain stationary. Knowing the position of the combined center of mass, we can determine the positions of the person and barge when the person is at the right-hand end of the barge.

Solution

(a) Before the person starts running, the total linear momentum of the person and the barge in the horizontal direction is zero, so it must be zero after he starts running. Letting v_B be the value of the barge's velocity *to the left* while the person is running (Fig. a), we obtain

$$m_P v_P + m_B(-v_B) = 0,$$

so the velocity of the barge while he runs is

$$v_B = \left(\frac{m_P}{m_B}\right)v_P.$$

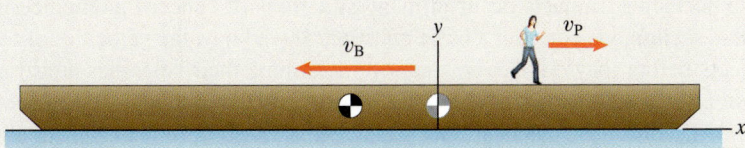

(a) Velocities of the person and barge.

(b) Let the origin of the coordinate system in Fig. b be the original horizontal position of the centers of mass of the barge and the person, and let x_B be the position of the barge's center of mass *to the left of the origin*. When the person has stopped at the right-hand end of the barge, the combined center of mass must still beat $x = 0$. That is,

$$\frac{x_P m_P + (-x_B)m_B}{m_P + m_B} = 0.$$

Solving this equation together with the relation $x_P + x_B = L/2$, we obtain

$$x_P = \frac{m_B L}{2(m_P + m_B)}, \qquad x_B = \frac{m_P L}{2(m_P + m_B)}.$$

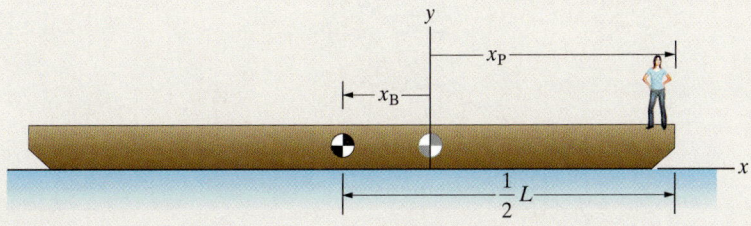

(b) Positions after the person has stopped running.

Critical Thinking

This example is a well-known illustration of the power of momentum methods. Notice that we were able to determine the velocity of the barge and the final positions of the person and barge even though we did not know the complicated time dependence of the horizontal forces they exert on each other.

16.3 Impacts

In machines that perform stamping or forging operations, dies impact against workpieces. Mechanical printers create images by impacting metal elements against the paper and platen. Vehicles impact each other intentionally, as when railroad cars are rolled against each other to couple them, and unintentionally in accidents. Impacts occur in many situations of concern in engineering. In this section, we consider a basic question: If we know the velocities of two objects before they collide, how can we determine their velocities afterward? In other words, what is the effect of the impact on the motions of the objects?

If colliding objects are not subjected to external forces, their total linear momentum must be the same before and after the impact. Even when they are subjected to external forces, the force of the impact is often so large, and its duration so brief, that the effect of external forces on the motions of the objects during the impact is negligible. Suppose that objects A and B with velocities $\mathbf{v}_A$ and $\mathbf{v}_B$ collide, and let $\mathbf{v}'_A$ and $\mathbf{v}'_B$ be their velocities after the impact (Fig. 16.9a). If the effects of external forces are negligible, then the total linear momentum of the system composed of A and B is conserved:

$$m_A\mathbf{v}_A + m_B\mathbf{v}_B = m_A\mathbf{v}'_A + m_B\mathbf{v}'_B. \tag{16.8}$$

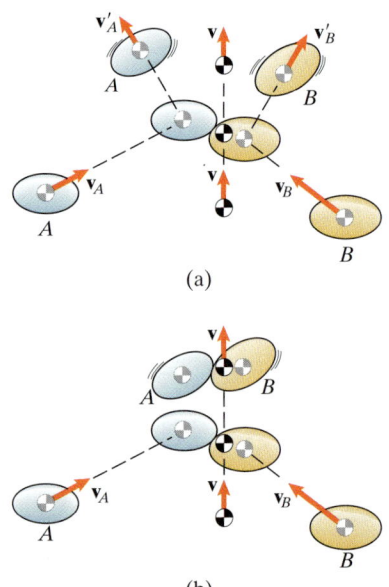

(a)

(b)

Figure 16.9
(a) Velocities of A and B before and after the impact, and the velocity $\mathbf{v}$ of their center of mass.
(b) A perfectly plastic impact.

Furthermore, the velocity $\mathbf{v}$ of the center of mass of A and B is the same before and after the impact. Thus, from Eq. (16.7),

$$\mathbf{v} = \frac{m_A\mathbf{v}_A + m_B\mathbf{v}_B}{m_A + m_B}. \tag{16.9}$$

If A and B adhere and remain together after they collide, they are said to undergo a *perfectly plastic impact*. Equation (16.9) gives the velocity of the center of mass of the object they form after the impact (Fig. 16.9b). A remarkable feature of this result is that we determine the velocity following the impact *without considering the physical nature of the impact*.

If A and B do not adhere, linear momentum conservation alone does not provide enough equations to determine their velocities after the impact. We first consider the case in which they travel along the same straight line before and after they collide.

Direct Central Impacts

Suppose that the centers of mass of A and B travel along the same straight line with velocities v_A and v_B before their impact (Fig. 16.10a). Let R be the magnitude of the force A and B exert on each other during the impact (Fig. 16.10b). We assume

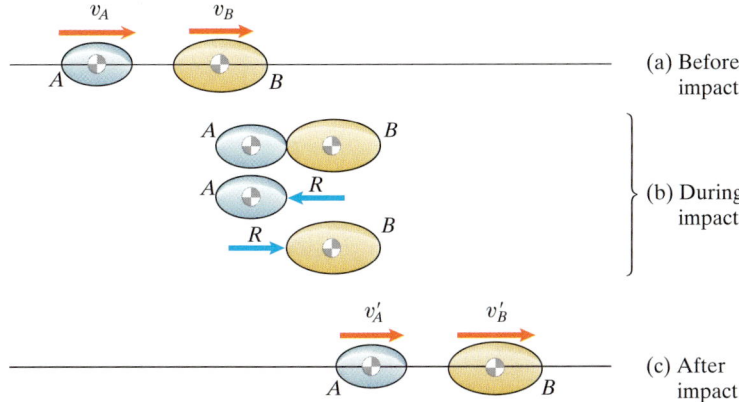

(a) Before impact

(b) During impact

(c) After impact

Figure 16.10
(a) Objects A and B traveling along the same straight line.
(b) During the impact, they exert a force R on each other.
(c) They travel along the same straight line after the central impact.

that the contacting surfaces are oriented so that R is parallel to the line along which the two objects travel and is directed toward their centers of mass. This condition, called *direct central impact*, means that A and B continue to travel along the same straight line after their impact (Fig. 16.10c). If the effects of external forces during the impact are negligible, the total linear momentum of the objects is conserved:

$$m_A v_A + m_B v_B = m_A v'_A + m_B v'_B. \tag{16.10}$$

However, we need another equation to determine the velocities v'_A and v'_B. To obtain it, we must consider the impact in more detail.

Let t_1 be the time at which A and B first come into contact (Fig. 16.11a). As a result of the impact, they will deform and their centers of mass will continue to approach each other. At a time t_C, their centers of mass will have reached their nearest proximity (Fig. 16.11b). At this time, the relative velocity of the two centers of mass is zero, so they have the same velocity. We denote it by v_C. The objects then begin to move apart and separate at a time t_2 (Fig. 16.11c). We apply the principle of impulse and momentum to A during the intervals of time from t_1 to the time of closest approach t_C and from t_C to t_2:

$$\int_{t_1}^{t_C} -R \, dt = m_A v_C - m_A v_A, \tag{16.11}$$

$$\int_{t_C}^{t_2} -R \, dt = m_A v'_A - m_A v_C. \tag{16.12}$$

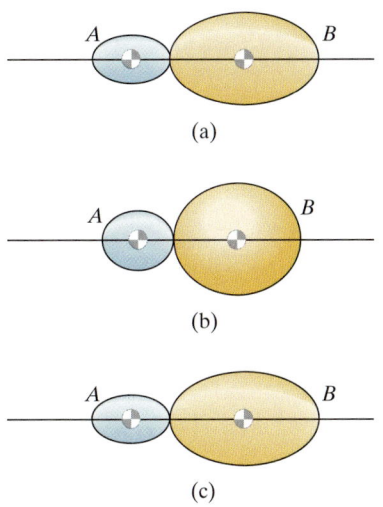

(a)

(b)

(c)

Figure 16.11
(a) First contact, $t = t_1$.
(b) Closest approach, $t = t_C$.
(c) End of contact, $t = t_2$.

Then we apply the principle to B for the same intervals of time:

$$\int_{t_1}^{t_C} R \, dt = m_B v_C - m_B v_B, \tag{16.13}$$

$$\int_{t_C}^{t_2} R \, dt = m_B v_B' - m_B v_C. \tag{16.14}$$

As a result of the impact, part of the objects' kinetic energy can be lost due to a variety of mechanisms, including permanent deformation and generation of heat and sound. As a consequence, the impulse they impart to each other during the "restitution" phase of the impact from t_C to t_2 is, in general, smaller than the impulse they impart from t_1 to t_C. The ratio of these impulses is called the *coefficient of restitution*:

$$e = \frac{\int_{t_C}^{t_2} R \, dt}{\int_{t_1}^{t_C} R \, dt}. \tag{16.15}$$

The value of e depends on the properties of the objects as well as their velocities and orientations when they collide, and it can be determined only by experiment or by a detailed analysis of the deformations of the objects during the impact.

If we divide Eq. (16.12) by Eq. (16.11) and divide Eq. (16.14) by Eq. (16.13), we can express the resulting equations in the forms

$$(v_C - v_A)e = v_A' - v_C$$

and

$$(v_C - v_B)e = v_B' - v_C.$$

Subtracting the first equation from the second, we obtain

$$e = \frac{v_B' - v_A'}{v_A - v_B}. \tag{16.16}$$

Thus, the coefficient of restitution is related in a simple way to the relative velocities of the objects before and after the impact. If e is known, Eq. (16.16) can be used together with the equation of conservation of linear momentum, Eq. (16.10), to determine v_A' and v_B'.

If $e = 0$, Eq. (16.16) indicates that $v_B' = v_A'$. The objects remain together after the impact, and the impact is perfectly plastic. If $e = 1$, it can be shown that the total kinetic energy is the same before and after the impact:

$$\tfrac{1}{2}m_A v_A^2 + \tfrac{1}{2}m_B v_B^2 = \tfrac{1}{2}m_A(v_A')^2 + \tfrac{1}{2}m_B(v_B')^2 \quad (\text{when } e = 1).$$

An impact in which kinetic energy is conserved is called *perfectly elastic*. Although this is sometimes a useful approximation, energy is lost in any impact in which material objects come into contact. If a collision can be heard, kinetic energy has been converted into sound. Permanent deformations of the colliding objects after the impact also represent losses of kinetic energy.

Oblique Central Impacts

We can extend the procedure used to analyze direct central impacts to the case in which the objects approach each other at an oblique angle. Suppose that A and B approach with arbitrary velocities $\mathbf{v}_A$ and $\mathbf{v}_B$ (Fig. 16.12) and that the forces they exert on each other during their impact are parallel to the x axis and point toward their centers of mass. No forces are exerted on A and B in the y or z directions, so their velocities in those directions are unchanged by the impact:

$$(\mathbf{v}'_A)_y = (\mathbf{v}_A)_y, \qquad (\mathbf{v}'_B)_y = (\mathbf{v}_B)_y,$$

$$(\mathbf{v}'_A)_z = (\mathbf{v}_A)_z, \qquad (\mathbf{v}'_B)_z = (\mathbf{v}_B)_z. \qquad (16.17)$$

In the x direction, linear momentum is conserved:

$$m_A(\mathbf{v}_A)_x + m_B(\mathbf{v}_B)_x = m_A(\mathbf{v}'_A)_x + m_B(\mathbf{v}'_B)_x. \qquad (16.18)$$

By the same analysis we used to arrive at Eq. (16.16), the x components of velocity satisfy the relation

$$e = \frac{(\mathbf{v}'_B)_x - (\mathbf{v}'_A)_x}{(\mathbf{v}_A)_x - (\mathbf{v}_B)_x}. \qquad (16.19)$$

We can analyze an oblique central impact in which an object A hits a stationary object B if friction is negligible. Suppose that B is constrained so that it cannot move relative to the inertial reference frame. For example, in Fig. 16.13, A strikes a wall B that is fixed relative to the earth. The y and z components of A's velocity are unchanged, because friction is neglected and the impact exerts no force in those directions. The x component of A's velocity after the impact is given by Eq. (16.19) with B's velocity equal to zero:

$$(\mathbf{v}'_A)_x = -e(\mathbf{v}_A)_x.$$

In summary, the velocity of the common center of mass of two objects A and B after they collide can be determined from Eq. (16.9). In an oblique central impact, in terms of the coordinate system shown in Fig. 16.12, the y and z components of the velocities of the two objects are unchanged, and the x components of the velocities after the impact can be determined using Eqs. (16.18) and (16.19).

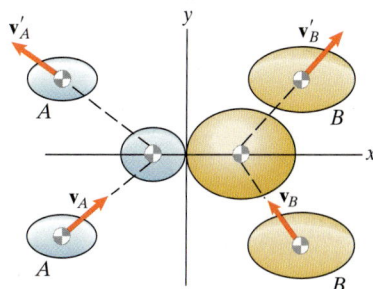

Figure 16.12
An oblique central impact.

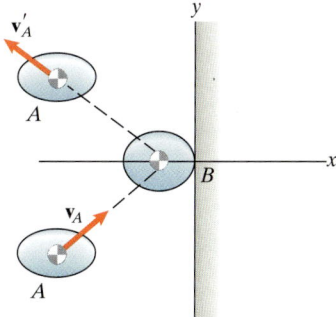

Figure 16.13
Impact with a stationary object.

Study Questions

1. If two objects collide, what can you infer about the motion of their common center of mass?

2. What are perfectly plastic and perfectly elastic impacts?

3. What is the definition of the coefficient of restitution?

4. If the coefficient of restitution is known for a direct central impact, what equations are used to determine the velocities of the objects after their collision?

Example 16.5   Analyzing an Impact

The 4-kg masses A and B in Fig. 16.14 slide on the smooth horizontal bar. Determine their velocities after they collide if (a) they are coated with Velcro® and stick together and (b) their coefficient of restitution is $e = 0.8$.

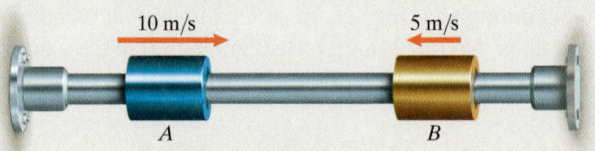

Figure 16.14

Strategy

(a) If the masses stick together, they have the same velocity after their collision. We can determine the velocity from conservation of linear momentum.

(b) Knowing the coefficient of restitution, we can determine the velocity of each mass after the collision by using conservation of linear momentum together with the definition of the coefficient of restitution, Eq. (16.16).

Solution

(a) The velocities of the masses before the impact are $v_A = 10$ m/s and $v_B = -5$ m/s. Let v' be their common velocity after the impact. Conservation of linear momentum requires that

$$m_A v_A + m_B v_B = (m_A + m_B)v':$$

$$(4\,\text{kg})(10\,\text{m/s}) + (4\,\text{kg})(-5\,\text{m/s}) = (4\,\text{kg} + 4\,\text{kg})v'.$$

Solving, we obtain $v' = 2.5$ m/s. The connected masses move toward the right at 2 m/s after the impact.

(b) Let v'_A and v'_B be the velocities of the masses after the impact. Conservation of linear momentum requires that

$$m_A v_A + m_B v_B = m_A v'_A + m_B v'_B:$$

$$(4\,\text{kg})(10\,\text{m/s}) + (4\,\text{kg})(-5\,\text{m/s}) = (4\,\text{kg})v'_A + (4\,\text{kg})v'_B.$$

From Eq. (16.16),

$$e = \frac{v'_B - v'_A}{v_A - v_B}:$$

$$0.8 = \frac{v'_B - v'_A}{10\,\text{m/s} - (-5\,\text{m/s})}.$$

We now have two equations in v'_A and v'_B. Solving them, we obtain $v'_A = -3.5$ m/s and $v'_B = 8.5$ m/s. Mass A moves toward the left at 3.5 m/s and mass B moves toward the right at 8.5 m/s after the impact.

Critical Thinking

This example demonstrates the most common types of impact problems. If the masses stick together, there is only one velocity to determine after the impact, so it can be obtained using conservation of linear momentum alone. If the masses do not stick together, determining both velocities after the impact requires another equation. The coefficient of restitution supplies the additional equation. Even if impacting objects do not stick together, the velocity of their common center of mass after the impact is equal to the velocity the objects would have if they did stick together. In part (b), the velocity of the common center of mass after the impact is

$$\frac{m_A v'_A + m_B v'_B}{m_A + m_B} = \frac{(4 \text{ kg})(-3.5 \text{ m/s}) + (4 \text{ kg})(8.5 \text{ m/s})}{4 \text{ kg} + 4 \text{ kg}} = 2.5 \text{ m/s},$$

which is the velocity we obtained in part (a).

Example 16.6   Applying Momentum Methods to Spacecraft Docking

The *Apollo* command-service module (*A*) attempts to dock with the *Soyuz* capsule (*B*), July 15, 1975 (Fig. 16.15). Their masses are $m_A = 18$ Mg and $m_B = 6.6$ Mg. The *Soyuz* is stationary relative to the reference frame shown, and the command-service module approaches with velocity $\mathbf{v}_A = 0.2\mathbf{i} + 0.03\mathbf{j} - 0.02\mathbf{k}$ (m/s).
(a) If the first attempt at docking is successful, what is the velocity of the center of mass of the combined vehicles afterward?
(b) If the first attempt is unsuccessful and the coefficient of restitution of the resulting impact is $e = 0.95$, what are the velocities of the two spacecraft after the impact?

Figure 16.15
(A) Apollo command-service module (B) Soyuz capsule.

Strategy

(a) If the docking is successful, the impact is perfectly plastic, and we can use Eq. (16.9) to determine the velocity of the center of mass of the combined object after the impact.

(b) By assuming an oblique central impact with the forces exerted by the docking collars parallel to the x axis, we can use Eqs. (16.18) and (16.19) to determine the velocities of both spacecraft after the impact.

Solution

(a) From Eq. (16.9), the velocity of the center of mass of the combined vehicles is

$$\mathbf{v} = \frac{m_A \mathbf{v}_A + m_B \mathbf{v}_B}{m_A + m_B}$$

$$= \frac{(18 \text{ Mg})[0.2\mathbf{i} + 0.03\mathbf{j} - 0.02\mathbf{k} \text{ (m/s)}] + 0}{18 \text{ Mg} + 6.6 \text{ Mg}}$$

$$= 0.146\mathbf{i} + 0.0220\mathbf{j} - 0.0146\mathbf{k} \text{ (m/s)}.$$

(b) The y and z components of the velocities of both spacecraft are unchanged. To determine the x components, we first use conservation of linear momentum, Eq. (16.18).

$$m_A(\mathbf{v}_A)_x + m_B(\mathbf{v}_B)_x = m_A(\mathbf{v}'_A)_x + m_B(\mathbf{v}'_B)_x:$$

$$(18 \text{ Mg})(0.2 \text{ m/s}) + 0 = (18 \text{ Mg})(\mathbf{v}'_A)_x + (6.6 \text{ Mg})(\mathbf{v}'_B)_x.$$

We then use the coefficient of restitution, Eq. (16.19), to obtain

$$e = \frac{(\mathbf{v}'_B)_x - (\mathbf{v}'_A)_x}{(\mathbf{v}_A)_x - (\mathbf{v}_B)_x}:$$

$$0.95 = \frac{(\mathbf{v}'_B)_x - (\mathbf{v}'_A)_x}{0.2 \text{ m/s} - 0}.$$

Solving these two equations yields $(\mathbf{v}'_A)_x = 0.0954$ m/s and $(\mathbf{v}'_B)_x = 0.285$ m/s, so the velocities of the spacecraft after the impact are

$$\mathbf{v}'_A = 0.0954\mathbf{i} + 0.03\mathbf{j} - 0.02\mathbf{k} \text{ (m/s)},$$

$$\mathbf{v}'_B = 0.285\mathbf{i} \text{ (m/s)}.$$

Critical Thinking

Why are calculations of this kind useful? Analytical simulations of the impact of the two spacecraft as they docked were used in designing the docking mechanisms and also in training the astronauts who performed the docking maneuver.

Problems

16.43 A girl weighing 80 lb stands at rest on a 325-lb floating platform. She starts running at 10 ft/s *relative to the water* and runs off the end of the platform. Neglect the horizontal force exerted on the floating platform by the water.

(a) After she starts running, what is the velocity of the platform relative to the water?

(b) While she is running, what is the velocity of the common center of mass of the girl and the platform relative to the water?

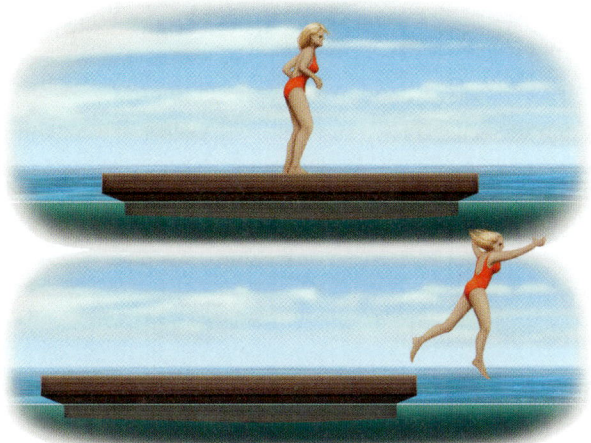

Problem 16.43

16.44 Two railroad cars with weights $W_A = 120,000$ lb and $W_B = 70,000$ lb collide and become coupled together. Car A is full, and car B is half full, of carbolic acid. When the cars collide, the acid in B sloshes back and forth violently.

(a) Immediately after the impact, what is the velocity of the common center of mass of the two cars?

(b) When the sloshing in B has subsided, what is the velocity of the two cars?

16.45 The weights of the railroad cars are $W_A = 120,000$ lb and $W_B = 70,000$ lb. The railroad track has a constant slope of 0.2 degrees upward toward the right. If the cars are 6 ft apart at the instant shown, what is the velocity of their common center of mass immediately after they become coupled together?

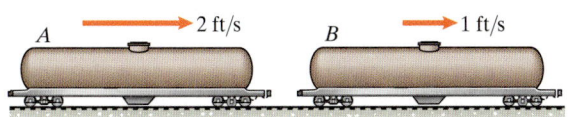

Problems 16.44/16.45

16.46 The 400-kg satellite S traveling at 7 km/s is hit by a 1-kg meteor M traveling at 12 km/s. The meteor is embedded in the satellite by the impact. Determine the magnitude of the velocity of their common center of mass after the impact and the angle β between the path of the center of mass and the original path of the satellite.

16.47 The 400-kg satellite S traveling at 7 km/s is hit by a 1-kg meteor M. The meteor is embedded in the satellite by the impact. What would the magnitude of the velocity of the meteor need to be to cause the angle β between the original path of the satellite and the path of the center of mass of the combined satellite and meteor after the impact to be 0.5°? What is the magnitude of the velocity of the center of mass after the impact?

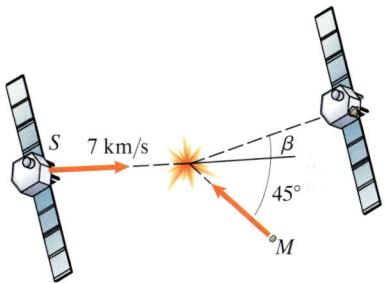

Problems 16.46/16.47

16.48 A 68-kg astronaut is initially stationary at the left side of an experiment module within an orbiting space shuttle. The 105,000-kg shuttle's center of mass is 4 m to the astronaut's right. He launches himself toward the center of mass at 1 m/s *relative to the shuttle*. He travels 8 m relative to the shuttle before bringing himself to rest at the opposite wall of the experiment module.

(a) What is the change in the magnitude of the shuttle's velocity relative to its original velocity while the astronaut is in motion?

(b) What is the change in the magnitude of the shuttle's velocity relative to its original velocity after his "flight"?

(c) Where is the shuttle's center of mass relative to the astronaut after his "flight"?

Problem 16.48

16.49 An 80-lb boy sitting in a stationary 20-lb wagon wants to simulate rocket propulsion by throwing bricks out of the wagon. Neglect horizontal forces on the wagon's wheels. If the boy has three bricks weighing 10 lb each and throws them with a horizontal velocity of 10 ft/s relative to the wagon, determine the velocity he attains (a) if he throws the bricks one at a time and (b) if he throws them all at once.

Problem 16.49

16.50 The catapult, designed to throw a line to ships in distress, throws a 2-kg projectile. The mass of the catapult is 36 kg, and it rests on a smooth surface. If the velocity of the projectile *relative to the earth* as it leaves the tube is 50 m/s at $\theta_0 = 30°$ relative to the horizontal, what is the resulting velocity of the catapult toward the left?

16.51 The catapult, which has a mass of 36 kg and throws a 2-kg projectile, rests on a smooth surface. The velocity of the projectile *relative to the catapult* as it leaves the tube is 50 m/s at $\theta_0 = 30°$ relative to the horizontal. What is the resulting velocity of the catapult toward the left?

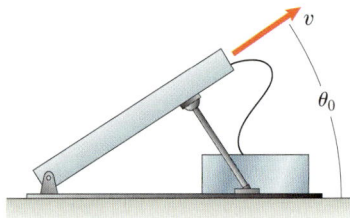

Problems 16.50/16.51

16.52 A bullet weighing 0.07 oz [16 oz (ounces) = 1 lb] strikes a stationary 10-lb block of wood with horizontal velocity $v = 2000$ ft/s and becomes embedded in it. The coefficient of kinetic friction between the block and the floor is $\mu_k = 0.12$. What distance does the block slide on the floor as a result of the impact?

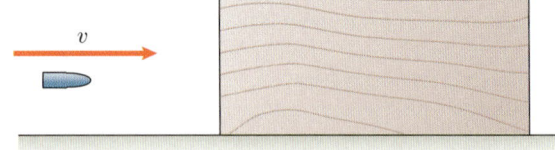

Problem 16.52

16.53 A 28-g bullet hits a suspended 45-kg block of wood and becomes embedded in it. The angle through which the wires supporting the block rotate as a result of the impact is measured and determined to be 7°. What was the bullet's velocity?

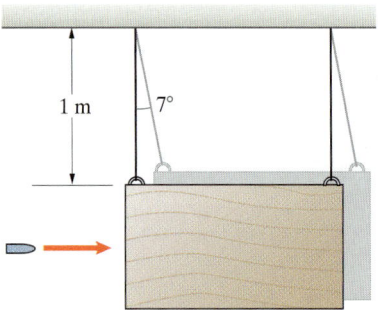

Problem 16.53

16.54 The overhead conveyor drops the 12-kg package A into the 1.6-kg carton B. The package is tacky and sticks to the bottom of the carton. If the coefficient of friction between the carton and the horizontal conveyor is $\mu_k = 0.2$, what distance does the carton slide after the impact?

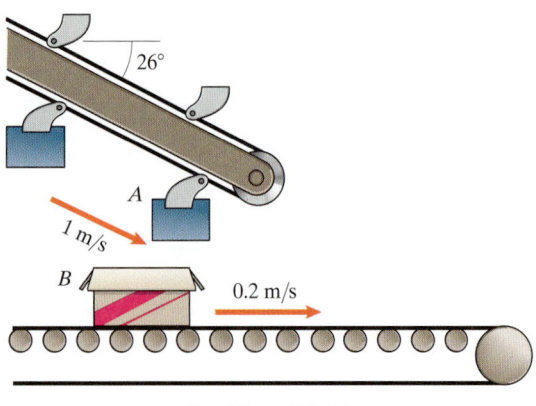

Problem 16.54

16.55 A 12,000-lb bus collides with a 2800-lb car. The velocity of the bus before the collision is $\mathbf{v}_B = 18\mathbf{i}$ (ft/s) and the velocity of the car is $\mathbf{v}_C = 33\mathbf{j}$ (ft/s). The two vehicles become entangled and remain together after the collision. The coefficient of kinetic friction between the vehicles' tires and the road is $\mu_k = 0.6$.

(a) What is the velocity of the common center of mass of the two vehicles immediately after the collision?

(b) Determine the approximate final position of the common center of mass of the vehicles relative to its position when the collision occurred. (Assume that the tires skid, not roll, on the road.)

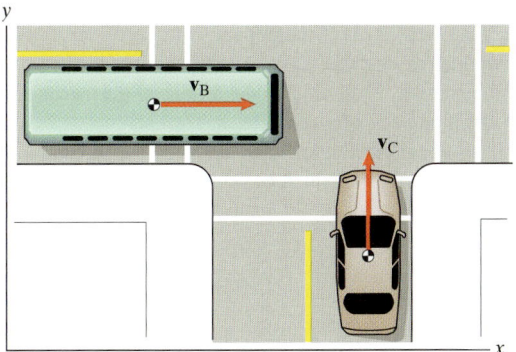

Problem 16.55

16.56 The velocity of the 200-kg astronaut A relative to the space station is $40\mathbf{i} + 30\mathbf{j}$ (mm/s). The velocity of the 300-kg structural member B relative to the station is $-20\mathbf{i} + 30\mathbf{j}$ (mm/s). When they approach each other, the astronaut grasps and clings to the structural member.

(a) What is the velocity of their common center of mass when they arrive at the station?

(b) Determine the approximate position at which they contact the station.

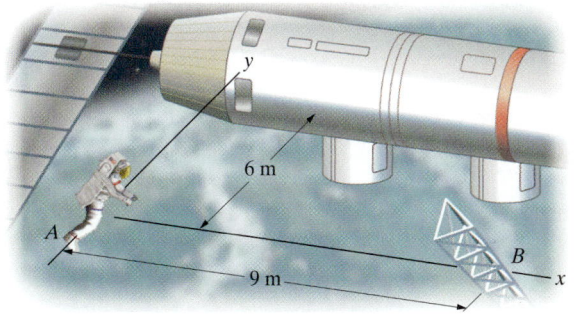

Problem 16.56

16.57 Objects A and B with weights $W_A = 5$ lb and $W_B = 8$ lb undergo a direct central impact. Before the impact, A is moving with velocity $v_A = 10$ ft/s and B is stationary. Determine the velocities of A and B after the impact if it is (a) perfectly plastic $(e = 0)$; (b) perfectly elastic $(e = 1)$.

16.58 The weights of object A and B are $W_A = 5$ lb and $W_B = 8$ lb. Object A undergoes a direct central impact with the stationary object B. The velocity v_A of A before the impact and the coefficient of restitution e of the impact are unknown. After the impact, measurements indicate that A is moving at 1.6 ft/s toward the right and B is moving at 2.8 ft/s toward the right. Determine v_A and e.

Problems 16.57/16.58

16.59 Objects A and B with velocities $v_A = 6$ m/s and $v_B = 3$ m/s undergo a direct central impact. The masses of the objects are $m_A = 2$ kg and $m_B = 4$ kg. Draw a graph of the velocity of the object A after the impact as a function of the coefficient of restitution from $e = 0$ to $e = 1$.

Problem 16.59

16.60 The 8-kg mass A and the mass B slide on the smooth horizontal bar with the velocities shown. After they collide, A is moving to the left at 1 m/s and B is moving to the right at 3 m/s. Determine the mass of B and the coefficient of restitution of the impact.

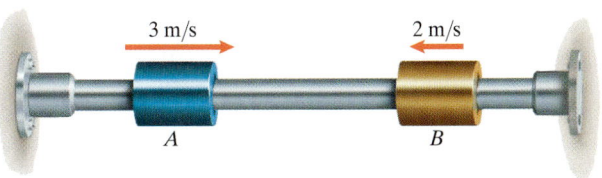

Problem 16.60

16.61 Two cars with energy-absorbing bumpers collide head-on. Their speeds are $v_A = 3$ mi/h and $v_B = 2$ mi/h in the directions shown. The weights of the cars are $W_A = 2800$ lb and $W_B = 4400$ lb. If the coefficient of restitution of the impact is $e = 0.2$, what are the velocities of the cars immediately after the collision?

16.62 The duration of the impact of the cars in Problem 16.61 depends on the designs of their energy-absorbing bumpers. What minimum duration is necessary if you want the magnitude of the average acceleration to which the occupants of either car are subjected to be no greater than 5 g's (161 ft/s^2)?

Strategy: Apply Eq. (16.2) to each car.

Problems 16.61/16.62

16.63 The balls are of equal mass m. Balls B and C are connected by an unstretched spring and are stationary. Ball A moves toward ball B with velocity v_A. The impact of A with B is perfectly elastic ($e = 1$).
(a) What is the velocity of the common center of mass of balls B and C immediately after the impact?
(b) What is the velocity of the common center of mass of B and C at time t after the impact?

16.64 In Problem 16.63, what is the maximum compressive force in the spring as a result of the impact?

16.65* The balls are of equal mass m. Balls B and C are connected by an unstretched spring and are stationary. Ball A moves toward ball B with velocity v_A. The impact of A with B is perfectly elastic ($e = 1$). Suppose that you interpret this as an impact between ball A and an "object" D consisting of the connected balls B and C.

(a) What is the coefficient of restitution of the impact between A and D?
(b) If you consider the total energy after the impact to be the sum of the kinetic energies $\frac{1}{2}m(v'_A)^2 + \frac{1}{2}(2m)(v'_D)^2$, where v'_D is the velocity of the center of mass of D after the impact, how much energy is "lost" as a result of the impact?
(c) How much energy is actually lost as a result of the impact? (This problem is an interesting model for one of the mechanisms of energy loss in impacts between objects. The energy "loss" calculated in part (b) is transformed into "internal energy"—the vibrational motions of B and C relative to their common center of mass.)

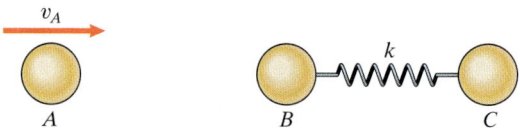

Problems 16.63–16.65

16.66 Suppose you investigate an accident in which a 1300-kg car A struck a parked 1200-kg car B. All four of B's wheels were locked, and skid marks indicate that B slid 2 m after the impact. If you estimate the coefficient of friction between B's tires and the road to be $\mu_k = 0.8$ and the coefficient of restitution of the impact to be $e = 0.4$, what was A's velocity just before the impact? (Assume that only one impact occurred.)

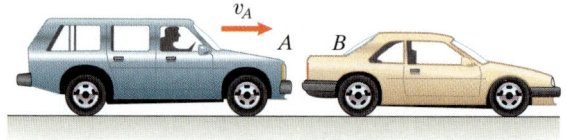

Problem 16.66

16.67 When the player releases the ball from rest at a height of 5 ft above the floor, it bounces to a height of 3.5 ft. If he throws the ball downward, releasing it at 3 ft above the floor, how fast would he need to throw it so that it would bounce to a height of 12 ft?

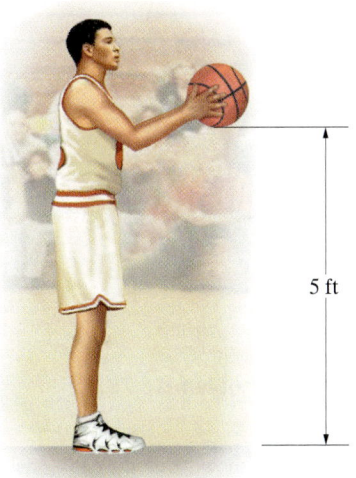

5 ft

Problem 16.67

16.68 The 0.45-kg soccer ball is 1 m above the ground when it is kicked upward at 12 m/s. If the coefficient of restitution between the ball and the ground is $e = 0.6$, what maximum height above the ground does the ball reach on its first bounce?

16.69 The 0.45-kg soccer ball is stationary just before it is kicked upward at 12 m/s. If the impact lasts 0.02 s, what average force is exerted on the ball by the player's foot?

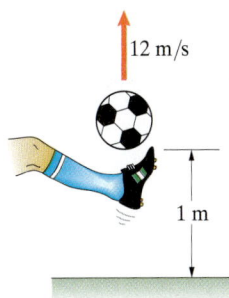

12 m/s

1 m

Problems 16.68/16.69

16.70 By making measurements directly from the photograph of the bouncing golf ball, estimate the coefficient of restitution.

16.71 If you throw the golf ball in Problem 16.70 horizontally at 2 ft/s and release it 4 ft above the surface, what is the distance between the first two bounces?

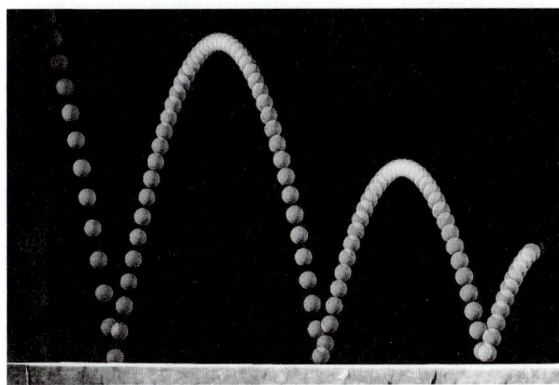

Problems 16.70/16.71

16.72 In a forging operation, the 100-lb weight is lifted into position 1 and released from rest. It falls and strikes a workpiece in position 2. If the weight is moving at 15 ft/s immediately before the impact and the coefficient of restitution is $e = 0.3$, what is the velocity of the weight immediately after the impact?

16.73 The 100-lb weight is released from rest in position 1. The spring constant is $k = 120$ lb/ft, and the springs are unstretched in position 1. If the coefficient of restitution of the impact of the weight with the workpiece in position 2 is $e = 0.2$, what is the velocity of the weight immediately after the impact?

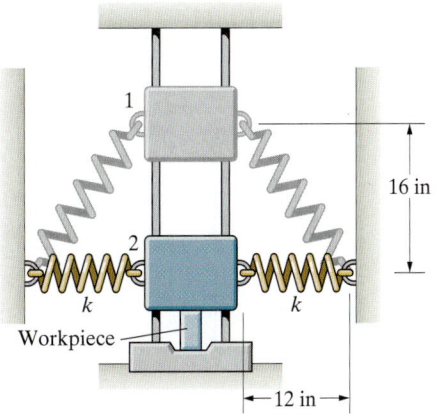

1

16 in

2

k k

Workpiece

←12 in→

Problems 16.72/16.73

16.74* A bioengineer studying helmet design uses an experimental apparatus that launches a 2.4-kg helmet containing a 2-kg model of the human head against a rigid surface at 6 m/s. The head, suspended within the helmet, is not immediately affected by the impact of the helmet with the surface and continues to move to the right at 6 m/s, so the head then undergoes an impact with the helmet. If the coefficient of restitution of the helmet's impact with the surface is 0.85 and the coefficient of restitution of the subsequent impact of the head with the helmet is 0.15, what is the velocity of the head after its initial impact with the helmet?

16.75* (a) If the duration of the impact of the head with the helmet in Problem 16.74 is 0.004 s, what is the magnitude of the average force exerted on the head by the impact?
(b) Suppose that the simulated head alone strikes the rigid surface at 6 m/s, the coefficient of restitution is 0.5, and the duration of the impact is 0.0002 s. What is the magnitude of the average force exerted on the head by the impact?

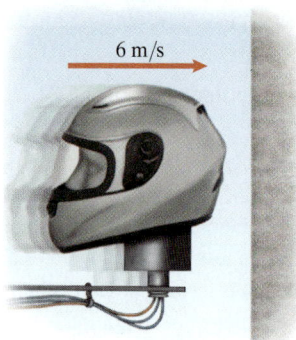

6 m/s

Problems 16.74/16.75

16.76 Two small balls, each of mass $m = 0.12$ kg, hang from strings of length $L = 1$ m. The left ball is released from rest with $\theta = 30°$. As a result of the initial collision, the right ball swings through a maximum angle of 25°. Determine the coefficient of restitution.

16.77 If the duration of the collision in Problem 16.76 is 0.008 s, what is the magnitude of the average force the balls exert on each other?

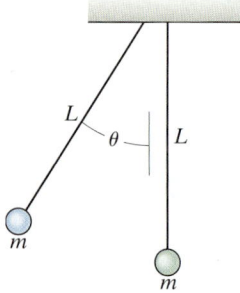

Problems 16.76/16.77

16.78 The 3-kg object A and 8-kg object B undergo an oblique central impact. The coefficient of restitution is $e = 0.8$. Before the impact, $\mathbf{v}_A = 10\mathbf{i} + 4\mathbf{j} + 8\mathbf{k}$ (m/s) and $\mathbf{v}_B = -2\mathbf{i} - 6\mathbf{j} + 5\mathbf{k}$ (m/s). What are the velocities of A and B after the impact?

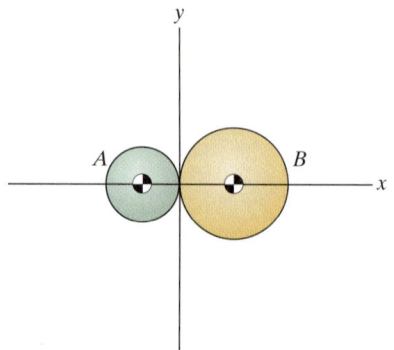

Problem 16.78

16.79 A baseball bat (shown with the bat's axis perpendicular to the page) strikes a thrown baseball. Before their impact, the velocity of the baseball is $\mathbf{v}_b = 132(\cos 45°\mathbf{i} + \cos 45°\mathbf{j})$ (ft/s) and the velocity of the bat is $\mathbf{v}_B = 60(-\cos 45°\mathbf{i} - \cos 45°\mathbf{j})$ (ft/s). Neglect the change in the velocity of the bat due to the direct central impact. The coefficient of restitution is $e = 0.2$. What is the ball's velocity after the impact? Assume that the baseball and the bat are moving horizontally. Does the batter achieve a potential hit or a foul ball?

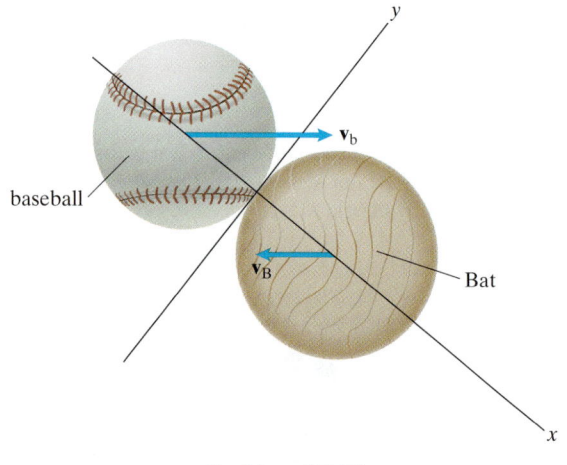

Problem 16.79

16.80 The cue gives the cue ball *A* a velocity parallel to the *y* axis. The cue ball hits the eight ball *B* and knocks it straight into the corner pocket. If the magnitude of the velocity of the cue ball just before the impact is 2 m/s and the coefficient of restitution is $e = 1$, what are the velocity vectors of the two balls just after the impact? (The balls are of equal mass.)

16.81 In Problem 16.80, what are the velocity vectors of the two balls just after the impact if the coefficient of restitution is $e = 0.9$?

Problems 16.80/16.81

16.82 If the coefficient of restitution is the same for both impacts, show that the cue ball's path after two banks is parallel to its original path.

Problem 16.82

16.83 The velocity of the 170-g hockey puck is $\mathbf{v_P} = 10\mathbf{i} - 4\mathbf{j}$ (m/s). If you neglect the change in the velocity $\mathbf{v_S} = v_S\mathbf{j}$ of the stick resulting from the impact, and if the coefficient of restitution is $e = 0.6$, what should v_S be to send the puck toward the goal?

16.84 In Problem 16.83, if the stick responds to the impact the way an object with the same mass as the puck would and the coefficient of restitution is $e = 0.6$, what should v_S be to send the puck toward the goal?

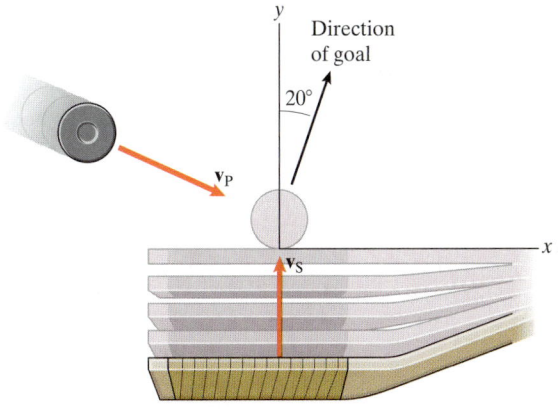

Problems 16.83/16.84

16.4 Angular Momentum

In this section we derive a result, analogous to the principle of impulse and momentum, that relates the integral of a moment with respect to time to the change in a quantity called the angular momentum.

Principle of Angular Impulse and Momentum

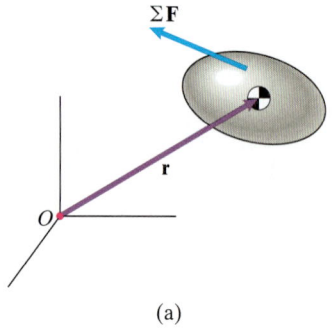

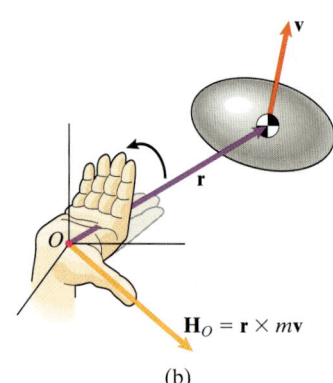

Figure 16.16
(a) The position vector and the total external force on an object.
(b) The angular momentum vector and the right-hand rule for determining its direction.

We describe the position of an object relative to an inertial reference frame with origin O by the position vector $\mathbf{r}$ from O to the object's center of mass (Fig. 16.16a). Recall that we obtained the very useful principle of work and energy by taking the dot product of Newton's second law with the velocity. Here we obtain another useful result by taking the cross product of Newton's second law with the position vector. This procedure gives us a relation between the moment of the external forces about O and the object's motion.

We take the cross product of Newton's second law with $\mathbf{r}$:

$$\mathbf{r} \times \Sigma\mathbf{F} = \mathbf{r} \times m\mathbf{a} = \mathbf{r} \times m\frac{d\mathbf{v}}{dt}. \tag{16.20}$$

Notice that the derivative of the quantity $\mathbf{r} \times m\mathbf{v}$ with respect to time is

$$\frac{d}{dt}(\mathbf{r} \times m\mathbf{v}) = \underbrace{\left(\frac{d\mathbf{r}}{dt} \times m\mathbf{v}\right)}_{= \, 0} + \left(\mathbf{r} \times m\frac{d\mathbf{v}}{dt}\right).$$

(The first term on the right side is zero because $d\mathbf{r}/dt = \mathbf{v}$ and the cross product of parallel vectors is zero.) Using this result, we can write Eq. (16.20) as

$$\mathbf{r} \times \Sigma\mathbf{F} = \frac{d\mathbf{H}_O}{dt}, \tag{16.21}$$

where the vector

$$\mathbf{H}_O = \mathbf{r} \times m\mathbf{v} \tag{16.22}$$

is called the *angular momentum* about O (Fig. 16.16b). If we interpret the angular momentum as the moment of the linear momentum of the object about point O, Eq. (16.21) states that the moment $\mathbf{r} \times \Sigma\mathbf{F}$ equals the rate of change of the moment of momentum about point O. *If the moment is zero during an interval of time,* $\mathbf{H}_O$ *is constant.*

Integrating Eq. (16.21) with respect to time, we obtain

$$\int_{t_1}^{t_2} (\mathbf{r} \times \Sigma\mathbf{F}) \, dt = (\mathbf{H}_O)_2 - (\mathbf{H}_O)_1. \tag{16.23}$$

The integral on the left is called the *angular impulse*, and the equation itself is called the *principle of angular impulse and momentum*: The angular impulse applied to an object during an interval of time is equal to the change in the object's angular momentum. If we know the moment $\mathbf{r} \times \Sigma\mathbf{F}$ as a function of time, we can determine the change in the angular momentum. The dimensions of the angular impulse and angular momentum are $(\text{mass}) \times (\text{length})^2/(\text{time})$.

Central-Force Motion

If the total force acting on an object remains directed toward a point that is fixed relative to an inertial reference frame, the object is said to be in *central-force motion*. The fixed point is called the *center* of the motion. Orbits are the most familiar instances of central-force motion. For example, the gravitational force on an earth satellite remains directed toward the center of the earth.

If we place the reference point O at the center of the motion (Fig. 16.17a), the position vector $\mathbf{r}$ is parallel to the total force, so $\mathbf{r} \times \Sigma\mathbf{F}$ equals zero. Therefore, Eq. (16.23) indicates that in central-force motion, an object's angular momentum is conserved:

$$\mathbf{H}_O = \text{constant}. \tag{16.24}$$

In plane central-force motion, we can express $\mathbf{r}$ and $\mathbf{v}$ in cylindrical coordinates (Fig. 16.17b):

$$\mathbf{r} = r\mathbf{e}_r, \qquad \mathbf{v} = v_r\mathbf{e}_r + v_\theta\mathbf{e}_\theta.$$

Substituting these expressions into Eq. (16.22), we obtain the angular momentum:

$$\mathbf{H}_O = (r\mathbf{e}_r) \times m(v_r\mathbf{e}_r + v_\theta\mathbf{e}_\theta) = mrv_\theta\mathbf{e}_z.$$

From this expression we see that in plane central-force motion, *the product of the radial distance from the center of the motion and the transverse component of the velocity is constant:*

$$rv_\theta = \text{constant}. \tag{16.25}$$

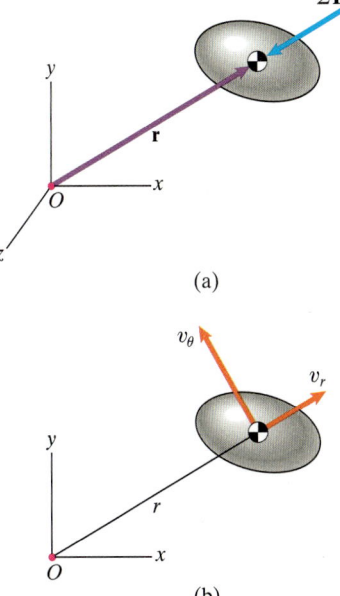

Figure 16.17
(a) Central-force motion.
(b) Expressing the position and velocity in cylindrical coordinates.

Study Questions

1. What is the definition of the angular momentum?
2. What is the principle of angular impulse and momentum?
3. What condition is necessary for angular momentum to be conserved?
4. In central-force motion, what can be inferred about the angular momentum?

Example 16.7 **Applying Angular Impulse and Momentum**

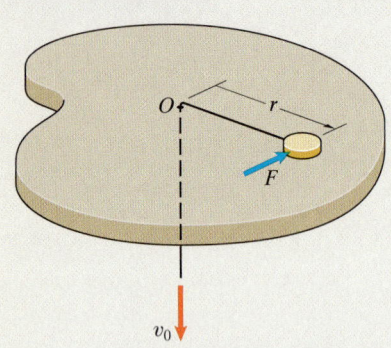

Figure 16.18

A disk of mass m attached to a string slides on a smooth horizontal table under the action of a constant transverse force F (Fig. 16.18). The string is drawn through a hole in the table at O at constant velocity v_0. At $t = 0$, $r = r_0$ and the transverse velocity of the disk is zero. What is the disk's velocity as a function of time?

Strategy

By expressing r as a function of time, we can determine the moment of the force on the disk about O as a function of time. The angular momentum of the disk depends on its velocity, so we can apply the principle of angular impulse and momentum to obtain information about the velocity as a function of time.

Solution

The radial position as a function of time is $r = r_0 - v_0 t$. In terms of polar coordinates (Fig. a), the moment about O of the forces on the disk is

$$\mathbf{r} \times \Sigma \mathbf{F} = r\mathbf{e}_r \times (-T\mathbf{e}_r + F\mathbf{e}_\theta) = F(r_0 - v_0 t)\mathbf{e}_z,$$

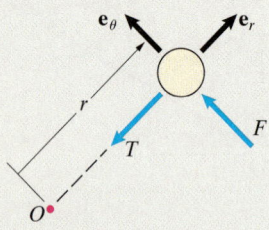

(a) Expressing the moment in terms of polar coordinates.

where T is the tension in the string. The angular momentum at time t is

$$\mathbf{H}_O = \mathbf{r} \times m\mathbf{v} = r\mathbf{e}_r \times m(v_r\mathbf{e}_r + v_\theta\mathbf{e}_\theta)$$

$$= mv_\theta(r_0 - v_0 t)\mathbf{e}_z.$$

Substituting these expressions into the principle of angular impulse and momentum, we get

$$\int_{t_1}^{t_2} (\mathbf{r} \times \Sigma \mathbf{F})\, dt = (\mathbf{H}_O)_2 - (\mathbf{H}_O)_1:$$

$$\int_0^t F(r_0 - v_0 t)\mathbf{e}_z\, dt = mv_\theta(r_0 - v_0 t)\mathbf{e}_z - \mathbf{0}.$$

Evaluating the integral, we obtain the transverse component of velocity as a function of time:

$$v_\theta = \frac{[r_0 t - (1/2)v_0 t^2]F}{(r_0 - v_0 t)m}.$$

The disk's velocity as a function of time is

$$\mathbf{v} = -v_0\mathbf{e}_r + \frac{[r_0 t - (1/2)v_0 t^2]F}{(r_0 - v_0 t)m}\mathbf{e}_\theta.$$

Critical Thinking

You can recognize that this example is a contrived situation. In fact, useful engineering applications of the principle of angular impulse and momentum in the form given by Eq. (16.23) are not common. In contrast, conservation of angular momentum is often useful. (See Example 16.8 and Problems 16.86–16.98.)

Example 16.8   Applying Conservation of Momentum and Energy to a Satellite

When an earth satellite is at perigee (the point at which it is nearest to the earth), the magnitude of its velocity is $v_P = 7000$ m/s and its distance from the center of the earth is $r_P = 10{,}000$ km (Fig. 16.19). What are the magnitude of the velocity v_A and the distance r_A of the satellite from the earth at apogee (the point at which it is farthest from the earth)? The radius of the earth is $R_E = 6370$ km.

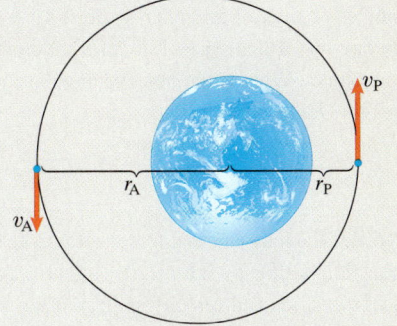

Figure 16.19

Strategy
Because the satellite undergoes central-force motion about the center of the earth, the product of its distance from the center of the earth and the transverse component of its velocity is constant. This gives us one equation relating v_A and r_A. We can obtain a second equation relating v_A and r_A by using conservation of energy.

Solution
From Eq. (16.25), conservation of angular momentum requires that

$$r_A v_A = r_P v_P.$$

From Eq. (15.27), the potential energy of the satellite in terms of its distance from the center of the earth is

$$V = -\frac{mgR_E^2}{r}.$$

The sum of the kinetic and potential energies at apogee and perigee must be equal:

$$\frac{1}{2}mv_A^2 - \frac{mgR_E^2}{r_A} = \frac{1}{2}mv_P^2 - \frac{mgR_E^2}{r_P}.$$

Substituting $r_A = r_P v_P / v_A$ into this equation and rearranging terms, we obtain

$$(v_A - v_P)\left(v_A + v_P - \frac{2gR_E^2}{r_P v_P}\right) = 0.$$

This equation yields the trivial solution $v_A = v_P$ and also the solution for the velocity at apogee:

$$v_A = \frac{2gR_E^2}{r_P v_P} - v_P.$$

Substituting the values of g, R_E, r_P, and v_P, we obtain $v_A = 4370$ m/s and $r_A = 16{,}000$ km.

Critical Thinking
In this example, the satellite's velocity and its distance from the center of the earth at perigee were known. Notice that with this information, you can use Eq. (16.25) to determine the transverse component of the satellite's velocity v_θ at *any* given radial position. You can use conservation of energy to determine the magnitude of the satellite's velocity at the same radial position, which means that you can also determine the radial component of the satellite's velocity v_r. (See Problem 16.87.)

Problems

16.85 At the instant shown ($t_1 = 0$), the position of the 2-kg object's center of mass is $\mathbf{r} = 6\mathbf{i} + 4\mathbf{j} + 2\mathbf{k}$ (m) and its velocity is $\mathbf{v} = -16\mathbf{i} + 8\mathbf{j} - 12\mathbf{k}$ (m/s). No external forces act on the object. What is the object's angular momentum about the origin O at $t_2 = 1$ s?

16.86 The total external force on the 2-kg object is given as a function of time by $\Sigma \mathbf{F} = 2t\mathbf{i} + 4\mathbf{j}$ (N). At time $t_1 = 0$, the object's position and velocity are $\mathbf{r} = \mathbf{0}$ and $\mathbf{v} = \mathbf{0}$.
(a) Use Newton's second law to determine the object's velocity $\mathbf{v}$ and position $\mathbf{r}$ as functions of time.
(b) By integrating $\mathbf{r} \times \Sigma \mathbf{F}$ with respect to time from $t_1 = 0$ to $t_2 = 6$ s, determine the angular impulse about O exerted on the object during this interval of time.
(c) Use the results of part (a) to determine the change in the object's angular momentum from $t_1 = 0$ to $t_2 = 6$ s.

16.87 A satellite is in the elliptic earth orbit shown. Its velocity at perigee A is 8640 m/s. The radius of the earth is 6370 km.
(a) Use conservation of angular momentum to determine the magnitude of the satellite's velocity at apogee C.
(b) Use conservation of energy to determine the magnitude of the velocity at C.

16.88 For the satellite in Problem 16.87, determine the magnitudes of the radial velocity v_r and transverse velocity v_θ at B.

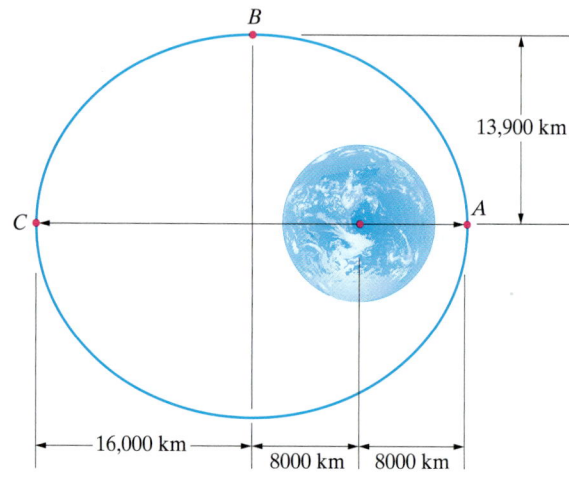

Problems 16.87/16.88

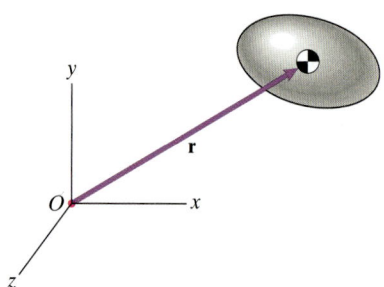

Problems 16.85/16.86

16.89 The bar rotates *in the horizontal plane* about a smooth pin at the origin. The 2-kg sleeve A slides on the smooth bar, and the mass of the bar is negligible in comparison to the mass of the sleeve. The spring constant is $k = 40$ N/m, and the spring is unstretched when $r = 0$. At $t = 0$, the radial position of the sleeve is $r = 0.2$ m and the angular velocity of the bar is $\omega_0 = 6$ rad/s. What is the angular velocity of the bar when $r = 0.25$ m?

16.90 The bar rotates *in the horizontal plane* about a smooth pin at the origin. The 2-kg sleeve A slides on the smooth bar, and the mass of the bar is negligible in comparison to the mass of the sleeve. The spring constant is $k = 40$ N/m, and the spring is unstretched when $r = 0$. At $t = 0$, the radial position of the sleeve is $r = 0.2$ m, its radial velocity is $v_r = 0$, and the angular velocity of the bar is $\omega_0 = 6$ rad/s. What are the angular velocity of the bar and the radial velocity of the sleeve when $r = 0.25$ m?

16.91 A 2-kg disk slides on a smooth horizontal table and is connected to an elastic cord whose tension is $T = 6r$ N, where r is the radial position of the disk in meters. If the disk is at $r = 1$ m and is given an initial velocity of 4 m/s in the transverse direction, what are the magnitudes of the radial and transverse components of its velocity when $r = 2$ m?

16.92 In Problem 16.91, determine the maximum value of r reached by the disk.

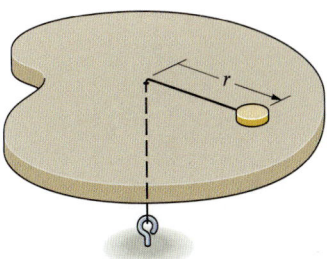

Problems 16.91/16.92

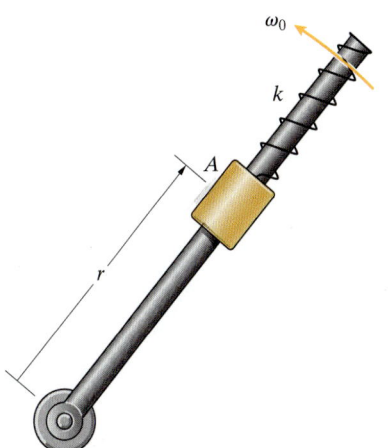

Problems 16.89/16.90

16.93 A 1-kg disk slides on a smooth horizontal table and is attached to a string that passes through a hole in the table.

(a) If the mass moves in a circular path of constant radius $r = 1$ m with a velocity of 2 m/s, what is the tension T?
(b) Starting from the initial condition described in part (a), the tension T is increased in such a way that the string is pulled through the hole at a constant rate until $r = 0.5$ m. Determine the value of T as a function of r while this is taking place.

16.94 In Problem 16.93, how much work is done on the mass in pulling the string through the hole as described in part (b)?

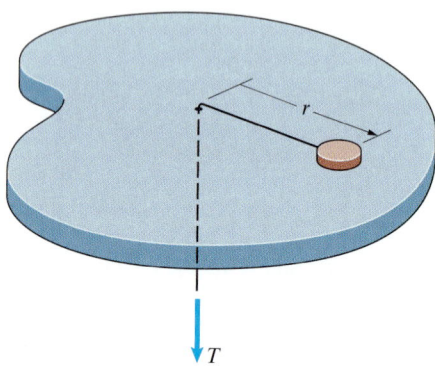

Problems 16.93/16.94

16.95 Two gravity research satellites ($m_A = 250$ kg, $m_B = 50$ kg) are tethered by a cable. The satellites and cable rotate with angular velocity $\omega_0 = 0.25$ revolution per minute. Ground controllers order satellite A to slowly unreel 6 m of additional cable. What is the angular velocity afterward?

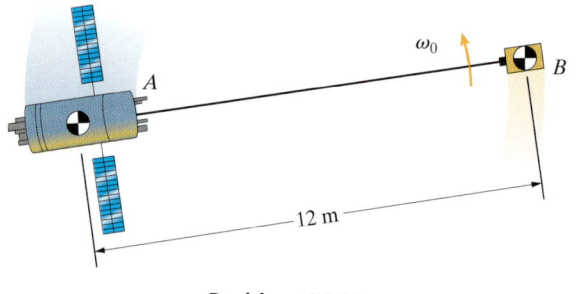

Problem 16.95

16.96 The astronaut moves in the x–y plane at the end of a 10-m tether attached to a large space station at O. The total mass of the astronaut and his equipment is 120 kg.

(a) What is the astronaut's angular momentum about O before the tether becomes taut?
(b) What is the magnitude of the component of his velocity perpendicular to the tether immediately after the tether becomes taut?

16.97 The astronaut moves in the x–y plane at the end of a 10-m tether attached to a large space station at O. The total mass of the astronaut and his equipment is 120 kg. The coefficient of restitution of the "impact" that occurs when he comes to the end of the tether is $e = 0.8$. What are the x and y components of his velocity immediately after the tether becomes taut?

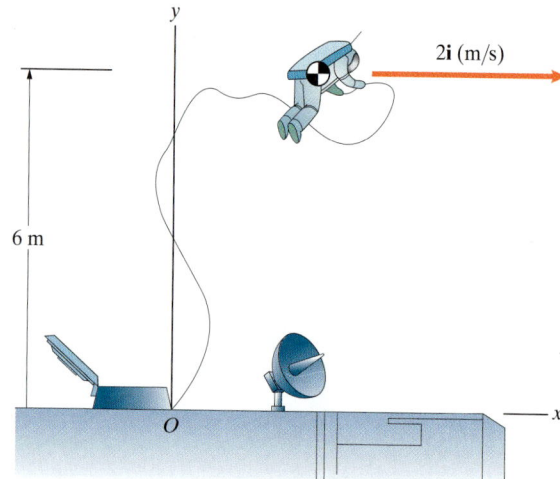

Problems 16.96/16.97

16.98 A ball suspended from a string that goes through a hole in the ceiling at O moves with velocity v_A in a horizontal circular path of radius r_A. The string is then drawn through the hole until the ball moves with velocity v_B in a horizontal circular path of radius r_B. Use the principle of angular impulse and momentum to show that $r_A v_A = r_B v_B$.

    Strategy: Let $\mathbf{e}$ be a unit vector that is perpendicular to the ceiling. Although this is not a central-force problem—the ball's weight does not point toward O—you can show that $\mathbf{e} \cdot (\mathbf{r} \times \Sigma\mathbf{F}) = 0$, so that $\mathbf{e} \cdot \mathbf{H}_O$ is conserved.

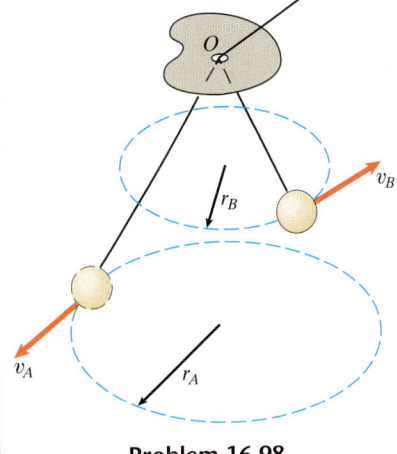

Problem 16.98

16.5 Mass Flows

In this section, we use conservation of linear momentum to determine the force exerted on an object as a result of emitting or absorbing a continuous flow of mass. The resulting equation applies to a variety of situations, including determining the thrust of a rocket and calculating the forces exerted on objects by flows of liquids or granular materials.

    Suppose that an object of mass m and velocity $\mathbf{v}$ is subjected to no external forces (Fig. 16.20a) and emits an element of mass Δm_f with velocity $\mathbf{v}_f$ *relative to the object* (Fig. 16.20b). We denote the new velocity of the object by $\mathbf{v} + \Delta\mathbf{v}$. The linear momentum of the object before the element of mass is emitted equals the total linear momentum of the object and the element afterward:

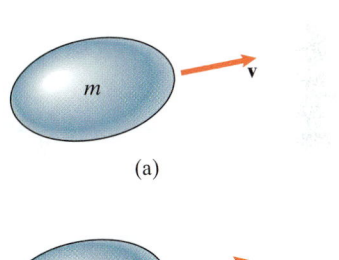

(a)

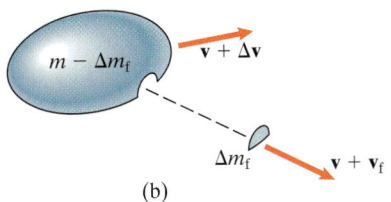

(b)

$$mv = (m - \Delta m_f)(\mathbf{v} + \Delta\mathbf{v}) + \Delta m_f(\mathbf{v} + \mathbf{v}_f).$$

Evaluating the products and simplifying, we obtain

$$m\Delta\mathbf{v} + \Delta m_f\mathbf{v}_f - \Delta m_f\Delta\mathbf{v} = 0. \tag{16.26}$$

Figure 16.20
An object's mass and velocity (a) before and (b) after emitting an element of mass.

Now we assume that instead of shedding a discrete element of mass, the object emits a continuous flow of mass and that Δm_f is the amount emitted in an interval of time Δt. We divide Eq. (16.26) by Δt and write the resulting equation as

$$m\frac{\Delta \mathbf{v}}{\Delta t} + \frac{\Delta m_f}{\Delta t}\mathbf{v}_f - \frac{\Delta m_f}{\Delta t}\frac{\Delta \mathbf{v}}{\Delta t}\Delta t = 0.$$

Taking the limit of this equation as $\Delta t \to 0$, we obtain

$$-\frac{dm_f}{dt}\mathbf{v}_f = m\mathbf{a},$$

where $\mathbf{a}$ is the acceleration of the object's center of mass and the term dm_f/dt is the *mass flow rate*—the rate at which mass flows from the object. Comparing this equation with Newton's second law, we conclude that a flow of mass *from* an object exerts a force

$$\mathbf{F}_f = -\frac{dm_f}{dt}\mathbf{v}_f \tag{16.27}$$

on the object. The force is proportional to the mass flow rate and to the magnitude of the relative velocity of the flow, and its direction is *opposite* to the direction of the relative velocity. Conversely, a flow of mass *to* an object exerts a force in the *same* direction as the relative velocity.

Example 16.9   Force Resulting from a Mass Flow

The rocket sled shown in Fig. 16.21 is slowed by a water brake after its motor has burned out. There is a trough of water between the tracks where the sled is brought to rest. A tube from the sled extends into the water with its open end pointing forward, so that water enters the tube in the direction parallel to the x axis as the sled moves forward. The other end of the tube points upward, so that the water flows out in the direction parallel to the y axis. If v is the sled's velocity, the water enters the tube with velocity v relative to the sled and flows out with the same velocity. The mass flow rate of water through the tube is $\rho v A$, where $\rho = 1000 \text{ kg/m}^3$ is the mass density of the water and $A = 0.01 \text{ m}^2$ is the cross-sectional area of the tube. At an instant when $v = 300 \text{ m/s}$, what forces are exerted on the sled by the flows of water entering and leaving it?

Strategy
We can use Eq. (16.27) to determine the forces exerted by the flows of water entering and leaving the sled.

Figure 16.21

Solution

Relative to the sled, the velocity vector of the water entering it is $\mathbf{v}_f = -v\mathbf{i}$. Because water is *entering* the sled, $dm_f/dt = -\rho vA$. The force exerted on the sled is

$$\mathbf{F}_f = -\frac{dm_f}{dt}\mathbf{v}_f = -(-\rho vA)(-v\mathbf{i}) = -\rho v^2\, A\mathbf{i}$$

$$= -(1000 \text{ kg/m}^3)(300 \text{ m/s})^2(0.01 \text{ m}^2)\mathbf{i}$$

$$= -900{,}000\,\mathbf{i} \text{ (N)}.$$

The water entering the sled subjects it to a braking force of 900 kN (202,000 lb).

The water leaves the sled with relative velocity $\mathbf{v}_f = v\mathbf{j}$, and the mass flow rate is $dm_f/dt = \rho vA$. The force exerted on the sled is

$$\mathbf{F}_f = -\frac{dm_f}{dt}\mathbf{v}_f = -(\rho vA)(v\mathbf{j}) = -\rho v^2\, A\mathbf{j}$$

$$= -(1000 \text{ kg/m}^3)(300 \text{ m/s})^2(0.01 \text{ m}^2)\mathbf{j}$$

$$= -900{,}000\mathbf{j} \text{ (N)}.$$

The water leaving the sled subjects it to a downward force of 900 kN.

Critical Thinking

As this example illustrates, you must use care in determining the signs of the terms in Eq. (16.27). In deriving it, we assumed that dm_f/dt is the rate at which mass is *leaving* an object. If mass is entering the object instead, dm_f/dt is negative. The term $\mathbf{v}_f$ is the velocity of the flow of mass *relative to the object*. The sled's velocity in terms of the reference frame shown in Fig. 16.21 is $v\mathbf{i}$, so the velocity of the stationary water entering the sled *relative to the sled* is $-v\mathbf{i}$.

Example 16.10 Thrust of a Rocket

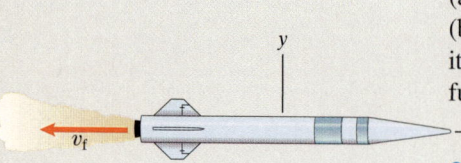

Figure 16.22

The classic example of a force created by a mass flow is the rocket. The one in Fig. 16.22 has a uniform, constant exhaust velocity v_f parallel to the x axis.
(a) What force is exerted on the rocket by the mass flow of its exhaust?
(b) If the force determined in part (a) is the only force acting on the rocket, and it starts from rest with an initial mass m_0, determine the rocket's velocity as a function of its mass m.

Strategy
(a) Equation (16.27) gives the force exerted on the rocket in terms of the exhaust velocity and the mass flow rate of fuel.
(b) We can use Newton's second law to obtain an equation for the rocket's velocity as a function of its mass.

Solution
(a) In terms of the coordinate system in Fig. 16.22, the velocity vector of the exhaust is $\mathbf{v}_f = -v_f\mathbf{i}$. From Eq. (16.27), the force exerted on the rocket is

$$\mathbf{F}_f = -\frac{dm_f}{dt}\mathbf{v}_f = \frac{dm_f}{dt}v_f\mathbf{i},$$

where dm_f/dt is the mass flow rate of the rocket's fuel. The force exerted on the rocket by its exhaust is toward the right, opposite to the direction of the flow of the exhaust.
(b) Newton's second law applied to the rocket is

$$\Sigma F_x = \frac{dm_f}{dt}v_f = m\frac{dv_x}{dt},$$

where m is the rocket's mass. The mass flow rate of fuel is the rate at which the rocket's mass is being consumed. Therefore, the rate of change of the mass of the rocket is

$$\frac{dm}{dt} = -\frac{dm_f}{dt}.$$

Using this expression, we can write Newton's second law as

$$dv_x = -v_f\frac{dm}{m}.$$

Because the exhaust velocity v_f is constant, we can integrate the preceding equation to determine the velocity of the rocket as a function of its mass:

$$\int_0^{v_x} dv_x = -v_f\int_{m_0}^{m}\frac{dm}{m}.$$

The result is

$$v_x = v_f \ln\left(\frac{m_0}{m}\right).$$

Critical Thinking
The velocity attained by the rocket is determined by the exhaust velocity and the amount of mass expended. Thus, a rocket can gain more velocity by expending more of its mass. However, notice that increasing the ratio m_0/m from 10 to 100 increases the velocity attained by only a factor of two. In contrast, increasing the exhaust velocity results in a proportional increase in the rocket's velocity. Rocket

engineers use fuels such as liquid oxygen and liquid hydrogen because they produce a relatively large exhaust velocity. This objective has also led to research on rocket engines that use electromagnetic fields to accelerate charged particles of fuel to large velocities.

Example 16.11   Force Resulting from a Mass Flow

A horizontal stream of water with velocity v_0 and mass flow rate dm_f/dt hits a plate that deflects the water in the horizontal plane through an angle θ (Fig. 16.23). Assume that the magnitude of the velocity of the water when it leaves the plate is approximately equal to v_0. What force is exerted on the plate by the water?

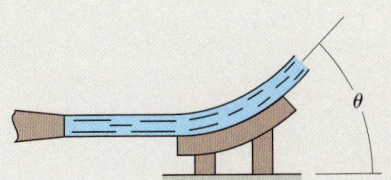

Figure 16.23

Strategy
We can determine the force exerted on the plate by treating the part of the stream in contact with the plate as an object with mass flows entering and leaving it.

Solution
In Fig. a, we draw the free-body diagram of the part of the stream in contact with the plate. Streams of mass with velocity v_0 enter and leave this "object," and $\mathbf{F_P}$ is the force exerted on the stream by the plate. We wish to determine the force $-\mathbf{F_P}$ exerted on the plate by the stream. First we consider the departing stream of water. The mass flow rate of water leaving the free-body diagram must be equal to the mass flow rate entering. In terms of the coordinate system shown, the velocity of the departing stream is

$$\mathbf{v_f} = v_0 \cos \theta \, \mathbf{i} + v_0 \sin \theta \, \mathbf{j}.$$

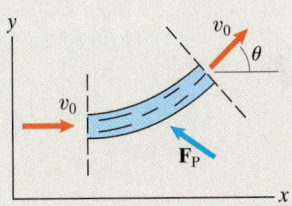

(a) Free-body diagram of the stream.

Let $\mathbf{F_D}$ be the force exerted on the object by the departing stream. From Eq. (16.27),

$$\mathbf{F_D} = -\frac{dm_f}{dt}\mathbf{v_f} = -\frac{dm_f}{dt}(v_0 \cos \theta \, \mathbf{i} + v_0 \sin \theta \, \mathbf{j}).$$

The velocity of the entering stream is $\mathbf{v_f} = v_0 \mathbf{i}$. Since this flow is entering the object rather than leaving it, the resulting force $\mathbf{F_E}$ is in the same direction as the relative velocity:

$$\mathbf{F_E} = \frac{dm_f}{dt}\mathbf{v_f} = \frac{dm_f}{dt}v_0 \mathbf{i}.$$

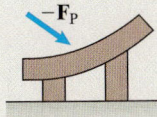

(b) Force exerted on the plate.

The sum of the forces on the free-body diagram must equal zero:

$$\mathbf{F_D} + \mathbf{F_E} + \mathbf{F_P} = \mathbf{0}.$$

Hence, the force exerted on the plate by the water is (Fig. b)

$$-\mathbf{F_P} = \mathbf{F_D} + \mathbf{F_E} = \frac{dm_f}{dt}v_0[(1 - \cos \theta)\mathbf{i} - \sin \theta \, \mathbf{j}].$$

Critical Thinking
This simple example affords an insight into how turbine blades and airplane wings can create forces by deflecting streams of liquid or gas (Fig. c).

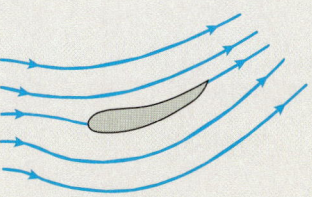

(c) Pattern of moving fluid around an airplane wing.

Design Example 16.12 | Jet Engines

In a turbojet engine (Fig. 16.24), a mass flow rate dm_c/dt of inlet air enters the compressor with velocity v_i. The air is mixed with fuel and ignited in the combustion chamber. The mixture then flows through the turbine, which powers the compressor. The exhaust, with a mass flow rate equal to that of the air plus the mass flow rate of the fuel $(dm_c/dt + dm_f/dt)$, exits at a high exhaust velocity v_e, exerting a large force on the engine. Suppose that $dm_c/dt = 13.5$ kg/s and $dm_f/dt = 0.130$ kg/s. The inlet air velocity is $v_i = 120$ m/s and the exhaust velocity is $v_e = 480$ m/s. What is the engine's thrust?

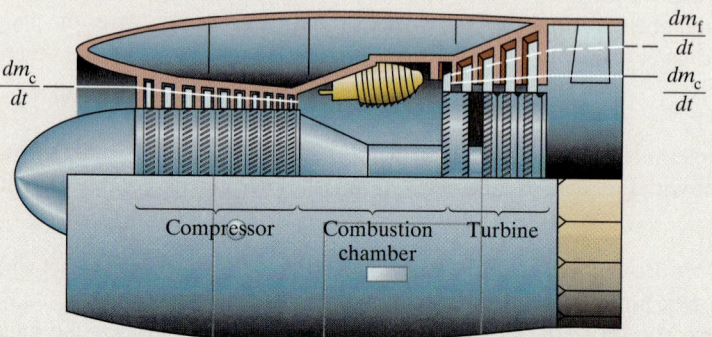

Figure 16.24

Strategy

We can determine the engine's thrust by using Eq. (16.27). We must include both the force exerted by the engine's exhaust and the force exerted by the mass flow of air entering the compressor to determine the net thrust.

Solution

The engine's exhaust exerts a force to the left equal to the product of the mass flow rate of the fuel–air mixture and the exhaust velocity. The inlet air exerts a force to the right equal to the product of the mass flow rate of the inlet air and the inlet velocity. The engine's thrust (the net force to the left) is

$$T = \left(\frac{dm_c}{dt} + \frac{dm_f}{dt} \right) v_e - \frac{dm_c}{dt} v_i$$

$$= (13.5 + 0.130)(480) - (13.5)(120)$$

$$= 4920 \text{ N}.$$

Design Issues

The jet engine was developed in Europe in the years just prior to World War II. Although the turbojet engine in Fig. 16.24 was a successful design that dominated both military and commercial aviation for many years, it has the drawback that it consumes a relatively large amount of fuel. During the last 30 years, the fan-jet engine, shown in Fig. 16.25, has become the most commonly used design, particularly for commercial airplanes. Part of its thrust is provided by air that is accelerated by the fan. The ratio of the mass flow rate of air entering the fan, dm_b/dt, to the mass flow rate of air entering the compressor, dm_c/dt, is called the *bypass ratio*.

The force exerted by a jet engine's exhaust equals the product of the mass flow rate and the exhaust velocity. In the fan-jet engine, the air passing through the fan is not heated by the combustion of fuel and therefore has a higher density than the exhaust of the turbojet engine. As a result, the fan-jet engine can provide a given thrust with a lower average exhaust velocity. Since the work that must be expended to create the thrust depends on the kinetic energy of the exhaust, the fan-jet engine creates thrust more efficiently.

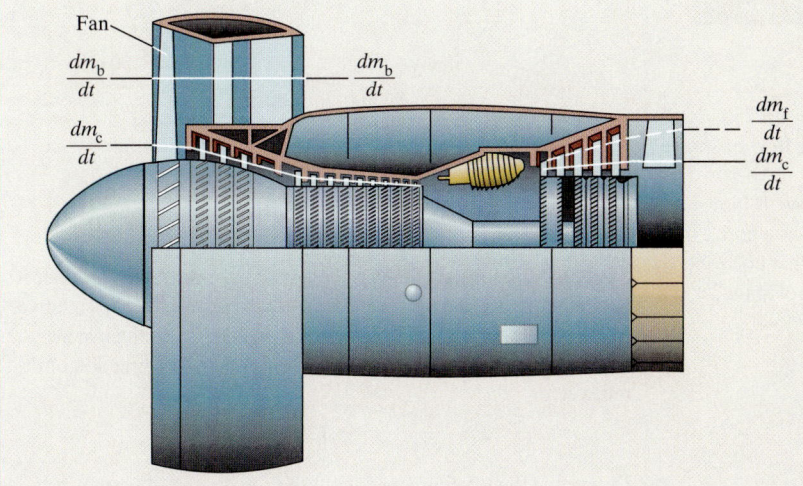

Figure 16.25
A fan-jet engine. Part of the entering mass flow of air is accelerated by the fan and does not enter the compressor.

Problems

16.99 The Cheverton fire-fighting and rescue boat can pump 3.8 kg/s of water from each of its two pumps at a velocity of 44 m/s. If both pumps point in the same direction, what total force do they exert on the boat?

Problem 16.99

16.100 The mass flow rate of water through the nozzle is 1.6 slugs/s. Determine the magnitude of the horizontal force exerted on the truck by the flow of water.

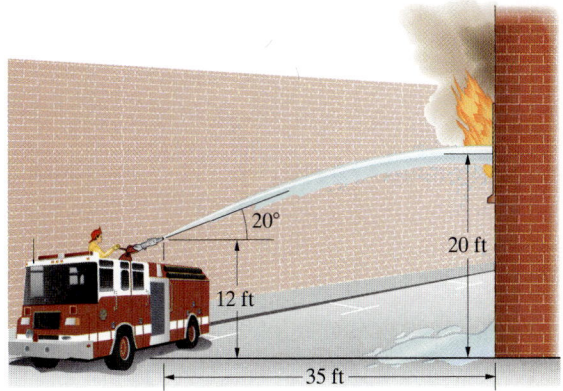

Problem 16.100

16.101 The front-end loader moves at a constant speed of 2 mi/h scooping up iron ore. The constant horizontal force exerted on the loader by the road is 400 lb. What weight of iron ore is scooped up in 3 s?

Problem 16.101

16.102 The snowblower moves at 1 m/s and scoops up 750 kg/s of snow. Determine the force exerted by the entering flow of snow.

16.103 The snowblower scoops up 750 kg/s of snow. It blows the snow out the side at 45° above the horizontal from a port 2 m above the ground and the snow lands 20 m away. What horizontal force is exerted on the blower by the departing flow of snow?

Problems 16.102/16.103

16.104 A nozzle ejects a stream of water horizontally at 40 m/s with a mass flow rate of 30 kg/s, and the stream is deflected in the horizontal plane by a plate. Determine the force exerted on the plate by the stream in cases (a), (b), and (c).

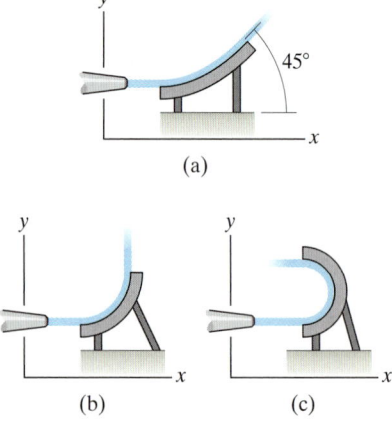

Problem 16.104

16.105* A stream of water with velocity $80\mathbf{i}$ (m/s) and a mass flow rate of 6 kg/s strikes a turbine blade moving with constant velocity $20\mathbf{i}$ (m/s).

(a) What force is exerted on the blade by the water?
(b) What is the magnitude of the velocity of the water as it leaves the blade?

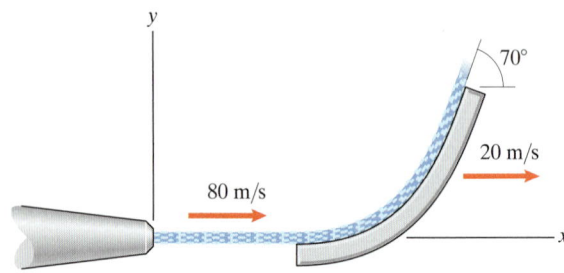

Problem 16.105

16.106 At the instant shown, the nozzle A of the lawn sprinkler is located at $(0.1, 0, 0)$ m. Water exits each nozzle at 8 m/s relative to the nozzle with a mass flow rate of 0.22 kg/s. At the instant shown, the flow relative to the nozzle at A is in the direction of the unit vector

$$\mathbf{e} = \frac{1}{\sqrt{3}}\mathbf{i} - \frac{1}{\sqrt{3}}\mathbf{j} + \frac{1}{\sqrt{3}}\mathbf{k}.$$

Determine the total moment about the z axis exerted on the sprinkler by the flows from all four nozzles.

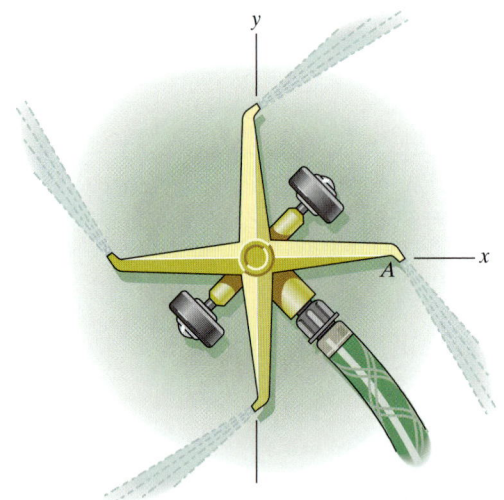

Problem 16.106

16.107 A 45-kg/s flow of gravel exits the chute at 2 m/s and falls onto a conveyer moving at 0.3 m/s. Determine the components of the force exerted on the conveyer by the flow of gravel if $\theta = 0$.

16.108 Solve Problem 16.107 if $\theta = 30°$.

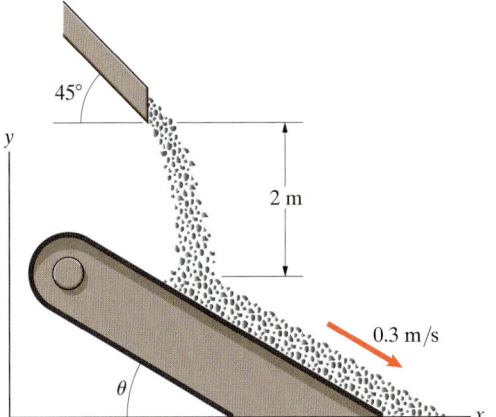

Problems 16.107/16.108

16.109 Suppose that you are designing a toy car that will be propelled by water that squirts from an internal tank at 10 ft/s relative to the car. The total weight of the car and its water "fuel" is to be 2 lb. If you want the car to achieve a maximum speed of 12 ft/s, what part of the total weight must be water?

Problem 16.109

16.110 The rocket consists of a 1000-kg payload and a 9000-kg booster. Eighty percent of the booster's mass is fuel, and its exhaust velocity is 1200 m/s. If the rocket starts from rest and external forces are neglected, what velocity will it attain?

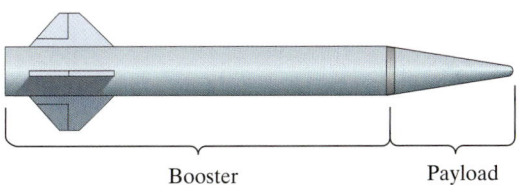

Booster   Payload

Problem 16.110

16.111* The rocket consists of a 1000-kg payload and a booster. The booster has two stages whose total mass is 9000 kg. Eighty percent of the mass of each stage is fuel, and the exhaust velocity of each stage is 1200 m/s. When the fuel of stage 1 is expended, it is discarded and the motor of stage 2 is ignited. Assume that the rocket starts from rest and neglect external forces. Determine the velocity attained by the rocket if the masses of the stages are $m_1 = 6000$ kg and $m_2 = 3000$ kg. Compare your result to the answer to Problem 16.110.

1   2   Payload

Problem 16.111

16.112 A rocket of initial mass m_0 takes off straight up. Its exhaust velocity v_f and the mass flow rate of its engine $\dot{m}_f = dm_f/dt$ are constant. Show that, during the initial part of the flight, when aerodynamic drag is negligible, the rocket's upward velocity as a function of time is

$$v = v_f \ln\left(\frac{m_0}{m_0 - \dot{m}_f t}\right) - gt.$$

Problem 16.112

16.113 The mass of the rocket sled in Example 16.9 is 440 kg. Assuming that the only significant force acting on the sled in the direction of its motion is the force exerted by the flow of water entering it, what distance is required for the sled to decelerate from 300 m/s to 100 m/s?

16.114* Suppose that you grasp the end of a chain that weighs 3 lb/ft and lift it straight up off the floor at a constant speed of 2 ft/s.
(a) Determine the upward force F you must exert as a function of the height s.
(b) How much work do you do in lifting the top of the chain to $s = 4$ ft?

    Strategy: Treat the part of the chain you have lifted as an object that is gaining mass.

16.115* Solve Problem 16.114, assuming that you lift the end of the chain straight up off the floor with a constant acceleration of 2 ft/s^2.

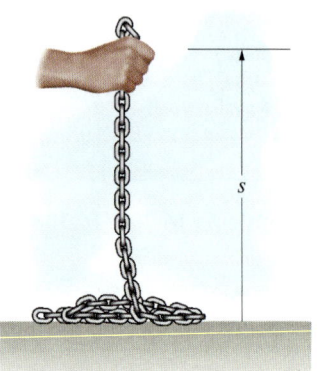

Problems 16.114/16.115

16.116* It has been suggested that a heavy chain could be used to gradually stop an airplane that rolls past the end of the runway. A hook attached to the end of the chain engages the plane's nose wheel, and the plane drags an increasing length of the chain as it rolls. Let m be the airplane's mass and v_0 its initial velocity, and let ρ_L be the mass per unit length of the chain. Neglecting friction and aerodynamic drag, what is the airplane's velocity as a function of s?

16.117* In Problem 16.116, the frictional force exerted on the chain by the ground would actually dominate other forces as the distance s increases. If the coefficient of kinetic friction between the chain and the ground is μ_k and you neglect all forces except the frictional force, what is the airplane's velocity as a function of s?

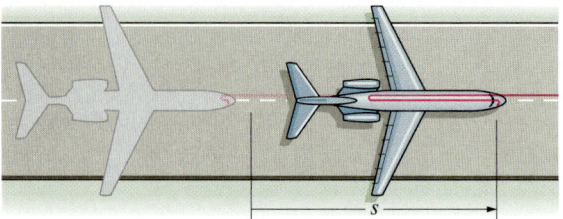

Problems 16.116/16.117

Design Experience

Problems 16.118–16.122 are related to Design Example 16.12.

16.118 The turbojet engine in Fig. 16.24 is being operated on a test stand. The mass flow rate of air entering the compressor is 13.5 kg/s and the mass flow rate of fuel is 0.13 kg/s. The effective velocity of the air entering the compressor is zero, and the exhaust velocity is 500 m/s. What is the thrust of the engine?

16.119 The turbojet engine in Fig. 16.24 is in an airplane flying at 400 km/h. The mass flow rate of air entering the compressor is 13.5 kg/s and the mass flow rate of fuel is 0.13 kg/s. The effective velocity of the air entering the inlet is equal to the airplane's velocity, and the exhaust velocity (relative to the airplane) is 500 m/s. What is the thrust of the engine?

16.120 A turbojet engine's thrust reverser causes the exhaust to exit the engine at 20° from the engine centerline. The mass flow rate of air entering the compressor is 44 kg/s, and the air enters at 60 m/s. The mass flow rate of fuel is 1.5 kg/s, and the exhaust velocity is 370 m/s. What braking force does the engine exert on the airplane?

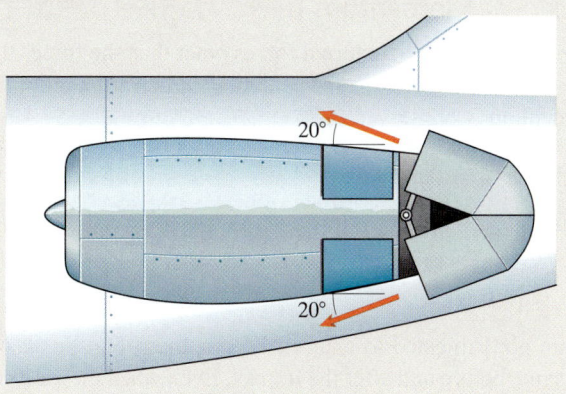

Problem 16.120

16.121 The 13,600-kg airplane is moving at 400 km/h. The total mass flow rate of air entering the compressors of the plane's turbo-jet engines is 280 kg/s, and the total mass flow rate of fuel is 2.6 kg/s. The effective velocity of the air entering the compressors is equal to the airplane's velocity, and the exhaust velocity is 480 m/s. The ratio of the lift force L to the drag force D is 6, and the z component of the airplane's acceleration is zero. What is the x component of the acceleration?

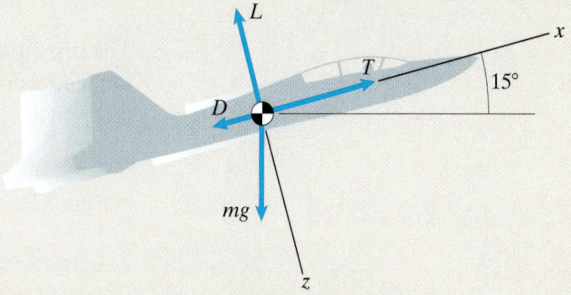

Problem 16.121

16.122 The fan-jet engine in Fig. 16.25 is similar to the Pratt and Whitney JT9D-3A engine used on early models of the Boeing 747. When the airplane begins its takeoff run, the velocity of the air entering the compressor and fan is negligible. A mass flow rate of 38.5 slugs/s enters the fan and is accelerated to 885 ft/s. A mass flow rate of 7.7 slugs/s enters the compressor. The mass flow rate of fuel is 0.23 slugs/s, and the exhaust velocity is 1190 ft/s. (a) What is the bypass ratio? (b) What is the thrust of the engine? (c) If the airplane weighs 500,000 lb, what is its initial acceleration? (It has four engines.)

CHAPTER SUMMARY

Principle of Impulse and Momentum

The linear impulse applied to an object during an interval of time is equal to the change in the object's linear momentum:

$$\int_{t_1}^{t_2} \Sigma \mathbf{F} \, dt = m\mathbf{v}_2 - m\mathbf{v}_1. \tag{16.1}$$

This result can also be expressed in terms of the average of the total force with respect to time:

$$(t_2 - t_1)\Sigma \mathbf{F}_{\mathrm{av}} = m\mathbf{v}_2 - m\mathbf{v}_1. \tag{16.2}$$

The principle of impulse and momentum can also be expressed in terms of the tangential component of the total force and the linear momentum along the object's path:

$$\int_{t_1}^{t_2} \Sigma F_t \, dt = mv_2 - mv_1. \tag{16.3}$$

The average of the tangential component of the total force with respect to time is related to the change in the linear momentum along the path by

$$(t_2 - t_1)\Sigma F_{t\,\mathrm{av}} = mv_2 - mv_1. \tag{16.4}$$

Conservation of Linear Momentum

If objects A and B are not subjected to external forces other than the forces they exert on each other (or if the effects of other external forces are negligible), their total linear momentum is conserved:

$$m_A\mathbf{v}_A + m_B\mathbf{v}_B = \text{constant}. \tag{16.6}$$

Also, the velocity of the common center of mass of A and B is constant.

Impacts

If colliding objects are not subjected to external forces, their total linear momentum must be the same before and after the impact. Even when they are subjected to external forces, the force of the impact is often so large, and its duration so brief, that the effect of the external forces on their motions during the impact is negligible.

If objects A and B adhere and remain together after they collide, they are said to undergo a *perfectly plastic impact*. The velocity of their common center of mass before and after the impact is given by

$$\mathbf{v} = \frac{m_A\mathbf{v}_A + m_B\mathbf{v}_B}{m_A + m_B}. \tag{16.9}$$

Central Impacts

In a *direct central impact* (Fig. a), linear momentum is conserved, or

$$m_A v_A + m_B v_B = m_A v_A' + m_B v_B', \tag{16.10}$$

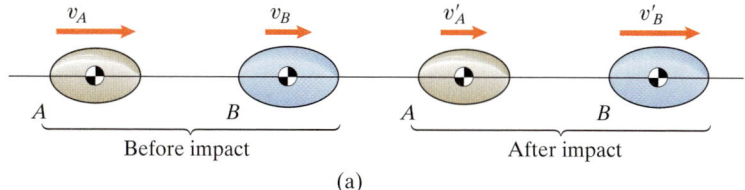

(a)

and the velocities are related by the *coefficient of restitution*:

$$e = \frac{v'_B - v'_A}{v_A - v_B}. \tag{16.16}$$

If $e = 0$, the impact is perfectly plastic. If $e = 1$, the total kinetic energy is conserved and the impact is called *perfectly elastic*.

In an *oblique central impact* (Fig. b), the components of velocity in the y and z directions are unchanged by the impact:

$$(\mathbf{v}'_A)_y = (\mathbf{v}_A)_y, \qquad (\mathbf{v}'_B)_y = (\mathbf{v}_B)_y,$$

$$(\mathbf{v}'_A)_z = (\mathbf{v}_A)_z, \qquad (\mathbf{v}'_B)_z = (\mathbf{v}_B)_z. \tag{16.17}$$

In the x direction, linear momentum is conserved, or

$$m_A(\mathbf{v}_A)_x + m_B(\mathbf{v}_B)_x = m_A(\mathbf{v}'_A)_x + m_B(\mathbf{v}'_B)_x, \tag{16.18}$$

and the velocity components are related by the coefficient of restitution:

$$e = \frac{(\mathbf{v}'_B)_x - (\mathbf{v}'_A)_x}{(\mathbf{v}_A)_x - (\mathbf{v}_B)_x}. \tag{16.19}$$

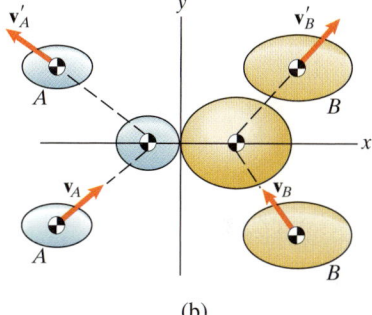

(b)

Principle of Angular Impulse and Momentum

For a fixed point O, the angular impulse applied to an object during an interval of time is equal to the change in the object's angular momentum about O; that is,

$$\int_{t_1}^{t_2} (\mathbf{r} \times \Sigma\mathbf{F}) \, dt = (\mathbf{H}_O)_2 - (\mathbf{H}_O)_1, \tag{16.23}$$

where the angular momentum is

$$\mathbf{H}_O = \mathbf{r} \times m\mathbf{v} \tag{16.22}$$

Central-Force Motion

If the total force acting on an object remains directed toward a fixed point, the object is said to be in *central-force motion*, and its angular momentum about the fixed point is conserved:

$$\mathbf{H}_O = \text{constant.} \tag{16.24}$$

In plane central-force motion, the product of the radial distance and the transverse component of the velocity is constant:

$$rv_\theta = \text{constant.} \tag{16.25}$$

Mass Flows

A flow of mass *from* an object with velocity $\mathbf{v}_f$ *relative to the object* exerts a force

$$\mathbf{F}_f = -\frac{dm_f}{dt}\mathbf{v}_f \qquad (16.27)$$

on the object, where dm_f/dt is the *mass flow rate*. The direction of the force is opposite to the direction of the relative velocity. A flow of mass *to* an object exerts a force in the same direction as the relative velocity.

Review Problems

16.123 The total external force on a 10-kg object is constant and equal to $90\mathbf{i} - 60\mathbf{j} + 20\mathbf{k}$ (N). At $t = 2$ s, the object's velocity is $-8\mathbf{i} + 6\mathbf{j}$ (m/s).

(a) What impulse is applied to the object from $t = 2$ s to $t = 4$ s?
(b) What is the object's velocity at $t = 4$ s?

16.124 The total external force on an object is $\mathbf{F} = 10t\mathbf{i} + 60\mathbf{j}$ (lb). At $t = 0$, the object's velocity is $\mathbf{v} = 20\mathbf{j}$ (ft/s). At $t = 12$ s, the x component of its velocity is 48 ft/s.

(a) What impulse is applied to the object from $t = 0$ to $t = 6$ s?
(b) What is the object's velocity at $t = 6$ s?

16.125 An aircraft arresting system is used to stop airplanes whose braking systems fail. The system stops a 47,500-kg airplane moving at 80 m/s in 9.15 s.

(a) What impulse is applied to the airplane during the 9.15 s?
(b) What is the average deceleration to which the passengers are subjected?

Problem 16.125

16.126 The 1895 Austrian 150-mm howitzer had a 1.94-m barrel, possessed a muzzle velocity of 300 m/s, and fired a 38-kg shell. If the shell took 0.013 s to travel the length of the barrel, what average force was exerted on the shell?

16.127 An athlete throws a shot weighing 16 lb. When he releases it, the shot is 7 ft above the ground and its components of velocity are $v_x = 31$ ft/s and $v_y = 26$ ft/s.

(a) Suppose the athlete accelerates the shot from rest in 0.8 s, and assume as a first approximation that the force $\mathbf{F}$ he exerts on the shot is constant. Use the principle of impulse and momentum to determine the x and y components of $\mathbf{F}$.
(b) What is the horizontal distance from the point where he releases the shot to the point where it strikes the ground?

Problem 16.127

16.128 The 6000-lb pickup truck A moving at 40 ft/s collides with the 4000-lb car B moving at 30 ft/s.

(a) What is the magnitude of the velocity of their common center of mass after the impact?

(b) Treat the collision as a perfectly plastic impact. How much kinetic energy is lost?

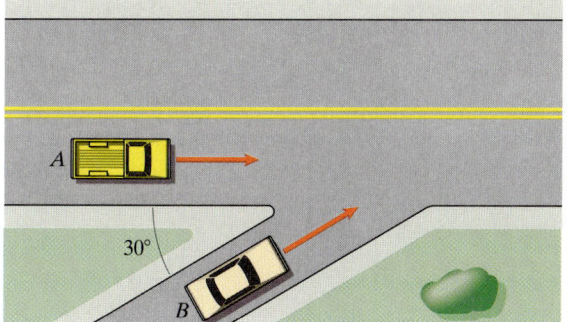

Problem 16.128

16.129 Two hockey players ($m_A = 80$ kg, $m_B = 90$ kg) converging on the puck at $x = 0$, $y = 0$ become entangled and fall. Before the collision, $v_A = 9\mathbf{i} + 4\mathbf{j}$ (m/s) and $v_B = -3\mathbf{i} + 6\mathbf{j}$ (m/s). If the coefficient of kinetic friction between the players and the ice is $\mu_k = 0.1$, what is their approximate position when they stop sliding?

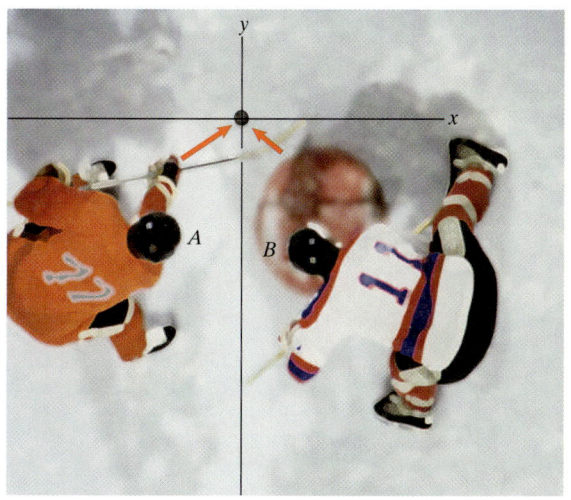

Problem 16.129

16.130 The cannon weighed 400 lb, fired a cannonball weighing 10 lb, and had a muzzle velocity of 200 ft/s. For the 10° elevation angle shown, determine (a) the velocity of the cannon after it was fired and (b) the distance the cannonball traveled. (Neglect drag.)

Problem 16.130

16.131 A 1-kg ball moving horizontally at 12 m/s strikes a 10-kg block. The coefficient of restitution of the impact is $e = 0.6$, and the coefficient of kinetic friction between the block and the inclined surface is $\mu_k = 0.4$. What distance does the block slide before stopping?

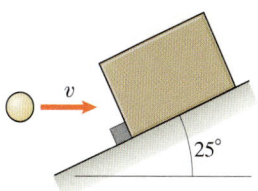

Problem 16.131

16.132 A Peace Corps volunteer designs the simple device shown for drilling water wells in remote areas. A 70-kg "hammer," such as a section of log or a steel drum partially filled with concrete, is hoisted to $h = 1$ m and allowed to drop onto a protective cap on the section of pipe being pushed into the ground. The combined mass of the cap and section of pipe is 20 kg. Assume that the coefficient of restitution is nearly zero.

(a) What is the velocity of the cap and pipe immediately after the impact?

(b) If the pipe moves 30 mm downward when the hammer is dropped, what resistive force was exerted on the pipe by the ground? (Assume that the resistive force is constant during the motion of the pipe.)

Hammer

h

Problem 16.132

16.133 A tugboat (mass = 40,000 kg) and a barge (mass = 160,000 kg) are stationary with a slack hawser connecting them. The tugboat accelerates to 2 knots (1 knot = 1852 m/h) before the hawser becomes taut. Determine the velocities of the tugboat and the barge just after the hawser becomes taut (a) if the "impact" is perfectly plastic ($e = 0$) and (b) if the "impact" is perfectly elastic ($e = 1$). Neglect the forces exerted by the water and the tugboat's engines.

16.134 In Problem 16.133, determine the magnitude of the impulsive force exerted on the tugboat in the two cases if the duration of the "impact" is 4 s. Neglect the forces exerted by the water and the tugboat's engines during this period.

Problems 16.133/16.134

16.135 The 10-kg mass A is moving at 5 m/s when it is 1 m from the stationary 10-kg mass B. The coefficient of kinetic friction between the floor and the two masses is $\mu_k = 0.6$, and the coefficient of restitution of the impact is $e = 0.5$. Determine how far B moves from its initial position as a result of the impact.

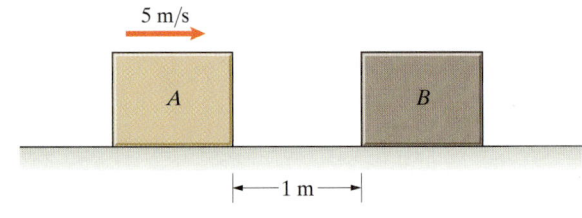

5 m/s

A

B

$\leftarrow$ 1 m $\rightarrow$

Problem 16.135

16.136 The kinetic coefficients of friction between the 5-kg crates A and B and the inclined surface are 0.1 and 0.4, respectively. The coefficient of restitution between the crates is $e = 0.8$. If the crates are released from rest in the positions shown, what are the magnitudes of their velocities immediately after they collide?

16.137 Solve Problem 16.136 if crate A has a velocity of 0.2 m/s down the inclined surface and crate B is at rest when the crates are in the positions shown.

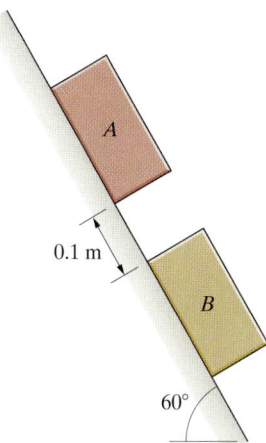

Problems 16.136/16.137

16.138 A small object starts from rest at A and slides down the smooth ramp. The coefficient of restitution of the impact of the object with the floor is $e = 0.8$. At what height above the floor does the object hit the wall?

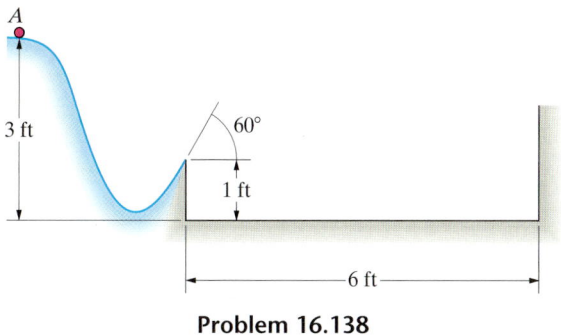

Problem 16.138

16.139 The cue gives the cue ball A a velocity of magnitude 3 m/s. The angle $\beta = 0$ and the coefficient of restitution of the impact of the cue ball and the eight ball B is $e = 1$. If the magnitude of the eight ball's velocity after the impact is 0.9 m/s, what was the coefficient of restitution of the cue ball's impact with the cushion? (The balls are of equal mass.)

16.140 What is the solution to Problem 16.139 if the angle $\beta = 10°$?

16.141 What is the solution to Problem 16.139 if the angle $\beta = 15°$ and the coefficient of restitution of the impact between the two balls is $e = 0.9$?

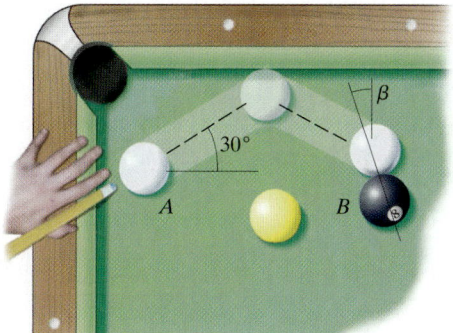

Problems 16.139–16.141

16.142 A ball is given a horizontal velocity of 3 m/s at 2 m above the smooth floor. Determine the distance D between the ball's first and second bounces if the coefficient of restitution is $e = 0.6$.

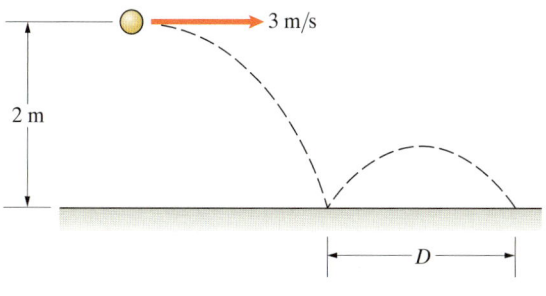

Problem 16.142

16.143* A basketball dropped on the floor from a height of 4 ft rebounds to a height of 3 ft. In the layup shot shown, the magnitude of the ball's velocity is 5 ft/s, and the angles between its velocity vector and the positive coordinate axes are $\theta_x = 42°$, $\theta_y = 68°$, and $\theta_z = 124°$ just before it hits the backboard. What are the magnitude of its velocity and the angles between its velocity vector and the positive coordinate axes just after the ball hits the backboard?

16.144* In Problem 16.143, the basketball's diameter is 9.5 in, the coordinates of the center of the basket rim are $x = 0$, $y = 0$, $z = 12$ in, and the backboard lies in the $x–y$ plane. Determine the x and y coordinates of the point where the ball must hit the backboard so that the center of the ball passes through the center of the basket rim.

Problems 16.143/16.144

16.145 A satellite at $r_0 = 10{,}000$ mi from the center of the earth is given an initial velocity $v_0 = 20{,}000$ ft/s in the direction shown. Determine the magnitude of the transverse component of the satellite's velocity when $r = 20{,}000$ mi. (The radius of the earth is 3960 mi.)

16.146 In Problem 16.145, determine the magnitudes of the radial and transverse components of the satellite's velocity when $r = 15{,}000$ mi.

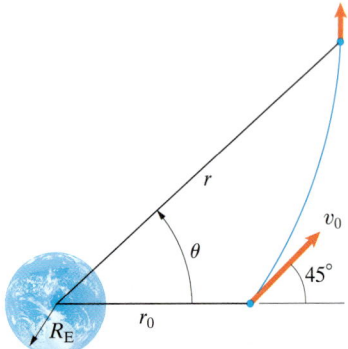

Problems 16.145/16.146

16.147 The snow is 2 ft deep and weighs 20 lb/ft^3, the snowplow is 8 ft wide, and the truck travels at 5 mi/h. What force does the snow exert on the truck?

Problem 16.147

16.148 An empty 55-lb drum, 3 ft in diameter, stands on a set of scales. Water begins pouring into the drum at 1200 lb/min from 8 ft above the bottom of the drum. The weight density of water is approximately 62.4 lb/ft^3. What do the scales read 40 s after the water starts pouring?

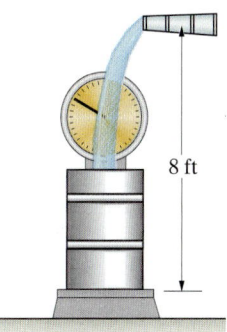

8 ft

Problem 16.148

16.149 The ski boat's jet propulsive system draws water in at A and expels it at B at 80 ft/s relative to the boat. Assume that the water drawn in enters with no horizontal velocity relative to the surrounding water. The maximum mass flow rate of water through the engine is 2.5 slugs/s. Hydrodynamic drag exerts a force on the boat of magnitude $1.5v$ lb, where v is the boat's velocity in feet per second. Neglecting aerodynamic drag, what is the ski boat's maximum velocity?

16.150 The ski boat in Problem 16.149 weighs 2800 lb. The mass flow rate of water through the engine is 2.5 slugs/s, and the craft starts from rest at $t = 0$. Determine the boat's velocity at (a) $t = 20$ s and (b) $t = 60$ s.

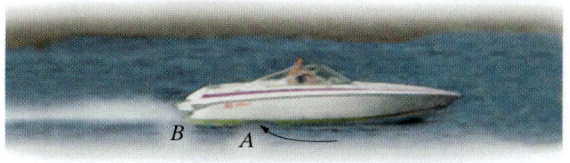

Problems 16.149/16.150

16.151* A crate of mass m slides across a smooth floor pulling a chain from a stationary pile. The mass per unit length of the chain is ρ_L. If the velocity of the crate is v_0 when $s = 0$, what is its velocity as a function of s?

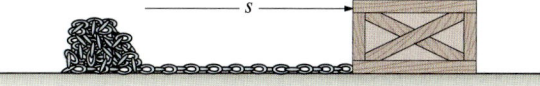

Problem 16.151

Design Project

Design and carry out experiments to measure the coefficients of restitution when (a) a table tennis ball; (b) a tennis ball; and (c) a soccer ball or basketball strike a rigid surface. Investigate the repeatability of your results. Determine how sensitive your results are to the velocity of the ball. Write a brief report describing your experiments, discussing possible sources of error, and presenting your results. Also comment on possible reasons for the differences in your results for the three kinds of balls.

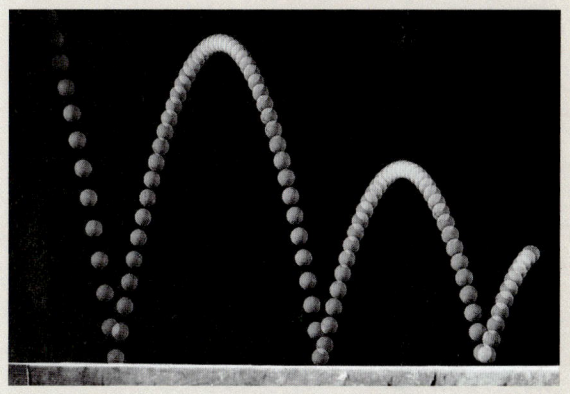

Planar Kinematics of Rigid Bodies

If the external forces acting on an object are known, Newton's second law can be used to determine the motion of the object's center of mass without considering any angular motion of the object about its center of mass. In many situations, however, angular motion must also be considered. The rotational motions of some objects are central to their functions, as in the cases of gears, wheels, generators, and turbines. In this chapter we analyze motions of objects, including their rotational motions.

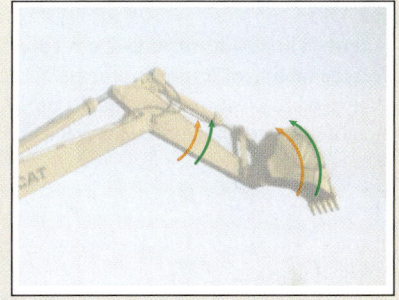

◀ To design the excavator, the engineer needed to analyze the translational and rotational motions of its members.

17.1 Rigid Bodies and Types of Motion

If a brick is thrown (Fig. 17.1a), we can determine the motion of its center of mass without having to be concerned about its rotational motion. The only significant force is the weight of the brick, and Newton's second law determines the acceleration of its center of mass. But suppose that the brick is standing on the floor, it is tipped over (Fig. 17.1b), and we want to determine the motion of its center of mass as it falls. In this case, the brick is subjected to its weight and also to a force exerted by the floor. We cannot determine either the force exerted by the floor or the motion of the brick's center of mass without also considering its rotational motion.

Figure 17.1
(a) A thrown brick—its rotation doesn't affect the motion of its center of mass.
(b) A tipped brick—its rotation and the motion of its center of mass are interrelated.

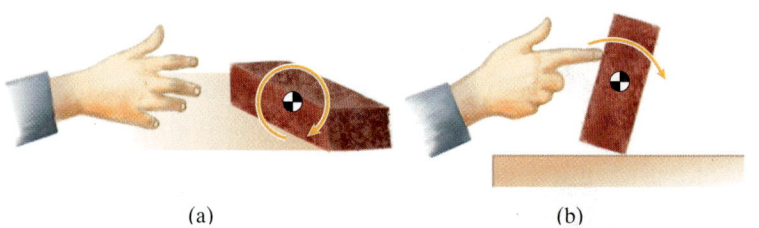

(a) (b)

Before we can analyze such motions, we must consider how to describe them. A brick is an example of an object whose motion can be described by treating the object as a rigid body. A *rigid body* is an idealized model of an object that does not deform, or change shape. The precise definition is that the distance between every pair of points of a rigid body remains constant. Although any object does deform as it moves, if its deformation is sufficiently small, we can approximate its motion by modeling it as a rigid body. For example, in normal use a twirler's baton (Fig. 17.2a) can be modeled as a rigid body, but a fishing rod (Fig. 17.2b) cannot.

(a) (b)

Figure 17.2
(a) A baton can be modeled as a rigid body.
(b) Under normal use, a fishing rod is too flexible to be modeled as a rigid body.

Describing the motion of a rigid body requires a reference frame (coordinate system) relative to which the motions of the points of the rigid body and its angular motion are measured. In many situations, it is convenient to use a reference frame that is fixed with respect to the earth. For example, we would use such an *earth-fixed* reference frame to describe the motion of the center of mass and the angular motion of the brick in Fig. 17.1. In the paragraphs that follow, we discuss some types of rigid-body motions relative to a given reference frame that occur frequently in applications.

Translation   If a rigid body in motion relative to a given reference frame does not rotate, it is said to be in *translation* (Fig. 17.3a). For example, the child's swing in Fig. 17.3b is designed so that the horizontal bar to which the seats are attached is in translation. Although each point of the horizontal bar moves in a circular path, the bar does not rotate. It remains horizontal, making it easier for the child to ride safely. Every point of a rigid body in translation has the same velocity and acceleration, so we describe the motion of the rigid body completely if we describe the motion of a single point.

(a)

(b)

Figure 17.3
(a) An object in translation does not rotate.
(b) The translating part of the swing on which the child sits remains level.

Rotation about a Fixed Axis   After translation, the simplest type of rigid-body motion is rotation about an axis that is fixed relative to a given reference frame (Fig. 17.4a). Each point of the rigid body on the axis is stationary, and each point not on the axis moves in a circular path about the axis as the rigid body rotates. The rotor of an electric motor (Fig. 17.4b) is an example of an object rotating about a fixed axis. The motion of a ship's propeller relative to the ship is rotation about a fixed axis. We discuss this type of motion in more detail in the next section.

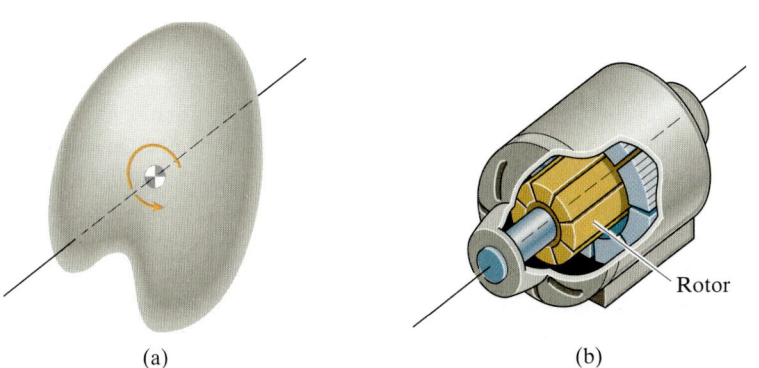

(a)

(b)

Rotor

Figure 17.4
(a) A rigid body rotating about a fixed axis.
(b) Relative to the frame of an electric motor, the rotor rotates about a fixed axis.

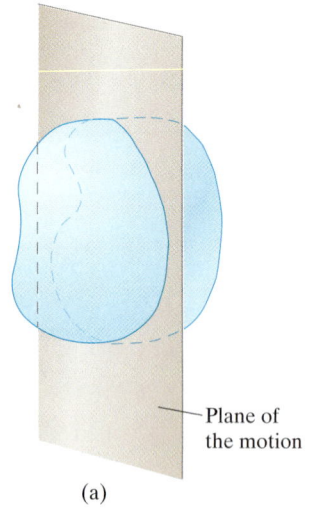

Plane of
the motion

(a)

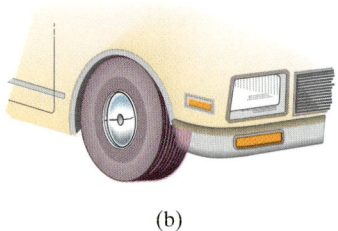

(b)

Figure 17.5
(**a**) A rigid body intersected by a fixed
     plane.
(**b**) A wheel undergoing planar motion.

Planar Motion   Consider a plane that is fixed relative to a given reference frame and a rigid body intersected by the plane (Fig. 17.5a). If the rigid body undergoes a motion in which the points intersected by the plane remain in the plane, the body is said to be in *two-dimensional*, or *planar*, motion. We refer to the fixed plane as the plane of the motion. Rotation of a rigid body about a fixed axis is a special case of planar motion. As another example, when a car moves in a straight path, its wheels are in planar motion (Fig. 17.5b).

The components of an internal combustion engine illustrate these types of motion (Fig. 17.6). Relative to a reference frame that is fixed with respect to the engine, the pistons translate within the cylinders. The connecting rods are in general planar motion, and the crankshaft rotates about a fixed axis.

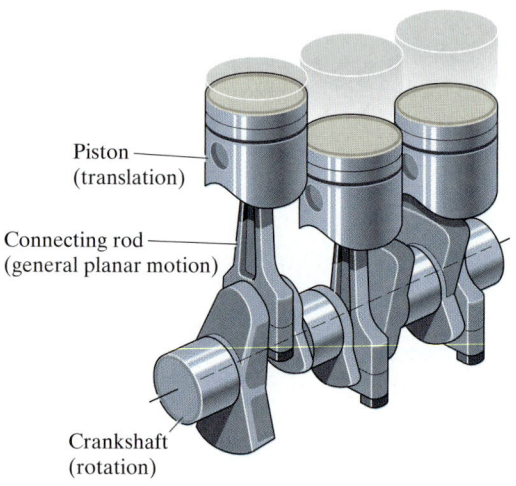

Piston
(translation)

Connecting rod
(general planar motion)

Crankshaft
(rotation)

Figure 17.6
Translation, rotation about a fixed axis, and planar motion in an automobile engine.

In the next section we begin our analysis of rigid-body motion with a discussion of rotation about a fixed axis. We then consider general planar motion, which combines rotation about an axis with translational motion, and derive expressions that relate relative velocities and accelerations of points of a rigid body to its angular velocity and angular acceleration. Using these relations, we analyze particular examples of general planar motion of a rigid body, including rolling, and also motions of connected rigid bodies.

17.2 Rotation about a Fixed Axis

By considering the rotation of an object about an axis that is fixed relative to a given reference frame, we can introduce some of the concepts of rigid-body motion in a familiar context. In this type of motion, each point of the rigid body moves in a circular path around the fixed axis, so we can analyze the motions of points using results developed in Chapter 13.

In Fig. 17.7, we show a rigid body rotating about a fixed axis and introduce two lines perpendicular to the axis. The reference line is fixed, and the body-fixed line rotates with the rigid body. The angle θ between the reference line and the body-fixed line describes the position, or orientation, of the rigid body about the fixed axis. The rigid body's *angular velocity*, or rate of rotation, and its *angular acceleration* are

$$\omega = \frac{d\theta}{dt}, \qquad \alpha = \frac{d\omega}{dt} = \frac{d^2\theta}{dt^2}. \tag{17.1}$$

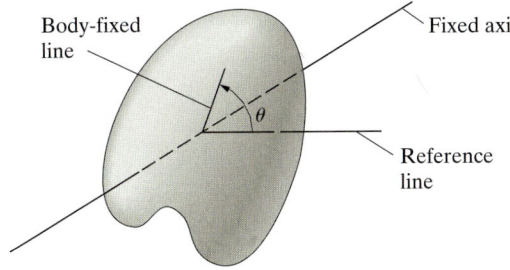

Body-fixed line

Fixed axis

θ

Reference line

Figure 17.7
Specifying the orientation of an object rotating about a fixed axis.

Each point of the object not on the fixed axis moves in a circular path about the axis. Using our knowledge of the motion of a point in a circular path, we can relate the velocity and acceleration of a point to the object's angular velocity and angular acceleration. In Fig. 17.8, we view the object in the direction parallel to the fixed axis. The velocity of a point at a distance r from the fixed axis is tangent to the point's circular path (Fig. 17.8a) and is given in terms of the angular velocity of the object by

$$v = r\omega. \tag{17.2}$$

A point has components of acceleration tangential and normal to its circular path (Fig. 17.8b). In terms of the angular velocity and angular acceleration of the object, the components of acceleration are

$$a_t = r\alpha, \qquad a_n = \frac{v^2}{r} = r\omega^2. \tag{17.3}$$

With these relations, we can analyze problems involving objects rotating about fixed axes. For example, suppose that we know the angular velocity ω_A and angular

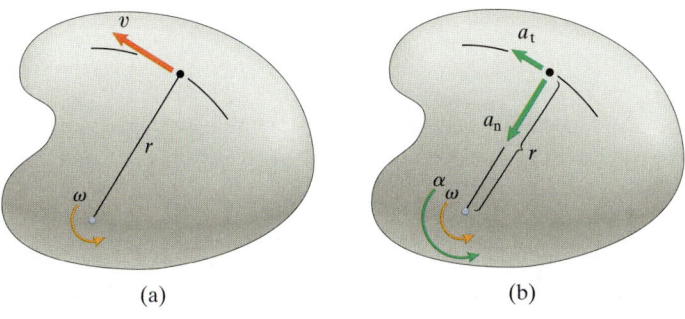

(a)

(b)

Figure 17.8
(a) Velocity and (b) acceleration of a point of a rigid body rotating about a fixed axis.

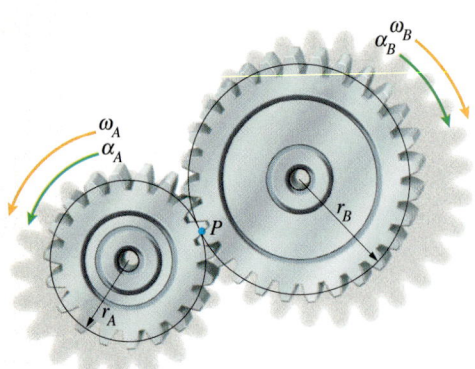

Figure 17.9
Relating the angular velocities and angular accelerations of meshing gears.

acceleration α_A of the gear in Fig. 17.9 relative to a particular reference frame, and we want to determine ω_B and α_B. The velocities of the gears must be equal at P, because there is no relative motion between them in the tangential direction at P. Therefore, $r_A\omega_A = r_B\omega_B$, and we find that the angular velocity of gear B is

$$\omega_B = \left(\frac{r_A}{r_B}\right)\omega_A.$$

By taking the derivative of this equation with respect to time, we determine the angular acceleration of gear B:

$$\alpha_B = \left(\frac{r_A}{r_B}\right)\alpha_A.$$

From this result, we see that the tangential components of the accelerations of the gears at P are equal: $r_A\alpha_A = r_B\alpha_B$. However, the normal components of the accelerations of the gears at P are different in direction and, if the gears have different radii, are different in magnitude as well. The normal component of the acceleration of gear A at P points toward the center of gear A, and its magnitude is $r_A\omega_A^2$. The normal component of the acceleration of gear B at P points toward the center of gear B, and its magnitude is $r_B\omega_B^2 = (r_A/r_B)(r_A\omega_A^2)$.

Study Questions

1. What is the definition of a rigid body?
2. What is meant by two-dimensional, or planar, motion of a rigid body?
3. If a rigid body rotates about a fixed axis, what can you infer about the path of each point not on the axis?
4. If a rigid body is rotating about a fixed axis with angular velocity ω and angular acceleration α, what can you infer about the velocity and acceleration of a point at a distance r from the axis?

Example 17.1   Objects Rotating about Fixed Axes

Figure 17.10

Gear A of the winch in Fig. 17.10 turns gear B, raising the hook H. If gear A starts from rest at $t = 0$ and its clockwise angular acceleration is $\alpha_A = 0.2t$ rad/s^2, what vertical distance has the hook H risen and what is its velocity at $t = 10$ s?

Strategy

By equating the tangential components of acceleration of gears A and B at their point of contact, we can determine the angular acceleration of gear B. Then we can integrate twice to obtain the angular velocity of gear B and the angle through which it has turned at $t = 10$ s.

Solution

The tangential acceleration of the point of contact of the two gears (Fig. a) is

$$a_t = (0.05 \text{ m})(0.2t \text{ rad/s}^2) = (0.2 \text{ m})(\alpha_B).$$

Therefore, the angular acceleration of gear B is

$$\alpha_B = \frac{d\omega_B}{dt} = \frac{(0.05 \text{ m})(0.2t \text{ rad/s}^2)}{(0.2 \text{ m})} = 0.05t \text{ rad/s}^2.$$

Integrating this equation yields

$$\int_0^{\omega_B} d\omega_B = \int_0^t 0.05t \, dt,$$

and we obtain the angular velocity of gear B:

$$\omega_B = \frac{d\theta_B}{dt} = 0.025t^2 \text{ rad/s}.$$

Integrating again, we obtain the angle through which gear B has turned:

$$\theta_B = 0.00833t^3 \text{ rad}.$$

At $t = 10$ s, $\theta_B = 8.33$ rad. The amount of cable wound around the drum, which is the distance the hook H has risen, is the product of θ_B and the radius of the drum: $(8.33 \text{ rad})(0.1 \text{ m}) = 0.833$ m.

At $t = 10$ s, $\omega_B = 2.5$ rad/s. The velocity of a point on the rim, which equals the velocity of the hook H (Fig. b), is

$$v_H = (0.1 \text{ m})(2.5 \text{ rad/s}) = 0.25 \text{ m/s}.$$

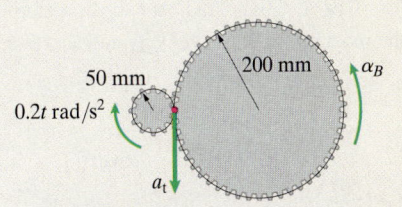

(a) The tangential accelerations of the gears are equal at their point of contact.

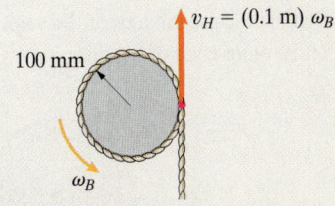

(b) Determining the hook's velocity.

Critical Thinking

Why are the calculations in this example useful? By knowing the angular accelerations of the gears as a function of time, we were able to determine their angular motions and the motion of the hook H. Engineers use analyses of this kind to design and evaluate the behavior of machines.

Problems

17.1 At the instant shown, the disk's angular velocity is 2 rad/s counterclockwise and its angular acceleration is 6 rad/s^2 counterclockwise. What are the magnitudes of the velocity and acceleration of point A?

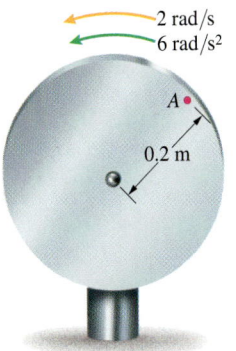

Problem 17.1

17.2 The angle θ is given as a function of time by $\theta = 0.3t + 0.018t^3$ rad. At $t = 4$ s, determine θ in degrees and the magnitudes of the velocity and acceleration of point A.

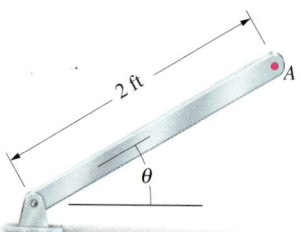

Problem 17.2

17.3 The weight A starts from rest at $t = 0$ and falls with a constant acceleration of 2 m/s^2.

(a) What is the magnitude of the disk's angular velocity at $t = 1$ s?
(b) What are the magnitudes of the velocity and acceleration of a point at the outer edge of the disk at $t = 1$ s?

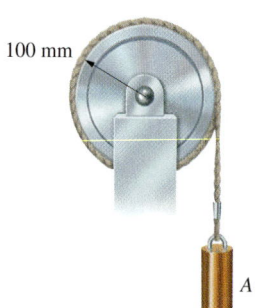

Problem 17.3

17.4 At the instant shown, the left disk has an angular velocity of 3 rad/s counterclockwise and an angular acceleration of 1 rad/s^2 clockwise.

(a) What are the angular velocity and angular acceleration of the right disk? (Assume that there is no relative motion between the disks at their point of contact.)
(b) What are the magnitudes of the velocity and acceleration of point A?

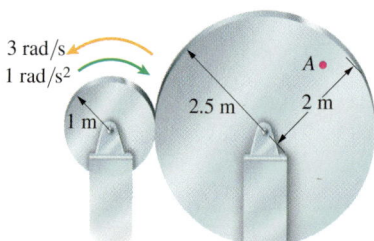

Problem 17.4

17.5 The angular velocity of the left disk is given as a function of time by $\omega_A = 4 + 0.2t$ rad/s.

(a) What are the angular velocities ω_B and ω_C at $t = 5$ s?
(b) Through what angle does the right disk turn from $t = 0$ to $t = 5$ s?

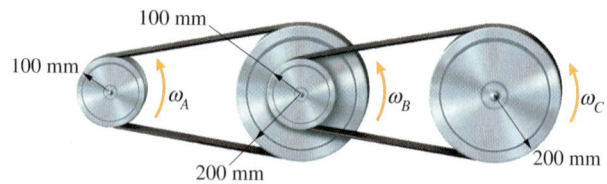

Problem 17.5

17.6 (a) If the bicycle's 120-mm sprocket wheel rotates through one revolution, through how many revolutions does the 45-mm gear turn? (b) If the angular velocity of the sprocket wheel is 1 rad/s, what is the angular velocity of the gear?

17.7 The rear wheel of the bicycle has a 330-mm radius and is rigidly attached to the 45-mm gear. It the rider turns the pedals, which are rigidly attached to the 120-mm sprocket wheel, at one revolution per second, what is the bicycle's velocity?

Problems 17.6/17.7

17.8 Relative to the given coordinate system, the disk rotates about the origin with a constant counterclockwise angular velocity of 10 rad/s. What are the x and y components of the velocity and acceleration of points A and B at the instant shown?

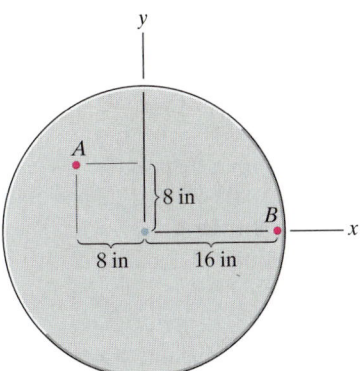

Problem 17.8

17.9 The Corvette is traveling at a constant speed of 65 mi/h. Its tires are 28 in in diameter.

(a) What is the angular velocity of the car's wheels?
(b) Relative to a nonrotating reference frame with its origin at the center of a wheel, what is the magnitude of the acceleration of a point of the wheel at a radial distance of 10 in from the center of the wheel?

17.10 The driver of the Corvette in Problem 17.9 suddenly applies the brakes, subjecting the car to a constant deceleration of 8 ft/s². Assume that the tires continue to roll, not skid, on the road surface.

(a) What is the resulting angular acceleration of the car's wheels?
(b) Consider a point of one of the wheels that is at a radial distance of 10 in from the center of the wheel. Immediately after the brakes are applied, relative to a nonrotating reference frame with its origin at the center of the wheel, what are the magnitudes of the tangential and normal components of the acceleration of the point?

Problems 17.9/17.10

17.11 If the bar has a counterclockwise angular velocity of 8 rad/s and a clockwise angular acceleration of 40 rad/s², what are the magnitudes of the accelerations of points A and B?

17.12 If the magnitudes of the velocity and acceleration of point A of the rotating bar are $|\mathbf{v}_A| = 3$ m/s and $|\mathbf{a}_A| = 28$ m/s², what are $|\mathbf{v}_B|$ and $|\mathbf{a}_B|$?

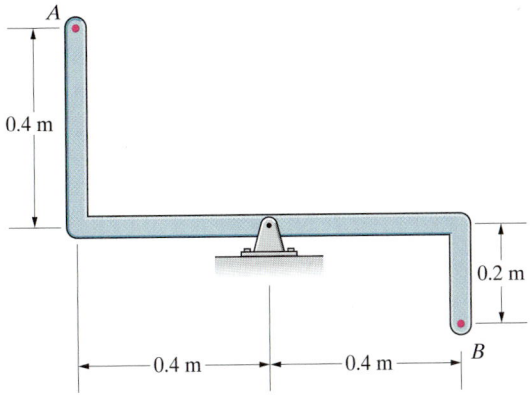

Problems 17.11/17.12

17.13 A disk of radius $R = 0.5$ m rolls on a horizontal surface. The relationship between the horizontal distance x the center of the disk moves and the angle β through which the disk rotates is $x = R\beta$. Suppose that the center of the disk is moving to the right with a constant velocity of 2 m/s.

(a) What is the disk's angular velocity?
(b) Relative to a nonrotating reference frame with its origin at the center of the disk, what are the magnitudes of the velocity and acceleration of a point on the edge of the disk?

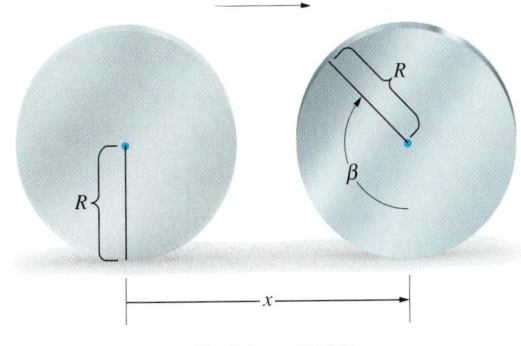

Problem 17.13

17.3 General Motions: Velocities

Each point of a rigid body in translation undergoes the same motion. Each point of a rigid body rotating about a fixed axis undergoes circular motion about the axis. To analyze more complicated motions that combine translation and rotation, we must develop equations that relate the relative motions of points of a rigid body to its angular motion.

Relative Velocities

In Fig. 17.11a, we view a rigid body from a perspective perpendicular to the plane of its motion. Points A and B are points of the rigid body contained in the latter plane, and O is the origin of a given reference frame. The position of A relative to B, $\mathbf{r}_{A/B}$, is related to the positions of A and B relative to O by

$$\mathbf{r}_A = \mathbf{r}_B + \mathbf{r}_{A/B}.$$

Taking the derivative of this equation with respect to time, we obtain

$$\mathbf{v}_A = \mathbf{v}_B + \mathbf{v}_{A/B}, \tag{17.4}$$

where $\mathbf{v}_A$ and $\mathbf{v}_B$ are the velocities of A and B relative to the given reference frame and $\mathbf{v}_{A/B} = d\mathbf{r}_{A/B}/dt$ is the velocity of A relative to B. (*When we simply speak of the velocity of a point, we will mean its velocity relative to the given reference frame.*)

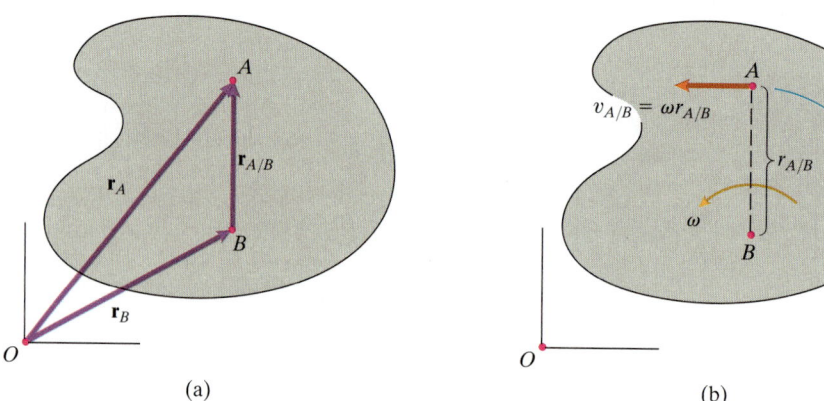

(a)

(b)

(c)

Figure 17.11
(a) A rigid body in planar motion.
(b) The velocity of A relative to B.
(c) The velocity of A is the sum of its velocity relative to B and the velocity of B.

We can show that $\mathbf{v}_{A/B}$ is related in a simple way to the rigid body's angular velocity. Since A and B are points of the rigid body, the distance between them, $r_{A/B} = |\mathbf{r}_{A/B}|$, is constant. That means that, relative to B, A moves in a circular path as the rigid body rotates. That velocity of A relative to B is therefore tangent to the circular path and equal to the product of $r_{A/B}$ and the angular velocity ω of the rigid body (Fig. 17.11b). From Eq. (17.4), the velocity of A is the sum of the velocity of B and the velocity of A relative to B (Fig. 17.11c). This result can be used to relate velocities of points of a rigid body in planar motion when the angular velocity of the body is known.

For example, in Fig. 17.12a, we show a circular disk of radius R rolling with counterclockwise angular velocity ω on a stationary plane surface. Saying that the surface is stationary means that we are describing the motion of the disk in terms of a reference frame that is fixed with respect to the surface. By *rolling*, we mean that the velocity of the disk relative to the surface is zero at the point of contact C. The velocity of the center B of the disk relative to C is illustrated in Fig. 17.12b. Since $\mathbf{v}_C = \mathbf{0}$, the velocity of B in terms of the fixed coordinate system shown is

$$\mathbf{v}_B = \mathbf{v}_C + \mathbf{v}_{B/C} = -R\omega\mathbf{i}.$$

This result is very useful: *The magnitude of the velocity of the center of a round object rolling on a stationary plane surface equals the product of the radius and the magnitude of the angular velocity.*

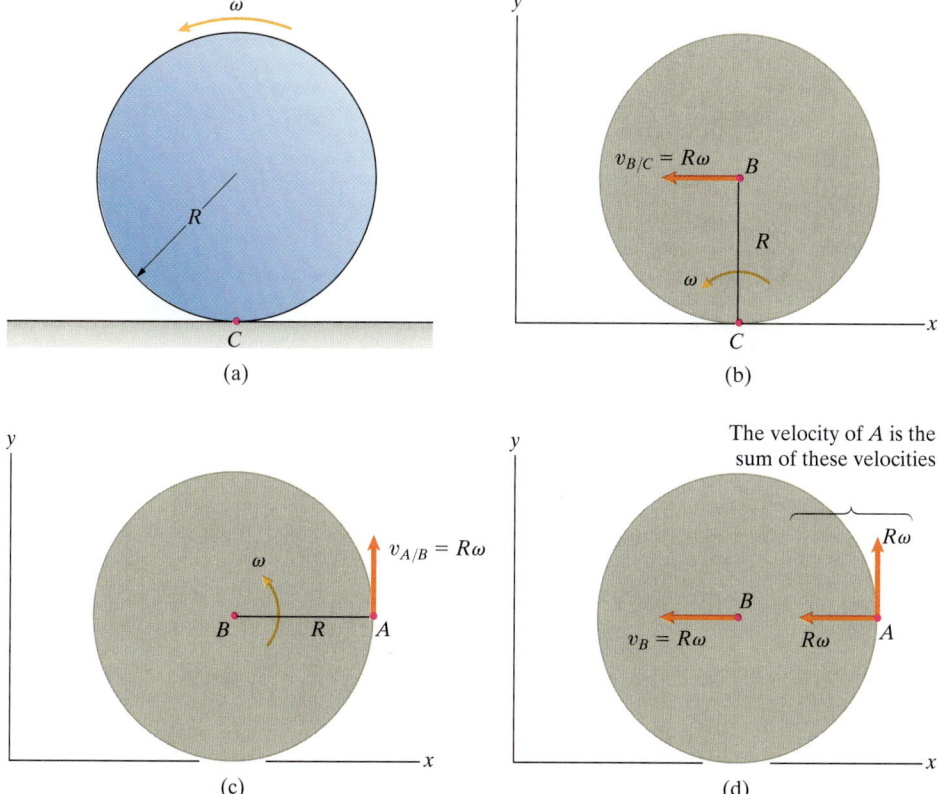

Figure 17.12
(a) A disk rolling with angular velocity ω.
(b) The velocity of the center B relative to C.
(c) The velocity of A relative to B.
(d) The velocity of A equals the sum of the velocity of B and the velocity of A relative to B.

We can determine the velocity of any other point of the disk in the same way. Figure 17.12c shows the velocity of a point A relative to point B. The velocity of A is the sum of the velocity of B and the velocity of A relative to B (Fig. 17.12d):

$$\mathbf{v}_A = \mathbf{v}_B + \mathbf{v}_{A/B} = -R\omega\mathbf{i} + R\omega\mathbf{j}.$$

The Angular Velocity Vector

We can express the rate of rotation of a rigid body as a vector. *Euler's theorem* states that a rigid body constrained to rotate about a fixed point B can move between any two positions by a single rotation about some axis through B. Suppose that we choose an arbitrary point B of a rigid body that is undergoing an arbitrary motion at a time t. Euler's theorem allows us to express the rigid body's change in position relative to B during an interval of time from t to $t + dt$ as a single rotation through an angle $d\theta$ about some axis. At time t, the rigid body's rate of rotation about the axis is its angular velocity $\omega = d\theta/dt$, and the axis about which the body rotates is called the *instantaneous axis of rotation*.

The *angular velocity vector*, denoted by $\boldsymbol{\omega}$, specifies both the direction of the instantaneous axis of rotation and the angular velocity. The angular velocity vector is defined to be parallel to the instantaneous axis of rotation (Fig. 17.13a), and its magnitude is the rate of rotation, the absolute value of ω. Its direction is related to the direction of the rigid body's rotation through a right-hand rule: Pointing the thumb of the right hand in the direction of $\boldsymbol{\omega}$, the fingers curl around $\boldsymbol{\omega}$ in the direction of the rotation (Fig. 17.13b).

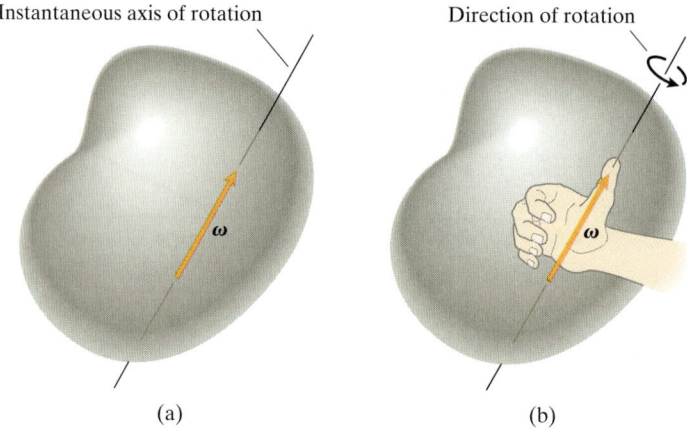

Instantaneous axis of rotation

Direction of rotation

(a)   (b)

Figure 17.13
(a) An angular velocity vector.
(b) Right-hand rule for the direction of the vector.

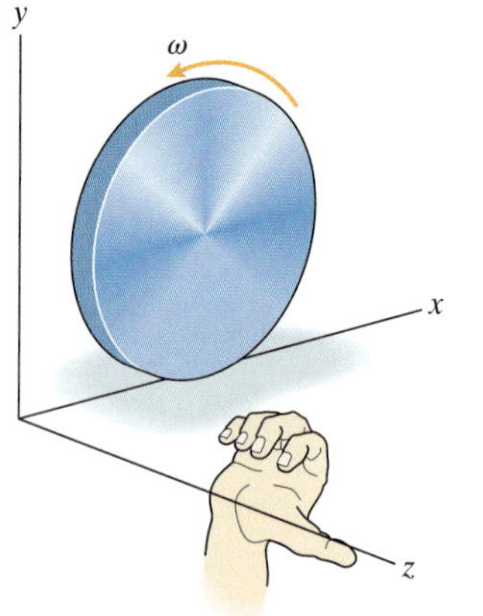

Figure 17.14
Determining the direction of the angular velocity vector of a rolling disk.

For example, the axis of rotation of the rolling disk in Fig. 17.12 is parallel to the z axis, so the angular velocity vector of the disk is parallel to the z axis and its magnitude is ω. Curling the fingers of the right hand around the z axis in the direction of the rotation, the thumb points in the positive z direction (Fig. 17.14). The angular velocity vector of the disk is $\boldsymbol{\omega} = \omega\mathbf{k}$.

The angular velocity vector allows us to express the results of the previous section in a convenient form. Let A and B be points of a rigid body with angular velocity $\boldsymbol{\omega}$ (Fig. 17.15a). We can show that the velocity of A relative to B is

$$\mathbf{v}_{A/B} = \frac{d\mathbf{r}_{A/B}}{dt} = \boldsymbol{\omega} \times \mathbf{r}_{A/B}. \tag{17.5}$$

At the present instant, relative to B, point A is moving in a circular path of radius $|\mathbf{r}_{A/B}| \sin \beta$, where β is the angle between the vectors $\mathbf{r}_{A/B}$ and $\boldsymbol{\omega}$ (Fig. 17.15b). The magnitude of the velocity of A relative to B is equal to the product of the radius of the circular path and the angular velocity of the rigid body; that is, $|\mathbf{v}_{A/B}| = (|\mathbf{r}_{A/B}| \sin \beta)|\boldsymbol{\omega}|$. The right-hand side of this equation is the magnitude of the cross product of $\mathbf{r}_{A/B}$ and $\boldsymbol{\omega}$. In addition, $\mathbf{v}_{A/B}$ is perpendicular both to $\boldsymbol{\omega}$ and $\mathbf{r}_{A/B}$. But is $\mathbf{v}_{A/B}$ equal to $\boldsymbol{\omega} \times \mathbf{r}_{A/B}$ or $\mathbf{r}_{A/B} \times \boldsymbol{\omega}$? Notice in Fig. 17.15b that, pointing the fingers of the right hand in the direction of $\boldsymbol{\omega}$ and closing them toward $\mathbf{r}_{A/B}$, the thumb points in the direction of the velocity of A relative to B, so $\mathbf{v}_{A/B} = \boldsymbol{\omega} \times \mathbf{r}_{A/B}$. Substituting Eq. (17.5) into Eq. (17.4), we obtain an equation for the relation between the velocities of two points of a rigid body in terms of its angular velocity:

$$\mathbf{v}_A = \mathbf{v}_B + \underbrace{\boldsymbol{\omega} \times \mathbf{r}_{A/B}}_{\mathbf{v}_{A/B}}. \tag{17.6}$$

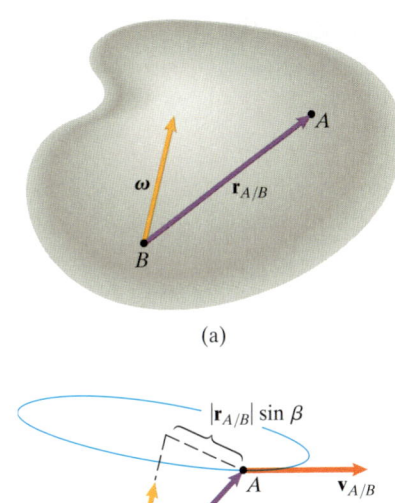

(a)

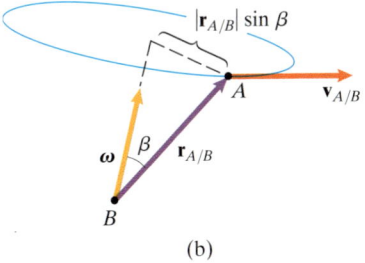

(b)

Figure 17.15
(a) Points A and B of a rotating rigid body.
(b) A is moving in a circular path relative to B.

If the angular velocity vector and the velocity of one point of a rigid body are known, Eq. (17.6) can be used to determine the velocity of any other point of the rigid body. Returning to the example of a disk of radius R rolling with angular velocity ω (Fig. 17.16), let us use Eq. (17.6) to determine the velocity of point A. The velocity of the center of the disk is given in terms of the angular velocity by $\mathbf{v}_B = -R\omega\mathbf{i}$, the disk's angular velocity vector is $\boldsymbol{\omega} = \omega\mathbf{k}$, and the position vector of A relative to the center is $\mathbf{r}_{A/B} = R\mathbf{i}$. The velocity of point A is

$$\mathbf{v}_A = \mathbf{v}_B + \boldsymbol{\omega} \times \mathbf{r}_{A/B} = -R\omega\mathbf{i} + (\omega\mathbf{k}) \times (R\mathbf{i})$$

$$= -R\omega\mathbf{i} + R\omega\mathbf{j}.$$

Compare this result with the velocity of point A shown in Fig. 17.12d.

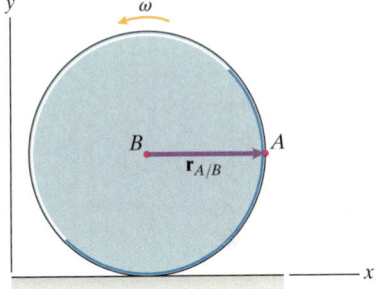

Figure 17.16
A rolling disk and the position vector of A relative to B.

Study Questions

1. What is meant when an object is said to be rolling on a surface?
2. How is the angular velocity vector of a rigid body defined?
3. If you know the direction of the angular velocity vector of a rigid body, how do you determine the direction of rotation of the body?
4. If you know the angular velocity vector and the velocity of one point of a rigid body, how can you determine the velocity of a different point of the rigid body?

Example 17.2	Determining Velocities and Angular Velocities

Bar AB in Fig. 17.17 rotates with a clockwise angular velocity of 10 rad/s. Determine the angular velocity of bar BC and the velocity of point C.

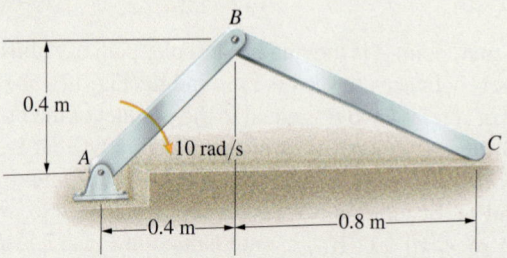

Figure 17.17

Strategy

Bar AB rotates about the fixed point A with a known angular velocity, so we can determine the velocity of B. Then, by expressing the horizontal velocity of C in terms of the velocity of B and the angular velocity of bar BC, we can obtain two equations in terms of the velocity of C and the angular velocity of bar BC.

Solution

Because the angular velocity of bar AB is given and point A is stationary, we can use Eq. (17.6) to determine the velocity of point B. From Fig. a, the position vector of B relative to A is $\mathbf{r}_{B/A} = 0.4\mathbf{i} + 0.4\mathbf{j}$ (m). The angular velocity vector of bar AB is $\boldsymbol{\omega}_{AB} = -10\mathbf{k}$ (rad/s), so the velocity of B is

$$\mathbf{v}_B = \mathbf{v}_A + \boldsymbol{\omega} \times \mathbf{r}_{B/A}$$

$$= \mathbf{0} + \begin{bmatrix} \mathbf{i} & \mathbf{j} & \mathbf{k} \\ 0 & 0 & -10 \\ 0.4 & 0.4 & 0 \end{bmatrix}$$

$$= 4\mathbf{i} - 4\mathbf{j} \ (\text{m/s}).$$

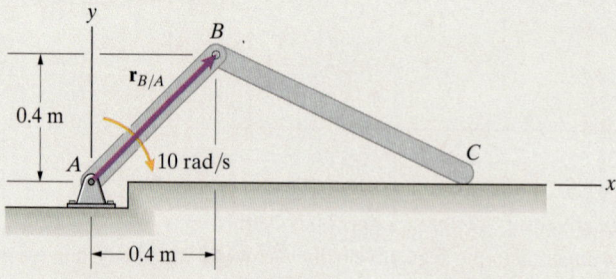

(a) The position vector of B relative to A.

We now use Eq. (17.6) to express the velocity of point C in terms of the velocity of point B. Let ω_{BC} be the counterclockwise angular velocity of bar BC (Fig. b), so that the angular velocity vector of bar BC is $\boldsymbol{\omega}_{BC} = \omega_{BC}\mathbf{k}$. Because point C is moving horizontally (Fig. b), we can write its velocity as $\mathbf{v}_C = v_C\mathbf{i}$. Therefore,

$$\mathbf{v}_C = \mathbf{v}_B + \boldsymbol{\omega}_{BC} \times \mathbf{r}_{C/B}:$$

$$v_C\mathbf{i} = 4\mathbf{i} - 4\mathbf{j} + \begin{bmatrix} \mathbf{i} & \mathbf{j} & \mathbf{k} \\ 0 & 0 & \omega_{BC} \\ 0.8 & -0.4 & 0 \end{bmatrix}$$

$$= (4 + 0.4\omega_{BC})\mathbf{i} - (4 - 0.8\omega_{BC})\mathbf{j}.$$

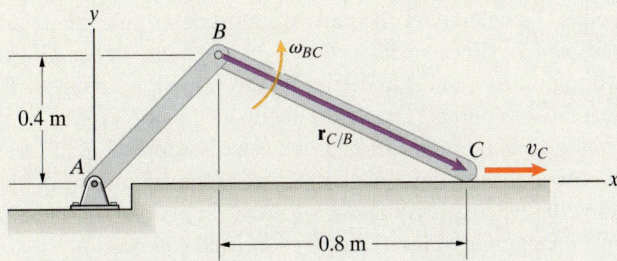

(b) The angular velocity of bar BC, the velocity of C, and the position vector of C relative to B.

Equating the $\mathbf{i}$ and $\mathbf{j}$ components in this equation, we obtain

$$v_C = 4 + 0.4\omega_{BC},$$

$$0 = 4 - 0.8\omega_{BC}.$$

Solving, we get $v_C = 6$ m/s and $\omega_{BC} = 5$ rad/s. The angular velocity of bar BC is $\boldsymbol{\omega}_{BC} = 5\mathbf{k}$ (rad/s), and the velocity of point C is $\mathbf{v}_C = 6\mathbf{i}$ (m/s).

Critical Thinking

Why did we express the velocity of point C as $\mathbf{v}_C = v_C\mathbf{i}$? Doing so introduces into the solution the constraint imposed on the motion of bar BC by the horizontal surface. This critical step states that the vertical component of the velocity of point C is zero.

Example 17.3 Analysis of a Linkage

Bar AB in Fig. 17.18 rotates with a clockwise angular velocity of 10 rad/s. What is the vertical velocity v_R of the rack of the rack-and-pinion gear?

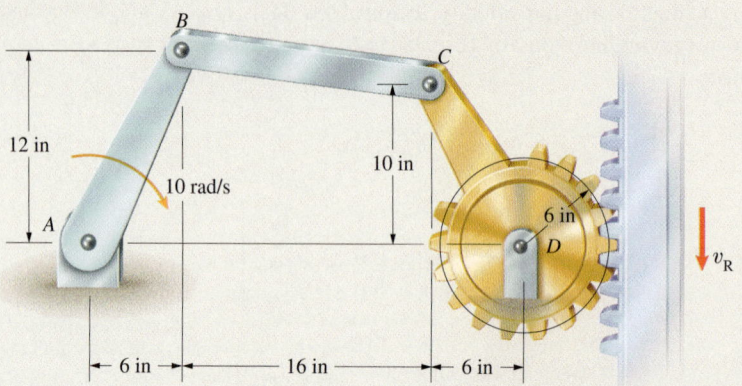

Figure 17.18

Strategy

To determine the velocity of the rack, we must determine the angular velocity of the member CD. Since we know the angular velocity of bar AB, we can apply Eq. (17.6) to points A and B to determine the velocity of point B. Then we can apply Eq. (17.6) to points C and D to obtain an equation for $\mathbf{v}_C$ in terms of the angular velocity of the member CD. We can also apply Eq. (17.6) to points B and C to obtain an equation for $\mathbf{v}_C$ in terms of the angular velocity of bar BC. By equating the two expressions for $\mathbf{v}_C$, we will obtain a vector equation in two unknowns: the angular velocities of bars BC and CD.

Solution

We first apply Eq. (17.6) to points A and B (Fig. a). In terms of the coordinate system shown, the position vector of B relative to A is $\mathbf{r}_{B/A} = 0.5\mathbf{i} + \mathbf{j}$ (ft), and the angular velocity vector of bar AB is $\boldsymbol{\omega}_{AB} = -10\mathbf{k}$ (rad/s). The velocity of B is

$$\mathbf{v}_B = \mathbf{v}_A + \boldsymbol{\omega}_{AB} \times \mathbf{r}_{B/A} = \mathbf{0} + \begin{vmatrix} \mathbf{i} & \mathbf{j} & \mathbf{k} \\ 0 & 0 & -10 \\ 0.5 & 1 & 0 \end{vmatrix}$$

$$= 10\mathbf{i} - 5\mathbf{j} \text{ (ft/s)}.$$

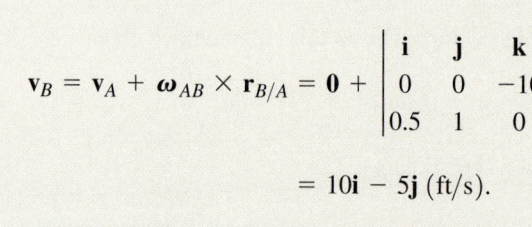

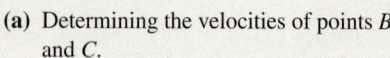

(a) Determining the velocities of points B and C.

We now apply Eq. (17.6) to points C and D. Let ω_{CD} be the unknown angular velocity of member CD (Fig. a). The position vector of C relative to D is $\mathbf{r}_{C/D} = -0.500\mathbf{i} + 0.833\mathbf{j}$ (ft), and the angular velocity vector of member CD is $\boldsymbol{\omega}_{CD} = -\omega_{CD}\mathbf{k}$. The velocity of C is

$$\mathbf{v}_C = \mathbf{v}_D + \boldsymbol{\omega}_{CD} \times \mathbf{r}_{C/D} = \mathbf{0} + \begin{vmatrix} \mathbf{i} & \mathbf{j} & \mathbf{k} \\ 0 & 0 & -\omega_{CD} \\ -0.500 & 0.833 & 0 \end{vmatrix}$$

$$= 0.833\omega_{CD}\mathbf{i} + 0.500\omega_{CD}\mathbf{j}.$$

Now we apply Eq. (17.6) to points B and C (Fig. b). We denote the unknown angular velocity of bar BC by ω_{BC}. The position vector of C relative to B is $\mathbf{r}_{C/B} = 1.333\mathbf{i} - 0.167\mathbf{j}$ (ft), and the angular velocity vector of bar BC is $\boldsymbol{\omega}_{BC} = \omega_{BC}\mathbf{k}$. Expressing the velocity of C in terms of the velocity of B, we obtain

$$\mathbf{v}_C = \mathbf{v}_B + \boldsymbol{\omega}_{BC} \times \mathbf{r}_{C/B} = \mathbf{v}_B + \begin{vmatrix} \mathbf{i} & \mathbf{j} & \mathbf{k} \\ 0 & 0 & \omega_{BC} \\ 1.333 & -0.167 & 0 \end{vmatrix}$$

$$= \mathbf{v}_B + 0.167\omega_{BC}\mathbf{i} + 1.333\omega_{BC}\mathbf{j}.$$

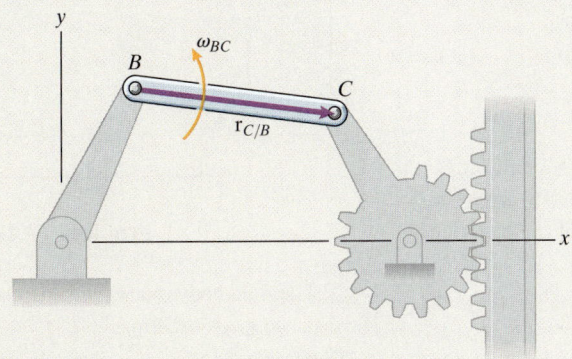

(b) Expressing the velocity of point C in terms of the velocity of point B.

Substituting our expressions for $\mathbf{v}_B$ and $\mathbf{v}_C$ into this equation, we get

$$0.833\omega_{CD}\mathbf{i} + 0.500\omega_{CD}\mathbf{j} = 10\mathbf{i} - 5\mathbf{j} + 0.167\omega_{BC}\mathbf{i} + 1.333\omega_{BC}\mathbf{j}.$$

Equating the $\mathbf{i}$ and $\mathbf{j}$ components yields two equations in terms of ω_{BC} and ω_{CD}:

$$0.833\omega_{CD} = 10 + 0.167\omega_{BC},$$

$$0.500\omega_{CD} = -5 + 1.333\omega_{BC}.$$

Solving these equations, we obtain $\omega_{BC} = 8.92$ rad/s and $\omega_{CD} = 13.78$ rad/s.

The vertical velocity of the rack is equal to the velocity of the gear where it contacts the rack:

$$v_R = (0.5 \text{ ft})\omega_{CD} = (0.5)(13.78) = 6.89 \text{ ft/s}.$$

Critical Thinking

How did we know the sequence of steps that would determine the velocity of the rack and pinion gear? The Strategy section may give you the impression that there is only one method of solution and that it should be obvious. Neither of these things is true. But most problems of this kind can be solved by repeatedly applying Eq. (17.6) until enough equations have been obtained to determine what you need to know. Just remember that Eq. (17.6) *applies only to two points of the same rigid body.* For example, in this case we could apply Eq. (17.6) to points B and C, but we could not have applied it to points A and C.

Problems

17.14 The turbine rotates relative to the coordinate system at 30 rad/s about a fixed axis coincident with the x axis. What is its angular velocity vector?

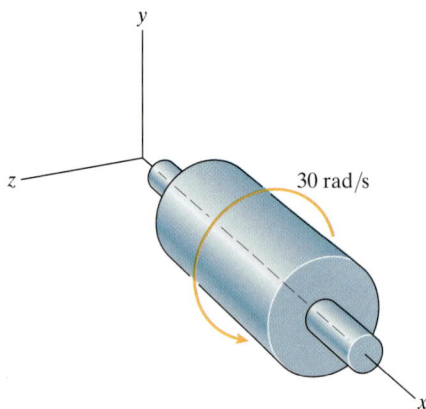

Problem 17.14

17.15 The rectangular plate swings in the x–y plane from arms of equal length. What is the angular velocity vector of (a) the rectangular plate and (b) the bar AB?

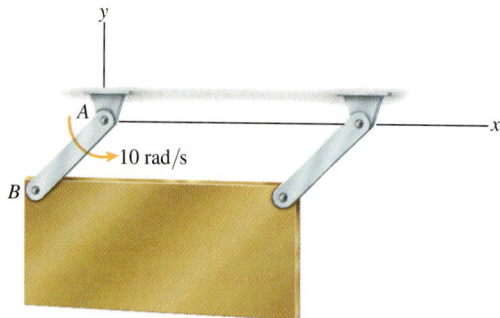

Problem 17.15

17.16 Bar OQ is rotating in the clockwise direction at 4 rad/s. What are the angular velocity vectors of the bars OQ and PQ?

Strategy: Notice that if you know the angular velocity of bar OQ, you also know the angular velocity of bar PQ.

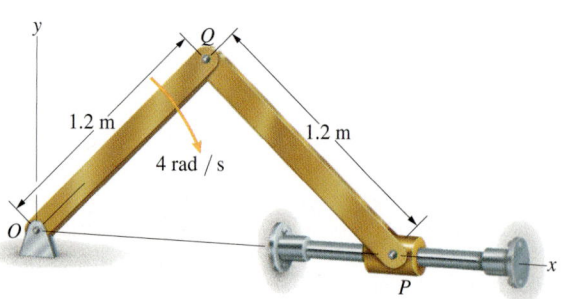

Problem 17.16

17.17 A disk of radius $R = 0.5$ m rolls on a horizontal surface. The relationship between the horizontal distance x the center of the disk moves and the angle β through which the disk rotates is $x = R\beta$. Suppose that the center of the disk is moving to the right with a constant velocity of 2 m/s.

(a) What is the disk's angular velocity?
(b) What is the disk's angular velocity vector?

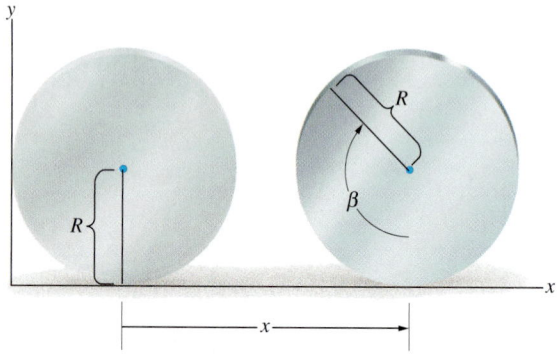

Problem 17.17

17.18 The rigid body rotates with angular velocity $\omega = 12$ rad/s. The distance $r_{A/B} = 0.4$ m.

(a) Determine the x and y components of the velocity of A relative to B by representing the velocity as shown in Fig. 17.11b.
(b) What is the angular velocity vector of the rigid body?
(c) Use Eq. (17.5) to determine the velocity of A relative to B.

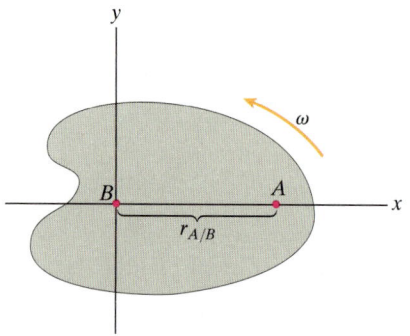

Problem 17.18

17.19 The bar is rotating in the counterclockwise direction at 8 rad/s. Use Eq. (17.5) to determine the velocity of A relative to B.

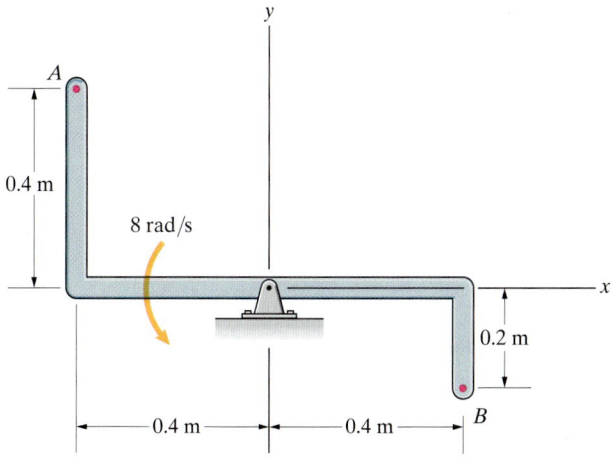

Problem 17.19

17.20 The bracket rotates about the shaft O with a counterclockwise angular velocity of 20 rad/s. By using Eq. (17.5), determine (a) the velocity of A relative to B and (b) the velocity of B relative to A.

17.21 The bracket rotates about the shaft O with a counterclockwise angular velocity of 20 rad/s.
(a) By applying Eq. (17.6) to point A and the fixed point O, determine the velocity of A.
(b) By using the result of part (a) and applying Eq. (17.6) to points A and B, determine the velocity of B.

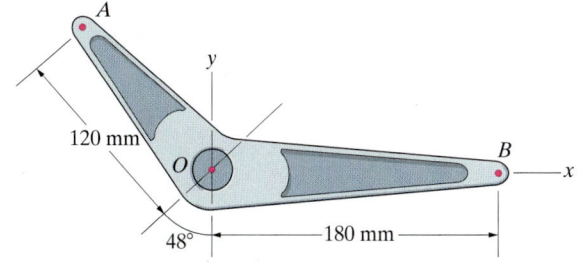

Problems 17.20/17.21

17.22 Determine the x and y components of the velocity of point A.

17.23 If the angular velocity of the bar is constant, what are the x and y components of the velocity of point A 0.1 s after the instant shown?

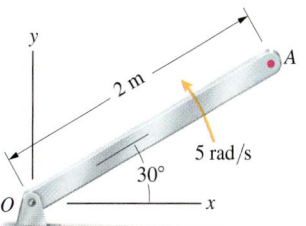

Problems 17.22/17.23

17.24 The disk is rotating about the z axis at 50 rad/s in the clockwise direction. Determine the x and y components of the velocities of points A, B, and C.

17.25 If the magnitude of the velocity of point A relative to point B is 4 m/s, what is the magnitude of the disk's angular velocity?

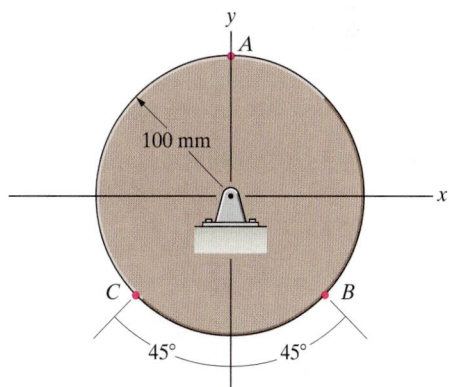

Problems 17.24/17.25

17.26 The Corvette is traveling at a constant speed of 65 mi/h. Its tires are 28 in in diameter.
(a) What is the angular velocity vector of one of the car's wheels?
(b) What is the velocity of the center of the wheel?
(c) Use Eq. (17.6) to determine the velocity of the highest point of the wheel's tire.

Problem 17.26

17.27 Point *A* of the rolling disk is moving to the right. The magnitude of the velocity of point *C* relative to point *B* is 8 m/s. What is the velocity of point *D*?

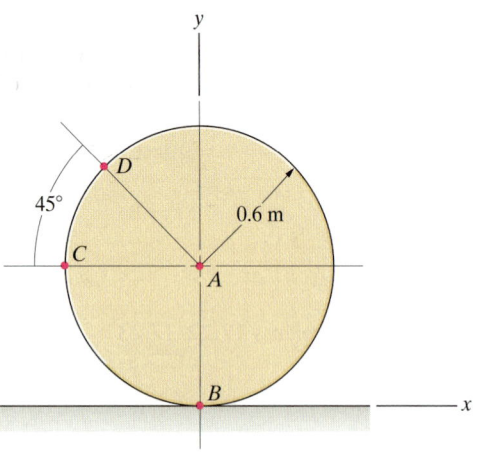

Problem 17.27

17.28 The helicopter is in planar motion in the *x*–*y* plane. At the instant shown, the position of the craft's center of mass, *G*, is $x = 2$ m, $y = 2.5$ m, and its velocity is $\mathbf{v}_G = 12\mathbf{i} + 4\mathbf{j}$ (m/s). The position of point *T*, where the tail rotor is mounted, is $x = -3.5$ m, $y = 4.5$ m. The helicopter's angular velocity is 0.2 rad/s clockwise. What is the velocity of point *T*?

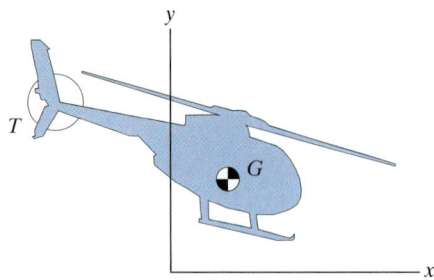

Problem 17.28

17.29 The bar *AB* is rotating in the counterclockwise direction at 3 rad/s. Point *A* is moving in the direction of the unit vector $0.966\mathbf{i} - 0.259\mathbf{j}$, and point *B* is moving in the direction of the unit vector $0.766\mathbf{i} + 0.643\mathbf{j}$. Determine the velocities of points *A* and *B*.

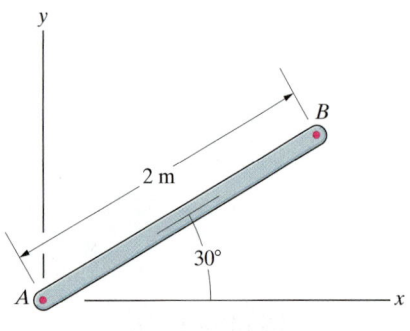

Problem 17.29

17.30 Points *A* and *B* of the 2-m bar slide on the plane surfaces. Point *B* is moving to the right at 3 m/s. What is the velocity of the midpoint *G* of the bar?

Strategy: First apply Eq. (17.6) to points *A* and *B* to determine the bar's angular velocity. Then apply Eq. (17.6) to points *B* and *G*.

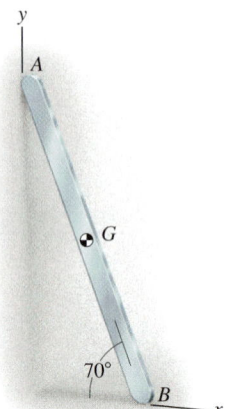

Problem 17.30

17.31 Point *B* is moving to the right at 4 in/s. Determine the velocity of point *A* and the angular velocity of the bar *AB*.

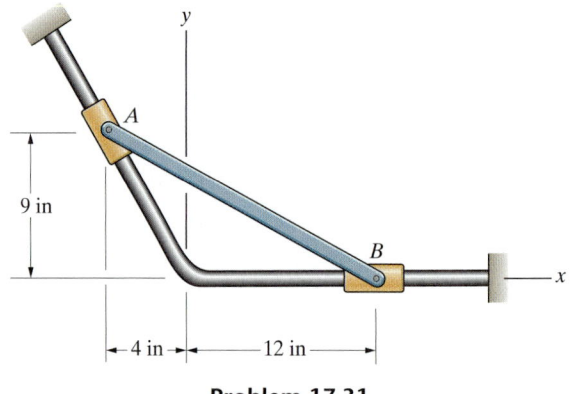

Problem 17.31

17.32 If $\theta = 45°$ and the sleeve *P* is moving to the right at 2 m/s, what are the angular velocities of bars *OQ* and *PQ*?

17.33 If the sleeve *P* is moving to the right at 2 m/s and bar *PQ* is rotating counterclockwise at 1 rad/s, what is the angle θ?

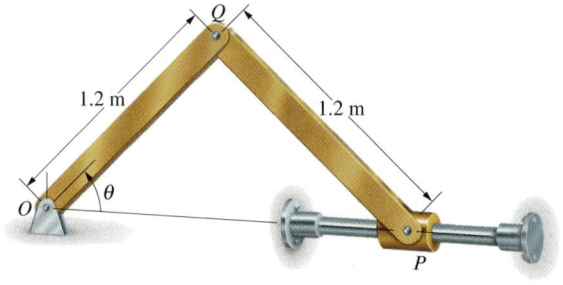

Problems 17.32/17.33

17.34 Bar *AB* rotates in the counterclockwise direction at 6 rad/s. Determine the angular velocity of bar *BD* and the velocity of point *D*.

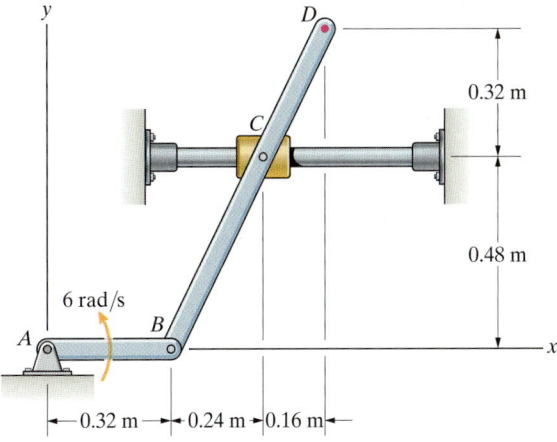

Problem 17.34

17.35 The crank *AB* is rotating in the clockwise direction at 2000 rpm (revolutions per minute). What is the velocity of the piston *C* at the instant shown?

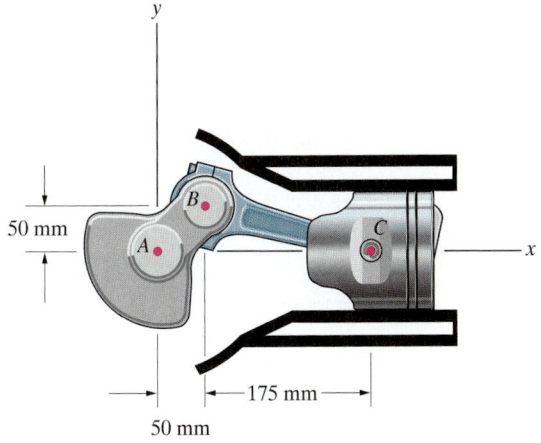

Problem 17.35

17.36 Bar *AB* rotates at 10 rad/s in the counterclockwise direction. Determine the angular velocity of bar *CD*.

Strategy: Since the angular velocity of the bar *AB* is known, the velocity of *B* can be determined. Apply Eq. (17.6) to points *B* and *C* to obtain an equation for $\mathbf{v}_C$ in terms of the angular velocity of bar *BC*, and then apply Eq. (17.6) to points *C* and *D* to obtain an equation for $\mathbf{v}_C$ in terms of the angular velocity of bar *CD*. By equating the two expressions, you will obtain a vector equation in two unknowns: the angular velocities of bars *BC* and *CD*.

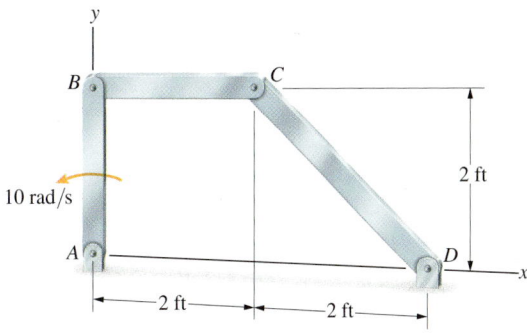

Problem 17.36

17.37 Bar *AB* rotates at 12 rad/s in the clockwise direction. Determine the angular velocities of bars *BC* and *CD*.

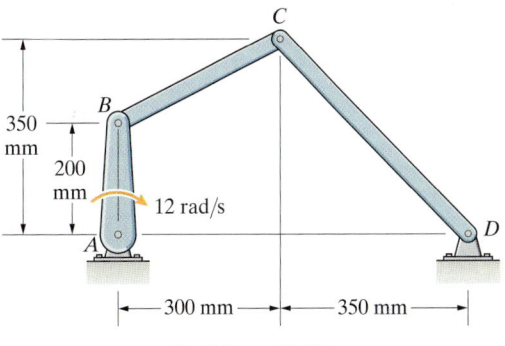

Problem 17.37

17.38 Bar *AB* rotates at 4 rad/s in the counterclockwise direction. Determine the angular velocities of bars *BC* and *CD*.

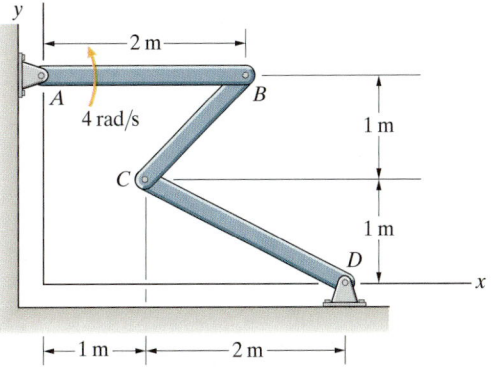

Problem 17.38

17.39 Bar AB rotates at 2 rad/s in the counterclockwise direction. Determine the velocity of the midpoint G of bar BC.

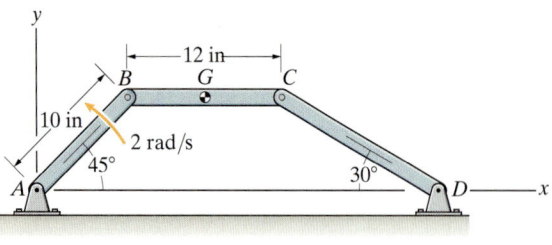

Problem 17.39

17.40 Bar AB rotates at 10 rad/s in the counterclockwise direction. Determine the velocity of point E.

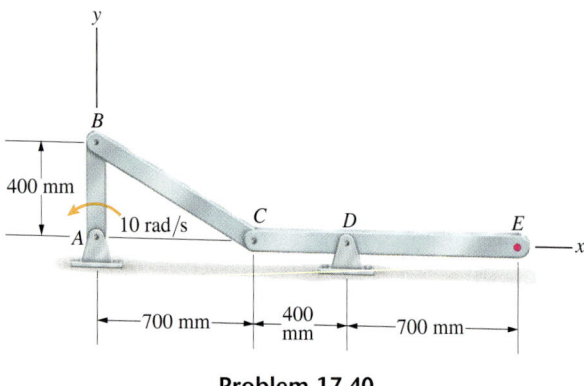

Problem 17.40

17.41 Bar AB rotates at 4 rad/s in the counterclockwise direction. Determine the velocity of point C.

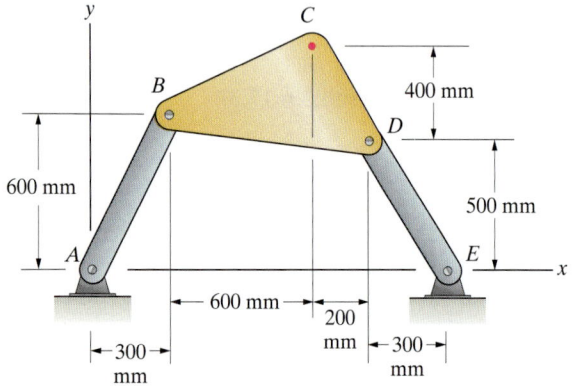

Problem 17.41

17.42 The upper grip and jaw of the pliers ABC is stationary. The lower grip DEF is rotating a 0.2 rad/s in the clockwise direction. At the instant shown, what is the angular velocity of the lower jaw CFG?

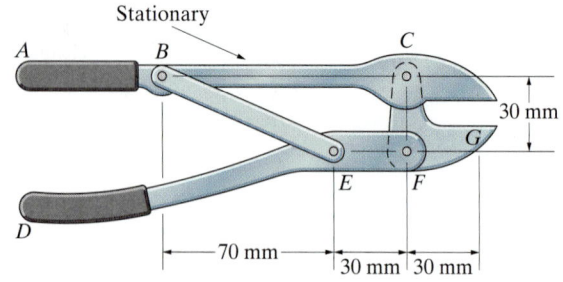

Problem 17.42

17.43 The horizontal member ADE supporting the scoop is stationary. If the link BD is rotating in the clockwise direction at 1 rad/s, what is the angular velocity of the scoop?

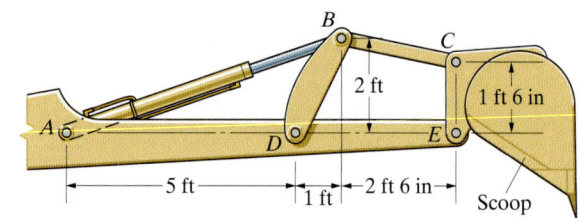

Problem 17.43

17.44 The diameter of the disk is 1 m, and the length of bar AB is 1 m. The disk is rolling, and point B slides on the plane surface. Determine the angular velocity of bar AB and the velocity of point B.

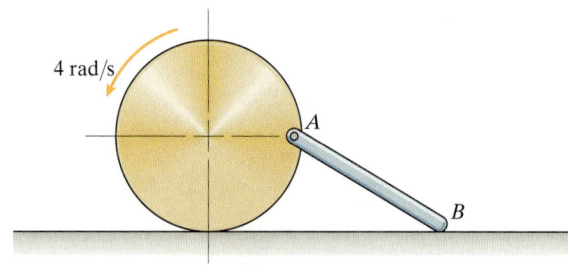

Problem 17.44

17.45 A motor rotates the circular disk mounted at A, moving the saw back and forth. (The saw is supported by a horizontal slot so that point C moves horizontally.) The radius AB is 4 in, and the link BC is 14 in long. In the position shown, $\theta = 45°$ and the link BC is horizontal. If the angular velocity of the disk is one revolution per second counterclockwise, what is the velocity of the saw?

17.46 In Problem 17.45, if the angular velocity of the disk is one revolution per second counterclockwise and $\theta = 270°$, what is the velocity of the saw?

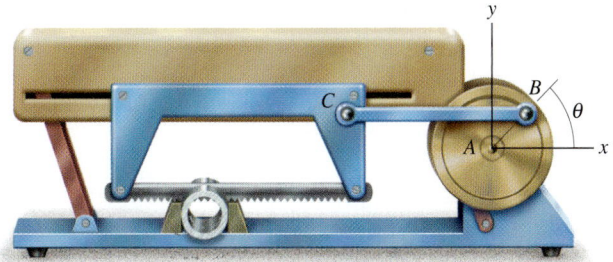

Problems 17.45/17.46

17.47 The disks roll on the plane surface. The angular velocity of the left disk is 2 rad/s in the clockwise direction. What is the angular velocity of the right disk?

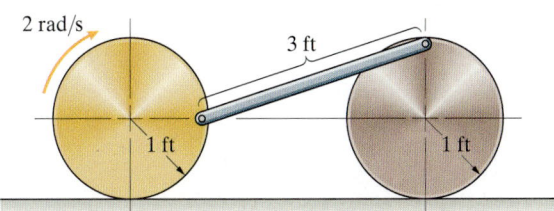

Problem 17.47

17.48 The disk rolls on the curved surface. The bar rotates at 10 rad/s in the counterclockwise direction. Determine the velocity of point A.

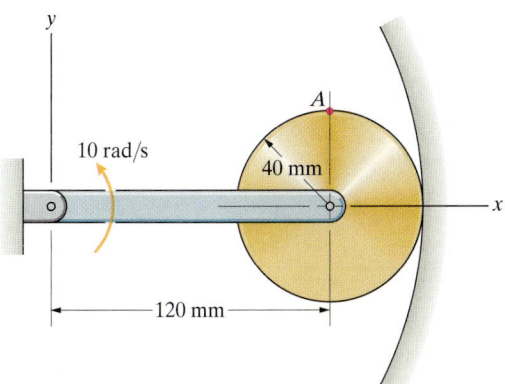

Problem 17.48

17.49 If $\omega_{AB} = 2$ rad/s and $\omega_{BC} = 4$ rad/s, what is the velocity of point C, where the excavator's bucket is attached?

17.50 If $\omega_{AB} = 2$ rad/s, what clockwise angular velocity ω_{BC} will cause the vertical component of the velocity of point C to be zero? What is the resulting velocity of point C?

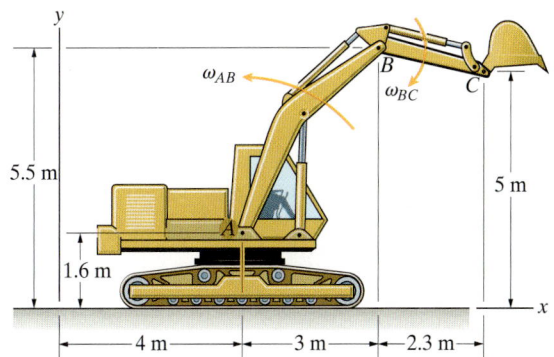

Problems 17.49/17.50

17.51 The motorcycle's rear wheel is rolling on the ground (the velocity of its point of contact with the ground is zero) at 500 rpm, the wheel's radius is 280 mm, and the body of the motorcycle is rotating in the clockwise direction at 6 rad/s. Determine the velocity of the center of mass G.

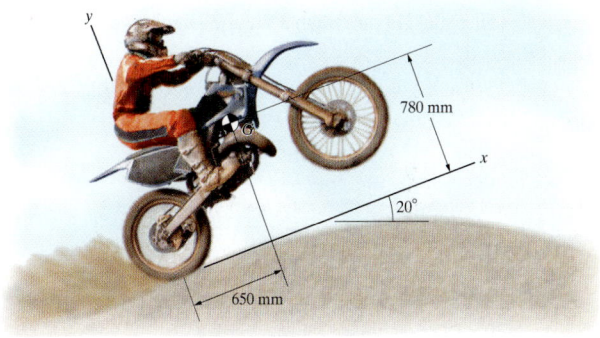

Problem 17.51

17.52 An athlete exercises his arm by raising the mass m. The shoulder joint A is stationary. The distance AB is 300 mm, and the distance BC is 400 mm. At the instant shown, $\omega_{AB} = 1$ rad/s and $\omega_{BC} = 2$ rad/s. How fast is the mass m rising?

17.53 The distance AB is 12 in, the distance BC is 16 in, $\omega_{AB} = 0.6$ rad/s, and the mass m is rising at 24 in/s. What is the angular velocity ω_{BC}?

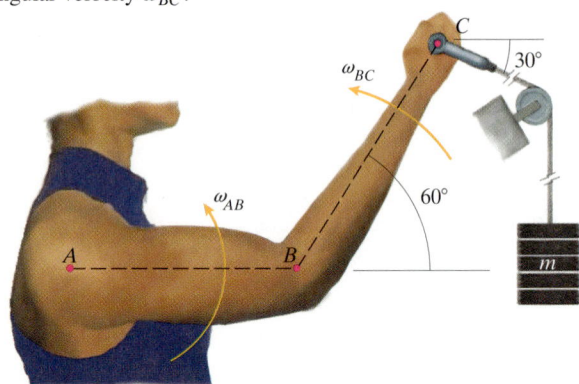

Problems 17.52/17.53

17.54 Points B and C are in the x–y plane. The angular velocity vectors of the arms AB and BC are $\omega_{AB} = -0.2\mathbf{k}$ (rad/s) and $\omega_{BC} = 0.4\mathbf{k}$ (rad/s). What is the velocity of point C?

17.55 If the velocity of point C of the robotic arm is $\mathbf{v}_C = -0.15\mathbf{i} + 0.42\mathbf{j}$ (m/s), what are the angular velocities of arms AB and BC?

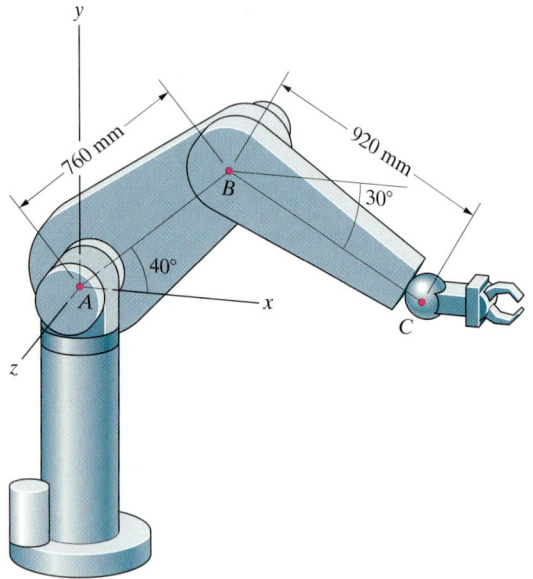

Problems 17.54/17.55

17.56 The link AB of the robot's arm is rotating at 2 rad/s in the counterclockwise direction, the link BC is rotating at 3 rad/s in the clockwise direction, and the link CD is rotating at 4 rad/s in the counterclockwise direction. What is the velocity of point D?

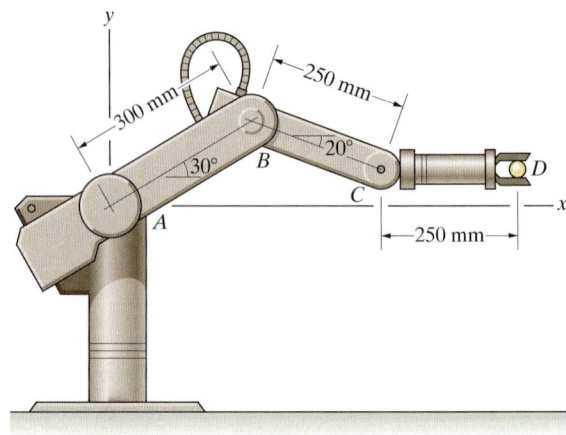

Problem 17.56

17.57 The person squeezes the grips of the shears, causing the angular velocities shown. What is the resulting angular velocity of the jaw BD?

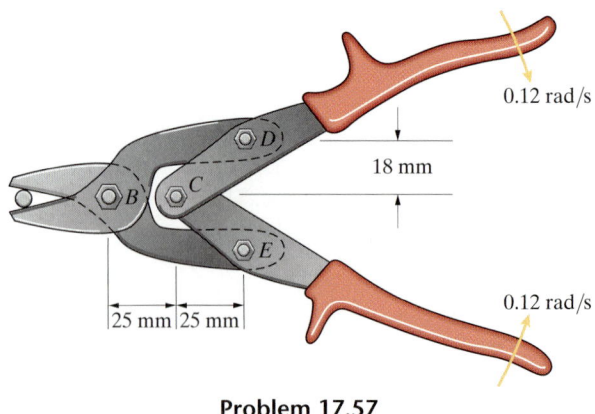

Problem 17.57

17.58 Determine the velocity v_W and the angular velocity of the small pulley.

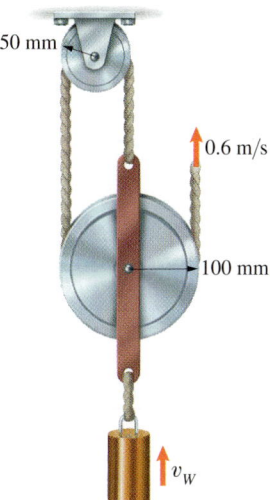

Problem 17.58

17.59 Determine the velocity of the block and the angular velocity of the small pulley.

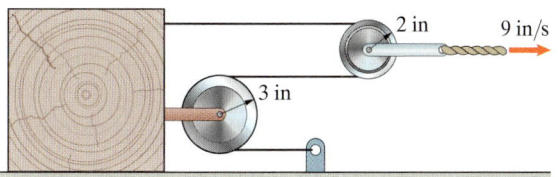

Problem 17.59

17.60 The device shown is used in the semiconductor industry to polish silicon wafers. The wafers are placed on the faces of the carriers. The outer and inner rings are then rotated, causing the wafers to move and rotate against an abrasive surface. If the outer ring rotates in the clockwise direction at 7 rpm and the inner ring rotates in the counterclockwise direction at 12 rpm, what is the angular velocity of the carriers?

17.61 Suppose that the outer ring rotates in the clockwise direction at 5 rpm and you want the centerpoints of the carriers to remain stationary during the polishing process. What is the necessary angular velocity of the inner ring?

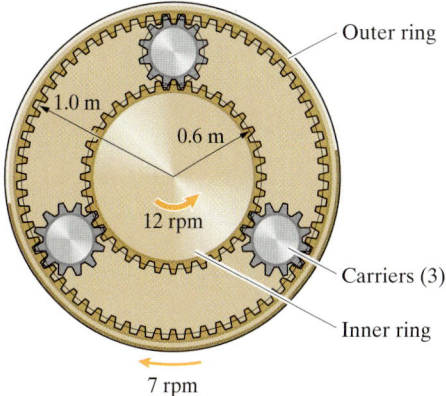

Problems 17.60/17.61

17.62 The ring gear is fixed and the hub and planet gears are bonded together. The connecting rod rotates in the counterclockwise direction at 60 rpm. Determine the angular velocity of the sun gear and the magnitude of the velocity of point A.

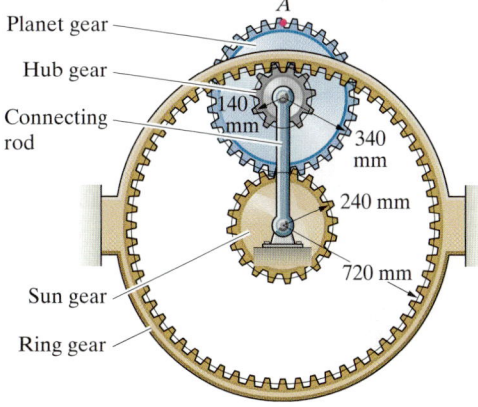

Problem 17.62

17.63 The large gear is fixed. Bar AB has a counterclockwise angular velocity of 2 rad/s. What are the angular velocities of bars CD and DE?

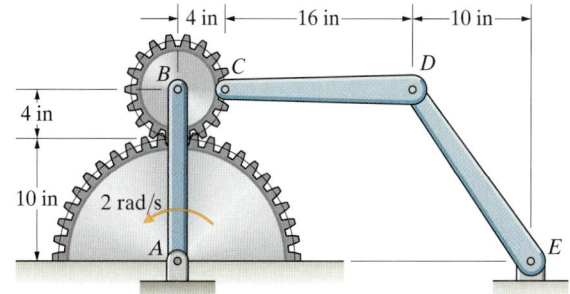

Problem 17.63

Instantaneous Centers

A point of a rigid body whose velocity is zero at a given instant is called an *instantaneous center*. "Instantaneous" means that the point may have zero velocity *only* at the instant under consideration, although we also refer to a fixed point, such as a point of a fixed axis about which a rigid body rotates, as an instantaneous center.

When we know the location of an instantaneous center of a rigid body in two-dimensional motion and we know its angular velocity, the velocities of other points are easy to determine. For example, suppose that point C in Fig. 17.19a is the instantaneous center of a rigid body in plane motion with angular velocity ω. Relative to C, a point A moves in a circular path. The velocity of A relative to C is tangent to the circular path and equal to the product of the distance from C to A and the angular velocity. But since C is stationary at the present instant, the velocity of A relative to C is the velocity of A. At this instant, every point of the rigid body rotates about C (Fig. 17.19b).

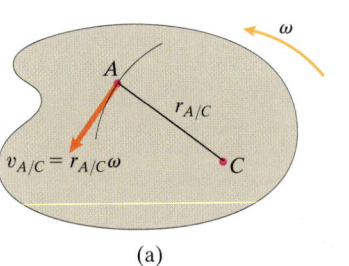 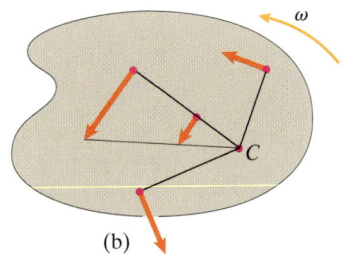

(a)   (b)

Figure 17.19
(a) An instantaneous center C and a different point A.
(b) Every point is rotating about the instantaneous center.

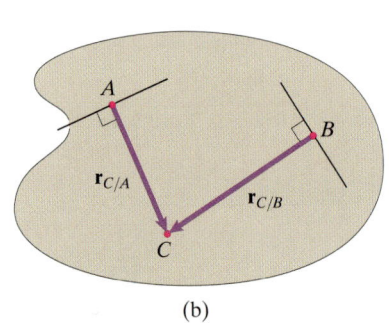

Direction of motion of A Direction of motion of B

Instantaneous center

(a)

(b)

Figure 17.20
(a) Locating the instantaneous center in planar motion.
(b) Proving that $\mathbf{v}_C = \mathbf{0}$.

The instantaneous center of a rigid body in planar motion can often be located by a simple procedure. Suppose that the directions of the motions of two points A and B are known and are not parallel (Fig. 17.20a). If we draw lines through A and B perpendicular to their directions of motion, then the point C where the lines intersect is the instantaneous center. To show that this is true, let us express the velocity of C in terms of the velocity of A (Fig. 17.20b):

$$\mathbf{v}_C = \mathbf{v}_A + \boldsymbol{\omega} \times \mathbf{r}_{C/A}.$$

The vector $\boldsymbol{\omega} \times \mathbf{r}_{C/A}$ is perpendicular to $\mathbf{r}_{C/A}$, so this equation indicates that $\mathbf{v}_C$ is parallel to the direction of motion of A. We also express the velocity of C in terms of the velocity of B:

$$\mathbf{v}_C = \mathbf{v}_B + \boldsymbol{\omega} \times \mathbf{r}_{C/B}.$$

The vector $\boldsymbol{\omega} \times \mathbf{r}_{C/B}$ is perpendicular to $\mathbf{r}_{C/B}$, so this equation indicates that $\mathbf{v}_C$ is parallel to the direction of motion of B. We have shown that the component of $\mathbf{v}_C$ perpendicular to the direction of motion of A is zero and that the component of $\mathbf{v}_C$ perpendicular to the direction of motion of B is zero, so $\mathbf{v}_C = \mathbf{0}$.

An instantaneous center may not be a point of the rigid body (Fig. 17.21a). This simply means that at the instant in question, the rigid body is rotating about an external point. It is helpful to *imagine* extending the rigid body so that it includes the instantaneous center (Fig. 17.21b). The velocity of point C of the extended body would be zero at the instant under consideration.

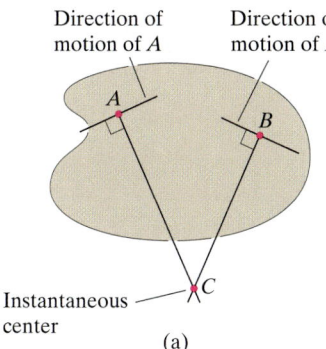 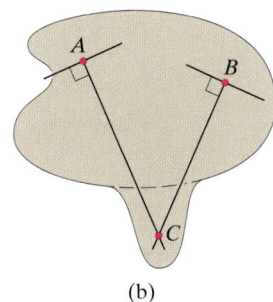

(a)

(b)

Figure 17.21
(a) An instantaneous center external to the rigid body.
(b) A hypothetical extended body. Point C would be stationary.

Notice in Fig. 17.21a that if the directions of motion of A and B are changed so that the lines perpendicular to their directions of motion become parallel, C moves to infinity. In that case, the rigid body is in pure translation, with an angular velocity of zero.

Returning once again to our example of a disk of radius R rolling with angular velocity ω (Fig. 17.22a), the point C in contact with the floor is stationary at the instant shown—it is the instantaneous center of the disk. Therefore, the velocity of any other point is perpendicular to the line from C to the point, and its magnitude equals the product of ω and the distance from C to the point. In terms of the coordinate system given in Fig. 17.22b, the velocity of point A is

$$\mathbf{v}_A = -\sqrt{2}R\omega \cos 45°\mathbf{i} + \sqrt{2}R\omega \sin 45°\mathbf{j}$$

$$= -R\omega\mathbf{i} + R\omega\mathbf{j}.$$

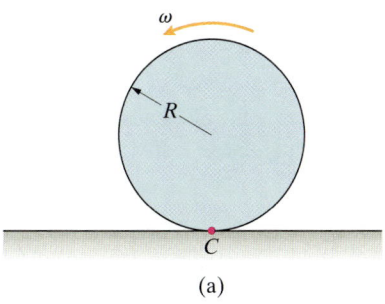 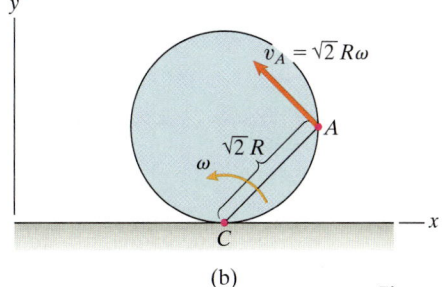

(a)

(b)

Figure 17.22
(a) Point C is the instantaneous center of the rolling disk.
(b) Determining the velocity of point A.

Study Questions

1. If a point of a rigid body is an instantaneous center, what is its velocity?

2. If you know both the angular velocity of a rigid body in planar motion and the position of the body's instantaneous center, what can you infer about the velocity of a different point of the rigid body?

3. If you know the directions of motion of two points of a rigid body in planar motion and these directions are not parallel, how can you determine the position of the instantaneous center of the body?

4. Where is the instantaneous center of a rolling object?

Example 17.4 | Linkage Analysis by Instantaneous Centers

Bar AB in Fig. 17.23 rotates with a counterclockwise angular velocity of 10 rad/s. What are the angular velocities of bars BC and CD?

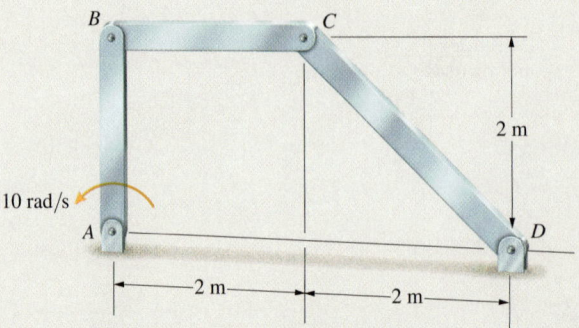

Figure 17.23

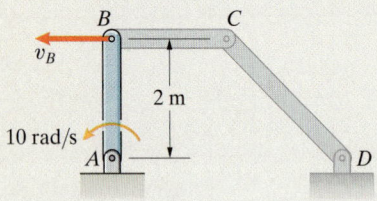

(a) Determining v_B.

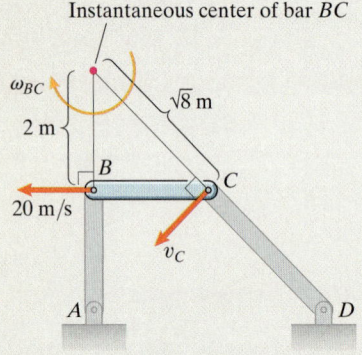

(b) Determining ω_{BC} and v_C.

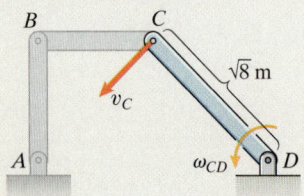

(c) Determining ω_{CD}.

Strategy

Because bars AB and CD rotate about fixed axes, we know the directions of motion of points B and C and so can locate the instantaneous center of bar BC. Beginning with bar AB (because we know its angular velocity), we can use the instantaneous centers of the bars to determine both the velocities of the points where they are connected and their angular velocities.

Solution

The velocity of B due to the rotation of bar AB about A (Fig. a) is

$$v_B = (2 \text{ m})(10 \text{ rad/s}) = 20 \text{ m/s}.$$

Drawing lines perpendicular to the directions of motion of B and C, we locate the instantaneous center of bar BC (Fig. b). The velocity of B is equal to the product of its distance from the instantaneous center of bar BC and the angular velocity ω_{BC}:

$$v_B = 20 \text{ m/s} = (2 \text{ m})\omega_{BC}.$$

Hence, $\omega_{BC} = 10$ rad/s. (Notice that bar BC rotates in the clockwise direction.) Using the instantaneous center of bar BC and its angular velocity ω_{BC}, we can determine the velocity of point C:

$$v_C = \left(\sqrt{8} \text{ m}\right)\omega_{BC} = 10\sqrt{8} \text{ m/s}.$$

Our last step is to use the velocity of point C to determine the angular velocity of bar CD about point D (Fig. c). We have

$$v_C = 10\sqrt{8} \text{ m/s} = \left(\sqrt{8} \text{ m}\right)\omega_{CD},$$

so that $\omega_{CD} = 10$ rad/s counterclockwise.

Critical Thinking

In this example, the use of instantaneous centers greatly simplified determining the angular velocities of bars BC and CD in comparison to our previous approach. However, notice that the lengths and positions of the bars made it very easy for us to locate the instantaneous center of bar BC. If the geometry is too complicated, the use of instantaneous centers can be impractical.

Problems

17.64 If the bar has a clockwise angular velocity of 10 rad/s and $v_A = 20$ m/s, what are the coordinates of the instantaneous center of the bar, and what is the value of v_B?

17.65 If $v_A = 24$ m/s and $v_B = 36$ m/s, what are the coordinates of the instantaneous center of the bar, and what is its angular velocity?

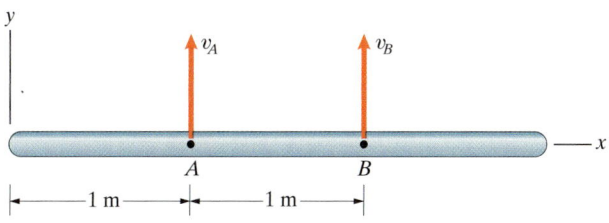

Problems 17.64/17.65

17.66 The velocity of point O of the bat is $\mathbf{v}_O = -6\mathbf{i} - 14\mathbf{j}$ (ft/s), and the bat rotates about the z axis with a counterclockwise angular velocity of 4 rad/s. What are the x and y coordinates of the bat's instantaneous center?

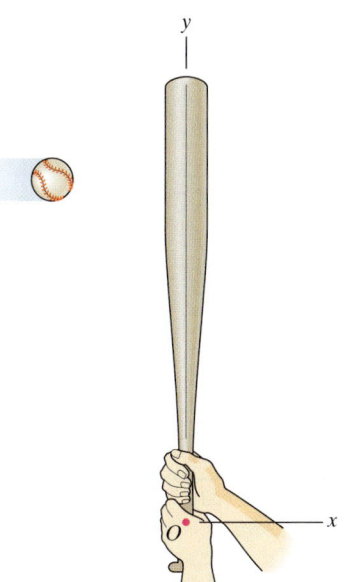

Problem 17.66

17.67 Points A and B of the 1-m bar slide on the plane surfaces. The velocity of B is $\mathbf{v}_B = 2\mathbf{i}$ (m/s).

(a) What are the coordinates of the instantaneous center of the bar?

(b) Use the instantaneous center to determine the velocity of A.

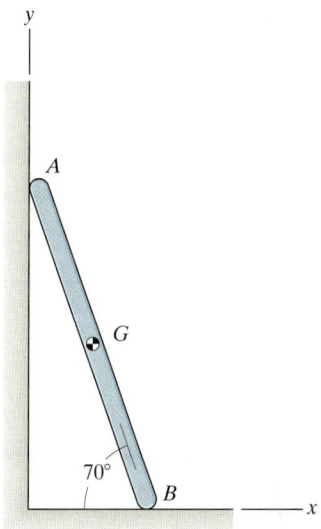

Problem 17.67

17.68 The bar is in two-dimensional motion in the x–y plane. The velocity of point A is $\mathbf{v}_A = 8\mathbf{i}$ (ft/s), and B is moving in the direction parallel to the bar. Determine the velocity of B (a) by using Eq. (17.6) and (b) by using the instantaneous center of the bar.

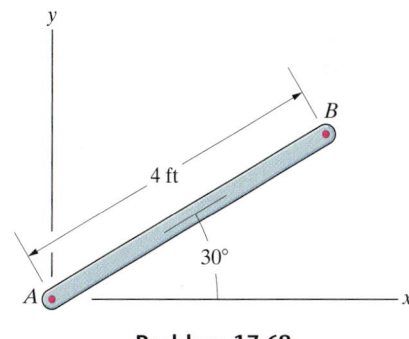

Problem 17.68

17.69 Point A of the bar is moving at 8 m/s in the direction of the unit vector $0.966\mathbf{i} - 0.259\mathbf{j}$, and point B is moving in the direction of the unit vector $0.766\mathbf{i} + 0.643\mathbf{j}$.

(a) What are the coordinates of the bar's instantaneous center?

(b) What is the bar's angular velocity?

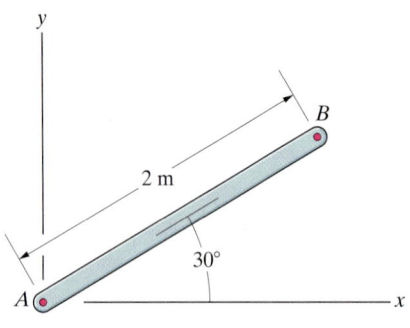

Problem 17.69

17.70 Points A and B of the 4-ft bar slide on the plane surfaces. Point B is sliding down the slanted surface at 2 ft/s.

(a) What are the coordinates of the instantaneous center of the bar?

(b) Use the instantaneous center to determine the velocity of A.

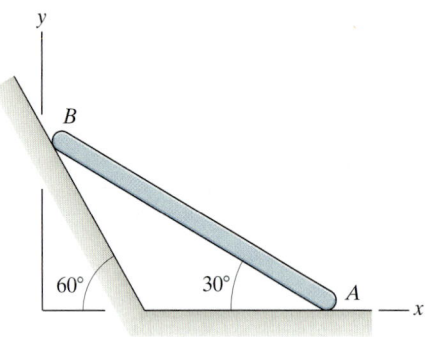

Problem 17.70

17.71 Use instantaneous centers to determine the horizontal velocity of B.

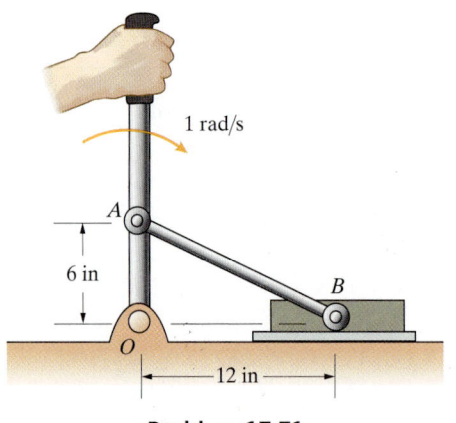

Problem 17.71

17.72 When the mechanism in Problem 17.71 is in the position shown here, use instantaneous centers to determine the horizontal velocity of B.

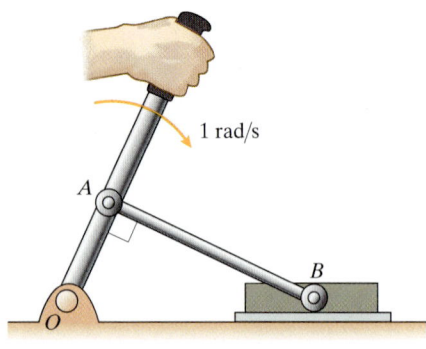

Problem 17.72

17.73 The angle $\theta = 45°$, and bar OQ is rotating in the counterclockwise direction at 0.2 rad/s. Use instantaneous centers to determine the velocity of the sleeve P.

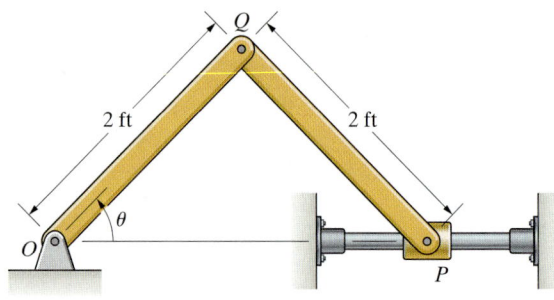

Problem 17.73

17.74 Bar AB is rotating in the counterclockwise direction at 5 rad/s. The disk rolls on the horizontal surface. Determine the angular velocity of bar BC.

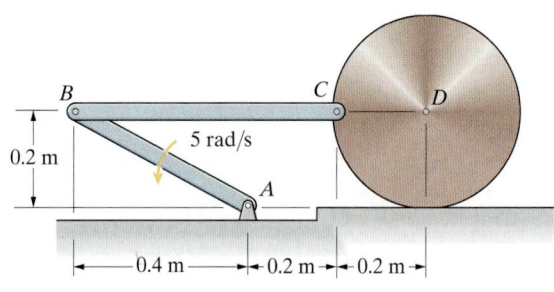

Problem 17.74

17.75 Bar *AB* rotates at 6 rad/s in the clockwise direction. Use instantaneous centers to determine the angular velocity of bar *BC*.

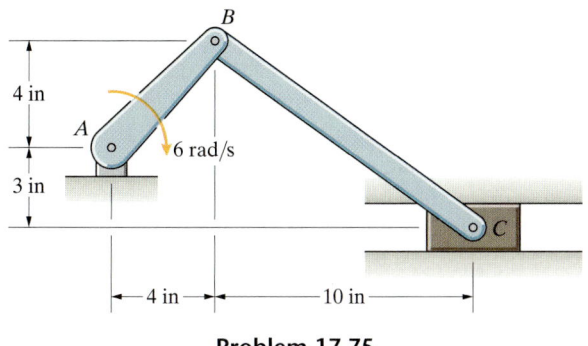

Problem 17.75

17.76 The crank *AB* is rotating in the clockwise direction at 2000 rpm (revolutions per minute).

(a) At the instant shown, what are the coordinates of the instantaneous center of the connecting rod *BC*?

(b) Use instantaneous centers to determine the angular velocity of the connecting rod *BC* at the instant shown.

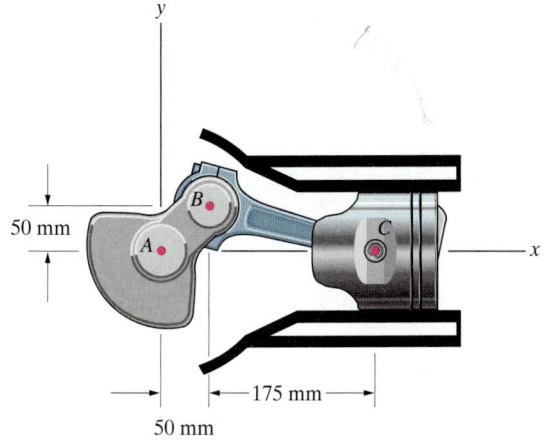

Problem 17.76

17.77 The disks roll on the plane surface. The left disk rotates at 2 rad/s in the clockwise direction. Use instantaneous centers to determine the angular velocities of the bar and the right disk.

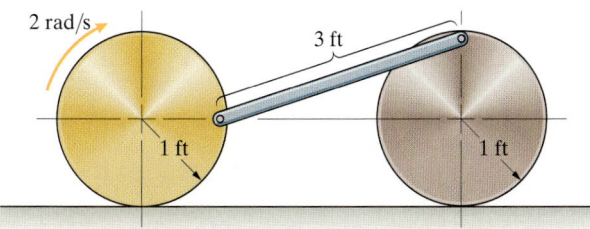

Problem 17.77

17.78 Bar *AB* rotates at 12 rad/s in the clockwise direction. Use instantaneous centers to determine the angular velocities of bars *BC* and *CD*.

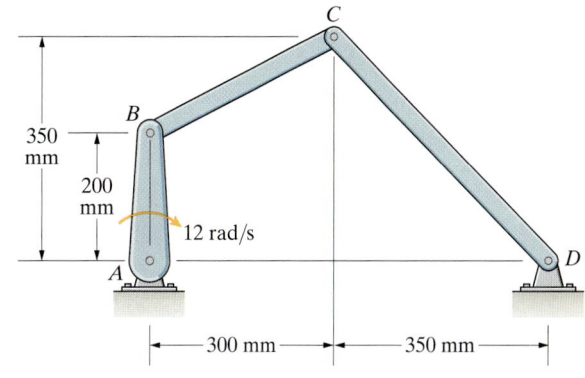

Problem 17.78

17.79 The horizontal member *ADE* supporting the scoop is stationary. The link *BD* is rotating in the clockwise direction at 1 rad/s. Use instantaneous centers to determine the angular velocity of the scoop.

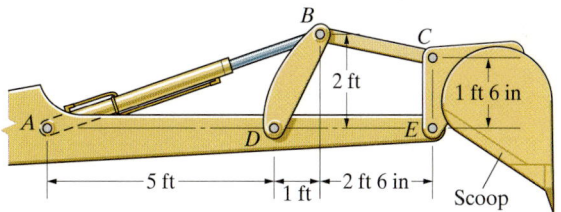

Problem 17.79

17.80 The disk is in planar motion. The directions of the velocities of points *A* and *B* are shown. The velocity of point *A* is $v_A = 2$ m/s.

(a) What are the coordinates of the disk's instantaneous center?

(b) Determine the velocity v_B and the disk's angular velocity.

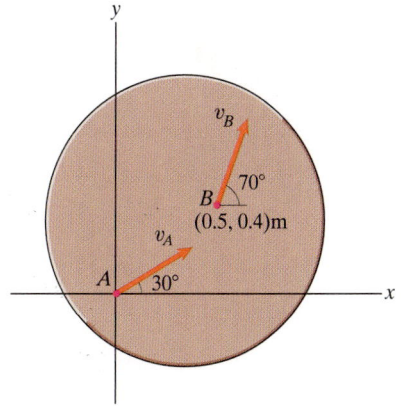

Problem 17.80

17.4 General Motions: Accelerations

In Chapter 18, we will be concerned with determining the motion of a rigid body when we know the external forces and couples acting on it. The governing equations are expressed in terms of the acceleration of the center of mass of the rigid body and its angular acceleration. To solve such problems, we need the relationship between the accelerations of points of a rigid body and its angular acceleration. In this section, we extend the methods we have used to analyze velocities of points of rigid bodies to accelerations.

Consider points A and B of a rigid body in planar motion relative to a given reference frame (Fig. 17.24a). Their velocities are related by

$$\mathbf{v}_A = \mathbf{v}_B + \mathbf{v}_{A/B}.$$

Taking the derivative of this equation with respect to time, we obtain

$$\mathbf{a}_A = \mathbf{a}_B + \mathbf{a}_{A/B},$$

where $\mathbf{a}_A$ and $\mathbf{a}_B$ are the accelerations of A and B relative to the reference frame and $\mathbf{a}_{A/B}$ is the acceleration of A relative to B. (*When we simply speak of the acceleration of a point, we will mean its acceleration relative to the given reference frame.*) Because A moves in a circular path relative to B as the rigid body rotates, $\mathbf{a}_{A/B}$ has normal and tangential components (Fig. 17.24b). The tangential

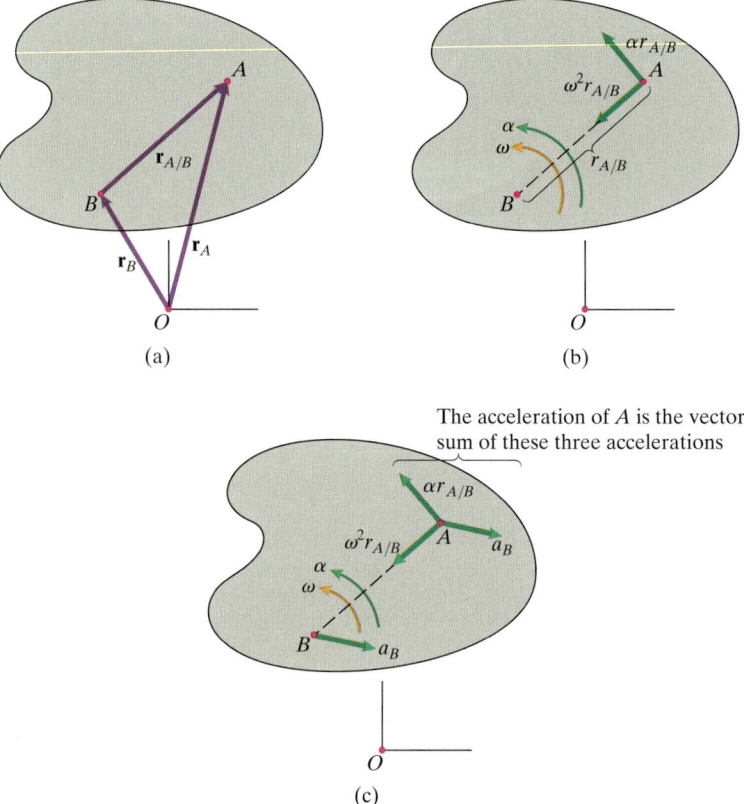

Figure 17.24
(a) Points A and B of a rigid body in planar motion and the position vector of A relative to B.
(b) Components of the acceleration of A relative to B.
(c) The acceleration of A.

component equals the product of the distance $r_{A/B} = |\mathbf{r}_{A/B}|$ and the angular acceleration α of the rigid body. The normal component points toward the center of the circular path, and its magnitude is $|\mathbf{v}_{A/B}|^2/r_{A/B} = \omega^2 r_{A/B}$. The acceleration of A equals the sum of the acceleration of B and the acceleration of A relative to B (Fig. 17.24c).

For example, let us consider a circular disk of radius R rolling on a stationary plane surface. The disk has counterclockwise angular velocity ω and counterclockwise angular acceleration α (Fig. 17.25a). The disk's center B is moving in a straight line with velocity $R\omega$, toward the left if ω is positive. Therefore, the acceleration of B is $d/dt(R\omega) = R\alpha$ and is toward the left if α is positive (Fig. 17.25b). In other words, *the magnitude of the acceleration of the center of a round object rolling on a stationary plane surface is the product of the radius and the angular acceleration.*

Now that we know the acceleration of the disk's center, let us determine the acceleration of the point C that is in contact with the surface. Relative to B, C moves in a circular path of radius R with angular velocity ω and angular acceleration α. The tangential and normal components of the acceleration of C relative to B are shown in Fig. 17.25c. The acceleration of C is the sum of the acceleration of B and the acceleration of C relative to B (Fig. 17.25d). In terms of the coordinate system shown,

$$\mathbf{a}_C = \mathbf{a}_B + \mathbf{a}_{C/B} = -R\alpha\mathbf{i} + R\alpha\mathbf{i} + R\omega^2\mathbf{j}$$

$$= R\omega^2\mathbf{j}.$$

The acceleration of point C parallel to the surface is zero, but C does have an acceleration normal to the surface.

Expressing the acceleration of a point A relative to a point B in terms of A's circular path about B as we have done is useful for visualizing and understanding the relative acceleration. However, just as we did in the case of the relative velocity, we can obtain $\mathbf{a}_{A/B}$ in a form more convenient for applications by using the angular velocity vector $\boldsymbol{\omega}$. The velocity of A relative to B is given in terms of $\boldsymbol{\omega}$ by Eq. (17.5):

$$\mathbf{v}_{A/B} = \boldsymbol{\omega} \times \mathbf{r}_{A/B}.$$

Taking the derivative of this equation with respect to time, we obtain

$$\mathbf{a}_{A/B} = \frac{d\boldsymbol{\omega}}{dt} \times \mathbf{r}_{A/B} + \boldsymbol{\omega} \times \mathbf{v}_{A/B}$$

$$= \frac{d\boldsymbol{\omega}}{dt} \times \mathbf{r}_{A/B} + \boldsymbol{\omega} \times (\boldsymbol{\omega} \times \mathbf{r}_{A/B}).$$

We next define the *angular acceleration vector* $\boldsymbol{\alpha}$ to be the rate of change of the angular velocity vector:

$$\boldsymbol{\alpha} = \frac{d\boldsymbol{\omega}}{dt}. \tag{17.7}$$

Then the acceleration of A relative to B is

$$\mathbf{a}_{A/B} = \boldsymbol{\alpha} \times \mathbf{r}_{A/B} + \boldsymbol{\omega} \times (\boldsymbol{\omega} \times \mathbf{r}_{A/B}).$$

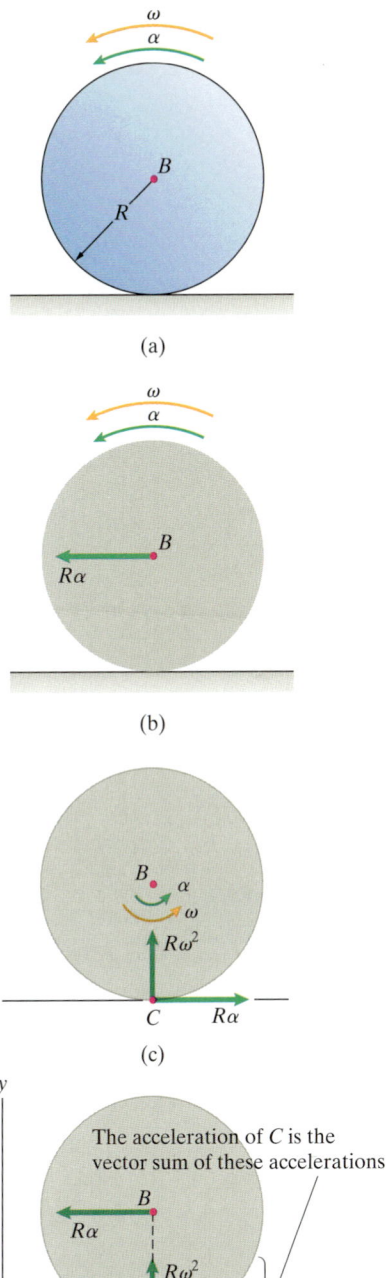

(a)

(b)

(c)

(d)

Figure 17.25
(a) A disk rolling with angular velocity ω and angular acceleration α.
(b) Acceleration of the center B.
(c) Components of the acceleration of C relative to B.
(d) The acceleration of C.

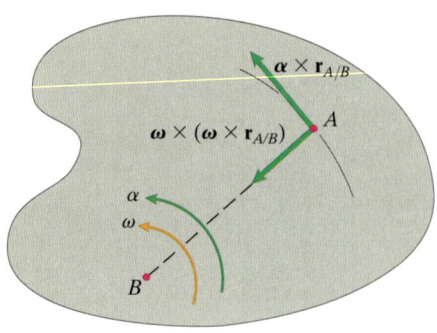

Figure 17.26
Vector components of the acceleration of A relative to B in planar motion.

Using this expression, we can write equations relating the velocities and accelerations of two points of a rigid body in terms of its angular velocity and angular acceleration:

$$\mathbf{v}_A = \mathbf{v}_B + \boldsymbol{\omega} \times \mathbf{r}_{A/B}, \tag{17.8}$$

$$\mathbf{a}_A = \mathbf{a}_B + \boldsymbol{\alpha} \times \mathbf{r}_{A/B} + \boldsymbol{\omega} \times (\boldsymbol{\omega} \times \mathbf{r}_{A/B}). \tag{17.9}$$

In the case of planar motion, the term $\boldsymbol{\alpha} \times \mathbf{r}_{A/B}$ in Eq. (17.9) is the tangential component of the acceleration of A relative to B, and $\boldsymbol{\omega} \times (\boldsymbol{\omega} \times \mathbf{r}_{A/B})$ is the normal component (Fig. 17.26). Therefore, for planar motion, we can write Eq. (17.9) in the simpler form

$$\mathbf{a}_A = \mathbf{a}_B + \boldsymbol{\alpha} \times \mathbf{r}_{A/B} - \omega^2 \mathbf{r}_{A/B}. \qquad \text{planar motion} \tag{17.10}$$

Study Questions

1. If a rigid body is in planar motion with angular velocity ω and angular acceleration α, what can you infer about the acceleration of a point A of the rigid body relative to a point B?
2. What is the definition of the angular acceleration vector?
3. If you know the angular velocity and angular acceleration vectors and the acceleration of one point of a rigid body, how can you determine the acceleration of a different point?

Example 17.5 **Acceleration of a Point**

The rolling disk in Fig. 17.27 has counterclockwise angular velocity ω and counterclockwise angular acceleration α. What is the acceleration of point A?

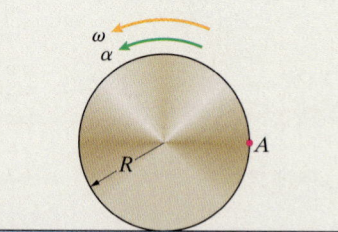

Figure 17.27

Strategy

We know that the magnitude of the acceleration of the center of the disk is the product of the radius and the angular acceleration. Therefore, we can express the acceleration of A as the sum of the acceleration of the center of the disk and the acceleration of A relative to the center. We will do so first by using vector diagrams as shown in Fig. 17.24c and then by using Eq. (17.10).

Solution

First Method In terms of the coordinate system in Fig. a, the acceleration of the center B is $\mathbf{a}_B = -\alpha R\mathbf{i}$. A's motion in a circular path of radius R relative to B results in the tangential and normal components of relative acceleration shown in Fig. b:

$$\mathbf{a}_{A/B} = -\omega^2 R\mathbf{i} + \alpha R\mathbf{j}.$$

Therefore, the acceleration of A is

$$\mathbf{a}_A = \mathbf{a}_B + \mathbf{a}_{A/B} = -\alpha R\mathbf{i} - \omega^2 R\mathbf{i} + \alpha R\mathbf{j}$$

$$= (-\alpha R - \omega^2 R)\mathbf{i} + \alpha R\mathbf{j}.$$

Second Method The angular acceleration vector of the disk is $\boldsymbol{\alpha} = \alpha\mathbf{k}$, and the position of A relative to B is $\mathbf{r}_{A/B} = R\mathbf{i}$ (Fig. c). From Eq. (17.10), the acceleration of A is

$$\mathbf{a}_A = \mathbf{a}_B + \boldsymbol{\alpha} \times \mathbf{r}_{A/B} - \omega^2 \mathbf{r}_{A/B}$$

$$= -\alpha R\mathbf{i} + (\alpha\mathbf{k}) \times (R\mathbf{i}) - \omega^2(R\mathbf{i})$$

$$= (-\alpha R - \omega^2 R)\mathbf{i} + \alpha R\mathbf{j}.$$

Critical Thinking

We were able to determine the acceleration of point A of the rolling disk because we could express the acceleration of the center B in terms of the disk's angular acceleration. You will find it very useful to remember that the magnitude of the velocity of the center of a round rolling object is equal to the product of its radius and its angular velocity, and the magnitude of the acceleration of the center is equal to the product of its radius and its angular acceleration.

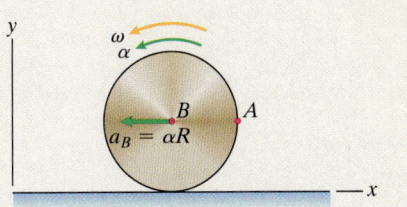

(a) Acceleration of the center of the disk.

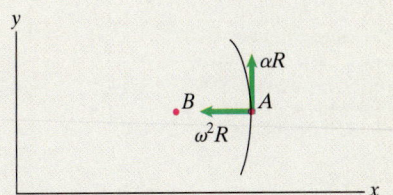

(b) Components of the acceleration of A relative to B.

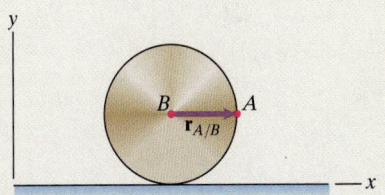

(c) Position of A relative to B.

Example 17.6   Angular Accelerations of Members of a Linkage

Bar AB in Fig. 17.28 has a counterclockwise angular velocity of 10 rad/s and a clockwise angular acceleration of 300 rad/s². What are the angular accelerations of bars BC and CD?

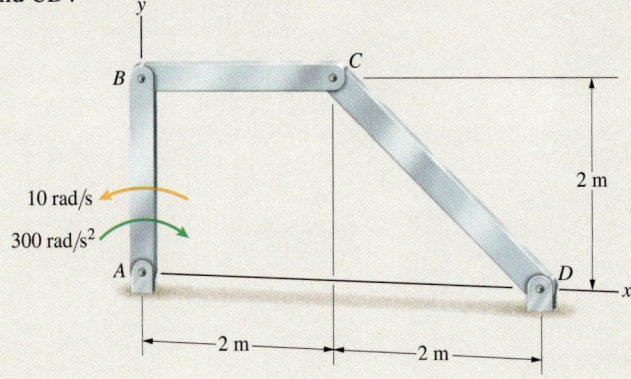

Figure 17.28

Strategy

Since we know the angular velocity of bar AB, we can determine the velocity of point B. Then we can apply Eq. (17.8) to points C and D to obtain an equation for $\mathbf{v}_C$ in terms of the angular velocity of bar CD. We can also apply Eq. (17.8) to points B and C to obtain an equation for $\mathbf{v}_C$ in terms of the angular velocity of bar BC. By equating the two expressions for $\mathbf{v}_C$, we will obtain a vector equation in two unknowns: the angular velocities of bars BC and CD. Then, by following the same sequence of steps, but using Eq. (17.10), we can obtain the angular accelerations of bars BC and CD.

Solution

The velocity of B is (Fig. a)

$$\mathbf{v}_B = \mathbf{v}_A + \boldsymbol{\omega}_{AB} \times \mathbf{r}_{B/A}$$

$$= \mathbf{0} + (10\mathbf{k}) \times (2\mathbf{j})$$

$$= -20\mathbf{i} \; (\text{m/s}).$$

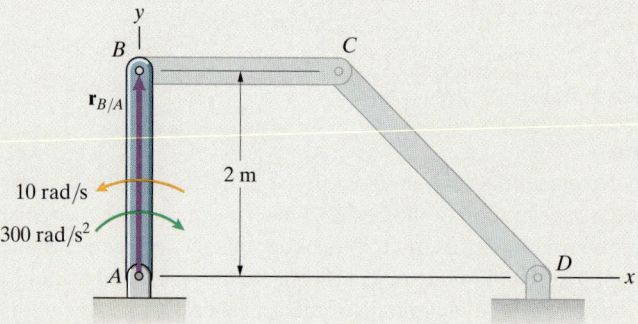

(a) Determining the motion of B.

Let ω_{CD} be the unknown angular velocity of bar CD (Fig. b). The velocity of C in terms of the velocity of D is

$$\mathbf{v}_C = \mathbf{v}_D + \boldsymbol{\omega}_{CD} \times \mathbf{r}_{C/D}$$

$$= \mathbf{0} + \begin{vmatrix} \mathbf{i} & \mathbf{j} & \mathbf{k} \\ 0 & 0 & \omega_{CD} \\ -2 & 2 & 0 \end{vmatrix}$$

$$= -2\omega_{CD}\mathbf{i} - 2\omega_{CD}\mathbf{j}.$$

Denoting the angular velocity of bar BC by ω_{BC} (Fig. c), we obtain the velocity of C in terms of the velocity of B:

$$\mathbf{v}_C = \mathbf{v}_B + \boldsymbol{\omega}_{BC} \times \mathbf{r}_{C/B}$$

$$= -20\mathbf{i} + (\omega_{BC}\mathbf{k}) \times (2\mathbf{i})$$

$$= -20\mathbf{i} + 2\omega_{BC}\mathbf{j}.$$

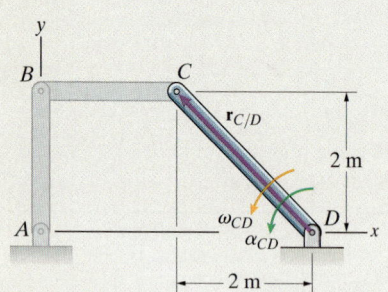

(b) Determining the motion of C in terms of the angular motion of bar CD.

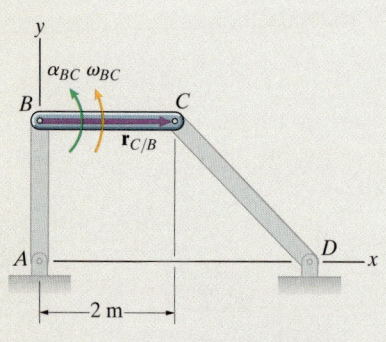

(c) Determining the motion of C in terms of the angular motion of bar BC.

Equating our two expressions for $\mathbf{v}_C$ yields

$$-2\omega_{CD}\mathbf{i} - 2\omega_{CD}\mathbf{j} = -20\mathbf{i} + 2\omega_{BC}\mathbf{j}.$$

Equating the $\mathbf{i}$ and $\mathbf{j}$ components, we obtain $\omega_{CD} = 10\ \text{rad/s}$ and $\omega_{BC} = -10\ \text{rad/s}$.

We can use the same sequence of steps to determine the angular accelerations. The acceleration of B is (Fig. a)

$$\mathbf{a}_B = \mathbf{a}_A + \boldsymbol{\alpha}_{AB} \times \mathbf{r}_{B/A} - \omega_{AB}^2 \mathbf{r}_{B/A}$$

$$= \mathbf{0} + (-300\mathbf{k}) \times (2\mathbf{j}) - (10)^2(2\mathbf{j})$$

$$= 600\mathbf{i} - 200\mathbf{j}\ (\text{m/s}^2).$$

The acceleration of C in terms of the acceleration of D is (Fig. b)

$$\mathbf{a}_C = \mathbf{a}_D + \boldsymbol{\alpha}_{CD} \times \mathbf{r}_{C/D} - \omega_{CD}^2 \mathbf{r}_{C/D}$$

$$= \mathbf{0} + \begin{vmatrix} \mathbf{i} & \mathbf{j} & \mathbf{k} \\ 0 & 0 & \alpha_{CD} \\ -2 & 2 & 0 \end{vmatrix} - (10)^2(-2\mathbf{i} + 2\mathbf{j})$$

$$= (200 - 2\alpha_{CD})\mathbf{i} - (200 + 2\alpha_{CD})\mathbf{j}.$$

The acceleration of C in terms of the acceleration of B is (Fig. c)

$$\mathbf{a}_C = \mathbf{a}_B + \boldsymbol{\alpha}_{BC} \times \mathbf{r}_{C/B} - \omega_{BC}^2 \mathbf{r}_{C/B}$$

$$= 600\mathbf{i} - 200\mathbf{j} + (\alpha_{BC}\mathbf{k}) \times (2\mathbf{i}) - (-10)^2(2\mathbf{i})$$

$$= 400\mathbf{i} - (200 - 2\alpha_{BC})\mathbf{j}.$$

Equating the expressions for $\mathbf{a}_C$, we obtain

$$(200 - 2\alpha_{CD})\mathbf{i} - (200 + 2\alpha_{CD})\mathbf{j} = 400\mathbf{i} - (200 - 2\alpha_{BC})\mathbf{j}.$$

Equating $\mathbf{i}$ and $\mathbf{j}$ components, we obtain the angular accelerations $\alpha_{BC} = 100\ \text{rad/s}^2$ and $\alpha_{CD} = -100\ \text{rad/s}^2$.

Critical Thinking

To determine the angular accelerations of a set of pinned rigid bodies, it is usually necessary to determine their angular velocities first, because the angular velocity appears in Eqs. (17.9) and (17.10). But as this example demonstrates, this initial step provides you with a guide for completing the solution. Once you have found a sequence of steps using Eq. (17.8) for determining the angular velocities, the same sequence of steps using Eq. (17.9) or (17.10) will determine the angular accelerations.

Problems

17.81 The rigid body rotates about the z axis with counterclockwise angular velocity $\omega = 4$ rad/s and counterclockwise angular acceleration $\alpha = 2$ rad/s^2. The distance $r_{A/B} = 0.6$ m.

(a) What are the rigid body's angular velocity and angular acceleration vectors?

(b) Determine the acceleration of point A relative to point B first by using Eq. (17.9) and then by using Eq. (17.10).

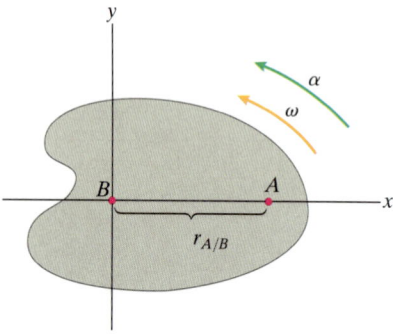

Problem 17.81

17.82 The bar rotates with a counterclockwise angular velocity of 5 rad/s and a counterclockwise angular acceleration of 30 rad/s^2. Determine the acceleration of A (a) by using Eq. (17.9) and (b) by using Eq. (17.10).

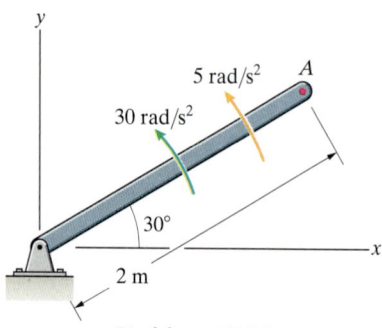

Problem 17.82

17.83 The bar rotates with a counterclockwise angular velocity of 20 rad/s and a counterclockwise angular acceleration of 6 rad/s^2.

(a) By applying Eq. (17.10) to point A and the fixed point O, determine the acceleration of A.

(b) By using the result of part (a) and applying Eq. (17.10) to points A and B, determine the acceleration of B.

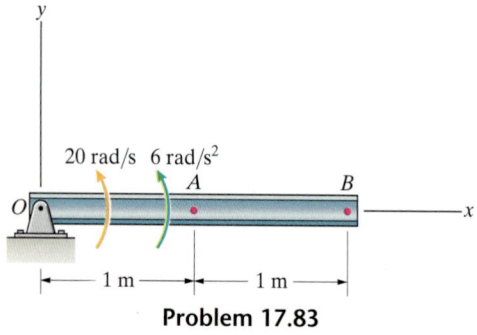

Problem 17.83

17.84 The helicopter is in planar motion in the x–y plane. At the instant shown, the position of its center of mass G is $x = 2$ m, $y = 2.5$ m, its velocity is $\mathbf{v}_G = 12\mathbf{i} + 4\mathbf{j}$ (m/s), and its acceleration is $\mathbf{a}_G = 2\mathbf{i} + 3\mathbf{j}$ (m/s^2). The position of point T where the tail rotor is mounted is $x = -3.5$ m, $y = 4.5$ m. The helicopter's angular velocity is 0.2 rad/s clockwise, and its angular acceleration is 0.1 rad/s^2 counterclockwise. What is the acceleration of point T?

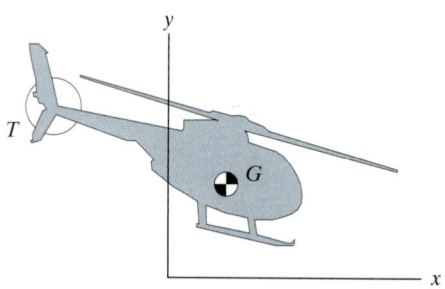

Problem 17.84

17.85 The disk rolls on the plane surface. The velocity of point A is 6 m/s to the right, and the acceleration of A is 20 m/s^2 to the right.

(a) What is the angular acceleration vector of the disk?

(b) Determine the accelerations of points B, C, and D.

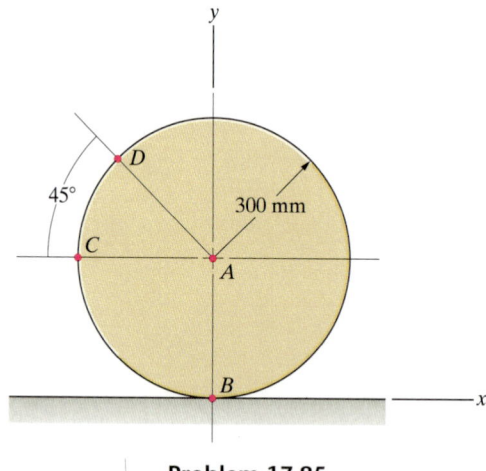

Problem 17.85

17.86 The disk rolls on the circular surface with a constant clockwise angular velocity of 1 rad/s. What are the accelerations of points A and B?

Strategy: Begin by determining the acceleration of the center of the disk. Notice that the center moves in a circular path and the magnitude of its velocity is constant.

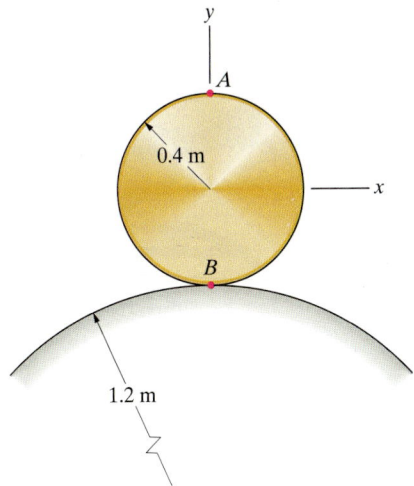

Problem 17.86

17.87 The endpoints of the bar slide on the plane surfaces. Show that the acceleration of the midpoint G is related to the bar's angular velocity and angular acceleration by

$$\mathbf{a}_G = \tfrac{1}{2}L[(\alpha \cos \theta - \omega^2 \sin \theta)\mathbf{i} - (\alpha \sin \theta + \omega^2 \cos \theta)\mathbf{j}].$$

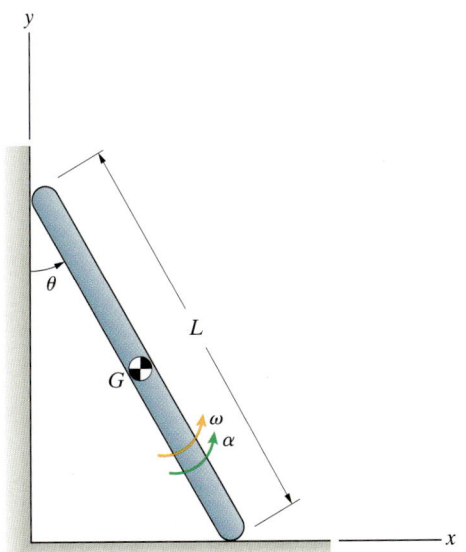

Problem 17.87

17.88 The angular velocity and angular acceleration of bar AB are $\omega_{AB} = 2$ rad/s and $\alpha_{AB} = 10$ rad/s². The dimensions of the rectangular plate are 12 in × 24 in. What are the angular velocity and angular acceleration of the rectangular plate?

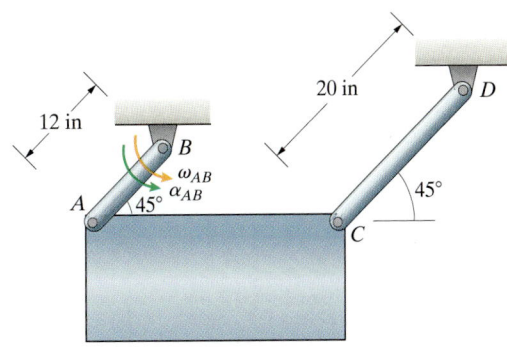

Problem 17.88

17.89 The ring gear is stationary, and the sun gear has an angular acceleration of 10 rad/s² in the counterclockwise direction. Determine the angular acceleration of the planet gears.

17.90 The sun gear has a counterclockwise angular velocity of 4 rad/s and a clockwise angular acceleration of 12 rad/s². What is the magnitude of the acceleration of the centerpoints of the planet gears?

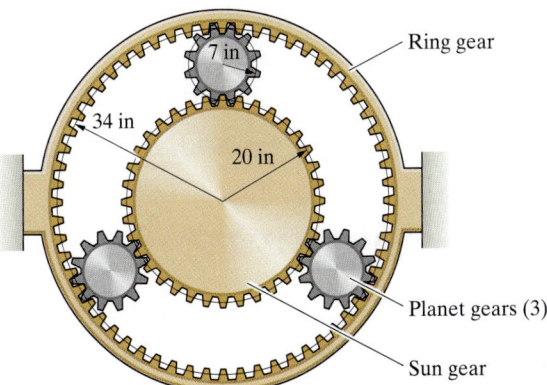

Problems 17.89/17.90

17.91 The 1-m-diameter disk rolls, and point B of the 1-m-long bar slides, on the plane surface. Determine the angular acceleration of the bar and the acceleration of point B.

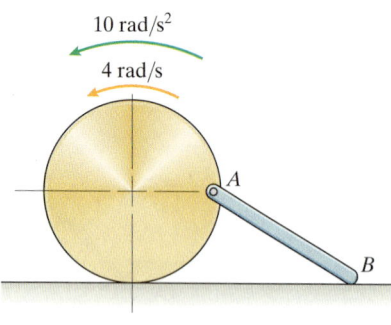

10 rad/s²

4 rad/s

A

B

Problem 17.91

17.92 If $\theta = 45°$ and sleeve P is moving to the right with a constant velocity of 2 m/s, what are the angular accelerations of bars OQ and PQ?

17.93 If $\theta = 50°$ and bar OQ has a constant clockwise angular velocity of 1 rad/s, what is the acceleration of sleeve P?

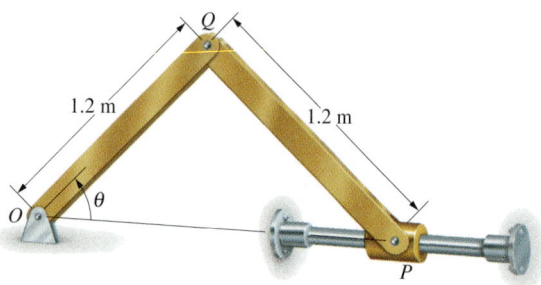

Q

1.2 m

1.2 m

O

θ

P

Problems 17.92/17.93

17.94 The angle $\theta = 60°$, and bar OQ has a constant counterclockwise angular velocity of 2 rad/s. What is the angular acceleration of bar PQ?

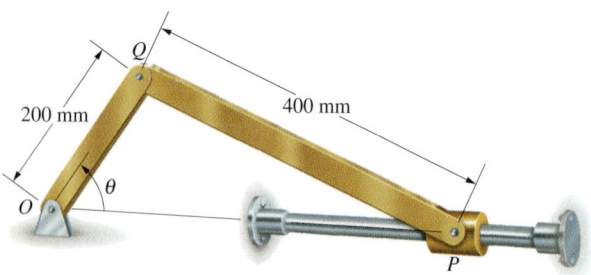

Q

200 mm

400 mm

O

θ

P

Problem 17.94

17.95 The crank AB is rotating in the clockwise direction with a constant angular velocity of 2000 rpm (revolutions per minute). What is the acceleration of the piston C at the instant shown?

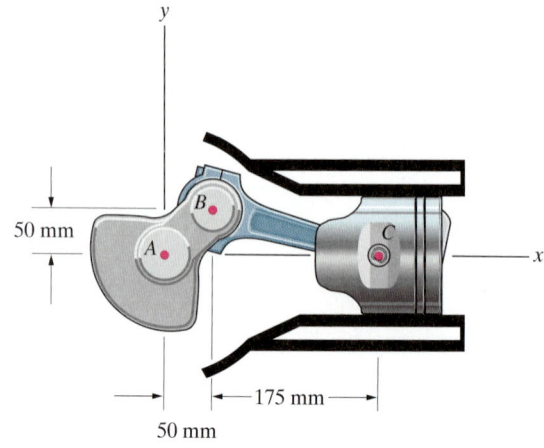

y

B

50 mm

A

C

x

175 mm

50 mm

Problem 17.95

17.96 The angular velocity and angular acceleration of bar AB are $\omega_{AB} = 4$ rad/s and $\alpha_{AB} = -6$ rad/s². Determine the angular accelerations of bars BC and CD.

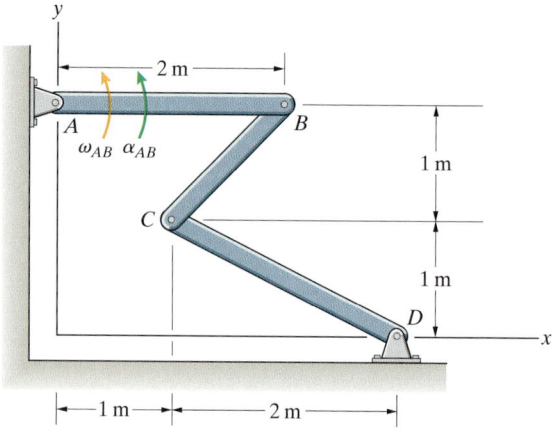

y

2 m

A

B

$\omega_{AB}\ \alpha_{AB}$

1 m

C

1 m

D

x

1 m

2 m

Problem 17.96

17.97 The angular velocity and angular acceleration of bar AB are $\omega_{AB} = 2$ rad/s and $\alpha_{AB} = 8$ rad/s^2. What is the acceleration of point D?

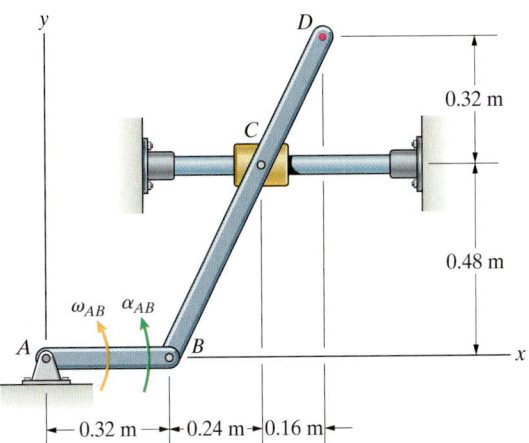

Problem 17.97

17.98 If $\omega_{AB} = 6$ rad/s and $\alpha_{AB} = 20$ rad/s^2, what are the velocity and acceleration of point C?

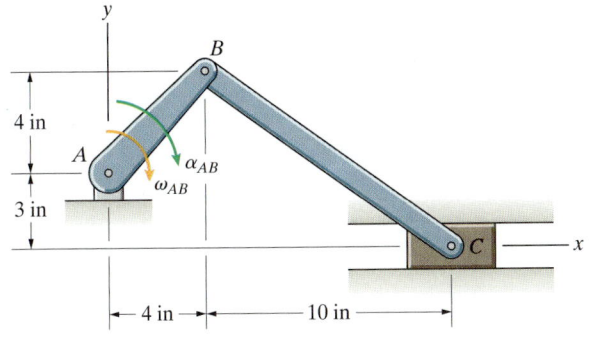

Problem 17.98

17.99 The angular velocity and angular acceleration of bar AB are $\omega_{AB} = 5$ rad/s and $\alpha_{AB} = 10$ rad/s^2. Determine the angular acceleration of bar BC.

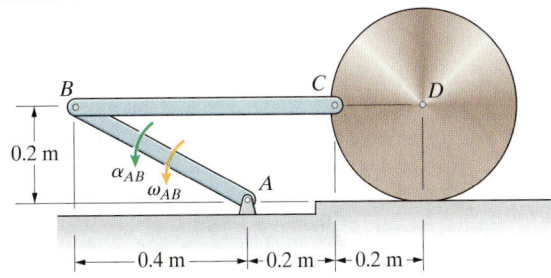

Problem 17.99

17.100 A motor rotates the circular disk mounted at A, moving the saw back and forth. (The saw is supported by a horizontal slot so that point C moves horizontally.) The radius AB is 4 in, and the link BC is 14 in long. In the position shown, $\theta = 45°$ and the link BC is horizontal. If the disk has a constant angular velocity of one revolution per second counterclockwise, what is the acceleration of the saw?

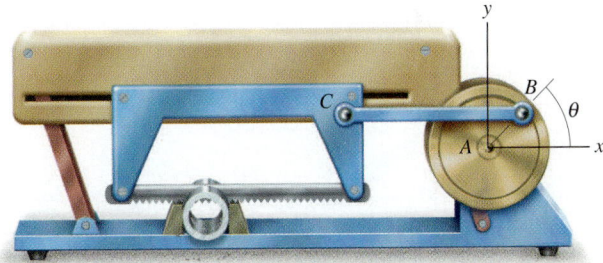

Problem 17.100

17.101 If $\omega_{AB} = 2$ rad/s, $\alpha_{AB} = 2$ rad/s^2, $\omega_{BC} = 1$ rad/s, and $\alpha_{BC} = 4$ rad/s^2, what is the acceleration of point C where the scoop of the excavator is attached?

17.102 If the velocity of point C of the excavator is $\mathbf{v}_C = 4\mathbf{i}$ (m/s) and is constant, what are ω_{AB}, α_{AB}, ω_{BC}, and α_{BC}?

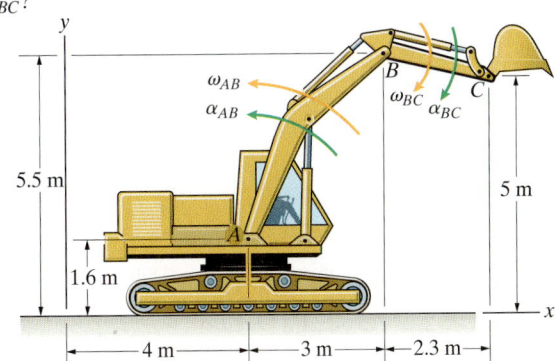

Problems 17.101/17.102

17.103 Bar AB rotates in the counterclockwise direction with a constant angular velocity of 10 rad/s. What is the acceleration of the midpoint G of bar CD?

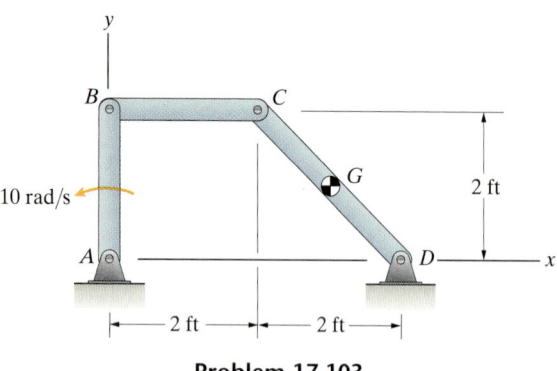

Problem 17.103

17.104 At the instant shown, bar AB has no angular velocity, but has a counterclockwise angular acceleration of 10 rad/s². Determine the acceleration of point E.

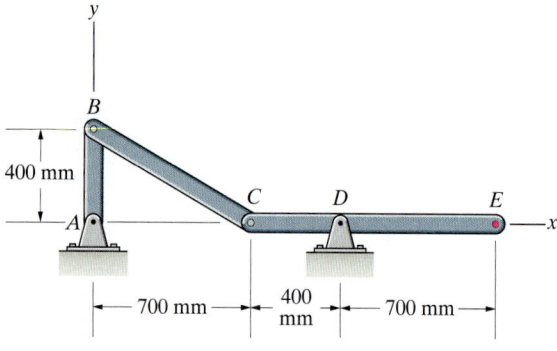

Problem 17.104

17.105 If $\omega_{AB} = 12$ rad/s and $\alpha_{AB} = 100$ rad/s², what are the angular accelerations of bars BC and CD?

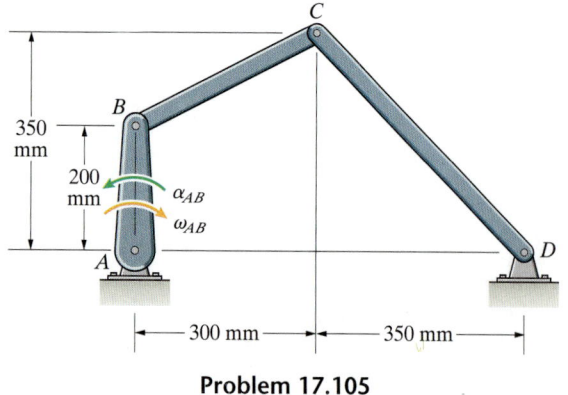

Problem 17.105

17.106 If $\omega_{AB} = 4$ rad/s counterclockwise and $\alpha_{AB} = 12$ rad/s² counterclockwise, what is the acceleration of point C?

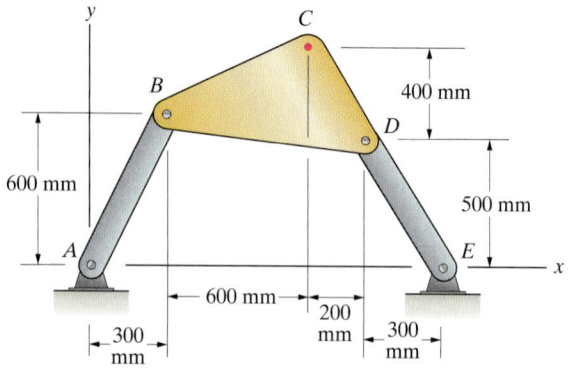

Problem 17.106

17.107 The angular velocities and angular accelerations of the grips of the shears are shown. What is the resulting angular acceleration of the jaw BD?

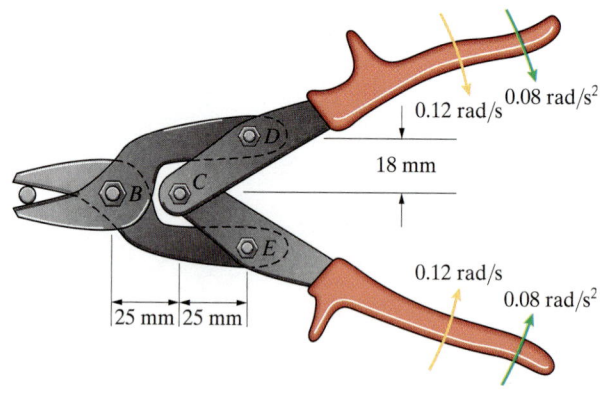

Problem 17.107

17.108 If arm *AB* has a constant clockwise angular velocity of 0.8 rad/s, arm *BC* has a constant clockwise angular velocity of 0.2 rad/s, and arm *CD* remains vertical, what is the acceleration of part *D*?

17.109 If arm *AB* has a constant clockwise angular velocity of 0.8 rad/s and you want *D* to have zero velocity and acceleration, what are the necessary angular velocities and angular accelerations of arms *BC* and *CD*?

17.110 If you want arm *CD* to remain vertical and you want part *D* to have velocity $\mathbf{v}_D = 1.0\mathbf{i}$ (m/s) and zero acceleration, what are the necessary angular velocities and angular accelerations of arms *AB* and *BC*?

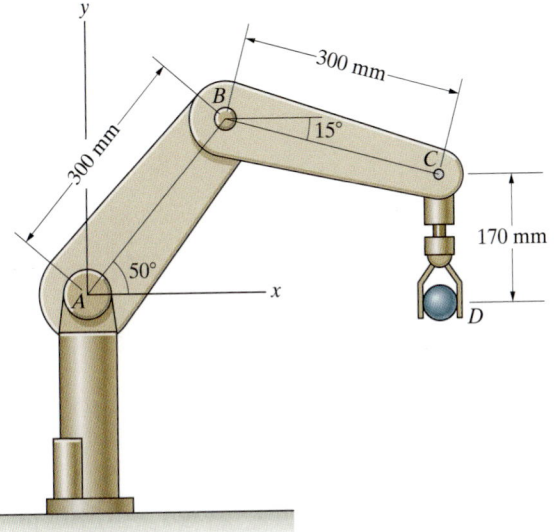

Problems 17.108–17.110

17.111 Link *AB* of the robot's arm is rotating with a constant counterclockwise angular velocity of 2 rad/s, and link *BC* is rotating with a constant clockwise angular velocity of 3 rad/s. Link *CD* is rotating at 4 rad/s in the counterclockwise direction and has a counterclockwise angular acceleration of 6 rad/s². What is the acceleration of point *D*?

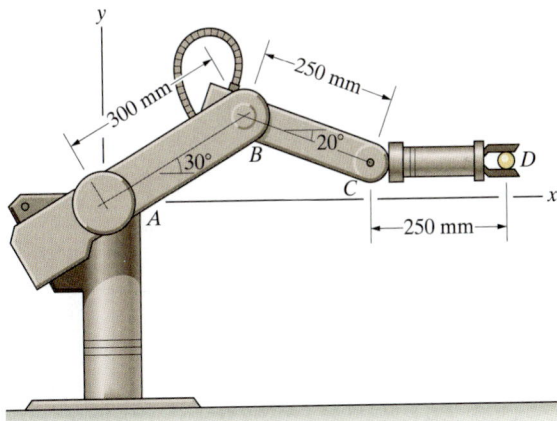

Problem 17.111

17.112 The upper grip and jaw of the pliers *ABC* is stationary. The lower grip *DEF* is rotating in the clockwise direction with a constant angular velocity of 0.2 rad/s. At the instant shown, what is the angular acceleration of the lower jaw *CFG*?

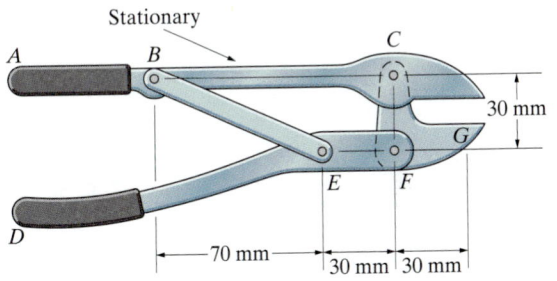

Problem 17.112

17.113 The horizontal member *ADE* supporting the scoop is stationary. If link *BD* has a clockwise angular velocity of 1 rad/s and a counterclockwise angular acceleration of 2 rad/s², what is the angular acceleration of the scoop?

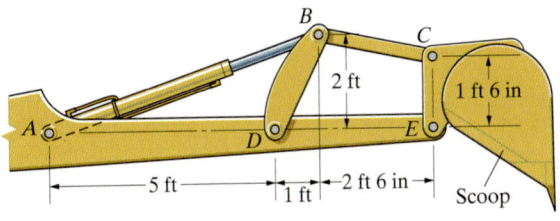

Problem 17.113

17.114 The ring gear is fixed, and the hub and planet gears are bonded together. The connecting rod has a counterclockwise angular acceleration of 10 rad/s². Determine the angular accelerations of the planet and sun gears.

17.115 The connecting rod has a counterclockwise angular velocity of 4 rad/s and a clockwise angular acceleration of 12 rad/s². Determine the magnitude of the acceleration of point A.

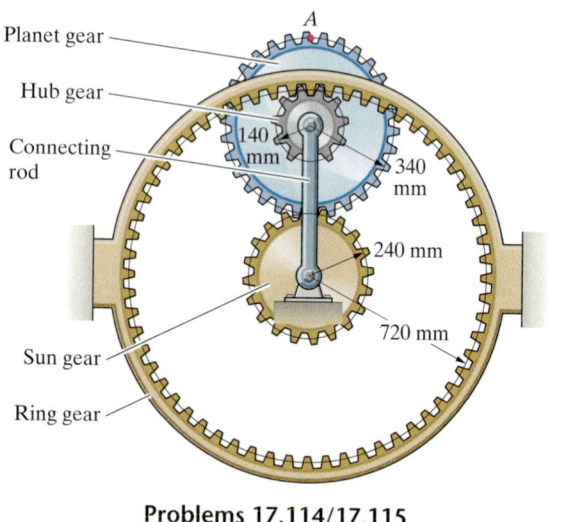

Problems 17.114/17.115

17.116 The large gear is fixed. The angular velocity and angular acceleration of bar AB are $\omega_{AB} = 2$ rad/s and $\alpha_{AB} = 4$ rad/s². Determine the angular accelerations of bars CD and DE.

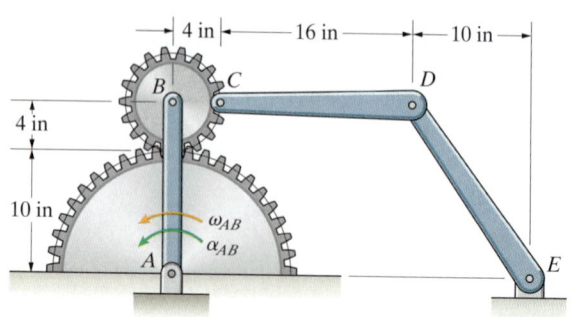

Problem 17.116

17.5 Sliding Contacts

In this section we consider a type of problem that is superficially similar to those we have discussed previously in this chapter, but which requires a different method of solution. For example, in Fig. 17.29, pin A of the bar connected at C slides in a slot in the bar connected at B. Suppose that we know the angular velocity of the bar connected at B, and we want to determine the angular velocity of bar AC. We cannot use the equation $\mathbf{v}_A = \mathbf{v}_B + \boldsymbol{\omega} \times \mathbf{r}_{A/B}$ to express the velocity of pin A in terms of the angular velocity of the bar fixed at B, because we derived that equation under the assumption that A and B are points of the same rigid body. Here pin A moves relative to the bar connected at B as it slides along the slot. This is an example of a *sliding contact* between rigid bodies. To solve this type of problem, we must rederive Eqs. (17.8), (17.9), and (17.10) without making the assumption that A is a point of the rigid body.

To describe the motion of a point that moves relative to a given rigid body, it is convenient to use a reference frame that moves with the rigid body. We say that such a reference frame is *body-fixed*. In Fig. 17.30, we introduce a body-fixed reference frame *xyz* with its origin at a point B of the rigid body, in addition to the *primary reference frame* with origin O. (The primary reference frame is the reference frame relative to which we are describing the motion of the

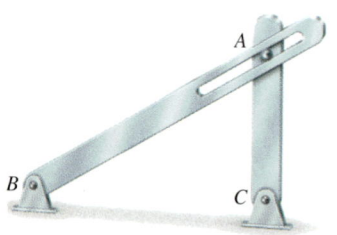

Figure 17.29
Linkage with a sliding contact.

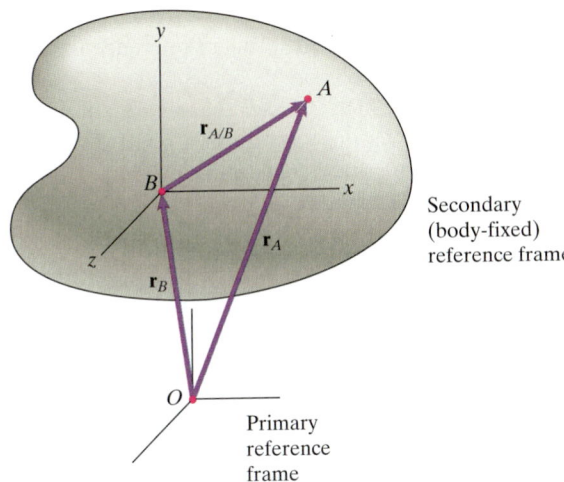

Figure 17.30
A point B of a rigid body, a body-fixed secondary reference frame, and an arbitrary point A.

rigid body.) We do not assume A to be a point of the rigid body. The position of A relative to O is

$$\mathbf{r}_A = \mathbf{r}_B + \underbrace{x\mathbf{i} + y\mathbf{j} + z\mathbf{k}}_{\mathbf{r}_{A/B}},$$

where x, y, and z are the coordinates of A in terms of the body-fixed reference frame. Our next step is to take the derivative of this expression with respect to time in order to obtain an equation for the velocity of A. In doing so, we recognize that the unit vectors $\mathbf{i}$, $\mathbf{j}$, and $\mathbf{k}$ are not constant, because they rotate with the body-fixed reference frame:

$$\mathbf{v}_A = \mathbf{v}_B + \frac{dx}{dt}\mathbf{i} + x\frac{d\mathbf{i}}{dt} + \frac{dy}{dt}\mathbf{j} + y\frac{d\mathbf{j}}{dt} + \frac{dz}{dt}\mathbf{k} + z\frac{d\mathbf{k}}{dt}.$$

Now, what are the derivatives of the unit vectors? In Section 17.3, we showed that if $\mathbf{r}_{P/B}$ is the position of a point P of a rigid body relative to another point B of the same rigid body, $d\mathbf{r}_{P/B}/dt = \mathbf{v}_{P/B} = \boldsymbol{\omega} \times \mathbf{r}_{P/B}$. Since we can regard the unit vector $\mathbf{i}$ as the position vector of a point P of the rigid body (Fig. 17.31), its derivative is $d\mathbf{i}/dt = \boldsymbol{\omega} \times \mathbf{i}$. Applying the same argument to the unit vectors $\mathbf{j}$ and $\mathbf{k}$, we obtain

$$\frac{d\mathbf{i}}{dt} = \boldsymbol{\omega} \times \mathbf{i}, \qquad \frac{d\mathbf{j}}{dt} = \boldsymbol{\omega} \times \mathbf{j}, \qquad \frac{d\mathbf{k}}{dt} = \boldsymbol{\omega} \times \mathbf{k}.$$

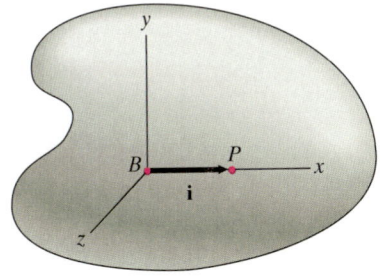

Figure 17.31
Interpreting $\mathbf{i}$ as the position vector of a point P relative to B.

Using these expressions, we can write the velocity of point A as

$$\mathbf{v}_A = \mathbf{v}_B + \underbrace{\mathbf{v}_{A\,\text{rel}} + \boldsymbol{\omega} \times \mathbf{r}_{A/B}}_{\mathbf{v}_{A/B}}, \tag{17.11}$$

where

$$\mathbf{v}_{A\,\text{rel}} = \frac{dx}{dt}\mathbf{i} + \frac{dy}{dt}\mathbf{j} + \frac{dz}{dt}\mathbf{k} \tag{17.12}$$

is the velocity of A relative to the body-fixed reference frame. That is, $\mathbf{v}_A$ is the velocity of A relative to the primary reference frame, and $\mathbf{v}_{A\,\text{rel}}$ is the velocity of A relative to the rigid body.

Equation (17.11) expresses the velocity of a point A as the sum of three terms (Fig. 17.32): the velocity of a point B of the rigid body, the velocity $\boldsymbol{\omega} \times \mathbf{r}_{A/B}$ of A relative to B due to the rotation of the rigid body, and the velocity $\mathbf{v}_{A\,\text{rel}}$ of A relative to the rigid body.

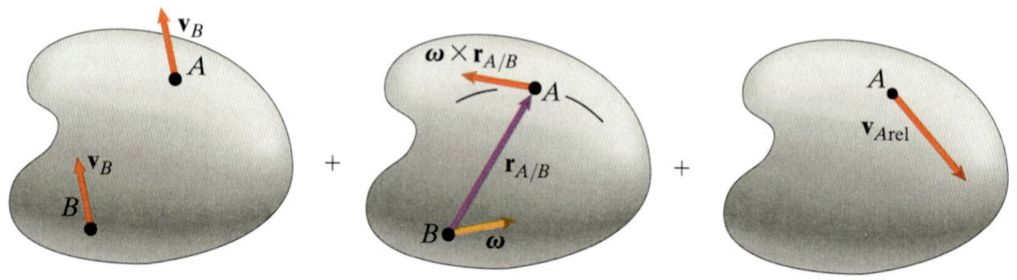

Figure 17.32
Expressing the velocity of A in terms of the velocity of a point B of the rigid body.

To obtain an equation for the acceleration of point A, we take the derivative of Eq. (17.11) with respect to time and use Eq. (17.12). The result is (see Problem 17.142)

$$\mathbf{a}_A = \mathbf{a}_B + \underbrace{\mathbf{a}_{A\,\text{rel}} + 2\boldsymbol{\omega} \times \mathbf{v}_{A\,\text{rel}} + \boldsymbol{\alpha} \times \mathbf{r}_{A/B} + \boldsymbol{\omega} \times (\boldsymbol{\omega} \times \mathbf{r}_{A/B})}_{\mathbf{a}_{A/B}}, \qquad (17.13)$$

where

$$\mathbf{a}_{A\,\text{rel}} = \frac{d^2 x}{dt^2}\mathbf{i} + \frac{d^2 y}{dt^2}\mathbf{j} + \frac{d^2 z}{dt^2}\mathbf{k} \qquad (17.14)$$

is the acceleration of A relative to the body-fixed reference frame. That is, $\mathbf{a}_A$ is the acceleration of A relative to the primary reference frame, and $\mathbf{a}_{A\,\text{rel}}$ is the acceleration of A relative to the rigid body.

In the case of planar motion, we can express Eq. (17.13) in the simpler form

$$\mathbf{a}_A = \mathbf{a}_B + \underbrace{\mathbf{a}_{A\,\text{rel}} + 2\boldsymbol{\omega} \times \mathbf{v}_{A\,\text{rel}} + \boldsymbol{\alpha} \times \mathbf{r}_{A/B} - \omega^2 \mathbf{r}_{A/B}}_{\mathbf{a}_{A/B}}. \qquad (17.15)$$

In summary, $\mathbf{v}_A$ and $\mathbf{a}_A$ are the velocity and acceleration of point A relative to the primary reference frame—the reference frame relative to which the rigid body's motion is being described. The terms $\mathbf{v}_{A\,\text{rel}}$ and $\mathbf{a}_{A\,\text{rel}}$ are the velocity and acceleration of point A relative to the body-fixed reference frame. That is, they are the velocity and acceleration measured by an observer moving with the rigid body (Fig. 17.33). If A is a point of the rigid body, then $\mathbf{v}_{A\,\text{rel}}$ and $\mathbf{a}_{A\,\text{rel}}$ are zero, and Eqs. (17.11) and (17.13) are identical to Eqs. (17.8) and (17.9).

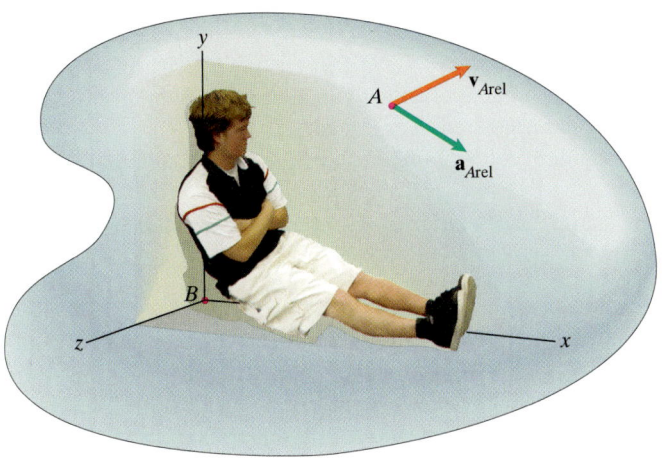

Figure 17.33
Imagine yourself to be stationary relative to the rigid body.

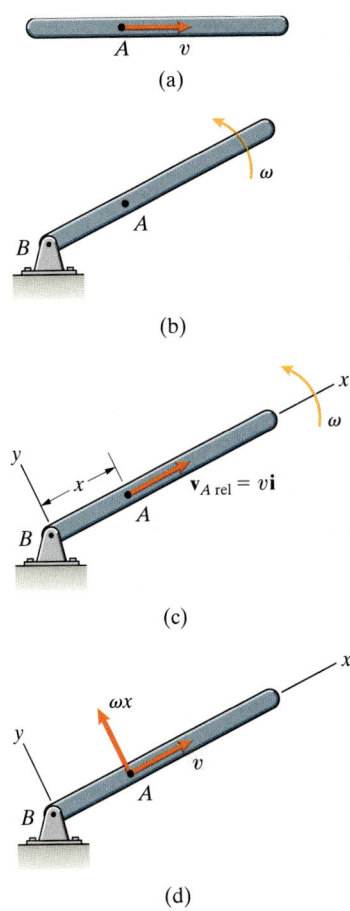

We can illustrate these concepts with a simple example. Figure 17.34(a) shows a point A moving with velocity v parallel to the axis of a bar. (Imagine that A is a bug walking along the bar.) Suppose that at the same time, the bar is rotating about a fixed point B with a constant angular velocity ω relative to an earth-fixed reference frame (Fig. 17.34b). We will use Eq. (17.11) to determine the velocity of A relative to the earth-fixed reference frame.

Let the coordinate system in Fig. 17.34(c) be fixed with respect to the bar, and let x be the present position of A. In terms of this body-fixed reference frame, the angular velocity vector of the bar (and the reference frame) relative to the primary earth-fixed reference frame is $\boldsymbol{\omega} = \omega\mathbf{k}$. Relative to the body-fixed reference frame, point A moves along the x axis with velocity v, so $\mathbf{v}_{A\,\text{rel}} = v\mathbf{i}$. From Eq. (17.11), the velocity of A relative to the earth-fixed reference frame is

$$\mathbf{v}_A = \mathbf{v}_B + \mathbf{v}_{A\,\text{rel}} + \boldsymbol{\omega} \times \mathbf{r}_{A/B}$$

$$= \mathbf{0} + v\mathbf{i} + (\omega\mathbf{k}) \times (x\mathbf{i})$$

$$= v\mathbf{i} + \omega x\mathbf{j}.$$

Relative to the earth-fixed reference frame, A has a component of velocity parallel to the bar and also a perpendicular component due to the bar's rotation (Fig. 17.34d). Although $\mathbf{v}_A$ is the velocity of A relative to the earth-fixed reference frame, notice that it is expressed in components that are in terms of the body-fixed reference frame.

Figure 17.34
(a) A point moving along a bar.
(b) The bar is rotating.
(c) A body-fixed reference frame.
(d) Components of $\mathbf{v}_A$.

Study Questions

1. What is meant by a body-fixed reference frame?

2. How are the primary and secondary reference frames defined?

3. What is the definition of the term $\mathbf{v}_{A\,\text{rel}}$ in Eq. (17.11)?

4. What is the definition of the term $\mathbf{a}_{A\,\text{rel}}$ in Eq. (17.13)?

Example 17.7   Linkage with a Sliding Contact

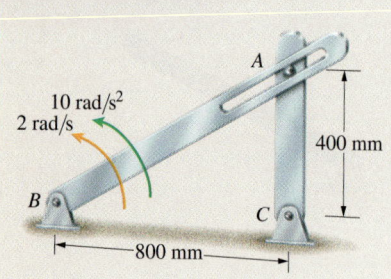

Figure 17.35

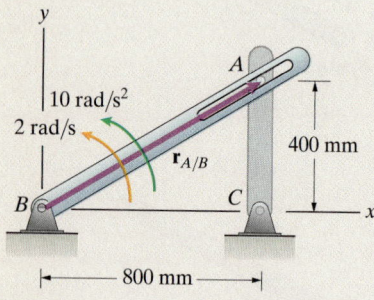

(a) A body-fixed coordinate system and the position vector of pin A relative to B.

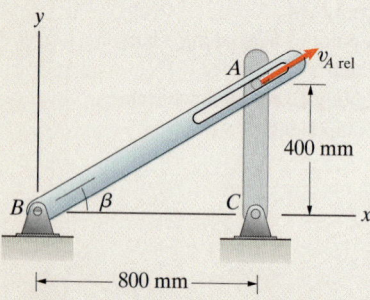

(b) The velocity of pin A relative to the body-fixed coordinate system.

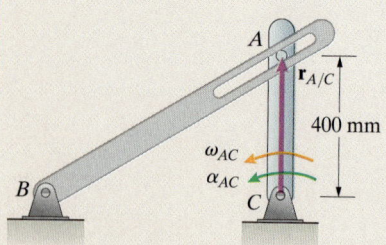

(c) The position vector of A relative to C.

Bar AB in Fig. 17.35 has a counterclockwise angular velocity of 2 rad/s and a counterclockwise angular acceleration of 10 rad/s^2.

(a) Determine the angular velocity of bar AC and the velocity of pin A relative to the slot in bar AB.

(b) Determine the angular acceleration of bar AC and the acceleration of pin A relative to the slot in bar AB.

Strategy

By employing a secondary reference frame that is fixed with respect to the slotted bar, we can use Eq. (17.11) to express the velocity of pin A in terms of its velocity relative to the slot and the known angular velocity of bar AB. Pins A and C are both points of bar AC, so we can express $\mathbf{v}_A$ in terms of the angular velocity of bar AC in the usual way. By equating the resulting expressions for $\mathbf{v}_A$, we will obtain a vector equation in terms of the velocity of A relative to the slot and the angular velocity of bar AC. Then, by following the same sequence of steps, but this time using Eq. (17.15), we can obtain the acceleration of A relative to the slot and the angular acceleration of bar AC.

Solution

(a) Let the coordinate system in Fig. a be body fixed with respect to the slotted bar. Applying Eq. (17.11) to points A and B, we find that the velocity of A is

$$\mathbf{v}_A = \mathbf{v}_B + \mathbf{v}_{A\,rel} + \boldsymbol{\omega}_{AB} \times \mathbf{r}_{A/B}$$

$$= \mathbf{0} + \mathbf{v}_{A\,rel} + \begin{vmatrix} \mathbf{i} & \mathbf{j} & \mathbf{k} \\ 0 & 0 & 2 \\ 0.8 & 0.4 & 0 \end{vmatrix}.$$

The velocity of pin A *relative to the body-fixed coordinate system* is parallel to the slot (Fig. b). Therefore, we can express that velocity as

$$\mathbf{v}_{A\,rel} = v_{A\,rel} \cos \beta \mathbf{i} + v_{A\,rel} \sin \beta \mathbf{j}, \tag{1}$$

where $\beta = \arctan (0.4/0.8)$. Substituting this expression into our equation for $\mathbf{v}_A$, we obtain

$$\mathbf{v}_A = (v_{A\,rel} \cos \beta - 0.8)\mathbf{i} + (v_{A\,rel} \sin \beta + 1.6)\mathbf{j}. \tag{2}$$

Let ω_{AC} be the angular velocity of bar AC (Fig. c). Expressing the velocity of A in terms of the velocity of C, we get

$$\mathbf{v}_A = \mathbf{v}_C + \boldsymbol{\omega}_{AC} \times \mathbf{r}_{A/C}$$

$$= \mathbf{0} + (\omega_{AC}\mathbf{k}) \times (0.4\mathbf{j})$$

$$= -0.4\omega_{AC}\mathbf{i}. \tag{3}$$

Notice that there is no relative velocity term in this equation, because A is a point of bar AC. Equating Eqs. (2) and (3), we obtain

$$(v_{A\,rel} \cos \beta - 0.8)\mathbf{i} + (v_{A\,rel} \sin \beta + 1.6)\mathbf{j} = -0.4\omega_{AC}\mathbf{i}.$$

Equating $\mathbf{i}$ and $\mathbf{j}$ components yields the two equations

$$v_{A\,rel} \cos \beta - 0.8 = -0.4\omega_{AC},$$

$$v_{A\,rel} \sin \beta + 1.6 = 0.$$

Solving these equations, we obtain $v_{A \text{ rel}} = -3.58$ m/s and $\omega_{AC} = 10$ rad/s. At this instant, pin A is moving relative to the slot at 3.58 m/s toward B. The vector

$$\mathbf{v}_{A \text{ rel}} = -3.58(\cos \beta \mathbf{i} + \sin \beta \mathbf{j}) = -3.2\mathbf{i} - 1.6\mathbf{j} \text{ (m/s)}.$$

(b) Applying Eq. (17.15) to bar AB (Fig. b) yields the acceleration of A:

$$\mathbf{a}_A = \mathbf{a}_B + \mathbf{a}_{A \text{ rel}} + 2\boldsymbol{\omega}_{AB} \times \mathbf{v}_{A \text{ rel}} + \boldsymbol{\alpha}_{AB} \times \mathbf{r}_{A/B} - \omega_{AB}^2 \mathbf{r}_{A/B}$$

$$= \mathbf{0} + \mathbf{a}_{A \text{ rel}} + 2 \begin{vmatrix} \mathbf{i} & \mathbf{j} & \mathbf{k} \\ 0 & 0 & 2 \\ -3.2 & -1.6 & 0 \end{vmatrix} + \begin{vmatrix} \mathbf{i} & \mathbf{j} & \mathbf{k} \\ 0 & 0 & 10 \\ 0.8 & 0.4 & 0 \end{vmatrix}$$

$$- (2)^2(0.8\mathbf{i} + 0.4\mathbf{j}).$$

The acceleration of A relative to the body-fixed coordinate system is parallel to the slot (Fig. d), so we can write it in the same way we did $\mathbf{v}_{A \text{ rel}}$:

$$\mathbf{a}_{A \text{ rel}} = a_{A \text{ rel}} \cos \beta \mathbf{i} + a_{A \text{ rel}} \sin \beta \mathbf{j}. \tag{4}$$

Substituting this expression into our equation for $\mathbf{a}_A$ gives

$$\mathbf{a}_A = (a_{A \text{ rel}} \cos \beta - 0.8)\mathbf{i} + (a_{A \text{ rel}} \sin \beta - 6.4)\mathbf{j}. \tag{5}$$

Expressing the acceleration of A in terms of the acceleration of C (Fig. c), we obtain

$$\mathbf{a}_A = \mathbf{a}_C + \boldsymbol{\alpha}_{AC} \times \mathbf{r}_{A/C} - \omega_{AC}^2 \mathbf{r}_{A/C}$$

$$= \mathbf{0} + (\alpha_{AC}\mathbf{k}) \times (0.4\mathbf{j}) - (10)^2(0.4\mathbf{j})$$

$$= -0.4\alpha_{AC}\mathbf{i} - 40\mathbf{j}. \tag{6}$$

Equating Eqs. (5) and (6), we get

$$(a_{A \text{ rel}} \cos \beta - 0.8)\mathbf{i} + (a_{A \text{ rel}} \sin \beta - 6.4)\mathbf{j} = -0.4\alpha_{AC}\mathbf{i} - 40\mathbf{j}.$$

Equating $\mathbf{i}$ and $\mathbf{j}$ components yields the equations

$$a_{A \text{ rel}} \cos \beta - 0.8 = -0.4\alpha_{AC},$$

$$a_{A \text{ rel}} \sin \beta - 6.4 = -40.$$

Solving these equations, we obtain $a_{A \text{ rel}} = -75.1$ m/s^2 and $\alpha_{AC} = 170$ rad/s^2. At the given instant, pin A is accelerating relative to the slot at 75.1 m/s^2 toward B.

Critical Thinking

This example illustrates a typical sliding contact between rigid bodies. Because the pin A is not a point of the slotted bar, we applied Eqs. (17.11) and (17.15) to points A and B instead of Eqs. (17.8) and (17.10). To do so, we introduced a secondary coordinate system x–y that was fixed with respect to the slotted bar. That means that *the slotted bar is stationary with respect to the body-fixed coordinate system*. Since the pin A slides in the slot, *A moves in a straight line relative to the body-fixed coordinate system*. We therefore knew that the velocity $\mathbf{v}_{A \text{ rel}}$ and acceleration $\mathbf{a}_{A \text{ rel}}$ of A relative to the body-fixed coordinate system were parallel to the slot and could be expressed in the forms given by Eqs. (1) and (4). That was the key to the solution. By using these expressions, we were able to determine the angular velocity and angular acceleration of bar AC and also the terms $v_{A \text{ rel}}$ and $a_{A \text{ rel}}$, which told us the velocity and acceleration of the pin A relative to the slot.

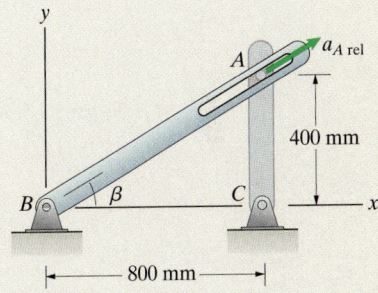

(d) The acceleration of pin A relative to the body-fixed coordinate system.

Example 17.8   Bar Sliding Relative to a Support

The collar at B in Fig. 17.36 slides along the circular bar, causing pin B to move at constant speed v_0 in a circular path of radius R. Bar BC slides in the collar at A. At the instant shown, determine the angular velocity and angular acceleration of bar BC.

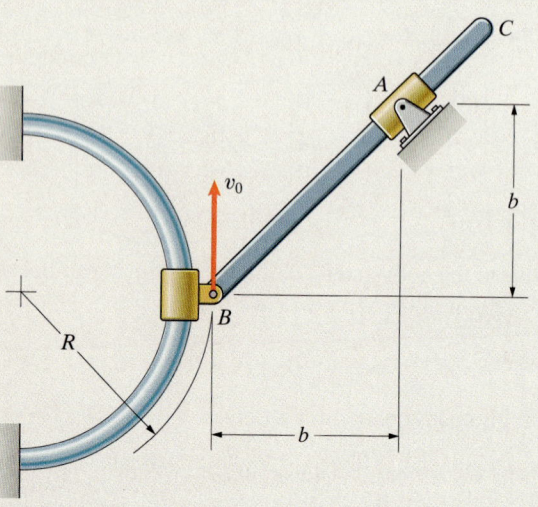

Figure 17.36

Strategy

We will use a secondary reference frame with its origin at B that is body fixed with respect to bar BC. By using Eqs. (17.11) and (17.15) to express the velocity and acceleration of the stationary pin A in terms of the velocity and acceleration of pin B, we can determine the angular velocity and angular acceleration of bar BC.

Solution

Angular Velocity   Let the angular velocity and angular acceleration of bar BC, which are also the angular velocity and angular acceleration of the body-fixed coordinate system, be ω_{BC} and α_{BC} (Fig. a). The velocity of the stationary pin

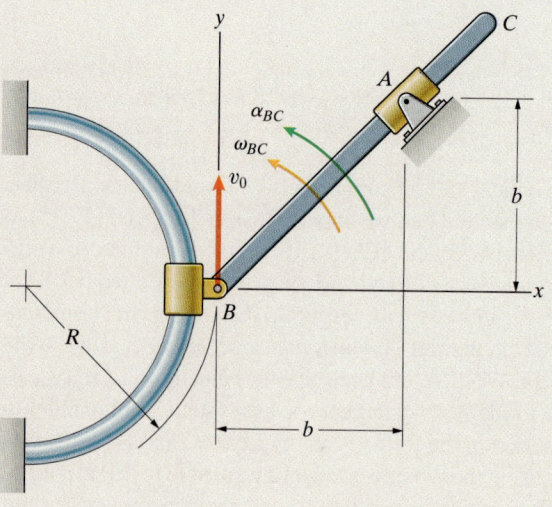

(a) A reference frame fixed with respect to bar BC.

A is zero. From Eq. (17.11),

$$\mathbf{v}_A = \mathbf{0} = \mathbf{v}_B + \mathbf{v}_{A\,\text{rel}} + \boldsymbol{\omega} \times \mathbf{r}_{A/B}, \tag{1}$$

where $\mathbf{v}_{A\,\text{rel}}$ is the velocity of A relative to the body-fixed coordinate system and $\boldsymbol{\omega} = \omega_{BC}\mathbf{k}$ is the angular velocity vector of the coordinate system. The velocity of pin B is $\mathbf{v}_B = v_0\mathbf{j}$. The velocity of the stationary pin A relative to the body-fixed coordinate system is parallel to the bar (Fig. b), so we can express it in the form

$$\mathbf{v}_{A\,\text{rel}} = v_{A\,\text{rel}}\cos 45°\mathbf{i} + v_{A\,\text{rel}}\sin 45°\mathbf{j} \tag{2}$$

and write Eq. (1) as

$$\mathbf{0} = v_0\mathbf{j} + v_{A\,\text{rel}}\cos 45°\mathbf{i} + v_{A\,\text{rel}}\sin 45°\mathbf{j}$$

$$+ \begin{vmatrix} \mathbf{i} & \mathbf{j} & \mathbf{k} \\ 0 & 0 & \omega_{BC} \\ b & b & 0 \end{vmatrix}.$$

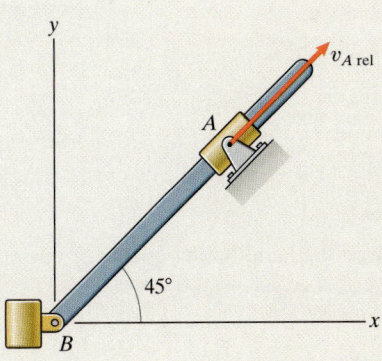

(b) Direction of the velocity of the fixed pin A relative to the body-fixed coordinate system.

From the $\mathbf{i}$ and $\mathbf{j}$ components of this equation, we obtain

$$v_{A\,\text{rel}}\cos 45° - b\omega_{BC} = 0,$$

$$v_0 + v_{A\,\text{rel}}\sin 45° + b\omega_{BC} = 0.$$

Solving these equations, we determine that the velocity of pin A relative to the body-fixed coordinate system is

$$\mathbf{v}_{A\,\text{rel}} = v_{A\,\text{rel}}\cos 45°\mathbf{i} + v_{A\,\text{rel}}\sin 45°\mathbf{j}$$

$$= -\frac{v_0}{2}\mathbf{i} - \frac{v_0}{2}\mathbf{j}$$

and the angular velocity of bar BC is

$$\omega_{BC} = -\frac{v_0}{2b}.$$

Angular Acceleration   The acceleration of pin A is zero. From Eq. (17.15),

$$\mathbf{a}_A = \mathbf{0} = \mathbf{a}_B + \mathbf{a}_{A\,\text{rel}} + 2\boldsymbol{\omega} \times \mathbf{v}_{A\,\text{rel}} + \boldsymbol{\alpha} \times \mathbf{r}_{A/B} - \omega^2 \mathbf{r}_{A/B}. \tag{3}$$

The acceleration of pin B is $\mathbf{a}_B = -(v_0^2/R)\mathbf{i}$. The acceleration of pin A relative to the body-fixed coordinate system is parallel to the bar (Fig. c). We can therefore express it as

$$\mathbf{a}_{A\,\text{rel}} = a_{A\,\text{rel}} \cos 45°\mathbf{i} + a_{A\,\text{rel}} \sin 45°\mathbf{j} \tag{4}$$

and write Eq. (3) as

$$\mathbf{0} = -\frac{v_0^2}{R}\mathbf{i} + a_{A\,\text{rel}} \cos 45°\mathbf{i} + a_{A\,\text{rel}} \sin 45°\mathbf{j}$$

$$+ 2 \begin{vmatrix} \mathbf{i} & \mathbf{j} & \mathbf{k} \\ 0 & 0 & \omega_{BC} \\ -v_0/2 & -v_0/2 & 0 \end{vmatrix} + \begin{vmatrix} \mathbf{i} & \mathbf{j} & \mathbf{k} \\ 0 & 0 & \alpha_{BC} \\ b & b & 0 \end{vmatrix}$$

$$- \omega_{BC}^2(b\mathbf{i} + b\mathbf{j}).$$

From the $\mathbf{i}$ and $\mathbf{j}$ components of this equation, we obtain

$$-\frac{v_0^2}{R} + a_{A\,\text{rel}} \cos 45° + v_0\omega_{BC} - b\alpha_{BC} - b\omega_{BC}^2 = 0,$$

$$a_{A\,\text{rel}} \sin 45° - v_0\omega_{BC} + b\alpha_{BC} - b\omega_{BC}^2 = 0.$$

Solving these equations, we determine that the angular acceleration of bar BC is

$$\alpha_{BC} = -\frac{v_0^2}{2b}\left(\frac{1}{R} + \frac{1}{b}\right).$$

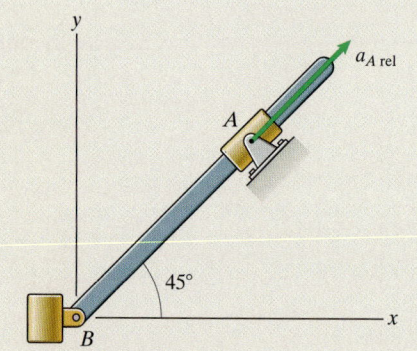

(c) Direction of the acceleration of the fixed pin A relative to the body-fixed coordinate system.

Critical Thinking
In this example, the bar BC slides relative to its support at A. Notice that the fixed support A moves relative to the coordinate system that is body-fixed with respect to the bar BC. Because the bar BC is stationary with respect to the body-fixed coordinate system, we knew that the velocity and acceleration of the support A relative to the body-fixed coordinate system is parallel to bar BC. That was why we could express them in the forms given by Eqs. (2) and (4).

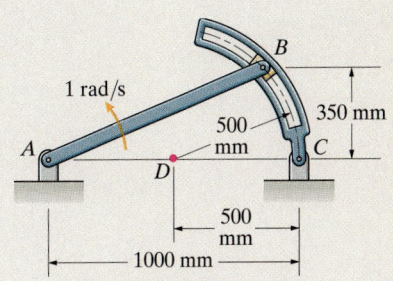

Example 17.9 | **Analysis of a Sliding Contact**

Bar *AB* in Fig. 17.37 rotates with a constant counterclockwise angular velocity of 1 rad/s. Block *B* slides in a circular slot in the curved bar *BC*. At the instant shown, the center of the circular slot is at *D*. Determine the angular velocity and angular acceleration of bar *BC*.

Figure 17.37

Strategy

Since we know the angular velocity of bar *AB*, we can determine the velocity of point *B*. Because *B* is not a point of bar *BC*, we must apply Eq. (17.11) to points *B* and *C*. By equating our expressions for $\mathbf{v}_B$, we can solve for the angular velocity of bar *BC*. Then, by following the same sequence of steps, but this time using Eq. (17.15), we can determine the angular acceleration of bar *BC*.

Solution

To determine the velocity of *B*, we express it in terms of the velocity of *A* and the angular velocity of bar *AB*: $\mathbf{v}_B = \mathbf{v}_A + \boldsymbol{\omega}_{AB} \times \mathbf{r}_{B/A}$. In terms of the coordinate system shown in Fig. a, the position vector of *B* relative to *A* is

$$\mathbf{r}_{B/A} = (0.500 + 0.500 \cos \beta)\mathbf{i} + 0.350\mathbf{j} = 0.857\mathbf{i} + 0.350\mathbf{j} \text{ (m)},$$

where $\beta = \arcsin(350/500) = 44.4°$. Therefore, the velocity of *B* is

$$\mathbf{v}_B = \mathbf{v}_A + \boldsymbol{\omega}_{AB} \times \mathbf{r}_{B/A} = 0 + \begin{vmatrix} \mathbf{i} & \mathbf{j} & \mathbf{k} \\ 0 & 0 & 1 \\ 0.857 & 0.350 & 0 \end{vmatrix}$$

$$= -0.350\mathbf{i} + 0.857\mathbf{j} \text{ (m/s)}. \quad (1)$$

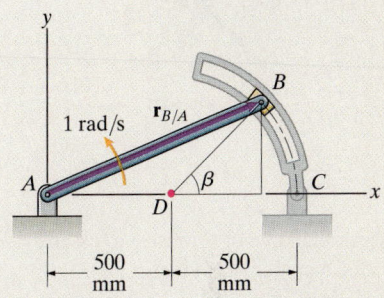

(a) Determining the velocity of point *B*.

To apply Eq. (17.11) to points *B* and *C*, we introduce a parallel secondary coordinate system that rotates with the curved bar (Fig. b). The velocity of *B* is

$$\mathbf{v}_B = \mathbf{v}_C + \mathbf{v}_{B\,\text{rel}} + \boldsymbol{\omega}_{BC} \times \mathbf{r}_{B/C}. \quad (2)$$

The position vector of *B* relative to *C* is

$$\mathbf{r}_{B/C} = -(0.500 - 0.500 \cos \beta)\mathbf{i} + 0.350\mathbf{j} = -0.143\mathbf{i} + 0.350\mathbf{j} \text{ (m)}.$$

Relative to the body-fixed coordinate system, point *B* moves in a circular path about point *D* (Fig. c). In terms of the angle β, the vector

$$\mathbf{v}_{B\,\text{rel}} = -v_{B\,\text{rel}} \sin \beta \mathbf{i} + v_{B\,\text{rel}} \cos \beta \mathbf{j}.$$

We substitute the preceding expressions for $\mathbf{r}_{B/C}$ and $\mathbf{v}_{B\,\text{rel}}$ into Eq. (2), obtaining

$$\mathbf{v}_B = -v_{B\,\text{rel}} \sin \beta \mathbf{i} + v_{B\,\text{rel}} \cos \beta \mathbf{j} + \begin{vmatrix} \mathbf{i} & \mathbf{j} & \mathbf{k} \\ 0 & 0 & \omega_{BC} \\ -0.143 & 0.350 & 0 \end{vmatrix}.$$

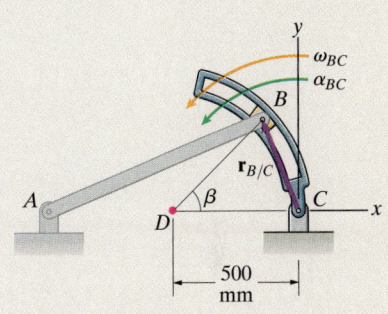

(b) A coordinate system fixed with respect to the curved bar.

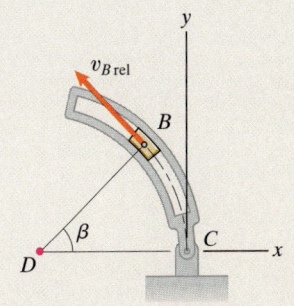

(c) The velocity of *B* relative to the body-fixed coordinate system.

Equating this expression for $\mathbf{v}_B$ to its value given in Eq. (1) yields the two equations

$$-v_{B\,\text{rel}} \sin \beta - 0.350\omega_{BC} = -0.350,$$

$$v_{B\,\text{rel}} \cos \beta - 0.143\omega_{BC} = 0.857.$$

Solving these equations, we obtain $v_{B\,\text{rel}} = 1.0$ m/s and $\omega_{BC} = -1.0$ rad/s.

We follow the same sequence of steps to determine the angular acceleration of bar BC. The acceleration of point B is

$$\mathbf{a}_B = \mathbf{a}_A + \boldsymbol{\alpha}_{AB} \times \mathbf{r}_{B/A} - \omega_{AB}^2 \mathbf{r}_{B/A}$$

$$= \mathbf{0} + \mathbf{0} - (1)^2(0.857\mathbf{i} + 0.350\mathbf{j})$$

$$= -0.857\mathbf{i} - 0.350\mathbf{j} \ (\text{m/s}^2). \tag{3}$$

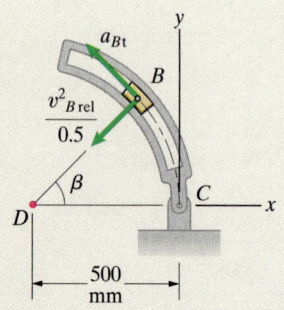

(d) Acceleration of B relative to the body-fixed coordinate system.

Because the motion of point B relative to the body-fixed coordinate system is a circular path about point D, there is a tangential component of acceleration, which we denote a_{Bt}, and a normal component of acceleration $v_{B\,\text{rel}}^2/(0.5 \text{ m})$. These components are shown in Fig. d. In terms of the angle β, the vector

$$\mathbf{a}_{B\,\text{rel}} = -a_{Bt} \sin \beta \mathbf{i} + a_{Bt} \cos \beta \mathbf{j}$$

$$- (v_{B\,\text{rel}}^2/0.5) \cos \beta \mathbf{i} - (v_{B\,\text{rel}}^2/0.5) \sin \beta \mathbf{j}.$$

Applying Eq. (17.15) to points B and C, we obtain the acceleration of B:

$$\mathbf{a}_B = \mathbf{a}_C + \mathbf{a}_{B\,\text{rel}} + 2\boldsymbol{\omega}_{BC} \times \mathbf{v}_{B\,\text{rel}}$$

$$+ \boldsymbol{\alpha}_{BC} \times \mathbf{r}_{B/C} - \omega_{BC}^2 \mathbf{r}_{B/C}$$

$$= \mathbf{0} - a_{Bt} \sin \beta \mathbf{i} + a_{Bt} \cos \beta \mathbf{j}$$

$$- [(1)^2/0.5] \cos \beta \mathbf{i} - [(1)^2/0.5] \sin \beta \mathbf{j}$$

$$+ 2 \begin{vmatrix} \mathbf{i} & \mathbf{j} & \mathbf{k} \\ 0 & 0 & -1 \\ -(1)\sin\beta & (1)\cos\beta & 0 \end{vmatrix}$$

$$+ \begin{vmatrix} \mathbf{i} & \mathbf{j} & \mathbf{k} \\ 0 & 0 & \alpha_{BC} \\ -0.143 & 0.350 & 0 \end{vmatrix} - (-1)^2(-0.143\mathbf{i} + 0.350\mathbf{j}).$$

Equating this expression for $\mathbf{a}_B$ to its value given in Eq. (3) yields the two equations

$$-a_{Bt} \sin \beta - 0.350\alpha_{BC} + 0.143 = -0.857,$$

$$a_{Bt} \cos \beta - 0.143\alpha_{BC} - 0.350 = -0.350.$$

Solving we obtain $a_{Bt} = 0.408 \text{ m/s}^2$ and $\alpha_{BC} = 2.040 \text{ rad/s}^2$.

Critical Thinking

This example was distinguished by the fact that the slotted bar has the shape of a circular arc. As a result, the block B moved in a circular path relative to the coordinate system that is fixed with respect to the circular bar. The direction of the velocity $\mathbf{v}_{B\,\text{rel}}$ was tangential to the circular path. However, as a result of the curved path, we knew that $\mathbf{a}_{B\,\text{rel}}$ would have components tangential and normal to the path. Furthermore, the magnitude of the normal component could be written in terms of the magnitude of the velocity and the radius of the circular path. Therefore, once $\mathbf{v}_{B\,\text{rel}}$ was known, only the tangential component of $\mathbf{a}_{B\,\text{rel}}$ needed to be determined.

Problems

17.117 The bar rotates with a counterclockwise angular velocity of 10 rad/s and sleeve A slides at 4 ft/s relative to the bar. Use Eq. (17.11) and the body-fixed coordinate system shown to determine the velocity of A.

17.118 The bar rotates with a constant counterclockwise angular velocity of 10 rad/s and sleeve A slides at a constant velocity of 4 ft/s relative to the bar. Use Eq. (17.15) to determine the acceleration of A.

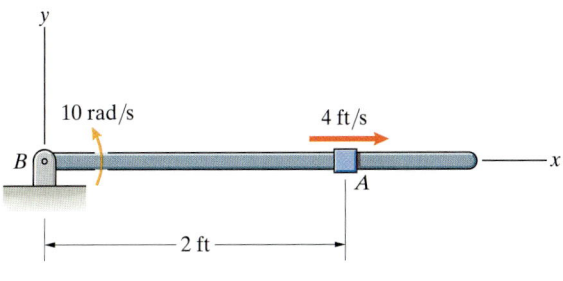

Problems 17.117/17.118

17.119 Sleeve C slides at 1 m/s relative to bar BD. Use the body-fixed coordinate system shown to determine the velocity of C.

17.120 The angular accelerations of the two bars are zero and sleeve C slides at a constant velocity of 1 m/s relative to bar BD. What is the acceleration of sleeve C?

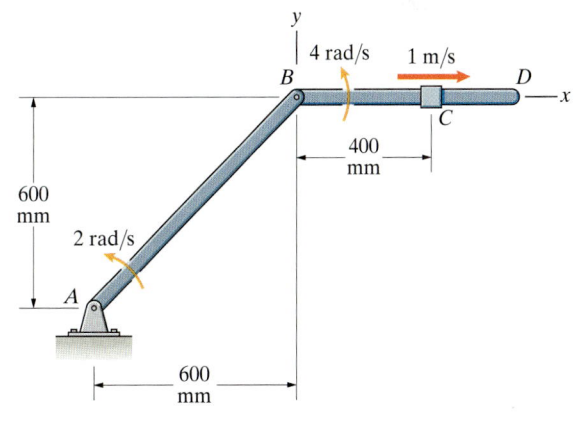

Problems 17.119/17.120

17.121 Bar *AC* has an angular velocity of 2 rad/s in the counterclockwise direction that is decreasing at 4 rad/s². The pin at *C* slides in the slot in bar *BD*.

(a) Determine the angular velocity of bar *BD* and the velocity of the pin relative to the slot.

(b) Determine the angular acceleration of bar *BD* and the acceleration of the pin relative to the slot.

17.122 The velocity of pin *C* relative to the slot is 21 in/s upward and is decreasing at 42 in/s². What are the angular velocity and angular acceleration of bar *AC*?

17.123 What should the angular velocity and acceleration of bar *AC* be if you want the angular velocity and acceleration of bar *BD* to be 4 rad/s counterclockwise and 24 rad/s² counterclockwise, respectively?

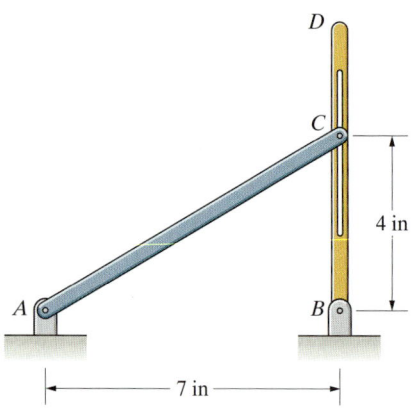

Problems 17.121–17.123

17.124 Bar *AB* has an angular velocity of 4 rad/s in the clockwise direction. What is the velocity of pin *B* relative to the slot?

17.125 Bar *AB* has an angular velocity of 4 rad/s in the clockwise direction and an angular acceleration of 10 rad/s² in the counterclockwise direction. What is the acceleration of pin *B* relative to the slot?

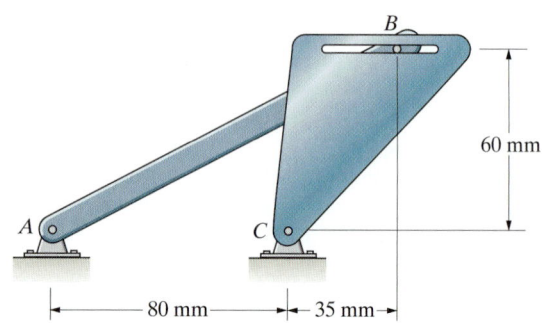

Problems 17.124/17.125

17.126 The hydraulic actuator *BC* of the crane is extending (increasing in length) at a rate of 0.2 m/s. At the instant shown, what is the angular velocity of the crane's boom *AD*?

    Strategy: Use Eq. (17.8) to write the velocity of point *C* in terms of the velocity of point *A*, and use Eq. (17.11) to write the velocity of point *C* in terms of the velocity of point *B*. Then equate your two expressions for the velocity of point *C*.

17.127 The hydraulic actuator *BC* of the crane is extending (increasing in length) at a constant rate of 0.2 m/s. At the instant shown, what is the angular acceleration of the crane's boom *AD*?

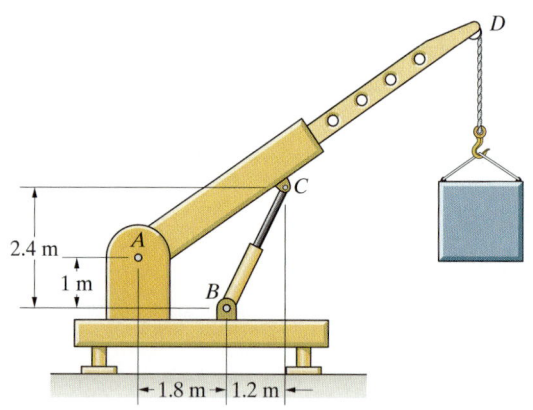

Problems 17.126/17.127

17.128 The angular velocity $\omega_{AC} = 5°$ per second. Determine the angular velocity of the hydraulic actuator BC and the rate at which the actuator is extending.

17.129 The angular velocity $\omega_{AC} = 5°$ per second and the angular acceleration $\alpha_{AC} = -2°$ per second squared. Determine the angular acceleration of the hydraulic actuator BC and the rate of change of the actuator's rate of extension.

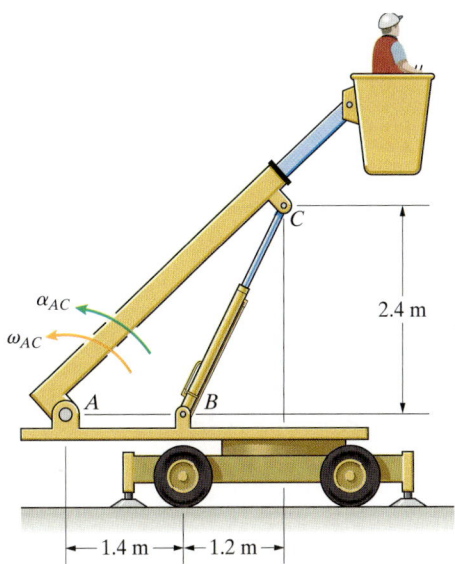

Problems 17.128/17.129

17.130 The sleeve at A slides upward at a constant velocity of 10 m/s. Bar AC slides through the sleeve at B. Determine the angular velocity of bar AC and the velocity at which the bar slides relative to the sleeve at B.

17.131 The sleeve at A slides upward at a constant velocity of 10 m/s. Determine the angular acceleration of bar AC and the rate of change of the velocity at which the bar slides relative to the sleeve at B.

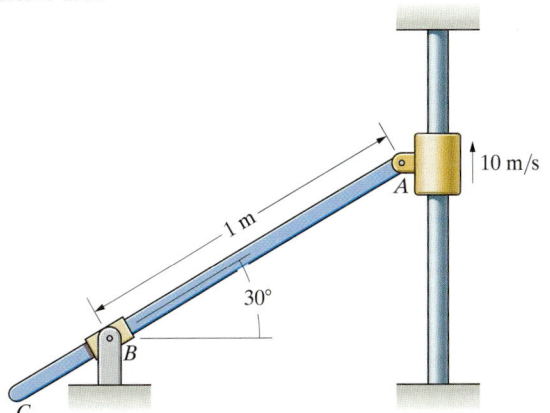

Problems 17.130/17.131

17.132 Block A slides up the inclined surface at 2 ft/s. Determine the angular velocity of bar AC and the velocity of point C.

17.133 Block A slides up the inclined surface at a constant velocity of 2 ft/s. Determine the angular acceleration of bar AC and the acceleration of point C.

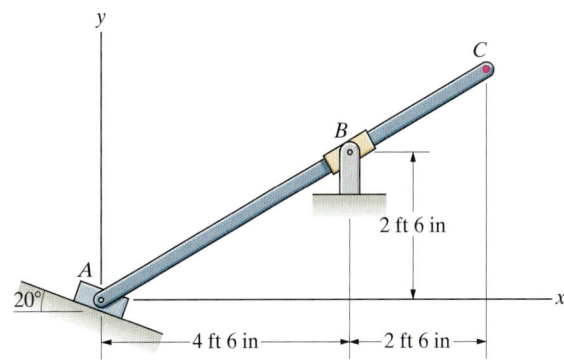

Problems 17.132/17.133

17.134 The angular velocity of the scoop is 1 rad/s clockwise. Determine the rate at which the hydraulic actuator AB is extending.

17.135 The angular velocity of the scoop is 1 rad/s clockwise and its angular acceleration is zero. Determine the rate of change of the rate at which the hydraulic actuator AB is extending.

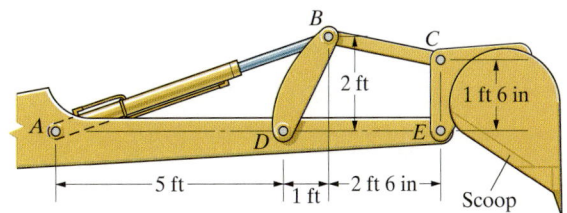

Problems 17.134/17.135

17.136 Suppose that the curved bar in Example 17.9 rotates with a counterclockwise angular velocity of 2 rad/s.

(a) What is the angular velocity of bar AB?
(b) What is the velocity of block B relative to the slot?

17.137 Suppose that the curved bar in Example 17.9 has a clockwise angular velocity of 4 rad/s and a counterclockwise angular acceleration of 10 rad/s^2. What is the angular acceleration of bar AB?

17.138* The disk rolls on the plane surface with a counterclockwise angular velocity of 10 rad/s. Bar AB slides on the surface of the disk at A. Determine the angular velocity of bar AB.

17.139* The disk rolls on the plane surface with a constant counterclockwise angular velocity of 10 rad/s. Determine the angular acceleration of bar AB.

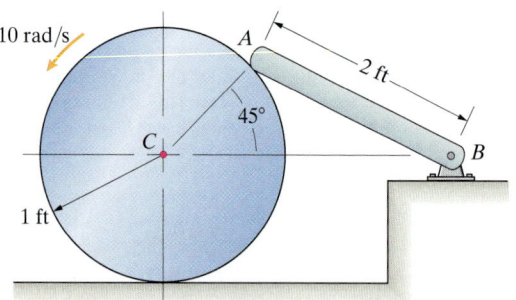

Problems 17.138/17.139

17.140* Bar BC rotates with a counterclockwise angular velocity of 2 rad/s. A pin at B slides in a circular slot in the rectangular plate. Determine the angular velocity of the plate and the velocity at which the pin slides relative to the circular slot.

17.141* Bar BC rotates with a constant counterclockwise angular velocity of 2 rad/s. Determine the angular acceleration of the plate.

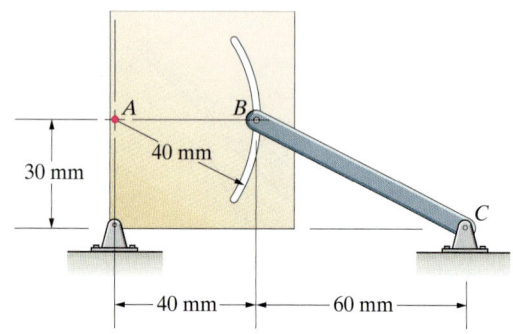

Problems 17.140/17.141

17.142* By taking the derivative of Eq. (17.11) with respect to time and using Eq. (17.12), derive Eq. (17.13).

17.6 Moving Reference Frames

In this section we revisit the subjects of Chapters 13 and 14—the motion of a point and Newton's second law. In many situations, it is convenient to describe the motion of a point by using a secondary reference frame that moves relative to some primary reference frame. For example, to measure the motion of a point relative to a moving vehicle, we would choose a secondary reference frame that is fixed with respect to the vehicle. Here we show how the velocity and acceleration of a point relative to a primary reference frame are related to their values relative to a moving secondary reference frame. We also discuss how to apply Newton's second law using moving secondary reference frames when the primary reference frame is inertial. In Chapter 14 we mentioned the example of playing tennis on the deck of a cruise ship. If the ship translates with constant velocity, we can use the equation $\Sigma\mathbf{F} = m\mathbf{a}$ expressed in terms of a reference frame fixed with respect to the ship to analyze the ball's motion. We cannot do so if the ship is turning or changing its speed. However, we can apply the second law using reference frames that accelerate and rotate relative to an inertial reference frame by properly accounting for the acceleration and rotation. We explain how this is done in this section.

Motion of a Point Relative to a Moving Reference Frame

Equations (17.11) and (17.13) give the velocity and acceleration of an arbitrary point A relative to a point B of a rigid body in terms of a body-fixed secondary reference frame:

$$\mathbf{v}_A = \mathbf{v}_B + \mathbf{v}_{A\,\text{rel}} + \boldsymbol{\omega} \times \mathbf{r}_{A/B}, \tag{17.16}$$

$$\mathbf{a}_A = \mathbf{a}_B + \mathbf{a}_{A\,\text{rel}} + 2\boldsymbol{\omega} \times \mathbf{v}_{A\,\text{rel}} + \boldsymbol{\alpha} \times \mathbf{r}_{A/B}$$
$$+ \boldsymbol{\omega} \times (\boldsymbol{\omega} \times \mathbf{r}_{A/B}). \tag{17.17}$$

But these equations do not require us to assume that the secondary reference frame is connected to some rigid body. They apply to any reference frame having a moving origin B and rotating with angular velocity $\boldsymbol{\omega}$ and angular acceleration $\boldsymbol{\alpha}$ relative to a primary reference frame (Fig. 17.38). The terms $\mathbf{v}_A$ and $\mathbf{a}_A$ are the velocity and acceleration of A relative to the primary reference frame. The terms $\mathbf{v}_{A\,\text{rel}}$ and $\mathbf{a}_{A\,\text{rel}}$ are the velocity and acceleration of A relative to the secondary reference frame. That is, they are the velocity and acceleration measured by an observer moving with the secondary reference frame (Fig. 17.39).

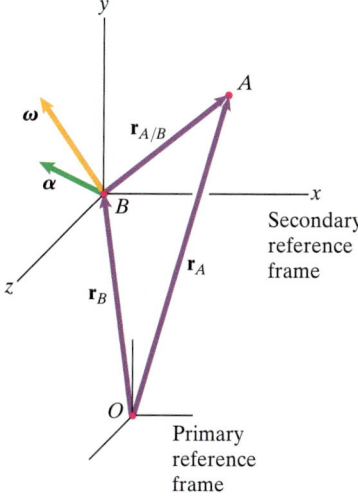

Figure 17.38
A secondary reference frame with origin B and an arbitrary point A.

When the velocity and acceleration of a point A relative to a moving secondary reference frame are known, we can use Eqs. (17.16) and (17.17) to determine the velocity and acceleration of A relative to the primary reference frame. There will also be situations in which the velocity and acceleration of A relative to the primary reference frame will be known and we will want to use Eqs. (17.16) and (17.17) to determine the velocity and acceleration of A relative to a moving secondary reference frame.

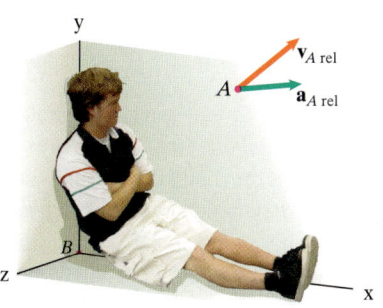

Figure 17.39
Imagine yourself to be stationary relative to the secondary reference frame.

Example 17.10	A Rotating Secondary Reference Frame

The merry-go-round in Fig. 17.40 rotates with constant counterclockwise angular velocity ω. Suppose that you are in the center at B and observe the motion of a second person A using a coordinate system that rotates with the merry-go-round. Consider two cases.

Case 1 Person A is not on the merry-go-round, but stands on the ground next to it. At the instant shown, what are her velocity and acceleration relative to your coordinate system?

Case 2 Person A is on the edge of the merry-go-round and moves with it. What are her velocity and acceleration relative to the earth?

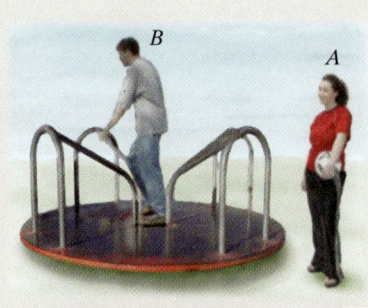

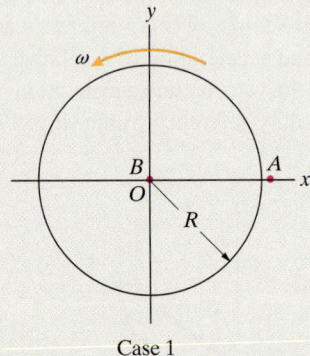

Case 1

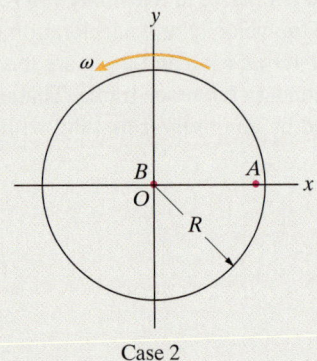

Case 2

Figure 17.40

Strategy

This simple example clarifies the distinction between the terms $\mathbf{v}_A$, $\mathbf{a}_A$ and the terms $\mathbf{v}_{A\,\text{rel}}$, $\mathbf{a}_{A\,\text{rel}}$ in Eqs. (17.16) and (17.17). We choose the coordinate system that rotates with the merry-go-round as the secondary reference frame and let the primary reference frame be fixed with respect to the earth. In case 1, A's velocity and acceleration relative to the earth, $\mathbf{v}_A$ and $\mathbf{a}_A$, are known: She is standing still. We can use Eqs. (17.16) and (17.17) to determine $\mathbf{v}_{A\,\text{rel}}$ and $\mathbf{a}_{A\,\text{rel}}$, which are her velocity and acceleration relative to your rotating coordinate system. In case 2, $\mathbf{v}_{A\,\text{rel}}$ and $\mathbf{a}_{A\,\text{rel}}$ are known: A is stationary relative to your coordinate system. We can use Eqs. (17.16) and (17.17) to determine $\mathbf{v}_A$ and $\mathbf{a}_A$.

Solution

Case 1 A is standing on the ground, so her velocity relative to the earth is $\mathbf{v}_A = \mathbf{0}$. The angular velocity vector of your coordinate system is $\boldsymbol{\omega} = \omega\mathbf{k}$, and at the instant shown $\mathbf{r}_{A/B} = R\mathbf{i}$. From Eq. (17.16),

$$\mathbf{v}_A = \mathbf{v}_B + \mathbf{v}_{A\,\text{rel}} + \boldsymbol{\omega} \times \mathbf{r}_{A/B}:$$

$$\mathbf{0} = \mathbf{0} + \mathbf{v}_{A\,\text{rel}} + (\omega\mathbf{k}) \times (R\mathbf{i}).$$

We find that $\mathbf{v}_{A\,\text{rel}} = -\omega R\mathbf{j}$. Although A is stationary relative to the earth, $\mathbf{v}_{A\,\text{rel}}$ is not zero. What does this term represent? As you sit at the center of the merry-go-round, you see A moving around you in a circular path. *Relative to your rotating coordinate system*, A moves in a circular path of radius R in the clockwise direction with a velocity of constant magnitude ωR. At the instant shown, A's velocity relative to your coordinate system is $-\omega R\mathbf{j}$.

You know that a point moving in a circular path of radius R with velocity v has a normal component of acceleration equal to v^2/R. Relative to your coordinate system, person A moves in a circular path of radius R with velocity ωR. Therefore, *relative to your coordinate system, A has a normal component of acceleration* $(\omega R)^2/R = \omega^2 R$. At the instant shown, the normal acceleration points in the negative x direction. We conclude that A's acceleration relative to your coordinate system is $\mathbf{a}_{A\,\text{rel}} = -\omega^2 R\mathbf{i}$.

We can confirm this result with Eq. (17.17). A's acceleration relative to the earth is $\mathbf{a}_A = \mathbf{0}$. The angular velocity vector of the coordinate system is constant, so $\boldsymbol{\alpha} = \mathbf{0}$. From Eq. (17.17),

$$\mathbf{a}_A = \mathbf{a}_B + \mathbf{a}_{A\,\text{rel}} + 2\boldsymbol{\omega} \times \mathbf{v}_{A\,\text{rel}} + \boldsymbol{\alpha} \times \mathbf{r}_{A/B} + \boldsymbol{\omega} \times (\boldsymbol{\omega} \times \mathbf{r}_{A/B}):$$

$$\mathbf{0} = \mathbf{0} + \mathbf{a}_{A\,\text{rel}} + 2(\omega\mathbf{k}) \times (-\omega R\mathbf{j}) + \mathbf{0} + (\omega\mathbf{k}) \times [(\omega\mathbf{k}) \times (R\mathbf{i})].$$

Solving this equation for $\mathbf{a}_{A\,\text{rel}}$, we obtain $\mathbf{a}_{A\,\text{rel}} = -\omega^2 R\mathbf{i}$. A's velocity and acceleration relative to your coordinate system are shown in Fig. a.

Case 2 *Relative to your coordinate system, A is stationary, so* $\mathbf{v}_{A\,\text{rel}} = \mathbf{0}$ and $\mathbf{a}_{A\,\text{rel}} = \mathbf{0}$. From Eq. (17.16), A's velocity relative to the earth is

$$\mathbf{v}_A = \mathbf{v}_B + \mathbf{v}_{A\,\text{rel}} + \boldsymbol{\omega} \times \mathbf{r}_{A/B} = \mathbf{0} + \mathbf{0} + (\omega\mathbf{k}) \times (R\mathbf{i})$$

$$= \omega R\mathbf{j}.$$

In this case, A is moving in a circular path of radius R with a velocity of constant magnitude ωR relative to the earth.

From Eq. (17.17), A's acceleration relative to the earth is

$$\mathbf{a}_A = \mathbf{a}_B + \mathbf{a}_{A\,\text{rel}} + 2\boldsymbol{\omega} \times \mathbf{v}_{A\,\text{rel}} + \boldsymbol{\alpha} \times \mathbf{r}_{A/B} + \boldsymbol{\omega} \times (\boldsymbol{\omega} \times \mathbf{r}_{A/B})$$

$$= \mathbf{0} + \mathbf{0} + \mathbf{0} + \mathbf{0} + (\omega\mathbf{k}) \times [(\omega\mathbf{k}) \times (R\mathbf{i})]$$

$$= -\omega^2 R\mathbf{i}.$$

This is A's acceleration relative to the earth due to her circular motion. A's velocity and acceleration relative to the earth are shown in Fig. b.

Critical Thinking

This example demonstrates the two principal uses of Eqs. (17.16) and (17.17). In some situations, you will know the velocity and acceleration relative to a primary reference frame and you can use Eqs. (17.16) and (17.17) to determine the velocity and acceleration relative to a secondary reference frame. In case 1 of this example, person A is stationary relative to the primary (earth-fixed) reference frame, and her velocity and acceleration relative to the rotating coordinate system were determined. In other situations, you will know the velocity and acceleration of an object relative to a secondary reference frame, and you can use Eqs. (17.16) and (17.17) to determine the velocity and acceleration relative to the underlying primary reference frame. In case 2 of this example, person A is stationary relative to the rotating reference frame, and his velocity and acceleration relative to the earth-fixed reference frame were determined. Although this example used a very simple context, in Example 17.11 we use exactly the same approach to analyze the motion of a helicopter relative to an earth-fixed reference frame and a secondary reference frame that is body-fixed with respect to a moving ship.

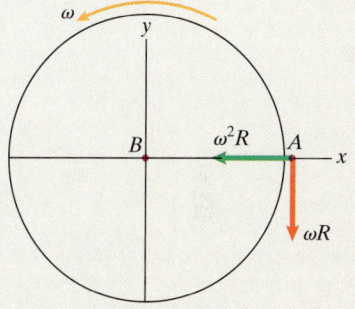

(a) The velocity and acceleration of A relative to the rotating coordinate system in case 1.

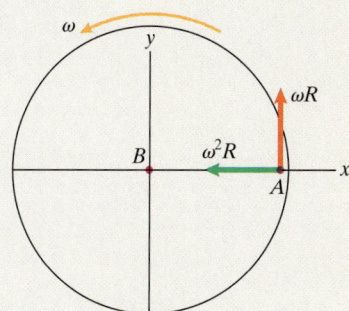

(b) The velocity and acceleration of A relative to the earth in case 2.

Example 17.11 | A Reference Frame Fixed with Respect to a Ship

At the instant shown, the ship B in Fig. 17.41 is moving north at a constant speed of 15.0 m/s relative to the earth and is turning toward the west at a constant rate of 5.0° per second. Relative to the ship's body-fixed coordinate system, its radar indicates that the position, velocity, and acceleration of the helicopter A are

$$\mathbf{r}_{A/B} = 420.0\mathbf{i} + 236.2\mathbf{j} + 212.0\mathbf{k} \text{ (m)},$$

$$\mathbf{v}_{A \text{ rel}} = -53.5\mathbf{i} + 2.0\mathbf{j} + 6.6\mathbf{k} \text{ (m/s)},$$

and

$$\mathbf{a}_{A \text{ rel}} = 0.4\mathbf{i} - 0.2\mathbf{j} - 13.0\mathbf{k} \text{ (m/s}^2).$$

What are the helicopter's velocity and acceleration relative to the earth?

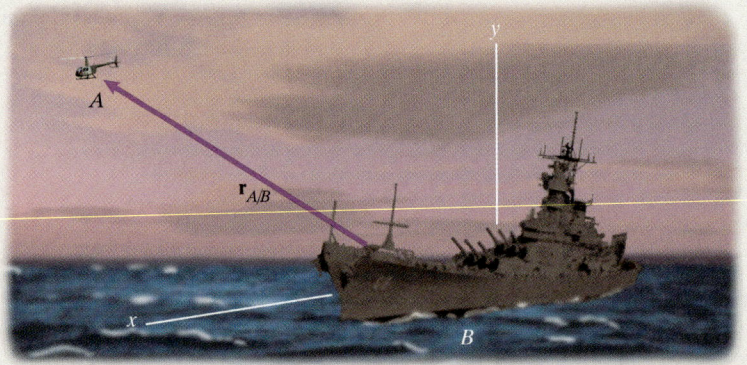

Figure 17.41

Strategy

We are given the ship's velocity and enough information to determine its acceleration, angular velocity, and angular acceleration relative to the earth. We also know the position, velocity, and acceleration of the helicopter relative to the secondary body-fixed coordinate system. Therefore, we can use Eqs. (17.16) and (17.17) to determine the helicopter's velocity and acceleration relative to the earth.

Solution

In terms of the body-fixed coordinate system, the ship's velocity is $\mathbf{v}_B = 15.0\mathbf{i}$ (m/s). The ship's angular velocity due to its rate of turning is $\omega = (5.0/180)\pi = 0.0873$ rad/s. The ship is rotating about the y axis. Pointing the arc of the fingers of the right hand around the y axis in the direction of the ship's rotation, we find that the thumb points in the positive y direction, so the ship's angular velocity vector is $\boldsymbol{\omega} = 0.0873\mathbf{j}$ (rad/s). The helicopter's velocity relative to the earth is

$$\mathbf{v}_A = \mathbf{v}_B + \mathbf{v}_{A \text{ rel}} + \boldsymbol{\omega} \times \mathbf{r}_{A/B}$$

$$= 15.0\mathbf{i} + (-53.5\mathbf{i} + 2.0\mathbf{j} + 6.6\mathbf{k}) + \begin{vmatrix} \mathbf{i} & \mathbf{j} & \mathbf{k} \\ 0 & 0.0873 & 0 \\ 420.0 & 236.2 & 212.0 \end{vmatrix}$$

$$= -20.0\mathbf{i} + 2.0\mathbf{j} - 30.1\mathbf{k} \text{ (m/s)}.$$

We can determine the ship's acceleration by expressing it in terms of normal and tangential components in the form given by Eq. (13.37) (Fig. a):

$$\mathbf{a}_B = \frac{dv}{dt}\mathbf{e}_\mathrm{t} + v\frac{d\theta}{dt}\mathbf{e}_\mathrm{n} = \mathbf{0} + (15)(0.0873)\mathbf{e}_\mathrm{n}$$

$$= 1.31\mathbf{e}_\mathrm{n} \ (\mathrm{m/s^2}).$$

The z axis is perpendicular to the ship's path and points toward the convex side of the path (Fig. b). Therefore, in terms of the body-fixed coordinate system, the ship's acceleration is $\mathbf{a}_B = -1.31\mathbf{k} \ (\mathrm{m/s^2})$. The ship's angular velocity vector is constant, so $\boldsymbol{\alpha} = \mathbf{0}$. The helicopter's acceleration relative to the earth is

$$\mathbf{a}_A = \mathbf{a}_B + \mathbf{a}_{A\,\mathrm{rel}} + 2\boldsymbol{\omega} \times \mathbf{v}_{A\,\mathrm{rel}} + \boldsymbol{\alpha} \times \mathbf{r}_{A/B}$$

$$+ \ \boldsymbol{\omega} \times (\boldsymbol{\omega} \times \mathbf{r}_{A/B})$$

$$= -1.31\mathbf{k} + (0.4\mathbf{i} - 0.2\mathbf{j} - 13.0\mathbf{k}) + 2\begin{vmatrix} \mathbf{i} & \mathbf{j} & \mathbf{k} \\ 0 & 0.0873 & 0 \\ -53.5 & 2.0 & 6.6 \end{vmatrix}$$

$$+ \ \mathbf{0} + (0.0873\mathbf{j}) \times \begin{vmatrix} \mathbf{i} & \mathbf{j} & \mathbf{k} \\ 0 & 0.0873 & 0 \\ 420.0 & 236.2 & 212.0 \end{vmatrix}$$

$$= -1.65\mathbf{i} - 0.20\mathbf{j} - 6.59\mathbf{k} \ (\mathrm{m/s^2}).$$

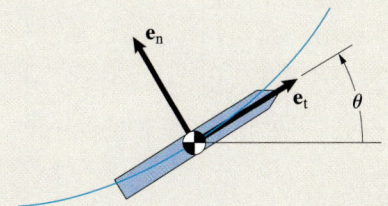

(a) Determining the ship's acceleration.

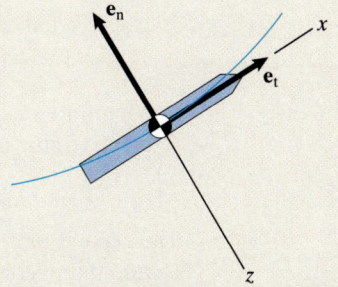

(b) Correspondence between the normal and tangential components and the body-fixed coordinate system.

Critical Thinking

Notice the substantial differences between the helicopter's velocity and acceleration relative to the earth and the values the ship measures using its body-fixed coordinate system. The ship's instruments, like those used on any moving vehicle, intrinsically make measurements relative to a body-fixed reference frame. Equations (17.16) and (17.17) must be used to transform measurements of velocity and acceleration into values relative to other reference frames.

Example 17.12 | An Earth-Fixed Reference Frame

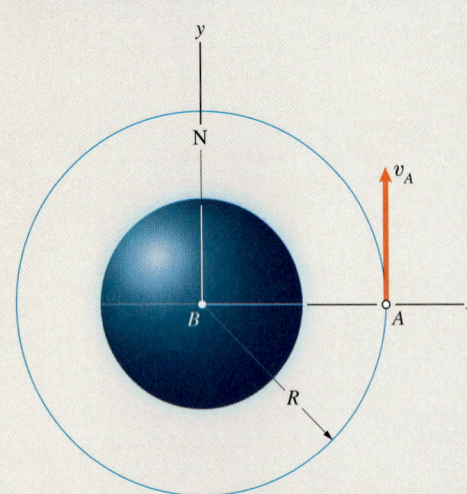

Figure 17.42

The satellite A shown in Fig. 17.42 is in a circular polar orbit (an orbit that intersects the earth's axis of rotation). Relative to a nonrotating primary reference frame with its origin at the center of the earth, the satellite moves in a circular path of radius R with a velocity of constant magnitude v_A. At the present instant, the satellite is above the equator. The secondary earth-fixed reference frame shown is oriented with the y axis in the direction of the north pole and the x axis in the direction of the satellite. What are the satellite's velocity and acceleration relative to the earth-fixed reference frame? Let ω_E be the angular velocity of the earth.

Strategy

We are given enough information to determine the satellite's velocity and acceleration $\mathbf{v}_A$ and $\mathbf{a}_A$ relative to the nonrotating primary reference frame and the angular velocity vector $\boldsymbol{\omega}$ of the secondary reference frame. We can therefore use Eqs. (17.16) and (17.17) to determine the satellite's velocity and acceleration $\mathbf{v}_{A\,\text{rel}}$ and $\mathbf{a}_{A\,\text{rel}}$ relative to the earth-fixed reference frame.

Solution

At the present instant, the satellite's velocity and acceleration relative to a nonrotating primary reference frame with its origin at the center of the earth are $\mathbf{v}_A = v_A \mathbf{j}$ and $\mathbf{a}_A = -(v_A^2/R)\mathbf{i}$. The angular velocity vector of the earth points north (confirm this by using the right-hand rule), so the angular velocity of the earth-fixed reference frame is $\boldsymbol{\omega} = \omega_E \mathbf{j}$. From Eq. (17.16),

$$\mathbf{v}_A = \mathbf{v}_B + \mathbf{v}_{A\,\text{rel}} + \boldsymbol{\omega} \times \mathbf{r}_{A/B}:$$

$$v_A \mathbf{j} = \mathbf{0} + \mathbf{v}_{A\,\text{rel}} + \begin{vmatrix} \mathbf{i} & \mathbf{j} & \mathbf{k} \\ 0 & \omega_E & 0 \\ R & 0 & 0 \end{vmatrix}.$$

Solving for $\mathbf{v}_{A\,\text{rel}}$, we find that the satellite's velocity relative to the earth-fixed reference frame is

$$\mathbf{v}_{A\,\text{rel}} = v_A \mathbf{j} + R\omega_E \mathbf{k}.$$

The second term on the right side of this equation is the satellite's velocity toward the west relative to the rotating earth-fixed reference frame.

From Eq. (17.17),

$$\mathbf{a}_A = \mathbf{a}_B + \mathbf{a}_{A\,\text{rel}} + 2\boldsymbol{\omega} \times \mathbf{v}_{A\,\text{rel}} + \boldsymbol{\alpha} \times \mathbf{r}_{A/B} + \boldsymbol{\omega} \times (\boldsymbol{\omega} \times \mathbf{r}_{A/B}):$$

$$-\frac{v_A^2}{R}\mathbf{i} = \mathbf{0} + \mathbf{a}_{A\,\text{rel}} + 2\begin{vmatrix} \mathbf{i} & \mathbf{j} & \mathbf{k} \\ 0 & \omega_E & 0 \\ 0 & v_A & R\omega_E \end{vmatrix} + \mathbf{0} + \begin{vmatrix} \mathbf{i} & \mathbf{j} & \mathbf{k} \\ 0 & \omega_E & 0 \\ 0 & 0 & -R\omega_E \end{vmatrix}.$$

Solving for $\mathbf{a}_{A\,\text{rel}}$, we find the satellite's acceleration relative to the earth-fixed reference frame:

$$\mathbf{a}_{A\,\text{rel}} = -\left(\frac{v_A^2}{R} + \omega_E^2 R\right)\mathbf{i}.$$

Critical Thinking

In this example, we assumed that the motion of the satellite was known relative to a nonrotating primary reference frame with its origin at the center of the earth, and used Eqs. (17.16) and (17.17) to determine its velocity and acceleration relative to a secondary earth-fixed reference frame. The inverse of this procedure is more common in practice. Ground-based measuring instruments measure velocity and acceleration relative to an earth-fixed reference frame, and Eqs. (17.16) and (17.17) are used to determine their values relative to other reference frames. In the next section, we show why a nonrotating reference frame with its origin at the center of the earth is frequently used as the primary reference frame.

Inertial Reference Frames

We say that a reference frame is *inertial* if it can be used to apply Newton's second law in the form $\Sigma \mathbf{F} = m\mathbf{a}$. Why can an earth-fixed reference frame be treated as inertial in many situations, even though it both accelerates and rotates? How can Newton's second law be applied using a reference frame that is fixed with respect to a ship or airplane? We are now in a position to answer these questions.

Earth-Centered, Nonrotating Reference Frame   We begin by showing why a nonrotating reference frame with its origin at the center of the earth can be assumed to be inertial for the purpose of describing motions of objects near the earth. Figure 17.43a shows a hypothetical nonaccelerating, nonrotating reference frame with origin O and a secondary nonrotating, *earth-centered reference frame*. The earth (and therefore the earth-centered reference frame) accelerates due to the gravitational attractions of the sun, moon, etc. We denote the earth's acceleration by the vector $\mathbf{g}_B$.

Suppose that we want to determine the motion of an object A of mass m (Fig. 17.43b). A is also subject to the gravitational attractions of the sun, moon, etc., and we denote the resulting gravitational acceleration by the vector $\mathbf{g}_A$. The vector $\Sigma \mathbf{F}$ is the sum of all other external forces acting on A, including the gravitational force exerted on it by the earth. The total external force acting on A is $\Sigma \mathbf{F} + m\mathbf{g}_A$. We can apply Newton's second law to A, using our hypothetical inertial reference frame:

$$\Sigma \mathbf{F} + m\mathbf{g}_A = m\mathbf{a}_A. \tag{17.18}$$

Here, $\mathbf{a}_A$ is the acceleration of A relative to O. Since the earth-centered reference frame does not rotate, we can use Eq. (17.17) to write $\mathbf{a}_A$ as

$$\mathbf{a}_A = \mathbf{a}_B + \mathbf{a}_{A\,\text{rel}},$$

where $\mathbf{a}_{A\,\text{rel}}$ is the acceleration of A relative to the earth-centered reference frame. Using this relation and our definition of the earth's acceleration $\mathbf{a}_B = \mathbf{g}_B$ in Eq. (17.18), we obtain

$$\Sigma \mathbf{F} = m\mathbf{a}_{A\,\text{rel}} + m(\mathbf{g}_B - \mathbf{g}_A). \tag{17.19}$$

If the object A is on or near the earth, its gravitational acceleration $\mathbf{g}_A$ due to the attraction of the sun, etc., is very nearly equal to the earth's gravitational acceleration $\mathbf{g}_B$. If we neglect the difference, Eq. (17.19) becomes

$$\Sigma \mathbf{F} = m\mathbf{a}_{A\,\text{rel}}. \tag{17.20}$$

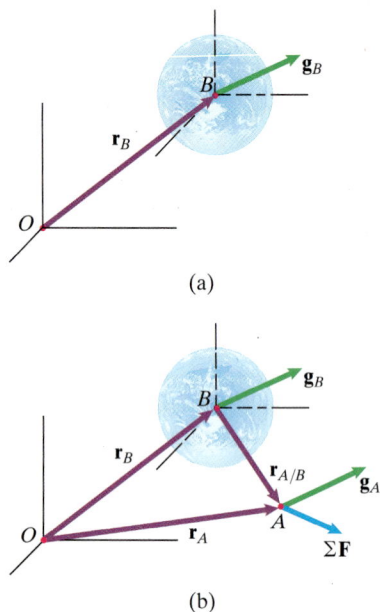

(a)

(b)

Figure 17.43
(a) An inertial reference frame and a nonrotating reference frame with its origin at the center of the earth.
(b) Determining the motion of an object A.

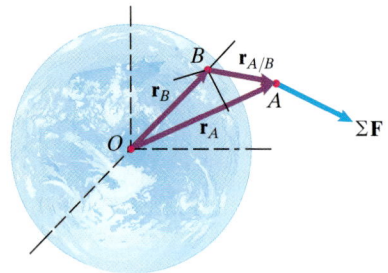

Figure 17.44
An earth-centered, nonrotating reference frame (origin O), an earth-fixed reference frame (origin B), and an object A.

Thus, we can apply Newton's second law using a nonrotating, earth-centered reference frame. Even though this reference frame accelerates, virtually the same gravitational acceleration acts on the object. Notice that this argument does not hold if the object is not near the earth.

Earth-Fixed Reference Frame For many applications, the most convenient reference frame is a local, *earth-fixed reference frame.* Why can we usually assume that an earth-fixed reference frame is inertial? Figure 17.44 shows a nonrotating reference frame with its origin O at the center of the earth, and a secondary earth-fixed reference frame with its origin at a point B. Since we can assume that the earth-centered, nonrotating reference frame is inertial, we can write Newton's second law for an object A of mass m as

$$\Sigma \mathbf{F} = m\mathbf{a}_A, \tag{17.21}$$

where $\mathbf{a}_A$ is A's acceleration relative to O. The earth-fixed reference frame rotates with the angular velocity of the earth, which we denote by $\boldsymbol{\omega}_E$. We can use Eq. (17.17) to write Eq. (17.21) in the form

$$\Sigma \mathbf{F} = m\mathbf{a}_{A\,\text{rel}} + m[\mathbf{a}_B + 2\boldsymbol{\omega}_E \times \mathbf{v}_{A\,\text{rel}}$$
$$+ \boldsymbol{\omega}_E \times (\boldsymbol{\omega}_E \times \mathbf{r}_{A/B})], \tag{17.22}$$

where $\mathbf{a}_{A\,\text{rel}}$ is A's acceleration relative to the earth-fixed reference frame. If we can neglect the terms in brackets on the right side of Eq. (17.22), we can take the earth-fixed reference frame to be inertial. Let us consider each term. (Recall from the definition of the cross product that $|\mathbf{U} \times \mathbf{V}| = |\mathbf{U}||\mathbf{V}| \sin \theta$, where θ is the angle between the two vectors. Therefore, the magnitude of the cross product is bounded by the product of the magnitudes of the vectors.)

- The term $\boldsymbol{\omega}_E \times (\boldsymbol{\omega}_E \times \mathbf{r}_{A/B})$: The earth's angular velocity ω_E is approximately one revolution per day $= 7.27 \times 10^{-5}$ rad/s. Therefore, the magnitude of this term is bounded by $\omega_E^2|\mathbf{r}_{A/B}| = (5.29 \times 10^{-9})|\mathbf{r}_{A/B}|$. For example, if the distance $|\mathbf{r}_{A/B}|$ from the origin of the earth-fixed reference frame to the object A is 10,000 m, this term is no larger than 5.3×10^{-5} m/s^2.
- The term $\mathbf{a}_B$: This term is the acceleration of the origin B of the earth-fixed reference frame relative to the center of the earth. B moves in a circular path due to the earth's rotation. If B lies on the earth's surface, $\mathbf{a}_B$ is bounded by $\omega_E^2 R_E$, where R_E is the radius of the earth. Using the value $R_E = 6370$ km, we find that $\omega_E^2 R_E = 0.0337$ m/s^2. This value is too large to neglect for many purposes. However, under normal circumstances, $\mathbf{a}_B$ is accounted for as a part of the local value of the acceleration due to gravity.
- The term $2\boldsymbol{\omega}_E \times \mathbf{v}_{A\,\text{rel}}$: This term is called the *Coriolis acceleration.* Its magnitude is bounded by $2\omega_E|\mathbf{v}_{A\,\text{rel}}| = (1.45 \times 10^{-4})|\mathbf{v}_{A\,\text{rel}}|$. For example, if the magnitude of the velocity of A relative to the earth-fixed reference frame is 10 m/s, this term is no larger than 1.45×10^{-3} m/s^2.

We see that in most applications, the terms in brackets in Eq. (17.22) can be neglected. However, in some cases this is not possible. The Coriolis acceleration becomes significant if an object's velocity relative to the earth is large, and even very small accelerations become significant if an object's motion must be predicted over a large period of time. In such cases, we can still use Eq. (17.22) to determine the motion, but must retain the significant terms. When this is done, the terms in brackets are usually moved to the left side:

$$\Sigma \mathbf{F} - m\mathbf{a}_B - 2m\boldsymbol{\omega}_E \times \mathbf{v}_{A\,\text{rel}} - m\boldsymbol{\omega}_E \times (\boldsymbol{\omega}_E \times \mathbf{r}_{A/B}) = m\mathbf{a}_{A\,\text{rel}}. \tag{17.23}$$

Written in this way, the equation has the usual form of Newton's second law, except that the left side contains additional "forces." (We use quotation marks because the quantities these terms represent are not forces, but arise from the motion of the earth-fixed reference frame.)

Coriolis Effects   The term $-2m\boldsymbol{\omega}_E \times \mathbf{v}_{A\,rel}$ in Eq. (17.23) is called the Coriolis force. It explains a number of physical phenomena that exhibit different behaviors in the northern and southern hemispheres. The earth's angular velocity vector $\boldsymbol{\omega}_E$ points north. When an object in the northern hemisphere that is moving tangent to the earth's surface travels north (Fig. 17.45a), the cross product $\boldsymbol{\omega}_E \times \mathbf{v}_{A\,rel}$ points west (Fig. 17.45b). Therefore, the Coriolis force points east; it causes an object moving north to turn to the right (Fig. 17.45c). If the object is moving south, the direction of $\mathbf{v}_{A\,rel}$ is reversed and the Coriolis force points west; its effect is to cause the object moving south to turn to the right (Fig. 17.45c). For example, in the northern hemisphere, winds converging on a center of low pressure tend to rotate about that center in the counterclockwise direction (Fig. 17.46a).

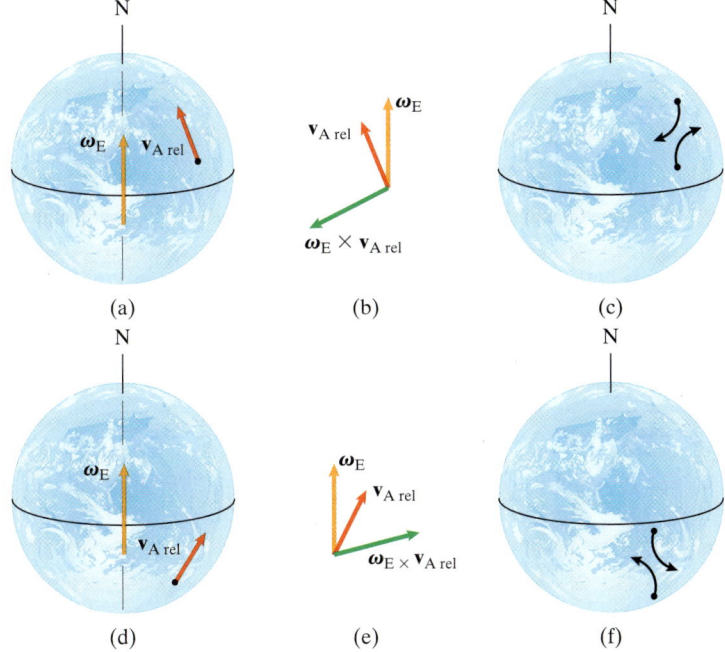

Figure 17.45
(a) An object in the northern hemisphere moving north.
(b) Cross product of the earth's angular velocity with the object's velocity.
(c) Effects of the Coriolis force in the northern hemisphere.
(d) An object in the southern hemisphere moving north.
(e) Cross product of the earth's angular velocity with the object's velocity.
(f) Effects of the Coriolis force in the southern hemisphere.

When an object in the southern hemisphere travels north (Fig. 17.45d), the cross product $\boldsymbol{\omega}_E \times \mathbf{v}_{A\,rel}$ points east (Fig. 17.45e). The Coriolis force points west and tends to cause the object to turn to the left (Fig. 17.45f). If the object is moving south, the Coriolis force points east and tends to cause the object to turn to the left (Fig. 17.45f). In the southern hemisphere, winds converging on a center of low pressure tend to rotate about that center in the clockwise direction (Fig. 17.46b).

Arbitrary Reference Frame   How can we analyze an object's motion relative to a reference frame that undergoes an arbitrary motion, such as a reference frame fixed with respect to a moving vehicle? Suppose that the primary

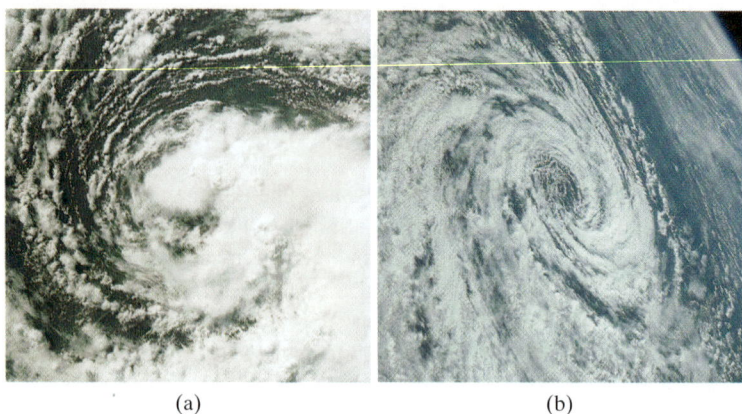

Figure 17.46
Storms in the (a) northern and (b) southern hemispheres.

(a)   (b)

reference frame with its origin at O in Fig. 17.47 is inertial and the secondary reference frame with its origin at B undergoes an arbitrary motion with angular velocity $\boldsymbol{\omega}$ and angular acceleration $\boldsymbol{\alpha}$. We can write Newton's second law for an object A of mass m as

$$\Sigma\mathbf{F} = m\mathbf{a}_A, \tag{17.24}$$

where $\mathbf{a}_A$ is A's acceleration relative to O. We use Eq. (17.17) to write Eq. (17.24) in the form

$$\Sigma\mathbf{F} - m[\mathbf{a}_B + 2\boldsymbol{\omega} \times \mathbf{v}_{A\,\text{rel}} + \boldsymbol{\alpha} \times \mathbf{r}_{A/B} + \boldsymbol{\omega} \times (\boldsymbol{\omega} \times \mathbf{r}_{A/B})] = m\mathbf{a}_{A\,\text{rel}}, \tag{17.25}$$

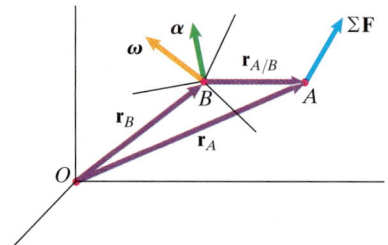

Figure 17.47
An inertial reference frame (origin O) and a reference frame undergoing an arbitrary motion (origin B).

where $\mathbf{a}_{A\,\text{rel}}$ is A's acceleration relative to the secondary reference frame. This is Newton's second law expressed in terms of a secondary reference frame undergoing an arbitrary motion relative to an inertial primary reference frame. If the forces acting on A and the secondary reference frame's motion are known, Eq. (17.25) can be used to determine $\mathbf{a}_{A\,\text{rel}}$.

Example 17.13   Inertial Reference Frames

Suppose that you and a friend play tennis on the deck of a cruise ship (Fig. 17.48), and you use the ship-fixed coordinate system with origin B to analyze the motion of the ball A. At the instant shown, the ball's position and velocity relative to the ship-fixed coordinate system are $\mathbf{r}_{A/B} = 5\mathbf{i} + 2\mathbf{j} + 12\mathbf{k}$ (m) and $\mathbf{v}_{A\,\text{rel}} = \mathbf{i} - 2\mathbf{j} + 7\mathbf{k}$ (m/s). The mass of the ball is 0.056 kg. Ignore the aerodynamic force on the ball for this example. As a result of the ship's motion, the acceleration of the origin B relative to the earth is $\mathbf{a}_B = 1.10\mathbf{i} + 0.07\mathbf{k}$ (m/s^2), and the angular velocity of the ship-fixed coordinate system is constant and equal to $\boldsymbol{\omega} = 0.1\mathbf{j}$ (rad/s). Use Newton's second law to determine the ball's acceleration relative to the ship-fixed coordinate system, (a) assuming that the ship-fixed coordinate system is inertial and (b) not assuming that the ship-fixed coordinate system is inertial, but assuming that a local earth-fixed coordinate system is inertial.

Strategy

In part (a), we know the ball's mass and the external forces acting on it, so we can simply apply Newton's second law to determine the acceleration. In part (b), we can express Newton's second law in the form given by Eq. (17.25), which applies to a coordinate system undergoing an arbitrary motion relative to an inertial reference frame.

Figure 17.48

Solution

(a) If the ship-fixed coordinate system is assumed to be inertial, we can apply Newton's second law to the ball in the form $\Sigma \mathbf{F} = m\mathbf{a}_{A\,\text{rel}}$. The only external force on the ball is its weight,

$$-(0.056)(9.81)\mathbf{j} = 0.056\mathbf{a}_{A\,\text{rel}},$$

so we obtain $\mathbf{a}_{A\,\text{rel}} = -9.81\mathbf{j} \ (\text{m/s}^2)$.

(b) If we treat the earth-fixed reference frame as inertial, we can solve Eq. (17.25) for the ball's acceleration relative to the ship-fixed reference frame:

$$\mathbf{a}_{A\,\text{rel}} = \frac{1}{m}\Sigma\mathbf{F} - \mathbf{a}_B - 2\boldsymbol{\omega} \times \mathbf{v}_{A\,\text{rel}} - \boldsymbol{\alpha} \times \mathbf{r}_{A/B} - \boldsymbol{\omega} \times (\boldsymbol{\omega} \times \mathbf{r}_{A/B})$$

$$= \frac{1}{0.056}[-(0.056)(9.81)\mathbf{j}] - (1.10\mathbf{i} + 0.07\mathbf{k}) - 2\begin{bmatrix} \mathbf{i} & \mathbf{j} & \mathbf{k} \\ 0 & 0.1 & 0 \\ 1 & -2 & 7 \end{bmatrix}$$

$$- \mathbf{0} - (0.1\mathbf{j}) \times \begin{bmatrix} \mathbf{i} & \mathbf{j} & \mathbf{k} \\ 0 & 0.1 & 0 \\ 5 & 2 & 12 \end{bmatrix}$$

$$= -2.45\mathbf{i} - 9.81\mathbf{j} + 0.25\mathbf{k} \ (\text{m/s}^2).$$

Critical Thinking

This example demonstrates the care you must exercise in applying Newton's second law. When we treated the ship-fixed coordinate system as inertial, Newton's second law did not correctly predict the ball's acceleration relative to the coordinate system, because the effects of the motion of the coordinate system on the ball's relative motion were not accounted for.

Problems

17.143 The x–y coordinate system is body-fixed with respect to the bar. The angle θ is given as a function of time by $\theta = 0.16t$ rad. The x coordinate of the sleeve A is given as a function of time by $x = 1.2 + 0.018t^2$ m. Use Eq. (17.16) to determine the velocity of the sleeve A at $t = 5$ s relative to a nonrotating reference frame with its origin at B. (Although you are determining the velocity of A relative to a nonrotating reference frame, your answer will be expressed in components in terms of the body-fixed reference frame.)

17.144 Determine the acceleration of the sleeve A in Problem 17.143 at $t = 5$ s relative to a nonrotating reference frame with its origin at B.

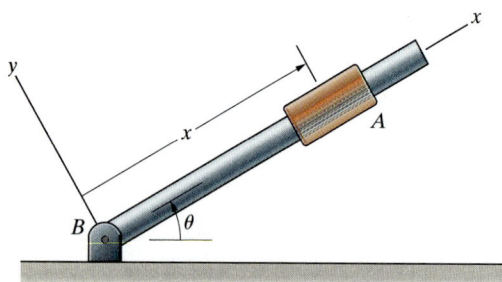

Problems 17.143/17.144

17.145 The metal plate is attached to a fixed ball-and-socket support at O. The pin A slides in a slot in the plate. At the instant shown, $x_A = 1$ m, $dx_A/dt = 2$ m/s, and $d^2x_A/dt^2 = 0$, and the plate's angular velocity and angular acceleration are $\omega = 2\mathbf{k}$ (rad/s) and $\alpha = 0$. What are the x, y, and z components of the velocity and acceleration of A relative to a nonrotating reference frame with its origin at O?

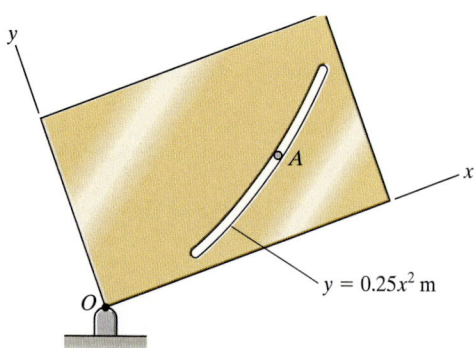

Problems 17.145/17.146

17.146 The pin A slides in a slot in the plate. Suppose that at the instant shown, $x_A = 1$ m, $dx_A/dt = -3$ m/s, $d^2x_A/dt^2 = 4$ m/s², and the plate's angular velocity and angular acceleration are $\omega = -4\mathbf{j} + 2\mathbf{k}$ (rad/s) and $\alpha = 3\mathbf{i} - 6\mathbf{j}$ (rad/s²). What are the x, y, z components of the velocity and acceleration of A relative to a nonrotating reference frame that is stationary with respect to O?

17.147 The coordinate system shown is fixed relative to the ship B. At the given instant, the ship is sailing north at 10 ft/s relative to the earth and its angular velocity is 0.02 rad/s clockwise. The airplane is flying east at 400 ft/s relative to the earth, and its position relative to the ship is $\mathbf{r}_{A/B} = 2000\mathbf{i} + 2000\mathbf{j} + 1000\mathbf{k}$ (ft). If the ship uses its radar to measure the plane's velocity relative to the ship's body-fixed coordinate system, what is the result?

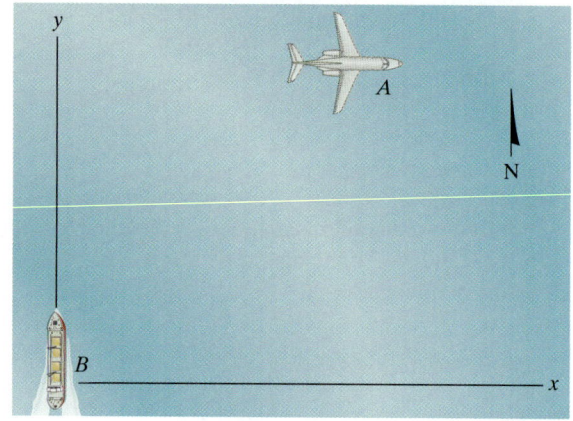

Problem 17.147

17.148 The space shuttle is attempting to recover a satellite for repair. At the current time, the satellite's position relative to a coordinate system fixed to the shuttle is $50\mathbf{i}$ (m). The rate gyros on the shuttle indicate that its current angular velocity is $0.05\mathbf{j} + 0.03\mathbf{k}$ (rad/s). The shuttle pilot measures the velocity of the satellite relative to the body-fixed coordinate system and determines it to be $-2\mathbf{i} - 1.5\mathbf{j} + 2.5\mathbf{k}$ (rad/s). What are the x, y, and z components of the satellite's velocity relative to a nonrotating coordinate system with its origin fixed to the shuttle's center of mass?

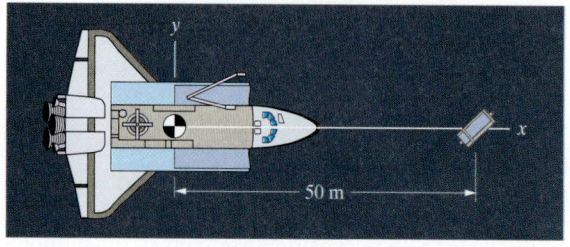

Problem 17.148

17.149 The train on the circular track is traveling at a constant speed of 50 ft/s in the direction shown. The train on the straight track is traveling at 20 ft/s in the direction shown and is increasing its speed at 2 ft/s². Determine the velocity of passenger A that passenger B observes relative to the given coordinate system, which is fixed to the car in which B is riding.

17.150 In Problem 17.149, determine the acceleration of passenger A that passenger B observes relative to the coordinate system fixed to the car in which B is riding.

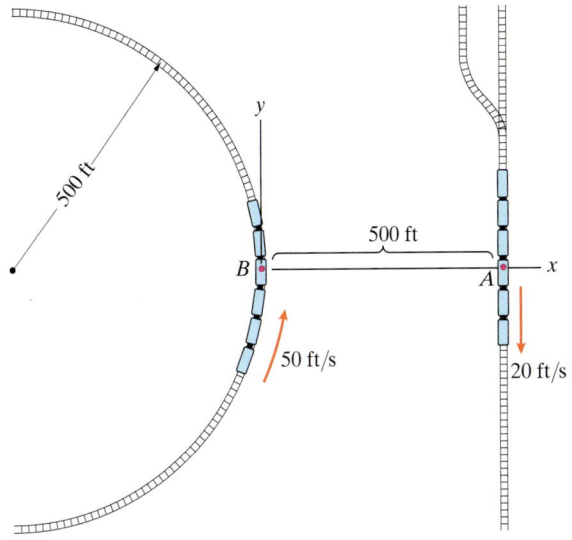

Problems 17.149/17.150

17.151 The satellite A is in a circular polar orbit (a circular orbit that intersects the earth's axis of rotation). The radius of the orbit is R, and the magnitude of the satellite's velocity relative to a nonrotating reference frame with its origin at the center of the earth is v_A. At the instant shown, the satellite is above the equator. An observer B on the earth directly below the satellite measures its motion using the earth-fixed coordinate system shown. What are the velocity and acceleration of the satellite relative to B's earth-fixed coordinate system? The radius of the earth is R_E and the angular velocity of the earth is ω_E.

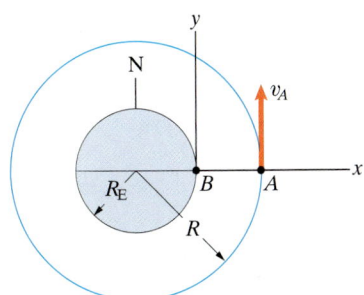

Problem 17.151

17.152 A car A at north latitude L drives north on a north–south highway with constant speed v. The earth's radius is R_E, and the earth's angular velocity is ω_E. (The earth's angular velocity vector points north.) The coordinate system is earth fixed, and the x axis passes through the car's position at the instant shown. Determine the car's velocity and acceleration (a) relative to the earth-fixed coordinate system and (b) relative to a nonrotating reference frame with its origin at the center of the earth.

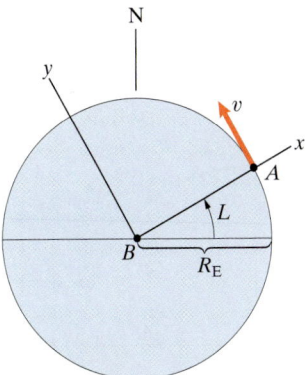

Problem 17.152

17.153 The airplane B conducts flight tests of a missile. At the instant shown, the airplane is traveling at 200 m/s relative to the earth in a circular path of 2000-m radius *in the horizontal plane*. The coordinate system is fixed relative to the airplane. The x axis is tangent to the plane's path and points forward. The y axis points out the plane's right side, and the z axis points out the bottom of the plane. The plane's bank angle (the inclination of the z axis from the vertical) is constant and equal to 20°. *Relative to the airplane's coordinate system*, the pilot measures the missile's position and velocity and determines them to be $\mathbf{r}_{A/B} = 1000\mathbf{i}$ (m) and $\mathbf{v}_{A/B} = 100.0\mathbf{i} + 94.0\mathbf{j} + 34.2\mathbf{k}$ (m/s).

(a) What are the x, y, and z components of the airplane's angular velocity vector?

(b) What are the x, y, and z components of the missile's velocity relative to the earth?

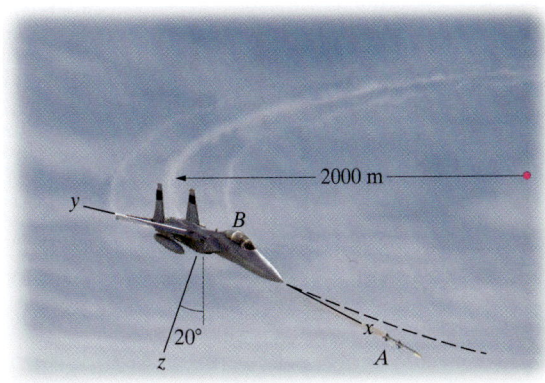

Problem 17.153

17.154 To conduct experiments related to long-term spaceflight, engineers construct a laboratory on earth that rotates about the vertical axis at B with a constant angular velocity ω of one revolution every 6 s. They establish a laboratory-fixed coordinate system with its origin at B and the z axis pointing upward. An engineer holds an object stationary relative to the laboratory at point A, 3 m from the axis of rotation, and releases it. At the instant he drops the object, determine its acceleration relative to the laboratory-fixed coordinate system, (a) assuming that the laboratory-fixed coordinate system is inertial and (b) not assuming that the laboratory-fixed coordinate system is inertial, but assuming that an earth-fixed coordinate system with its origin at B is inertial.

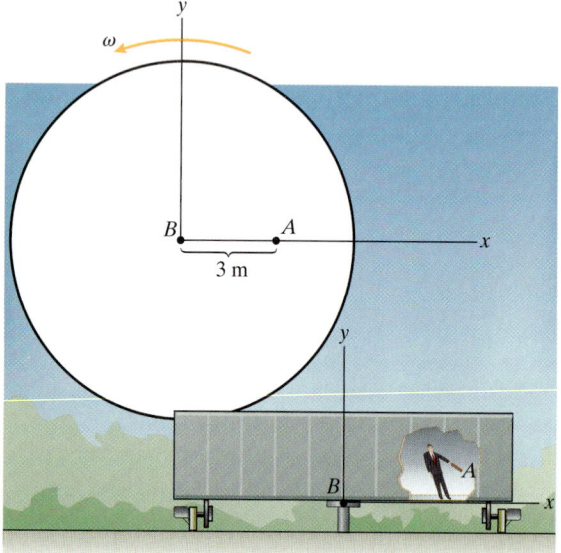

Problem 17.154

17.155 The disk rotates *in the horizontal plane* about a fixed shaft at the origin with constant angular velocity $\omega = 10$ rad/s. The 2-kg slider A moves in a smooth slot in the disk. The spring is unstretched when $x = 0$ and its constant is $k = 400$ N/m. Determine the acceleration of A relative to the body-fixed coordinate system when $x = 0.4$ m.

Strategy: Use Eq. (17.25) to express Newton's second law for the slider in terms of the body-fixed coordinate system.

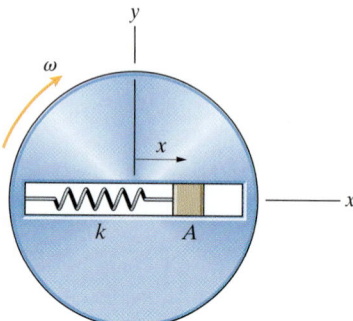

Problem 17.155

17.156* Engineers conduct flight tests of a rocket at 30° north latitude. They measure the rocket's motion using an earth-fixed coordinate system with the x axis pointing upward and the y axis directed northward. At a particular instant, the mass of the rocket is 4000 kg, the velocity of the rocket relative to the engineers' coordinate system is $2000\mathbf{i} + 2000\mathbf{j}$ (m/s), and the sum of the forces exerted on the rocket by its thrust, weight, and aerodynamic forces is $400\mathbf{i} + 400\mathbf{j}$ (N). Determine the rocket's acceleration relative to the engineers' coordinate system, (a) assuming that their earth-fixed coordinate system is inertial and (b) not assuming that their earth-fixed coordinate system is inertial.

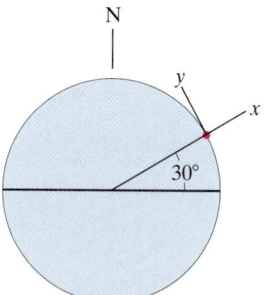

Problem 17.156

17.157* Consider a point A on the surface of the earth at north latitude L. The radius of the earth is R_E, and the earth's angular velocity is ω_E. A plumb bob suspended just above the ground at point A will hang at a small angle β relative to the vertical because of the earth's rotation. Show that β is related to the latitude by

$$\tan \beta = \frac{\omega_E^2 R_E \sin L \cos L}{g - \omega_E^2 R_E \cos^2 L}.$$

Strategy: Using the earth-fixed coordinate system shown, express Newton's second law in the form given by Eq. (17.22).

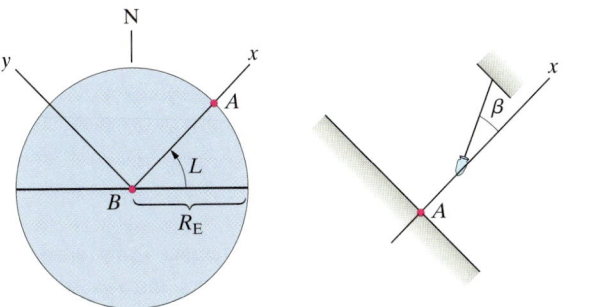

Problem 17.157

17.158* Suppose that a space station is in orbit around the earth and two astronauts on the station toss a ball back and forth. They observe that the ball appears to travel between them in a straight line at constant velocity.

(a) Write Newton's second law for the ball as it travels between the astronauts in terms of a nonrotating coordinate system with its origin fixed to the station. What is the term $\Sigma \mathbf{F}$? Use the equation you wrote to explain the behavior of the ball observed by the astronauts.

(b) Write Newton's second law for the ball as it travels between the astronauts in terms of a nonrotating coordinate system with its origin fixed to the center of the earth. What is the term $\Sigma \mathbf{F}$? Explain the difference between this equation arid the one you obtained in part (a).

CHAPTER SUMMARY

In this chapter we analyzed the motions of rigid bodies, showing how the velocities and accelerations of their points are related to their angular velocities and angular accelerations. In doing so, we did not consider the forces and couples causing the motions. In Chapter 18, we will use Newton's second law to determine the equations of motion for rigid bodies in planar motion. By drawing a free-body diagram of a rigid body, we will relate the acceleration of its center of mass and its angular acceleration to the forces and couples acting on it.

Types of Motion

If a rigid body in motion does not rotate relative to a given reference frame, it is said to be in *translation*. If the points of a rigid body intersected by a fixed plane remain in that plane, the rigid body is said to undergo *planar* motion. Rotation about a fixed axis is a special case of planar motion.

Relative Velocities and Accelerations

The *angular velocity vector* $\boldsymbol{\omega}$ of a rigid body is parallel to the axis of rotation of the body, and its magnitude $|\boldsymbol{\omega}|$ is the body's rate of rotation. If the thumb of the right hand points in the direction of $\boldsymbol{\omega}$, the fingers curl around $\boldsymbol{\omega}$ in the direction of the rotation. The *angular acceleration vector* $\boldsymbol{\alpha} = d\boldsymbol{\omega}/dt$ is the rate of change of the angular velocity vector of the body.

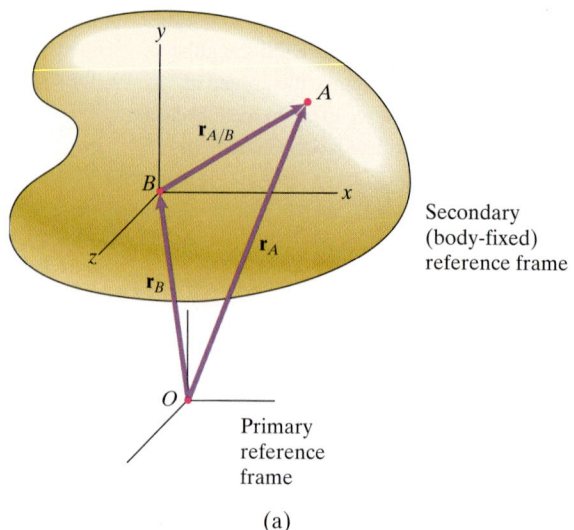

(a)

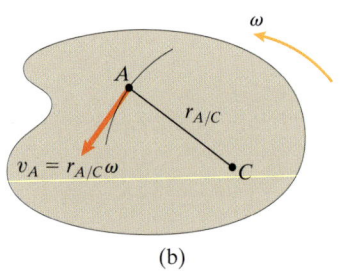

(b)

Consider a point B of a rigid body, a body-fixed reference frame with its origin at B, and an arbitrary point A (Fig. a). The velocities $\mathbf{v}_A$ and $\mathbf{v}_B$ of the points relative to the primary reference frame are related by

$$\mathbf{v}_A = \mathbf{v}_B + \mathbf{v}_{A\,\text{rel}} + \boldsymbol{\omega} \times \mathbf{r}_{A/B}, \tag{17.11}$$

where $\mathbf{v}_{A\,\text{rel}}$ is the velocity of A relative to the body-fixed reference frame. If A is a point of the rigid body, $\mathbf{v}_{A\,\text{rel}}$ is zero.

The accelerations $\mathbf{a}_A$ and $\mathbf{a}_B$ of the points relative to the primary reference frame are related by

$$\mathbf{a}_A = \mathbf{a}_B + \mathbf{a}_{A\,\text{rel}} + 2\boldsymbol{\omega} \times \mathbf{v}_{A\,\text{rel}} + \boldsymbol{\alpha} \times \mathbf{r}_{A/B}$$
$$+ \boldsymbol{\omega} \times (\boldsymbol{\omega} \times \mathbf{r}_{A/B}), \tag{17.13}$$

where $\mathbf{a}_{A\,\text{rel}}$ is the acceleration of A relative to the body-fixed reference frame. In planar motion, the term $\boldsymbol{\omega} \times (\boldsymbol{\omega} \times \mathbf{r}_{A/B})$ can be written in the simpler form $-\omega^2 \mathbf{r}_{A/B}$. If A is a point of the rigid body, $\mathbf{a}_{A\,\text{rel}}$ is zero.

Instantaneous Centers

A point of a rigid body whose velocity is zero at a given instant (or a point exterior to the rigid body whose velocity would be zero if the rigid body were extended to include that point—see Fig. 17.21) is called an *instantaneous center*. Consider a rigid body in planar motion, and suppose that C is an instantaneous center. The velocity of a point A is perpendicular to the line from C to A, and its magnitude is the product of the distance from C to A and the angular velocity (Fig. b).

If the directions of motion of two points A and B of a rigid body in planar motion are not parallel, lines drawn through A and B perpendicular to their directions of motion intersect at the instantaneous center (Fig. c).

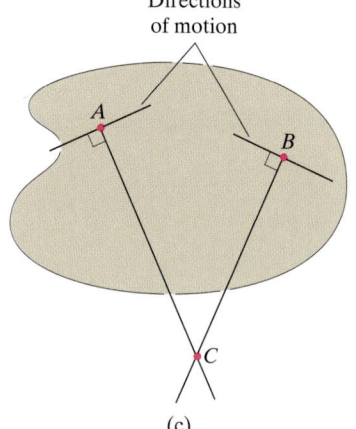

(c)

Moving Reference Frames

Consider a point A and a reference frame, with origin B, that rotates with angular velocity $\boldsymbol{\omega}$ and angular acceleration $\boldsymbol{\alpha}$ relative to a primary reference frame (Fig. d). The velocities of A and B relative to the primary reference frame are related by

$$\mathbf{v}_A = \mathbf{v}_B + \mathbf{v}_{A\,\text{rel}} + \boldsymbol{\omega} \times \mathbf{r}_{A/B}, \tag{17.16}$$

where $\mathbf{v}_{A\,\text{rel}}$ is the velocity of A relative to the secondary reference frame. The accelerations of A and B relative to the primary reference frame are related by

$$\mathbf{a}_A = \mathbf{a}_B + \mathbf{a}_{A\,\text{rel}} + 2\boldsymbol{\omega} \times \mathbf{v}_{A\,\text{rel}}$$
$$+ \boldsymbol{\alpha} \times \mathbf{r}_{A/B} + \boldsymbol{\omega} \times (\boldsymbol{\omega} \times \mathbf{r}_{A/B}), \tag{17.17}$$

where $\mathbf{a}_{A\,\text{rel}}$ is the acceleration of A relative to the secondary reference frame.

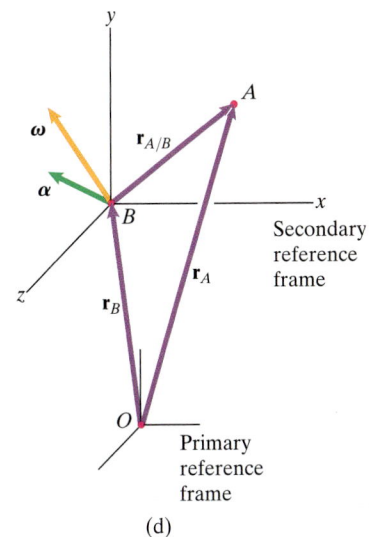

(d)

Review Problems

17.159 If $\theta = 60°$ and bar OQ rotates in the counterclockwise direction at 5 rad/s, what is the angular velocity of bar PQ?

17.160 If $\theta = 55°$ and the sleeve P is moving to the left at 2 m/s, what are the angular velocities of bars OQ and PQ?

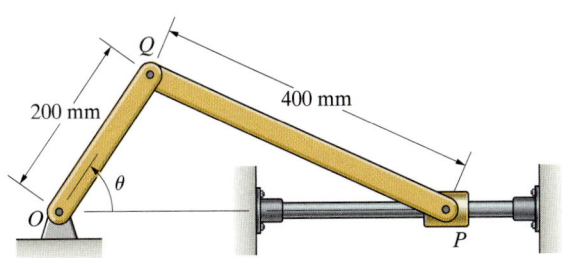

Problems 17.159/17.160

17.161 Determine the vertical velocity v_H of the hook and the angular velocity of the small pulley.

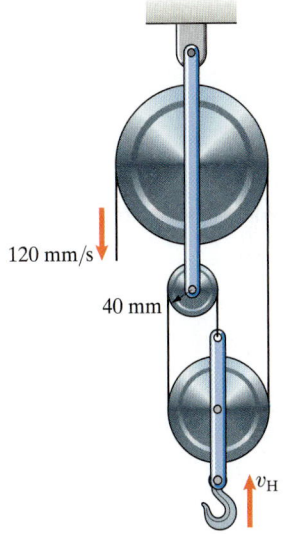

Problem 17.161

17.162 If the crankshaft AB is turning in the counterclockwise direction at 2000 rpm, what is the velocity of the piston?

17.163 If the piston is moving with velocity $\mathbf{v}_C = 20\mathbf{j}$ (ft/s), what are the angular velocities of the crankshaft AB and the connecting rod BC?

17.164 If the piston is moving with velocity $\mathbf{v}_C = 20\mathbf{j}$ (ft/s) and its acceleration is zero, what are the angular accelerations of the crankshaft AB and the connecting rod BC?

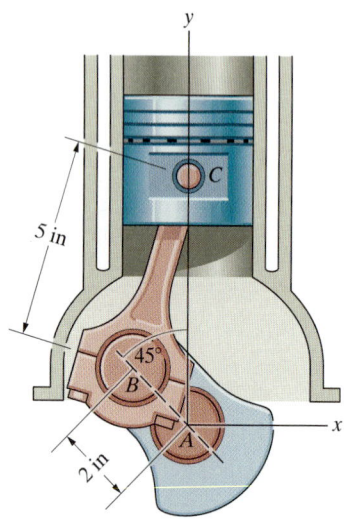

Problems 17.162–17.164

17.165 Bar AB rotates at 6 rad/s in the counterclockwise direction. Use instantaneous centers to determine the angular velocity of bar BCD and the velocity of point D.

17.166 Bar AB rotates with a constant angular velocity of 6 rad/s in the counterclockwise direction. Determine the acceleration of point D.

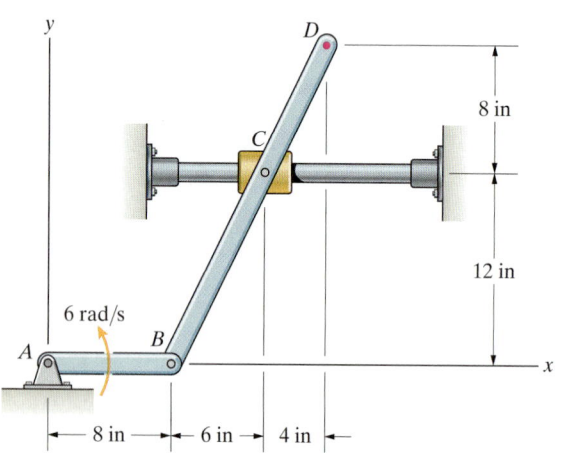

Problems 17.165/17.166

17.167 Point C is moving to the right at 20 in/s. What is the velocity of the midpoint G of bar BC?

17.168 Point C is moving to the right with a constant velocity of 20 in/s. What is the acceleration of the midpoint G of bar BC?

17.169 If the velocity of point C is $\mathbf{v}_C = 1.0\mathbf{i}$ (in/s), what are the angular velocity vectors of arms AB and BC?

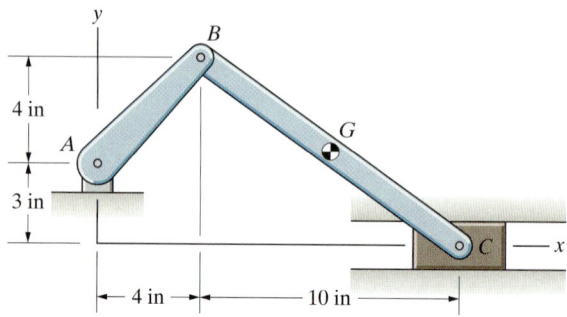

Problems 17.167–17.169

17.170 Points B and C are in the x–y plane. The angular velocity vectors of arms AB and BC are $\boldsymbol{\omega}_{AB} = -0.5\mathbf{k}$ (rad/s) and $\boldsymbol{\omega}_{BC} = -2.0\mathbf{k}$ (rad/s). Determine the velocity of point C.

17.171 If the velocity of point C is $\mathbf{v}_C = 1.0\mathbf{i}$ (m/s), what are the angular velocity vectors of arms AB and BC?

17.172 The angular velocity vectors of arms AB and BC are $\boldsymbol{\omega}_{AB} = -0.5\mathbf{k}$ (rad/s) and $\boldsymbol{\omega}_{BC} = 2.0\mathbf{k}$ (rad/s), and their angular acceleration vectors are $\boldsymbol{\alpha}_{AB} = 1.0\mathbf{k}$ (rad/s²) and $\boldsymbol{\alpha}_{BC} = 1.0\mathbf{k}$ (rad/s²). What is the acceleration of point C?

17.173 The velocity of point C is $\mathbf{v}_C = 1.0\mathbf{i}$ (m/s) and $\mathbf{a}_C = \mathbf{0}$. What are the angular velocity and angular acceleration vectors of arm BC?

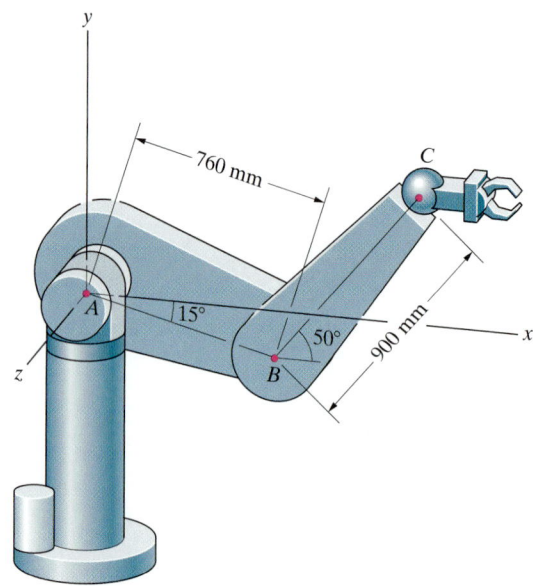

Problems 17.170–17.173

17.174 The crank AB has a constant clockwise angular velocity of 200 rpm. What are the velocity and acceleration of the piston P?

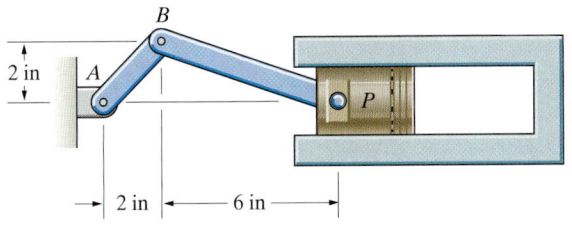

Problem 17.174

17.175 Bar AB has a counterclockwise angular velocity of 10 rad/s and a clockwise angular acceleration of 20 rad/s². Determine the angular acceleration of bar BC and the acceleration of point C.

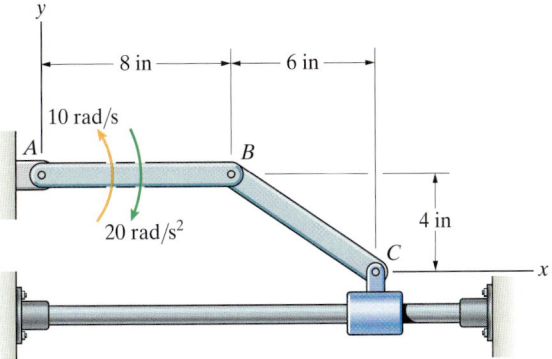

Problem 17.175

17.176 The angular velocity of arm AC is 1 rad/s counterclockwise. What is the angular velocity of the scoop?

17.177 The angular velocity of arm AC is 2 rad/s counterclockwise and its angular acceleration is 4 rad/s² clockwise. What is the angular acceleration of the scoop?

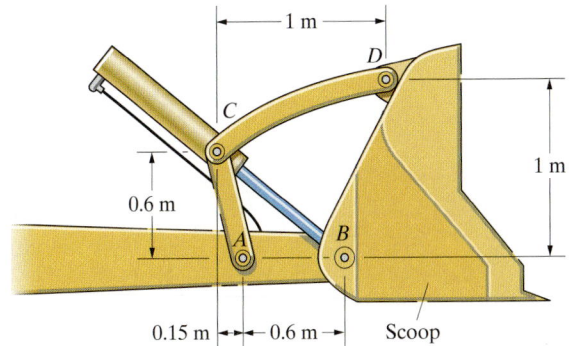

Problems 17.176/17.177

17.178 If you want to program the robot so that, at the instant shown, the velocity of point D is $\mathbf{v}_D = 0.2\mathbf{i} + 0.8\mathbf{j}$ (m/s) and the angular velocity of arm CD is 0.3 rad/s counterclockwise, what are the necessary angular velocities of arms AB and BC?

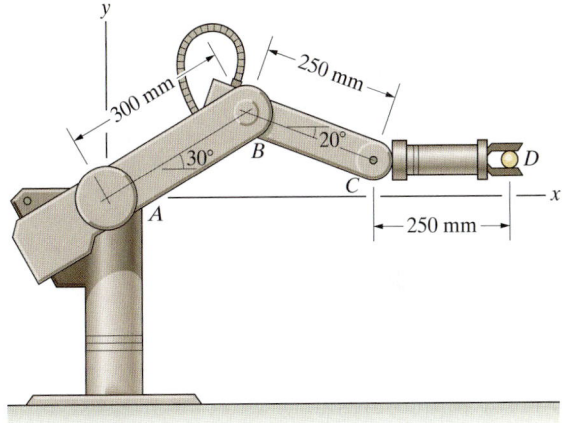

Problem 17.178

17.179 The ring gear is stationary, and the sun gear rotates at 120 rpm in the counterclockwise direction. Determine the angular velocity of the planet gears and the magnitude of the velocity of their centerpoints.

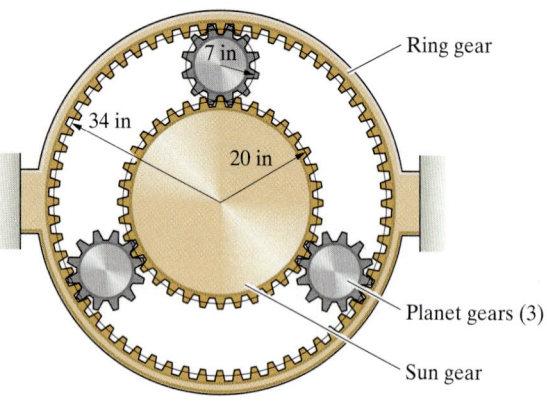

Problem 17.179

17.180 Arm AB is rotating at 10 rad/s in the clockwise direction. Determine the angular velocity of arm BC and the velocity at which the arm slides relative to the sleeve at C.

17.181 Arm AB is rotating with an angular velocity of 10 rad/s and an angular acceleration of 20 rad/s^2, both in the clockwise direction. Determine the angular acceleration of arm BC.

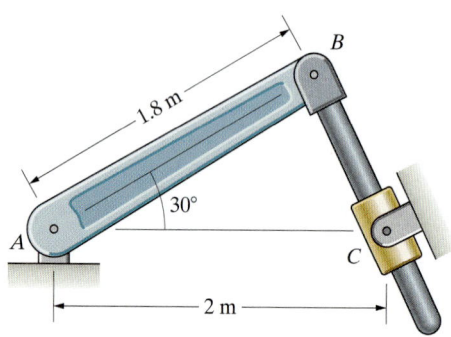

Problems 17.180/17.181

17.182 Arm AB is rotating with a constant counterclockwise angular velocity of 10 rad/s. Determine the vertical velocity and acceleration of the rack R of the rack-and-pinion gear.

17.183 The rack R of the rack-and-pinion gear is moving upward with a constant velocity of 10 ft/s. What are the angular velocity and angular acceleration of bar BC?

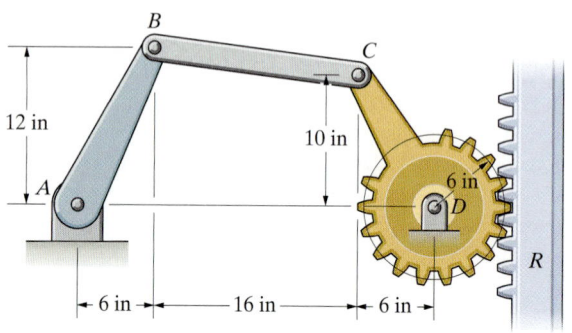

Problems 17.182/17.183

17.184 Bar AB has a constant counterclockwise angular velocity of 2 rad/s. The 1-kg collar C slides on the smooth horizontal bar. At the instant shown, what is the tension in the cable BC?

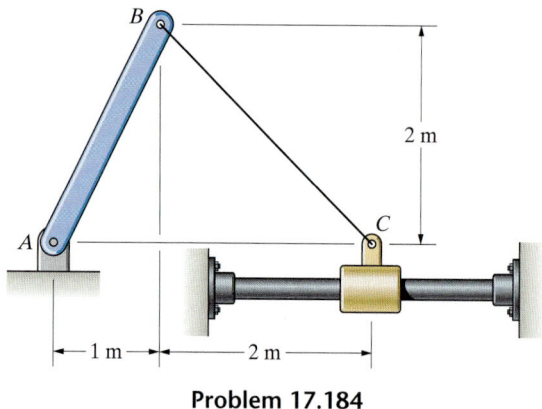

Problem 17.184

17.185 An athlete exercises his arm by raising the 8-kg mass m. The shoulder joint A is stationary. The distance AB is 300 mm, the distance BC is 400 mm, and the distance from C to the pulley is 340 mm. The angular velocities $\omega_{AB} = 1.5$ rad/s and $\omega_{BC} = 2$ rad/s are constant. What is the tension in the cable?

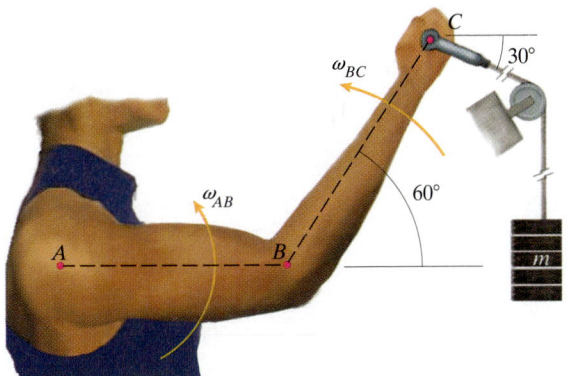

Problem 17.185

17.186 The hydraulic actuator BC of the crane is extending (increasing in length) at a constant rate of 0.2 m/s. When the angle $\beta = 35°$, what is the angular velocity of the crane's boom AD?

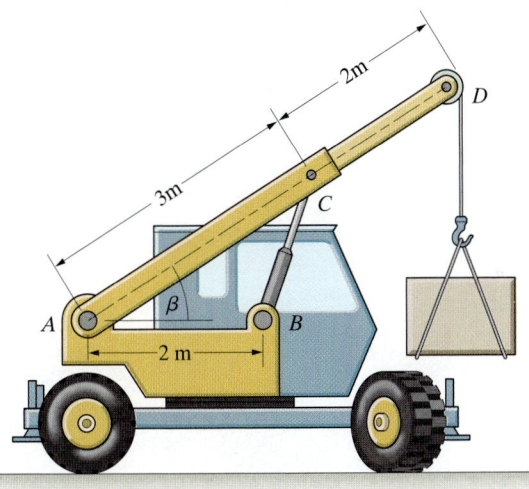

Problem 17.186

17.187 The coordinate system shown is fixed relative to the ship B. The ship uses its radar to measure the position of a stationary buoy A and determines it to be $400\mathbf{i} + 200\mathbf{j}$ (m). The ship also measures the velocity of the buoy relative to its body-fixed coordinate system and determines it to be $2\mathbf{i} - 8\mathbf{j}$ (m/s). What are the ship's velocity and angular velocity relative to the earth? (Assume that the ship's velocity is in the direction of the y axis.)

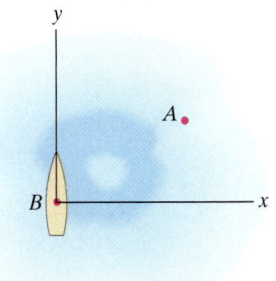

Problem 17.187

Design Project

Bar AB rotates about the fixed point A with constant angular velocity ω_0.

(a) Determine the lengths of the bars AB and BC so that as bar AB rotates, the collar C moves back and forth between the positions D and E.

(b) Draw graphs of the velocity and acceleration of the collar C as functions of the angular position of bar AB.

(c) Suppose that a design constraint is that the magnitude of the acceleration of collar C must not exceed 200 m/s^2. What is the maximum allowable value of ω_0?

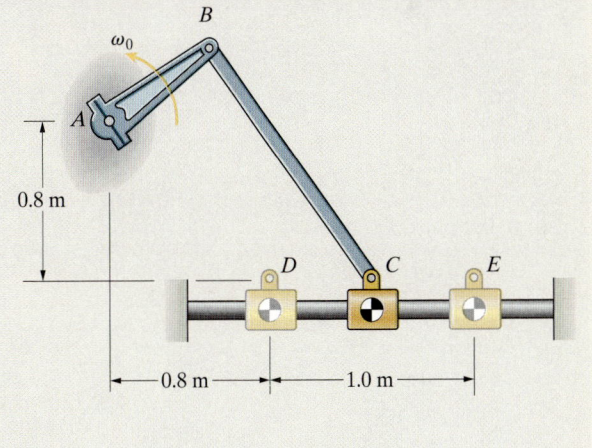

Planar Dynamics of Rigid Bodies

In Chapter 17 we analyzed planar motions of rigid bodies without considering the forces and couples causing them. In this chapter we derive planar equations of angular motion for a rigid body. By drawing the free-body diagram of an object and applying the equations of motion, we can determine both the acceleration of its center of mass and its angular acceleration in terms of the forces and couples to which it is subjected.

◄ Electric motors exert torque on the wind-tunnel fan, accelerating it to its operating angular velocity.

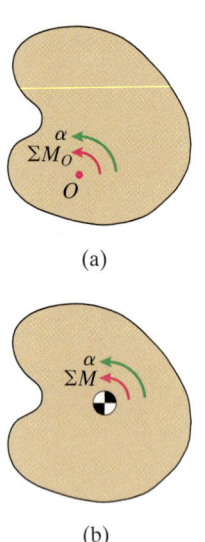

(a)

(b)

Figure 18.1
(a) A rigid body rotating about a fixed axis O.
(b) A rigid body in general planar motion.

18.1 Preview of the Equations of Motion

The two-dimensional equations of angular motion for a rigid body are simple in form, but their derivations are rather involved. To provide help in following the derivations, we first summarize the equations in this section.

The equations governing the planar motion of a rigid body are Newton's second law,

$$\Sigma \mathbf{F} = m\mathbf{a},$$

which states that the sum of the external forces acting on the body equals the product of its mass and the acceleration of its center of mass, and an equation of angular motion. If the rigid body rotates about a fixed axis O (Fig. 18.1a), the sum of the moments about the axis due to external forces and couples acting on the body is related to its angular acceleration by

$$\Sigma M_O = I_O \alpha,$$

where I_O is the moment of inertia of the rigid body about O. Just as an object's mass determines the acceleration resulting from the forces acting on it, its moment of inertia I_O about a fixed axis determines the angular acceleration resulting from the sum of the moments about the axis.

In general planar motion (Fig. 18.1b), the sum of the moments about the center of mass of a rigid body is related to its angular acceleration by

$$\Sigma M = I\alpha,$$

where I is the moment of inertia of the rigid body about its center of mass. If the external forces and couples acting on a rigid body in planar motion are known, we can use these equations to determine the acceleration of the center of mass of the body and its angular acceleration.

18.2 Momentum Principles for a System of Particles

Our derivations of the equations of motion for rigid bodies are based on principles governing the motion of a system of particles. We summarize these general and important principles in this section.

Force–Linear Momentum Principle

We begin by showing that the sum of the external forces on a system of particles equals the rate of change of the total linear momentum of the system. Let us consider a system of N particles. We denote the mass of the ith particle by m_i and denote its position vector relative to the origin O of an inertial reference frame by $\mathbf{r}_i$ (Fig. 18.2). Let $\mathbf{f}_{ij}$ be the force exerted on the ith particle by the jth particle, and let the external force on the ith particle (i.e., the total force exerted by objects other than the system of particles we are considering) be $\mathbf{f}_i^E$. Newton's second law states that the total force acting on the ith particle equals the product of its mass and the rate of change of its linear momentum; that is,

$$\sum_j \mathbf{f}_{ij} + \mathbf{f}_i^E = \frac{d}{dt}(m_i \mathbf{v}_i), \tag{18.1}$$

where $\mathbf{v}_i = d\mathbf{r}_i/dt$ is the velocity of the ith particle. Writing this equation for each particle of the system and summing from $i = 1$ to N, we obtain

$$\sum_i \sum_j \mathbf{f}_{ij} + \sum_i \mathbf{f}_i^E = \frac{d}{dt} \sum_i m_i \mathbf{v}_i. \tag{18.2}$$

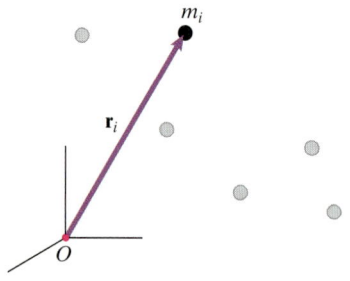

Figure 18.2
A system of particles. The vector $\mathbf{r}_i$ is the position vector of the ith particle.

The first term on the left side of this equation is the sum of the internal forces on the system of particles. As a consequence of Newton's third law ($\mathbf{f}_{ij} + \mathbf{f}_{ji} = \mathbf{0}$), that term equals zero:

$$\sum_i \sum_j \mathbf{f}_{ij} = \mathbf{f}_{12} + \mathbf{f}_{21} + \mathbf{f}_{13} + \mathbf{f}_{31} + \cdots = \mathbf{0}.$$

The second term on the left side of Eq. (18.2) is the sum of the external forces on the system. Denoting it by $\Sigma\mathbf{F}$, we conclude that the sum of the external forces on the system equals the rate of change of its total linear momentum:

$$\Sigma\mathbf{F} = \frac{d}{dt} \sum_i m_i \mathbf{v}_i. \tag{18.3}$$

Let m be the sum of the masses of the particles:

$$m = \sum_i m_i.$$

The position of the center of mass of the system is

$$\mathbf{r} = \frac{\sum_i m_i \mathbf{r}_i}{m}, \tag{18.4}$$

so the velocity of the center of mass is

$$\mathbf{v} = \frac{d\mathbf{r}}{dt} = \frac{\sum_i m_i \mathbf{v}_i}{m}.$$

By using this expression, we can write Eq. (18.3) as

$$\Sigma\mathbf{F} = \frac{d}{dt}(m\mathbf{v}).$$

The total external force on a system of particles thus equals the rate of change of the product of the total mass of the system and the velocity of its center of mass. Since any object or collection of objects, including a rigid body, can be modeled as a system of particles, this result is one of the most general and elegant in mechanics. Furthermore, if the total mass m is constant, we obtain

$$\Sigma\mathbf{F} = m\mathbf{a},$$

where $\mathbf{a} = d\mathbf{v}/dt$ is the acceleration of the center of mass. We see that the total external force equals the product of the total mass and the acceleration of the center of mass.

Moment–Angular Momentum Principles

We now obtain relations between the sum of the moments due to the forces acting on a system of particles and the rate of change of the total angular momentum of the system. In Fig. 18.3, $\mathbf{r}_i$ is the position vector of the ith particle of a system of particles, $\mathbf{r}$ is the position vector of the center of mass of the system, and $\mathbf{R}_i$ is the position vector of the ith particle relative to the center of mass. These vectors are related by

$$\mathbf{r}_i = \mathbf{r} + \mathbf{R}_i. \tag{18.5}$$

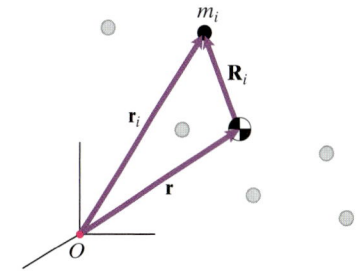

Figure 18.3
The vector $\mathbf{R}_i$ is the position vector of the ith particle relative to the center of mass.

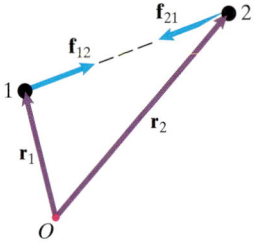

Figure 18.4
Particles 1 and 2 and the forces they exert on each other. If the forces act along the line between the particles, their total moment about O is zero.

We take the cross product of Newton's second law for the ith particle, Eq. (18.1), with the position vector $\mathbf{r}_i$ and sum from $i = 1$ to N, writing the resulting equation in the form

$$\sum_i \sum_j \mathbf{r}_i \times \mathbf{f}_{ij} + \sum_i \mathbf{r}_i \times \mathbf{f}_i^E = \frac{d}{dt} \sum_i \mathbf{r}_i \times m_i \mathbf{v}_i. \qquad (18.6)$$

The first term on the left side of this equation is the sum of the moments about O due to the forces exerted on the particles by the other particles of the system. This term vanishes if we assume that the mutual forces exerted by each pair of particles are not only equal and opposite, but directed along the straight line between the particles. For example, consider particles 1 and 2 in Fig. 18.4. If the forces the particles exert on each other are directed along the line between the particles, we can write the moment about O as

$$\mathbf{r}_1 \times \mathbf{f}_{12} + \mathbf{r}_1 \times \mathbf{f}_{21} = \mathbf{r}_1 \times (\mathbf{f}_{12} + \mathbf{f}_{21}) = \mathbf{0}. \qquad (18.7)$$

The second term on the left side of Eq. (18.6) is the sum of the moments about O due to external forces, which we denote by $\Sigma \mathbf{M}_O$. We write Eq. (18.6) as

$$\Sigma \mathbf{M}_O = \frac{d\mathbf{H}_O}{dt}, \qquad (18.8)$$

where

$$\mathbf{H}_O = \sum_i \mathbf{r}_i \times m_i \mathbf{v}_i \qquad (18.9)$$

is the total angular momentum about O. The sum of the moments about O is equal to the rate of change of the total angular momentum about O. By using Eqs. (18.4) and (18.5), we can write Eq. (18.8) as

$$\Sigma \mathbf{M}_O = \frac{d}{dt}(\mathbf{r} \times m\mathbf{v} + \mathbf{H}), \qquad (18.10)$$

where

$$\mathbf{H} = \sum_i \mathbf{R}_i \times m_i \frac{d\mathbf{R}_i}{dt} \qquad (18.11)$$

is the total angular momentum of the system about the center of mass.

We also need to determine the relation between the sum of the moments about the center of mass, which we denote by $\Sigma \mathbf{M}$, and $\mathbf{H}$. We can obtain this relation by letting the fixed point O be coincident with the center of mass at the present instant. In that case, $\Sigma \mathbf{M}_O = \Sigma \mathbf{M}$ and $\mathbf{r} = \mathbf{0}$, and we see from Eq. (18.10) that

$$\Sigma \mathbf{M} = \frac{d\mathbf{H}}{dt}. \qquad (18.12)$$

The sum of the moments about the center of mass is equal to the rate of change of the total angular momentum about the center of mass. The angular momenta about point O and about the center of mass are related by (Fig. 18.5)

$$\mathbf{H}_O = \mathbf{H} + \mathbf{r} \times m\mathbf{v}. \tag{18.13}$$

18.3 Derivation of the Equations of Motion

We now derive the equations of motion for a rigid body in planar motion. We have shown that the total external force on any object equals the product of the mass of the object and the acceleration of its center of mass:

$$\Sigma \mathbf{F} = m\mathbf{a}.$$

This equation, which we refer to as Newton's second law, describes the motion of the center of mass of a rigid body. To derive the equations of angular motion, we consider first rotation about a fixed axis and then general planar motion.

Rotation about a Fixed Axis

Let O be a point that is stationary relative to an inertial reference frame, and let L_O be a nonrotating line through O. Suppose that a rigid body rotates about L_O. In terms of a coordinate system with its z axis aligned with L_O (Fig. 18.6a), we can express the rigid body's angular velocity vector as $\boldsymbol{\omega} = \omega\mathbf{k}$, and the velocity of the ith particle of the rigid body is

$$\frac{d\mathbf{r}_i}{dt} = \boldsymbol{\omega} \times \mathbf{r}_i = \omega\mathbf{k} \times \mathbf{r}_i.$$

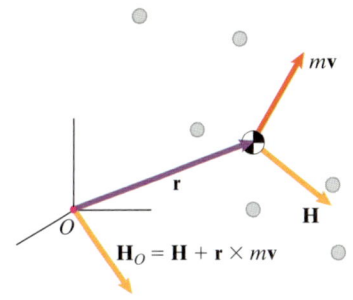

Figure 18.5
The angular momentum about O equals the sum of the angular momentum about the center of mass and the angular momentum about O due to the velocity of the center of mass.

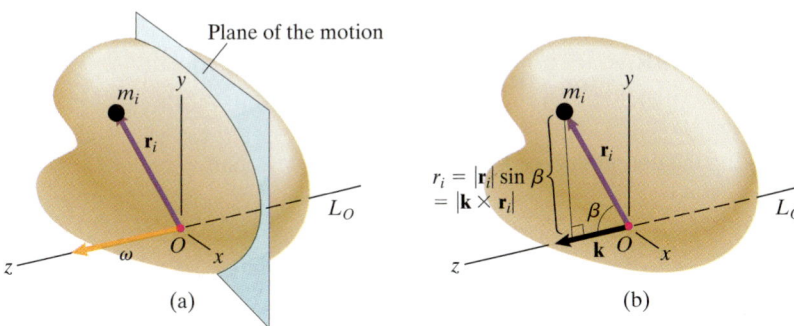

(a)   (b)

Figure 18.6
(a) A coordinate system with the z axis aligned with the axis of rotation, L_O.
(b) The magnitude of $\mathbf{k} \times \mathbf{r}_i$ is the perpendicular distance from the axis of rotation to m_i.

Let $\Sigma M_O = \Sigma\mathbf{M}_O \cdot \mathbf{k}$ denote the sum of the moments about L_O. By taking the dot product of Eq. (18.8) with $\mathbf{k}$, we obtain

$$\Sigma M_O = \frac{dH_O}{dt}, \tag{18.14}$$

where

$$H_O = \mathbf{H}_O \cdot \mathbf{k} = \sum_i [\mathbf{r}_i \times m_i(\omega\mathbf{k} \times \mathbf{r}_i)] \cdot \mathbf{k} \tag{18.15}$$

is the angular momentum about L_O. Using the identity $\mathbf{U} \cdot (\mathbf{V} \times \mathbf{W}) = (\mathbf{U} \times \mathbf{V}) \cdot \mathbf{W}$, we can write Eq. (18.15) as

$$H_O = \sum_i m_i(\mathbf{k} \times \mathbf{r}_i) \cdot (\mathbf{k} \times \mathbf{r}_i)\omega = \sum_i m_i|\mathbf{k} \times \mathbf{r}_i|^2\omega. \tag{18.16}$$

In Fig. 18.6b, we show that $|\mathbf{k} \times \mathbf{r}_i|$ is the perpendicular distance from L_O to the ith particle, which we denote by r_i. Using the definition of the moment of inertia of the rigid body about L_O,

$$I_O = \sum_i m_i r_i^2,$$

we can write Eq. (18.16) as

$$H_O = I_O \omega.$$

Substituting this expression into Eq. (18.14), we obtain the equation of angular motion for a rigid body rotating about a fixed axis. The sum of the moments about the fixed axis equals the product of the moment of inertia about the fixed axis and the angular acceleration:

$$\Sigma M_O = I_O \alpha. \tag{18.17}$$

General Planar Motion

Figure 18.7a shows the plane of the motion of a rigid body in general planar motion. Point O is a fixed point contained in the plane, L_O is the line through O that is perpendicular to the plane, and L is the line parallel to L_O that passes through the center of mass of the rigid body. In terms of the coordinate system shown, we can express the rigid body's angular velocity vector as $\boldsymbol{\omega} = \omega \mathbf{k}$, and the velocity of the ith particle of the rigid body relative to the center of mass is

$$\frac{d\mathbf{R}_i}{dt} = \boldsymbol{\omega} \times \mathbf{R}_i = \omega \mathbf{k} \times \mathbf{R}_i.$$

By taking the dot product of Eq. (18.10) with $\mathbf{k}$, we obtain

$$\Sigma M_O = \frac{d}{dt}[(\mathbf{r} \times m\mathbf{v}) \cdot \mathbf{k} + H], \tag{18.18}$$

where

$$H = \mathbf{H} \cdot \mathbf{k} = \sum_i [\mathbf{R}_i \times m_i(\omega \mathbf{k} \times \mathbf{R}_i)] \cdot \mathbf{k}$$

is the angular momentum about L. Using the same identity we applied to Eq. (18.15), we can write this equation for H as

$$H = \sum_i m_i(\mathbf{k} \times \mathbf{R}_i) \cdot (\mathbf{k} \times \mathbf{R}_i)\omega = \sum_i m_i |\mathbf{k} \times \mathbf{R}_i|^2 \omega.$$

The term $|\mathbf{k} \times \mathbf{R}_i| = r_i$ is the perpendicular distance from L to the ith particle (Fig. 18.7b). In terms of the moment of inertia of the rigid body about L,

$$I = \sum_i m_i r_i^2,$$

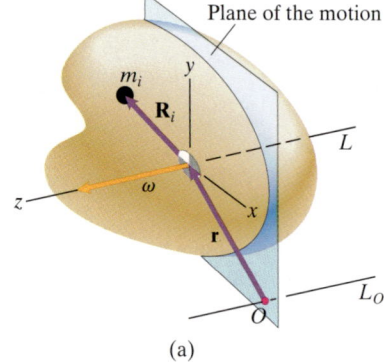

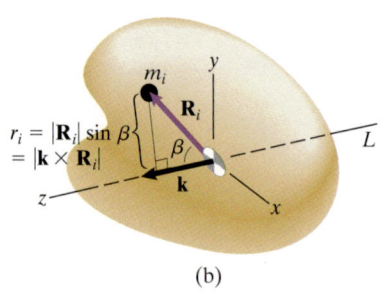

Figure 18.7
(a) A coordinate system with the z axis aligned with L.
(b) The magnitude of $\mathbf{k} \times \mathbf{R}_i$ is the perpendicular distance from L to m_i.

the rigid body's angular momentum about L is

$$H = I\omega.$$

Substituting this expression into Eq. (18.18), we obtain

$$\Sigma M_O = \frac{d}{dt}[(\mathbf{r} \times m\mathbf{v}) \cdot \mathbf{k} + I\omega] = (\mathbf{r} \times m\mathbf{a}) \cdot \mathbf{k} + I\alpha. \quad (18.19)$$

With this equation we can obtain the relation between the sum of the moments about L, which we denote by ΣM, and the angular acceleration. If we let the fixed axis L_O be coincident with L at the present instant, then $\Sigma M_O = \Sigma M$ and $\mathbf{r} = \mathbf{0}$, and from Eq. (18.19) we obtain the equation of angular motion for a rigid body in general planar motion. The sum of the moments about the center of mass equals the product of the moment of inertia about the center of mass and the angular acceleration:

$$\Sigma M = I\alpha. \quad (18.20)$$

18.4 Applications

We have seen that the equations of motion for a rigid body in planar motion include Newton's second law,

$$\Sigma \mathbf{F} = m\mathbf{a}, \quad (18.21)$$

where $\mathbf{a}$ is the acceleration of the center of mass of the body, and an equation relating the moments due to forces and couples to the angular acceleration of the body. If the rigid body rotates about a fixed axis O, the total moment about O equals the product of the moment of inertia about O and the angular acceleration:

$$\Sigma M_O = I_O\alpha. \quad (18.22)$$

In *any* planar motion, the total moment about the center of mass equals the product of the moment of inertia about the center of mass and the angular acceleration:

$$\Sigma M = I\alpha. \quad (18.23)$$

Of course, this equation applies to the case of rotation about a fixed axis, but for that type of motion it is often more convenient to use Eq. (18.22).

These equations can be used to obtain information about an object's motion, or to determine the values of unknown forces or couples acting on the object, or both. Doing so typically involves three steps:

1. *Draw the free-body diagram.* Isolate the object and identify the external forces and couples acting on it.
2. *Apply the equations of motion.* Write equations of motion suitable for the type of motion, choosing an appropriate coordinate system for applying Newton's second law. For example, if the center of mass moves in a circular path, it may be advantageous to use normal and tangential components.

3. *Determine kinematic relationships.* If necessary, supplement the equations of motion with relationships between the acceleration of the center of mass and the angular acceleration of the object.

As we show in the sections that follow, the best approach depends in part on the type of motion involved.

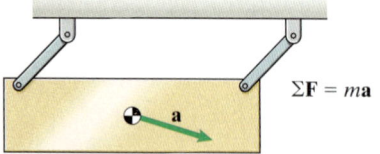

Translation

$\Sigma \mathbf{F} = m\mathbf{a}$

If a rigid body is in translation (Fig. 18.8), only Newton's second law is required to determine its motion. There is no rotational motion to determine. Nevertheless, it may be necessary to apply the angular equation of motion to determine unknown forces or couples. Since $\alpha = 0$, Eq. (18.23) states that the total moment *about the center of mass* equals zero:

Figure 18.8
A rigid body in translation. There is no rotational motion to determine.

$$\Sigma M = 0.$$

Example 18.1 **Translating Object**

The mass of the airplane in Fig. 18.9 is $m = 250,000$ kg, and the thrust of the engines during the plane's takeoff roll is $T = 700$ kN. Determine the acceleration of the airplane and the normal forces exerted on its wheels at A and B. Neglect the horizontal forces exerted on its wheels.

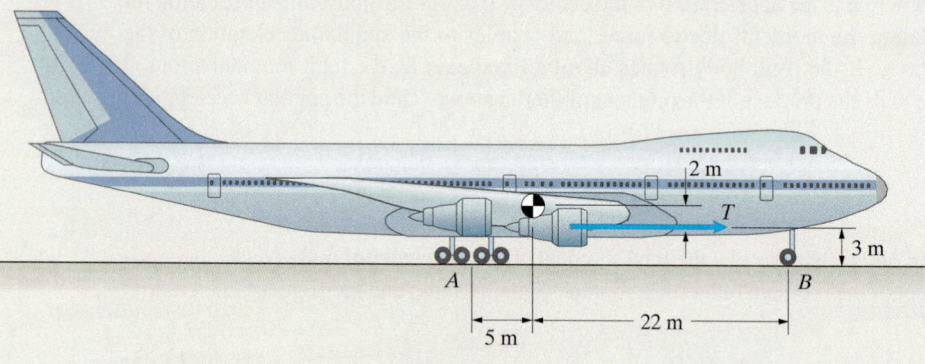

Figure 18.9

Strategy
The airplane is in translation during its takeoff roll, so the sum of the moments about its center of mass equals zero. Using this condition and Newton's second law, we can determine the acceleration of the plane and the normal forces exerted on its wheels.

Solution
Draw the Free-Body Diagram We draw the free-body diagram in Fig. a, showing the airplane's weight and the normal forces A and B exerted on the wheels.

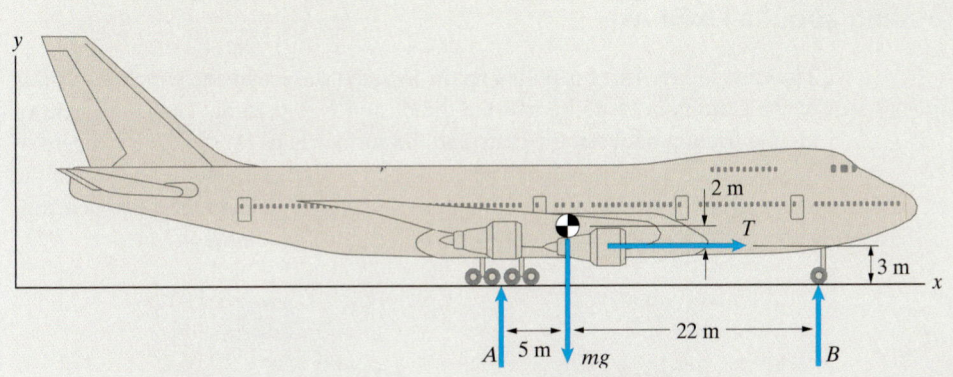

(a) Free-body diagram of the airplane.

Apply the Equations of Motion In terms of the coordinate system in Fig. a, Newton's second law is

$$\Sigma F_x = T = ma_x,$$

$$\Sigma F_y = A + B - mg = 0.$$

From the first equation, the airplane's acceleration is

$$a_x = \frac{T}{m} = \frac{700{,}000 \text{ N}}{250{,}000 \text{ kg}} = 2.8 \text{ m/s}^2.$$

The equation of angular motion is

$$\Sigma M = (2 \text{ m})T + (22 \text{ m})B - (5 \text{ m})A = 0.$$

Solving this equation together with the second equation we obtained from Newton's second law yields $A = 2050$ kN and $B = 402$ kN.

Critical Thinking

When an object is in equilibrium, the sum of the moments about any point due to the external forces and couples acting on it is zero. When a translating rigid body is not in equilibrium, you know only that the sum of the moments *about the center of mass* is zero. It would be instructive to try reworking this example by assuming that the sum of the moments about A or B is zero. You will not obtain the correct values for the normal forces exerted on the wheels.

Rotation about a Fixed Axis

In the case of rotation about a fixed axis (Fig. 18.10), only Eq. (18.22) is required to determine the rotational motion, although it may be necessary to use Newton's second law to determine unknown forces or couples.

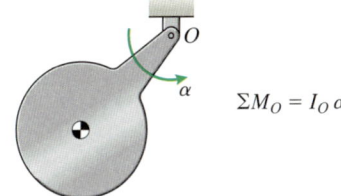

$$\Sigma M_O = I_O \alpha$$

Figure 18.10
A rigid body rotating about O. You need only the equation of angular motion about O to determine the angular acceleration of the body.

Example 18.2 | **Object Rotating about a Fixed Axis**

The crate in Fig. 18.11 is pulled up the inclined surface by the winch. The mass of the crate is $m = 45$ kg and the dimension $b = 0.15$ m. The coefficient of kinetic friction between the crate and the surface is $\mu_k = 0.4$. The moment of inertia of the drum on which the cable is being wound, including the part of the cable already wound on the drum, is $I_A = 4$ kg-m². If the motor exerts a couple $M = 50$ N-m on the drum, what is the crate's acceleration?

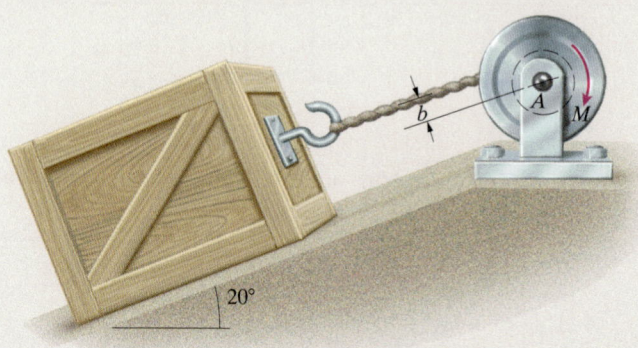

Figure 18.11

Strategy
We will draw separate free-body diagrams of the crate and drum and apply the equations of motion to them individually. The drum rotates about a fixed axis, so we can use the equation of angular motion about the axis to determine its angular acceleration. To complete the solution, we must determine the relationship between the crate's acceleration and the drum's angular acceleration.

Solution
Draw the Free-Body Diagrams We draw the free-body diagrams in Fig. a, showing the equal forces exerted on the crate and the drum by the cable.

Apply the Equations of Motion We denote the crate's acceleration up the inclined surface by a_x and the *clockwise* angular acceleration of the drum by α (Fig. b). Newton's second law for the crate yields the equations

$$\Sigma F_x = T - mg \sin 20° - \mu_k N = ma_x,$$

$$\Sigma F_y = N - mg \cos 20° = 0.$$

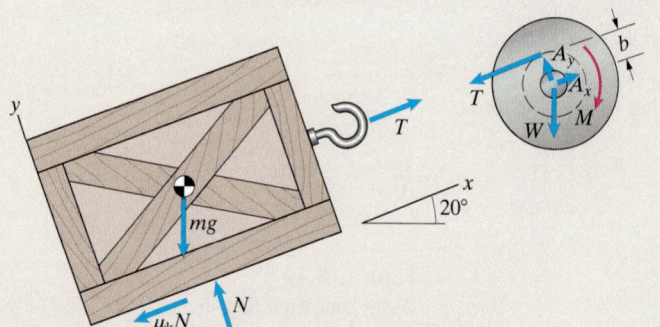

(a) Free-body diagrams of the crate and the drum.

(b) The crate's acceleration and the angular acceleration of the drum.

The equation of angular motion for the drum is

$$\Sigma M_A = M - bT = I_A\alpha.$$

From the equations of motion, we have obtained three equations in four unknowns: T, N, a_x, and α. Eliminating T and N from these equations, we get one equation in terms of a_x and α:

$$M - bmg(\sin 20° + \mu_k \cos 20°) - bma_x = I_A\alpha. \qquad (1)$$

To complete the solution, we must determine the kinematic relationship between a_x and α.

Determine Kinematic Relationship   The tangential component of acceleration of the drum at the point where the cable begins winding onto it is equal to the crate's acceleration (Fig. b):

$$a_x = b\alpha.$$

Solving this equation together with Eq. (1), we find that the crate's acceleration is

$$a_x = \frac{M - bmg(\sin 20° + \mu_k \cos 20°)}{bm + I_A/b}$$

$$= \frac{(50 \text{ N-m}) - (0.15 \text{ m})(45 \text{ kg})(9.81 \text{ m/s}^2)(\sin 20° + 0.4 \cos 20°)}{(0.15 \text{ m})(45 \text{ kg}) + (4 \text{ kg-m}^2)/(0.15 \text{ m})}$$

$$= 0.0737 \text{ m/s}^2.$$

Critical Thinking

Why did we define the angular acceleration of the drum to be positive in the clockwise direction? That choice meant that a positive acceleration a_x of the crate up the inclined surface corresponded to a positive angular acceleration α of the drum, and they were related by $a_x = b\alpha$. Notice that our equation of angular motion for the drum states that the sum of the *clockwise* moments about A equals the product of the moment of inertia about A and the angular acceleration. Although we felt that this choice simplified the presentation of this particular example, we could have followed the usual convention and defined the angular acceleration of the drum to be positive in the counterclockwise direction. In that case, the acceleration of the crate up the inclined surface would be related to the drum's angular acceleration by $a_x = -b\alpha$, and the equation of angular motion for the drum would state that the sum of the *counterclockwise* moments about A equals the product of the moment of inertia about A and the angular acceleration. You can confirm that this results in the same solution for the crate's acceleration.

Example 18.3 Bar Rotating about a Fixed Axis

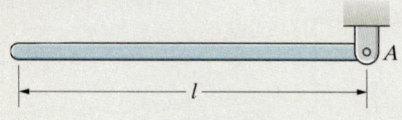

Figure 18.12

The slender bar of mass m in Fig. 18.12 is released from rest in the horizontal position shown. Determine the bar's angular acceleration and the force exerted on the bar by the support A at that instant.

Strategy
Since the bar rotates about a fixed point, we can use Eq. (18.22) to determine its angular acceleration. The advantage of using this equation instead of Eq. (18.23) is that the unknown reactions at A will not appear in the equation of angular motion. Once we know the angular acceleration, we can determine the acceleration of the center of mass and use Newton's second law to obtain the reactions at A.

Solution
Draw the Free-Body Diagram   In Fig. a, we draw the free-body diagram of the bar, showing the reactions at the pin support.

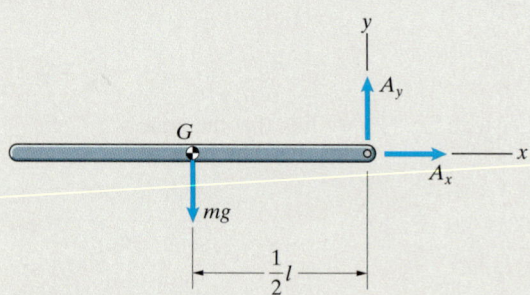

(a) Free-body diagram of the bar.

Apply the Equations of Motion   Let the acceleration of the center of mass G of the bar be $\mathbf{a}_G = a_x\mathbf{i} + a_y\mathbf{j}$, and let its counterclockwise angular acceleration be α (Fig. b). Newton's second law for the bar is

$$\Sigma F_x = A_x = ma_x,$$

$$\Sigma F_y = A_y - mg = ma_y.$$

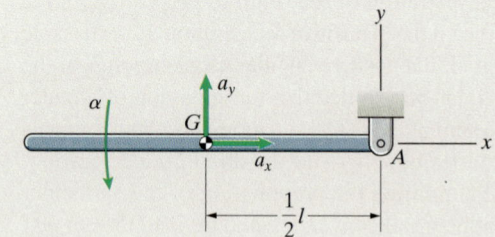

(b) The angular acceleration and components of the acceleration of the center of mass.

The equation of angular motion about the fixed point A is

$$\Sigma M_A = \left(\frac{1}{2}l\right)mg = I_A\alpha. \tag{1}$$

The moment of inertia of a slender bar about its center of mass is $I = \frac{1}{12}ml^2$. (See Appendix C.) From the parallel-axis theorem, the moment of inertia of the bar about A is

$$I_A = I + d^2 m = \frac{1}{12}ml^2 + \left(\frac{1}{2}l\right)^2 m = \frac{1}{3}ml^2.$$

Substituting this expression into Eq. (1), we obtain the angular acceleration:

$$\alpha = \frac{\frac{1}{2}mgl}{\frac{1}{3}ml^2} = \frac{3}{2}\frac{g}{l}.$$

Determine Kinematic Relationships   To determine the reactions A_x and A_y, we need to determine the acceleration components a_x and a_y. We can do so by expressing the acceleration of G in terms of the acceleration of A:

$$\mathbf{a}_G = \mathbf{a}_A + \boldsymbol{\alpha} \times \mathbf{r}_{G/A} - \omega^2 \mathbf{r}_{G/A}.$$

At the instant the bar is released, its angular velocity $\omega = 0$. Also, $\mathbf{a}_A = \mathbf{0}$, so we obtain

$$\mathbf{a}_G = a_x\mathbf{i} + a_y\mathbf{j} = (\alpha\mathbf{k}) \times \left(-\frac{1}{2}l\mathbf{i}\right) = -\frac{1}{2}l\alpha\mathbf{j}.$$

Equating $\mathbf{i}$ and $\mathbf{j}$ components, we get

$$a_x = 0,$$

$$a_y = -\frac{1}{2}l\alpha = -\frac{3}{4}g.$$

Substituting these acceleration components into Newton's second law, we find that the reactions at A at the instant the bar is released are

$$A_x = 0,$$

$$A_y = mg + m\left(-\frac{3}{4}g\right) = \frac{1}{4}mg.$$

Critical Thinking

We could have determined the kinematic relationship between the bar's angular acceleration and the acceleration of the center of mass G in a less formal way. Because G describes a circular path about A, we know that the tangential component of its acceleration equals the product of the radial distance from A to G and the angular acceleration of the bar. Due to the directions in which we defined a_y and α to be positive, $a_y = -(l/2)\alpha$. The normal component of the acceleration of G is equal to the square of its velocity divided by the radius of its circular path. The velocity equals zero at the instant the bar is released, so $a_x = 0$.

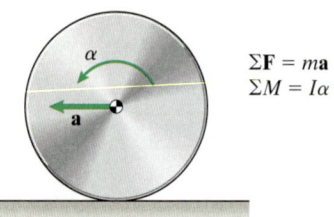

$$\Sigma \mathbf{F} = m\mathbf{a}$$
$$\Sigma M = I\alpha$$

Figure 18.13
A rigid body in planar motion. You must apply both Newton's second law and the equation of angular motion about the center of mass.

General Planar Motion

If a rigid body undergoes general planar motion, both Newton's second law and the equation of angular motion are required to determine its motion. If the motion of the center of mass and the rotational motion are not independent—the rolling disk in Fig. 18.13 is an example—there will be more unknown quantities than equations of motion. In such cases, it is necessary to obtain additional equations by deriving kinematic relationships between the acceleration of the center of mass and the angular acceleration.

Example 18.4 | **Bar in General Planar Motion**

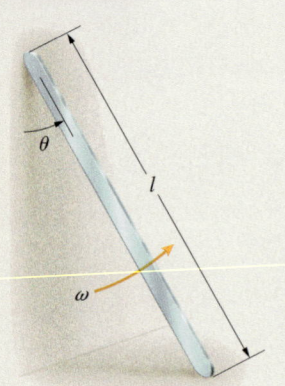

Figure 18.14

The slender bar of mass m in Fig. 18.14 slides on the smooth floor and wall and has counterclockwise angular velocity ω at the instant shown. What is the bar's angular acceleration?

Strategy

We must draw the free-body diagram of the bar and apply Newton's second law and the equation of angular motion. This will result in three equations in terms of the two components of acceleration of the bar's center of mass, the bar's angular acceleration, and the unknown forces exerted on the bar by the floor and the wall. Two kinematic relations will be needed to complete the solution.

Solution

Draw the Free-Body Diagram We draw the free-body diagram in Fig. a, showing the bar's weight and the normal forces exerted by the floor and wall.

Apply the Equations of Motion Writing the acceleration of the center of mass G as $\mathbf{a}_G = a_x\mathbf{i} + a_y\mathbf{j}$ (Fig. b), we obtain from Newton's second law

$$\Sigma F_x = P = ma_x,$$

$$\Sigma F_y = N - mg = ma_y.$$

Let α be the bar's counterclockwise angular acceleration (Fig. b). The equation of angular motion is

$$\Sigma M = N\left(\frac{1}{2}l \sin \theta\right) - P\left(\frac{1}{2}l \cos \theta\right) = I\alpha,$$

where I is the moment of inertia of the bar about its center of mass. We have three equations of motion in terms of five unknowns: P, N, a_x, a_y, and α. To complete the solution, we must relate the acceleration of the center of mass of the bar to its angular acceleration.

Determine Kinematic Relationships Although we don't know the accelerations of the endpoints A and B (Fig. c), we know that A moves horizontally and B moves vertically. We can use this information to obtain the needed relations between the acceleration of the center of mass and the angular acceleration.

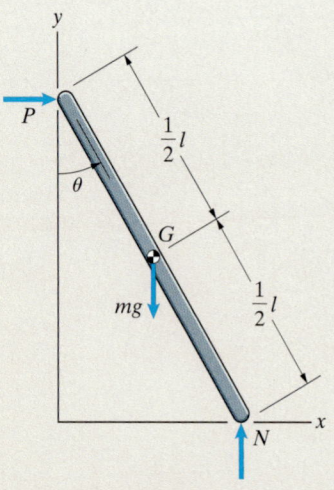

(a) Free-body diagram of the bar.

Expressing the acceleration of A as $\mathbf{a}_A = a_A\mathbf{i}$, we can write the acceleration of the center of mass as (Figs. b and c)

$$\mathbf{a}_G = \mathbf{a}_A + \boldsymbol{\alpha} \times \mathbf{r}_{G/A} - \omega^2\mathbf{r}_{G/A}:$$

$$a_x\mathbf{i} + a_y\mathbf{j} = a_A\mathbf{i} + \begin{vmatrix} \mathbf{i} & \mathbf{j} & \mathbf{k} \\ 0 & 0 & \alpha \\ -\dfrac{1}{2}l\sin\theta & \dfrac{1}{2}l\cos\theta & 0 \end{vmatrix} - \omega^2\left(-\frac{1}{2}l\sin\theta\mathbf{i} + \frac{1}{2}l\cos\theta\mathbf{j}\right).$$

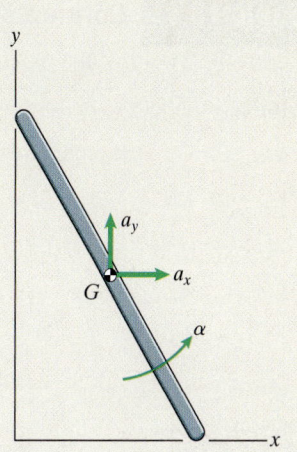

Taking advantage of the fact that $\mathbf{a}_A$ has no $\mathbf{j}$ component, we equate the $\mathbf{j}$ components in this equation, obtaining

$$a_y = -\frac{1}{2}l(\alpha\sin\theta + \omega^2\cos\theta).$$

Now we express the acceleration of B as $\mathbf{a}_B = a_B\mathbf{j}$ and write the acceleration of the center of mass as

$$\mathbf{a}_G = \mathbf{a}_B + \boldsymbol{\alpha} \times \mathbf{r}_{G/B} - \omega^2\mathbf{r}_{G/B}:$$

$$a_x\mathbf{i} + a_y\mathbf{j} = a_B\mathbf{j} + \begin{vmatrix} \mathbf{i} & \mathbf{j} & \mathbf{k} \\ 0 & 0 & \alpha \\ \dfrac{1}{2}l\sin\theta & -\dfrac{1}{2}l\cos\theta & 0 \end{vmatrix} - \omega^2\left(\frac{1}{2}l\sin\theta\mathbf{i} - \frac{1}{2}l\cos\theta\mathbf{j}\right).$$

We equate the $\mathbf{i}$ components in this equation, obtaining

$$a_x = \frac{1}{2}l(\alpha\cos\theta - \omega^2\sin\theta).$$

With these two kinematic relationships, we have five equations in five unknowns. Solving them for the angular acceleration and using the relation $I = \frac{1}{12}ml^2$ for the bar's moment of inertia (Appendix C), we obtain

$$\alpha = \frac{3}{2}\frac{g}{l}\sin\theta.$$

(b) Components of the acceleration of the center of mass and the angular acceleration.

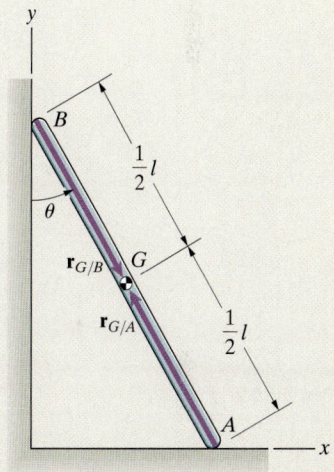

(c) Position vectors of G relative to the endpoints A and B.

Critical Thinking

Why did we express the acceleration of the center of mass G in terms of the acceleration of point A and then express the acceleration of G in terms of the acceleration of point B? One answer is pragmatic. Doing so resulted in equations that allowed us to solve the problem. But notice that those steps introduced into our formulation the effects of the floor and wall on the bar's motion. By writing the acceleration of A as $\mathbf{a}_A = a_A\mathbf{i}$, we introduced the constraint that A moves horizontally, and by writing the acceleration of B as $\mathbf{a}_B = a_B\mathbf{j}$, we introduced the constraint that B moves vertically.

Example 18.5 | **Connected Rigid Bodies**

The slender bar in Fig. 18.15 has mass m and is pinned at A to a metal block of mass m_B that rests on a smooth, level surface. The system is released from rest in the position shown. What is the bar's angular acceleration at the instant of release?

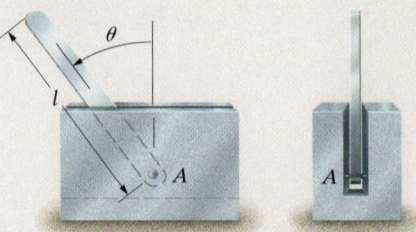

Figure 18.15

Strategy

We must draw free-body diagrams of the bar and the block and apply the equations of motion to them individually. To complete the solution, we must also relate the acceleration of the bar's center of mass and its angular acceleration to the acceleration of the block.

Solution

Draw the Free-Body Diagrams We draw the free-body diagrams of the bar and block in Fig. a. Notice the opposite forces they exert on each other where they are pinned together.

Apply the Equations of Motion Writing the acceleration of the center of mass of the bar as $\mathbf{a}_G = a_x\mathbf{i} + a_y\mathbf{j}$ (Fig. b), we have, from Newton's second law,

$$\Sigma F_x = A_x = ma_x$$

and

$$\Sigma F_y = A_y - mg = ma_y.$$

Letting α be the counterclockwise angular acceleration of the bar (Fig. b), its equation of angular motion is

$$\Sigma M = A_x\left(\frac{1}{2}l\cos\theta\right) + A_y\left(\frac{1}{2}l\sin\theta\right) = I\alpha.$$

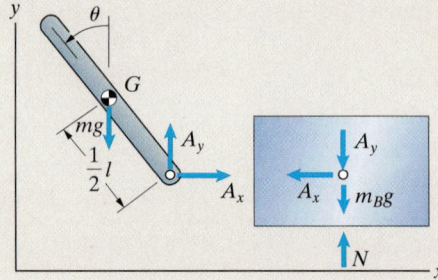

(a) Free-body diagrams of the bar and block.

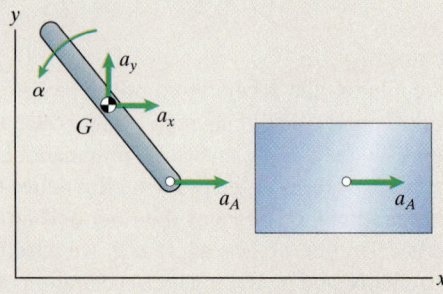

(b) Definitions of the accelerations.

We express the block's acceleration as $a_A\mathbf{i}$ (Fig. b) and write Newton's second law for the block:

$$\Sigma F_x = -A_x = m_B a_A,$$

$$\Sigma F_y = N - A_y - m_B g = 0.$$

Determine Kinematic Relationships To relate the bar's motion to that of the block, we express the acceleration of the bar's center of mass in terms of the acceleration of point A (Figs. b and c):

$$\mathbf{a}_G = \mathbf{a}_A + \boldsymbol{\alpha} \times \mathbf{r}_{G/A} - \omega^2 \mathbf{r}_{G/A},$$

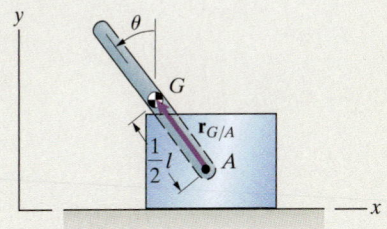

$$a_x\mathbf{i} + a_y\mathbf{j} = a_A\mathbf{i} + \begin{vmatrix} \mathbf{i} & \mathbf{j} & \mathbf{k} \\ 0 & 0 & \alpha \\ -\dfrac{1}{2}l\sin\theta & \dfrac{1}{2}l\cos\theta & 0 \end{vmatrix} - \mathbf{0}.$$

(c) Position vector of G relative to A.

Equating $\mathbf{i}$ and $\mathbf{j}$ components, we obtain

$$a_x = a_A - \frac{1}{2}l\alpha\cos\theta$$

and

$$a_y = -\frac{1}{2}l\alpha\sin\theta.$$

We have five equations of motion and two kinematic relations in terms of seven unknowns: A_x, A_y, N, a_x, a_y, α, and a_A. Solving them for the angular acceleration and using the relation $I = \frac{1}{12}ml^2$ for the bar's moment of inertia, we obtain

$$\alpha = \frac{\frac{3}{2}(g/l)\sin\theta}{1 - \frac{3}{4}\big[m/(m + m_B)\big]\cos^2\theta}.$$

Critical Thinking

This example is a simple case of a very important type of problem. The approach we used is applicable to analyzing planar motions of a broad class of machines consisting of interconnected moving parts. The free-body diagrams of the individual parts are first drawn, including the forces and couples the parts exert on each other. The equations of motion are written for each part. The solution is completed by determining kinematic relationships. In this example, the block is constrained to move horizontally. The block and bar have the same acceleration at the pin A, and the acceleration of the center of mass G of the bar is related to the bar's angular acceleration and the acceleration of point A because the bar is pinned at A. As this example illustrates, determining kinematic relationships is usually the most challenging part of the solution.

Example 18.6 | Wheel With an Offset Center of Mass

The drive wheel in Fig. 18.16 rolls on the horizontal track. The wheel is subjected to a downward force F_A by its axle A and a horizontal force F_C by the connecting rod. The mass of the wheel is m and the moment of inertia about its center of mass is I. The center of mass G is offset a distance b from the wheel's center. At the instant shown, the wheel has a counterclockwise angular velocity ω. What is the wheel's angular acceleration?

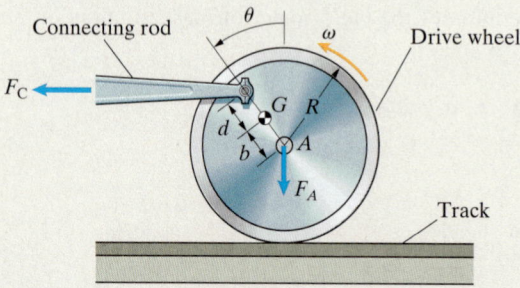

Figure 18.16

Strategy

By drawing the free-body diagram of the wheel and applying Newton's second law and the equation of angular motion, we can obtain three equations in terms of the two components of acceleration of the wheel's center of mass, the wheel's angular acceleration, and the unknown normal and friction forces exerted on the wheel by the floor. We must obtain two kinematic relations to complete the solution.

Solution

Draw the Free-Body Diagram We draw the free-body diagram of the drive wheel in Fig. a, showing its weight and the normal and frictional forces exerted by the track.

Apply the Equations of Motion Writing the acceleration of the center of mass G as $\mathbf{a}_G = a_x\mathbf{i} + a_y\mathbf{j}$ (Fig. b), we have, from Newton's second law,

$$\Sigma F_x = f - F_C = ma_x$$

and

$$\Sigma F_y = N - F_A - mg = ma_y.$$

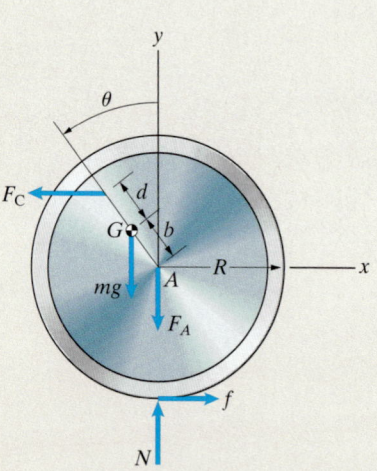

(a) Free-body diagram of the wheel.

Remember that we must express the equation of angular motion in terms of the sum of the moments about the center of mass, G, *not the center of the wheel*. The equation of angular motion is (Figs. a and b)

$$\Sigma M = F_C(d\cos\theta) - F_A(b\sin\theta) + N(b\sin\theta) + f(b\cos\theta + R)$$

$$= I\alpha.$$

We have three equation of motion in terms of five unknowns: N, f, a_x, a_y, and α. To complete the solution, we must relate the acceleration of the wheel's center of mass to its angular acceleration.

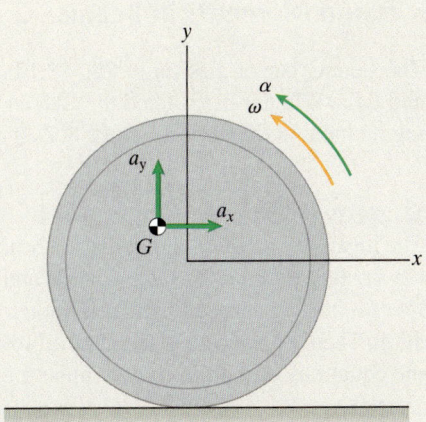

(b) Components of the acceleration of G, the angular velocity, and the angular acceleration.

Determine Kinematic Relationships   The acceleration of the center A of the rolling wheel is $\mathbf{a}_A = -R\alpha\mathbf{i}$. By expressing the acceleration of the center of mass $\mathbf{a}_G$ in terms of $\mathbf{a}_A$ (Figs. b and c), we obtain relations between the components of $\mathbf{a}_G$ and α.

$$\mathbf{a}_G = \mathbf{a}_A + \boldsymbol{\alpha} \times \mathbf{r}_{G/A} - \omega^2 \mathbf{r}_{G/A}:$$

$$a_x\mathbf{i} + a_y\mathbf{j} = -R\alpha\mathbf{i} + \begin{vmatrix} \mathbf{i} & \mathbf{j} & \mathbf{k} \\ 0 & 0 & \alpha \\ -b\sin\theta & b\cos\theta & 0 \end{vmatrix} - \omega^2(-b\sin\theta\mathbf{i} + b\cos\theta\mathbf{j}).$$

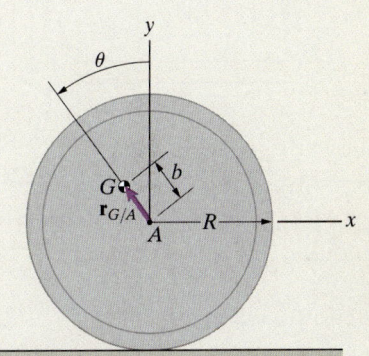

(c) Position vector of G relative to A.

We equate the $\mathbf{i}$ and $\mathbf{j}$ components in this equation, obtaining

$$a_x = -R\alpha - b\alpha\cos\theta + b\omega^2\sin\theta$$

and

$$a_y = -b\alpha\sin\theta - b\omega^2\cos\theta.$$

With these two kinematic relationships, we have five equations in five unknowns. Solving them for the angular acceleration, we obtain

$$\alpha = \frac{F_C(R + b\cos\theta + d\cos\theta) + mgb\sin\theta + mbR\omega^2\sin\theta}{m(b^2 + 2bR\cos\theta + R^2) + I}.$$

Critical Thinking

How did we know that expressing the acceleration of the center of mass G in terms of the acceleration of the center A of the wheel would result in the kinematic relations we needed to complete the solution? We wanted to relate the components of the acceleration of G to the wheel's angular acceleration and knew that the acceleration of A could be expressed in terms of the wheel's angular acceleration. You will often find that obtaining appropriate kinematic relations requires some trial and error.

Design Example 18.7 **Internal Forces and Moments in Beams**

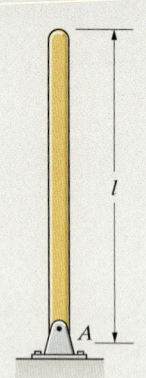

Figure 18.17

The slender bar of mass m in Fig. 18.17 starts from rest in the position shown and falls. When it has rotated through an angle θ, what is the maximum bending moment in the bar and where does it occur?

Strategy

The internal forces and moments in a beam subjected to two-dimensional loading are the axial force P, shear force V, and bending moment M (Fig. a). We must first use the equation of angular motion to determine the bar's angular acceleration. Then we can cut the bar at an arbitrary distance x from one end and apply the equations of motion to determine the bending moment as a function of x.

Solution

The moment of inertia of the bar about A is

$$I_A = I + d^2 m = \frac{1}{12} m l^2 + \left(\frac{1}{2} l\right)^2 m = \frac{1}{3} m l^2.$$

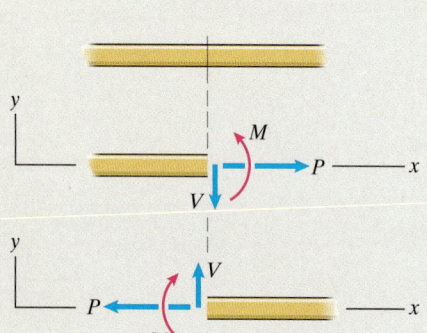

(a) The axial force, shear force, and bending moment in a beam.

When the bar has rotated through an angle θ (Fig. b), the total moment about A is $\Sigma M_A = mg\left(\frac{1}{2} l \sin \theta\right)$. Point A is fixed, so we can write the equation of angular motion as

$$\Sigma M_A = I_A \alpha:$$

$$\frac{1}{2} mgl \sin \theta = \frac{1}{3} m l^2 \alpha.$$

Solving for the angular acceleration, we obtain

$$\alpha = \frac{3}{2} \frac{g}{l} \sin \theta.$$

In Fig. c, we introduce a coordinate system, cut the bar at a distance x from the top, and draw the free-body diagram of the top part. The center of mass is at the midpoint, and we determine the mass of the free body by multiplying the bar's mass by the ratio of the length of the free body to that of the bar. Applying Newton's second law in the y direction, we obtain

$$\Sigma F_y = -V - \frac{x}{l} mg \sin \theta = \frac{x}{l} m a_y.$$

The moment of inertia of the free body about its center of mass is $\frac{1}{12}[(x/l)m]x^2$, so the equation of angular motion is

$$\Sigma M = I\alpha:$$

$$M - \left(\frac{1}{2} x\right) V = \frac{1}{12}\left(\frac{x}{l} m\right) x^2 \frac{3}{2} \frac{g}{l} \sin \theta.$$

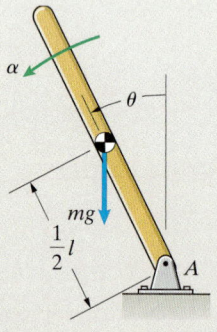

(b) Determining the moment about A.

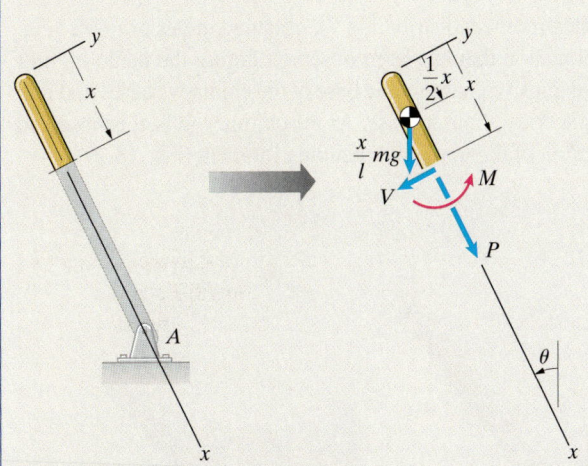

(c) Cutting the bar at an arbitrary distance x.

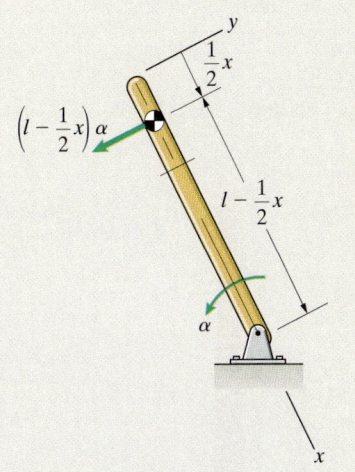

(d) Determining the acceleration of the center of mass of the free body.

The y component of the acceleration of the center of mass is equal to the product of its radial distance from A and the angular acceleration (Fig. d):

$$a_y = -\left(l - \frac{1}{2}x\right)\alpha = -\left(l - \frac{1}{2}x\right)\frac{3}{2}\frac{g}{l}\sin\theta.$$

Using this expression, we can solve the two equations of motion for V and M in terms of θ. The solution for M is

$$M = \frac{1}{4}mgl\sin\theta\left(\frac{x}{l}\right)^2\left(1 - \frac{x}{l}\right). \tag{1}$$

The bending moment equals zero at both ends of the bar. Taking the derivative of this expression with respect to x and equating it to zero to determine where M is a maximum, we obtain $x = \frac{2}{3}l$. Substituting this value of x into Eq. (1), we obtain the maximum bending moment:

$$M_{max} = \frac{1}{27}mgl\sin\theta.$$

The distribution of M is shown in Fig. 18.18.

Design Issues

To design a member of a structure, engineers must consider both the external and internal forces and moments to which the member will be subjected. In the case of a beam, they must determine the distributions of the axial force P, shear force V, and bending moment M as the first step in determining whether the beam will support its design loads without failing. If they know the external loads and reactions, and if the beam is in equilibrium, they can apply the equilibrium equations to determine the internal forces and moment at a given cross section. But in many situations, a beam will not be in equilibrium. It could be a member of a structure, such as the internal frame of an airplane, that is accelerating, or it could be a connecting rod in an internal combustion engine. In such cases, the maximum internal forces and moments can far exceed the values predicted by a static analysis, and the procedure we describe in this example must be used.

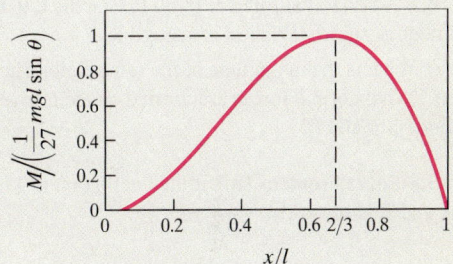

Figure 18.18
Distribution of the bending moment in a falling bar.

The dynamic bending moment distribution we obtained in this example (Fig. 18.18) explains a phenomenon that has been observed during the demolition of masonry chimneys. An explosive charge at the base of the chimney causes it to fall, initially rotating as a rigid body about its base. As the chimney falls, it is observed to fracture near the location of the maximum bending moment (Fig. 18.19).

Figure 18.19
A falling chimney fractures as it falls due to the bending moment to which it is subjected.

Problems

18.1 A horizontal force $F = 30$ lb is applied to the 230-lb refrigerator as shown. Friction is negligible.

(a) What is the magnitude of the refrigerator's acceleration?
(b) What normal forces are exerted on the refrigerator by the floor at A and B?

18.2 Solve Problem 18.1 if the coefficient of kinetic friction at A and B is $\mu_k = 0.1$.

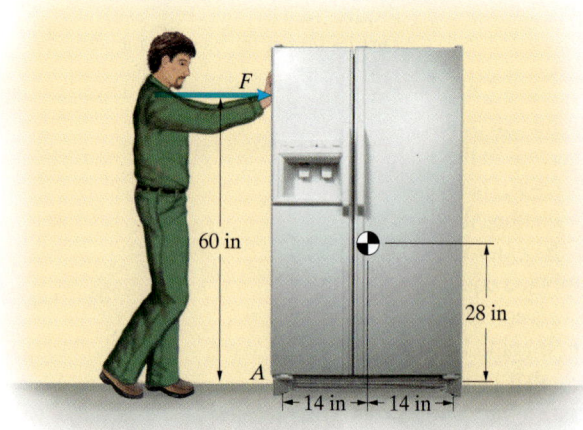

Problems 18.1/18.2

18.3 As the airplane begins its takeoff run, the normal forces exerted on its tires by the runway at A and B are $N_A = 720$ lb and $N_B = 1660$ lb. What is the magnitude of the airplane's acceleration?

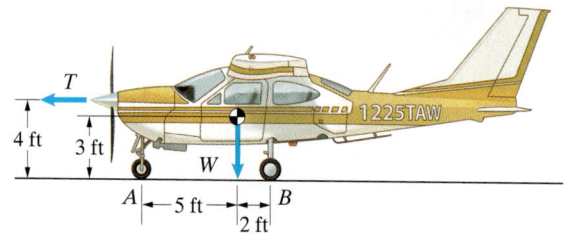

Problem 18.3

18.4 The mass of the Boeing 747 is 300,000 kg. As it begins its takeoff run, its four engines exert a total horizontal thrust $T = 670$ kN. Neglect horizontal forces exerted on the tires.

(a) What is the magnitude of the airplane's acceleration?
(b) What normal forces are exerted on the tires at A and B?

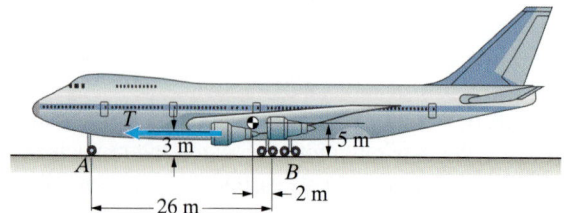

Problem 18.4

18.5 The crane moves to the right with constant acceleration, and the 800-kg load moves without swinging.

(a) What is the acceleration of the crane and load?
(b) What are the tensions in the cables attached at *A* and *B*?

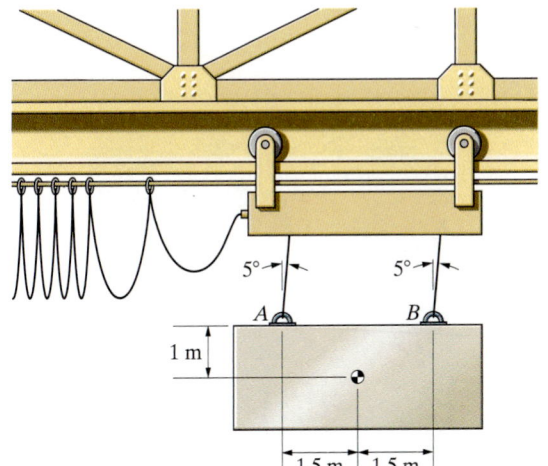

Problem 18.5

18.6 The total weight of the go-cart and driver is 240 lb. The location of their combined center of mass is shown. The rear drive wheels together exert a 24-lb horizontal force on the track. Neglect the horizontal forces exerted on the front wheels.

(a) What is the magnitude of the go-cart's acceleration?
(b) What normal forces are exerted on the tires at *A* and *B*?

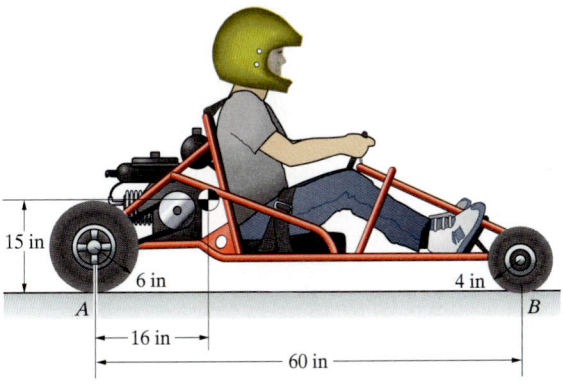

Problem 18.6

18.7 The total weight of the bicycle and rider is 160 lb. The location of their combined center of mass is shown. The dimensions shown are $b = 21$ in, $c = 16$ in, and $h = 38$ in. What is the largest acceleration the bicycle can have without the front wheel leaving the ground? Neglect the horizontal force exerted on the front wheel by the road.

Strategy: You want to determine the value of the acceleration that causes the normal force exerted on the front wheel by the road to equal zero.

18.8 The total mass of the bicycle and rider is 72 kg. The location of their combined center of mass is shown. The dimensions shown are $b = 530$ mm, $c = 400$ mm, and $h = 960$ mm. If the bicycle is traveling at 5 m/s and the rider engages the brakes, achieving the largest deceleration for which the rear wheel will not leave the ground, how long does it take the bicycle to stop, and what distance does it travel during that time?

Problems 18.7/18.8

18.9 The combined mass of the motorcycle and rider is 160 kg. The rear wheel exerts a 400-N horizontal force on the road, and you can neglect the horizontal force exerted on the road by the front wheel. Modeling the motorcycle and its wheels as a rigid body, determine (a) the motorcycle's acceleration and (b) the normal forces exerted on the road by the rear and front wheels.

660 mm

A |← 660 mm →| B

←——— 1500 mm ———→

Problem 18.9

18.10 The moment of inertia of the disk about O is 22 kg-m^2. At $t = 0$, the stationary disk is subjected to a constant 50 N-m torque.
(a) What is the magnitude of the disk's angular velocity at $t = 5$ s?
(b) Through how many revolutions does the disk rotate from $t = 0$ to $t = 5$ s?

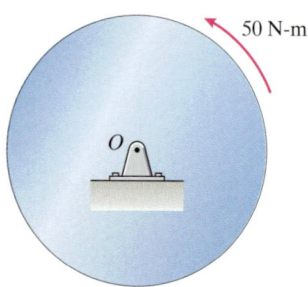

50 N-m

O

Problem 18.10

18.11 During extravehicular activity, an astronaut fires a thruster of his maneuvering unit, exerting a force $T = 14.2$ N for 1 s. It requires 60 s from the time the thruster is fired for him to rotate through one revolution. If you model the astronaut and maneuvering unit as a rigid body, what is the moment of inertia about their center of mass?

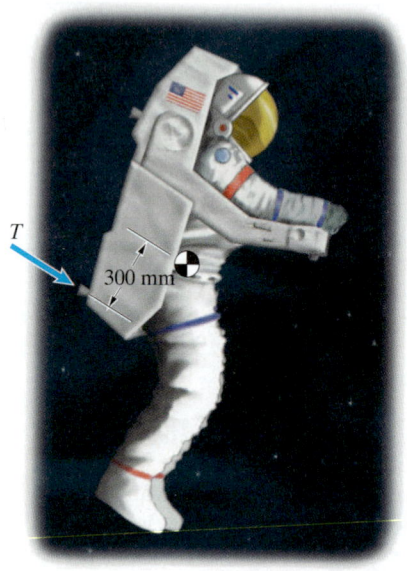

T

300 mm

Problem 18.11

18.12 The moment of inertia of the helicopter's rotor is 420 slug-ft^2. The rotor starts from rest at $t = 0$. The torque exerted on it by the engine is given as a function of time by $500 - 20t$ ft-lb.
(a) What is the magnitude of the rotor's angular velocity at $t = 10$ s?
(b) Through how many revolutions does the rotor rotate from $t = 0$ to $t = 10$ s?

Problem 18.12

18.13 The moment of inertia of the robot manipulator arm about the vertical y axis is 10 kg-m². The moment of inertia of the 14-kg casting held by the arm about the y' axis is 1.2 kg-m². The system is initially stationary. At t = 0, the arm is subjected to a torque about the y axis that is given as a function of time by 220 + 100t N-m. How long does it take the arm to undergo one revolution?

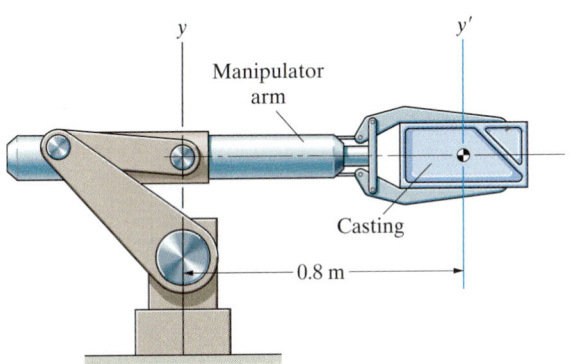

Problem 18.13

18.14 The moment of inertia of the wind-tunnel fan is 225 kg-m². The fan starts from rest. The torque exerted on it by the engine is given as a function of the angular velocity of the fan by T = 140 − 0.02ω² N-m.

(a) When the fan has turned 620 revolutions, what is its angular velocity in rpm (revolutions per minute)?
(b) What maximum angular velocity in rpm does the fan attain?

Strategy: By writing the equation of angular motion, determine the angular acceleration of the fan in terms of its angular velocity. Then use the chain rule:

$$\alpha = \frac{d\omega}{dt} = \frac{d\omega}{d\theta}\frac{d\theta}{dt} = \frac{d\omega}{d\theta}\omega.$$

Problem 18.14

18.15 The gears A and B can turn freely on their pin supports. Their moments of inertia are I_A = 0.002 kg-m² and I_B = 0.006 kg-m². They are initially stationary, and at t = 0 a constant couple M = 2 N-m is applied to gear B. How many revolutions has gear A turned at t = 4 s?

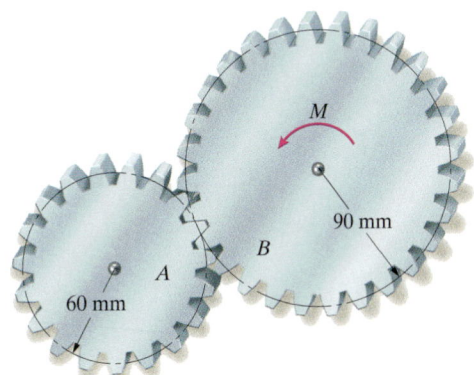

Problem 18.15

18.16 The disks A, B, and C do not slip relative to each other at their points of contact. Their masses are m_A = 4 kg, m_B = 16 kg, and m_C = 9 kg. They are initially stationary. At t = 0, a constant 10 N-m counterclockwise couple is applied to disk A. What is the angular velocity of disk C at t = 5 s?

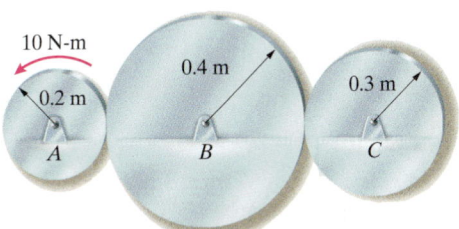

Problem 18.16

18.17 The moment of inertia of the pulley is 0.4 slug-ft². The 5-lb weight slides on the smooth horizontal surface. If the system starts from rest, determine how far to the right the 5-lb weight moves in 1 s in each case.

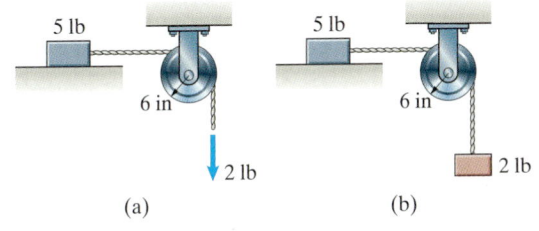

Problem 18.17

18.18 The masses of the slender bar and disk are 5 kg and 10 kg, respectively. The coefficient of kinetic friction between the disk and the horizontal surface is $\mu_k = 0.1$. Determine the disk's angular acceleration if it is subjected to (a) an 8 N-m counterclockwise couple; (b) an 8 N-m clockwise couple.

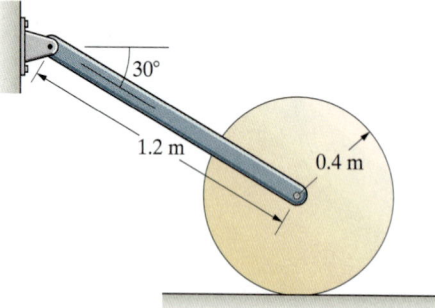

Problem 18.18

18.19 The 5-kg slender bar is released from rest in the horizontal position shown. Determine the magnitude of the bar's angular velocity when it has fallen to the vertical position.

Strategy: By drawing a free-body diagram of the bar when it has fallen through an arbitrary angle θ and using the equation of angular motion, determine the bar's angular acceleration as a function of θ. Then apply the chain rule:

$$\alpha = \frac{d\omega}{dt} = \frac{d\omega}{d\theta}\frac{d\theta}{dt} = \frac{d\omega}{d\theta}\omega.$$

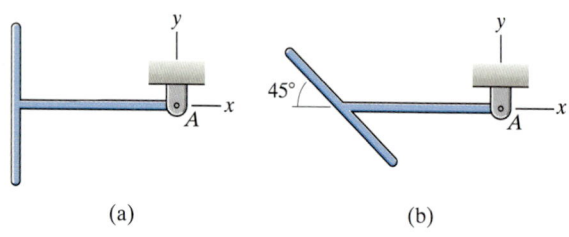

Problem 18.19

18.20 The objects consist of identical 3-ft, 10-lb bars welded together. If they are released from rest in the positions shown, what are their angular accelerations and what are the components of the reactions at A at that instant? (The y axes are vertical.)

(a)   (b)

Problem 18.20

18.21 The object consists of the 2-kg slender bar ABC welded to the 3-kg slender bar BDE. The y axis is vertical. The object is released from rest in the position shown. Determine its angular acceleration and the components of the force exerted on the object by the pin at D at the instant it is released.

18.22 Determine the angular acceleration and the components of the force exerted on the object in Problem 18.21 by the pin at D at the instant the bar BDE has fallen to the vertical position.

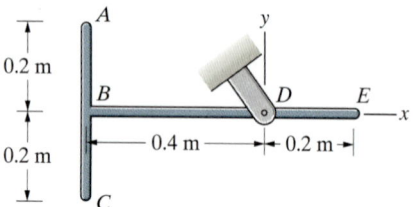

Problems 18.21/18.22

18.23 For what value of x is the horizontal bar's angular acceleration a maximum, and what is the maximum angular acceleration?

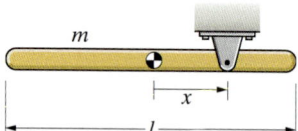

Problem 18.23

18.24 Model the arm ABC as a single rigid body. Its mass is 300 kg, and the moment of inertia about its center of mass is $I = 360$ kg-m². If point A is stationary and the angular acceleration of the arm is 0.6 rad/s² counterclockwise, what force does the hydraulic cylinder exert on the arm at B? (The arm is actuated by two hydraulic cylinders, one on each side of the vehicle. You are to determine the total force exerted by the two cylinders.)

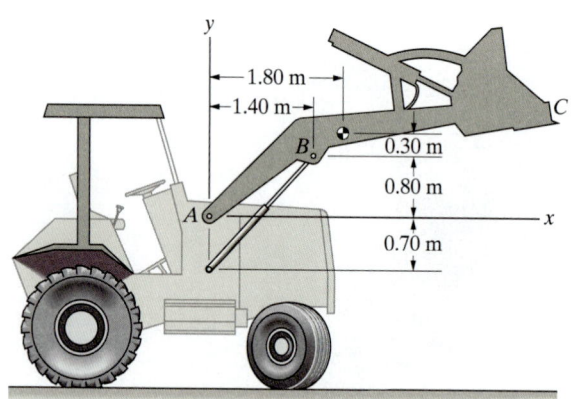

Problem 18.24

18.25 The mass of the truck's bed is 2500 kg and its moment of inertia about O is 78,000 kg-m^2. At the instant shown, the coordinates of the center of mass of the bed are $(3, 3.75)$ m and the coordinates of point B are $(4.5, 3.5)$ m. If the angular acceleration of the bed is 0.5 rad/s^2 in the clockwise direction, what is the magnitude of the force exerted on the bed at B by the hydraulic cylinder AB?

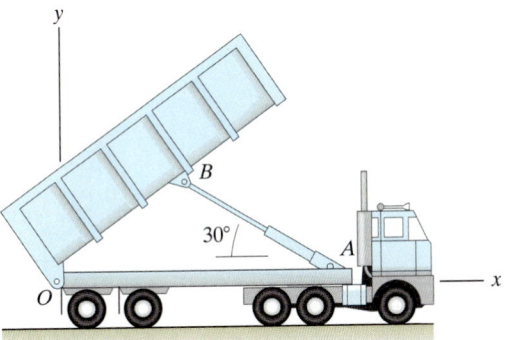

Problem 18.25

18.26 Arm BC has a mass of 12 kg and the moment of inertia about its center of mass is 3 kg-m^2. Point B is stationary and arm BC has a constant counterclockwise angular velocity of 2 rad/s. At the instant shown, what are the couple and the components of force exerted on arm BC at B?

18.27 Arm BC has a mass of 12 kg and the moment of inertia about its center of mass is 3 kg-m^2. At the instant shown, arm AB has a constant clockwise angular velocity of 2 rad/s and arm BC has a counterclockwise angular velocity of 2 rad/s and a clockwise angular acceleration of 4 rad/s^2. What are the couple and the components of force exerted on arm BC at B?

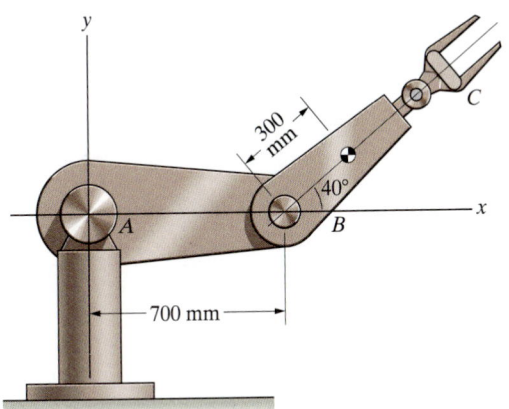

Problems 18.26/18.27

18.28 The space shuttle's attitude control engines exert two forces $F_f = 8$ kN and $F_r = 2$ kN. The force vectors and the center of mass G lie in the x–y plane of the inertial reference frame. The mass of the shuttle is 54,000 kg, and its moment of inertia about the axis through the center of mass that is parallel to the z axis is 4.5×10^6 kg-m^2. Determine the acceleration of the center of mass and the angular acceleration. (You can ignore the force exerted on the shuttle by its weight.)

18.29 In Problem 18.28, suppose that $F_f = 4$ kN and you want the shuttle's angular acceleration to be zero. Determine the necessary force F_r and the resulting acceleration of the center of mass.

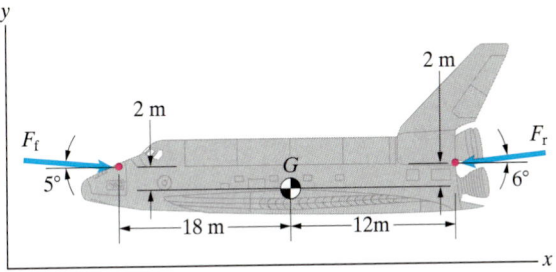

Problems 18.28/18.29

18.30 Points B and C lie in the x–y plane. The y axis is vertical. The center of mass of the 18-kg arm BC is at the midpoint of the line from B to C, and the moment of inertia of the arm about the axis through the center of mass that is parallel to the z axis is 1.5 kg-m^2. At the instant shown, the angular velocity and angular acceleration vectors of arm AB are $\omega_{AB} = 0.6\mathbf{k}$ (rad/s) and $\alpha_{AB} = -0.3\mathbf{k}$ (rad/s^2). The angular velocity and angular acceleration vectors of arm BC are $\omega_{BC} = 0.4\mathbf{k}$ (rad/s) and $\alpha_{BC} = 2\mathbf{k}$ (rad/s^2). Determine the force and couple exerted on arm BC at B.

18.31 Points B and C lie in the x–y plane. The y axis is vertical. The center of mass of the 18-kg arm BC is at the midpoint of the line from B to C, and the moment of inertia of the arm about the axis through the center of mass that is parallel to the z axis is 1.5 kg-m^2. At the instant shown, the angular velocity and angular acceleration vectors of arm AB are $\omega_{AB} = 0.6\mathbf{k}$ (rad/s) and $\alpha_{AB} = -0.3\mathbf{k}$ (rad/s^2). The angular velocity vector of arm BC is $\omega_{BC} = 0.4\mathbf{k}$ (rad/s). If you want to program the robot so that the angular acceleration of arm BC is zero at this instant, what couple must be exerted on arm BC at B?

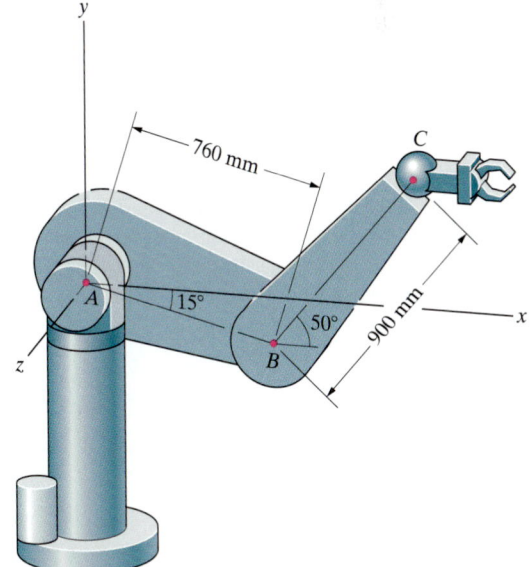

Problems 18.30/18.31

18.32 The 9000-kg airplane has just landed. At the instant shown, its angular velocity is zero. Its landing gear are rolling and contact the runway at $x = 10$ m. The frictional force on the wheels is negligible. The coordinates of the airplane's center of mass are $x = 10.50$, $y = 3.00$ m. The total aerodynamic force is $-26.8\mathbf{i} + 30.4\mathbf{j}$ (kN), and it effectively acts at the center of pressure located at $x = 10.75$, $y = 3.2$ m. The thrust $T = 4.40$ kN exerts no moment about the center of mass. The moment of inertia of the airplane about its center of mass is 75,000 kg-m². Determine the airplane's angular acceleration.

Strategy: Draw a free-body diagram of the airplane, including the normal force exerted on the landing gear. To relate the acceleration of the center of mass to the angular acceleration, use the fact that the acceleration of the airplane (treated as a rigid body) is horizontal at the point where the wheels contact the runway.

Problem 18.32

18.33 The radius of the 2-kg disk is $R = 80$ mm. It is released from rest on the rough inclined surface.

(a) How long does it take the disk to roll through one revolution?
(b) What minimum coefficient of static friction between the disk and the surface is necessary for the disk to roll instead of slipping when it is released?

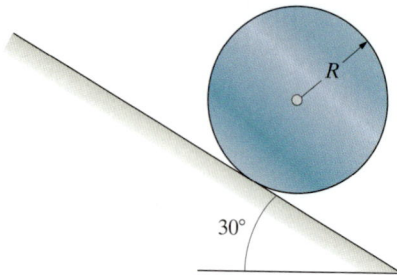

Problem 18.33

18.34 A thin ring and a circular disk, each of mass m and radius R, are released from rest on an inclined surface and allowed to roll a distance D. Determine the ratio of the times required to traverse the distance D in the two cases.

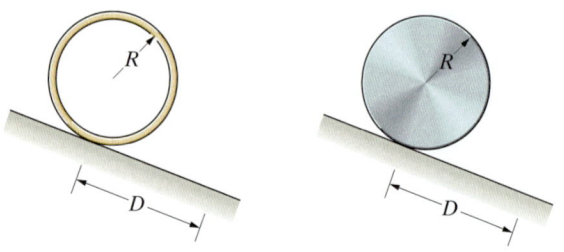

Problem 18.34

18.35 The stepped disk weighs 40 lb and its moment of inertia is $I = 0.2$ slug-ft². If the disk is released from rest, how long does it take its center to fall 3 ft? (Assume that the string remains vertical.)

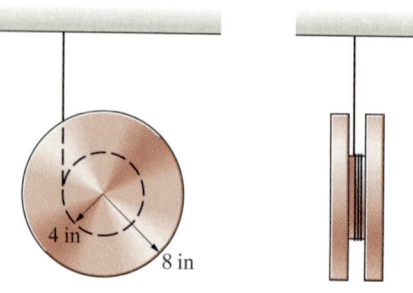

Problem 18.35

18.36 The radius of the pulley is $R = 100$ mm and its moment of inertia is $I = 0.1$ kg-m². The mass $m = 5$ kg. The spring constant is $k = 135$ N/m. The system is released from rest with the spring unstretched. Determine how fast the mass is moving when it has fallen a distance $x = 0.5$ m.

Strategy: Draw individual free-body diagrams of the mass and pulley, and use them to determine the acceleration a of the mass as a function of the distance x it has fallen. Then use the chain rule: $a = dv/dt = (dv/dx)(dx/dt) = (dv/dx)v$.

18.37 The radius of the pulley is $R = 100$ mm and its moment of inertia is $I = 0.1$ kg-m². The mass $m = 5$ kg. The spring constant is $k = 135$ N/m. The system is released from rest with the spring unstretched. What maximum velocity does the mass attain as it falls?

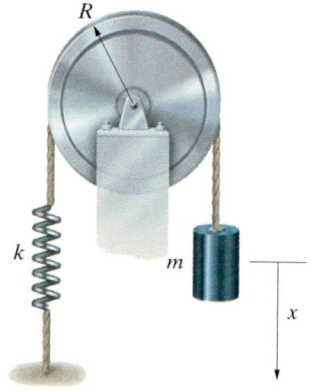

Problems 18.36/18.37

18.38 The mass of the disk is 45 kg and its radius is $R = 0.3$ m. The spring constant is $k = 600$ N/m. The disk is rolled to the left until the spring is compressed 0.5 m and released from rest.

(a) If you assume that the disk rolls, what is its angular acceleration at the instant it is released?

(b) What is the minimum coefficient of static friction for which the disk will not slip when it is released?

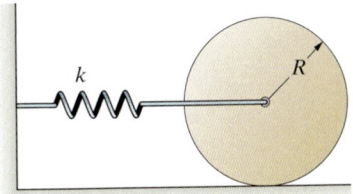

Problem 18.38

18.39 The disk weighs 12 lb and its radius is $R = 6$ in. The spring constant is $k = 3$ lb/ft. The disk is released from rest with the spring unstretched. Determine the magnitude of the velocity of the center of the disk when it has moved 2 ft from its initial position if (a) the inclined surface is smooth (friction is negligible); (b) the disk rolls on the surface.

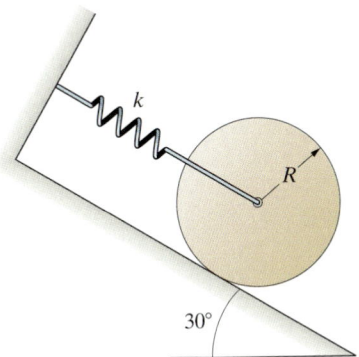

Problem 18.39

18.40 A 42-lb sphere with radius $R = 4$ in is placed on a horizontal surface with initial angular velocity $\omega_0 = 40$ rad/s. The coefficient of kinetic friction between the sphere and the surface is $\mu_k = 0.06$. What maximum velocity will the center of the sphere attain, and how long does it take to reach that velocity?

Strategy: The friction force exerted on the spinning sphere by the surface will cause the sphere to accelerate to the right. The friction force will also cause the sphere's angular velocity to decrease. The center of the sphere will accelerate until the sphere is rolling on the surface instead of slipping relative to it. Use the relation between the velocity of the center and the angular velocity of the sphere when it is rolling to determine when the sphere begins rolling.

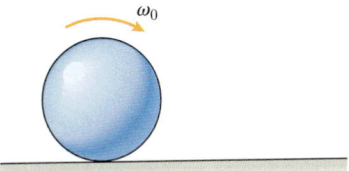

Problem 18.40

18.41 A soccer player kicks the ball to a teammate 8 m away. The ball leaves the player's foot moving parallel to the ground at 6 m/s with no angular velocity. The coefficient of kinetic friction between the ball and the grass is $\mu_k = 0.32$. How long does it take the ball to reach his teammate? The radius of the ball is 112 mm and its mass is 0.4 kg. Estimate the ball's moment of inertia by using the equation for a thin spherical shell: $I = \frac{2}{3} mR^2$.

Problem 18.41

18.42 The 100-kg cylindrical disk is at rest when the force F is applied to a cord wrapped around it. The static and kinetic coefficients of friction between the disk and the surface are 0.2. Determine the angular acceleration of the disk if (a) $F = 500$ N and (b) $F = 1000$ N.

Strategy: First solve the problem by assuming that the disk does not slip, but rolls on the surface. Determine the frictional force, and find out whether it exceeds the product of the coefficient of friction and the normal force. If it does, you must rework the problem assuming that the disk slips.

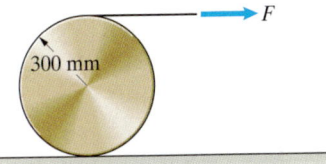

Problem 18.42

18.43 The ring gear is fixed. The mass and moment of inertia of the sun gear are $m_S = 320$ kg and $I_S = 40$ kg-m². The mass and moment of inertia of each planet gear are $m_P = 38$ kg and $I_P = 0.60$ kg-m². If a couple $M = 200$ N-m is applied to the sun gear, what is the latter's angular acceleration?

18.44 In Problem 18.43, what is the magnitude of the tangential force exerted on the sun gear by each planet gear at their points of contact when the 200 N-m couple is applied to the sun gear?

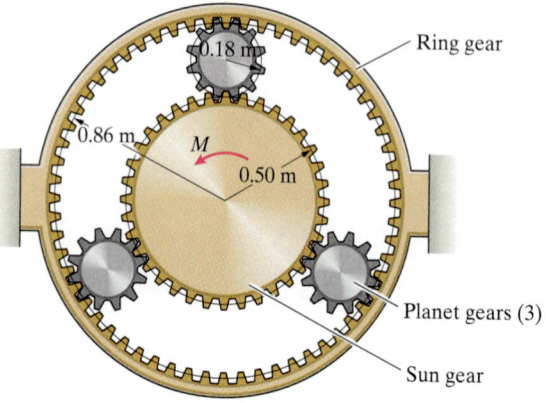

Problems 18.43/18.44

18.45 The 18-kg ladder is released from rest in the position shown. Model it as a slender bar and neglect friction. At the instant of release, determine (a) the angular acceleration of the ladder and (b) the normal force exerted on the ladder by the floor.

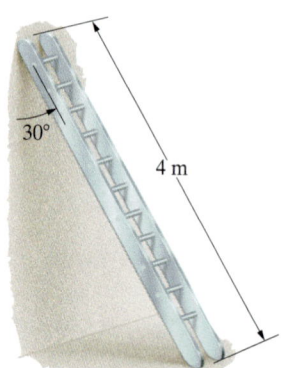

Problem 18.45

18.46 The 18-kg ladder is released from rest in the position shown. Model it as a slender bar and neglect friction. Determine its angular acceleration at the instant of release.

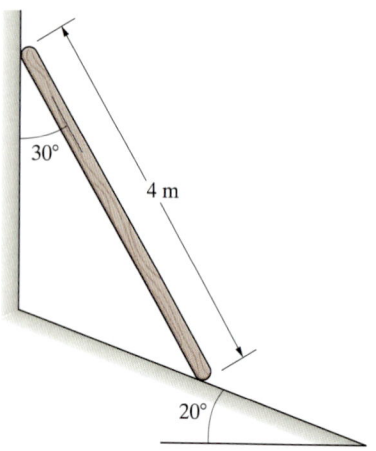

Problem 18.46

18.47 The 4-kg slender bar is released from rest on the rough surface in the position shown. Determine the minimum value of the coefficient of static friction for which the bar will not slip relative to the floor when it is released.

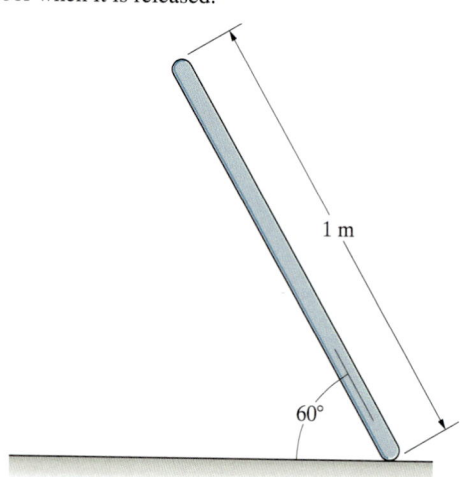

Problem 18.47

18.48 The masses of the bar and disk are 14 kg and 9 kg, respectively. The system is released from rest with the bar horizontal. Determine the bar's angular acceleration at that instant if (a) the bar and disk are welded together at A and (b) the bar and disk are connected by a smooth pin at A.

Strategy: In part (b), draw individual free-body diagrams of the bar and disk.

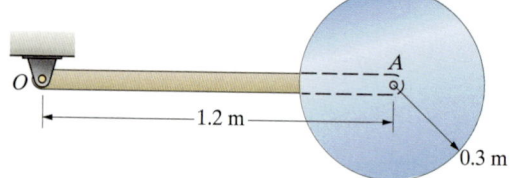

Problem 18.48

18.49 The 5-lb horizontal bar is connected to the 10-lb disk by a smooth pin at A. The system is released from rest in the position shown. What are the angular accelerations of the bar and disk at that instant?

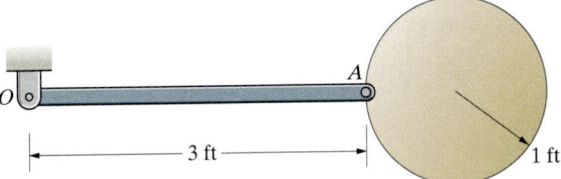

Problem 18.49

18.50 The 0.1-kg slender bar and 0.2-kg cylindrical disk are released from rest with the bar horizontal. The disk rolls on the curved surface. What is the bar's angular acceleration at the instant it is released?

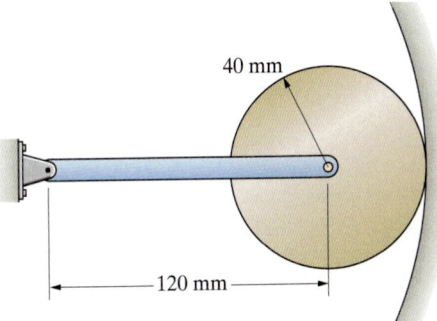

Problem 18.50

18.51 The mass of the suspended object A is 8 kg. The mass of the pulley is 5 kg, and its moment of inertia is 0.036 kg-m². If the force $T = 70$ N, what is the magnitude of the acceleration of A?

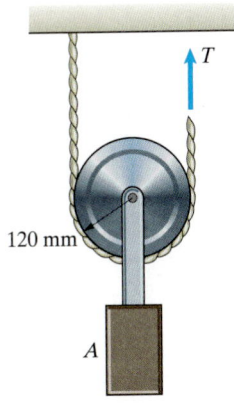

Problem 18.51

18.52 The suspended object A weighs 20 lb. The pulleys are identical, each weighing 10 lb and having moment of inertia 0.022 slug-ft². If the force $T = 15$ lb, what is the magnitude of the acceleration of A?

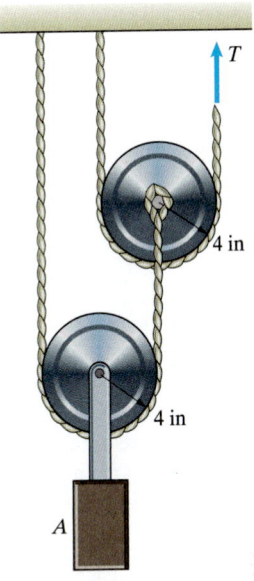

Problem 18.52

18.53 The 2-kg slender bar and 5-kg block are released from rest in the position shown. If friction is negligible, what is the block's acceleration at that instant?

18.54 The 2-kg slender bar and 5-kg block are released from rest in the position shown. What minimum coefficient of static friction between the block and the horizontal surface would be necessary for the block not to move when the system is released?

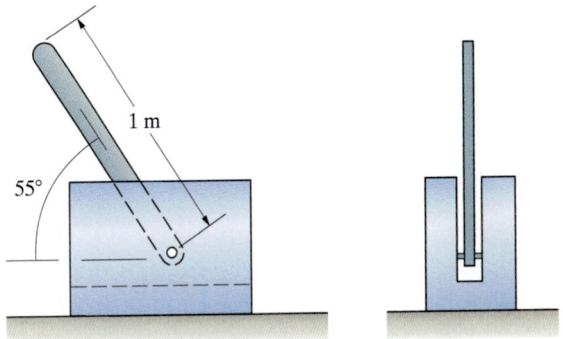

Problems 18.53/18.54

18.55 The 0.4-kg slender bar and 1-kg disk are released from rest in the position shown. If the disk rolls, what is the bar's angular acceleration at the instant of release?

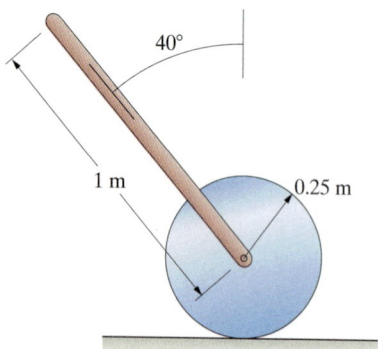

Problem 18.55

18.56 The masses of the slender bar and the crate are 9 kg and 36 kg, respectively. The crate rests on a smooth horizontal surface. If the system is stationary at the instant shown and a counterclockwise couple $M = 300$ N-m is applied to the bar, what is the resulting acceleration of the crate?

18.57 In Problem 18.56, determine the resulting acceleration of the crate if the coefficient of kinetic friction between the crate and the horizontal surface is $\mu_k = 0.2$.

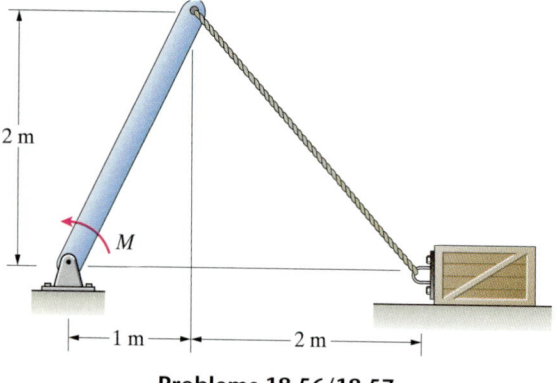

Problems 18.56/18.57

18.58 Bar AB rotates with a constant angular velocity of 6 rad/s in the counterclockwise direction. The slender bar BCD weighs 10 lb, and the collar that bar BCD is attached to at C weighs 2 lb. The y axis points upward. Neglecting friction, determine the components of the forces exerted on bar BCD by the pins at B and C at the instant shown.

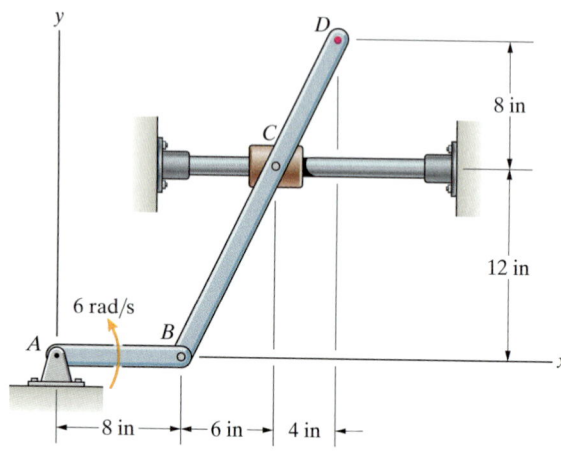

Problem 18.58

18.59 The masses of the slender bars AB and BC are 10 kg and 12 kg, respectively. The angular velocities of the bars are zero at the instant shown and the horizontal force $F = 150$ N. The horizontal surface is smooth. Determine the angular accelerations of the bars.

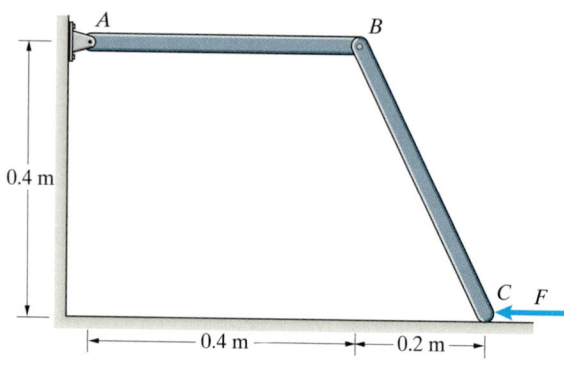

Problem 18.59

18.60 Let the total moment of inertia of the car's two rear wheels and axle be I_R, and let the total moment of inertia of the two front wheels be I_F. The radius of the tires is R, and the total mass of the car, including the wheels, is m. If the car's engine exerts a torque (couple) T on the rear wheels and the wheels do not slip, show that the car's acceleration is

$$a = \frac{RT}{R^2m + I_R + I_F}.$$

Strategy: Isolate the wheels and draw three free-body diagrams.

Problem 18.60

18.61 The combined mass of the motorcycle and rider is 160 kg. Each 9-kg wheel has a 330-mm radius and a moment of inertia $I = 0.8$ kg-m^2. The engine drives the rear wheel by exerting a couple on it. If the rear wheel exerts a 400-N horizontal force on the road and you do *not* neglect the horizontal force exerted on the road by the front wheel, determine (a) the motorcycle's acceleration and (b) the normal forces exerted on the road by the rear and front wheels. (The location of the center of mass of the motorcycle, *not including* its wheels, is shown.)

18.62 In Problem 18.61, if the front wheel lifts slightly off the road when the rider accelerates, determine (a) the motorcycle's acceleration and (b) the torque exerted by the engine on the rear wheel.

Problems 18.61/18.62

18.63 The moment of inertia of the vertical handle about O is 0.12 slug-ft^2. The object B weighs 15 lb and rests on a smooth surface. The weight of the bar AB is negligible (which means that you can treat the bar as a two-force member). If the person exerts a 0.2-lb horizontal force on the handle 15 in above O, what is the resulting angular acceleration of the handle?

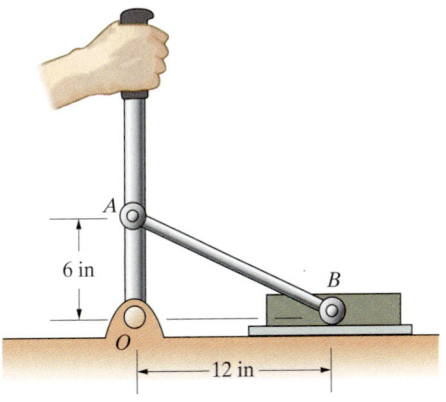

Problem 18.63

18.64 The bars are each 1 m in length and have a mass of 2 kg. They rotate *in the horizontal plane*. Bar AB rotates with a constant angular velocity of 4 rad/s in the counterclockwise direction. At the instant shown, bar BC is rotating in the counterclockwise direction at 6 rad/s. What is the angular acceleration of bar BC?

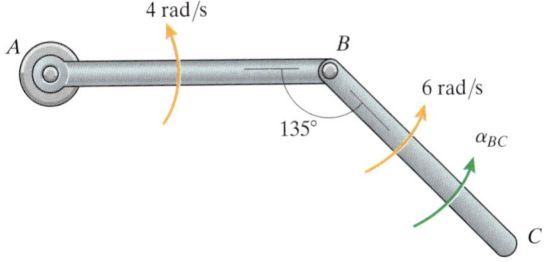

Problem 18.64

18.65 Bars OQ and PQ each weigh 6 lb. The weight of the collar P and the friction between the collar and the horizontal bar are negligible. If the system is released from rest with $\theta = 45°$, what are the angular accelerations of the two bars?

18.66 In Problem 18.65, what are the angular accelerations of the two bars if the collar P weighs 2 lb?

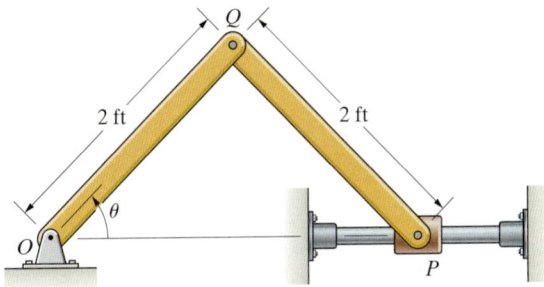

Problems 18.65/18.66

18.67 The 4-kg slender bar is pinned to 2-kg sliders at A and B. If friction is negligible and the system is released from rest in the position shown, what is the angular acceleration of the bar at that instant?

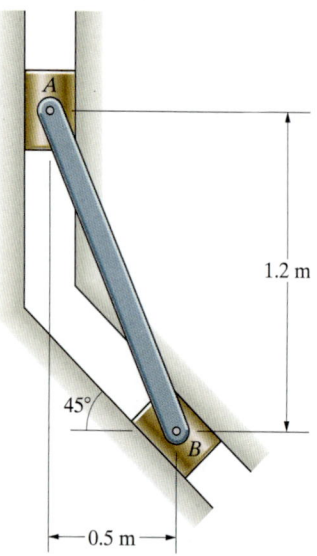

1.2 m

45°

0.5 m

Problem 18.67

18.68 The mass of the slender bar is m and the mass of the homogeneous disk is $4m$. The system is released from rest in the position shown. If the disk rolls and the friction between the bar and the horizontal surface is negligible, show that the disk's angular acceleration is $\alpha = 6g/95R$ counterclockwise.

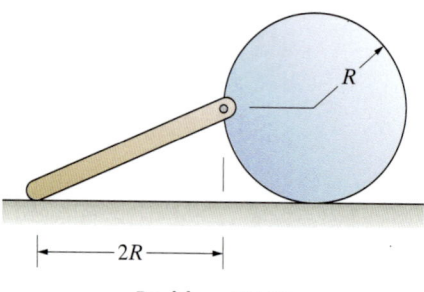

R

2R

Problem 18.68

18.69 Bar AB rotates *in the horizontal plane* with a constant angular velocity of 10 rad/s in the counterclockwise direction. The masses of the slender bars BC and CD are 3 kg and 4.5 kg, respectively. Determine the x and y components of the forces exerted on bar BC by the pins at B and C at the instant shown.

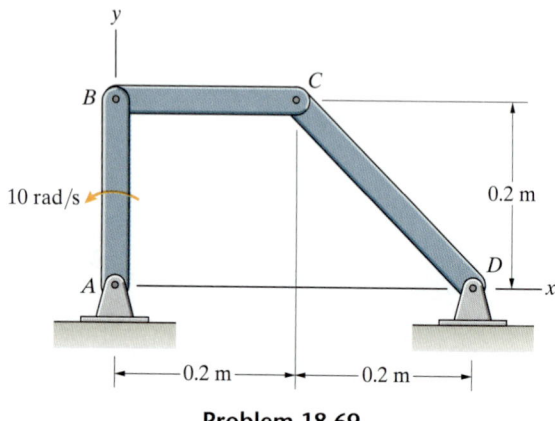

10 rad/s

0.2 m

0.2 m   0.2 m

Problem 18.69

18.70 The 2-kg bar rotates *in the horizontal plane* about the smooth pin. The 6-kg collar A slides on the smooth bar. At the instant shown, $r = 1.2$ m, $\omega = 0.4$ rad/s, and the collar is sliding outward at 0.5 m/s relative to the bar. If you neglect the moment of inertia of the collar (that is, treat the collar as a particle), what is the bar's angular acceleration?

Strategy: Draw individual free-body diagrams of the bar and the collar, and write Newton's second law for the collar in polar coordinates.

18.71 In Problem 18.70, suppose that the moment of inertia of the collar about its center of mass is 0.2 kg-m^2. Determine the angular acceleration of the bar and compare your answer with the answer to Problem 18.70.

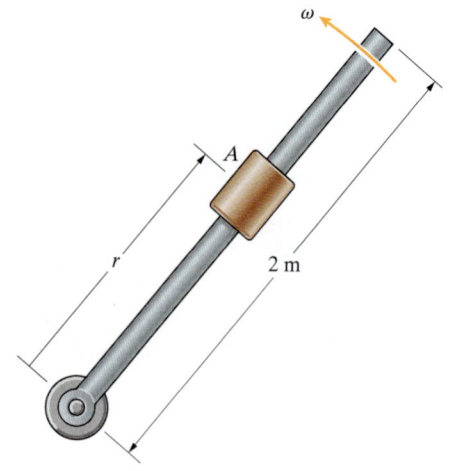

ω

A

r

2 m

Problems 18.70/18.71

Design Experience

Problems 18.72–18.78 are related to Design Example 18.7.

18.72 The 3-Mg rocket is accelerating upward at 2 g's. If you model it as a homogeneous bar, what is the magnitude of the axial force P at the midpoint?

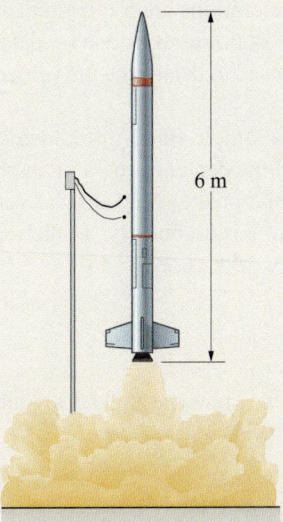

6 m

Problem 18.72

18.73 The 20-kg slender bar is attached to a vertical shaft at A and rotates *in the horizontal plane* with a constant angular velocity of 10 rad/s. What is the axial force P at the bar's midpoint?

18.74 The 20-kg slender bar is attached to a vertical shaft at A and rotates in the horizontal plane with a constant angular velocity of 10 rad/s. Draw a graph of the axial force in the bar as a function of x.

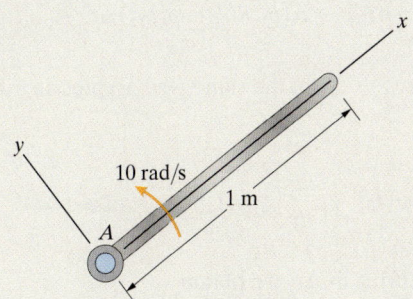

10 rad/s

1 m

A

x

y

Problems 18.73/18.74

18.75 The 100-lb slender bar AB has a built-in support at A. The y axis points upward. Determine the magnitudes of the shear force and bending moment at the bar's midpoint if (a) the support is stationary and (b) the support is accelerating upward at 10 ft/s^2.

18.76 For the bar in Problem 18.75, draw the shear force and bending moment diagrams for the two cases.

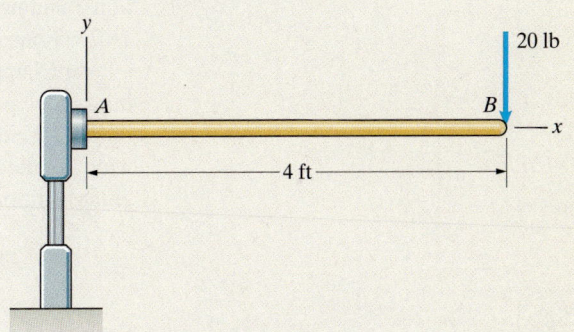

y

A

B

20 lb

x

4 ft

Problems 18.75/18.76

18.77 The 18-kg ladder is held in equilibrium in the position shown by the force F. Neglect friction and model the ladder as a slender bar.
(a) What are the axial force, shear force, and bending moment at the ladder's midpoint?
(b) If the force F is suddenly removed, what are the axial force, shear force, and bending moment at the ladder's midpoint at that instant?

18.78 For the ladder in Problem 18.77, draw the shear force and bending moment diagrams for the two cases.

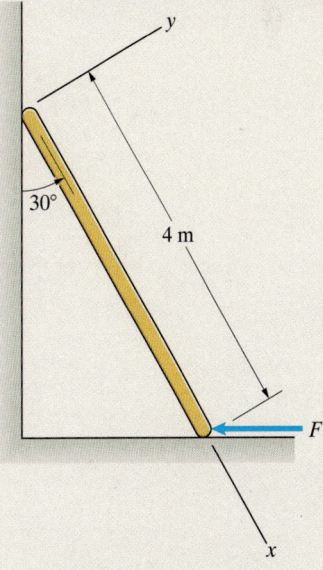

y

30°

4 m

F

x

Problems 18.77/18.78

18.5 Numerical Solutions

When the forces and couples acting on a rigid body are known, the equations of motion can be used to determine the acceleration of the center of mass and the angular acceleration of the body. In some situations, we can then integrate to obtain closed-form expressions for the velocity and position of the center of mass and for the angular velocity and angular position as functions of time. But if the functions describing the accelerations are too complicated, or the forces and couples are known in terms of continuous or analog data instead of equations, a numerical method must be used to determine the velocities and positions as functions of time.

In Chapter 14, we described a simple finite-difference method for determining the position and velocity of the center of mass of an object as functions of time. We can determine the angular position and angular velocity in the same way. Let the angular acceleration of a rigid body be a function of time, the angular position, and angular velocity of the body:

$$\alpha = \alpha(t, \theta, \omega).$$

Suppose that at a particular time t_0, we know the angle $\theta(t_0)$ and angular velocity $\omega(t_0)$. The angular acceleration at t_0 is

$$\frac{d\omega}{dt}(t_0) = \alpha(t_0, \theta(t_0), \omega(t_0)). \tag{18.24}$$

To determine the angular velocity at a time $t + \Delta t$, we express it as a Taylor series:

$$\omega(t_0 + \Delta t) = \omega(t_0) + \frac{d\omega}{dt}(t_0)\Delta t + \frac{1}{2}\frac{d^2\omega}{dt^2}(t_0)(\Delta t)^2 + \cdots.$$

By choosing a sufficiently small value of Δt, we can neglect terms of second and higher order in Δt and substitute Eq. (18.24) to obtain an approximation for the angular velocity at $t_0 + \Delta t$:

$$\omega(t_0 + \Delta t) = \omega(t_0) + \alpha(t_0, \theta(t_0), \omega(t_0))\Delta t. \tag{18.25}$$

We approximate the angle at $t_0 + \Delta t$ in the same way. Expressing it as a Taylor series yields

$$\theta(t_0 + \Delta t) = \theta(t_0) + \frac{d\theta}{dt}(t_0)\Delta t + \frac{1}{2}\frac{d^2\theta}{dt^2}(t_0)(\Delta t)^2 + \cdots,$$

and neglecting higher order terms in Δt, we obtain

$$\theta(t_0 + \Delta t) = \theta(t_0) + \omega(t_0)\Delta t. \tag{18.26}$$

With Eqs. (18.25) and (18.26), we can determine the approximate values of the angular velocity and position at $t_0 + \Delta t$. Using these values as initial conditions, we can repeat the procedure to determine the angular velocity and position at $t_0 + 2\Delta t$, and so forth.

COMPUTATIONAL MECHANICS

The following example and problems are designed to be worked with the use of a programmable calculator or computer.

Computational Example 18.8

The 18-kg ladder in Fig. 18.20 is released from rest in the position shown at $t = 0$. Neglecting friction, determine the angular position and angular velocity of the ladder as functions of time. Use time increments Δt of 0.1 s, 0.01 s, and 0.001 s.

Strategy
The initial steps—drawing the free-body diagram of the ladder, applying the equations of motion, and determining the angular acceleration—are presented in Example 18.4. The ladder's angular acceleration is

$$\alpha = \frac{3g}{2l} \sin \theta,$$

where θ is the angle between the ladder and the wall and l is the length of the ladder. With this expression, we can use Eqs. (18.25) and (18.26) to approximate the ladder's angular position and angular velocity as functions of time.

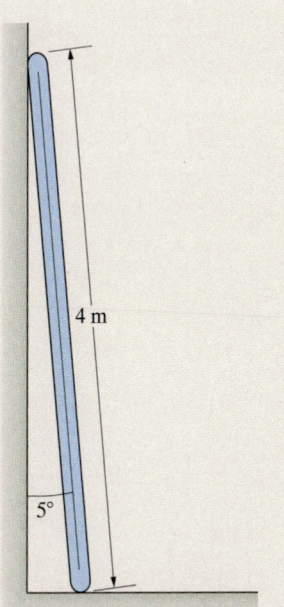

4 m

5°

Figure 18.20

Solution
The angular acceleration is

$$\alpha = \frac{3(9.81 \text{ m/s}^2)}{2(4 \text{ m})} \sin \theta = 3.68 \sin \theta \text{ rad/s}^2.$$

Let $\Delta t = 0.1$ s. At the initial time $t_0 = 0$, $\theta(t_0) = 5° = 0.0873$ rad and $\omega(t_0) = 0$. From Eq. (18.26), the angular position at time $t_0 + \Delta t = 0.1$ s is

$$\theta(t_0 + \Delta t) = \theta(t_0) + \omega(t_0)\Delta t:$$

$$\theta(0.1) = \theta(0) + \omega(0)\Delta t$$

$$= 0.0873 + (0)(0.1) = 0.0873 \text{ rad}.$$

From Eq. (18.25), the angular velocity is

$$\omega(t_0 + \Delta t) = \omega(t_0) + \alpha(t_0)\Delta t:$$

$$\omega(0.1) = 0 + [3.68 \sin(0.0873)](0.1) = 0.0321 \text{ rad/s}.$$

Using these values as the initial conditions for the next time step, we find that the angular position at $t = 0.2$ s is

$$\theta(0.2) = \theta(0.1) + \omega(0.1)\Delta t$$

$$= 0.0873 + (0.0321)(0.1) = 0.0905 \text{ rad},$$

and the angular velocity is

$$\omega(0.2) = \omega(0.1) + \alpha(0.1)\Delta t$$

$$= 0.0321 + [3.68 \sin(0.0873)](0.1) = 0.0641 \text{ rad/s}.$$

Continuing in this way, we obtain the following values for the first five time steps:

Time (s)	θ (rad)	ω (rad/s)
0.0	0.0873	0.0000
0.1	0.0873	0.0321
0.2	0.0905	0.0641
0.3	0.0969	0.0974
0.4	0.1066	0.1329
0.5	0.1199	0.1721

Figures 18.21 and 18.22 show the numerical solutions for the angular position and angular velocity obtained using $\Delta t = 0.1$ s, $\Delta t = 0.01$ s, and $\Delta t = 0.001$ s. Trials with smaller time intervals indicate that $\Delta t = 0.001$ s closely approximates the exact solution. We show the positions of the falling ladder at 0.2-s intervals in Fig. 18.23.

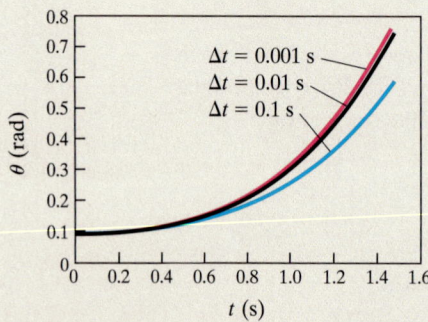

Figure 18.21
Numerical solutions for the ladder's angular position.

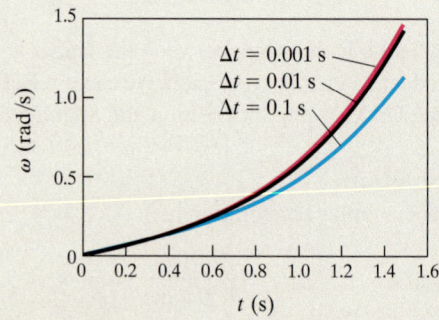

Figure 18.22
Numerical solutions for the ladder's angular velocity.

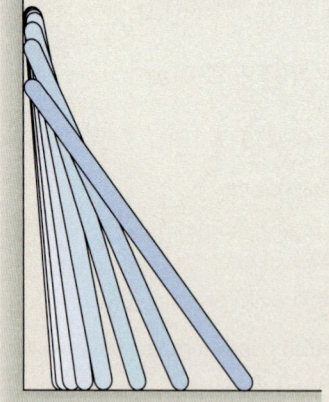

Figure 18.23
Position of the falling ladder at 0.2-s intervals from $t = 0$ to $t = 1.4$ s.

Critical Thinking

By using the chain rule, we can write the ladder's angular acceleration as

$$\alpha = \frac{d\omega}{dt} = \frac{d\omega}{d\theta}\omega = \frac{3g}{2l}\sin\theta.$$

Separating variables, we can integrate to determine the angular velocity as a function of the angular position:

$$\int_0^\omega \omega \, d\omega = \int_{5°}^\theta \frac{3g}{2l} \sin \theta \, d\theta.$$

We obtain

$$\omega = \sqrt{\left(\frac{3g}{l}\right)(\cos 5° - \cos \theta)}.$$

This closed-form result is compared with the graph of our numerical solution (using $\Delta t = 0.001$ s) in Fig. 18.24. The curves are indistinguishable.

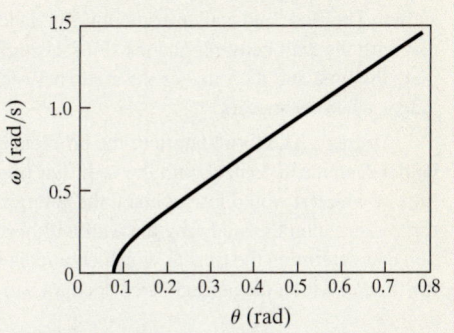

Figure 18.24
Analytical and numerical solutions for the ladder's angular velocity as a function of its angular position.

Computational Problems

18.79 Continue the calculations presented in Example 18.8, using $\Delta t = 0.1$ s, and determine the ladder's angular position and angular velocity at $t = 0.6$ s and $t = 0.7$ s.

18.80 The moment of inertia of the helicopter's rotor is 400 slug-ft². The helicopter starts from rest at $t = 0$, the engine exerts a constant torque of 500 ft-lb, and aerodynamic drag exerts a torque of magnitude $20\omega^2$ ft-lb, where ω is the rotor's angular velocity in radians per second. Using $\Delta t = 0.2$ s, determine the rotor's angular position and angular velocity for the first five time steps. Compare your results for the angular velocity with the closed-form solution.

18.81 In Problem 18.80, draw a graph of the rotor's angular velocity as a function of time from $t = 0$ to $t = 10$ s, comparing the closed-form solution, the numerical solution using $\Delta t = 1.0$ s and the numerical solution using $\Delta t = 0.2$ s.

18.82 The slender 10-kg bar is released from rest in the horizontal position shown. Using $\Delta t = 0.1$ s, determine the bar's angular position and angular velocity for the first five time steps.

18.83 The slender 10-kg bar is released from rest in the horizontal position shown. Determine the bar's angular position and angular velocity as functions of time from $t = 0$ to $t = 0.8$ s, using $\Delta t = 0.1$ s, $\Delta t = 0.01$ s, and $\Delta t = 0.001$ s. Draw the graphs of the angular velocity as a function of the angular position for these three cases, and compare them with the graph of the closed-form solution for the angular velocity as a function of the angular position.

18.84 The slender 10-kg bar is released from rest in the horizontal position shown. Suppose that the bar's pin support contains a damping device that exerts a resisting couple on the bar of magnitude $c\omega$ (N-m), where ω is the angular velocity in radians per second. Using $\Delta t = 0.001$ s, draw graphs of the bar's angular velocity as a function of time from $t = 0$ to $t = 0.8$ s for the cases $c = 0$, $c = 2$, $c = 4$, and $c = 8$.

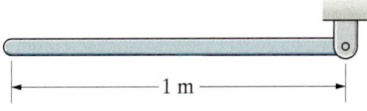

Problems 18.82–18.84

Problems 18.80/18.81

18.85 The 18-kg ladder is released from rest in the position shown. The floor and wall are smooth. The ladder will lose contact with the wall before it hits the floor. Using $\Delta t = 0.001$ s, estimate the time and the value of the angle between the wall and the ladder when this occurs.

Strategy: The formulation of the problem assumes that the ladder remains in contact with the wall. For times greater than the time at which it would lose contact, the solution for the normal force exerted on the ladder by the wall will become negative. So you can determine the time at which contact is lost by determining the time at which the normal force decreases to zero.

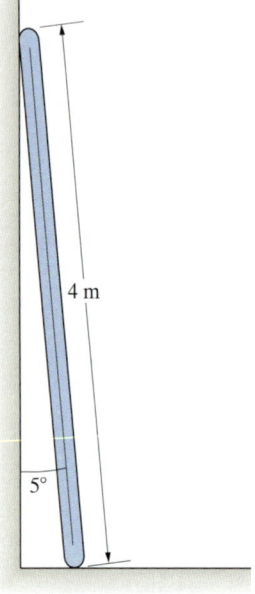

4 m

5°

Problem 18.85

18.86 A torsional spring exerts a counterclockwise couple $k\theta$ on the bar, where $k = 20$ N-m and θ is in radians. The 2-kg bar is 1 m long. At $t = 0$, the bar is released from rest in the horizontal position ($\theta = 0$). Using $\Delta t = 0.01$ s, determine the bar's angular position and angular velocity for the first five time steps.

18.87 Using a numerical solution with $\Delta t = 0.001$ s, estimate the maximum angle θ reached by the bar in Problem 18.86 when it is released from rest in the horizontal position. At what time after release does the maximum angle occur?

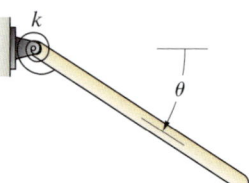

k

θ

Problems 18.86/18.87

Appendix: Moments of Inertia

When a rigid body is subjected to forces and couples, the rotational motion that results depends not only on the mass of the body, but also on how its mass is *distributed*. Although the two objects in Fig. 18.25 have the same mass, the angular accelerations caused by the couple M are different. This difference is reflected in the equation of angular motion $M = I\alpha$ through the moment of inertia, I. The object in Fig. 18.25a has a smaller moment of inertia about the axis L, so its angular acceleration is greater.

In deriving the equations of motion of a rigid body in Sections 18.2 and 18.3, we modeled the body as a finite number of particles and expressed its moment of inertia about an axis L_O as

$$I_O = \sum_i m_i r_i^2,$$

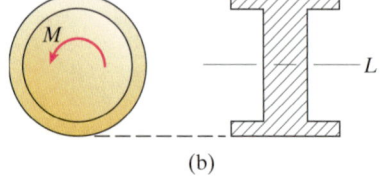

M

L

(a)

M

L

(b)

Figure 18.25
Objects of equal mass that have different moments of inertia about L.

where m_i is the mass of the ith particle and r_i is the perpendicular distance from L_O to the ith particle (Fig. 18.26a). To calculate moments of inertia of objects, it is often more convenient to model them as continuous distributions of mass and express the moment of inertia about L_O as

$$I_O = \int_m r^2 \, dm, \qquad (18.27)$$

where r is the perpendicular distance from L_O to the differential element of mass dm (Fig. 18.26b). When the axis passes through the center of mass of the object, we denote the axis by L and the moment of inertia about L by I.

The dimensions of the moment of inertia of an object are $(\text{mass}) \times (\text{length})^2$. Notice that the definition implies that its value must be positive.

Simple Objects

We begin by determining moments of inertia of some simple objects. Then, in the next section, we describe the parallel-axis theorem, which simplifies the task of determining moments of inertia of objects composed of combinations of simple parts.

Slender Bars   We will determine the moment of inertia of a straight slender bar about a perpendicular axis L through the center of mass of the bar (Fig. 18.27a). "Slender" means we assume that the bar's length is much greater than its width. Let the bar have length l, cross-sectional area A, and mass m. We assume that A is uniform along the length of the bar and that the material is homogeneous. Consider a differential element of the bar of length dr at a distance r from the center of mass (Fig. 18.27b). The element's mass is equal to the product of its volume and the mass density: $dm = \rho A \, dr$. Substituting this expression into Eq. (18.27), we obtain the moment of inertia of the bar about a perpendicular axis through its center of mass:

$$I = \int_m r^2 \, dm = \int_{-l/2}^{l/2} \rho A r^2 \, dr = \frac{1}{12} \rho A l^3.$$

The mass of the bar equals the product of the mass density and the volume of the bar $(m = \rho A l)$, so we can express the moment of inertia as

$$I = \frac{1}{12} m l^2. \qquad (18.28)$$

We have neglected the lateral dimensions of the bar in obtaining this result. That is, we treated the differential element of mass dm as if it were concentrated on the axis of the bar. As a consequence, Eq. (18.28) is an approximation for the moment of inertia of a bar. Later in this section, we will determine the moments of inertia for a bar of finite lateral dimension and show that Eq. (18.28) is a good approximation when the width of the bar is small in comparison to its length.

Thin Plates   Consider a homogeneous flat plate that has mass m and uniform thickness T. We will leave the shape of the cross-sectional area of the plate unspecified. Let a cartesian coordinate system be oriented so that the plate lies in the x–y plane (Fig. 18.28a). Our objective is to determine the moments of inertia of the plate about the x, y, and z axes.

We can obtain a differential element of volume of the plate by projecting an element of area dA through the thickness T of the plate (Fig. 18.28b). The resulting

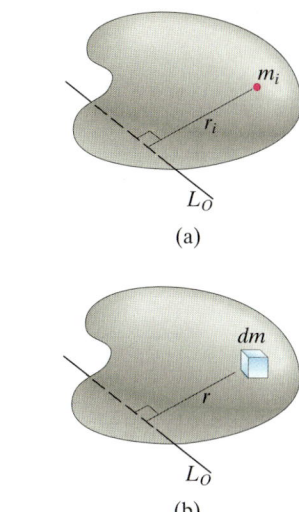

(a)

(b)

Figure 18.26
Determining the moment of inertia by modeling an object as (a) a finite number of particles and (b) a continuous distribution of mass.

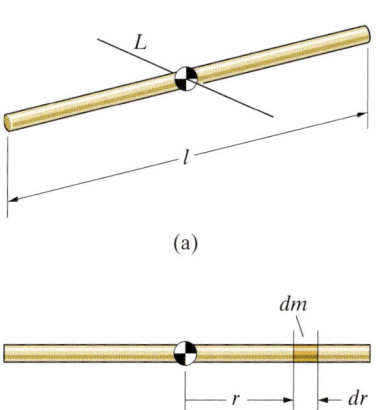

(a)

(b)

Figure 18.27
(a) A slender bar.
(b) A differential element of length dr.

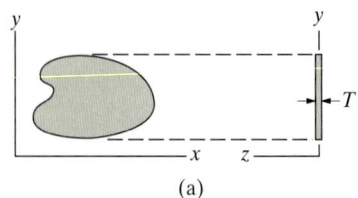

(a)

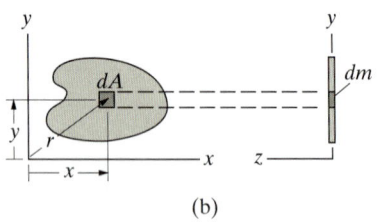

(b)

Figure 18.28
(a) A plate of arbitrary shape and uniform thickness T.
(b) An element of volume obtained by projecting an element of area dA through the plate.

volume is TdA. The mass of this element of volume is equal to the product of the mass density and the volume: $dm = \rho T \, dA$. Substituting this expression into Eq. (18.27), we obtain the moment of inertia of the plate about the z axis in the form

$$I_{z \text{ axis}} = \int_m r^2 \, dm = \rho T \int_A r^2 \, dA,$$

where r is the distance from the z axis to dA. Because the mass of the plate is $m = \rho T A$, where A is the cross-sectional area of the plate, the product $\rho T = m/A$. The integral on the right is the polar moment of inertia J_O of the cross-sectional area of the plate. Therefore, we can write the moment of inertia of the plate about the z axis as

$$I_{z \text{ axis}} = \frac{m}{A} J_O. \tag{18.29}$$

From Fig. 18.28b, we see that the perpendicular distance from the x axis to the element of area dA is the y coordinate of dA. Consequently, the moment of inertia of the plate about the x axis is

$$I_{x \text{ axis}} = \int_m y^2 \, dm = \rho T \int_A y^2 \, dA = \frac{m}{A} I_x, \tag{18.30}$$

where I_x is the moment of inertia of the cross-sectional area of the plate about the x axis. The moment of inertia of the plate about the y axis is

$$I_{y \text{ axis}} = \int_m x^2 \, dm = \rho T \int_A x^2 \, dA = \frac{m}{A} I_y, \tag{18.31}$$

where I_y is the moment of inertia of the cross-sectional area of the plate about the y axis.

Because the sum of the area moments of inertia I_x and I_y is equal to the polar moment of inertia J_O, the moment of inertia of the thin plate about the z axis is equal to the sum of its moments of inertia about the x and y axes:

$$I_{z \text{ axis}} = I_{x \text{ axis}} + I_{y \text{ axis}}. \tag{18.32}$$

Thus, we have expressed the moments of inertia of a thin homogeneous plate of uniform thickness in terms of the moments of inertia of the cross-sectional area of the plate. In fact, these results explain why the area integrals I_x, I_y, and J_O are called moments of inertia.

The use of the same terminology and similar symbols for moments of inertia of areas and moments of inertia of objects can be confusing, but is entrenched in engineering practice. The type of moment of inertia being referred to can be determined either from the context or from the units, $(\text{length})^4$ for moments of inertia of areas and $(\text{mass}) \times (\text{length})^2$ for moments of inertia of objects.

Example 18.9   Moment of Inertia of an L-Shaped Bar

Two homogeneous slender bars, each of length l, mass m, and cross-sectional area A, are welded together to form an L-shaped object (Fig. 18.29). Determine the moment of inertia of the object about the axis L_O through point O. (The axis L_O is perpendicular to the two bars.)

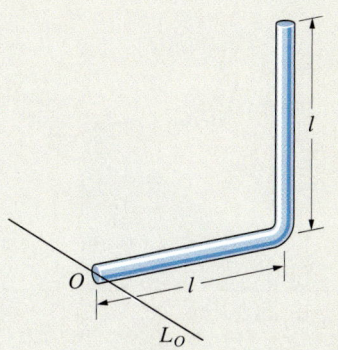

Figure 18.29

Strategy

Using the same integration procedure we used for a single bar, we can determine the moment of inertia of each bar about L_O and sum the results.

Solution

We orient a coordinate system with the z axis along L_O and the x axis colinear with bar 1 (Fig. a). The mass of the differential element of length dx of bar 1 is $dm = \rho A\, dx$. The moment of inertia of bar 1 about L_O is

$$(I_O)_1 = \int_m r^2\, dm = \int_0^l \rho A x^2\, dx = \frac{1}{3}\rho A l^3.$$

In terms of the mass of the bar, $m = \rho A l$, we can write this result as

$$(I_O)_1 = \frac{1}{3}ml^2.$$

The mass of the element of length dy of bar 2 shown in Fig. b is $dm = \rho A\, dy$. From the figure, we see that the perpendicular distance from L_O to the element is $r = \sqrt{l^2 + y^2}$. Therefore, the moment of inertia of bar 2 about L_O is

$$(I_O)_2 = \int_m r^2\, dm = \int_0^l \rho A (l^2 + y^2)\, dy = \frac{4}{3}\rho A l^3.$$

In terms of the mass of the bar, we obtain

$$(I_O)_2 = \frac{4}{3}ml^2.$$

The moment of inertia of the L-shaped object about L_O is

$$I_O = (I_O)_1 + (I_O)_2 = \frac{1}{3}ml^2 + \frac{4}{3}ml^2 = \frac{5}{3}ml^2.$$

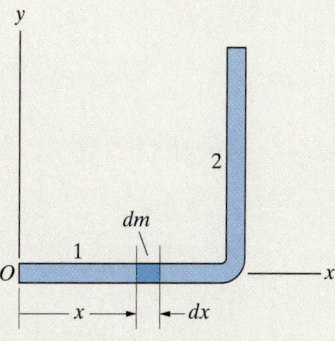

(a) Differential element of bar 1.

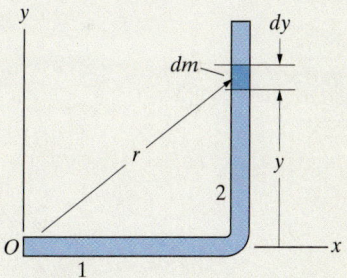

(b) Differential element of bar 2.

Critical Thinking

In this example we used integration to determine a moment of inertia of an object consisting of two straight bars. The same procedure could be applied to more complicated objects made of such bars, but it would obviously be cumbersome. Once we have used integration to determine a moment of inertia of a single bar, such as Eq. (18.28), it would be very convenient to be able to use that result to determine moments of inertia of composite objects made of bars without having to resort to integration. We show how this can be done in the next section.

Example 18.10 | Moments of Inertia of a Triangular Plate

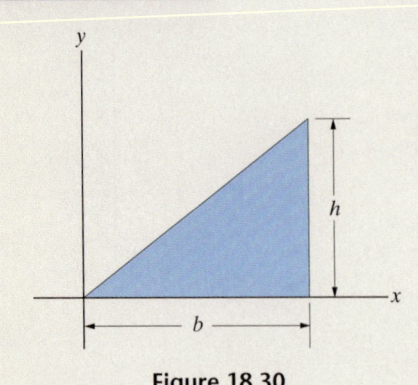

Figure 18.30

The thin, homogeneous plate in Fig. 18.30 is of uniform thickness and mass m. Determine its moments of inertia about the x, y, and z axes.

Strategy

The moments of inertia about the x and y axes are given by Eqs. (18.30) and (18.31) in terms of the moments of inertia of the cross-sectional area of the plate. We can determine the moment of inertia of the plate about the z axis from Eq. (18.32).

Solution

From Appendix B, the moments of inertia of the triangular area about the x and y axes are $I_x = \frac{1}{12}bh^3$ and $I_y = \frac{1}{4}hb^3$. Therefore, the moments of inertia of the plate about the x and y axes are

$$I_{x\,\text{axis}} = \frac{m}{A}I_x = \frac{m}{\frac{1}{2}bh}\left(\frac{1}{12}bh^3\right) = \frac{1}{6}mh^2$$

and

$$I_{y\,\text{axis}} = \frac{m}{A}I_y = \frac{m}{\frac{1}{2}bh}\left(\frac{1}{4}hb^3\right) = \frac{1}{2}mb^2.$$

The moment of inertia about the z axis is

$$I_{z\,\text{axis}} = I_{x\,\text{axis}} + I_{y\,\text{axis}} = m\left(\frac{1}{6}h^2 + \frac{1}{2}b^2\right).$$

Critical Thinking

As this example demonstrates, you can use the moments of inertia of areas tabulated in Appendix B to determine moments of inertia of thin homogeneous plates. For plates with more complicated shapes, you can use the methods for determining moments of inertia of composite areas.

Parallel-Axis Theorem

The parallel-axis theorem allows us to determine the moment of inertia of a composite object when we know the moments of inertia of its parts. Suppose that we know the moment of inertia I about an axis L through the center of mass of an object and we wish to determine its moment of inertia I_O about a parallel axis L_O (Fig. 18.31a). To determine I_O, we introduce parallel coordinate systems xyz and $x'y'z'$, with the z axis along L_O and the z' axis along L, as shown in Fig. 18.31b. (In this figure, the axes L_O and L are perpendicular to the page.) The origin O of the xyz coordinate system is contained in the $x'-y'$ plane. The terms d_x and d_y are the coordinates of the center of mass relative to the xyz coordinate system.

The moment of inertia of the object about L_O is

$$I_O = \int_m r^2\, dm = \int_m (x^2 + y^2)\, dm, \tag{18.33}$$

where r is the perpendicular distance from L_O to the differential element of mass dm and x, y are the coordinates of dm in the $x-y$ plane. The $x-y$ coordinates of dm are related to its $x'-y'$ coordinates by

$$x = x' + d_x$$

and

$$y = y' + d_y.$$

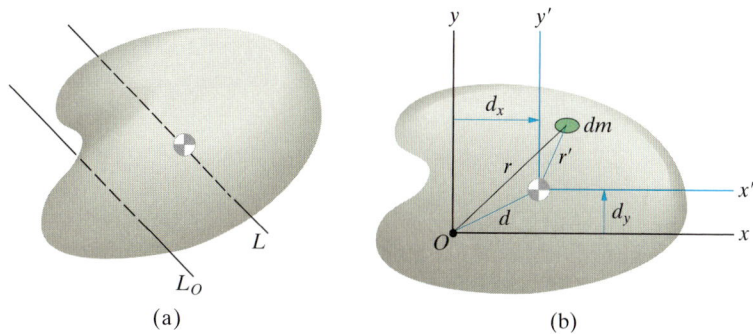

(a)   (b)

Figure 18.31
(a) An axis L through the center of mass of an object and a parallel axis L_O.
(b) The xyz and $x'y'z'$ coordinate systems.

Substituting these expressions into Eq. (18.33), we can write that equation as

$$I_O = \int_m [(x')^2 + (y')^2] \, dm + 2d_x \int_m x' \, dm + 2d_y \int_m y' \, dm$$

$$+ \int_m (d_x^2 + d_y^2) \, dm. \qquad (18.34)$$

Since $(x')^2 + (y')^2 = (r')^2$, where r' is the perpendicular distance from L to dm, the first integral on the right side of this equation is the moment of inertia I of the object about L. Recall that the x' and y' coordinates of the center of mass of the object relative to the $x'y'z'$ coordinate system are defined by

$$\overline{x}' = \frac{\int_m x' \, dm}{\int_m dm}, \quad \overline{y}' = \frac{\int_m y' \, dm}{\int_m dm}.$$

Because the center of mass of the object is at the origin of the $x'y'z'$ system, $\overline{x}' = 0$ and $\overline{y}' = 0$. Therefore, the integrals in the second and third terms on the right side of Eq. (18.34) are equal to zero. From Fig. 18.31(b), we see that $d_x^2 + d_y^2 = d^2$, where d is the perpendicular distance between the axes L and L_O. Therefore, we obtain

$$I_O = I + d^2 m. \qquad (18.35)$$

This is the *parallel-axis theorem* for moments of inertia of objects. Equation (18.35) relates the moment of inertia I of an object about an axis *through the center of mass* to its moment of inertia I_O about any parallel axis, where d is the perpendicular distance between the two axes and m is the mass of the object.

The parallel-axis theorem makes it possible to determine moments of inertia of composite objects. Determine the moment of inertia about a given axis L_O typically requires three steps:

1. *Choose the parts.* Try to divide the object into parts whose moments of inertia can easily be determined.

2. *Determine the moments of inertia of the parts.* Determine the moment of inertia of each part about the axis through its center of mass parallel to L_O. Then use the parallel-axis theorem to determine its moment of inertia about L_O.

3. *Sum the results.* Sum the moments of inertia of the parts (or subtract in the case of a hole or cutout) to obtain the moment of inertia of the composite object.

| **Example 18.11** | **Application of the Parallel-Axis Theorem** |

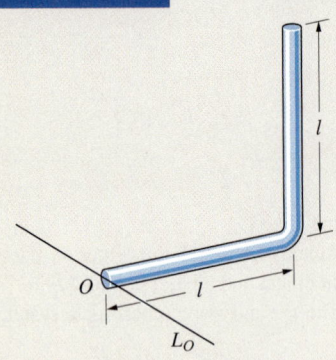

Figure 18.32

Two homogeneous, slender bars, each of length l and mass m, are welded together to form an L-shaped object (Fig. 18.32). Determine the moment of inertia of the object about the axis L_O through point O. (The axis L_O is perpendicular to the two bars.)

Strategy
The moment of inertia of a straight slender bar about a perpendicular axis through its center of mass is given by Eq. (18.28). We can use the parallel-axis theorem to determine the moments of inertia of the bars about the axis L_O and sum them to obtain the moment of inertia of the composite bar.

Solution
Choose the Parts   The parts are the two bars, which we call bar 1 and bar 2 (Fig. a).

Determine the Moments of Inertia of the Parts   From Appendix C, the moment of inertia of each bar about a perpendicular axis through its center of mass is $I = \frac{1}{12}ml^2$. The distance from L_O to the parallel axis through the center of mass of bar 1 is $\frac{1}{2}l$ (Fig. a). Therefore, the moment of inertia of bar 1 about L_O is

$$(I_O)_1 = I + d^2m = \frac{1}{12}ml^2 + \left(\frac{1}{2}l\right)^2 m = \frac{1}{3}ml^2.$$

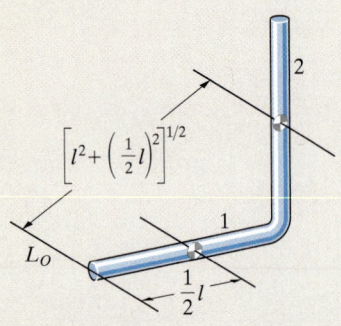

(a) The distances from L_O to parallel axes through the centers of mass of bars 1 and 2.

The distance from L_O to the parallel axis through the center of mass of bar 2 is $[l^2 + \left(\frac{1}{2}l\right)^2]^{1/2}$. The moment of inertia of bar 2 about L_O is

$$(I_O)_2 = I + d^2m = \frac{1}{12}ml^2 + \left[l^2 + \left(\frac{1}{2}l\right)^2\right]m = \frac{4}{3}ml^2.$$

Sum the Results   The moment of inertia of the L-shaped object about L_O is

$$I_O = (I_O)_1 + (I_O)_2 = \frac{1}{3}ml^2 + \frac{4}{3}ml^2 = \frac{5}{3}ml^2.$$

Critical Thinking
Compare this solution with Example 18.9, in which we used integration to determine the moment of inertia of the same object about L_O. We obtained the result much more easily with the parallel-axis theorem, but of course we needed to know the moments of inertia of the bars about the axes through their centers of mass.

Example 18.12   Moments of Inertia of a Composite Object

The object in Fig. 18.33 consists of a slender 3-kg bar welded to a thin, circular 2-kg disk. Determine the moment of inertia of the object about the axis L through its center of mass. (The axis L is perpendicular to the bar and disk.)

Strategy
We must locate the center of mass of the composite object and then apply the parallel-axis theorem. We can obtain the moments of inertia of the bar and disk from Appendix C.

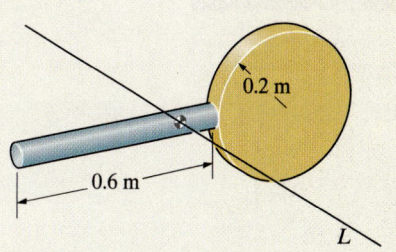

Figure 18.33

Solution

Choose the Parts   The parts are the bar and the disk. Introducing the coordinate system in Fig. a, we have, for the x coordinate of the center of mass of the composite object,

$$\bar{x} = \frac{\bar{x}_{\text{bar}} m_{\text{bar}} + \bar{x}_{\text{disk}} m_{\text{disk}}}{m_{\text{bar}} + m_{\text{disk}}}$$

$$= \frac{(0.3 \text{ m})(3 \text{ kg}) + (0.6 \text{ m} + 0.2 \text{ m})(2 \text{ kg})}{(3 \text{ kg}) + (2 \text{ kg})} = 0.5 \text{ m}.$$

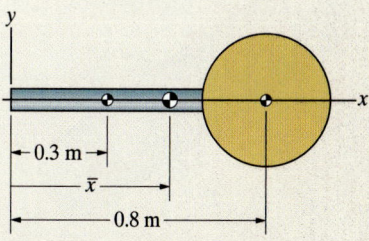

(a) The coordinate $\bar{x}$ of the center of mass of the object.

Determine the Moments of Inertia of the Parts   The distance from the center of mass of the bar to the center of mass of the composite object is 0.2 m (Fig. b). Therefore, the moment of inertia of the bar about L is

$$I_{\text{bar}} = \frac{1}{12}(3 \text{ kg})(0.6 \text{ m})^2 + (3 \text{ kg})(0.2 \text{ m})^2 = 0.210 \text{ kg-m}^2.$$

The distance from the center of mass of the disk to the center of mass of the composite object is 0.3 m (Fig. c). The moment of inertia of the disk about L is

$$I_{\text{disk}} = \frac{1}{2}(2 \text{ kg})(0.2 \text{ m})^2 + (2 \text{ kg})(0.3 \text{ m})^2 = 0.220 \text{ kg-m}^2.$$

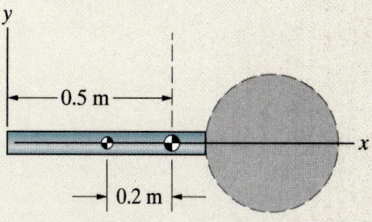

(b) Distance from L to the center of mass of the bar.

Sum the Results   The moment of inertia of the composite object about L is

$$I = I_{\text{bar}} + I_{\text{disk}} = 0.430 \text{ kg-m}^2.$$

Critical Thinking
This example demonstrates the most common procedure for determining moments of inertia of objects in engineering applications. Objects usually consist of assemblies of parts. The center of mass of each part and its moment of inertia about the axis through its center of mass must be determined. (It may be necessary to determine this information experimentally, or it is sometimes supplied by manufacturers of subassemblies.) Then the center of mass of the composite object is determined and the parallel axis theorem is used to determine the moment of inertia of each part about the axis through the center of mass of the composite object. Finally, the individual moments of inertia are summed to obtain the moment of inertia of the composite object.

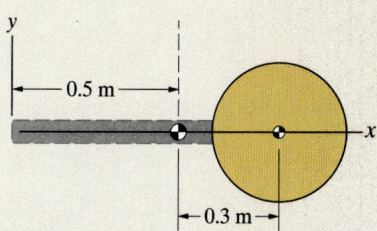

(c) Distance from L to the center of mass of the disk.

Example 18.13   **Moments of Inertia of a Homogeneous Cylinder**

The homogeneous cylinder in Fig. 18.34 has mass m, length l, and radius R. Determine the moments of inertia of the cylinder about the x, y, and z axes.

Strategy

We can determine the moments of inertia of the cylinder by an interesting application of the parallel-axis theorem. We use it to determine the moments of inertia about the x, y, and z axes of an infinitesimal element of the cylinder consisting of a disk of thickness dz. Then we integrate the results with respect to z to obtain the moments of inertia of the cylinder.

Solution

Consider an element of the cylinder of thickness dz at a distance z from the center of the cylinder (Fig. a). (You can imagine obtaining this element by "slicing" the cylinder perpendicular to its axis.) The mass of the element is equal to the product of the mass density and the volume of the element: $dm = \rho(\pi R^2\, dz)$. We obtain the moments of inertia of the element by using the values for a thin circular plate given in Appendix C. The moment of inertia about the z axis is

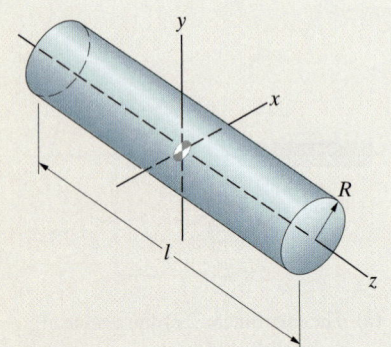

Figure 18.34

$$dI_{z \text{ axis}} = \frac{1}{2} dm\, R^2 = \frac{1}{2}(\rho \pi R^2\, dz) R^2.$$

We integrate this result with respect to z from $-l/2$ to $l/2$, thereby summing the moments of inertia of the infinitesimal disk elements that make up the cylinder. The result is the moment of inertia of the cylinder about the z axis:

$$I_{z \text{ axis}} = \int_{-l/2}^{l/2} \frac{1}{2} \rho \pi R^4\, dz = \frac{1}{2} \rho \pi R^4 l.$$

We can write this result in terms of the mass of the cylinder, $m = \rho(\pi R^2 l)$, as

$$I_{z \text{ axis}} = \frac{1}{2} m R^2.$$

The moment of inertia of the disk element about the x' axis is

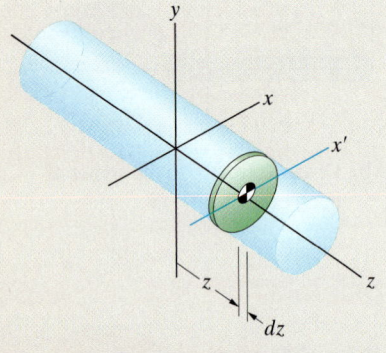

(a) A differential element of the cylinder in the form of a disk.

$$dI_{x' \text{ axis}} = \frac{1}{4} dm\, R^2 = \frac{1}{4}(\rho \pi R^2\, dz) R^2.$$

We use this result and the parallel-axis theorem to determine the moment of inertia of the element about the x axis:

$$dI_{x \text{ axis}} = dI_{x' \text{ axis}} + z^2\, dm = \frac{1}{4}(\rho \pi R^2\, dz) R^2 + z^2(\rho \pi R^2\, dz).$$

Integrating this expression with respect to z from $-l/2$ to $l/2$, we obtain the moment of inertia of the cylinder about the x axis:

$$I_{x \text{ axis}} = \int_{-l/2}^{l/2} \left(\frac{1}{4} \rho \pi R^4 + \rho \pi R^2 z^2 \right) dz = \frac{1}{4} \rho \pi R^4 l + \frac{1}{12} \rho \pi R^2 l^3.$$

In terms of the mass of the cylinder,

$$I_{x \text{ axis}} = \frac{1}{4}mR^2 + \frac{1}{12}ml^2.$$

Due to the symmetry of the cylinder,

$$I_{y \text{ axis}} = I_{x \text{ axis}}.$$

Critical Thinking

When the cylinder is very long in comparison to its width ($l \gg R$), the first term in the equation for $I_{x \text{ axis}}$ can be neglected, and we obtain the moment of inertia of a slender bar about a perpendicular axis, Eq. (18.28). On the other hand, when the radius of the cylinder is much greater than its length ($R \gg l$), the second term in the equation for $I_{x \text{ axis}}$ can be neglected, and we obtain the moment of inertia for a thin circular disk about an axis parallel to the disk. This indicates the sizes of the terms you neglect when you use the approximate expressions for the moments of inertia of a "slender" bar and a "thin" disk.

Problems

18.88 The axis L_O is perpendicular to both segments of the L-shaped slender bar. The mass of the bar is 6 kg and the material is homogeneous. Use integration to determine the moment of inertia of the bar about L_O.

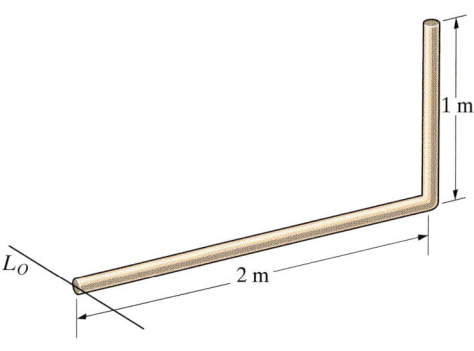

Problem 18.88

18.89 Two homogeneous slender bars, each of mass m and length l, are welded together to form the T-shaped object. Use integration to determine the moment of inertia of the object about the axis through point O that is perpendicular to the bars.

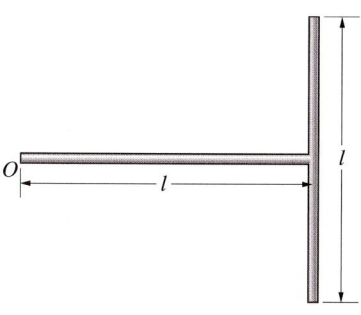

Problem 18.89

18.90 The slender bar lies in the x–y plane. Its mass is 6 kg and the material is homogeneous. Use integration to determine its moment of inertia about the z axis.

18.91 The slender bar lies in the x–y plane. Its mass is 6 kg and the material is homogeneous. Use integration to determine its moment of inertia about the y axis.

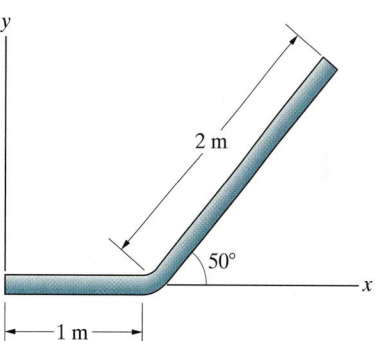

Problems 18.90/18.91

18.92 The homogeneous thin plate has mass $m = 12$ kg and dimensions $b = 1$ m and $h = 2$ m. Determine the moments of inertia of the plate about the x, y, and z axes.

Strategy: The moments of inertia of a thin plate of arbitrary shape are given by Eqs. (18.30)–(18.32) in terms of the moments of inertia of the cross-sectional area of the plate. You can obtain the moments of inertia of the triangular area from Appendix B.

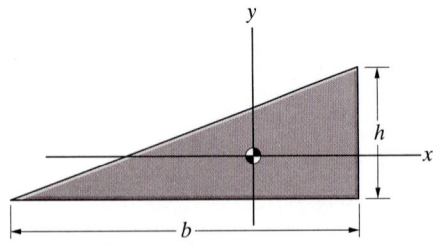

Problem 18.92

18.93 The brass washer is of uniform thickness and mass m.

(a) Determine its moments of inertia about the x and z axes.
(b) Let $R_i = 0$, and compare your results with the values given in Appendix C for a thin circular plate.

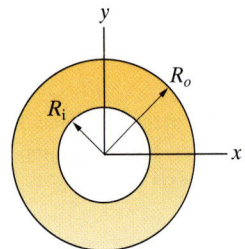

Problem 18.93

18.94 The homogeneous thin plate is of uniform thickness and weighs 20 lb. Determine its moment of inertia about the y axis.

18.95 Determine the moment of inertia of the 20-lb plate about the x axis.

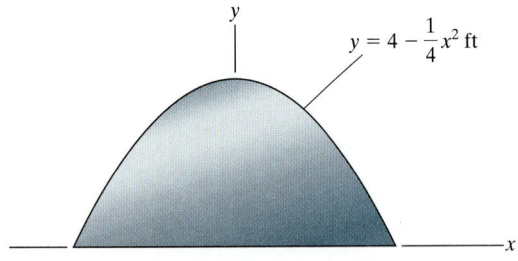

Problems 18.94/18.95

18.96 The mass of the object is 10 kg. Its moment of inertia about L_1 is 10 kg-m^2. What is its moment of inertia about L_2? (The three axes lie in the same plane.)

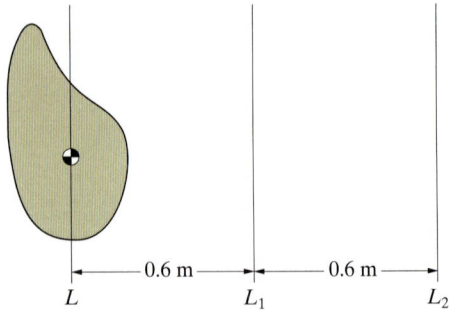

Problem 18.96

18.97 An engineer gathering data for the design of a maneuvering unit determines that the astronaut's center of mass is at $x = 1.01$ m, $y = 0.16$ m and that her moment of inertia about the z axis is 105.6 kg-m^2. The astronaut's mass is 81.6 kg. What is her moment of inertia about the z' axis through her center of mass?

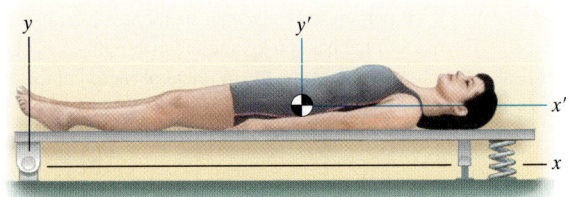

Problem 18.97

18.98 Two homogeneous slender bars, each of mass m and length l, are welded together to form the T-shaped object. Use the parallel-axis theorem to determine the moment of inertia of the object about the axis through point O that is perpendicular to the bars.

18.99 Use the parallel-axis theorem to determine the moment of inertia of the T-shaped object in Problem 18.98 about the axis through the center of mass of the object that is perpendicular to the two bars.

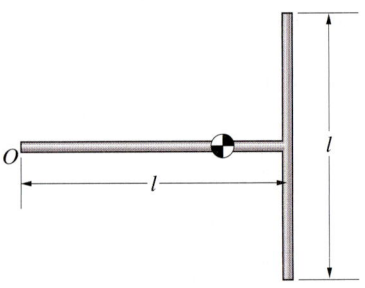

Problems 18.98/18.99

18.100 The mass of the homogeneous slender bar is 30 kg. Determine its moment of inertia about the z axis.

18.101 The mass of the homogeneous slender bar is 30 kg. Determine the moment of inertia of the bar about the z' axis through its center of mass.

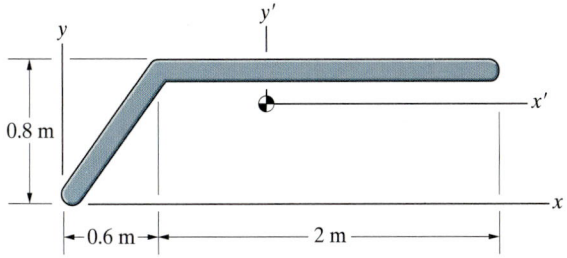

Problems 18.100/18.101

18.102 The homogeneous slender bar weighs 5 lb. Determine its moment of inertia about the z axis.

18.103 Determine the moment of inertia of the 5-lb bar about the z' axis through its center of mass.

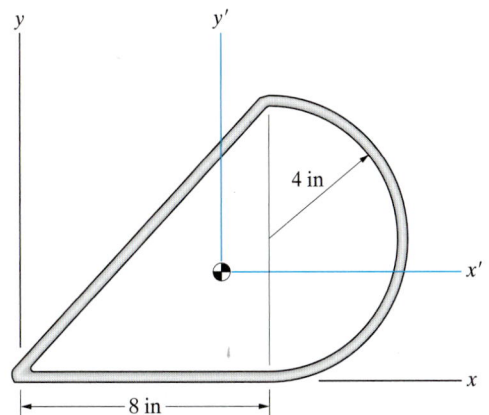

Problems 18.102/18.103

18.104 The rocket is used for atmospheric research. Its weight and its moment of inertia about the z axis through its center of mass (including its fuel) are 10,000 lb and 10,200 slug-ft^2, respectively. The rocket's fuel weighs 6000 lb, its center of mass is located at $x = -3$ ft, $y = 0$, $z = 0$, and the moment of inertia of the fuel about the axis through the fuel's center of mass parallel to the z axis is 2200 slug-ft^2. When the fuel is exhausted, what is the rocket's moment of inertia about the axis through its new center of mass parallel to the z axis?

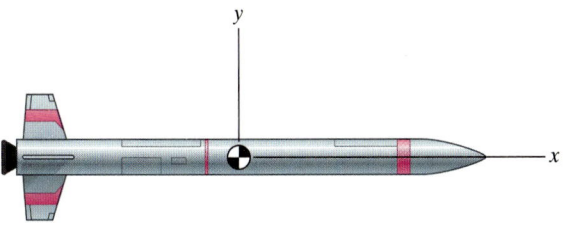

Problem 18.104

18.105 The mass of the homogeneous thin plate is 36 kg. Determine the moment of inertia of the plate about the x axis.

18.106 Determine the moment of inertia of the 36-kg plate about the z axis.

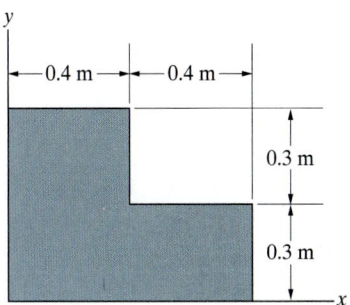

Problems 18.105/18.106

18.107 The mass of the homogeneous thin plate is 20 kg. Determine its moment of inertia about the x axis.

18.108 The mass of the homogeneous thin plate is 20 kg. Determine its moment of inertia about the y axis.

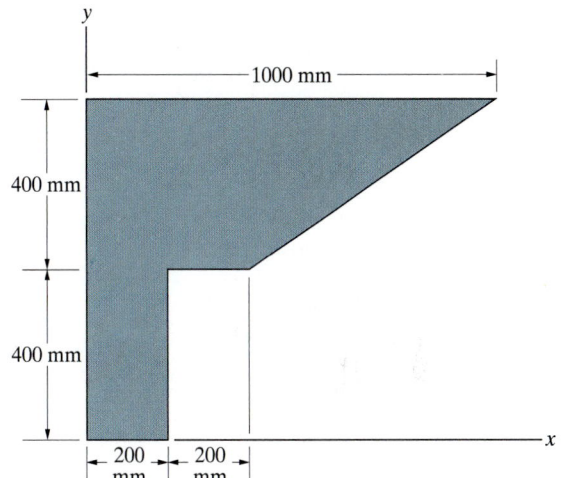

Problems 18.107/18.108

18.109 The thermal radiator (used to eliminate excess heat from a satellite) can be modeled as a homogeneous thin rectangular plate. The mass of the radiator is 5 slugs. Determine its moments of inertia about the x, y, and z axes.

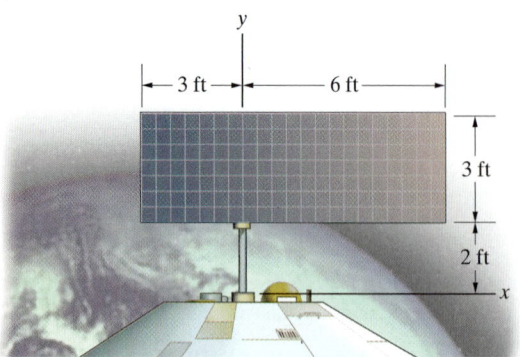

Problem 18.109

18.110 The mass of the homogeneous thin plate is 2 kg. Determine the moment of inertia of the plate about the axis through point O that is perpendicular to the plate.

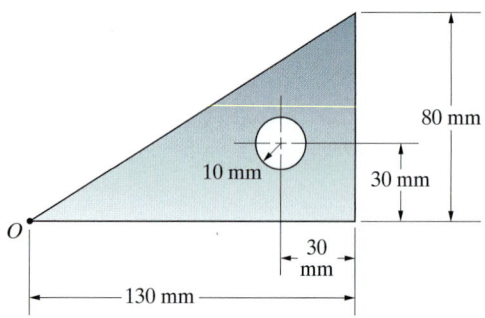

Problem 18.110

18.111 The homogeneous cone is of mass m. Determine its moment of inertia about the z axis, and compare your result with the value given in Appendix C.

  Strategy: Use the same approach we used in Example 18.13 to obtain the moments of inertia of a homogeneous cylinder.

18.112 Determine the moments of inertia of the homogeneous cone of mass m about the x and y axes, and compare your results with the values given in Appendix C.

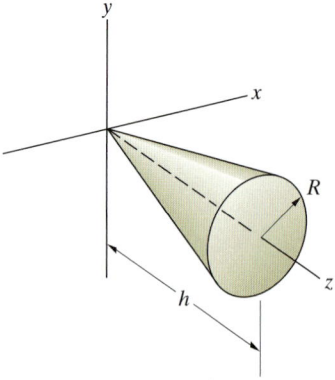

Problems 18.111/18.112

18.113 The homogeneous object has the shape of a truncated cone and consists of bronze with mass density $\rho = 8200$ kg/m^3. Determine the moment of inertia of the object about the z axis.

18.114 Determine the moment of inertia of the object in Problem 18.113 about the x axis.

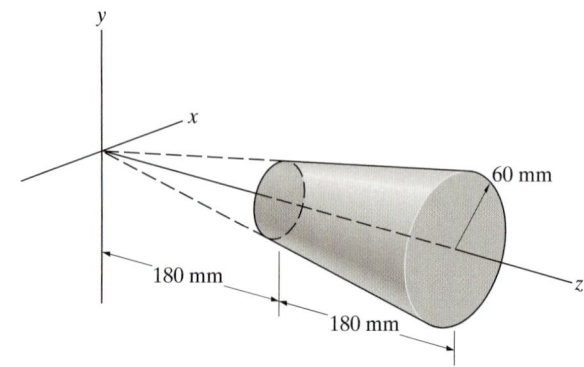

Problems 18.113/18.114

18.115 The homogeneous rectangular parallelepiped is of mass m. Determine its moments of inertia about the x, y, and z axes and compare your results with the values given in Appendix C.

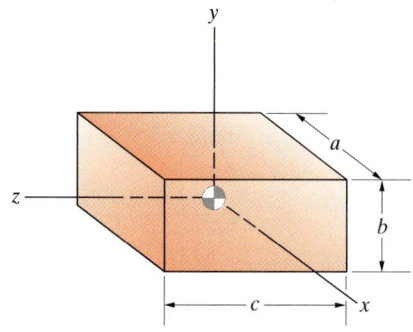

Problem 18.115

18.116 The sphere-capped cone consists of material with density 7800 kg/m³. The radius $R = 80$ mm. Determine its moment of inertia about the x axis.

18.117 Determine the moment of inertia of the sphere-capped cone described in Problem 18.116 about the y axis.

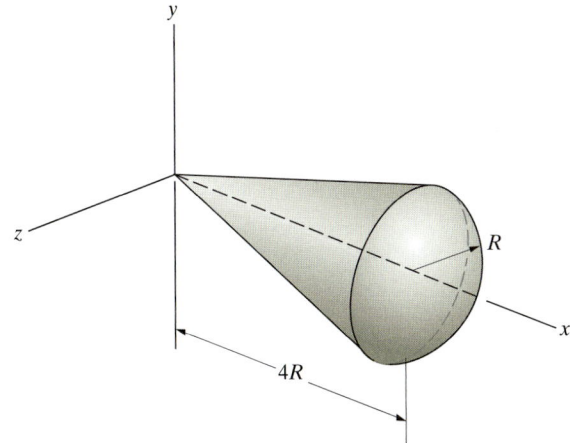

Problems 18.116/18.117

18.118 The circular cylinder is made of aluminum (Al) with density 2700 kg/m³ and iron (Fe) with density 7860 kg/m³. Determine its moment of inertia about the x' axis.

18.119 Determine the moment of inertia of the composite cylinder described in Problem 18.118 about the y' axis.

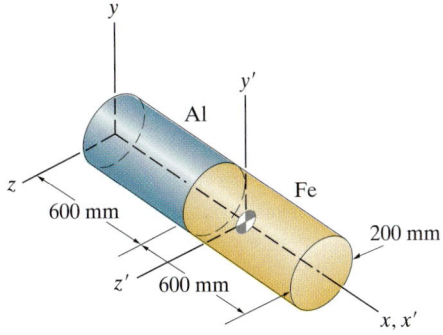

Problems 18.118/18.119

18.120 The homogeneous machine part is made of aluminum alloy with mass density $\rho = 2800$ kg/m³. Determine the moment of inertia of the part about the z axis.

18.121 Determine the moment of inertia of the machine part described in Problem 18.120 about the x axis.

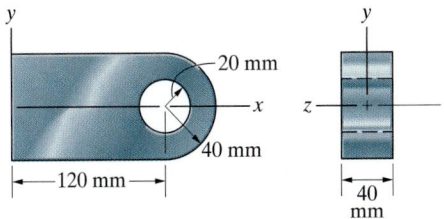

Problems 18.120/18.121

18.122 The object shown consists of steel of density $\rho = 7800$ kg/m³. Determine its moment of inertia about the axis L_O through point O.

18.123 Determine the moment of inertia of the object described in Problem 18.122 about the axis through the center of mass of the object parallel to L_O.

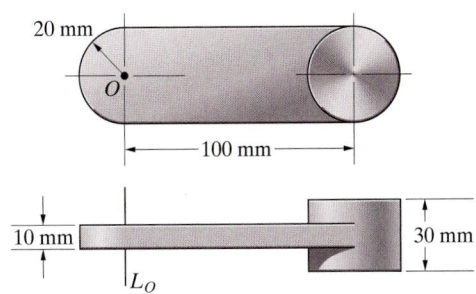

Problems 18.122/18.123

18.124 The thick plate consists of steel of density $\rho = 15$ slug/ft³. Determine the moment of inertia of the plate about the z axis.

18.125 Determine the moment of inertia of the object described in Problem 18.124 about the x axis.

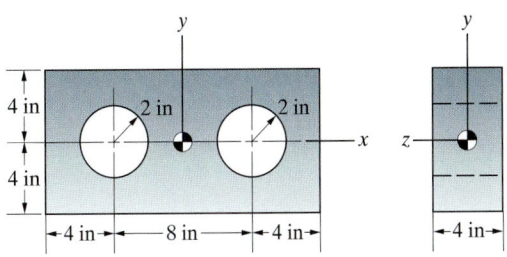

Problems 18.124/18.125

CHAPTER SUMMARY

In this chapter we derived the equations of planar motion for a rigid body. We used those equations together with the kinematics relationships developed in Chapter 17 to determine motions of rigid bodies resulting from the forces and couples acting on them. In Chapter 19, we will apply energy and momentum methods to the motions of rigid bodies and show that those methods can greatly simplify the solution of particular types of problems.

Moment–Angular-Momentum Relations

Let $\mathbf{r}_i$ be the position of the ith particle of a system of particles relative to a fixed point O, $\mathbf{r}$ the position of the center of mass of the system, and $\mathbf{R}_i$ the position of the ith particle relative to the center of mass. The sum of the moments due to external forces about O is equal to the rate of change of the total angular momentum about O:

$$\Sigma \mathbf{M}_O = \frac{d\mathbf{H}_O}{dt}. \tag{18.8}$$

The angular momentum about O is

$$\mathbf{H}_O = \sum_i \mathbf{r}_i \times m_i \mathbf{v}_i, \tag{18.9}$$

where m_i is the mass of the ith particle and $\mathbf{v}_i$ is its velocity. This relationship can also be written as

$$\Sigma \mathbf{M}_O = \frac{d}{dt}(\mathbf{r} \times m\mathbf{v} + \mathbf{H}), \tag{18.10}$$

where m is the total mass of the system of particles, $\mathbf{v}$ is the velocity of the center of mass, and

$$\mathbf{H} = \sum_i \mathbf{R}_i \times m_i \frac{d\mathbf{R}_i}{dt} \tag{18.11}$$

is the total angular momentum about the center of mass. The sum of the moments due to external forces about the center of mass is equal to the rate of change of the total angular momentum about the center of mass:

$$\Sigma \mathbf{M} = \frac{d\mathbf{H}}{dt}. \tag{18.12}$$

The angular momenta about point O and about the center of mass are related by

$$\mathbf{H}_O = \mathbf{H} + \mathbf{r} \times m\mathbf{v}. \tag{18.13}$$

Equations of Planar Motion

The equations of motion for a rigid body in planar motion include Newton's second law,

$$\Sigma \mathbf{F} = m\mathbf{a}, \tag{18.21}$$

where $\mathbf{a}$ is the acceleration of the center of mass of the body. If the rigid body rotates about a fixed point O, the total moment about O equals the product of the moment of inertia about O and the angular acceleration (Fig. a):

$$\Sigma M_O = I_O \alpha. \tag{18.22}$$

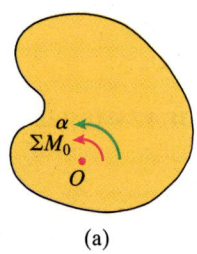

α
ΣM_0
O

(a)

In any planar motion, the total moment about the center of mass equals the product of the moment of inertia about the center of mass and the angular acceleration (Fig. b):

$$\Sigma M = I\alpha. \qquad (18.23)$$

If a rigid body is in translation, Newton's second law is sufficient to determine its motion. Nevertheless, the angular equation of motion may be needed to determine unknown forces or couples. Since $\alpha = 0$, the total moment about the center of mass equals zero. In the case of rotation about a fixed axis, Eq. (18.22) is sufficient to determine the rotational motion, although Newton's second law may be needed to determine unknown forces or couples. If a rigid body undergoes general planar motion, both Newton's second law and the equation of angular motion are needed.

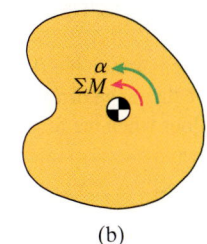

(b)

Moments of Inertia

The moment of inertia of an object about an axis L_O is

$$I_O = \int_m r^2 \, dm, \qquad (18.27)$$

where r is the perpendicular distance from L_O to the differential element of mass dm (Fig. c).

Let L be an axis through the center of mass of an object, and let L_O be a parallel axis. The moment of inertia, I_O, about L_O is given in terms of the moment of inertia, I, about L by the *parallel-axis theorem*

$$I_O = I + d^2 m, \qquad (18.35)$$

where m is the mass of the object and d is the distance between L and L_O.

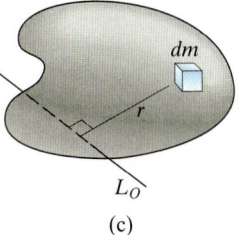

(c)

Review Problems

18.126 The airplane is at the beginning of its takeoff run. Its weight is 1000 lb, and the initial thrust T exerted by its engine is 300 lb. Assume that the thrust is horizontal, and neglect the tangential forces exerted on the wheels.

(a) If the acceleration of the airplane remains constant, how long will it take to reach its takeoff speed of 80 mi/h?
(b) Determine the normal force exerted on the forward landing gear at the beginning of the takeoff run.

18.127 The pulleys can turn freely on their pin supports. Their moments of inertia are $I_A = 0.002$ kg-m², $I_B = 0.036$ kg-m², and $I_C = 0.032$ kg-m². They are initially stationary, and at $t = 0$ a constant couple $M = 2$ N-m is applied to pulley A. What is the angular velocity of pulley C and how many revolutions has it turned at $t = 2$ s?

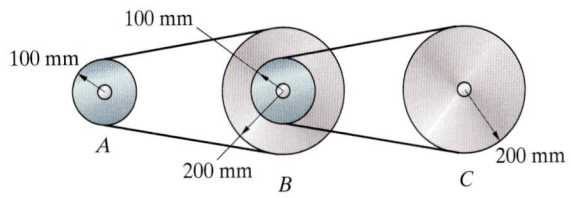

Problem 18.127

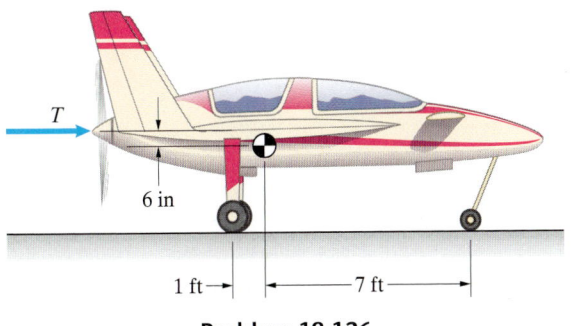

Problem 18.126

18.128 A 2-kg box is subjected to a 40-N horizontal force. Neglect friction.

(a) If the box remains on the floor, what is its acceleration?
(b) Determine the range of values of c for which the box will remain on the floor when the force is applied.

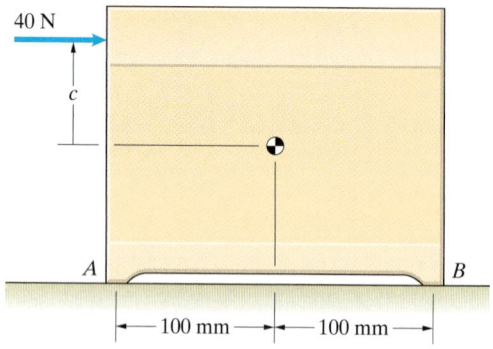

Problem 18.128

18.129 The slender 2-slug bar AB is 3 ft long. It is pinned to the cart at A and leans against it at B.

(a) If the acceleration of the cart is $a = 20$ ft/s^2, what normal force is exerted on the bar by the cart at B?
(b) What is the largest acceleration a for which the bar will remain in contact with the surface at B?

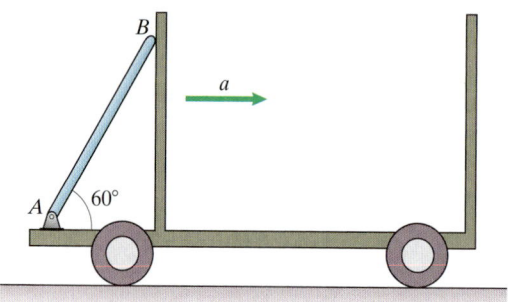

Problem 18.129

18.130 To determine a 4.5-kg tire's moment of inertia, an engineer lets the tire roll down an inclined surface. If it takes the tire 3.5 s to start from rest and roll 3 m down the surface, what is the tire's moment of inertia about its center of mass?

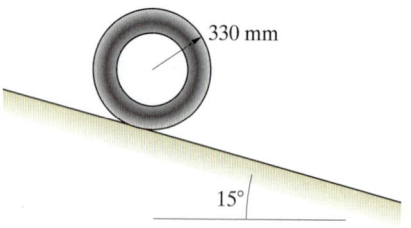

Problem 18.130

18.131 Pulley A weighs 4 lb, $I_A = 0.060$ slug-ft^2, and $I_B = 0.014$ slug-ft^2. If the system is released from rest, what distance does the 16-lb weight fall in 0.5 s?

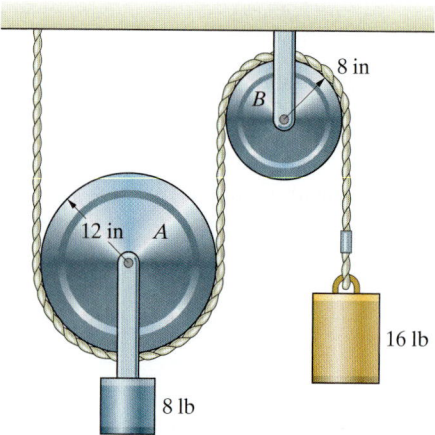

Problem 18.131

18.132 Model the excavator's arm *ABC* as a single rigid body. Its mass is 1200 kg, and the moment of inertia *about its center of mass* is $I = 3600$ kg-m^2. The angular velocity of the arm is zero and its angular acceleration is 1 rad/s^2 counterclockwise. What force does the vertical hydraulic cylinder exert on the arm at *B*?

18.133 Model the excavator's arm ABC as a single rigid body. Its mass is 1200 kg, and the moment of inertia *about its center of mass* is $I = 3600$ kg-m^2. The angular velocity of the arm is 2 rad/s counterclockwise and its angular acceleration is 1 rad/s^2 counterclockwise. What are the components of the force exerted on the arm at *A*?

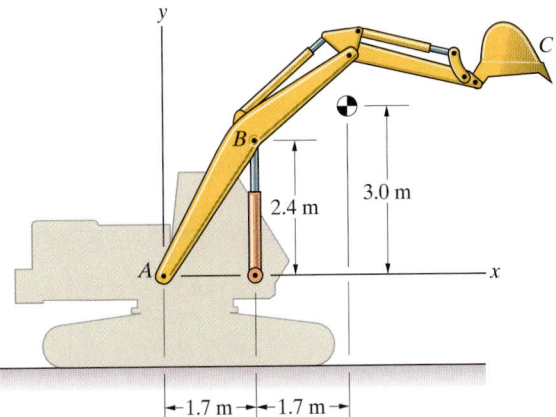

Problems 18.132/18.133

18.134 To decrease the angle of elevation of the stationary 200-kg ladder, the gears that raised it are disengaged, and a fraction of a second later a second set of gears that lower it are engaged. At the instant the gears that raised the ladder are disengaged, what is the ladder's angular acceleration and what are the components of force exerted on the ladder by its support at *O*? The moment of inertia of the ladder about *O* is $I_O = 14{,}000$ kg-m^2, and the coordinates of its center of mass at the instant the gears are disengaged are $\overline{x} = 3$ m, $\overline{y} = 4$ m.

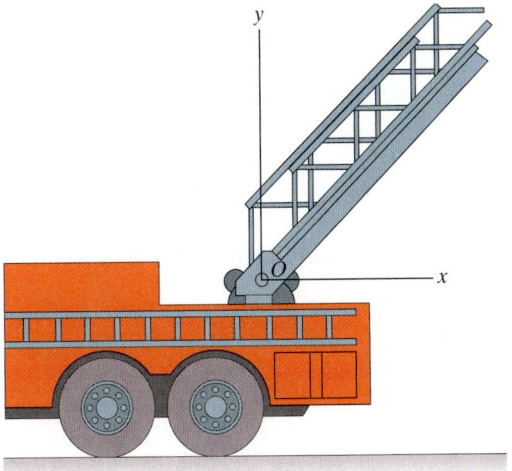

Problem 18.134

18.135 The slender bars each weigh 4 lb and are 10 in long. The homogeneous plate weighs 10 lb. If the system is released from rest in the position shown, what is the angular acceleration of the bars at that instant?

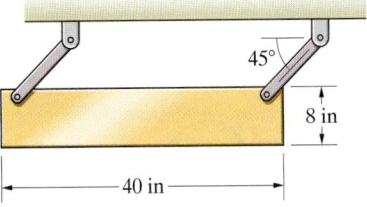

Problem 18.135

18.136 A slender bar of mass *m* is released from rest in the position shown. The static and kinetic coefficients of friction at the floor and wall have the same value μ. If the bar slips, what is its angular acceleration at the instant of release?

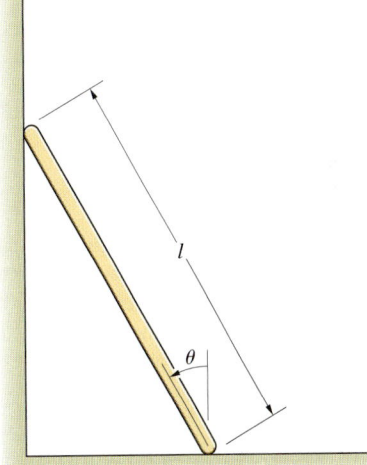

Problem 18.136

18.137 Each of the go-cart's front wheels weighs 5 lb and has a moment of inertia of 0.01 slug-ft². The two rear wheels and rear axle form a single rigid body weighing 40 lb and having a moment of inertia of 0.1 slug-ft². The total weight of the go-cart and driver is 240 lb. (The location of the center of mass of the go-cart and driver, *not including* the front wheels or the rear wheels and rear axle, is shown.) If the engine exerts a torque of 12 ft-lb on the rear axle, what is the go-cart's acceleration?

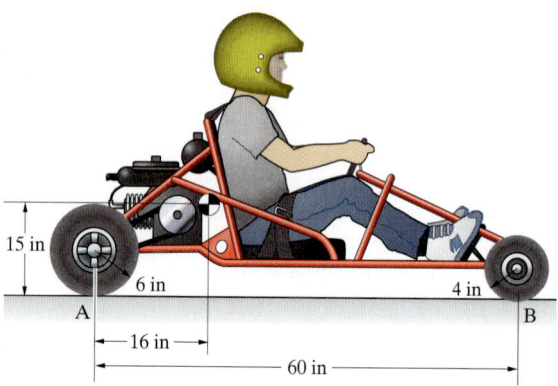

Problem 18.137

18.138 Bar *AB* rotates with a constant angular velocity of 10 rad/s in the counterclockwise direction. The masses of the slender bars *BC* and *CDE* are 2 kg and 3.6 kg, respectively. The *y* axis points upward. Determine the components of the forces exerted on bar *BC* by the pins at *B* and *C* at the instant shown.

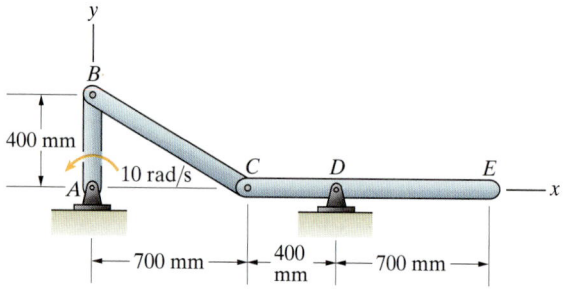

Problem 18.138

18.139 At the instant shown, the arms of the robotic manipulator have constant counterclockwise angular velocities $\omega_{AB} = -0.5$ rad/s, $\omega_{BC} = 2$ rad/s, and $\omega_{CD} = 4$ rad/s. The mass of arm *CD* is 10 kg, and its center of mass is at its midpoint. At this instant, what force and couple are exerted on arm *CD* at *C*?

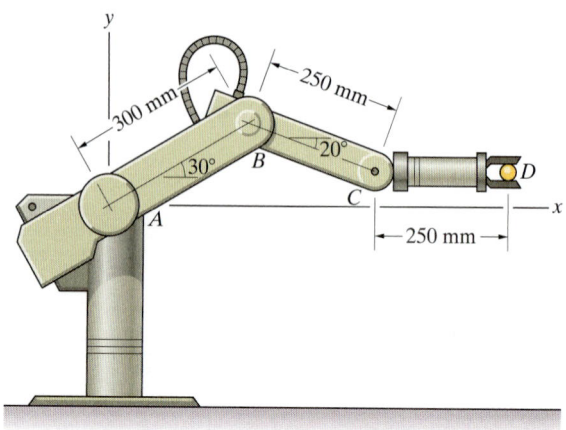

Problem 18.139

18.140 Each bar is 1 m in length and has a mass of 4 kg. The inclined surface is smooth. If the system is released from rest in the position shown, what are the angular accelerations of the bars at that instant?

18.141 Each bar is 1 m in length and has a mass of 4 kg. The inclined surface is smooth. If the system is released from rest in the position shown, what is the magnitude of the force exerted on bar *OA* by the support at *O* at that instant?

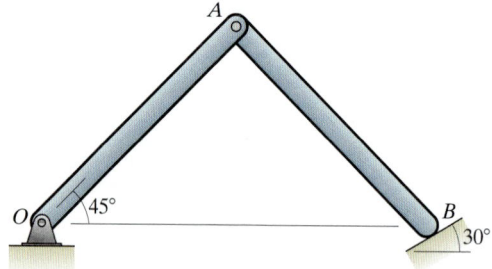

Problems 18.140/18.141

18.142 The fixed ring gear lies *in the horizontal plane*. The hub and planet gears are bonded together. The mass and moment of inertia of the combined hub and planet gears are $m_{HP} = 130$ kg and $I_{HP} = 130$ kg-m^2. The moment of inertia of the sun gear is $I_S = 60$ kg-m^2. The mass of the connecting rod is 5 kg, and it can be modeled as a slender bar. If a 1 kN-m counterclockwise couple is applied to the sun gear, what is the resulting angular acceleration of the bonded hub and planet gears?

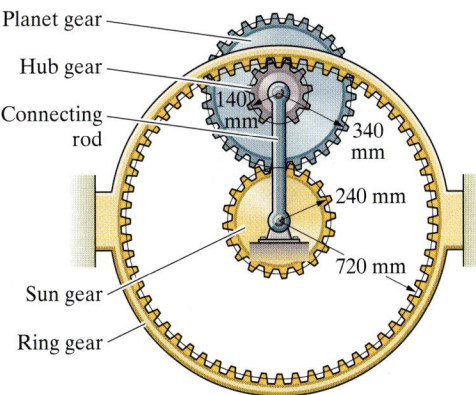

Planet gear
Hub gear
Connecting rod
140 mm
340 mm
240 mm
720 mm
Sun gear
Ring gear

Problem 18.142

18.143 The system is stationary at the instant shown. The net force exerted on the piston by the exploding fuel–air mixture and friction is 5 kN to the left. A clockwise couple $M = 200$ N-m acts on the crank AB. The moment of inertia of the crank about A is 0.0003 kg-m^2. The mass of the connecting rod BC is 0.36 kg, and its center of mass is 40 mm from B on the line from B to C. The connecting rod's moment of inertia about its center of mass is 0.0004 kg-m^2. The mass of the piston is 4.6 kg. What is the piston's acceleration? (Neglect the gravitational forces on the crank and connecting rod.)

18.144 If the crank AB in Problem 18.143 has a counterclockwise angular velocity of 2000 rpm at the instant shown, what is the piston's acceleration due to the couple M and the force exerted by the fuel–air mixture?

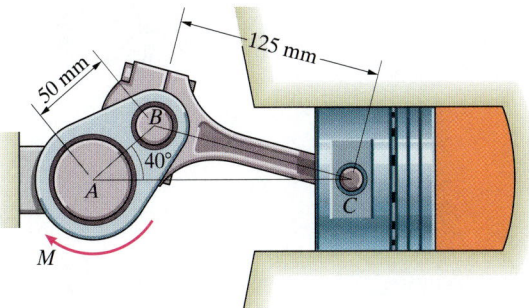

50 mm
125 mm
B
40°
A
C
M

Problems 18.143/18.144

Design Project

Investigate the effects of the coefficients of friction and the dimensions b, c, and h on the bicycle rider's ability to accelerate and decelerate. Consider a range of coefficients of friction that you think might encompass dry, wet, and icy road surfaces. In studying deceleration, consider the cases in which the brakes act on the front wheel only, the rear wheel only, and both wheels. Pay particular attention to the constraint that the bicycle's front and rear wheels should not leave the ground. Write a brief report presenting your analyses and making observations about their implications for bicycle design. Notice that the rider can alter the dimensions h, b, and c to some extent by changing the position of his upper body. Comment on how the rider can thereby affect the bicycle's performance with respect to acceleration and deceleration.

h

b

c

Energy and Momentum in Rigid-Body Dynamics

In Chapters 15 and 16 we demonstrated that energy and momentum methods are useful for solving particular types of problems in dynamics. If the forces on an object are known as functions of position, the principle of work and energy can be used to determine the change in the magnitude of the velocity of the object as it moves between two positions. If the forces are known as functions of time, the principle of impulse and momentum can be used to determine the change in the object's velocity during an interval of time. We now extend these methods to situations in which both the translational and rotational motions of objects must be considered.

◄ The wind exerts torque on the wind generators, performing work that is transformed into electrical energy. In this chapter we use energy and momentum methods to analyze motions of rigid bodies.

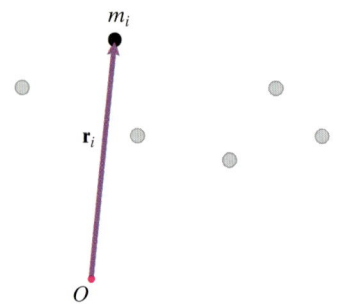

Figure 19.1
A system of particles. The vector $\mathbf{r}_i$ is the position vector of the ith particle.

19.1 Principle of Work and Energy

We will show that the work done on a rigid body by external forces and couples as it moves between two positions is equal to the change in its kinetic energy. To obtain this result, we adopt the same approach used in Chapter 18 to obtain the equations of motion for a rigid body. We derive the principle of work and energy for a system of particles and use it to deduce the principle for a rigid body.

Let m_i be the mass of the ith particle of a system of N particles. Let $\mathbf{r}_i$ be the position of the ith particle relative to a point O that is fixed with respect to an inertial reference frame (Fig. 19.1). We denote the sum of the kinetic energies of the particles by

$$T = \sum_i \tfrac{1}{2} m_i \mathbf{v}_i \cdot \mathbf{v}_i, \tag{19.1}$$

where $\mathbf{v}_i = d\mathbf{r}_i/dt$ is the velocity of the ith particle. Our objective is to relate the work done on the system of particles to the change in T. We begin with Newton's second law for the ith particle,

$$\sum_j \mathbf{f}_{ij} + \mathbf{f}_i^E = \frac{d}{dt}(m_i \mathbf{v}_i),$$

where $\mathbf{f}_{ij}$ is the force exerted on the ith particle by the jth particle and $\mathbf{f}_i^E$ is the external force on the ith particle. We take the dot product of this equation with $\mathbf{v}_i$ and sum from $i = 1$ to N:

$$\sum_i \sum_j \mathbf{f}_{ij} \cdot \mathbf{v}_i + \sum_i \mathbf{f}_i^E \cdot \mathbf{v}_i = \sum_i \mathbf{v}_i \cdot \frac{d}{dt}(m_i \mathbf{v}_i). \tag{19.2}$$

We can express the term on the right side of this equation as the rate of change of the total kinetic energy:

$$\sum_i \mathbf{v}_i \cdot \frac{d}{dt}(m_i \mathbf{v}_i) = \frac{d}{dt} \sum_i \frac{1}{2} m_i \mathbf{v}_i \cdot \mathbf{v}_i = \frac{dT}{dt}.$$

Multiplying Eq. (19.2) by dt yields

$$\sum_i \sum_j \mathbf{f}_{ij} \cdot d\mathbf{r}_i + \sum_i \mathbf{f}_i^E \cdot d\mathbf{r}_i = dT.$$

We integrate this equation, obtaining

$$\sum_i \sum_j \int_{(\mathbf{r}_i)_1}^{(\mathbf{r}_i)_2} \mathbf{f}_{ij} \cdot d\mathbf{r}_i + \sum_i \int_{(\mathbf{r}_i)_1}^{(\mathbf{r}_i)_2} \mathbf{f}_i^E \cdot d\mathbf{r}_i = T_2 - T_1. \tag{19.3}$$

The terms on the left side are the work done on the system by internal and external forces as the particles move from positions $(\mathbf{r}_i)_1$ to positions $(\mathbf{r}_i)_2$. We see that the work done by internal and external forces as a system of particles moves between two positions equals the change in the total kinetic energy of the system.

If the particles represent a rigid body, and we assume that the internal forces between each pair of particles are directed along the straight line between them, the work done by internal forces is zero. To show that this is true, we consider two particles of a rigid body designated 1 and 2 (Fig. 19.2). The sum of the forces the two particles exert on each other is zero $(\mathbf{f}_{12} + \mathbf{f}_{21} = \mathbf{0})$, so the rate at which the forces do work (the power) is

$$\mathbf{f}_{12} \cdot \mathbf{v}_1 + \mathbf{f}_{21} \cdot \mathbf{v}_2 = \mathbf{f}_{21} \cdot (\mathbf{v}_2 - \mathbf{v}_1).$$

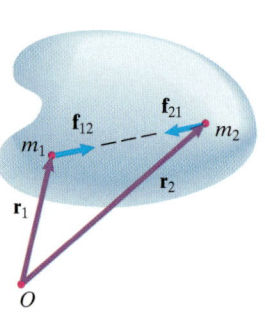

Figure 19.2
Particles 1 and 2 and the forces they exert on each other.

We can show that $\mathbf{f}_{21}$ is perpendicular to $\mathbf{v}_2 - \mathbf{v}_1$, and therefore the rate at which work is done by the internal forces between these two particles is zero. Because the particles are points of a rigid body, we can express their relative velocity in terms of the rigid body's angular velocity $\boldsymbol{\omega}$ as

$$\mathbf{v}_2 - \mathbf{v}_1 = \boldsymbol{\omega} \times (\mathbf{r}_2 - \mathbf{r}_1). \tag{19.4}$$

This equation shows that the relative velocity $\mathbf{v}_2 - \mathbf{v}_1$ is perpendicular to $\mathbf{r}_2 - \mathbf{r}_1$, which is the position vector from particle 1 to particle 2. Since the force $\mathbf{f}_{21}$ is parallel to $\mathbf{r}_2 - \mathbf{r}_1$, it is perpendicular to $\mathbf{v}_2 - \mathbf{v}_1$. We can repeat this argument for each pair of particles of the rigid body, so the total rate at which work is done by internal forces is zero. This implies that the work done by internal forces as a rigid body moves between two positions is zero. Notice that *if an object is not rigid, work can be done by internal forces.*

Therefore, in the case of a rigid body, the work done by internal forces in Eq. (19.3) vanishes. Denoting the work done by external forces by U_{12}, we obtain the principle of work and energy for a rigid body: *The work done by external forces and couples as a rigid body moves between two positions equals the change in the total kinetic energy of the body:*

$$U_{12} = T_2 - T_1. \tag{19.5}$$

We can also state this principle for a *system* of rigid bodies: *The work done by external and internal forces as a system of rigid bodies moves between two positions equals the change in the total kinetic energy of the system.*

19.2 Kinetic Energy

The kinetic energy of a rigid body can be expressed in terms of the velocity of the center of mass of the body and its angular velocity. We consider first general planar motion and then rotation about a fixed axis.

General Planar Motion

Let us model a rigid body as a system of particles, and let $\mathbf{R}_i$ be the position vector of the ith particle relative to the body's center of mass (Fig. 19.3). The position of the center of mass is

$$\mathbf{r} = \frac{\sum\limits_{i} m_i \mathbf{r}_i}{m},$$

where m is the mass of the rigid body. The position of the ith particle relative to O is related to its position relative to the center of mass by

$$\mathbf{r}_i = \mathbf{r} + \mathbf{R}_i, \tag{19.6}$$

and the vectors $\mathbf{R}_i$ satisfy the relation

$$\sum_{i} m_i \mathbf{R}_i = \mathbf{0}. \tag{19.7}$$

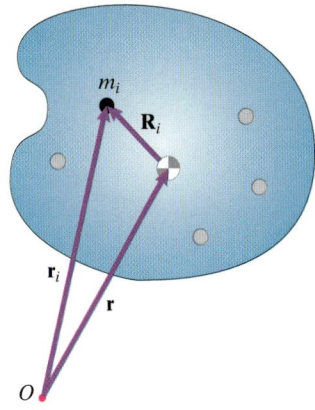

Figure 19.3
Representing a rigid body as a system of particles.

The kinetic energy of the rigid body is the sum of the kinetic energies of its particles, given by Eq. (19.1):

$$T = \sum_{i} \tfrac{1}{2} m_i \mathbf{v}_i \cdot \mathbf{v}_i. \tag{19.8}$$

By taking the derivative of Eq. (19.6) with respect to time, we obtain

$$\mathbf{v}_i = \mathbf{v} + \frac{d\mathbf{R}_i}{dt},$$

where $\mathbf{v}$ is the velocity of the center of mass. Substituting this expression into Eq. (19.8) and using Eq. (19.7), we obtain the kinetic energy of the rigid body in the form

$$T = \tfrac{1}{2}mv^2 + \sum_i \frac{1}{2}m_i \frac{d\mathbf{R}_i}{dt} \cdot \frac{d\mathbf{R}_i}{dt}, \tag{19.9}$$

where v is the magnitude of the velocity of the center of mass.

Let L be the axis through the center of mass that is perpendicular to the plane of the motion (Fig. 19.4a). In terms of the coordinate system shown, we can express the angular velocity vector as $\boldsymbol{\omega} = \omega\mathbf{k}$. The velocity of the ith particle relative to the center of mass is $d\mathbf{R}_i/dt = \omega\mathbf{k} \times \mathbf{R}_i$, so we can write Eq. (19.9) as

$$T = \tfrac{1}{2}mv^2 + \tfrac{1}{2}\left[\sum_i m_i(\mathbf{k} \times \mathbf{R}_i) \cdot (\mathbf{k} \times \mathbf{R}_i)\right]\omega^2. \tag{19.10}$$

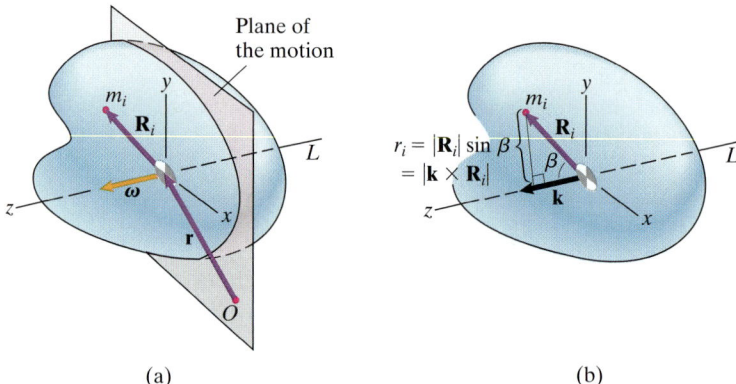

Figure 19.4
(a) A coordinate system with the z axis aligned with L.
(b) The magnitude of $\mathbf{k} \times \mathbf{R}_i$ is the perpendicular distance from L to m_i.

(a) (b)

The magnitude of the vector $\mathbf{k} \times \mathbf{R}_i$ is the perpendicular distance r_i from L to the ith particle (Fig. 19.4b), so the term in brackets in Eq. (19.10) is the moment of inertia of the body about L:

$$\sum_i m_i(\mathbf{k} \times \mathbf{R}_i) \cdot (\mathbf{k} \times \mathbf{R}_i) = \sum_i m_i|\mathbf{k} \times \mathbf{R}_i|^2 = \sum_i m_i r_i^2 = I.$$

Thus, we obtain the kinetic energy of a rigid body in general planar motion in the form

$$T = \tfrac{1}{2}mv^2 + \tfrac{1}{2}I\omega^2, \tag{19.11}$$

where m is the mass of the rigid body, v is the magnitude of the velocity of the center of mass, I is the moment of inertia about the axis L through the center of mass, and ω is the angular velocity. We see that the kinetic energy consists of two terms, the *translational kinetic energy* due to the velocity of the center of mass and the *rotational kinetic energy* due to the angular velocity (Fig. 19.5).

Fixed-Axis Rotation

An object rotating about a fixed axis is in general planar motion, and its kinetic energy is given by Eq. (19.11). But in this case there is another expression for

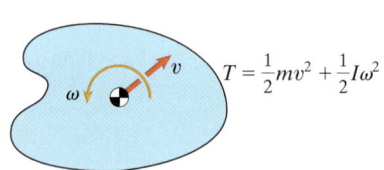

$$T = \frac{1}{2}mv^2 + \frac{1}{2}I\omega^2$$

Figure 19.5
Kinetic energy in general planar motion.

the kinetic energy that is often convenient. Suppose that a rigid body rotates with angular velocity ω about a fixed axis O. In terms of the distance d from O to the center of mass of the body, the velocity of the center of mass is $v = \omega d$ (Fig. 19.6a). From Eq. (19.11), the kinetic energy is

$$T = \tfrac{1}{2}m(\omega d)^2 + \tfrac{1}{2}I\omega^2 = \tfrac{1}{2}(I + d^2 m)\omega^2.$$

According to the parallel-axis theorem, the moment of inertia about O is $I_O = I + d^2 m$, so we obtain the kinetic energy of a rigid body rotating about a fixed axis O in the form (Fig. 19.6b)

$$T = \tfrac{1}{2}I_O\omega^2. \tag{19.12}$$

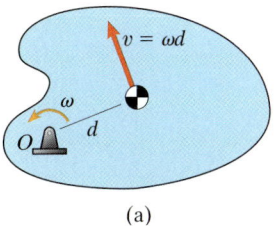 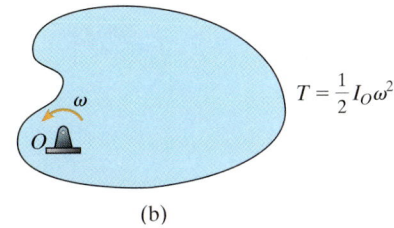

(a)   (b)

Figure 19.6
(a) Velocity of the center of mass.
(b) Kinetic energy of a rigid body rotating about a fixed axis.

19.3 Work and Potential Energy

The procedures for determining the work done by different types of forces and the expressions for the potential energies of forces discussed in Chapter 15 provide the essential tools for applying the principle of work and energy to a rigid body. The work done on a rigid body by a force $\mathbf{F}$ is given by

$$U_{12} = \int_{(\mathbf{r}_p)_1}^{(\mathbf{r}_p)_2} \mathbf{F} \cdot d\mathbf{r}_p, \tag{19.13}$$

where $\mathbf{r}_p$ is the position of the point of application of $\mathbf{F}$ (Fig. 19.7). If the point of application is stationary, or if its direction of motion is perpendicular to $\mathbf{F}$, no work is done.

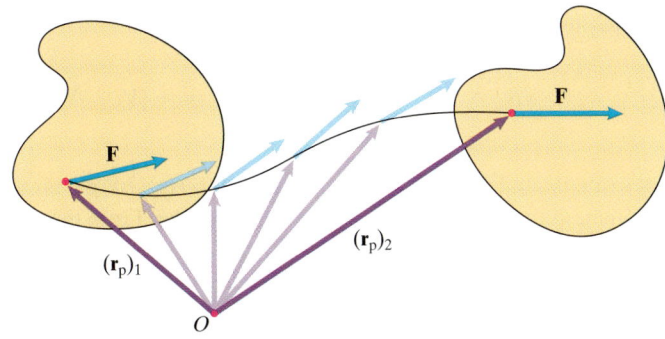

Figure 19.7
The work done by a force on a rigid body is determined by the path of the point of application of the force.

A force $\mathbf{F}$ is conservative if a potential energy V exists such that

$$\mathbf{F} \cdot d\mathbf{r}_p = -dV. \tag{19.14}$$

In terms of its potential energy, the work done by a conservative force $\mathbf{F}$ is

$$U_{12} = \int_{(\mathbf{r}_p)_1}^{(\mathbf{r}_p)_2} \mathbf{F} \cdot d\mathbf{r}_p = \int_{V_1}^{V_2} -dV = -(V_2 - V_1),$$

where V_1 and V_2 are the values of V at $(\mathbf{r}_p)_1$ and $(\mathbf{r}_p)_2$.

If a rigid body is subjected to a couple M (Fig. 19.8a), what work is done as the body moves between two positions? We can evaluate the work by representing the couple by forces (Fig. 19.8b) and determining the work done by the forces. If the rigid body rotates through an angle $d\theta$ in the direction of the couple (Fig. 19.8c), the work done by each force is $\left(\frac{1}{2} D \, d\theta\right) F$, so the total work is $DF \, d\theta = M \, d\theta$. Integrating this expression, we obtain the work done by a couple M as the rigid body rotates from θ_1 to θ_2 in the direction of M:

$$U_{12} = \int_{\theta_1}^{\theta_2} M \, d\theta. \tag{19.15}$$

Figure 19.8
(a) A rigid body subjected to a couple.
(b) An equivalent couple consisting of two forces: $DF = M$.
(c) Determining the work done by the forces.

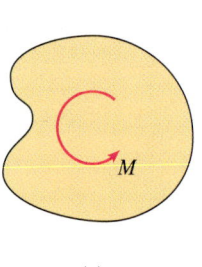
(a)

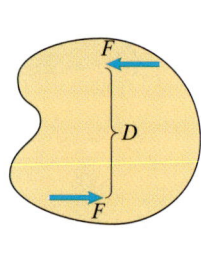
(b)

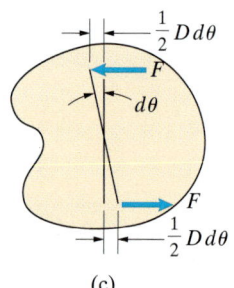
(c)

If M is constant , the work is simply the product of the couple and the angular displacement:

$$U_{12} = M(\theta_2 - \theta_1) \quad \text{(constant couple)}$$

A couple M is conservative if a potential energy V exists such that

$$M \, d\theta = -dV. \tag{19.16}$$

We can express the work done by a conservative couple in terms of its potential energy:

$$U_{12} = \int_{\theta_1}^{\theta_2} M \, d\theta = \int_{V_1}^{V_2} -dV = -(V_2 - V_1).$$

For example, in Fig. 19.9, a torsional spring exerts a couple on a bar that is proportional to the bar's angle of rotation: $M = -k\theta$. From the relation

$$M \, d\theta = -k\theta \, d\theta = -dV,$$

we see that the potential energy must satisfy the equation

$$\frac{dV}{d\theta} = k\theta.$$

(a)

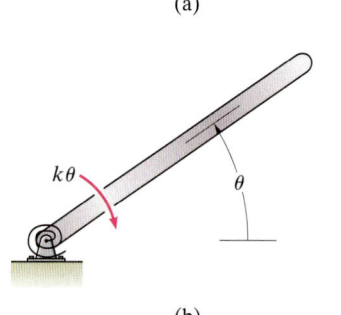
(b)

Figure 19.9
(a) A linear torsional spring connected to a bar.
(b) The spring exerts a couple of magnitude $k\theta$ in the direction opposite that of the bar's rotation.

Integrating this equation, we find that the potential energy of the torsional spring is

$$V = \tfrac{1}{2}k\theta^2. \tag{19.17}$$

If *all* the forces and couples that do work on a system of rigid bodies are conservative, we can express the total work done as the body moves between two positions 1 and 2 in terms of the total potential energy of the forces and couples:

$$U_{12} = V_1 - V_2.$$

Combining this relation with the principle of work and energy, Eq. (19.5), we conclude that the sum of the kinetic energy and the total potential energy is constant—energy is conserved:

$$T_1 + V_1 = T_2 + V_2 \tag{19.18}$$

If a system is subjected to both conservative and nonconservative forces, the principle of work and energy can be written in the form

$$T_1 + V_1 + U_{12} = T_2 + V_2 \tag{19.19}$$

The term U_{12} includes the work done by all nonconservative forces acting on the system as it moves from position 1 to position 2. If a force is conservative, there is a choice. The work it does can be calculated and included in U_{12}, *or* the force's potential energy can be included in V.

The results we have presented in Sections 19.1–19.3 can be used to relate changes in the translational and angular velocities of an object to a change in its position. This typically involves three steps:

1. *Identify the forces and couples that do work.* Use free-body diagrams to determine which external forces and couples do work.

2. *Apply the principle of work and energy or conservation of energy.* Either equate the total work done during a change in position to the change in the kinetic energy, or equate the sum of the kinetic and potential energies at two positions.

3. *Determine the kinematic relationships.* To complete the solution, it will often be necessary to obtain relations between velocities of points of rigid bodies and their angular velocities.

Study Questions

1. What is the principle of work and energy for a rigid body?

2. What is the kinetic energy of a rigid body in general planar motion?

3. How do you determine the work done by a couple acting on a rigid body in planar motion?

4. If all of the forces and couples that do work on a rigid body are conservative, what can you infer about the sum of the kinetic energy of the rigid body and the total potential energy?

Example 19.1 **Applying Work and Energy to a Rolling Disk**

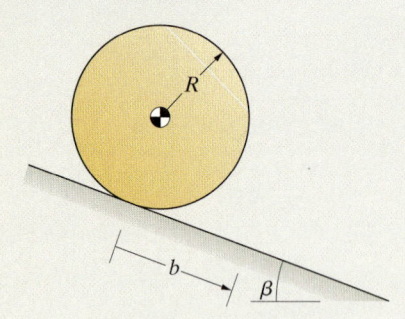

Figure 19.10

A disk of mass m and moment of inertia I is released from rest on an inclined surface (Fig. 19.10). Assuming that the disk rolls, what is the velocity of its center when it has moved a distance b?

Strategy

We can determine the velocity by equating the total work done as the disk rolls a distance b to the change in its kinetic energy.

Solution

Identify the Forces and Couples That Do Work   We draw the free-body diagram of the disk in Fig. a. The disk's weight does work as it rolls, but the normal force N and the friction force f do not. To explain why the friction force does no work, we can write the work done by a force $\mathbf{F}$ as

$$\int_{(\mathbf{r}_p)_1}^{(\mathbf{r}_p)_2} \mathbf{F} \cdot d\mathbf{r}_p = \int_{t_1}^{t_2} \mathbf{F} \cdot \frac{d\mathbf{r}_p}{dt} dt = \int_{t_1}^{t_2} \mathbf{F} \cdot \mathbf{v}_p \, dt,$$

where $\mathbf{v}_p$ is the velocity of the point of application of $\mathbf{F}$. Since the velocity of the point where f acts is zero as the disk rolls, the work done by f is zero.

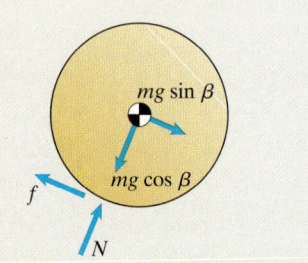

(a) Free-body diagram of the disk.

Apply Work and Energy   We can determine the work done by the weight by multiplying the component of the weight in the direction of the motion of the center of the disk by the distance b:

$$U_{12} = (mg \sin \beta)b.$$

Letting v and ω be the velocity of the center and the angular velocity of the disk when it has moved a distance b (Fig. b), we equate the work to the change in the disk's kinetic energy:

$$mgb \sin \beta = \tfrac{1}{2}mv^2 + \tfrac{1}{2}I\omega^2 - 0. \tag{1}$$

Determine the Kinematic Relationship   The angular velocity ω of the rolling disk is related to the velocity v by $\omega = v/R$. Substituting this relation into Eq. (1) and solving for v, we obtain

$$v = \sqrt{\frac{2gb \sin \beta}{1 + I/mR^2}}.$$

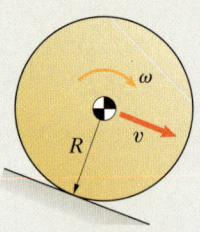

(b) Velocity of the center and the angular velocity when the disk has moved a distance b.

Critical Thinking

Suppose that the surface is smooth, so that the disk slides instead of rolling. In this case, the disk has no angular velocity, so Eq. (1) becomes

$$mgb \sin \beta = \tfrac{1}{2}mv^2 - 0,$$

and the velocity of the center of the disk is

$$v = \sqrt{2gb \sin \beta}.$$

The velocity is greater when the disk slides. You can see why by comparing the two expressions for the principle of work and energy. The work done by the disk's weight is the same in each case. When the disk rolls, part of the work increases the disk's translational kinetic energy, and part increases its rotational kinetic energy. When the disk slides, all of the work increases its translational kinetic energy.

Example 19.2 Applying Work and Energy to a Motorcycle

Each wheel of the motorcycle in Fig. 19.11 has mass $m_W = 9$ kg, radius $R = 330$ mm, and moment of inertia $I = 0.8$ kg-m^2. The combined mass of the rider and the motorcycle, not including the wheels, is $m_C = 142$ kg. The motorcycle starts from rest, and its engine exerts a constant couple $M = 140$ N-m on the rear wheel. Assume that the wheels do not slip. What horizontal distance b must the motorcycle travel to reach a velocity of 25 m/s?

Figure 19.11

Strategy

We can apply the principle of work and energy to the system consisting of the rider and the motorcycle, including its wheels, to determine the distance b.

Solution

Determining the distance b requires three steps.

Identify the Forces and Couples That Do Work We draw the free-body diagram of the system in Fig. a. The weights do no work because the motion is horizontal, and the forces exerted on the wheels by the road do no work because the velocity of their point of application is zero. (See Example 19.1.) Thus, no work is done by external forces and couples! However, work is done by the couple M exerted on the rear wheel by the engine (Fig. b). Although this is an internal couple for the system we are considering—the wheel exerts an opposite couple on the body of the motorcycle—net work is done because the wheel rotates whereas the body does not.

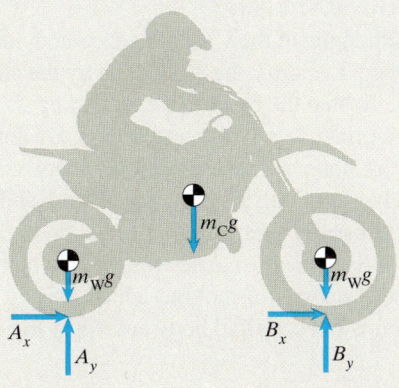

(a) Free-body diagram of the system.

(b) Isolating the rear wheel.

Apply Work and Energy   If the motorcycle moves a horizontal distance b, the wheels turn through an angle b/R rad, and the work done by the constant couple M is

$$U_{12} = M(\theta_2 - \theta_1) = M\left(\frac{b}{R}\right).$$

Let v be the motorcycle's velocity and ω the angular velocity of the wheels when the motorcycle has moved a distance b. The work equals the change in the total kinetic energy:

$$M\left(\frac{b}{R}\right) = \tfrac{1}{2}m_C v^2 + 2\left(\tfrac{1}{2}m_W v^2 + \tfrac{1}{2}I\omega^2\right) - 0. \tag{1}$$

Determine Kinematic Relationship   The angular velocity of the rolling wheels is related to the velocity v by $\omega = v/R$. Substituting this relation into Eq. (1) and solving for b, we obtain

$$b = \left(\tfrac{1}{2}m_C + m_W + \frac{I}{R^2}\right)\frac{Rv^2}{M}$$

$$= \left[\tfrac{1}{2}(142\text{ kg}) + (9\text{ kg}) + \frac{0.8\text{ kg-m}^2}{(0.33\text{ m})^2}\right]\frac{(0.33\text{ m})(25\text{ m/s})^2}{140\text{ N-m}}$$

$$= 129\text{ m}.$$

Critical Thinking

Although we drew separate free-body diagrams of the motorcycle and its rear wheel to clarify the work done by the couple exerted by the engine, notice that we treated the motorcycle, including its wheels, as a single system in applying the principle of work and energy. By doing so, we did not need to consider the work done by the internal forces between the motorcycle's body and its wheels. When applying the principle of work and energy to a system of rigid bodies, you will usually find it simplest to express the principle for the system as a whole. This is in contrast to determining the motion of a system of rigid bodies by using the equations of motion, which usually requires that you draw free-body diagrams of each rigid body and apply the equations to them individually.

Example 19.3   **Applying Conservation of Energy to a Linkage**

The slender bars AB and BC of the linkage in Fig. 19.12 have mass m and length l, and the collar C has mass m_C. A torsional spring at A exerts a clockwise couple $k\theta$ on bar AB. The system is released from rest in the position $\theta = 0$ and allowed to fall. Neglecting friction, determine the angular velocity $\omega = d\theta/dt$ of bar AB as a function of θ.

Strategy

The objective in this example—determining an angular velocity as a function of the position of the system—encourages an energy approach. We must first identify forces and couples that do work on the system. If they are conservative, we can apply conservation of energy to determine ω as a function of θ. If nonconservative forces do work on the system, we can apply the principle of work and energy.

Solution

Identify the Forces and Couples That Do Work   We draw the free-body diagram of the system in Fig. a. The forces and couples that do work—the weights of the bars and collar and the couple exerted by the torsional spring—are conservative. We can use conservation of energy and the kinematic relationships between the angular velocities of the bars and the velocity of the collar to determine ω as a function of θ.

Apply Conservation of Energy   We denote the center of mass of bar BC by G and the angular velocity of bar BC by ω_{BC} (Fig. b). The moment of inertia of each bar about its center of mass is $I = \frac{1}{12}ml^2$. Since bar AB rotates about the fixed point A, we can write its kinetic energy as

$$T_{\text{bar }AB} = \tfrac{1}{2}I_A\omega^2 = \tfrac{1}{2}\Big[I + \big(\tfrac{1}{2}l\big)^2 m\Big]\omega^2 = \tfrac{1}{6}ml^2\omega^2.$$

The kinetic energy of bar BC is

$$T_{\text{bar }BC} = \tfrac{1}{2}mv_G^2 + \tfrac{1}{2}I\omega_{BC}^2 = \tfrac{1}{2}mv_G^2 + \tfrac{1}{24}ml^2\omega_{BC}^2.$$

The kinetic energy of the collar C is

$$T_{\text{collar}} = \tfrac{1}{2}m_Cv_C^2.$$

Using the datum in Fig. (a), we obtain the potential energies of the weights:

$$V_{\text{bar }AB} + V_{\text{bar }BC} + V_{\text{collar}} = mg\big(\tfrac{1}{2}l\cos\theta\big) + mg\big(\tfrac{3}{2}l\cos\theta\big) + m_Cg(2l\cos\theta).$$

The potential energy of the torsional spring is given by Eq. (19.18):

$$V_{\text{spring}} = \tfrac{1}{2}k\theta^2.$$

We now have all the ingredients to apply conservation of energy. We equate the sum of the kinetic and potential energies at the position $\theta = 0$ to the sum of the kinetic and potential energies at an arbitrary value of θ:

$$T_1 + V_1 = T_2 + V_2:$$

$$0 + 2mgl + 2m_Cgl = \tfrac{1}{6}ml^2\omega^2 + \tfrac{1}{2}mv_G^2 + \tfrac{1}{24}ml^2\omega_{BC}^2 + \tfrac{1}{2}m_Cv_C^2$$
$$+ 2mgl\cos\theta + 2m_Cgl\cos\theta + \tfrac{1}{2}k\theta^2.$$

To determine ω from this equation, we must express the velocities v_G, v_C, and ω_{BC} in terms of ω.

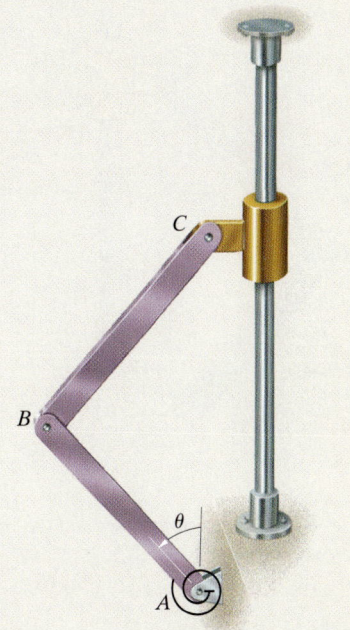

Figure 19.12

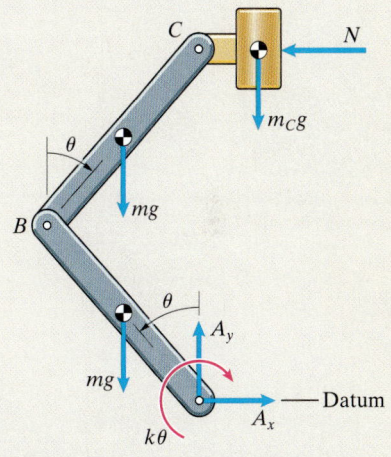

(a) Free-body diagram of the system.

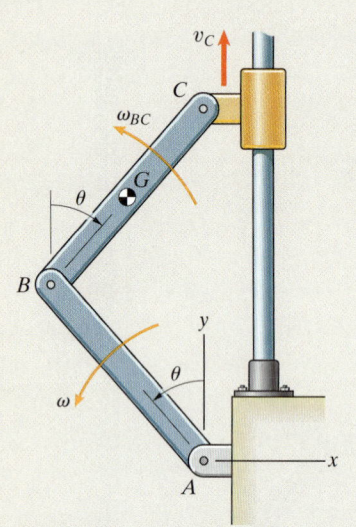

(b) Angular velocities of the bars and the velocity of the collar.

Determine Kinematic Relationships We can determine the velocity of point B in terms of ω and then express the velocity of point C in terms of the velocity of point B and the angular velocity ω_{BC}.

The velocity of B is

$$\mathbf{v}_B = \mathbf{v}_A + \boldsymbol{\omega}_{AB} \times \mathbf{r}_{B/A}$$

$$= \mathbf{0} + \begin{vmatrix} \mathbf{i} & \mathbf{j} & \mathbf{k} \\ 0 & 0 & \omega \\ -l\sin\theta & l\cos\theta & 0 \end{vmatrix}$$

$$= -l\omega\cos\theta\,\mathbf{i} - l\omega\sin\theta\,\mathbf{j}.$$

The velocity of C, expressed in terms of the velocity of B, is

$$v_C\mathbf{j} = \mathbf{v}_B + \boldsymbol{\omega}_{BC} \times \mathbf{r}_{C/B}$$

$$= -l\omega\cos\theta\,\mathbf{i} - l\omega\sin\theta\,\mathbf{j} + \begin{vmatrix} \mathbf{i} & \mathbf{j} & \mathbf{k} \\ 0 & 0 & \omega_{BC} \\ l\sin\theta & l\cos\theta & 0 \end{vmatrix}.$$

Equating $\mathbf{i}$ and $\mathbf{j}$ components, we obtain

$$\omega_{BC} = -\omega, \qquad v_C = -2l\omega\sin\theta.$$

(The minus signs indicate that the directions of the velocities are opposite to the directions we assumed in Fig. b.) Now that we know the angular velocity of bar BC in terms of ω, we can determine the velocity of its center of mass in terms of ω by expressing it in terms of $\mathbf{v}_B$:

$$\mathbf{v}_G = \mathbf{v}_B + \boldsymbol{\omega}_{BC} \times \mathbf{r}_{G/B}$$

$$= -l\omega\cos\theta\,\mathbf{i} - l\omega\sin\theta\,\mathbf{j} + \begin{vmatrix} \mathbf{i} & \mathbf{j} & \mathbf{k} \\ 0 & 0 & -\omega \\ \frac{1}{2}l\sin\theta & \frac{1}{2}l\cos\theta & 0 \end{vmatrix}$$

$$= -\frac{1}{2}l\omega\cos\theta\,\mathbf{i} - \frac{3}{2}l\omega\sin\theta\,\mathbf{j}.$$

Substituting these expressions for ω_{BC}, v_C, and $\mathbf{v}_G$ into our equation of conservation of energy and solving for ω, we obtain

$$\omega = \left[\frac{2gl(m + m_C)(1 - \cos\theta) - \frac{1}{2}k\theta^2}{\frac{1}{3}ml^2 + (m + 2m_C)l^2\sin^2\theta} \right]^{1/2}.$$

Critical Thinking

Newton's second law and the equation of angular motion for a rigid body can be applied to this example instead of conservation of energy. How can you decide what approach to use? The energy methods we have described are generally useful only when you can easily determine the work done by forces and couples acting on a system or their associated potential energies. When that is the case, an energy approach is often preferable. To apply Newton's second law and the equation of angular motion to this example, it would be necessary to draw individual free-body diagrams of the two bars and the collar C, thereby introducing into the formulation the forces exerted on the bars and collar at the pins connecting them. In contrast, we were able to apply conservation of energy to the system as a whole, greatly simplifying the solution.

19.4 Power

The work done on a rigid body by a force **F** during an infinitesimal displacement $d\mathbf{r}_\mathrm{p}$ of its point of application is

$$\mathbf{F} \cdot d\mathbf{r}_\mathrm{p}.$$

We obtain the power P transmitted to the rigid body—the rate at which work is done on it—by dividing this expression by the interval of time dt during which the displacement takes place. We obtain

$$P = \mathbf{F} \cdot \mathbf{v}_\mathrm{p}, \tag{19.20}$$

where $\mathbf{v}_\mathrm{p}$ is the velocity of the point of application of **F**.

Similarly, the work done on a rigid body in planar motion by a couple M during an infinitesimal rotation $d\theta$ in the direction of M is

$$M\, d\theta.$$

Dividing this expression by dt, we find that the power transmitted to the rigid body is the product of the couple and the angular velocity:

$$P = M\omega. \tag{19.21}$$

The total work done on a rigid body during an interval of time equals the change in kinetic energy of the body, so the total power transmitted equals the rate of change of the body's kinetic energy:

$$P = \frac{dT}{dt}.$$

The average with respect to time of the power during an interval of time from t_1 to t_2 is

$$P_{\mathrm{av}} = \frac{1}{t_2 - t_1} \int_{t_1}^{t_2} P\, dt = \frac{1}{t_2 - t_1} \int_{T_1}^{T_2} dT = \frac{T_2 - T_1}{t_2 - t_1}.$$

This expression shows that we can determine the average power transferred to or from a rigid body during an interval of time by dividing the change in kinetic energy of the body, or the total work done, by the interval of time:

$$P_{\mathrm{av}} = \frac{T_2 - T_1}{t_2 - t_1} = \frac{U_{12}}{t_2 - t_1}. \tag{19.22}$$

Problems

19.1 The rotating part of the medical centrifuge is turning at 6000 rpm (revolutions per minute). Its moment of inertia is $I = 0.2$ kg-m^2.

(a) What is its kinetic energy?

(b) How much work was done on it as it accelerated from rest to 6000 rpm?

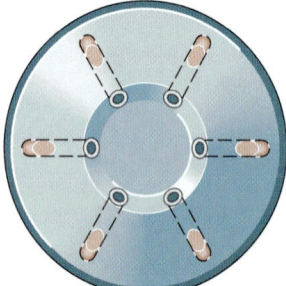

Problem 19.1

19.2 The 4-lb slender bar is 2 ft in length. At the instant shown, the velocity of the end A of the bar is $22\mathbf{i} + 14\mathbf{j}$ (ft/s) and the bar has a counterclockwise angular velocity of 16 rad/s. What is the bar's kinetic energy?

Strategy: You must use the given information to determine the velocity of the bar's center of mass.

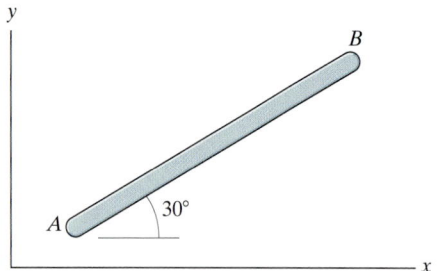

Problem 19.2

19.3 The 8-kg disk is at rest when the constant 10 N-m counterclockwise couple is applied.

(a) How much work has been done on the disk when it has rotated 5 revolutions?

(b) Use work and energy to determine the disk's angular velocity when it has rotated 5 revolutions.

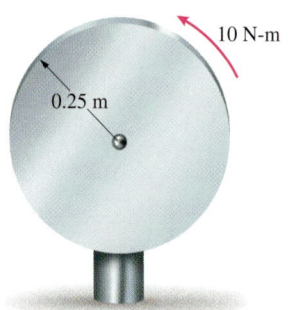

Problem 19.3

19.4 The space station's angular velocity is zero. Suppose that its reaction control system exerts a constant 5000 N-m couple on it about a particular axis and it undergoes planar motion about that axis. The moment of inertia of the station about the axis is $I = 1.5 \times 10^{10}$ kg-m^2.

(a) Use work and energy to determine the magnitude of the station's angular velocity when it has rotated through an angle of 2 degrees.

(b) What average power is transferred to the station while it rotates through an angle of 2 degrees?

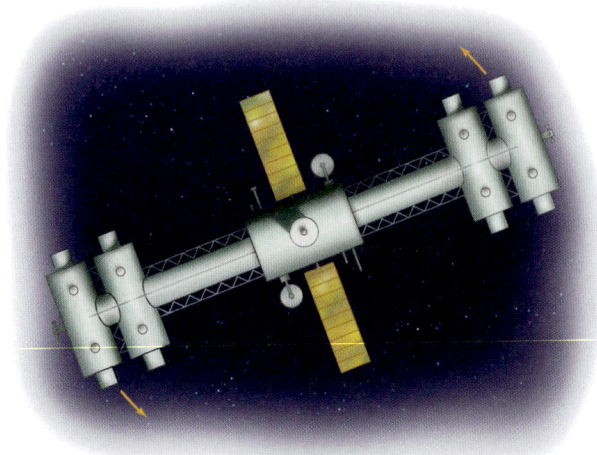

Problem 19.4

19.5 The helicopter's rotor starts from rest. Suppose that its engine exerts a constant 1200 ft-lb couple on the rotor and aerodynamic drag is negligible. The rotor's moment of inertia is $I = 400$ slug-ft^2.

(a) Use work and energy to determine the magnitude of the rotor's angular velocity when it has rotated through 5 revolutions.

(b) What average power is transferred to the rotor while it rotates through 5 revolutions?

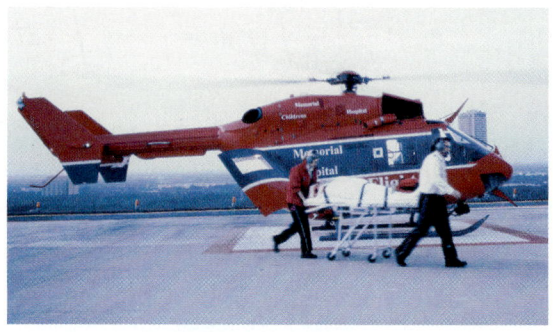

Problems 19.5/19.6

19.6 The helicopter's rotor starts from rest. The couple M exerted on the rotor by its engine and aerodynamic drag is given as a function of the angle θ (in radians) through which the rotor has turned by $M = 1200e^{-0.008\theta}$ ft-lb. The rotor's moment of inertia is $I = 400$ slug-ft^2.

(a) How much work has been done on the rotor when it has rotated through 5 revolutions?

(b) Use work and energy to determine the magnitude of the rotor's angular velocity when it has rotated through 5 revolutions.

19.7 During extravehicular activity, an astronaut's angular velocity is initially zero. She activates two thrusters of her maneuvering unit, exerting constant equal and opposite forces $T = 10$ N, until she has rotated through an angle of six degrees, then turns them off. The moment of inertia of the astronaut and her equipment about the axis through their center of mass perpendicular to the page is $I = 45$ kg-m^2.

(a) Use work and energy to determine her angular velocity when she has rotated through an angle of six degrees.

(b) How long does it take her to rotate through one-fourth of a revolution from her position when she activated the thrusters?

Problem 19.7

19.8 The 8-kg slender bar is released from rest in the horizontal position 1 and falls to position 2.

(a) Use work and energy to determine the magnitude of the bar's angular velocity when it is in position 2.

(b) Determine the x and y components of the force exerted on the bar by the pin support A when the bar is in position 2.

Strategy: To do part (b), draw the free-body diagram of the bar when it is in position 2 and apply Newton's second law and the equation of angular motion. Note that the center of mass of the bar undergoes circular motion.

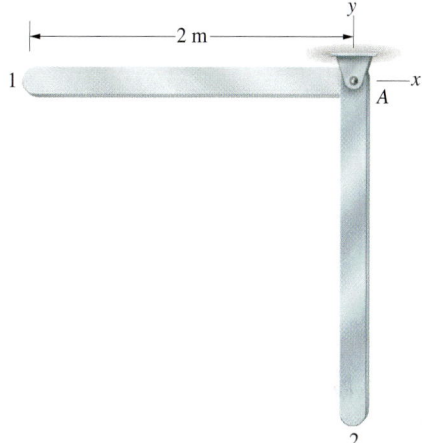

Problem 19.8

19.9 The 8-kg slender bar is released from rest in the horizontal position 1 and falls to position 2.

(a) Use work and energy to determine the magnitude of the bar's angular velocity when it is in position 2.

(b) Determine the x and y components of the force exerted on the bar by the pin support A when the bar is in position 2.

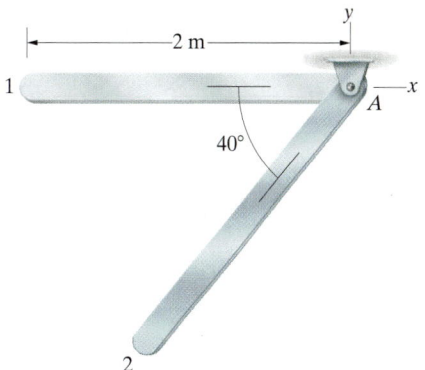

Problem 19.9

19.10 The object consists of an 8-lb slender bar welded to a 12-lb disk. It is released from rest in the horizontal position 1 and falls to position 2.

(a) Use work and energy to determine the magnitude of the object's angular velocity when it is in position 2.

(b) Determine the x and y components of the force exerted on the object by the pin support A when the bar is in position 2.

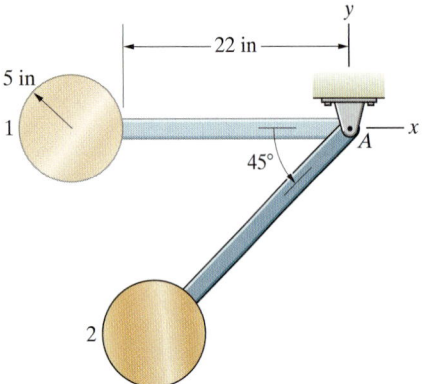

Problem 19.10

19.11 The slotted bar rotates *in the horizontal plane*. Its moment of inertia about its pinned end is 0.2 kg-m². The distance from the bar's pinned end to the fixed end of the spring is 0.3 m. The spring constant is $k = 960$ N/m, and the unstretched length of the spring is 0.25 m. As the follower slides on the smooth surface of the fixed cam, the radial distance r (in meters) is given as a function of the angle θ by

$$r = \frac{0.15}{1 + 0.5 \cos \theta}.$$

The mass of the slider is negligible. If the bar is released from rest in the position $\theta = 90°$ what is the magnitude of its angular velocity in the position $\theta = 0$?

19.12 If the slotted bar in Problem 19.11 is released from rest in the position $\theta = 180°$, what is the magnitude of its angular velocity in the position $\theta = 0$?

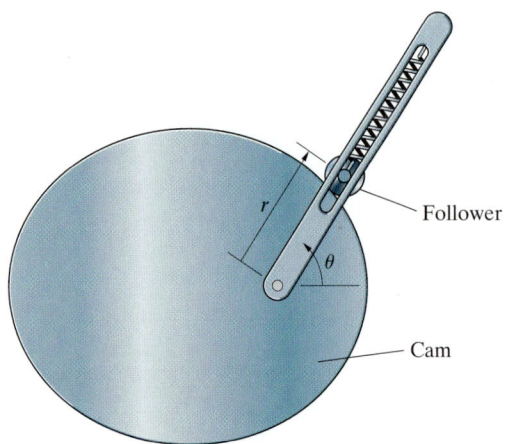

Problems 19.11/19.12

19.13 The 4-kg bar is released from rest in the horizontal position 1 and falls to position 2. The unstretched length of the spring is 0.6 m and the spring constant is $k = 20$ N/m. What is the magnitude of the bar's angular velocity when it is in position 2?

19.14* The 4-kg bar is released from rest in the horizontal position 1 and falls to position 2. The unstretched length of the spring is 0.6 m and the spring constant is $k = 20$ N/m. Determine the x and y components of the force exerted on the object by the pin support A when the bar is in position 2.

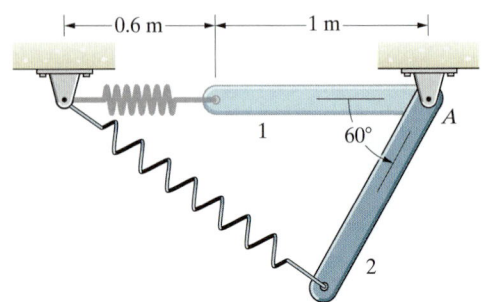

Problems 19.13/19.14

19.15 The moments of inertia of two gears that can turn freely on their pin supports are $I_A = 0.002$ kg-m² and $I_B = 0.006$ kg-m². The gears are at rest when a constant couple $M = 2$ N-m is applied to gear B. Neglecting friction, use the principle of work and energy to determine the angular velocities of the gears when gear A has turned 100 revolutions.

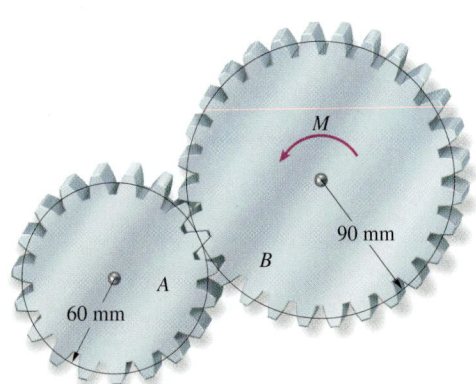

Problem 19.15

19.16 The moments of inertia of gears A and B are $I_A = 0.02$ kg-m^2 and $I_B = 0.09$ kg-m^2. Gear A is connected to a torsional spring with constant $k = 12$ N-m/rad. If gear B is given an initial counterclockwise angular velocity of 10 rad/s with the torsional spring unstretched, through what maximum counterclockwise angle does gear B rotate?

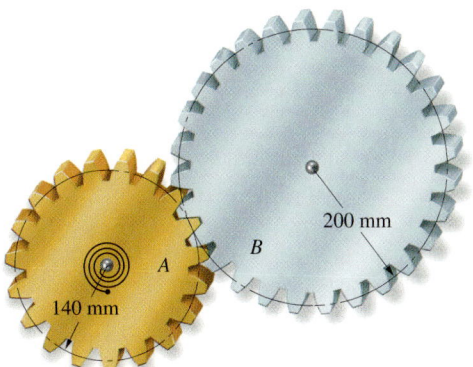

Problem 19.16

19.17 The moments of inertia of three pulleys that can turn freely on their pin supports are $I_A = 0.002$ kg-m^2, $I_B = 0.036$ kg-m^2, and $I_C = 0.032$ kg-m^2. The pulleys are stationary when a constant couple $M = 2$ N-m is applied to pulley A. What is the angular velocity of pulley A when it has turned 10 revolutions?

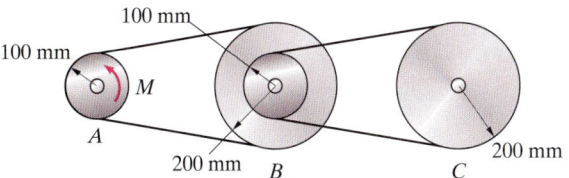

Problem 19.17

19.18 Model the arm ABC as a single rigid body. Its mass is 300 kg, and the moment of inertia about its center of mass is $I = 360$ kg-m^2. Starting from rest with its center of mass 2 m above the ground (position 1), arm ABC is pushed upward by the hydraulic cylinders. When it is in the position shown (position 2), the arm has a counterclockwise angular velocity of 1.4 rad/s. How much work do the hydraulic cylinders do on the arm in moving it from position 1 to position 2?

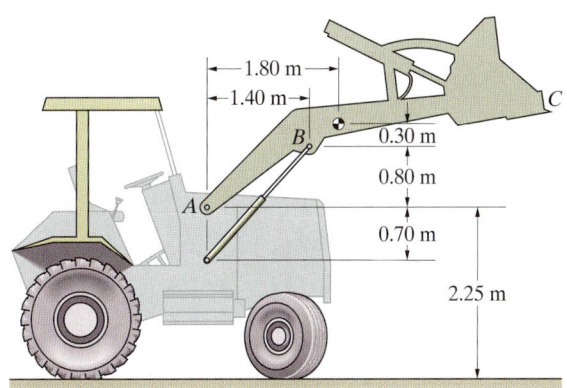

Problem 19.18

19.19 The mass of the homogeneous cylindrical disk is $m = 5$ kg and its radius is $R = 0.2$ m. The disk is stationary when a constant clockwise couple $M = 10$ N-m is applied to it. What is the velocity of the center of the disk when it has moved a distance $b = 0.4$ m? Determine the velocity in two ways:

(a) By drawing a free-body diagram of the disk and writing the equations of motion, determine the acceleration of the disk's center and use it to determine the velocity.

(b) Use work and energy.

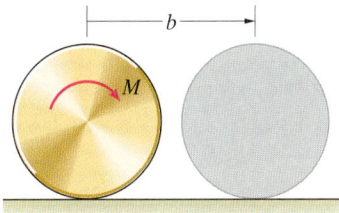

Problem 19.19

19.20 The mass of the homogeneous cylindrical disk is $m = 5$ kg and its radius is $R = 0.2$ m. The angle $\beta = 15°$. The disk is stationary when a constant clockwise couple $M = 10$ N-m is applied to it. What is the velocity of the center of the disk when it has moved a distance $b = 0.4$ m?

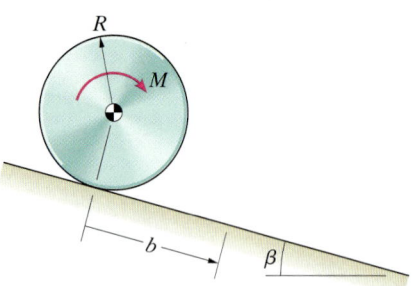

Problem 19.20

19.21 The mass of the stepped disk is 18 kg and its moment of inertia is 0.28 kg-m². If the disk is released from rest, what is its angular velocity when the center of the disk has fallen 1 m?

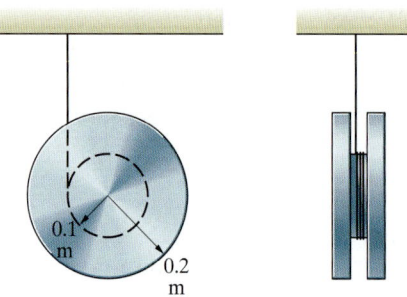

Problem 19.21

19.22 The 100-kg homogeneous cylindrical disk is at rest when the force $F = 500$ N is applied to a cord wrapped around it, causing the disk to roll. Use the principle of work and energy to determine the angular velocity of the disk when it has turned 1 revolution.

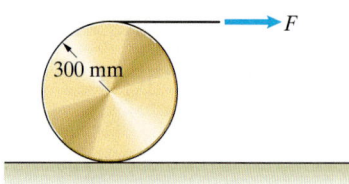

Problem 19.22

19.23 The 1-slug homogeneous cylindrical disk is given a clockwise angular velocity of 2 rad/s with the spring unstretched. The spring constant is $k = 3$ lb/ft. If the disk rolls, how far will its center move to the right?

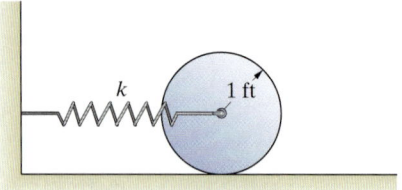

Problem 19.23

19.24 The 22-kg platen P rests on four roller bearings. The roller bearings can be modeled as 1-kg homogeneous cylinders with 30-mm radii. The platen is stationary and the spring $(k = 900$ N/m$)$ is unstretched when a constant horizontal force $F = 100$ N is applied as shown. What is the velocity of the platen when it has moved 200 mm to the right?

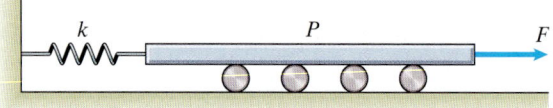

Problem 19.24

19.25 The two weights are released from rest. The 5-lb weight slides on the smooth horizontal surface. The moment of inertia of the pulley is $I = 0.02$ slug-ft². Determine the magnitude of the velocity of the 10-lb weight when it has fallen 2 ft.

19.26 The two weights are released from rest. The coefficient of kinetic friction between the 5-lb weight and the horizontal surface is $\mu_k = 0.4$. The moment of inertia of the pulley is $I = 0.02$ slug-ft². Determine the magnitude of the velocity of the 10-lb weight when it has fallen 2 ft.

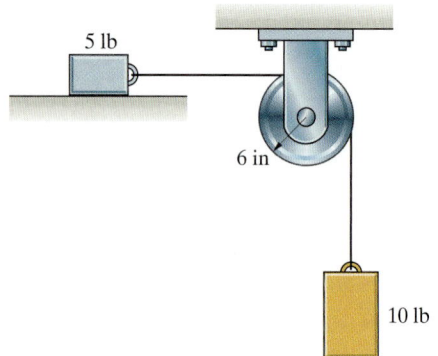

Problems 19.25/19.26

19.27 The total moment of inertia of the car's two rear wheels and axle is I_R, and the total moment of inertia of the two front wheels is I_F. The radius of the tires is R, and the total mass of the car, including the wheels, is m. The car is moving at velocity v_0 when the driver applies the brakes. If the car's brakes exert a constant retarding couple M on each wheel and the tires do not slip, determine the car's velocity as a function of the distance s from the point where the brakes are applied.

19.28 The total moment of inertia of the car's two rear wheels and axle is 0.24 kg-m^2. The total moment of inertia of the two front wheels is 0.2 kg-m^2. The radius of the tires is 0.3 m. The mass of the car, including the wheels, is 1480 kg. The car is moving at 100 km/h. If the car's brakes exert a constant retarding couple of 650 N-m on each wheel and the tires do not slip, what distance is required for the car to come to a stop? (Notice that this analysis neglects aerodynamic drag.)

Problems 19.27/19.28

19.29 The radius of the pulley is $R = 100$ mm and its moment of inertia is $I = 0.1$ kg-m^2. The mass $m = 5$ kg. The spring constant is $k = 135$ N/m. The system is released from rest with the spring unstretched. Determine how fast the mass is moving when it has fallen 0.5 m.

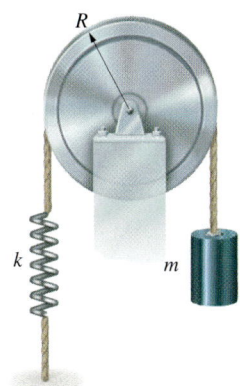

Problem 19.29

19.30 The masses of the bar and disk are 14 kg and 9 kg, respectively. The system is released from rest with the bar horizontal. Determine the angular velocity of the bar when it is vertical if the bar and disk are welded together at A.

19.31 The masses of the bar and disk are 14 kg and 9 kg, respectively. The system is released from rest with the bar horizontal. Determine the angular velocity of the bar when it is vertical if the bar and disk are connected by a smooth pin at A.

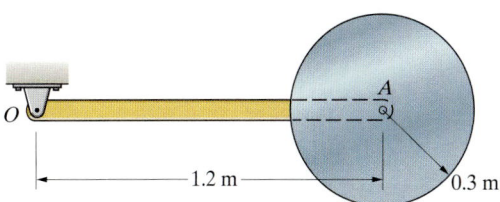

Problems 19.30/19.31

19.32 The 45-kg crate is pulled up the inclined surface by the winch. The coefficient of kinetic friction between the crate and the surface is $\mu_k = 0.4$. The moment of inertia of the drum on which the cable is being wound is $I_A = 4$ kg-m^2. The crate starts from rest, and the motor exerts a constant couple $M = 50$ N-m on the drum. Use the principle of work and energy to determine the magnitude of the velocity of the crate when it has moved 1 m.

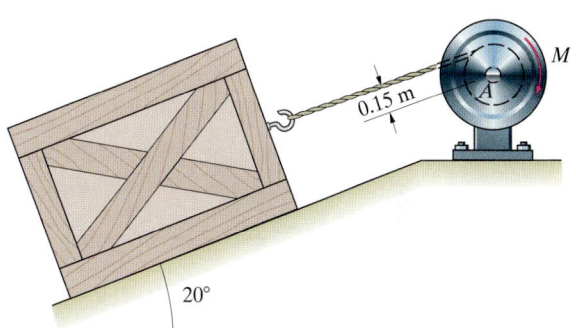

Problem 19.32

19.33 The 2-ft slender bars each weigh 4 lb, and the rectangular plate weighs 20 lb. If the system is released from rest in the position shown, what is the velocity of the plate when the bars are vertical?

Problem 19.33

19.34 The slender bar has mass $m = 8$ kg and length $l = 2$ m. A torsional spring with constant $k = 65$ N-m/rad is attached to the bar at the pin support. The spring is unstretched when the bar is vertical. The bar is released from rest with $\theta = 35°$.

(a) What is the magnitude of the bar's angular velocity when $\theta = 55°$?

(b) What maximum angle θ does the bar reach?

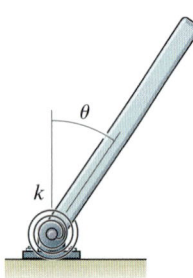

Problem 19.34

19.35 The mass of the suspended object A is 8 kg. The mass of the pulley is 5 kg, and its moment of inertia is 0.036 kg-m². If the force $T = 70$ N is applied to the stationary system, what is the magnitude of the velocity of A when it has risen 0.2 m?

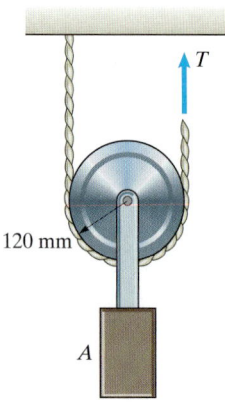

Problem 19.35

19.36 The mass of the left pulley is 7 kg, and its moment of inertia is 0.330 kg-m². The mass of the right pulley is 3 kg, and its moment of inertia is 0.054 kg-m². If the system is released from rest, how fast is the 18-kg mass moving when it has fallen 0.1 m?

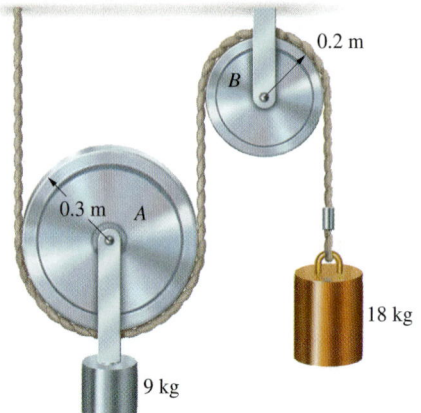

Problem 19.36

19.37 The 18-kg ladder is released from rest with $\theta = 10°$. The wall and floor are smooth. Modeling the ladder as a slender bar, use conservation of energy to determine the angular velocity of the bar when $\theta = 40°$.

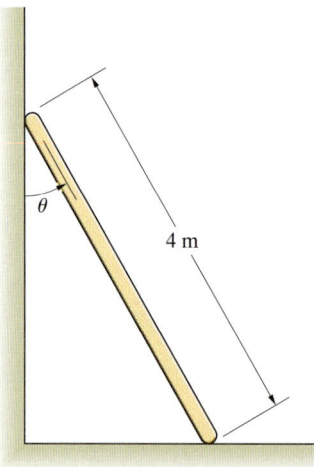

Problem 19.37

19.38 The 8-kg slender bar is released from rest with $\theta = 60°$. The horizontal surface is smooth. What is the bar's angular velocity when $\theta = 30°$.

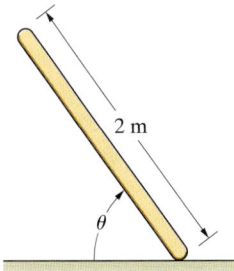

Problem 19.38

19.39 The mass and length of the bar are $m = 4$ kg and $l = 1.2$ m. The spring constant is $k = 180$ N/m. If the bar is released from rest in the position $\theta = 10°$, what is its angular velocity when it has fallen to $\theta = 20°$?

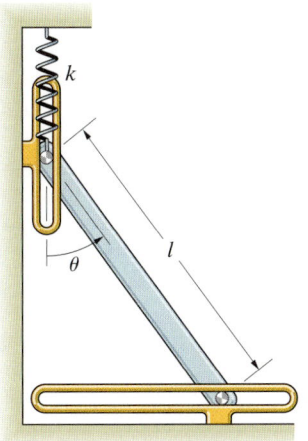

Problem 19.39

19.40 The 4-kg slender bar is pinned to a 2-kg slider at A and to a 4-kg homogeneous cylindrical disk at B. Neglect the friction force on the slider and assume that the disk rolls. If the system is released from rest with $\theta = 60°$, what is the bar's angular velocity when $\theta = 0$?

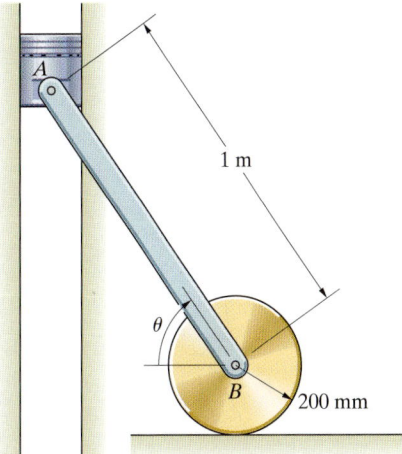

Problem 19.40

19.41* The sleeve P slides on the smooth horizontal bar. The mass of each bar is 4 kg and the mass of the sleeve P is 2 kg. If the system is released from rest with $\theta = 60°$, what is the magnitude of the velocity of the sleeve P when $\theta = 40°$?

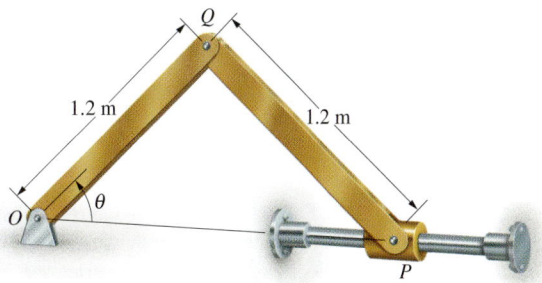

Problem 19.41

19.42* The system is in equilibrium in the position shown. The mass of the slender bar ABC is 6 kg, the mass of the slender bar BD is 3 kg, and the mass of the slider at C is 1 kg. The spring constant is $k = 200$ N/m. If a constant 100-N downward force is applied at A, what is the angular velocity of bar ABC when it has rotated 20° from its initial position?

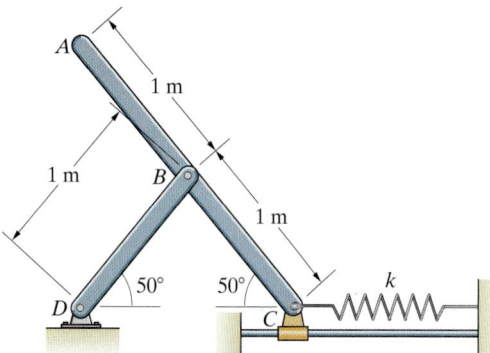

Problem 19.42

19.43* The masses of bars AB and BC are 5 kg and 3 kg, respectively. If the system is released from rest in the position shown, what are the angular velocities of the bars at the instant before the joint B hits the smooth floor?

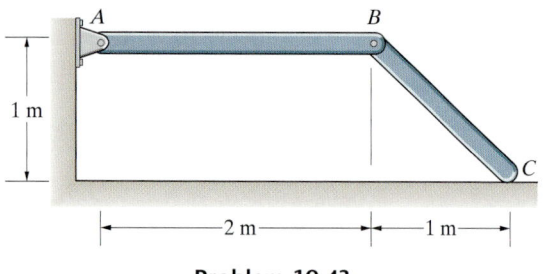

Problem 19.43

19.44* Bar AB weighs 5 lb. Each of the sleeves A and B weighs 2 lb. The system is released from rest in the position shown. What is the magnitude of the angular velocity of the bar when the sleeve B has moved 3 in to the right?

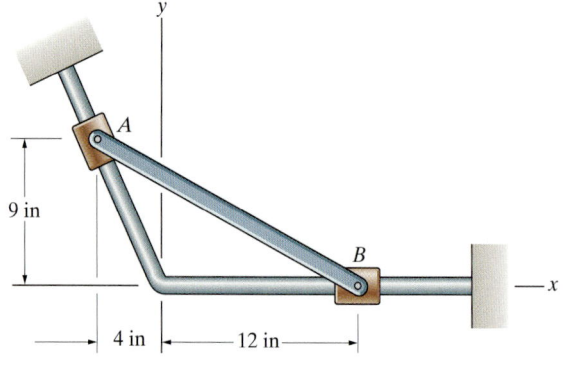

Problem 19.44

19.45* Each bar has a mass of 8 kg and a length of 1 m. The spring constant is $k = 100$ N/m, and the spring is unstretched when $\theta = 0$. If the system is released from rest with the bars vertical, what is the magnitude of the angular velocity of the bars when $\theta = 30°$?

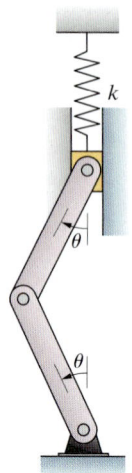

Problem 19.45

19.46* The system starts from rest with the crank AB vertical. A constant couple M exerted on the crank causes it to rotate in the clockwise direction, compressing the gas in the cylinder. Let s be the displacement (in meters) of the piston to the right relative to its initial position. The net force toward the left exerted on the piston by atmospheric pressure and the gas in the cylinder is $350/(1 - 10s)$ N. The moment of inertia of the crank about A is 0.0003 kg-m². The mass of the connecting rod BC is 0.36 kg, and the center of mass of the rod is at its midpoint. The connecting rod's moment of inertia about its center of mass is 0.0004 kg-m². The mass of the piston is 4.6 kg. If the clockwise angular velocity of the crank AB is 200 rad/s when it has rotated 90° from its initial position, what is M? (Neglect the work done by the weights of the crank and connecting rod.)

19.47* In Problem 19.46, if the system starts from rest with the crank AB vertical and the couple $M = 40$ N-m, what is the clockwise angular velocity of AB when it has rotated 45° from its initial position?

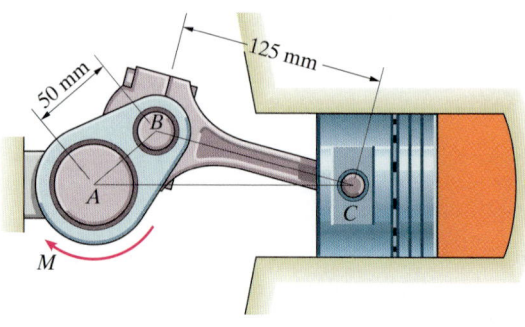

Problems 19.46/19.47

19.5 Principles of Impulse and Momentum

In this section, we review our discussion of the principle of linear impulse and momentum from Chapter 16 and then derive the principle of angular impulse and momentum for a rigid body. These principles relate time integrals of the forces and couples acting on a rigid body to changes in the velocity of its center of mass and its angular velocity.

Linear Momentum

Integrating Newton's second law with respect to time yields the principle of linear impulse and momentum for a rigid body:

$$\int_{t_1}^{t_2} \Sigma \mathbf{F} \, dt = m\mathbf{v}_2 - m\mathbf{v}_1. \tag{19.23}$$

Here, $\mathbf{v}_1$ and $\mathbf{v}_2$ are the velocities of the center of mass at the times t_1 and t_2 (Fig. 19.13). If the external forces acting on a rigid body are known as functions of time, this principle yields the change in the velocity of the center of mass of the body during an interval of time. In terms of the average of the total force from t_1 to t_2,

$$\Sigma \mathbf{F}_{av} = \frac{1}{t_2 - t_1} \int_{t_1}^{t_2} \Sigma \mathbf{F} \, dt,$$

we can write Eq. (19.23) as

$$(t_2 - t_1)\Sigma \mathbf{F}_{av} = m\mathbf{v}_2 - m\mathbf{v}_1. \tag{19.24}$$

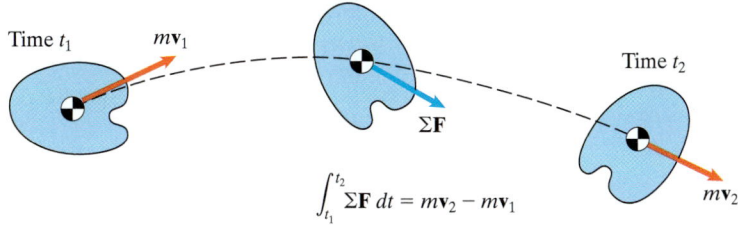

Figure 19.13
Principle of linear impulse and momentum.

This form of the principle of linear impulse and momentum is often useful when an object is subjected to impulsive forces. (See section 16.1.)

If the only forces acting on two rigid bodies A and B are the forces they exert on each other, or if other forces are negligible, the total linear momentum of A and B is conserved:

$$m_A\mathbf{v}_A + m_B\mathbf{v}_B = \text{constant}. \tag{19.25}$$

Angular Momentum

When momentum principles are applied to rigid bodies, it is often necessary to determine both the velocities of their centers of mass and their angular velocities. For this task, linear momentum principles alone are not sufficient. In this section, we derive angular momentum principles for a rigid body in planar motion.

Principles of Angular Impulse and Momentum   The total moment about the center of mass of a rigid body in planar motion equals the product of the moment of inertia of the body about its center of mass and the angular acceleration:

$$\Sigma M = I\alpha.$$

We can write this equation in the form

$$\Sigma M = \frac{dH}{dt}, \tag{19.26}$$

where

$$H = I\omega \tag{19.27}$$

is the rigid body's angular momentum about its center of mass. Integrating Eq. (19.26) with respect to time, we obtain one form of the principle of angular impulse and momentum:

$$\int_{t_1}^{t_2} \Sigma M \, dt = H_2 - H_1. \tag{19.28}$$

Here, H_1 and H_2 are the values of the angular momentum at the times t_1 and t_2. This equation says that angular impulse about the center of mass of the rigid body during the interval of time from t_1 to t_2 is equal to the change in the rigid body's angular momentum about its center of mass. If the total moment about the center of mass is known as a function of time, Eq. (19.28) can be used to determine the change in the angular velocity from t_1 to t_2.

We can derive another useful form of this principle: Let $\mathbf{r}$ be the position vector of the center of mass of the rigid body relative to a fixed point O (Fig. 19.14). In Chapter 18, we derived a relationship between the total moment about O due to external forces and couples and the rate of change of the rigid body's angular momentum about O:

$$\Sigma M_O = \frac{dH_O}{dt}, \tag{19.29}$$

where

$$H_O = (\mathbf{r} \times m\mathbf{v}) \cdot \mathbf{k} + I\omega. \tag{19.30}$$

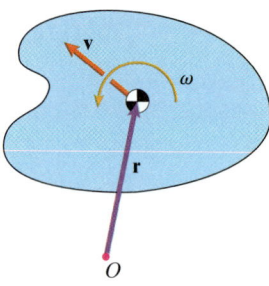

Figure 19.14
A rigid body in planar motion with velocity $\mathbf{v}$ and angular velocity ω.

Integrating Eq. (19.29) with respect to time, we obtain a second form of the principle of angular impulse and momentum:

$$\int_{t_1}^{t_2} \Sigma M_O \, dt = H_{O2} - H_{O1}. \tag{19.31}$$

The angular impulse about a fixed point O during the interval of time from t_1 to t_2 is equal to the change in the rigid body's angular momentum about O (Fig. 19.15).

The term $(\mathbf{r} \times m\mathbf{v}) \cdot \mathbf{k}$ in Eq. (19.30) is the rigid body's angular momentum about O due to the velocity of its center of mass. This term has the same form as the moment of a force, but with the linear momentum $m\mathbf{v}$ in place of the force. If we define ΣM_O and ω to be positive in the counterclockwise direction,

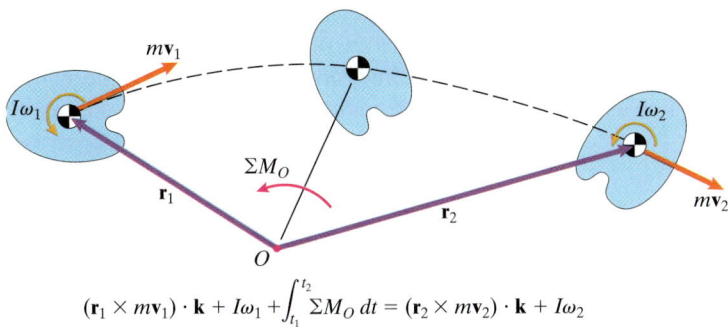

Figure 19.15
The impulse about O equals the change in the angular momentum about O.

$$(\mathbf{r}_1 \times m\mathbf{v}_1) \cdot \mathbf{k} + I\omega_1 + \int_{t_1}^{t_2} \Sigma M_O \, dt = (\mathbf{r}_2 \times m\mathbf{v}_2) \cdot \mathbf{k} + I\omega_2$$

the unit vector $\mathbf{k}$ points out of the page (Fig. 19.16) and $(\mathbf{r} \times m\mathbf{v}) \cdot \mathbf{k}$ is the counterclockwise "moment" of the linear momentum. The vector expression can be used to calculate this quantity, but it is often easier to use the fact that its magnitude is the product of the magnitude of the linear momentum and the perpendicular distance from point O to the line of action of the velocity. The "moment" is positive if it is counterclockwise (Fig. 19.17a) and negative if it is clockwise (Fig. 19.17b).

Impulsive Forces and Couples   The average of the moment about the center of mass from t_1 to t_2 is

$$\Sigma M_{\text{av}} = \frac{1}{t_2 - t_1} \int_{t_1}^{t_2} \Sigma M \, dt.$$

Using this equation, we can write Eq. (19.28) as

$$(t_2 - t_1) \Sigma M_{\text{av}} = H_2 - H_1. \qquad (19.32)$$

In the same way, we can express Eq. (19.31) in terms of the average moment about point O:

$$(t_2 - t_1)(\Sigma M_O)_{\text{av}} = H_{O2} - H_{O1}. \qquad (19.33)$$

When the average value of the moment and its duration are known, we can use Eq. (19.32) or Eq. (19.33) to determine the change in the angular momentum. These equations are often useful when a rigid body is subjected to impulsive forces and couples.

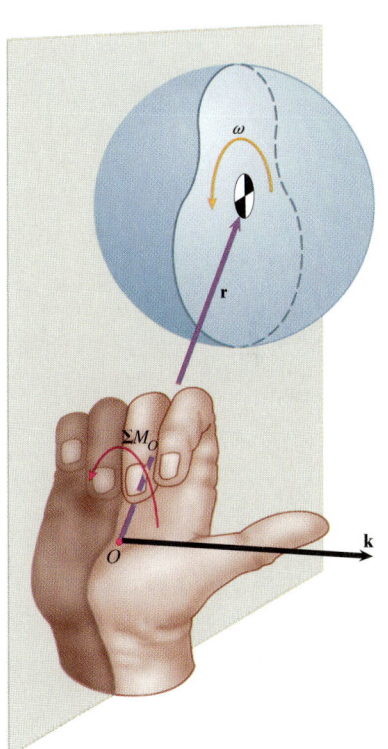

Figure 19.16
The direction of $\mathbf{k}$.

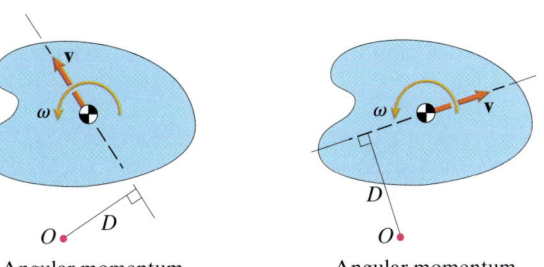

Angular momentum
$H_O = D(m|\mathbf{v}|) + I\omega$

(a)

Angular momentum
$H_O = -D(m|\mathbf{v}|) + I\omega$

(b)

Figure 19.17
Determining the angular momentum about O by calculating the "moment" of the linear momentum.

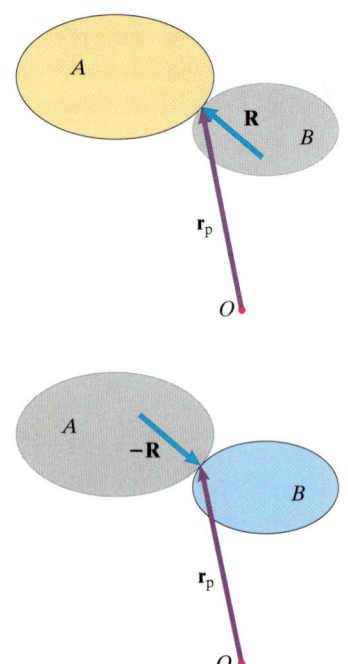

Figure 19.18
Rigid bodies A and B exerting forces on each other by contact.

Conservation of Angular Momentum   We can use Eq. (19.31) to obtain an equation of conservation of total angular momentum for two rigid bodies. Let A and B be rigid bodies in two-dimensional motion in the same plane, and suppose that they are subjected only to the forces and couples they exert on each other or that other forces and couples are negligible. Let M_{OA} be the moment about a fixed point O due to the forces and couples acting on A, and let M_{OB} be the moment about O due to the forces and couples acting on B. Under the same assumption we made in deriving the equations of motion—the forces between each pair of particles are directed along the line between the particles—the moment $M_{OB} = -M_{OA}$. For example, in Fig. 19.18, A and B exert forces on each other by contact. The resulting moments about O are $M_{OA} = (\mathbf{r}_p \times \mathbf{R}) \cdot \mathbf{k}$ and $M_{OB} = [\mathbf{r}_p \times (-\mathbf{R})] \cdot \mathbf{k} = -M_{OA}$.

We apply Eq. (19.31) to A and B for arbitrary times t_1 and t_2, obtaining

$$\int_{t_1}^{t_2} M_{OA}\, dt = H_{OA2} - H_{OA1}$$

and

$$\int_{t_1}^{t_2} M_{OB}\, dt = H_{OB2} - H_{OB1}.$$

Summing these equations, the terms on the left cancel, and we obtain

$$H_{OA1} + H_{OB1} = H_{OA2} + H_{OB2}.$$

We see that the total angular momentum of A and B about O is conserved:

$$H_{OA} + H_{OB} = \text{constant.} \tag{19.34}$$

Notice that this result holds even when A and B are subjected to significant external forces and couples if the total moment about O due to the external forces and couples is zero. The point O can sometimes be chosen so that this condition is satisfied. The result also applies to an arbitrary number of rigid bodies: Their total angular momentum about O is conserved if the total moment about O due to external forces and couples is zero.

Study Questions

1. If you know the total moment about the center of mass of a rigid body in planar motion during an interval of time, you can use a form of the principle of angular impulse and momentum to determine the change in the rigid body's angular velocity. Explain how.

2. Equation (19.30) is the angular momentum of a rigid body in planar motion about a fixed point O. You can evaluate the first term by calculating the "moment" of the linear momentum. How is this done?

3. If the total moment about a fixed point O due to the forces and couples acting on a rigid body is zero during an interval of time, what can you infer about the rigid body's angular momentum about O?

| Example 19.4 | **Principle of Angular Impulse and Momentum** |

Disk A in Fig. 19.19 initially has a counterclockwise angular velocity ω_0, and disk B is stationary. At $t = 0$, the disks are moved into contact. As a result of friction at the point of contact, the angular velocity of A decreases and the angular velocity of B increases until there is no slip between them. What are their final angular velocities ω_A and ω_B? The disks are supported at their centers of mass, and their moments of inertia are I_A, I_B.

Strategy

Since the disks rotate about fixed axes through their centers of mass while they are in contact, we can apply the principle of angular impulse and momentum in the form given in Eq. (19.28) to each disk. When there is no longer any slip between the disks, their velocities are equal at their point of contact. With this kinematic relationship and the equations we obtain with the principle of angular impulse and momentum, we can determine the final angular velocities.

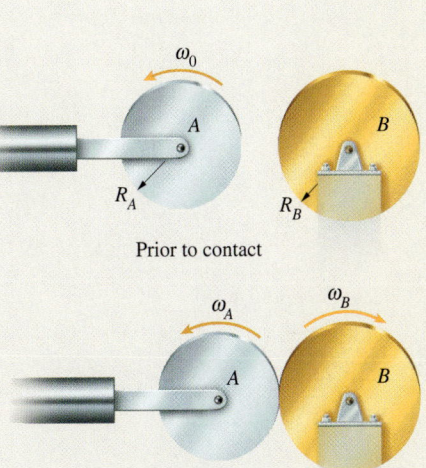

Prior to contact

Final angular velocities

Figure 19.19

Solution

We draw the free-body diagrams of the disks while slip occurs in Fig. a, showing the normal and frictional forces they exert on each other. Let t_f be the time at which slip ceases. We apply Eq. (19.28) to disk A for the interval of time from $t = 0$ to $t = t_f$:

$$\int_{t_1}^{t_2} \Sigma M \, dt = H_2 - H_1 = I\omega_2 - I\omega_1:$$

$$\int_0^{t_f} -R_A f \, dt = I_A \omega_A - I_A \omega_0. \tag{1}$$

We also apply Eq. (19.28) to disk B:

$$\int_{t_1}^{t_2} \Sigma M \, dt = H_2 - H_1 = I\omega_2 - I\omega_1:$$

$$\int_0^{t_f} -R_B f \, dt = -I_B \omega_B - 0. \tag{2}$$

(a) Free-body diagrams of the disks.

Notice that because ω_B is clockwise, $\omega_2 = -\omega_B$. We divide Eq. (1) by Eq. (2) and write the resulting equation as

$$\omega_A + \frac{R_A I_B}{R_B I_A} \omega_B = \omega_0.$$

When there is no slip, the velocities of the disks are equal at their point of contact:

$$R_A \omega_A = R_B \omega_B.$$

Solving these two equations, we obtain

$$\omega_A = \omega_0 \left[\frac{1}{1 + \dfrac{R_A^2 I_B}{R_B^2 I_A}} \right]$$

and

$$\omega_B = \omega_0 \left[\frac{R_A/R_B}{1 + \dfrac{R_A^2 I_B}{R_B^2 I_A}} \right].$$

If the disks have the same radius and moment of inertia, $\omega_A = \frac{1}{2}\omega_0$ and $\omega_B = \frac{1}{2}\omega_0$.

Critical Thinking

We did not know what force was exerted on disk A from the left to press it against disk B, which meant that we did not know the normal force N that the disks exerted on each other. The friction force f exerted on the disks while they were slipping relative to each other depends on N, so f was also unknown. It is the friction force f that causes the angular accelerations of the disks, so it is remarkable that we were able to determine their final angular velocities. Without knowing N, the time required for the disks to reach their final angular velocities cannot be determined, but the principle of angular impulse and momentum made it possible for us to obtain the final angular velocities.

Example 19.5 | Impulsive Force on a Rigid Body

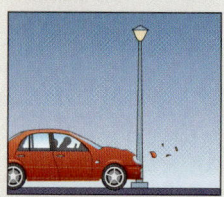

Figure 19.20

To help prevent injuries to passengers, engineers design a street light pole so that it shears off at ground level when struck by a vehicle (Fig. 19.20). From videotape of a test impact, the engineers estimate the angular velocity of the pole to be $\omega = 0.74$ rad/s and the horizontal velocity of its center of mass to be $v = 6.8$ m/s after the impact, and they estimate the duration of the impact to be $\Delta t = 0.01$ s. If the pole can be modeled as a 70-kg slender bar of length $l = 6$ m, the car strikes it at a height $h = 0.5$ m above the ground, and the couple exerted on the pole by its support during the impact is negligible, what average force was required to shear off the bolts supporting the pole?

Strategy

We will determine the average force by applying the principles of linear and angular impulse and momentum, expressed in terms of the average forces and moments exerted on the pole. We can apply the principle of angular impulse and momentum by using either Eq. (19.32) or Eq. (19.33). We will use Eq. (19.33) to demonstrate its use.

Solution

We draw the free-body diagram of the pole in Fig. a, where F is the average force exerted by the car and S is the average shearing force exerted on the pole by the bolts. Let m be the mass of the pole, and let v and ω be the velocity of its center of mass and its angular velocity at the end of the impact (Fig. b). From Eq. (19.24), the principle of linear impulse and momentum expressed in terms of the average horizontal force is

$$(t_2 - t_1)(\Sigma F_x)_{av} = (mv_x)_2 - (mv_x)_1:$$

$$\Delta t(F - S) = mv - 0. \tag{1}$$

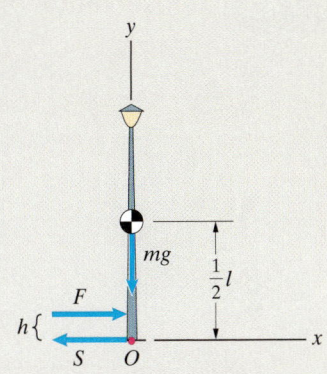

(a) Free-body diagram of the pole.

To apply the principle of angular impulse and momentum, we use Eq. (19.33), placing the fixed point O at the bottom of the pole (Figs. a and b). The pole's angular momentum about O at the end of the impact is

$$H_{O2} = [(\mathbf{r} \times m\mathbf{v}) \cdot \mathbf{k} + I\omega]_2 = \left[\left(\tfrac{1}{2}l\mathbf{j}\right) \times m(v\mathbf{i})\right] \cdot \mathbf{k} + I\omega$$

$$= -\tfrac{1}{2}lmv + I\omega.$$

We can also obtain this result by calculating the "moment" of the linear momentum about O and adding the term $I\omega$. The magnitude of the "moment" is the product of the magnitude of the linear momentum (mv) and the perpendicular distance from O to the line of action of the linear momentum $\left(\tfrac{1}{2}l\right)$, and it is negative because the "moment" is clockwise. (See Fig. 19.17.) From Eq. (19.33), we obtain

$$(t_2 - t_1)(\Sigma M_O)_{av} = H_{O2} - H_{O1}:$$

$$\Delta t(-hF) = -\tfrac{1}{2}lmv + I\omega - 0.$$

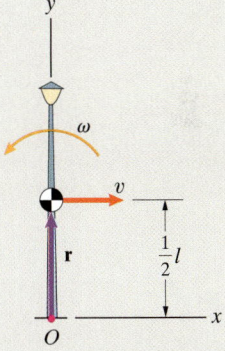

(b) Velocity and angular velocity at the end of the impact.

Solving this equation together with Eq. (1) for the average shear force S, we obtain

$$S = \frac{\left(\tfrac{1}{2}l - h\right)mv - I\omega}{h\,\Delta t}$$

$$= \frac{\left[\tfrac{1}{2}(6\text{ m}) - 0.5\text{ m}\right](70\text{ kg})(6.8\text{ m/s}) - \left[\tfrac{1}{12}(70\text{ kg})(6\text{ m})^2\right](0.74\text{ rad/s})}{(0.5\text{ m})(0.01\text{ s})}$$

$$= 207{,}000\text{ N}.$$

Critical Thinking

This example demonstrates both the power and the limitations of momentum methods. As the car collides with the light pole, the car's structure deforms, the light pole deforms, and the bolts supporting the pole fail. The time history of the force exerted on the car and pole by the impact hinges upon the details of these complicated phenomena. Using momentum methods and information about the motion of the pole after the impact, we were able to estimate the *average* value of the force, but we cannot determine its time history. To do so would require either more elaborate experiments or an analysis of the collision in which the deformations of the car and pole are modeled. This kind of trade-off occurs frequently in engineering. Often limited information about a phenomenon can be obtained quickly, as we do in this example, but more accurate and complete information could be obtained by investing the necessary time and resources. The question that must be answered is whether the additional investment is essential to achieve the required engineering objectives.

Example 19.6 **Conservation of Angular Momentum**

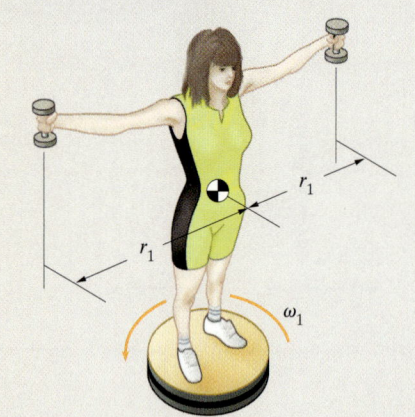

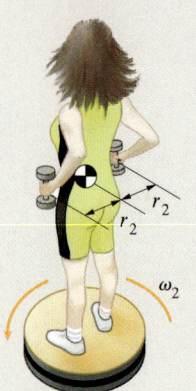

Figure 19.21

In a well-known demonstration of conservation of angular momentum, a person stands on a rotating platform holding a mass m in each hand (Fig. 19.21). The moment of inertia of the person and platform is $I_P = 0.4$ kg-m^2, the mass $m = 4$ kg, and the moment of inertia of each mass about the vertical axis through its center of mass is $I_M = 0.001$ kg-m^2. If the person's angular velocity with her arms extended to $r_1 = 0.6$ m is $\omega_1 = 1$ revolution per second, what is her angular velocity ω_2 when she pulls the masses inward to $r_2 = 0.2$ m? (You have observed skaters using this phenomenon to control their angular velocity in a spin by altering the positions of their arms.)

Strategy

If we neglect friction in the rotating platform, the total angular momentum of the person, platform, and masses about the vertical axis of rotation is conserved. We can use this condition to determine ω_2.

Solution

We begin by determining the angular momentum of one of the masses about the axis of rotation when the person's arms are extended. The magnitude of the velocity of this mass about the axis of rotation is $r_1\omega_1$, so the "moment" of its linear momentum about the axis of rotation is $r_1(mr_1\omega_1) = mr_1^2\omega_1$. The total angular momentum of the mass about the axis of rotation when the person's arms are extended is

$$(\mathbf{r} \times m\mathbf{v}) \cdot \mathbf{k} + I\omega = mr_1^2\omega_1 + I_M\omega_1.$$

Since the center of mass of the person and platform lie on the axis of rotation, their angular momentum about the axis of rotation is $I_P\omega_1$. Therefore, the combined angular momentum of the person, platform, and masses when her arms are extended is

$$H_{O1} = I_P\omega_1 + 2(mr_1^2\omega_1 + I_M\omega_1) = [I_P + 2(I_M + mr_1^2)]\omega_1.$$

By replacing ω_1 by ω_2 and r_1 by r_2 in this expression, we obtain the combined angular momentum when the person has pulled the masses inward. Angular momentum is conserved, so $H_{O1} = H_{O2}$.

$$[I_P + 2(I_M + mr_1^2)]\omega_1 = [I_P + 2(I_M + mr_2^2)]\omega_2:$$

$$\{0.4 \text{ kg-m}^2 + 2[0.001 \text{ kg-m}^2 + (4 \text{ kg})(0.6 \text{ m})^2]\}\omega_1$$

$$= \{0.4 \text{ kg-m}^2 + 2[0.001 \text{ kg-m}^2 + (4 \text{ kg})(0.2 \text{ m})^2]\}\omega_2.$$

Solving, we find that $\omega_2 = 4.55\omega_1 = 4.55$ revolutions per second.

Critical Thinking

The parallel-axis theorem states that if the person holds the masses at a distance r from the axis of rotation, the moment of inertia of each mass about the axis of rotation is $I_M + mr^2$. The term $I_P + 2(I_M + mr^2)$ that appears in our equation of conservation of angular momentum is simply the total moment of inertia of the person, the platform, and the two masses about the axis of rotation.

Problems

19.48 The moment of inertia of the disk about O is 22 kg-m^2. At $t = 0$, the stationary disk is subjected to a constant 50 N-m torque.

(a) Determine the angular impulse exerted on the disk from $t = 0$ to $t = 5$ s.
(b) What is the disk's angular velocity at $t = 5$ s?

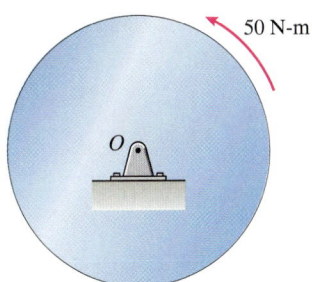

Problem 19.48

19.49 The moment of inertia of the jet engine's rotating assembly is 400 kg-m^2. The assembly starts from rest. At $t = 0$, the engine's turbine exerts a couple on it that is given as a function of time by $M = 6500 - 125t$ N-m.

(a) What is the magnitude of the angular impulse exerted on the assembly from $t = 0$ to $t = 20$ s?
(b) What is the magnitude of the angular velocity of the assembly (in rpm) at $t = 20$ s?

Problem 19.49

19.50 An astronaut fires a thruster of his maneuvering unit, exerting a force $T = 2(1 + t)$ N, where t is in seconds. The combined mass of the astronaut and his equipment is 122 kg, and the moment of inertia about their center of mass is 45 kg-m^2. Modeling the astronaut and his equipment as a rigid body, use the principle of angular impulse and momentum to determine how long it takes for his angular velocity to reach 0.1 rad/s.

19.51 The combined mass of the astronaut and his equipment is 122 kg, and the moment of inertia about their center of mass is 45 kg-m^2. The maneuvering unit exerts an impulsive force T of 0.2-s duration, giving him a counterclockwise angular velocity of 1 rpm.

(a) What is the average magnitude of the impulsive force?
(b) What is the magnitude of the resulting change in the velocity of the astronaut's center of mass?

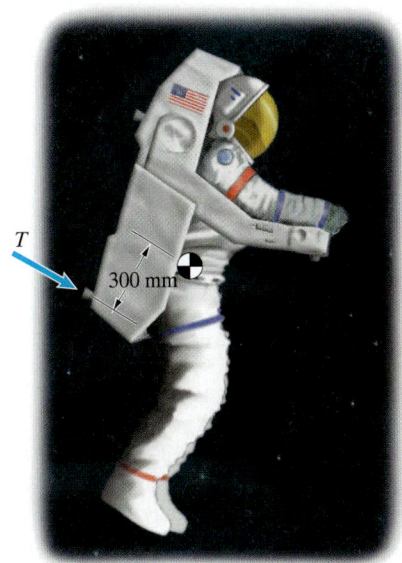

Problems 19.50/19.51

19.52 A flywheel attached to an electric motor is initially at rest. At $t = 0$, the motor exerts a couple $M = 200e^{-0.1t}$ N-m on the flywheel. The moment of inertia of the flywheel is 10 kg-m^2.

(a) What is the flywheel's angular velocity at $t = 10$ s?
(b) What maximum angular velocity will the flywheel attain?

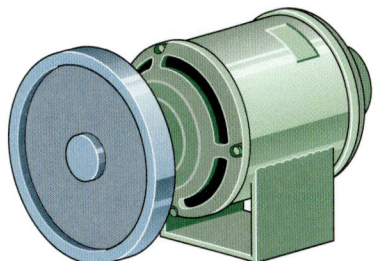

Problem 19.52

19.53 A main landing gear wheel of a Boeing 777 has a radius of 0.62 m and its moment of inertia is 24 kg-m^2. After the plane lands at 75 m/s, the skid marks of the wheel's tire is measured and determined to be 18 m in length. Determine the average friction force exerted on the wheel by the runway. Assume that the airplane's velocity is constant during the time the tire skids (slips) on the runway.

Problem 19.53

19.54 The force a club exerts on a 0.045-kg golf ball is shown. The ball is 42 mm in diameter and can be modeled as a homogeneous sphere. The club is in contact with the ball for 0.0006 s, and the magnitude of the velocity of the ball's center of mass after the ball is hit is 36 m/s. What is the magnitude of the ball's angular velocity after it is hit?

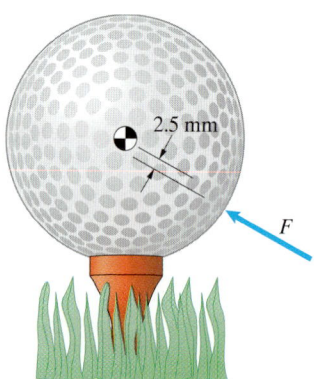

Problem 19.54

19.55 Disk A initially has a counterclockwise angular velocity $\omega_0 = 50$ rad/s. Disks B and C are initially stationary. At $t = 0$, disk A is moved into contact with disk B. Determine the angular velocities of the three disks when they have stopped slipping relative to each other. The masses of the disks are $m_A = 4$ kg, $m_B = 16$ kg, and $m_C = 9$ kg.

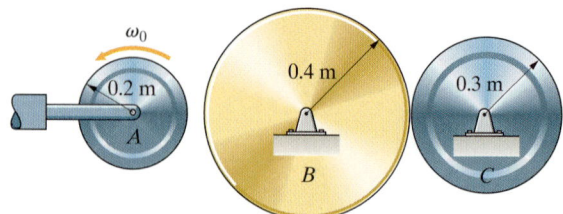

Problem 19.55

19.56* The suspended bars A and B each have a mass of 2 kg. The bars are stationary when a horizontal impulsive force F is applied to the bottom of bar B. The average value of the force is 1500 N and its duration is 0.0001 s. What are the angular velocities of the bars just after the impulsive force is applied?

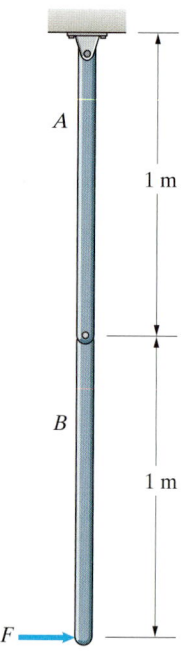

Problem 19.56

19.57 The force exerted on the cue ball by the cue is horizontal. Determine the value of h for which the ball rolls without slipping. (Assume that the frictional force exerted on the ball by the table is negligible.)

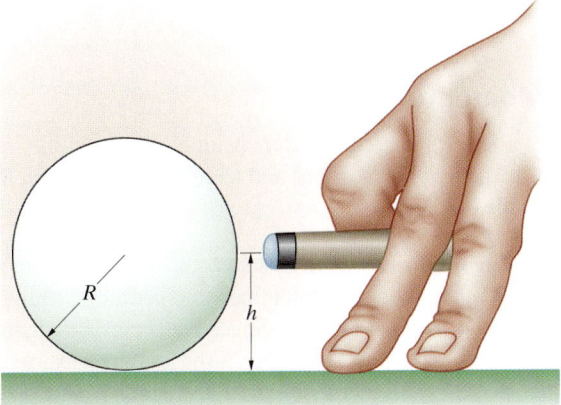

Problem 19.57

19.58 Two gravity research satellites $(m_A = 250 \text{ kg}, I_A = 350 \text{ kg-m}^2; m_B = 50 \text{ kg}, I_B = 16 \text{ kg-m}^2)$ are tethered by a cable. The satellites and cable rotate with angular velocity $\omega_0 = 0.25$ rpm. Ground controllers order satellite A to slowly unreel 6 m of additional cable. What is the angular velocity afterward?

19.59 Solve Problem 19.58 by treating the satellites as particles (that is, neglect their moments of inertia I_A and I_B), and compare your answer with that of Problem 19.58.

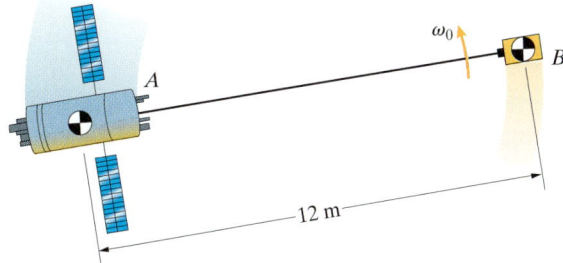

Problems 19.58/19.59

19.60 The 2-kg bar rotates *in the horizontal plane* about the smooth pin. The 6-kg collar A slides on the smooth bar. Assume that the moment of inertia of the collar A about its center of mass is negligible; that is, treat the collar as a particle. At the instant shown, the angular velocity of the bar is $\omega_0 = 60$ rpm and the distance from the pin to the collar is $r = 1.8$ m. Determine the bar's angular velocity when $r = 2.4$ m.

19.61 The 2-kg bar rotates *in the horizontal plane* about the smooth pin. The 6-kg collar A slides on the smooth bar. The moment of inertia of the collar A about its center of mass is 0.2 kg-m^2. At the instant shown, the angular velocity of the bar is $\omega_0 = 60$ rpm and the distance from the pin to the collar is $r = 1.8$ m. Determine the bar's angular velocity when $r = 2.4$ m and compare your answer to that of Problem 19.60.

19.62* The 2-kg bar rotates *in the horizontal plane* about the smooth pin. The 6-kg collar A slides on the smooth bar. The moment of inertia of the collar A about its center of mass is 0.2 kg-m^2. The spring is unstretched when $r = 0$, and the spring constant is $k = 10$ N/m. At the instant shown, the angular velocity of the bar is $\omega_0 = 2$ rad/s, the distance from the pin to the collar is $r = 1.8$ m, and the radial velocity of the collar is zero. Determine the radial velocity of the collar when $r = 2.4$ m.

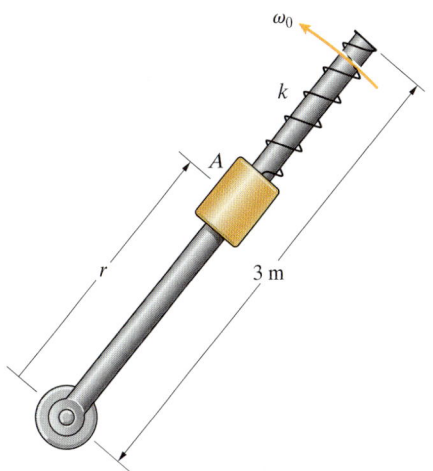

Problems 19.60–19.62

19.63 The circular bar is welded to the vertical shafts, which can rotate freely in bearings at A and B. Let I be the moment of inertia of the circular bar and shafts about the vertical axis. The circular bar has an initial angular velocity ω_0, and the mass m is released in the position shown with no velocity relative to the bar. Determine the angular velocity of the circular bar as a function of the angle β between the vertical and the position of the mass. Neglect the moment of inertia of the mass about its center of mass; that is, treat the mass as a particle.

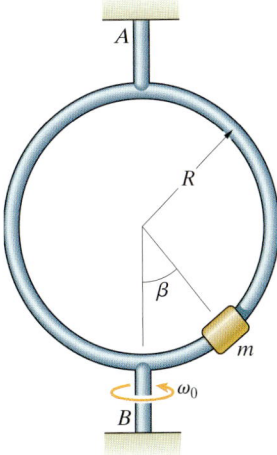

Problem 19.63

19.6 Impacts

In Chapter 16, we analyzed impacts between objects with the aim of determining the velocities of their centers of mass after the collision. We now discuss how to determine the velocities of the centers of mass *and the angular velocities* of rigid bodies after they collide.

Conservation of Momentum

Suppose that two rigid bodies *A* and *B*, in two-dimensional motion in the same plane, collide. What do the principles of linear and angular momentum tell us about their motions after the collision?

Linear Momentum   If other forces are negligible in comparison to the impact forces *A* and *B* exert on each other, then their total linear momentum is the same before and after the impact. But this result must be applied with care. For example, if one of the rigid bodies has a pin support (Fig. 19.22), the reactions exerted by the support cannot be neglected, and linear momentum is not conserved.

Angular Momentum   If other forces and couples are negligible in comparison to the impact forces and couples that *A* and *B* exert on each other, their total angular momentum about *any* fixed point *O* is the same before and after the impact. [See Eq. (19.34).] If, in addition, *A* and *B* exert forces on each other only at their point of impact *P*, and exert no couples on each other, the angular momentum about *P* of *each* rigid body is the same before and after the impact (Fig. 19.23). This result follows from the principle of angular impulse and momentum, Eq. (19.31), because the impact forces on *A* and *B* exert no moment about *P*. If one of the rigid bodies has a pin support at a point *O*, as in Fig. 19.22, their total angular momentum about *O* is the same before and after the impact.

Coefficient of Restitution

If two rigid bodies adhere and move as a single rigid body after colliding, their velocities and angular velocity can be determined by using momentum conservation and kinematic relationships alone. These relationships are not sufficient if the objects do not adhere. But some impacts of the latter type can be analyzed by also using the concept of the coefficient of restitution.

Let *P* be the point of contact of rigid bodies *A* and *B* during an impact (Fig. 19.24), and let their velocities at *P* be $\mathbf{v}_{AP}$ and $\mathbf{v}_{BP}$ just before the impact and $\mathbf{v}'_{AP}$ and $\mathbf{v}'_{BP}$ just afterward. The *x* axis is perpendicular to the contacting surfaces at *P*. If the frictional forces resulting from the impact are negligible, we can show that the components of the velocities normal to the surfaces at *P* are related to the coefficient of restitution *e* by

$$e = \frac{(\mathbf{v}'_{BP})_x - (\mathbf{v}'_{AP})_x}{(\mathbf{v}_{AP})_x - (\mathbf{v}_{BP})_x}. \tag{19.35}$$

To derive this result, we must consider the effects of the impact on the individual objects. Let t_1 be the time at which they first come into contact. The objects are not actually rigid, but will deform as a result of the collision. At a time t_C, the maximum deformation will occur, and the objects will begin a "recovery" phase in which they tend to resume their original shapes. Let t_2 be the time at which they separate.

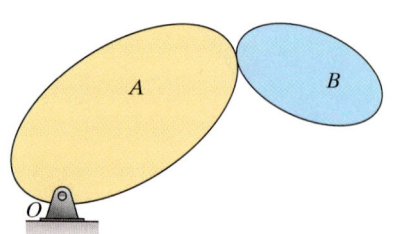

Figure 19.22
Rigid bodies *A* and *B* colliding. Because of the pin support, their total linear momentum is *not* conserved, but their total angular momentum about *O* is conserved.

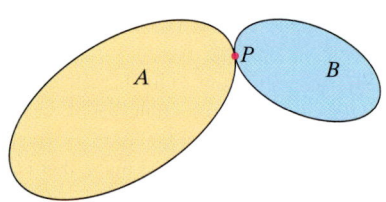

Figure 19.23
Rigid bodies *A* and *B* colliding at *P*. If forces are exerted only at *P*, the angular momentum of *A* about *P* and the angular momentum of *B* about *P* are each conserved.

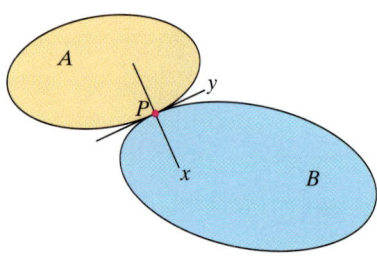

Figure 19.24
Rigid bodies *A* and *B* colliding at *P*. The *x* axis is perpendicular to the contacting surfaces.

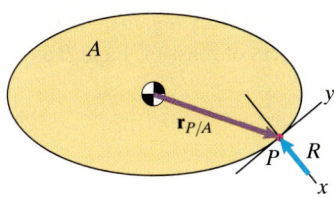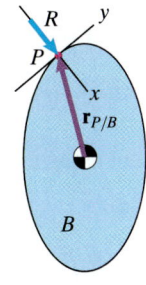

Figure 19.25
The normal force R resulting from the impact.

Our first step is to apply the principle of linear impulse and momentum to A and B for the intervals from t_1 to t_C and from t_C to t_2. Let R be the magnitude of the normal force exerted during the impact (Fig. 19.25). We denote the velocity of the center of mass of A at the times t_1, t_C, and t_2 by $\mathbf{v}_A$, $\mathbf{v}_{AC}$, and $\mathbf{v}'_A$, and denote the corresponding velocities of the center of mass of B by $\mathbf{v}_B$, $\mathbf{v}_{BC}$, and $\mathbf{v}'_B$. For A, we have

$$\int_{t_1}^{t_C} -R \, dt = m_A(\mathbf{v}_{AC})_x - m_A(\mathbf{v}_A)_x, \tag{19.36}$$

$$\int_{t_C}^{t_2} -R \, dt = m_A(\mathbf{v}'_A)_x - m_A(\mathbf{v}_{AC})_x. \tag{19.37}$$

For B, we have

$$\int_{t_1}^{t_C} R \, dt = m_B(\mathbf{v}_{BC})_x - m_B(\mathbf{v}_B)_x, \tag{19.38}$$

$$\int_{t_C}^{t_2} R \, dt = m_B(\mathbf{v}'_B)_x - m_B(\mathbf{v}_{BC})_x. \tag{19.39}$$

The coefficient of restitution is the ratio of the linear impulse during the recovery phase to the linear impulse during the deformation phase:

$$e = \frac{\displaystyle\int_{t_C}^{t_2} R \, dt}{\displaystyle\int_{t_1}^{t_C} R \, dt}. \tag{19.40}$$

If we divide Eq. (19.37) by Eq. (19.36) and divide Eq. (19.39) by Eq. (19.38), the resulting equations can be written as

$$(\mathbf{v}'_A)_x = -(\mathbf{v}_A)_x e + (\mathbf{v}_{AC})_x(1 + e),$$

$$(\mathbf{v}'_B)_x = -(\mathbf{v}_B)_x e + (\mathbf{v}_{BC})_x(1 + e). \tag{19.41}$$

We now apply the principle of angular impulse and momentum to A and B for the intervals of time from t_1 to t_C and from t_C to t_2. We denote the counterclockwise angular velocity of A at the times t_1, t_C, and t_2 by ω_A, ω_{AC}, and ω'_A and denote the corresponding angular velocities of B by ω_B, ω_{BC}, and ω'_B. We write the position vectors of P relative to the centers of mass of A and B as (Fig. 19.25)

$$\mathbf{r}_{P/A} = x_A \mathbf{i} + y_A \mathbf{j},$$

$$\mathbf{r}_{P/B} = x_B \mathbf{i} + y_B \mathbf{j}. \tag{19.42}$$

The moment about the center of mass of A due to the force exerted on A by the impact is $\mathbf{r}_{P/A} \times (-R\mathbf{i}) = y_A R \mathbf{k}$. From Eq. (19.28), we obtain the equations

$$\int_{t_1}^{t_C} y_A R \, dt = I_A \omega_{AC} - I_A \omega_A, \tag{19.43}$$

$$\int_{t_C}^{t_2} y_A R \, dt = I_A \omega'_A - I_A \omega_{AC}. \tag{19.44}$$

The corresponding equations for B are

$$\int_{t_1}^{t_C} -y_B R \, dt = I_B \omega_{BC} - I_B \omega_B, \tag{19.45}$$

$$\int_{t_C}^{t_2} -y_B R \, dt = I_B \omega'_B - I_B \omega_{BC}. \tag{19.46}$$

Dividing Eq. (19.44) by Eq. (19.43) and dividing Eq. (19.46) by Eq. (19.45), we can write the resulting equations as

$$\omega'_A = -\omega_A e + \omega_{AC}(1 + e),$$

$$\omega'_B = -\omega_B e + \omega_{BC}(1 + e). \tag{19.47}$$

By expressing the velocity of the point of A at P in terms of the velocity of the center of mass of A and the angular velocity of A, and expressing the velocity of the point of B at P in terms of the velocity of the center of mass of B and the angular velocity of B, we obtain

$$(\mathbf{v}_{AP})_x = (\mathbf{v}_A)_x - \omega_A y_A,$$

$$(\mathbf{v}'_{AP})_x = (\mathbf{v}'_A)_x - \omega'_A y_A,$$

$$(\mathbf{v}_{BP})_x = (\mathbf{v}_B)_x - \omega_B y_B,$$

$$(\mathbf{v}'_{BP})_x = (\mathbf{v}'_B)_x - \omega'_B y_B. \tag{19.48}$$

At time t_C, the x components of the velocities of the two objects are equal at P, which yields the relation

$$(\mathbf{v}_{AC})_x - \omega_{AC} y_A = (\mathbf{v}_{BC})_x - \omega_{BC} y_B. \tag{19.49}$$

From Eqs. (19.48),

$$\frac{(\mathbf{v}'_{BP})_x - (\mathbf{v}'_{AP})_x}{(\mathbf{v}_{AP})_x - (\mathbf{v}_{BP})_x} = \frac{(\mathbf{v}'_B)_x - \omega'_B y_B - (\mathbf{v}'_A)_x + \omega'_A y_A}{(\mathbf{v}_A)_x - \omega_A y_A - (\mathbf{v}_B)_x + \omega_B y_B}.$$

Substituting Eqs. (19.41) and (19.47) into this equation and collecting terms yields

$$\frac{(\mathbf{v}'_{BP})_x - (\mathbf{v}'_{AP})_x}{(\mathbf{v}_{AP})_x - (\mathbf{v}_{BP})_x} = e - \left[\frac{(\mathbf{v}_{AC})_x - \omega_{AC} y_A - (\mathbf{v}_{BC})_x + \omega_{BC} y_B}{(\mathbf{v}_A)_x - \omega_A y_A - (\mathbf{v}_B)_x + \omega_B y_B} \right] (e + 1).$$

From Eq. (19.49), the term in brackets vanishes, and we obtain the equation relating the normal components of the velocities at the point of contact to the coefficient of restitution:

$$e = \frac{(\mathbf{v}'_{BP})_x - (\mathbf{v}'_{AP})_x}{(\mathbf{v}_{AP})_x - (\mathbf{v}_{BP})_x}. \tag{19.50}$$

In arriving at this equation, we assumed that the contacting surfaces were smooth, so that *the collision exerts no force on A or B in the direction tangential to their contacting surfaces*.

  Although we derived Eq. (19.50) under the assumption that the motions of A and B are unconstrained, the relationship also holds if they are not—for example, if one of them is connected to a pin support.

  In summary, we see that the approach used for analyzing collisions of rigid bodies in planar motion depends on the type of collision. If other forces are negligible in comparison to the impact forces, the total linear momentum is conserved. If other forces and couples are negligible in comparison to the impact forces and couples, the total angular momentum about any fixed point is conserved. If, in addition, forces are exerted only at the point of impact P, the angular momentum about P of each rigid body is conserved. If one of the rigid bodies has a pin support at a point O, the total angular momentum about O is conserved. If the impact is assumed to exert no forces on the colliding objects in the direction tangential to their surface of contact, the coefficient of restitution e relates the normal components of the velocities at the point of contact through Eq. (19.50).

Study Questions

1. When two rigid bodies collide, you cannot always assume that their total linear momentum is conserved. Explain why.

2. If P is the point of impact of two colliding rigid bodies and external forces and couples are negligible, is the total angular momentum of the rigid bodies about P conserved? Is the angular momentum of each rigid body about P conserved?

3. If two rigid bodies collide, one of which is free and the other of which is attached to a fixed pin support at a point O, is the total angular momentum about O conserved? Is the angular momentum of each rigid body about O conserved?

4. If P is the point of impact of two colliding rigid bodies, is the coefficient of restitution defined in terms of the velocities of their centers of mass or in terms of their velocities at P?

Example 19.7   Impact of a Sphere and a Suspended Bar

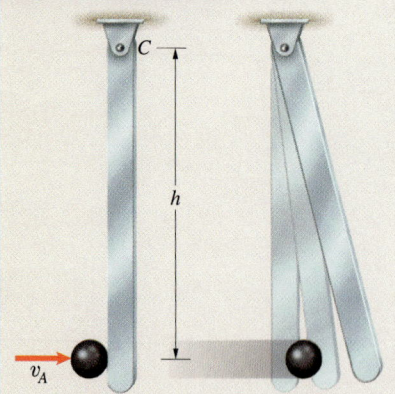

Figure 19.26

The homogeneous sphere in Fig. 19.26 is moving horizontally with velocity v_A and no angular velocity when it strikes the stationary slender bar. The sphere has mass m_A, and the bar has mass m_B and length l. The coefficient of restitution of the impact is e.
(a) What is the angular velocity of the bar after the impact?
(b) If the duration of the impact is Δt, what average horizontal force is exerted on the bar by the pin support C as a result of the impact?

Strategy

(a) From the definition of the coefficient of restitution, we can obtain an equation relating the horizontal velocity of the sphere and the velocity of the bar at the point of impact after the collision occurs. In addition, the total angular momentum of the sphere and bar about the pin C is conserved. With these two equations and kinematic relationships, we can determine the velocity of the sphere and the angular velocity of the bar after the impact.
(b) We can determine the average force exerted on the bar by the support by applying the principle of angular impulse and momentum to the bar.

Solution

(a) In Fig. a, we show the velocities just after the impact, where v'_{BP} is the bar's velocity at the point of impact. From the definition of the coefficient of restitution, Eq. (19.35), we obtain

$$e = \frac{v'_{BP} - v'_A}{v_A - 0}.$$

The equation of conservation of total angular momentum about C is

$$H_{CA} + H_{CB} = H'_{CA} + H'_{CB}:$$

$$(\mathbf{r}_A \times m_A\mathbf{v}_A) \cdot \mathbf{k} + I_A\omega_A + (\mathbf{r}_B \times m_B\mathbf{v}_B) \cdot \mathbf{k} + I_B\omega_B$$

$$= (\mathbf{r}_A \times m_A\mathbf{v}'_A) \cdot \mathbf{k} + I_A\omega'_A + (\mathbf{r}_B \times m_B\mathbf{v}'_B) \cdot \mathbf{k} + I_B\omega'_B,$$

$$(-h\mathbf{j} \times m_A v_A\mathbf{i}) \cdot \mathbf{k} = (-h\mathbf{j} \times m_A v'_A\mathbf{i}) \cdot \mathbf{k} + \left(-\tfrac{1}{2}l\mathbf{j} \times m_B v'_B\mathbf{i}\right) \cdot \mathbf{k} + I_B\omega'_B.$$

Carrying out the vector operations, we obtain

$$hm_A v_A = hm_A v'_A + \tfrac{1}{2}lm_B v'_B + I_B\omega'_B.$$

Notice in Fig. a that the velocities v'_B and v'_{BP} are related to the angular velocity of the bar ω'_B by

$$v'_B = \tfrac{1}{2}l\omega'_B, \qquad v'_{BP} = h\omega'_B.$$

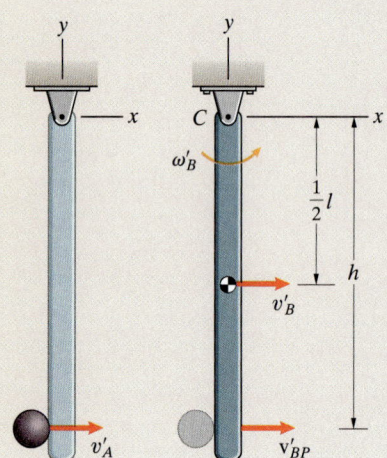

(a) Velocities of the sphere and bar after the impact.

We now have four equations in the four unknowns v'_A, v'_B, v'_{BP}, and ω'_B. Solving them for the angular velocity of the bar and using the relation $I_B = \frac{1}{12}m_B l^2$, we obtain

$$\omega'_B = \frac{(1+e)hm_A v_A}{h^2 m_A + \frac{1}{3}m_B l^2}.$$

(b) Let the forces on the free-body diagram of the bar in Fig. b represent the average forces exerted during the impact. We apply the principle of angular impulse and momentum, in the form given by Eq. (19.33), about the point of impact:

$$(t_2 - t_1)(\Sigma M_P)_{av} = H'_B - H_B:$$

$$(t_2 - t_1)(\Sigma M_P)_{av} =$$

$$[(\mathbf{r}_B \times m_B \mathbf{v}'_B) \cdot \mathbf{k} + I_B \omega'_B] - [(\mathbf{r}_B \times m_B \mathbf{v}_B) \cdot \mathbf{k} + I_B \omega_B],$$

$$\Delta t(-hC_x) = [(h - \tfrac{1}{2}l)\mathbf{j} \times m_B v'_B \mathbf{i}] \cdot \mathbf{k} + I_B \omega'_B - 0.$$

Solving for C_x yields

$$C_x = \frac{(h - \frac{1}{2}l)m_B v'_B - I_B \omega'_B}{h\,\Delta t}.$$

Using our solution for ω'_B from part (a) and the relation $v'_B = \frac{1}{2}l\omega'_B$, we obtain the average horizontal force exerted by the support:

$$C_x = \frac{(1+e)\left(\frac{1}{2}h - \frac{1}{3}l\right)lm_A m_B v_A}{\left(h^2 m_A + \frac{1}{3}m_B l^2\right)\Delta t}.$$

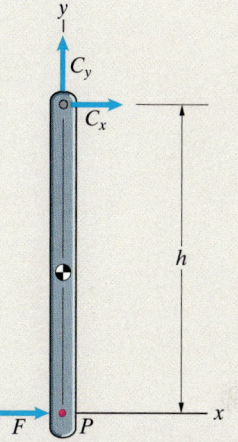

(b) Average forces exerted on the bar during the impact.

Critical Thinking

When the impact occurs, a force is exerted on the bar by the support at C. As a result, we could not assume that the total linear momentum of the sphere and bar was the same before and after the impact. But the force exerted by the support as a result of the impact exerts no moment about C, so we could assume that the total *angular* momentum of the sphere and bar was the same before and after the impact. That relationship, together with the coefficient of restitution, allowed us to determine both the velocity of the sphere and the angular velocity of the bars after the impact. Notice that the fact that the bar is pinned means that total linear momentum is not conserved, but it also reduces the number of unknowns, because the bar's angular velocity and the velocity of its center of mass are not independent.

Example 19.8 **Impact with a Fixed Obstacle**

The combined mass of the motorcycle and rider in Fig. 19.27 is $m = 170$ kg, and their combined moment of inertia about their center of mass is 22 kg-m^2. Following a jump, the motorcycle and rider are in the position shown just before the rear wheel contacts the ground. The velocity of their center of mass is of magnitude $|\mathbf{v}_G| = 8.8$ m/s, and their angular velocity is $\omega = 0.2$ rad/s. If the motorcycle and rider are modeled as a single rigid body and the coefficient of restitution of the impact is $e = 0.8$, what are the angular velocity ω' and velocity $\mathbf{v}'_G$ after the impact? Neglect the tangential component of force exerted on the motorcycle's wheel during the impact.

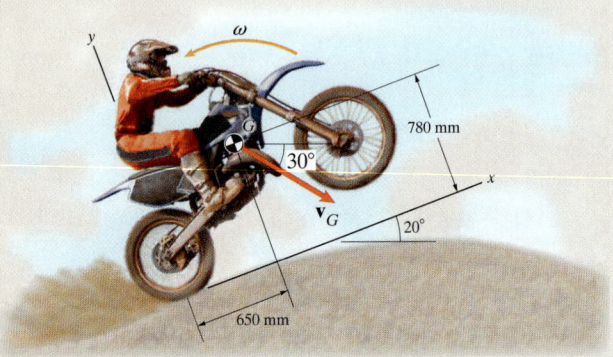

Figure 19.27

Strategy

Since the tangential component of force on the motorcycle's wheel during the impact is neglected, the component of the velocity of the center of mass parallel to the ground can be taken to be unchanged by the impact. The coefficient of restitution relates the motorcycle's velocity normal to the ground at the point of impact before the impact to its value after the impact. Also, the force of the impact exerts no moment about the point of impact, so the motorcycle's angular momentum about that point is conserved. (We assume the impact to be so brief that the angular impulse due to the weight is negligible.) With these three relations, we can determine the two components of the velocity of the center of mass and the angular velocity after the impact.

Solution

In Fig. a we align a coordinate system parallel and perpendicular to the ground at the point P where the impact occurs. Let the components of the velocity of the center of mass before and after the impact be $\mathbf{v}_G = v_x \mathbf{i} + v_y \mathbf{j}$ and $\mathbf{v}'_G = v'_x \mathbf{i} + v'_y \mathbf{j}$, respectively. The x and y components of the velocity are

$$v_x = 8.8 \cos 50° = 5.66 \text{ m/s}$$

and

$$v_y = -8.8 \sin 50° = -6.74 \text{ m/s}.$$

Because the component of the impact force tangential to the ground is neglected, the x component of the velocity of the center of mass is unchanged:

$$v'_x = v_x = 5.66 \text{ m/s}.$$

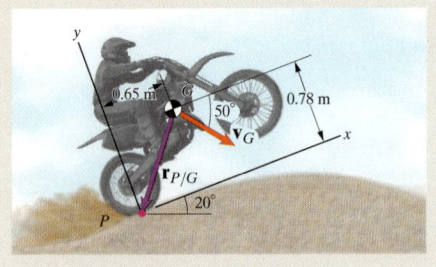

(a) Aligning the x axis of the coordinate system tangent to the ground at P.

We can express the y component of the wheel's velocity at P before the impact in terms of the velocity of the center of mass and the angular velocity (Fig. a) as

$$\mathbf{j} \cdot \mathbf{v}_P = \mathbf{j} \cdot (\mathbf{v}_G + \boldsymbol{\omega} \times \mathbf{r}_{P/G})$$

$$= \mathbf{j} \cdot \left\{ v_x \mathbf{i} + v_y \mathbf{j} + \begin{vmatrix} \mathbf{i} & \mathbf{j} & \mathbf{k} \\ 0 & 0 & \omega \\ -0.65 & -0.78 & 0 \end{vmatrix} \right\}$$

$$= v_y - 0.65\omega.$$

(Notice that this expression gives the y component of the velocity at P even though the wheel is spinning.) The y component of the wheel's velocity at P after the impact is

$$\mathbf{j} \cdot \mathbf{v}_P' = \mathbf{j} \cdot (\mathbf{v}_G' + \boldsymbol{\omega}' \times \mathbf{r}_{P/G})$$

$$= v_y' - 0.65\omega'.$$

The coefficient of restitution relates the y components of the wheel's velocity at P before and after the impact:

$$e = \frac{-(\mathbf{j} \cdot \mathbf{v}_P')}{\mathbf{j} \cdot \mathbf{v}_P} = \frac{-(v_y' - 0.65\omega')}{v_y - 0.65\omega}. \tag{1}$$

The force of the impact exerts no moment about P, so angular momentum about P is conserved:

$$H_P = H_P':$$

$$[(\mathbf{r}_{G/P} \times m\mathbf{v}_G) \cdot \mathbf{k} + I\omega] = [(\mathbf{r}_{G/P} \times m\mathbf{v}_G') \cdot \mathbf{k} + I\omega'],$$

$$\begin{vmatrix} \mathbf{i} & \mathbf{j} & \mathbf{k} \\ 0.65 & 0.78 & 0 \\ mv_x & mv_y & 0 \end{vmatrix} \cdot \mathbf{k} + I\omega = \begin{vmatrix} \mathbf{i} & \mathbf{j} & \mathbf{k} \\ 0.65 & 0.78 & 0 \\ mv_x' & mv_y' & 0 \end{vmatrix} \cdot \mathbf{k} + I\omega'.$$

Expanding the determinants and evaluating the dot products, we obtain

$$0.65mv_y - 0.78mv_x + I\omega = 0.65mv_y' - 0.78mv_x' + I\omega'. \tag{2}$$

Since we have already determined v_x', we can solve Eqs. (1) and (2) for v_y' and ω'. The results are

$$v_y' = -3.84 \text{ m/s}$$

and

$$\omega' = -14.4 \text{ rad/s}.$$

The velocity of the center of mass after the impact is $\mathbf{v}_G' = 5.66\mathbf{i} - 3.84\mathbf{j}$ m/s, and the angular velocity is 14.4 rad/s in the clockwise direction.

Critical Thinking

Although forces resulting from an impact are often so large that the effects of other forces can be neglected, that isn't always the case. In this example, we neglected the weight of the motorcycle and rider in determining their velocity and angular velocity following the impact of the rear wheel with the ground. Whenever there is doubt in engineering applications of momentum methods, such effects should be included in the analysis. In order to do so, the duration of the impact needs to be known or estimated so that the linear and angular impulses due to other forces can be evaluated.

Example 19.9	**Colliding Cars**

An engineer simulates a collision between two 1600-kg cars by modeling them as rigid bodies (Fig. 19.28). The moment of inertia of each car about its center of mass is 960 kg-m^2. The engineer assumes the contacting surfaces at P to be smooth and parallel to the x axis and assumes the coefficient of restitution to be $e = 0.2$. What are the angular velocities of the cars and the velocities of their centers of mass after the collision?

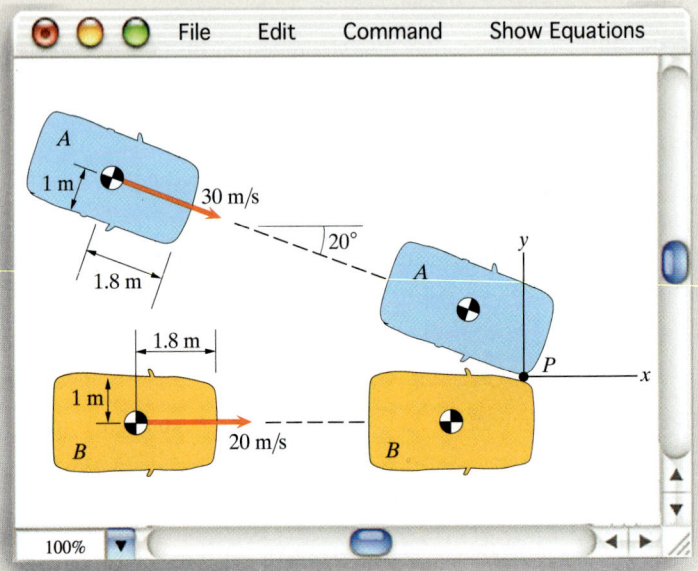

Figure 19.28

Strategy

Since the contacting surfaces are smooth, the x components of the velocities of the centers of mass are unchanged by the collision. The y components of the velocities must satisfy conservation of linear momentum, and the y components of the velocities at the point of impact before and after the impact are related by the coefficient of restitution. The force of the impact exerts no moment about P on either car, so the angular momentum of each car about P is conserved. From these conditions and kinematic relations between the velocities of the centers of mass and the velocities at P, we can determine the angular velocities and the velocities of the centers of mass after the impact.

Solution

The components of the velocities of the centers of mass before the impact are

$$\mathbf{v}_A = 30 \cos 20°\mathbf{i} - 30 \sin 20°\mathbf{j}$$

$$= 28.2\mathbf{i} - 10.3\mathbf{j} \ (\text{m/s})$$

and

$$\mathbf{v}_B = 20\mathbf{i} \ (\text{m/s}).$$

The x components of the velocities are unchanged by the impact:

$$v'_{Ax} = v_{Ax} = 28.2 \text{ m/s}, \qquad v'_{Bx} = v_{Bx} = 20 \text{ m/s}.$$

The y components of the velocities must satisfy conservation of linear momentum:

$$m_A v_{Ay} + m_B v_{By} = m_A v'_{Ay} + m_B v'_{By}. \tag{1}$$

Let the velocities of the two cars at P before the collision be $\mathbf{v}_{AP}$ and $\mathbf{v}_{BP}$. The coefficient of restitution $e = 0.2$ relates the y components of the velocities at P:

$$0.2 = \frac{v'_{BPy} - v'_{APy}}{v_{APy} - v_{BPy}}. \tag{2}$$

We can express the velocities at P after the impact in terms of the velocities of the centers of mass and the angular velocities after the impact (Fig. a). The position of P relative to the center of mass of car A is

$$\mathbf{r}_{P/A} = [(1.8)\cos 20° - (1)\sin 20°]\mathbf{i} - [(1.8)\sin 20° + (1)\cos 20°]\mathbf{j}$$

$$= 1.35\mathbf{i} - 1.56\mathbf{j} \text{ (m)}.$$

Therefore, we can express the velocity of point P of car A after the impact as

$$\mathbf{v}'_{AP} = \mathbf{v}'_A + \boldsymbol{\omega}'_A \times \mathbf{r}_{P/A}:$$

$$v'_{APx}\mathbf{i} + v'_{APy}\mathbf{j} = v'_{Ax}\mathbf{i} + v'_{Ay}\mathbf{j} + \begin{vmatrix} \mathbf{i} & \mathbf{j} & \mathbf{k} \\ 0 & 0 & \omega'_A \\ 1.35 & -1.56 & 0 \end{vmatrix}.$$

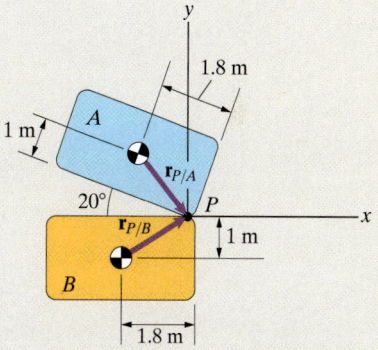

(a) Position vectors of P relative to the centers of mass.

Equating $\mathbf{i}$ and $\mathbf{j}$ components in this equation, we obtain

$$v'_{APx} = v'_{Ax} + 1.56\omega'_A,$$

$$v'_{APy} = v'_{Ay} + 1.35\omega'_A. \tag{3}$$

The position of P relative to the center of mass of car B is

$$\mathbf{r}_{P/B} = 1.8\mathbf{i} + \mathbf{j} \text{ (m)}.$$

We can express the velocity of point P of car B after the impact as

$$\mathbf{v}'_{BP} = \mathbf{v}'_B + \boldsymbol{\omega}'_B \times \mathbf{r}_{P/B}:$$

$$v'_{BPx}\mathbf{i} + v'_{BPy}\mathbf{j} = v'_{Bx}\mathbf{i} + v'_{By}\mathbf{j} + \begin{vmatrix} \mathbf{i} & \mathbf{j} & \mathbf{k} \\ 0 & 0 & \omega'_B \\ 1.8 & 1 & 0 \end{vmatrix}.$$

Equating $\mathbf{i}$ and $\mathbf{j}$ components yields

$$v'_{BPx} = v'_{Bx} - \omega'_B,$$

$$v'_{BPy} = v'_{By} + 1.8\omega'_B. \tag{4}$$

The angular momentum of car A about P is conserved:

$$H_{PA} = H'_{PA}:$$

$$[(\mathbf{r}_{A/P} \times m_A\mathbf{v}_A) \cdot \mathbf{k} + I_A\omega_A] = [(\mathbf{r}_{A/P} \times m_A\mathbf{v}'_A) \cdot \mathbf{k} + I_A\omega'_A],$$

$$\begin{vmatrix} \mathbf{i} & \mathbf{j} & \mathbf{k} \\ -1.35 & 1.56 & 0 \\ m_A v_{Ax} & m_A v_{Ay} & 0 \end{vmatrix} \cdot \mathbf{k} + 0 = \begin{vmatrix} \mathbf{i} & \mathbf{j} & \mathbf{k} \\ -1.35 & 1.56 & 0 \\ m_A v'_{Ax} & m_A v'_{Ay} & 0 \end{vmatrix} \cdot \mathbf{k} + I_A\omega'_A.$$

Expanding the determinants and evaluating the dot products, we obtain

$$-1.35 m_A v_{Ay} - 1.56 m_A v_{Ax}$$

$$= -1.35 m_A v'_{Ay} - 1.56 m_A v'_{Ax} + I_A\omega'_A. \tag{5}$$

The angular momentum of car B about P is also conserved,

$$H_{PB} = H'_{PB}:$$

$$[(\mathbf{r}_{B/P} \times m_B\mathbf{v}_B) \cdot \mathbf{k} + I_B\omega_B] = [(\mathbf{r}_{B/P} \times m_B\mathbf{v}'_B) \cdot \mathbf{k} + I_B\omega'_B],$$

$$\begin{vmatrix} \mathbf{i} & \mathbf{j} & \mathbf{k} \\ -1.8 & -1 & 0 \\ m_B v_{Bx} & 0 & 0 \end{vmatrix} \cdot \mathbf{k} + 0 = \begin{vmatrix} \mathbf{i} & \mathbf{j} & \mathbf{k} \\ -1.8 & -1 & 0 \\ m_B v'_{Bx} & m_B v'_{By} & 0 \end{vmatrix} \cdot \mathbf{k} + I_B\omega'_B.$$

From this equation, it follows that

$$m_B v_{Bx} = -1.8 m_B v'_{By} + m_B v'_{Bx} + I_B\omega'_B. \tag{6}$$

We can solve Eqs. (1)–(6) for $\mathbf{v}'_A$, $\mathbf{v}'_{AP}$, ω'_A, $\mathbf{v}'_B$, $\mathbf{v}'_{BP}$, and ω'_B. The results for the velocities of the centers of mass of the cars and their angular velocities are as follows:

$$\mathbf{v}'_A = 28.2\mathbf{i} - 9.08\mathbf{j} \text{ (m/s)}, \qquad \omega'_A = 2.65 \text{ rad/s},$$

$$\mathbf{v}'_B = 20.0\mathbf{i} - 1.18\mathbf{j} \text{ (m/s)}, \qquad \omega'_B = -3.54 \text{ rad/s}.$$

Critical Thinking

The total angular moment of the two cars about *any point* is the same before and after their collision, because we neglected the effects of horizontal forces other than the force due to their collision. But to determine their motions after the collision, we needed to use the fact that the angular momentum of *each* car about P is the same before and after the impact. That is true because the moment about P exerted on each car by the force of the collision is zero. Notice that we could not have assumed that the angular momentum of each car about any point is the same before and after the impact.

Problems

19.64 The 2-kg bar starts from rest in the vertical position and falls, striking the smooth surface at P. The coefficient of restitution of the impact is $e = 0.5$. When the bar rebounds, through what angle relative to the horizontal will it rotate?

Strategy: Use the coefficient of restitution to relate the bar's velocity at P just after the impact to its value just before the impact.

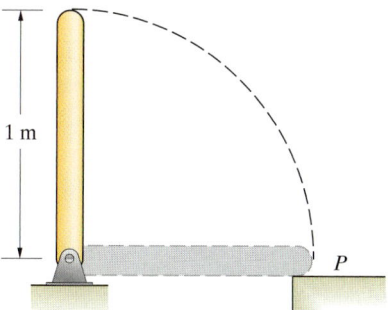

1 m

P

Problem 19.64

19.65 The 2-kg bar is released from rest above the fixed, smooth projection at A. If the coefficient of restitution of the impact is $e = 0.4$ and $b = 0.35$ m, what is the bar's angular velocity after the impact?

19.66 The 2-kg bar is released from rest above the fixed, smooth projection at A. The coefficient of restitution of the impact is $e = 0.4$. What value of the distance b would cause the velocity of the bar's center of mass to be zero immediately after the impact?

19.67 The 2-kg bar is released from rest above the fixed, smooth projection at A. The coefficient of restitution of the impact is $e = 0.4$ and $b = 0.35$ m. If the duration of the impact is 0.02 s, what is the magnitude of the average force exerted on the bar by the projection at A?

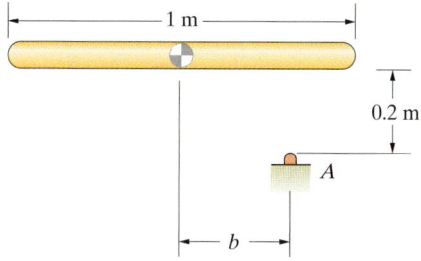

1 m

0.2 m

A

b

Problems 19.65–19.67

19.68 The mass of the ship is 544,000 kg, and the moment of inertia of the vessel about its center of mass is 4×10^8 kg-m². Wind causes the ship to drift sideways at 0.1 m/s and strike the stationary piling at P. The coefficient of restitution of the impact is $e = 0.2$. What is the ship's angular velocity after the impact?

19.69 In Problem 19.68, if the duration of the ship's impact with the piling is 10 s, what is the magnitude of the average force exerted on the ship by the impact?

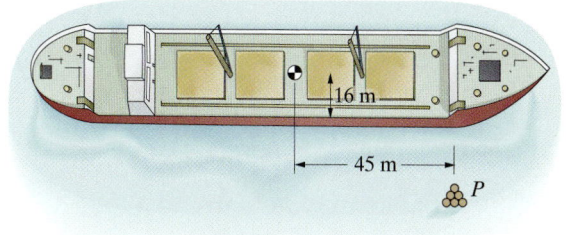

16 m

45 m

P

Problems 19.68/19.69

19.70 A 2-kg sphere A translating at 4 m/s strikes the end of a 5-kg slender bar B. The bar is pinned to a fixed support at O. What is the angular velocity of the bar after the impact if the sphere adheres to the bar?

19.71 In Problem 19.70, what is the angular velocity of the bar after the impact if the coefficient of restitution is $e = 0.8$?

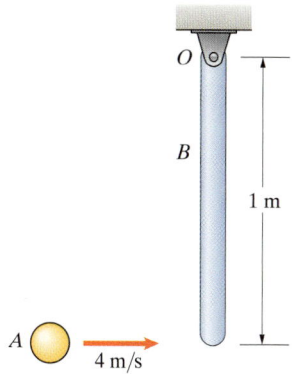

O

B

1 m

A

4 m/s

Problems 19.70/19.71

19.72 The 2-kg sphere A is moving at 10 m/s when it strikes the stationary unconstrained 4-kg bar. The coefficient of restitution is $e = 1$.

(a) What is the bar's angular velocity after the impact?
(b) Determine the total kinetic energy of the sphere and bar before and after the impact.

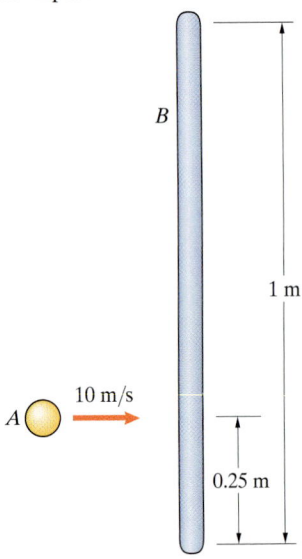

Problem 19.72

19.73* The suspended bars A and B each weigh 5 lb. The bars are stationary when the 1-lb sphere A moving at 10 ft/s strikes the end of bar B. What are the angular velocities of the bars after the impact if the sphere adheres to bar B?

19.74* The suspended bars A and B each weigh 5 lb. The bars are stationary when the 1-lb sphere A moving at 10 ft/s strikes the end of bar B. What are the angular velocities of the bars after the impact if the coefficient of restitution is $e = 0.5$.

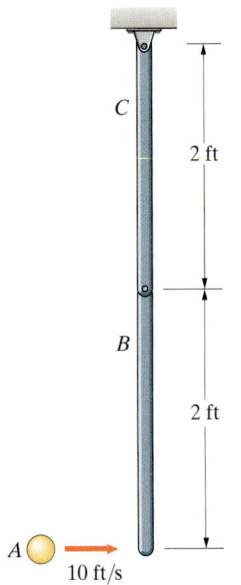

Problems 19.73/19.74

19.75 The 5-oz ball is translating with velocity $v_A = 80$ ft/s perpendicular to the bat just before impact. The player is swinging the 31-oz bat with angular velocity $\omega = 6\pi$ rad/s before the impact. Point C is the bat's instantaneous center both before and after the impact. The distances $b = 14$ in and $\bar{y} = 26$ in. The bat's moment of inertia about its center of mass is $I_B = 0.033$ slug-ft^2. The coefficient of restitution is $e = 0.6$, and the duration of the impact is 0.008 s. Determine the magnitude of the velocity of the ball after the impact and the average force A_x exerted on the bat by the player during the impact if (a) $d = 0$, (b) $d = 3$ in, and (c) $d = 8$ in.

19.76 In Problem 19.75, show that the force A_x is zero if $d = I_B/(m_B\bar{y})$, where m_B is the mass of the bat.

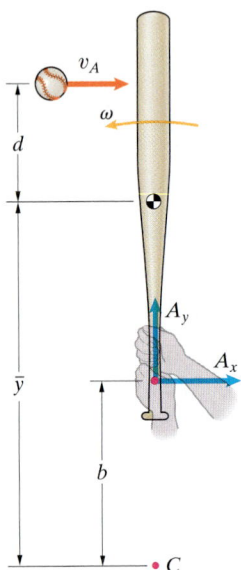

Problems 19.75/19.76

19.77 A slender bar of mass m is released from rest in the horizontal position at a height h above a peg (Fig. a). A small hook at the end of the bar engages the peg, and the bar swings from the peg (Fig. b). What minimum height h is necessary for the bar to swing 270° from its position when it engages the peg?

19.78 Is energy conserved in Problem 19.77? If not, how much energy is lost?

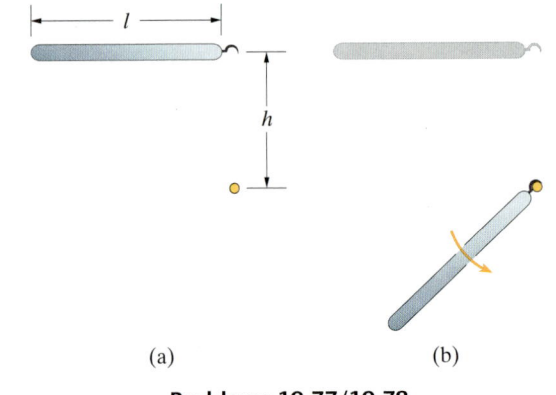

(a)   (b)

Problems 19.77/19.78

19.79 The 1-slug disk rolls at velocity $v = 10$ ft/s toward a 6-in step. The wheel remains in contact with the step and does not slip while rolling up onto it. What is the wheel's velocity once it is on the step?

19.80 The 1-slug disk rolls toward a 6-in step. The wheel remains in contact with the step and does not slip while rolling up onto it. What is the minimum velocity v the disk must have in order to climb up onto the step?

18 in

v

6 in

Problems 19.79/19.80

19.81 The length of the bar is 1 m and its mass is 2 kg. Just before the bar hits the floor, its angular velocity is $\omega = 0$ and its center of mass is moving downward at 4 m/s. If the end of the bar adheres to the floor, what is the bar's angular velocity after the impact?

19.82 The length of the bar is 1 m and its mass is 2 kg. Just before the bar hits the *smooth* floor, its angular velocity is $\omega = 0$ and its center of mass is moving downward at 4 m/s. If the coefficient of restitution of the impact is $e = 0.4$, what is the bar's angular velocity after the impact?

19.83 The length of the bar is 1 m and its mass is 2 kg. Just before the bar hits the *smooth* floor, it has angular velocity ω and its center of mass is moving downward at 4 m/s. The coefficient of restitution of the impact is $e = 0.4$. What value of ω would cause the bar to have no angular velocity after the impact?

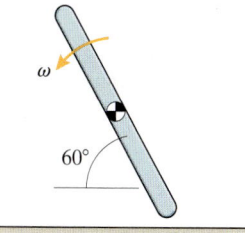

ω

60°

Problems 19.81–19.83

19.84 During her parallel-bars routine, the velocity of the 90-lb gymnast's center of mass is $4\mathbf{i} - 10\mathbf{j}$ (ft/s) and her angular velocity is zero just before she grasps the bar at A. In the position shown, her moment of inertia about her center of mass is 1.8 slug-ft^2. If she stiffens her shoulders and legs so that she can be modeled as a rigid body, what is the velocity of her center of mass and her angular velocity just after she grasps the bar?

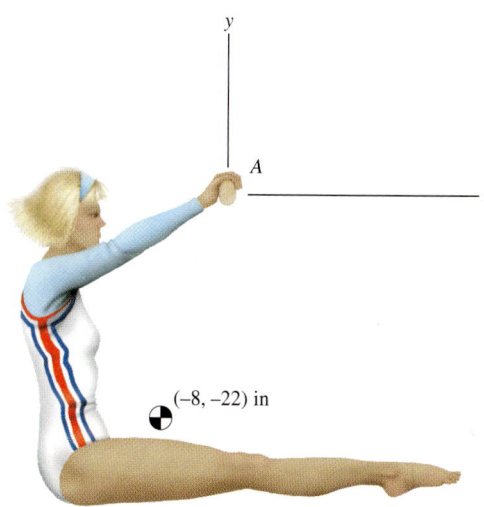

y

A

$(-8, -22)$ in

Problem 19.84

19.85 The 20-kg homogeneous rectangular plate is released from rest (Fig. a) and falls 200 mm before coming to the end of the string attached at the corner A (Fig. b). Assuming that the vertical component of the velocity of A is zero just after the plate reaches the end of the string, determine the angular velocity of the plate and the magnitude of the velocity of the corner B at that instant.

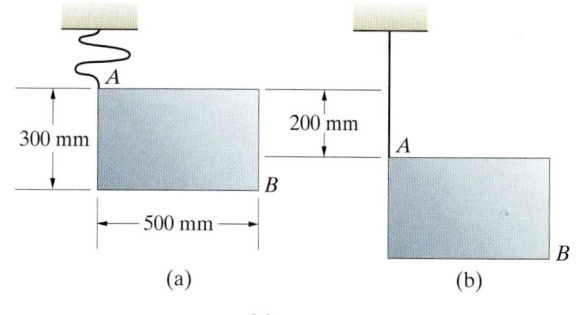

A

300 mm

B

500 mm

200 mm

A

B

(a)   (b)

Problem 19.85

19.86* The two bars A and B are each 2 m in length, and each has a mass of 4 kg. In Fig. a, bar A has no angular velocity and is moving to the right at 1 m/s, and bar B is stationary. If the bars bond together on impact (Fig. b), what is their angular velocity ω' after the impact?

19.87* In Problem 19.86, if the bars do not bond together on impact and the coefficient of restitution is $e = 0.8$, what are the angular velocities of the bars after the impact?

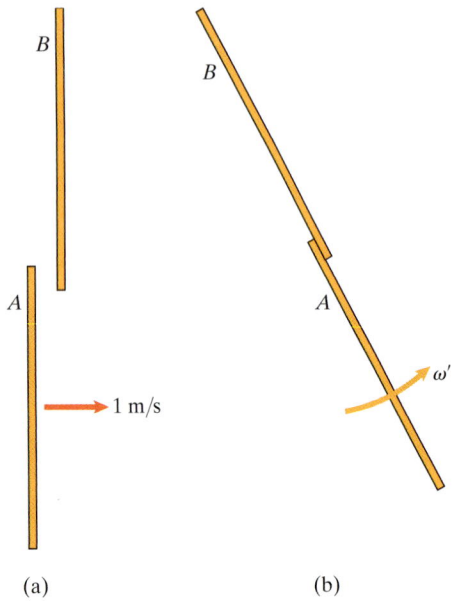

(a)   (b)

Problems 19.86/19.87

19.88* Two bars A and B are each 2 m in length, and each has a mass of 4 kg. In Fig. a, bar A has no angular velocity and is moving to the right at 1 m/s, and bar B is stationary. If the bars bond together on impact (Fig. b), what is their angular velocity ω' after the impact?

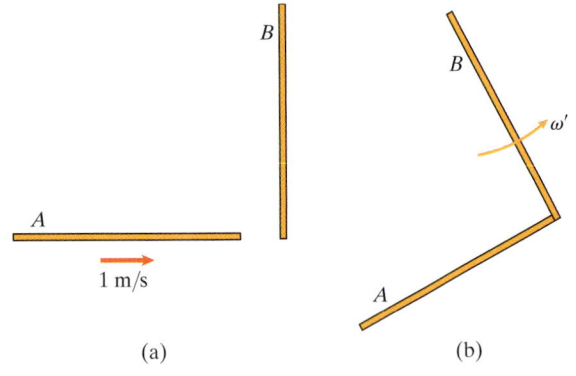

(a)   (b)

Problem 19.88

19.89* The horizontal velocity of the landing airplane is 50 m/s, its vertical velocity (rate of descent) is 2 m/s, and its angular velocity is zero. The mass of the airplane is 12 Mg, and the moment of inertia about its center of mass is 1×10^5 kg-m². When the rear wheels touch the runway, they remain in contact with it. Neglecting the horizontal force exerted on the wheels by the runway, determine the airplane's angular velocity just after it touches down.

19.90* Determine the angular velocity of the airplane in Problem 19.89 just after it touches down if its wheels don't stay in contact with the runway and the coefficient of restitution of the impact is $e = 0.4$.

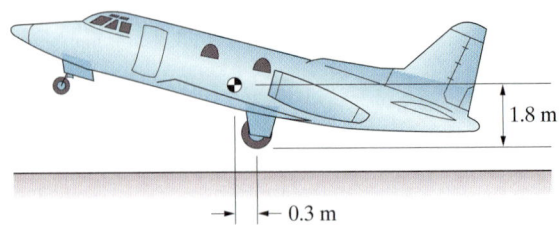

Problems 19.89/19.90

19.91* While attempting to drive on an icy street for the first time, a student skids his 1260-kg car (A) into the university president's stationary 2700-kg Rolls-Royce Corniche (B). The point of impact is P. Assume that the impacting surfaces are smooth and parallel to the y axis and that the coefficient of restitution of the impact is $e = 0.5$. The moments of inertia of the cars about their centers of mass are $I_A = 2400$ kg-m² and $I_B = 7600$ kg-m². Determine the angular velocities of the cars and the velocities of their centers of mass after the collision.

19.92* The student in Problem 19.91 claimed that he was moving at 5 km/h prior to the collision, but police estimate that the center of mass of the Rolls-Royce was moving at 1.7 m/s after the collision. What was the student's actual speed?

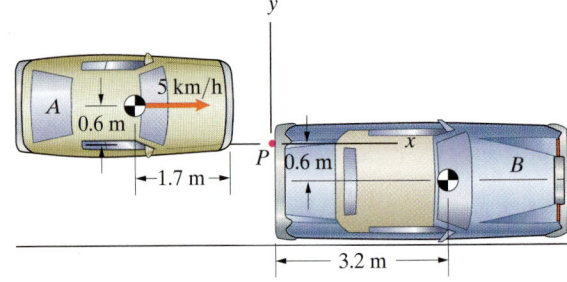

Problems 19.91/19.92

19.93 Each slender bar is 48 in long and weighs 20 lb. Bar A is released in the horizontal position shown. The bars are smooth, and the coefficient of restitution of their impact is $e = 0.8$. Determine the angle through which B swings afterward.

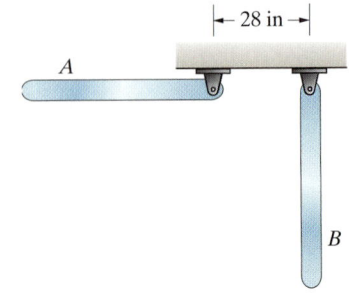

Problem 19.93

19.94* The *Apollo* CSM (*A*) approaches the *Soyuz* Space Station (*B*). The mass of the *Apollo* is $m_A = 18$ Mg, and the moment of inertia about the axis through its center of mass parallel to the z axis is $I_A = 114$ Mg-m^2. The mass of the *Soyuz* is $m_B = 6.6$ Mg, and the moment of inertia about the axis through its center of mass parallel to the z axis is $I_B = 70$ Mg-m^2. The *Soyuz* is stationary relative to the reference frame shown, and the CSM approaches with velocity $\mathbf{v}_A = 0.21\mathbf{i} + 0.05\mathbf{j}$ (m/s) and no angular velocity. What is the angular velocity of the attached spacecraft after docking?

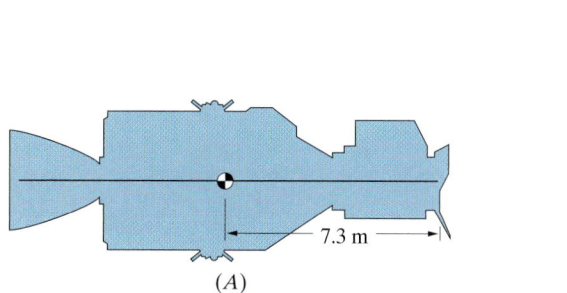

(A) (B)

Problem 19.94

CHAPTER SUMMARY

Work and Energy

The work done by external forces and couples as a rigid body moves between two positions is equal to the change in kinetic energy of the body:

$$U_{12} = T_2 - T_1. \tag{19.5}$$

The work done on a system of rigid bodies by external and internal forces and couples equals the change in the total kinetic energy.

The kinetic energy of a rigid body in general planar motion is

$$T = \tfrac{1}{2}mv^2 + \tfrac{1}{2}I\omega^2, \tag{19.11}$$

where v is the magnitude of the velocity of the center of mass of the body and I is the moment of inertia about the center of mass. If a rigid body rotates about a fixed axis O, its kinetic energy can also be expressed as

$$T = \tfrac{1}{2}I_O\omega^2. \tag{19.12}$$

The work done on a rigid body by a force $\mathbf{F}$ is

$$U_{12} = \int_{(\mathbf{r}_p)_1}^{(\mathbf{r}_p)_2} \mathbf{F} \cdot d\mathbf{r}_p, \tag{19.13}$$

where $\mathbf{r}_p$ is the position of the point of application of $\mathbf{F}$. If the point of application is stationary, or if its direction of motion is perpendicular to $\mathbf{F}$, no work is done.

The work done by a couple M on a rigid body in planar motion as the body rotates from θ_1 to θ_2 in the direction of M is

$$U_{12} = \int_{\theta_1}^{\theta_2} M \, d\theta. \tag{19.15}$$

A couple M is conservative if a potential energy V exists such that

$$M \, d\theta = -dV. \tag{19.16}$$

The potential energy of a linear torsional spring that exerts a couple $k\theta$ in the direction opposite to angular displacement θ of the spring (Fig. a) is $\frac{1}{2}k\theta^2$.

If all the forces and couples that do work on a system of rigid bodies are conservative, the sum of the kinetic energy and the total potential energy is constant:

$$T_1 + V_1 = T_2 + V_2 \tag{19.18}$$

If a system is subjected to both conservative and nonconservative forces, the principle of work and energy can be written in the form

$$T_1 + V_1 + U_{12} = T_2 + V_2 \tag{19.19}$$

The term U_{12} includes the work done by all nonconservative forces acting on the system. The work done by a conservative force can be calculated and included in U_{12}, *or* the force's potential energy can be included in V.

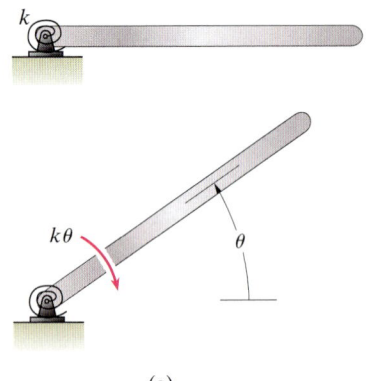

(a)

Power

The power transmitted to a rigid body by a force $\mathbf{F}$ is

$$P = \mathbf{F} \cdot \mathbf{v}_\mathrm{p}, \tag{19.20}$$

where $\mathbf{v}_\mathrm{p}$ is the velocity of the point of application of $\mathbf{F}$. The power transmitted to a rigid body in planar motion by a couple M is

$$P = M\omega. \tag{19.21}$$

The average power transferred to a rigid body during an interval of time is equal to the change in kinetic energy of the body, or the total work done during that time, divided by the interval of time:

$$P_\mathrm{av} = \frac{T_2 - T_1}{t_2 - t_1} = \frac{U_{12}}{t_2 - t_1}. \tag{19.22}$$

Impulse and Momentum

The principle of linear impulse and momentum states that the linear impulse applied to a rigid body during an interval of time is equal to the change in linear momentum of the body during that time:

$$\int_{t_1}^{t_2} \Sigma \mathbf{F} \, dt = m\mathbf{v}_2 - m\mathbf{v}_1. \tag{19.23}$$

This result can also be expressed in terms of the average of the total force with respect to time:

$$(t_2 - t_1)\Sigma \mathbf{F}_\mathrm{av} = m\mathbf{v}_2 - m\mathbf{v}_1. \tag{19.24}$$

If the only forces acting on two rigid bodies A and B are the forces they exert on each other, or if other forces are negligible, than the total linear momentum of the bodies is conserved:

$$m_A\mathbf{v}_A + m_B\mathbf{v}_B = \text{constant.} \tag{19.25}$$

The angular momentum about the center of mass of a rigid body in planar motion is

$$H = I\omega, \qquad (19.27)$$

where I is the moment of inertia of the body about its center of mass. One form of the principle of angular impulse and momentum states that the angular impulse about the center of mass during the interval from t_1 to t_2 is equal to the change in the rigid body's angular momentum about its center of mass during that interval:

$$\int_{t_1}^{t_2} \Sigma M \, dt = H_2 - H_1. \qquad (19.28)$$

The angular momentum of a rigid body about a fixed point O is

$$H_O = (\mathbf{r} \times m\mathbf{v}) \cdot \mathbf{k} + I\omega, \qquad (19.30)$$

where $\mathbf{r}$ is the position of the center of mass relative to O and $\mathbf{v}$ is the velocity of the center of mass. A second form of the principle of angular impulse and momentum states that the angular impulse about O during the interval from t_1 to t_2 is equal to the change in the rigid body's angular momentum about O during that interval:

$$\int_{t_1}^{t_2} \Sigma M_O \, dt = H_{O2} - H_{O1}. \qquad (19.31)$$

In terms of the averages of the total moments with respect to time, Eqs. (19.28) and (19.31) are

$$(t_2 - t_1)\Sigma M_{av} = H_2 - H_1 \qquad (19.32)$$

and

$$(t_2 - t_1)(\Sigma M_O)_{av} = H_{O2} - H_{O1}. \qquad (19.33)$$

If two rigid bodies A and B in planar motion are subjected only to internal forces and couples, or if the total moment due to external forces and couples about a fixed point O is zero, the total angular momentum of A and B about O is conserved:

$$H_{OA} + H_{OB} = \text{constant}. \qquad (19.34)$$

Impacts

Suppose that two rigid bodies A and B, in two-dimensional motion in the same plane, collide. If other forces and couples are negligible in comparison to the impact forces and couples that A and B exert on each other, then the total linear momentum of A and B and their total angular momentum about any fixed point O are conserved. If, in addition, A and B exert only forces on each other at their point of impact, P, the angular momentum about P of *each* rigid body is conserved. If one of the rigid bodies has a pin support at a point O, the total angular momentum of both bodies about that point is conserved.

Let P be the point of impact (Fig. b). The normal components of the velocities at P are related to the coefficient of restitution e by

$$e = \frac{(\mathbf{v}'_{BP})_x - (\mathbf{v}'_{AP})_x}{(\mathbf{v}_{AP})_x - (\mathbf{v}_{BP})_x}. \qquad (19.35)$$

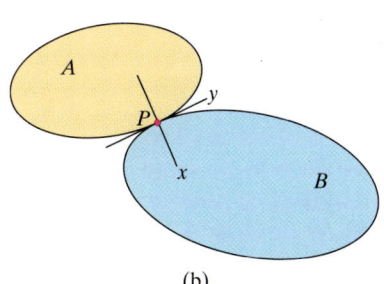

(b)

Review Problems

19.95 The moment of inertia of the pulley is 0.2 kg-m². The system is released from rest. Use the principle of work and energy to determine the velocity of the 10-kg cylinder when it has fallen 1 m.

19.96 The moment of inertia of the pulley is 0.2 kg-m². The system is released from rest. Use momentum principles to determine the velocity of the 10-kg cylinder 1 s after the system is released.

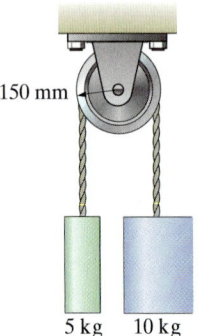

150 mm

5 kg   10 kg

Problems 19.95/19.96

19.97 Arm BC has a mass of 12 kg, and the moment of inertia about its center of mass is 3 kg-m². Point B is stationary. Arm BC is initially aligned with the (horizontal) x axis with zero angular velocity, and a constant couple M applied at B causes the arm to rotate upward. When it is in the position shown, its counterclockwise angular velocity is 2 rad/s. Determine M.

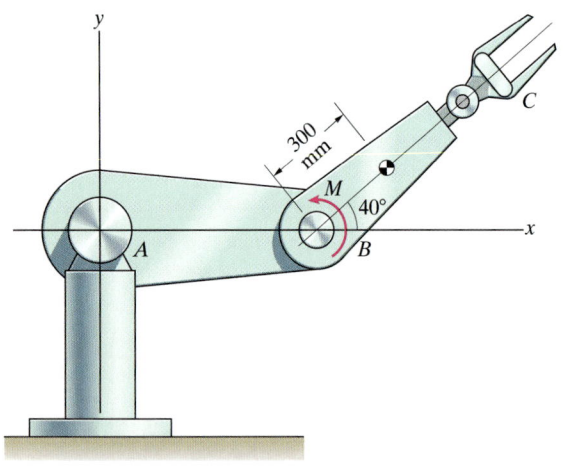

y

300 mm

M   40°

A   B   x

Problem 19.97

19.98 The cart is stationary when a constant force F is applied to it. What will the velocity of the cart be when it has rolled a distance b? The mass of the body of the cart is m_c, and each of the four wheels has mass m, radius R, and moment of inertia, I.

F

Problem 19.98

19.99 Each pulley has moment of inertia $I = 0.003$ kg-m², and the mass of the belt is 0.2 kg. If a constant couple $M = 4$ N-m is applied to the bottom pulley, what will its angular velocity be when it has turned 10 revolutions?

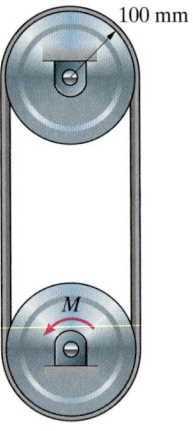

100 mm

M

Problem 19.99

19.100 The ring gear is fixed. The mass and moment of inertia of the sun gear are $m_S = 22$ slugs and $I_S = 4400$ slug-ft^2. The mass and moment of inertia of each planet gear are $m_P = 2.7$ slugs and $I_P = 65$ slug-ft^2. A couple $M = 600$ ft-lb is applied to the sun gear. Use work and energy to determine the angular velocity of the sun gear after it has turned 100 revolutions.

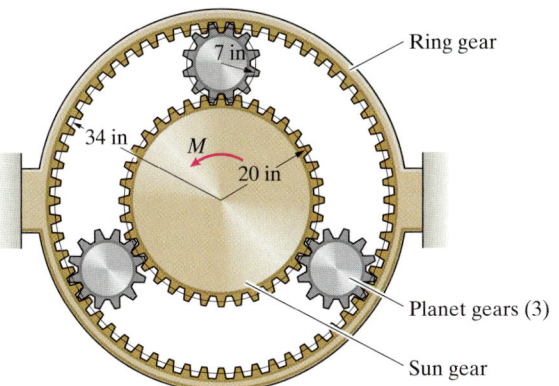

7 in
Ring gear
34 in
M
20 in
Planet gears (3)
Sun gear

Problem 19.100

19.101 The moments of inertia of gears A and B are $I_A = 0.014$ slug-ft^2 and $I_B = 0.100$ slug-ft^2. Gear A is connected to a torsional spring with constant $k = 0.2$ ft-lb/rad. If the spring is unstretched and the surface supporting the 5-lb weight is removed, what is the velocity of the weight when it has fallen 3 in?

19.102 Consider the system in Problem 19.101.
(a) What maximum distance does the 5-lb weight fall when the supporting surface is removed?
(b) What maximum velocity does the weight achieve?

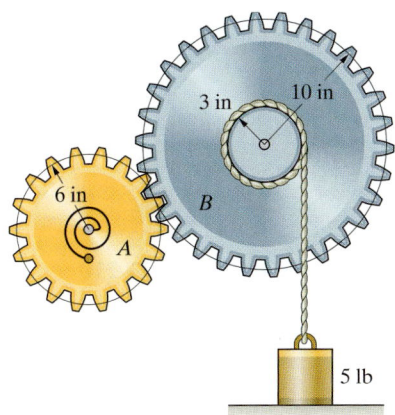

3 in
10 in
6 in
B
A
5 lb

Problems 19.101/19.102

19.103 Each of the go-cart's front wheels weighs 5 lb and has a moment of inertia of 0.01 slug-ft^2. The two rear wheels and rear axle form a single rigid body weighing 40 lb and having a moment of inertia of 0.1 slug-ft^2. The total weight of the rider and go-cart, including its wheels, is 240 lb. The go-cart starts from rest, its engine exerts a constant torque of 15 ft-lb on the rear axle, and its wheels do not slip. Neglecting friction and aerodynamic drag, how fast is the go-cart moving when it has traveled 50 ft?

19.104 Determine the maximum power and the average power transmitted to the go-cart in Problem 19.103 by its engine.

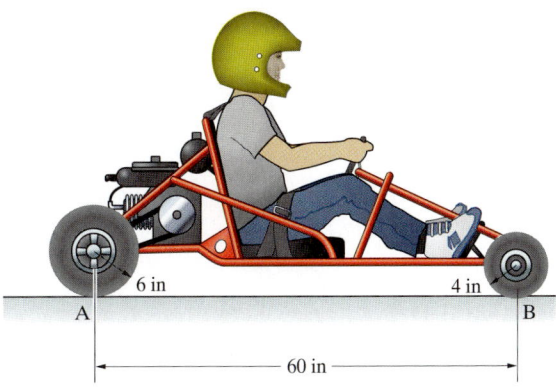

6 in
A
4 in
B
60 in

Problems 19.103/19.104

19.105 The system starts from rest with the 4-kg slender bar horizontal. The mass of the suspended cylinder is 10 kg. What is the angular velocity of the bar when it is in the position shown?

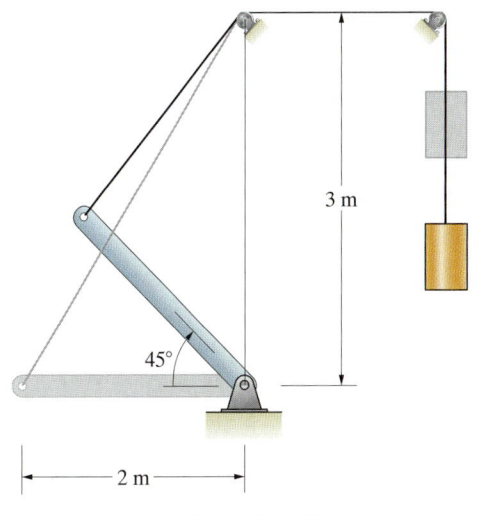

3 m
45°
2 m

Problem 19.105

19.106 The 0.1-kg slender bar and 0.2-kg cylindrical disk are released from rest with the bar horizontal. The disk rolls on the curved surface. What is the angular velocity of the bar when it is vertical?

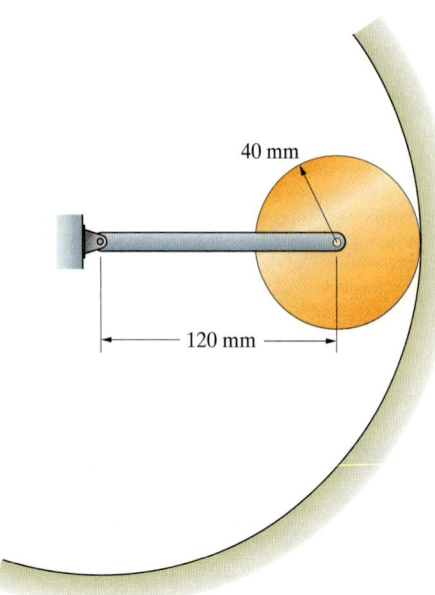

40 mm

120 mm

Problem 19.106

19.107 The slender bar of mass m is released from rest in the vertical position and allowed to fall. Neglecting friction and assuming that it remains in contact with the floor and wall, determine the bar's angular velocity as a function of θ.

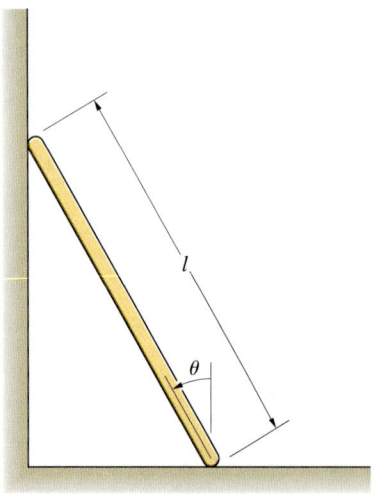

l

θ

Problem 19.107

19.108 The 4-kg slender bar is pinned to 2-kg sliders at A and B. If friction is negligible and the system starts from rest in the position shown, what is the bar's angular velocity when the slider at A has fallen 0.5 m?

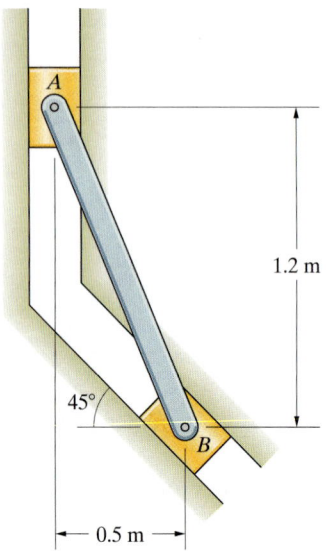

A

1.2 m

45°

B

0.5 m

Problem 19.108

19.109 The homogeneous hemisphere of mass m is released from rest in the position shown. If it rolls on the horizontal surface, what is its angular velocity when its flat surface is horizontal?

19.110 The homogeneous hemisphere of mass m is released from rest in the position shown. It rolls on the horizontal surface. What normal force is exerted on the hemisphere by the horizontal surface at the instant the flat surface of the hemisphere is horizontal?

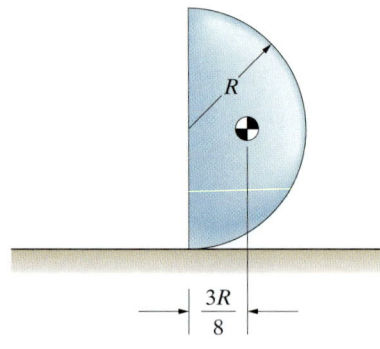

R

$\dfrac{3R}{8}$

Problems 19.109/19.110

19.111 The slender bar rotates freely *in the horizontal plane* about a vertical shaft at O. The bar weighs 20 lb and its length is 6 ft. The slider A weighs 2 lb. If the bar's angular velocity is $\omega = 10$ rad/s and the radial component of the velocity of A is zero when $r = 1$ ft, what is the angular velocity of the bar when $r = 4$ ft? (The moment of inertia of A about its center of mass is negligible; that is, treat A as a particle.)

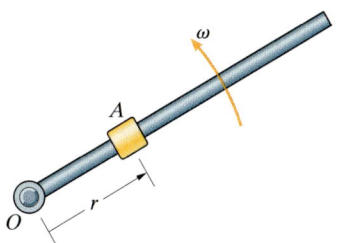

Problem 19.111

19.112 A satellite is deployed with angular velocity $\omega = 1$ rad/s (Fig. a). Two internally stored antennas that span the diameter of the satellite are then extended, and the satellite's angular velocity decreases to ω' (Fig. b). By modeling the satellite as a 500-kg sphere of 1.2-m radius and each antenna as a 10-kg slender bar, determine ω'.

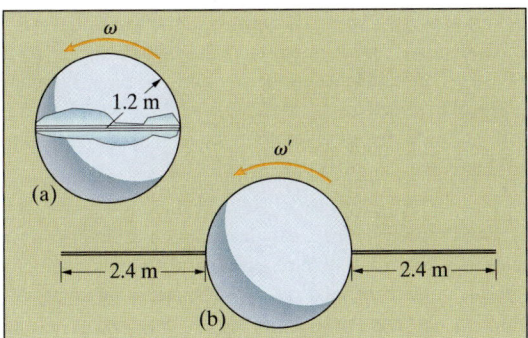

Problem 19.112

19.113 An engineer decides to control the angular velocity of a satellite by deploying small masses attached to cables. If the angular velocity of the satellite in configuration (a) is 4 rpm, determine the distance d in configuration (b) that will cause the angular velocity to be 1 rpm. The moment of inertia of the satellite is $I = 500$ kg-m^2 and each mass is 2 kg. (Assume that the cables and masses rotate with the same angular velocity as the satellite. Neglect the masses of the cables and the moments of inertia of the masses about their centers of mass.)

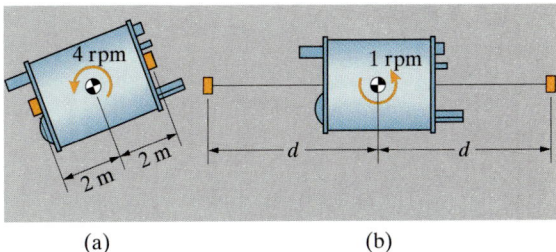

(a)   (b)

Problem 19.113

19.114 The homogeneous cylindrical disk of mass m rolls on the horizontal surface with angular velocity ω. If the disk does not slip or leave the slanted surface when it comes into contact with it, what is the angular velocity ω' of the disk immediately afterward?

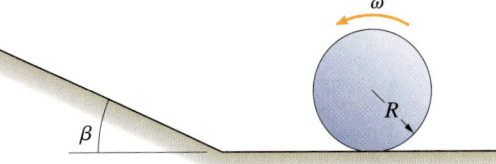

Problem 19.114

19.115 The 10-lb bar falls from rest in the vertical position and hits the smooth projection at B. The coefficient of restitution of the impact is $e = 0.6$, the duration of the impact is 0.1 s, and $b = 1$ ft. Determine the average force exerted on the bar at B as a result of the impact.

19.116 The 10-lb bar falls from rest in the vertical position and hits the smooth projection at B. The coefficient of restitution of the impact is $e = 0.6$ and the duration of the impact is 0.1 s. Determine the distance b for which the average force exerted on the bar by the support A as a result of the impact is zero.

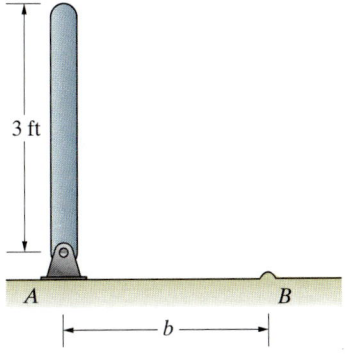

Problems 19.115/19.116

19.117 The 1-kg sphere A is moving at 2 m/s when it strikes the end of the 2-kg stationary slender bar B. If the velocity of the sphere after the impact is 0.8 m/s to the right, what is the coefficient of restitution?

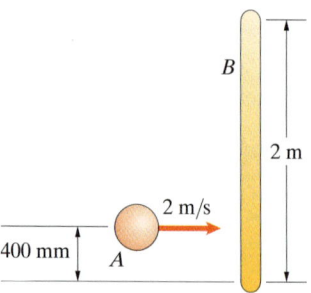

Problem 19.117

19.118 The slender bar is released from rest in the position shown in Fig. a and falls a distance $h = 1$ ft. When the bar hits the floor, its tip is supported by a depression and remains on the floor (Fig. b). The length of the bar is 1 ft and its weight is 4 oz. What is angular velocity ω of the bar just after it hits the floor?

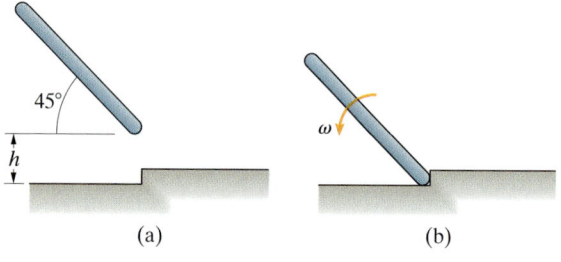

(a)   (b)

Problem 19.118

19.119 The slender bar is released from rest with $\theta = 45°$ and falls a distance $h = 1$ m onto the smooth floor. The length of the bar is 1 m and its mass is 2 kg. If the coefficient of restitution of the impact is $e = 0.4$, what is the angular velocity of the bar just after it hits the floor?

19.120 The slender bar is released from rest and falls a distance $h = 1$ m onto the smooth floor. The length of the bar is 1 m and its mass is 2 kg. The coefficient of restitution of the impact is $e = 0.4$. Determine the angle θ for which the angular velocity of the bar after it hits the floor is a maximum. What is the maximum angular velocity?

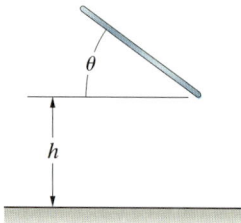

Problems 19.119/19.120

19.121 A nonrotating slender bar A moving with velocity v_0 strikes a stationary slender bar B. Each bar has mass m and length l. If the bars adhere when they collide, what is their angular velocity after the impact?

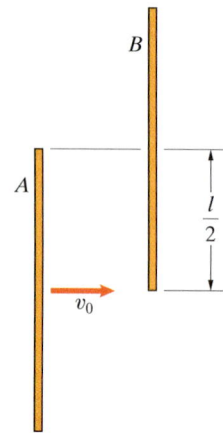

Problem 19.121

19.122 An astronaut translates toward a nonrotating satellite at $1.0\,\mathbf{i}$ (m/s) relative to the satellite. Her mass is 136 kg, and the moment of inertia about the axis through her center of mass parallel to the z axis is 45 kg-m². The mass of the satellite is 450 kg and its moment of inertia about the z axis is 675 kg-m². At the instant the astronaut attaches to the satellite and begins moving with it, the position of her center of mass is $(-1.8, -0.9, 0)$ m. The axis of rotation of the satellite after she attaches is parallel to the z axis. What is their angular velocity?

19.123 In Problem 19.122, suppose that the design parameters of the satellite's control system require that the angular velocity of the satellite not exceed 0.02 rad/s. If the astronaut is moving parallel to the x axis and the position of her center of mass when she attaches is $(-1.8, -0.9, 0)$ m, what is the maximum relative velocity at which she should approach the satellite?

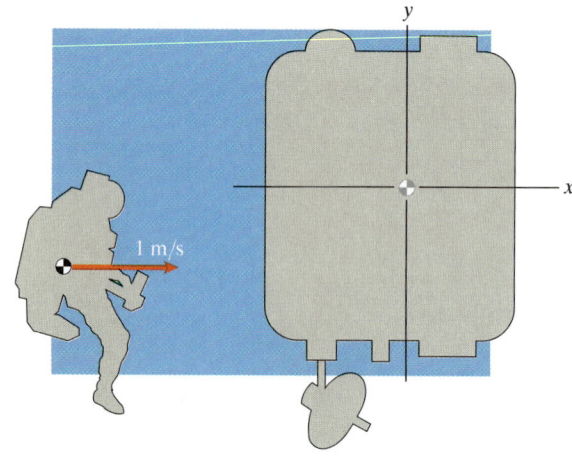

Problems 19.122/19.123

19.124 A 2800-lb car skidding on ice strikes a concrete abutment at 3 mi/h. The car's moment of inertia about its center of mass is 1800 slug-ft^2. Assume that the impacting surfaces are smooth and parallel to the y axis and that the coefficient of restitution of the impact is $e = 0.8$. What are the angular velocity of the car and the velocity of its center of mass after the impact?

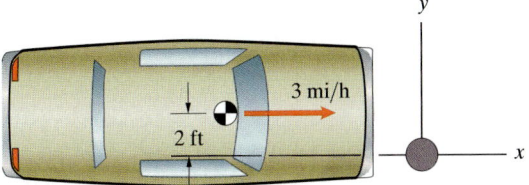

Problem 19.124

19.125 A 170-lb wide receiver jumps vertically to receive a pass and is stationary at the instant he catches the ball. At the same instant, he is hit at P by a 180-lb linebacker moving horizontally at 15 ft/s. The wide receiver's moment of inertia about his center of mass is 7 slug-ft^2. If you model the players as rigid bodies and assume that the coefficient of restitution is $e = 0$, what is the wide receiver's angular velocity immediately after the impact?

Problem 19.125

Design Project

Design and carry out experiments to determine the moments of inertia of (a) a homogeneous slender bar, such as a meter stick; and (b) a soccer ball or basketball. For the slender bar, compare your experimental values for the moments of inertia with the theoretical value $I = \frac{1}{12}ml^2$ for a slender bar of length l. For the ball, compare your experimental values with the theoretical value $I = \frac{2}{3}mR^2$ for a thin spherical shell of radius R. Investigate the repeatability of your experimental methods. Write a brief report describing your experiments, discussing possible sources of error, and presenting your results.

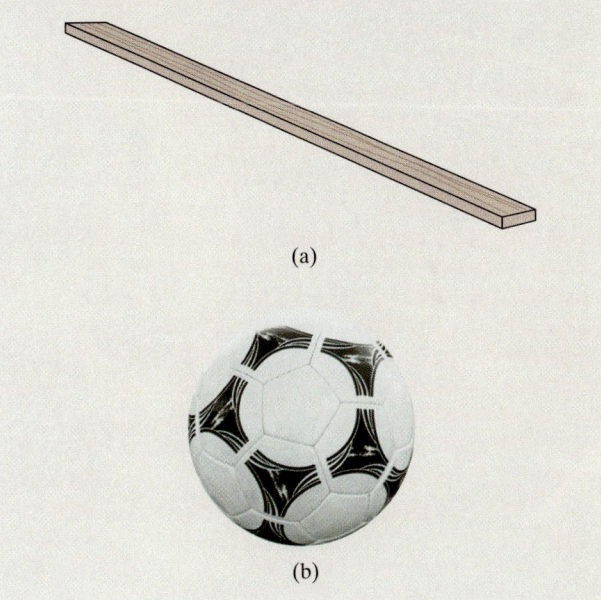

(a)

(b)

Three-Dimensional Kinematics and Dynamics of Rigid Bodies

For many engineering applications of dynamics, such as the design of airplanes and other vehicles, we must consider three-dimensional motion. After explaining how three-dimensional motion of a rigid body is described, we derive the equations of motion and use them to analyze simple motions. Finally, we introduce the Euler angles used to specify the orientation of a rigid body in three dimensions and express the equations of angular motion in terms of them.

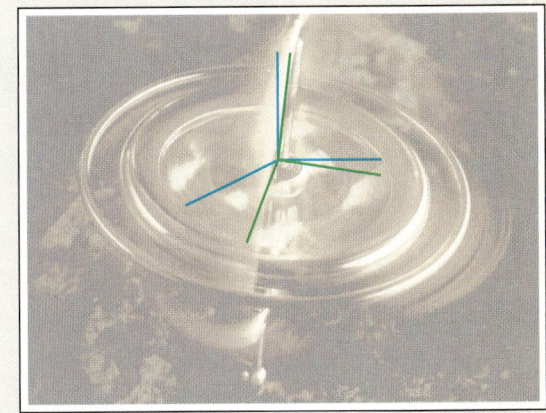

◄ Predicting the interesting behaviors of tops and gyroscopes requires a three-dimensional analysis. In this chapter we discuss three-dimensional motions of rigid bodies.

20.1 Kinematics

If a bicyclist rides in a straight path, the wheels undergo planar motion. But if the rider is turning, the motion of the wheels is three dimensional (Fig. 20.1a). Similarly, an airplane can remain in planar motion while it is in level flight or as it descends, climbs, or performs loops. But if it banks and turns, it is in three-dimensional motion (Fig. 20.1b). A spinning top may remain in planar motion for a brief period, rotating about a fixed vertical axis. But eventually, the top's axis begins to tilt and rotate. The top is then in three-dimensional motion and exhibits interesting, apparently gravity-defying behavior (Fig. 20.1c). In this section we begin the analysis of such motions by discussing the kinematics of rigid bodies in three-dimensional motion.

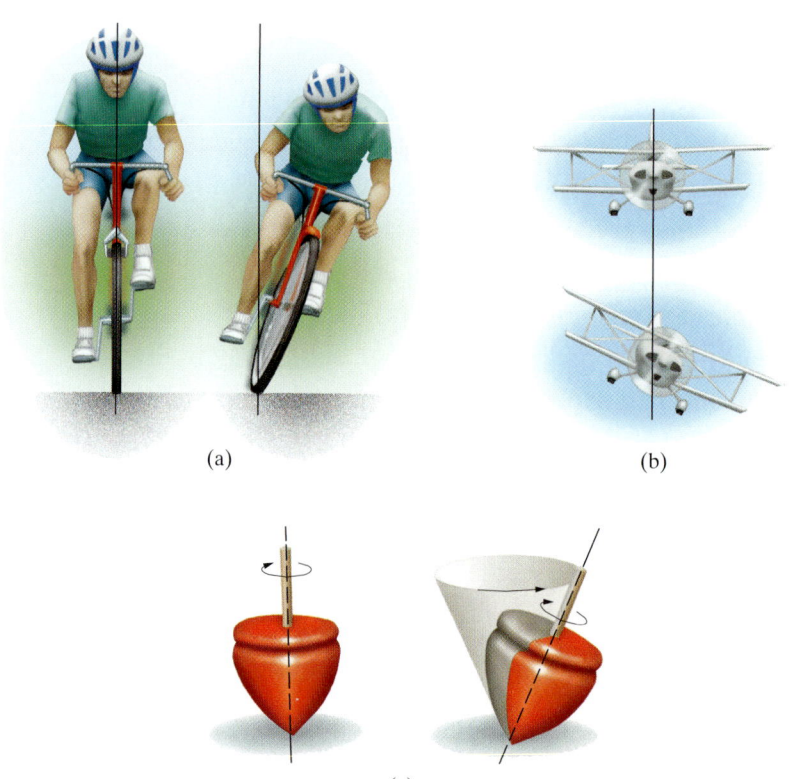

(a)

(b)

(c)

Figure 20.1

Examples of planar and three-dimensional motions.

Velocities and Accelerations

We have already discussed some of the concepts needed to describe the three-dimensional motion of a rigid body relative to a given reference frame. In Chapter 17, we showed that Euler's theorem implies that a rigid body undergoing any motion other than translation has an instantaneous axis of rotation. The direction of this axis at a particular instant and the rate at which the rigid body rotates about the axis are specified by the angular velocity vector $\boldsymbol{\omega}$.

We have also shown that a rigid body's velocity is completely specified by its angular velocity vector and the velocity of a single point of the body. For the rigid body and reference frame in Fig. 20.2, suppose that we know the angular velocity vector $\boldsymbol{\omega}$ and the velocity $\mathbf{v}_B$ of a point B. Then the velocity of *any* other point A of the body is given by Eq. (17.8):

$$\mathbf{v}_A = \mathbf{v}_B + \boldsymbol{\omega} \times \mathbf{r}_{A/B}. \tag{20.1}$$

A rigid body's acceleration is completely specified by its angular acceleration vector $\boldsymbol{\alpha} = d\boldsymbol{\omega}/dt$, its angular velocity vector, and the acceleration of a single point of the body. If we know $\boldsymbol{\alpha}, \boldsymbol{\omega}$, and the acceleration $\mathbf{a}_B$ of the point B in Fig. 20.2, the acceleration of *any* other point A of the rigid body is given by Eq. (17.9):

$$\mathbf{a}_A = \mathbf{a}_B + \boldsymbol{\alpha} \times \mathbf{r}_{A/B} + \boldsymbol{\omega} \times (\boldsymbol{\omega} \times \mathbf{r}_{A/B}). \tag{20.2}$$

Moving Reference Frames

The velocities and accelerations in Eqs. (20.1) and (20.2) are measured relative to the reference frame indicated in Fig. 20.2, which we will refer to as the *primary reference frame*. Although some situations require other choices, the most common primary reference frame used in engineering applications is one that is fixed relative to the earth. *When we do not state otherwise, you should assume that the primary reference frame is earth fixed.* We also use a *secondary reference frame* that moves relative to the primary reference frame. The secondary reference frame and its motion are chosen for convenience in describing the motion of a particular rigid body. In some situations, the secondary reference frame is defined to be fixed with respect to the rigid body. In other cases, it is advantageous to use a secondary reference frame that moves relative to the primary reference frame, but is not fixed with respect to the rigid body. (See Examples 20.1–20.3.)

Figure 20.3 shows a primary reference frame, a secondary reference frame *xyz*, and a rigid body. The angular velocity of the secondary reference frame relative to the primary reference frame is specified by the vector $\boldsymbol{\Omega}$, and the angular velocity of the rigid body relative to the primary reference frame is specified by the vector $\boldsymbol{\omega}$. If the secondary reference frame is fixed with respect to the rigid body, $\boldsymbol{\Omega} = \boldsymbol{\omega}$. If we express $\boldsymbol{\omega}$ *in terms of its components in the secondary reference frame* as

$$\boldsymbol{\omega} = \omega_x \mathbf{i} + \omega_y \mathbf{j} + \omega_z \mathbf{k},$$

the rigid body's angular acceleration vector relative to the primary reference frame is

$$\boldsymbol{\alpha} = \frac{d\boldsymbol{\omega}}{dt} = \frac{d\omega_x}{dt}\mathbf{i} + \omega_x \frac{d\mathbf{i}}{dt} + \frac{d\omega_y}{dt}\mathbf{j} + \omega_y \frac{d\mathbf{j}}{dt} + \frac{d\omega_z}{dt}\mathbf{k} + \omega_z \frac{d\mathbf{k}}{dt}. \tag{20.3}$$

The derivatives of $\mathbf{i}, \mathbf{j}$, and $\mathbf{k}$ can be expressed in terms of the angular velocity vector $\boldsymbol{\Omega}$ as (see Section 17.5)

$$\frac{d\mathbf{i}}{dt} = \boldsymbol{\Omega} \times \mathbf{i}, \qquad \frac{d\mathbf{j}}{dt} = \boldsymbol{\Omega} \times \mathbf{j}, \qquad \frac{d\mathbf{k}}{dt} = \boldsymbol{\Omega} \times \mathbf{k}.$$

Substituting these expressions into Eq. (20.3), we obtain the angular acceleration vector of the rigid body relative to the primary reference frame in the form

$$\boldsymbol{\alpha} = \frac{d\omega_x}{dt}\mathbf{i} + \frac{d\omega_y}{dt}\mathbf{j} + \frac{d\omega_z}{dt}\mathbf{k} + \boldsymbol{\Omega} \times \boldsymbol{\omega}. \tag{20.4}$$

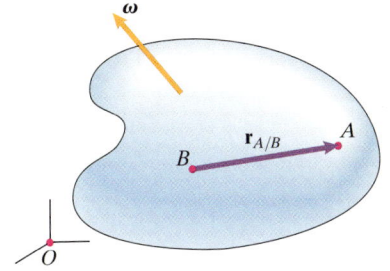

Figure 20.2
Points A and B of a rigid body. The velocity of A can be determined if the velocity of B and the rigid body's angular velocity vector $\boldsymbol{\omega}$ are known. The acceleration of A can be determined if the acceleration of B, the angular velocity vector, and the angular acceleration vector are known.

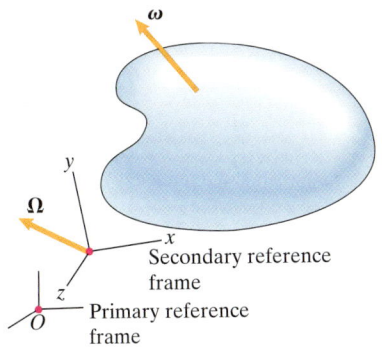

Figure 20.3
The primary and secondary reference frames. The vector $\boldsymbol{\Omega}$ is the angular velocity of the secondary reference frame relative to the primary reference frame. The vector $\boldsymbol{\omega}$ is the angular velocity of the rigid body relative to the primary reference frame.

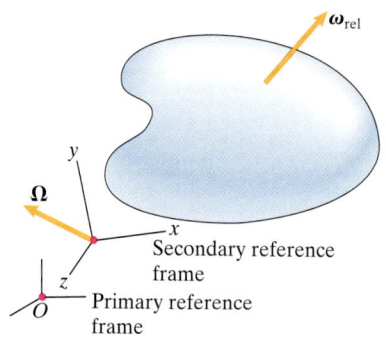

Figure 20.4

The vector $\boldsymbol{\omega}_{\text{rel}}$ is the rigid body's angular velocity relative to the secondary reference frame, and the vector $\boldsymbol{\Omega}$ is the angular velocity of the secondary reference frame relative to the primary reference frame. The rigid body's angular velocity vector relative to the primary reference frame is $\boldsymbol{\omega}_{\text{rel}} + \boldsymbol{\Omega}$.

Notice that, in general, the derivatives $d\omega_x/dt$, $d\omega_y/dt$, and $d\omega_z/dt$ are the components of $\boldsymbol{\alpha}$ only when $\boldsymbol{\Omega} = 0$ or when $\boldsymbol{\Omega} = \boldsymbol{\omega}$. Otherwise, the rigid body's angular acceleration vector relative to the primary reference frame must be determined from Eq. (20.4).

When the secondary reference frame is not fixed to the rigid body, it is often convenient to express the body's angular velocity vector $\boldsymbol{\omega}$ as the sum of the angular velocity vector $\boldsymbol{\Omega}$ of the secondary reference frame and the angular velocity vector $\boldsymbol{\omega}_{\text{rel}}$ of the rigid body relative to the secondary reference frame (Fig. 20.4):

$$\boldsymbol{\omega} = \boldsymbol{\Omega} + \boldsymbol{\omega}_{\text{rel}}. \tag{20.5}$$

Study Questions

1. If you know the velocity of a point B of a rigid body relative to a given reference frame, what additional information do you need to determine the velocity of a second point A of the rigid body relative to the given reference frame?

2. If you know the acceleration of a point B of a rigid body relative to a given reference frame, what additional information do you need to determine the acceleration of a second point A of the rigid body relative to the given reference frame?

3. Suppose that the angular velocity vector of a secondary reference frame relative to a primary reference frame is $\boldsymbol{\Omega}$. If the angular velocity vector of a rigid body relative to the secondary reference frame is $\boldsymbol{\omega}_{\text{rel}}$, what is the rigid body's angular velocity vector relative to the primary reference frame?

Example 20.1 | **Use of a Secondary Reference Frame**

The tire in Fig. 20.5 is rolling on the level surface. As the car turns, the midpoint B of the tire moves at 5 m/s in a circular path about the fixed point P (see the top view), and the tire remains perpendicular to the line from B to P.

(a) What is the tire's angular velocity vector $\boldsymbol{\omega}$ relative to an earth-fixed reference frame?

(b) Determine the velocity of point A, the rearmost point of the tire, at the instant shown.

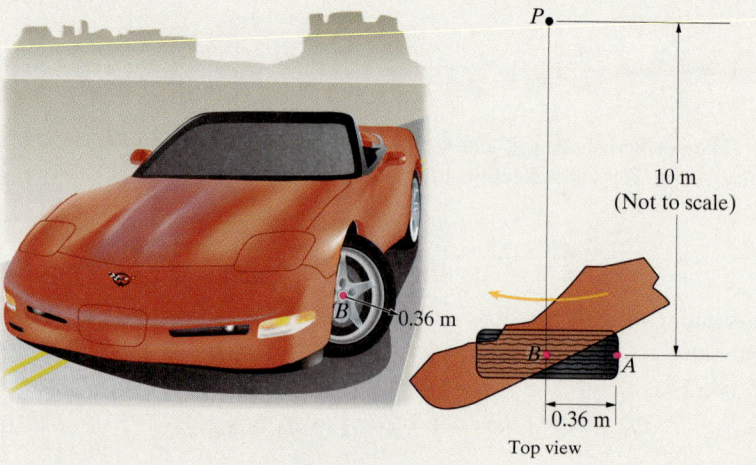

Figure 20.5

Strategy

(a) In Fig. a we introduce a secondary reference frame with its origin at B and its y axis along the line from B to P. We assume that the x axis *remains horizontal*. The motion of this reference frame is simple: It rotates about its z axis as the car turns. The motion of the tire relative to the secondary reference frame is also simple: It rotates about the y axis. By determining the angular velocity vector $\boldsymbol{\Omega}$ of the secondary reference frame and the angular velocity vector $\boldsymbol{\omega}_{\mathrm{rel}}$ of the tire relative to the secondary reference frame, we can use Eq. (20.5) to determine the tire's angular velocity vector relative to an earth-fixed reference frame.

(b) Knowing the velocity of point B and the tire's angular velocity vector $\boldsymbol{\omega}$, we can use Eq. (20.1) to determine the velocity of point A.

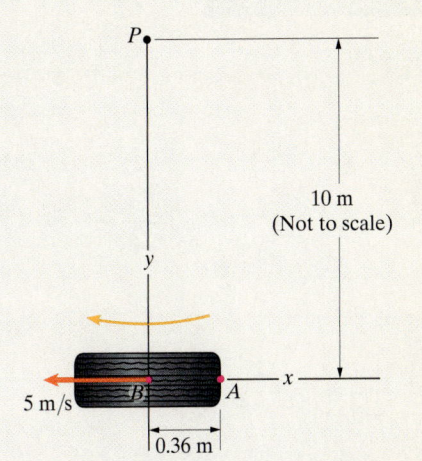

(a) A rotating secondary coordinate system. The y axis remains aligned with the line BP, and the x axis remains horizontal.

Solution

(a) The magnitude of the angular velocity of the line PB about P is $(5 \text{ m/s})/(10 \text{ m}) = 0.5 \text{ rad/s}$. The secondary coordinate system rotates about its z axis in the clockwise direction as viewed in Fig. a, so its angular velocity vector is

$$\boldsymbol{\Omega} = -0.5\mathbf{k} \; (\text{rad/s}).$$

The center of the tire moves at 5 m/s, so the magnitude of the tire's angular velocity about the y axis is $(5 \text{ m/s})/(0.36 \text{ m}) = 13.9 \text{ rad/s}$. The tire's angular velocity vector relative to the secondary coordinate system is

$$\boldsymbol{\omega}_{\mathrm{rel}} = -13.9\mathbf{j} \; (\text{rad/s}).$$

Therefore, the tire's angular velocity vector relative to an earth-fixed reference frame is

$$\boldsymbol{\omega} = \boldsymbol{\Omega} + \boldsymbol{\omega}_{\mathrm{rel}} = -13.9\mathbf{j} - 0.5\mathbf{k} \; (\text{rad/s}).$$

(b) The position vector of point A relative to point B is $\mathbf{r}_{A/B} = 0.36\mathbf{i} \; (\text{m})$. From Eq. (20.1), the velocity of point A is

$$\mathbf{v}_A = \mathbf{v}_B + \boldsymbol{\omega} \times \mathbf{r}_{A/B}$$

$$= -5\mathbf{i} + \begin{vmatrix} \mathbf{i} & \mathbf{j} & \mathbf{k} \\ 0 & -13.9 & -0.5 \\ 0.36 & 0 & 0 \end{vmatrix}$$

$$= -5\mathbf{i} - 0.18\mathbf{j} + 5\mathbf{k} \; (\text{m/s}).$$

Critical Thinking

Notice how using a secondary coordinate system rotating with the car simplified the determination of the tire's angular velocity vector. Although the tire's motion in space is quite complicated, its motion relative to the secondary coordinate system is simple.

Example 20.2 Angular Velocity and Angular Acceleration of a Rotating Disk

The disk shown in Fig. 20.6 is perpendicular to the horizontal part of the shaft and rotates relative to it with constant angular velocity ω_d. Relative to an earth-fixed reference frame, the shaft rotates about the vertical axis with constant angular velocity ω_0. Determine the angular velocity and angular acceleration vectors of the disk relative to the earth-fixed reference frame.

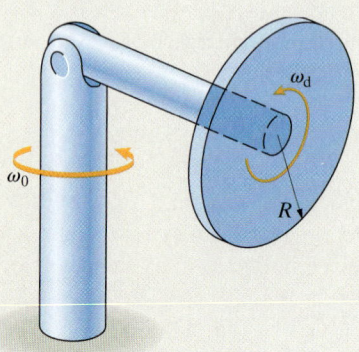

Figure 20.6

Strategy

The disk's motion relative to the earth-fixed reference frame is rather complicated. However, relative to a reference frame that is fixed with respect to the shaft, the disk simply rotates about a fixed axis with constant angular velocity. We will therefore introduce a secondary coordinate system that is fixed with respect to the shaft. The angular velocity vector we seek is the sum of the angular velocity vector of the secondary coordinate system and the disk's angular velocity vector relative to the secondary coordinate system. The disk's angular acceleration vector is given by Eq. (20.4).

Solution

We introduce the secondary coordinate system shown in Fig. a, which is fixed with respect to the shaft. The angular velocity vector of the secondary coordinate system relative to the earth-fixed reference frame is $\mathbf{\Omega} = \omega_0 \mathbf{j}$. The disk's angular velocity vector relative to the secondary coordinate system is $\boldsymbol{\omega}_{\mathrm{rel}} = \omega_d \mathbf{i}$. Therefore, the angular velocity vector of the disk relative to the earth-fixed reference frame is

$$\boldsymbol{\omega} = \mathbf{\Omega} + \boldsymbol{\omega}_{\mathrm{rel}} = \omega_d \mathbf{i} + \omega_0 \mathbf{j}.$$

Because ω_d and ω_0 are constants, we find from Eq. (20.4) that the disk's angular acceleration vector relative to the earth-fixed reference frame is

$$\boldsymbol{\alpha} = \mathbf{\Omega} \times \boldsymbol{\omega} = -\omega_0 \omega_d \mathbf{k}.$$

Critical Thinking

If the components of the disk's angular velocity vector are constants, how can the disk have an angular acceleration relative to the earth-fixed reference frame? Remember that ω_d and ω_0 are the components of $\boldsymbol{\omega}$ expressed in terms of the *secondary coordinate system*. In this example, the magnitude and direction of the vector $\boldsymbol{\omega}$ are constant relative to the secondary coordinate system, but $\boldsymbol{\omega}$ rotates relative to the earth-fixed reference frame.

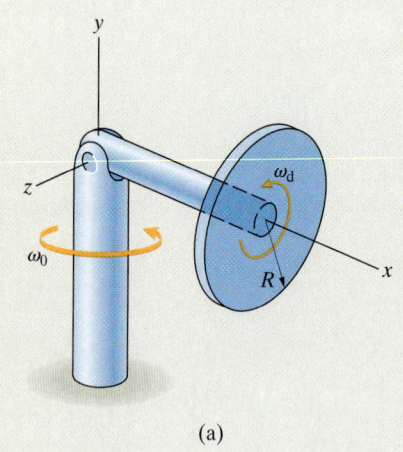

(a)

Example 20.3   Angular Velocity and Angular Acceleration of a Rolling Disk

The bent bar in Fig. 20.7 is rigidly attached to the vertical shaft, which rotates with constant angular velocity ω_0. The circular disk is pinned to the bent bar and rolls on the horizontal surface.
(a) Determine the disk's angular velocity vector $\boldsymbol{\omega}_{\text{disk}}$ and angular acceleration vector $\boldsymbol{\alpha}_{\text{disk}}$.
(b) Determine the velocity of point P, which is the uppermost point of the circular disk, at the present instant.

Strategy

(a) In this example, the primary reference frame is fixed with respect to the surface on which the disk rolls. To simplify our analysis of the disk's angular motion, we will use a secondary coordinate system that is fixed with respect to the bent bar. By applying Eq. (20.1) to the bent bar, we will determine the velocity of the center of the disk. We will determine the disk's angular velocity vector by recognizing that the velocity of the point of the disk in contact with the horizontal surface is zero. We can then use Eq. (20.4) to determine the disk's angular acceleration vector.
(b) Knowing the velocity of the center of the disk and the disk's angular velocity vector, we can apply Eq. (20.1) to the disk to determine the velocity of point P.

Solution

(a) Let the coordinate system in Fig. a be fixed with respect to the bent bar. The x axis coincides with the horizontal part of the bar, and the y axis coincides with the vertical shaft. The angular velocity vector $\boldsymbol{\omega}_{\text{bar}}$ of the bar and the angular velocity vector $\boldsymbol{\Omega}$ of the coordinate system are equal:

$$\boldsymbol{\omega}_{\text{bar}} = \boldsymbol{\Omega} = \omega_0 \mathbf{j}.$$

Let point B be the stationary origin of the coordinate system, and let point A be the center of the disk (Fig. a). The position vector of A relative to B is

$$\mathbf{r}_{A/B} = (h + b\cos\beta)\mathbf{i} - b\sin\beta\,\mathbf{j}.$$

From Eq. (20.1), the velocity of point A is

$$\mathbf{v}_A = \mathbf{v}_B + \boldsymbol{\omega}_{\text{bar}} \times \mathbf{r}_{A/B}$$

$$= \mathbf{0} + \begin{vmatrix} \mathbf{i} & \mathbf{j} & \mathbf{k} \\ 0 & \omega_0 & 0 \\ h + b\cos\beta & -b\sin\beta & 0 \end{vmatrix}$$

$$= -\omega_0(h + b\cos\beta)\mathbf{k}.$$

Because the coordinate system is fixed with respect to the bent bar, we can write the angular velocity vector of the disk relative to the coordinate system as (Fig. b)

$$\boldsymbol{\omega}_{\text{rel}} = \omega_{\text{rel}}\cos\beta\,\mathbf{i} - \omega_{\text{rel}}\sin\beta\,\mathbf{j}.$$

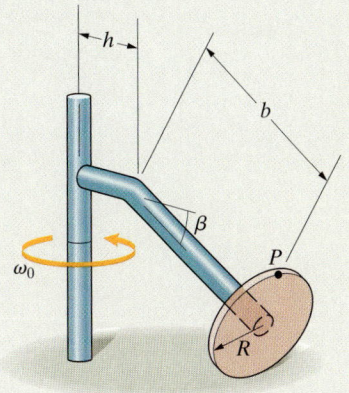

Figure 20.7

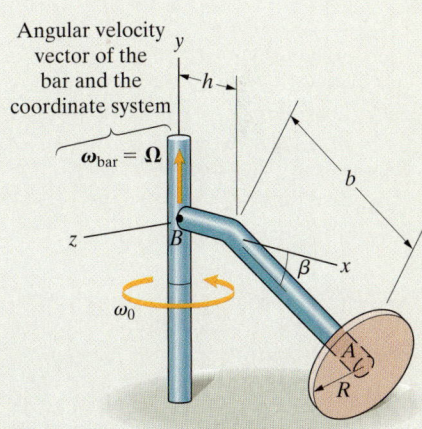

(a) A secondary coordinate system fixed to the bent bar.

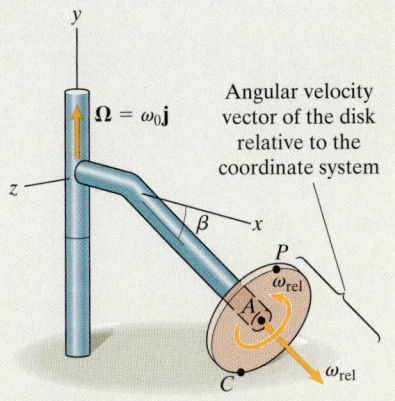

(b) Analyzing the motion of the disk.

Let point C in Fig. b be the point of the disk that is in contact with the surface. To determine ω_{rel}, we use the condition $\mathbf{v}_C = \mathbf{0}$. The position of C relative to A is

$$\mathbf{r}_{C/A} = -R \sin \beta \mathbf{i} - R \cos \beta \mathbf{j}.$$

Therefore,

$$\mathbf{v}_C = \mathbf{v}_A + \boldsymbol{\omega}_{disk} \times \mathbf{r}_{C/A}$$

$$= -\omega_0(h + b \cos \beta)\mathbf{k} + \begin{vmatrix} \mathbf{i} & \mathbf{j} & \mathbf{k} \\ \omega_{rel} \cos \beta & \omega_0 - \omega_{rel} \sin \beta & 0 \\ -R \sin \beta & -R \cos \beta & 0 \end{vmatrix} = \mathbf{0}.$$

Solving this equation for ω_{rel}, we obtain

$$\omega_{rel} = -\omega_0 \left(\frac{h}{R} + \frac{b}{R} \cos \beta - \sin \beta \right).$$

The angular velocity vector of the coordinate system is $\boldsymbol{\Omega} = \omega_0 \mathbf{j}$, so the disk's angular velocity vector is

$$\boldsymbol{\omega}_{disk} = \boldsymbol{\Omega} + \boldsymbol{\omega}_{rel} = \omega_{rel} \cos \beta \mathbf{i} + (\omega_0 - \omega_{rel} \sin \beta)\mathbf{j}.$$

Even though the components of $\boldsymbol{\omega}_{disk}$ are constants, we find from Eq. (20.4) that the disk's angular acceleration is not zero:

$$\boldsymbol{\alpha}_{disk} = \boldsymbol{\Omega} \times \boldsymbol{\omega}_{disk} = \begin{vmatrix} \mathbf{i} & \mathbf{j} & \mathbf{k} \\ 0 & \omega_0 & 0 \\ \omega_{rel} \cos \beta & \omega_0 - \omega_{rel} \sin \beta & 0 \end{vmatrix}$$

$$= -\omega_0 \omega_{rel} \cos \beta \mathbf{k}.$$

(b) The position vector of point P relative to the center of the disk is

$$\mathbf{r}_{P/A} = R \sin \beta \mathbf{i} + R \cos \beta \mathbf{j}.$$

Using Eq. (20.1) and our result for the velocity $\mathbf{v}_A$ of the center of the disk, we determine the velocity of point P:

$$\mathbf{v}_P = \mathbf{v}_A + \boldsymbol{\omega}_{disk} \times \mathbf{r}_{P/A}$$

$$= -\omega_0(h + b \cos \beta)\mathbf{k} + \begin{vmatrix} \mathbf{i} & \mathbf{j} & \mathbf{k} \\ \omega_{rel} \cos \beta & \omega_0 - \omega_{rel} \sin \beta & 0 \\ R \sin \beta & R \cos \beta & 0 \end{vmatrix}$$

$$= [-\omega_0(h + b \cos \beta + R \sin \beta) + \omega_{rel} R]\mathbf{k}$$

$$= -2\omega_0(h + b \cos \beta)\mathbf{k}.$$

Critical Thinking

In comparison to Example 20.2, this example was complicated by the fact that we didn't know the disk's angular velocity relative to the bar. But we knew the direction of its axis of rotation. Notice how we were able to use that information and the fact that the velocity of point C of the disk in contact with the surface is zero to determine ω_{rel}. That was the essential step in determining the disk's angular velocity vector relative to the primary reference frame.

Problems

20.1 The airplane's rate gyros indicate that relative to an earth-fixed reference frame, the airplane's angular velocity is $\boldsymbol{\omega} = -0.24\mathbf{i} + 0.62\mathbf{j} - 0.16\mathbf{k}$ (rad/s). The coordinates of point A of the airplane are $(4, 1, -0.5)$ m. What is the velocity vector of A relative to the airplane's center of mass?

Strategy: Apply Eq. (20.1) to point A and the center of mass.

20.2 The airplane's rate gyros indicate that relative to an earth-fixed reference frame, the airplane's angular velocity vector is $\boldsymbol{\omega} = -0.24\mathbf{i} + 0.62\mathbf{j} - 0.16\mathbf{k}$ (rad/s) and its angular acceleration is $\boldsymbol{\alpha} = 0.032\mathbf{i} - 0.140\mathbf{j} + 0.074\mathbf{k}$ (rad/s²). The coordinates of point A of the airplane are $(4, 1, -0.5)$ m. What is the acceleration of A relative to the airplane's center of mass?

20.3 The angular velocity vector of the rigid cube relative to the primary reference frame is $\boldsymbol{\omega} = 10\mathbf{i} + 8\mathbf{j} + 6\mathbf{k}$ (rad/s). The velocity of the center G of the cube relative to the primary reference frame is $\mathbf{v}_G = 4\mathbf{i} - 2\mathbf{j} + 3\mathbf{k}$ (m/s). What is the velocity of point A of the cube relative to the primary reference frame?

20.4 The *xyz* coordinate system is fixed with respect to the rigid cube. The angular velocity vector of the cube relative to the primary reference frame, $\boldsymbol{\omega} = 10\mathbf{i} + 8\mathbf{j} + 6\mathbf{k}$ (rad/s), is constant. The acceleration of the center G of the cube relative to the primary reference frame is zero. What is the acceleration of point A of the cube relative to the primary reference frame?

20.5 The origin of the *xyz* coordinate system is fixed to the center G of the rigid cube. Point G is fixed relative to the primary reference frame. The cube is rotating relative to the *xyz* system with constant angular velocity $\boldsymbol{\omega}_{\text{rel}} = -6\mathbf{i} + 2\mathbf{j} + 4\mathbf{k}$ (rad/s). The *xyz* system is rotating relative to the primary reference frame with constant angular velocity $\boldsymbol{\Omega} = 10\mathbf{i} + 8\mathbf{j} + 6\mathbf{k}$ (rad/s).

(a) What is the velocity of point A of the cube relative to the primary reference frame?

(b) What is the cube's angular acceleration vector relative to the primary reference frame?

(c) What is the acceleration of point A of the cube relative to the primary reference frame?

Problems 20.1/20.2

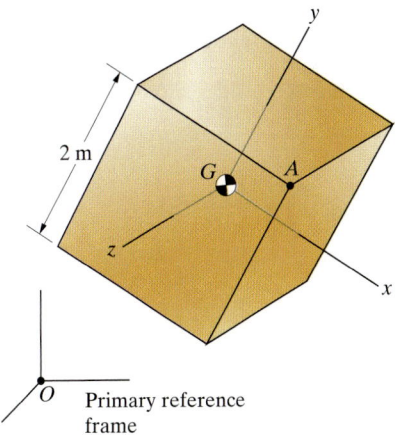

Problems 20.3–20.5

20.6 Relative to an earth-fixed reference frame, points A and B of the rigid parallelepiped are fixed and it rotates about the axis AB with an angular velocity of 30 rad/s. Determine the velocities of points C and D relative to the earth-fixed reference frame.

20.7 Relative to the xyz coordinate system shown, points A and B of the rigid parallelepiped are fixed and the parallelepiped rotates about the axis AB with an angular velocity of 30 rad/s. Relative to an earth-fixed reference frame, point A is fixed and the xyz coordinate system rotates with angular velocity $\mathbf{\Omega} = -5\mathbf{i} + 8\mathbf{j} + 6\mathbf{k}$ (rad/s). Determine the velocities of points C and D relative to the earth-fixed reference frame.

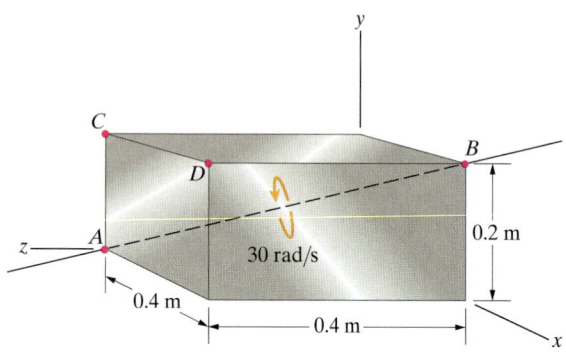

Problems 20.6/20.7

20.8 The disk of 0.2-m radius is supported by the vertical shaft. Relative to an earth-fixed reference frame, the shaft rotates about its axis with angular velocity $\omega_0 = 4$ rad/s. The disk rotates with angular velocity $\omega_d = 6$ rad/s relative to the shaft.

(a) What is the disk's angular velocity vector $\boldsymbol{\omega}$ relative to the earth-fixed reference frame?

(b) What is the velocity of point A of the disk relative to the earth-fixed reference frame?

Strategy: (a) Let the coordinate system shown be a secondary reference frame that is fixed with respect to the shaft. Determine the angular velocity vector $\mathbf{\Omega}$ of the secondary reference frame and the angular velocity vector $\boldsymbol{\omega}_{\text{rel}}$ of the disk relative to the secondary reference frame, and use Eq. (20.5). (b) Once $\boldsymbol{\omega}$ is known, you can use Eq. (20.1) to determine the velocity of point A. Notice that the center of the disk is a fixed point with respect to the earth-fixed reference frame.

20.9 The disk of 0.2-m radius is supported by the vertical shaft. Relative to an earth-fixed reference frame, the shaft rotates about its axis with constant angular velocity $\omega_0 = 4$ rad/s. The disk rotates with constant angular velocity $\omega_d = 6$ rad/s relative to the shaft.

(a) What is the disk's angular acceleration vector $\boldsymbol{\alpha}$ relative to the earth-fixed reference frame?

(b) What is the acceleration of point A relative to the earth-fixed reference frame?

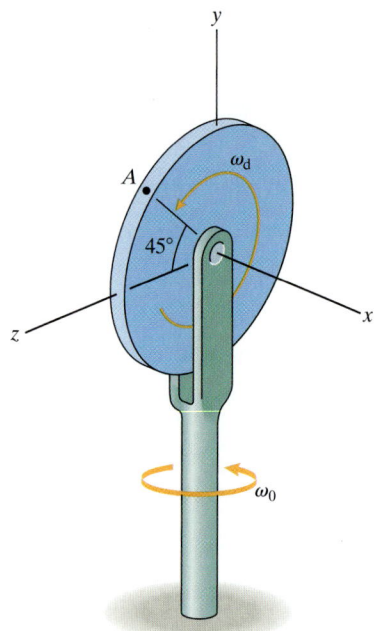

Problems 20.8/20.9

20.10 The radius of the disk is $R = 2$ ft. It is perpendicular to the horizontal part of the shaft and rotates relative to it with constant angular velocity $\omega_d = 36$ rad/s. Relative to an earth-fixed reference frame, the shaft rotates about the vertical axis with constant angular velocity $\omega_0 = 8$ rad/s.

(a) Determine the velocity relative to the earth-fixed reference frame of point P, which is the uppermost point of the disk.

(b) Determine the disk's angular acceleration vector α relative to the earth-fixed reference frame.

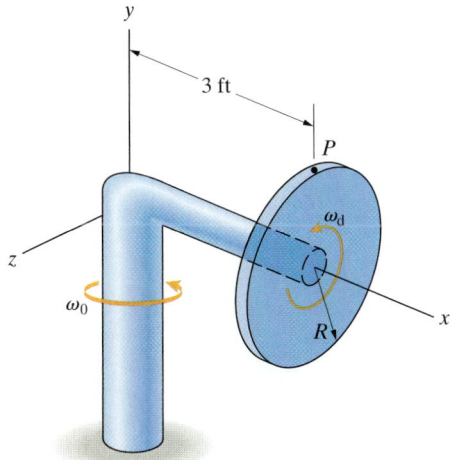

Problem 20.10

20.11 The vertical shaft supporting the dish antenna is rotating with angular velocity $\omega_0 = 0.2$ rad/s. The angle θ from the horizontal to the antenna's axis is 30° at the instant shown and is increasing at 15° per second.

(a) What is the antenna's angular velocity vector ω relative to an earth-fixed reference frame?

(b) What is the velocity of the point of the antenna with coordinates $(2, 2, -2)$ m relative to an earth-fixed reference frame?

Problem 20.11

20.12 The radius of the circular disk is $R = 0.2$ m, and $b = 0.3$ m. The disk rotates with angular velocity $\omega_d = 6$ rad/s relative to the horizontal bar. The horizontal bar rotates with angular velocity $\omega_b = 4$ rad/s relative to the vertical shaft, and the vertical shaft rotates with angular velocity $\omega_0 = 2$ rad/s relative to an earth-fixed reference frame. Assume that the secondary reference frame shown is fixed with respect to the horizontal bar.

(a) What is the angular velocity vector ω_{rel} of the disk relative to the secondary reference frame?

(b) Determine the velocity relative to the earth-fixed reference frame of point P, which is the uppermost point of the disk.

20.13 Solve Problem 20.12 assuming that the secondary reference frame is fixed with respect to the vertical shaft.

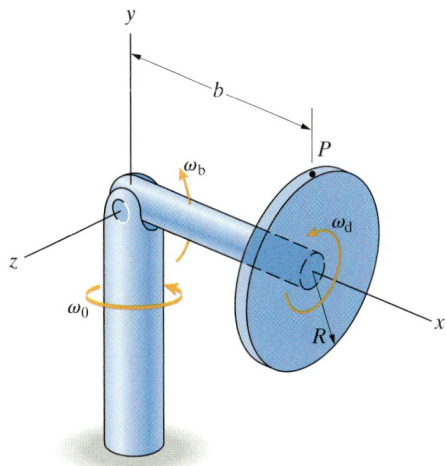

Problems 20.12/20.13

20.14 The bent bar is rigidly attached to the vertical shaft, which rotates with angular velocity ω_0 relative to an earth-fixed reference frame. The circular disk is pinned to the bent bar and rotates with angular velocity ω_d relative to the bar.

(a) Determine the disk's angular velocity vector $\boldsymbol{\omega}_{\text{disk}}$ relative to the earth-fixed reference frame.
(b) What is the velocity of point P, which is the uppermost point of the disk, relative to the earth-fixed reference frame?

20.15 The bent bar is rigidly attached to the vertical shaft, which rotates with constant angular velocity ω_0 relative to an earth-fixed reference frame. The circular disk is pinned to the bent bar and rotates with constant angular velocity ω_d relative to the bar.

(a) Determine the disk's angular acceleration vector $\boldsymbol{\alpha}_{\text{disk}}$ relative to the earth-fixed reference frame.
(b) What is the acceleration of point P, which is the uppermost point of the disk, relative to the earth-fixed reference frame?

20.16 Relative to a primary reference frame, the gyroscope's circular frame rotates about the vertical axis at 2 rad/s. The 60-mm diameter wheel rotates at 10 rad/s relative to the frame. Determine the velocities of points A and B relative to the primary reference frame.

20.17 Relative to a primary reference frame, the gyroscope's circular frame rotates about the vertical axis with a constant angular velocity of 2 rad/s. The 60-mm diameter wheel rotates with a constant angular velocity of 10 rad/s relative to the frame. Determine the accelerations of points A and B relative to the primary reference frame.

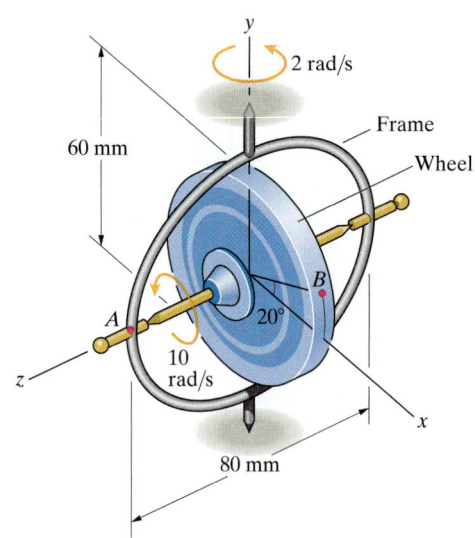

Problems 20.16/20.17

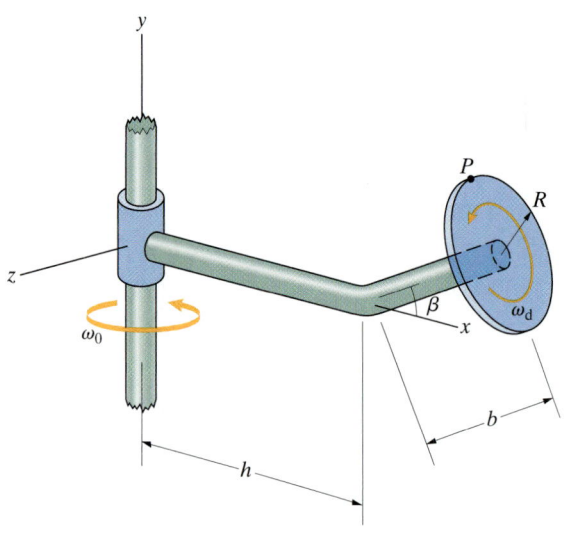

Problems 20.14/20.15

20.18 Relative to an earth-fixed reference frame, the manipulator rotates about the vertical axis with angular velocity $\omega_y = 0.1$ rad/s. The y axis of the secondary coordinate system remains vertical, and the x axis rotates with the manipulator, so that points A, B, and C remain in the x–y plane. The angular velocities of the arms AB and BC *relative to the secondary coordinate system* are $-0.2\mathbf{k}$ (rad/s) and $0.4\ \mathbf{k}$ (rad/s), respectively.

(a) What is the angular velocity vector $\boldsymbol{\omega}_{BC}$ of arm BC relative to the earth-fixed reference frame?

(b) What is the velocity of point C relative to the earth-fixed reference frame?

20.19 The angular velocity of the manipulator described in Problem 20.18 about the vertical axis is constant. The angular accelerations of the arms AB and BC relative to the secondary coordinate system are zero. What is the acceleration of point C relative to the earth-fixed reference frame?

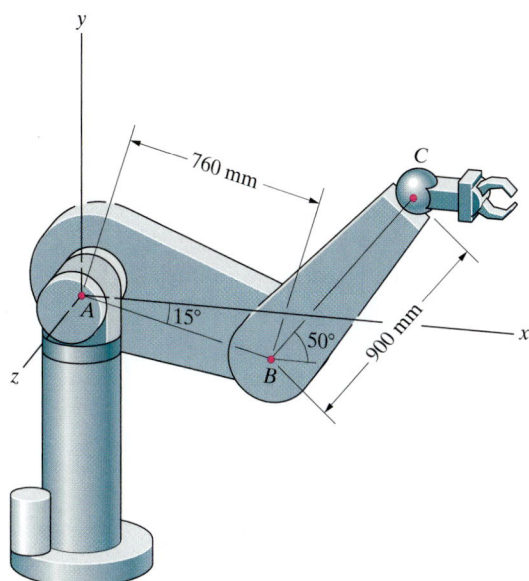

Problems 20.18/20.19

20.20* The cone rolls on the horizontal surface, which is fixed with respect to an earth-fixed reference frame. The x axis of the secondary reference frame remains coincident with the cone's axis, and the z axis remains horizontal. As the cone rolls, the z axis rotates in the horizontal plane with an angular velocity of 2 rad/s.

(a) What is the angular velocity vector $\boldsymbol{\Omega}$ of the secondary reference frame?

(b) What is the angular velocity vector $\boldsymbol{\omega}_{\mathrm{rel}}$ of the cone relative to the secondary reference frame?

Strategy: To solve part (b), use the fact that the velocity relative to the earth-fixed reference frame of points of the cone in contact with the surface is zero.

20.21* The cone rolls on the horizontal surface, which is fixed with respect to an earth-fixed reference frame. The x axis of the secondary reference frame remains coincident with the cone's axis, and the z axis remains horizontal. As the cone rolls, the z axis rotates in the horizontal plane with an angular velocity of 2 rad/s. Determine the velocity relative to the earth-fixed reference frame of the point of the base of the cone with coordinates $x = 0.4$ m, $y = 0$, $z = 0.2$ m.

20.22* The cone rolls on the horizontal surface, which is fixed with respect to an earth-fixed reference frame. The x axis of the secondary reference frame remains coincident with the cone's axis, and the z axis remains horizontal. As the cone rolls, the z axis rotates in the horizontal plane with a constant angular velocity of 2 rad/s. Determine the acceleration relative to the earth-fixed reference frame of the point of the base of the cone with coordinates $x = 0.4$ m, $y = 0$, $z = 0.2$ m.

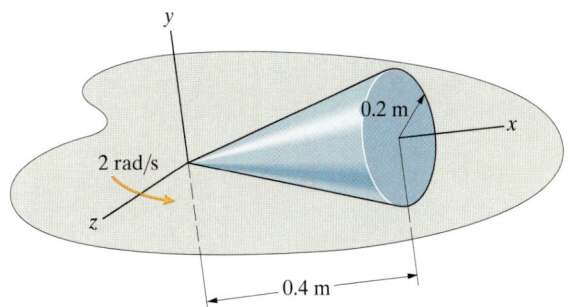

Problems 20.20–20.22

20.23* The radius and length of the cylinder are $R = 0.1$ m and $l = 0.4$ m. The horizontal surface is fixed with respect to an earth-fixed reference frame. One end of the cylinder rolls on the surface while its center, the origin of the secondary reference frame, remains stationary. The angle $\beta = 45°$. The z axis of the secondary reference frame remains coincident with the cylinder's axis, and the y axis remains horizontal. As the cylinder rolls, the y axis rotates in a horizontal plane with angular velocity $\omega_0 = 2$ rad/s.

(a) What is the angular velocity vector $\mathbf{\Omega}$ of the secondary reference frame?

(b) What is the angular velocity vector $\boldsymbol{\omega}_{\text{rel}}$ of the cylinder relative to the secondary reference frame?

20.24* The radius and length of the cylinder are $R = 0.1$ m and $l = 0.4$ m. The horizontal surface is fixed with respect to an earth-fixed reference frame. One end of the cylinder rolls on the surface while its center, the origin of the secondary reference frame, remains stationary. The angle $\beta = 45°$. The z axis of the secondary reference frame remains coincident with the cylinder's axis, and the y axis remains horizontal. As the cylinder rolls, the y axis rotates in a horizontal plane with angular velocity $\omega_0 = 2$ rad/s. Determine the velocity relative to the earth-fixed reference frame of the point of the upper end of the cylinder with coordinates $x = 0.1$ m, $y = 0$, $z = 0.2$ m.

20.25* The landing gear of the P-40 airplane used in World War II retracts by rotating 90° about the horizontal axis toward the rear of the airplane. As the wheel retracts, a linkage rotates the strut supporting the wheel 90° about the strut's longitudinal axis so that the wheel is horizontal in the retracted position. (Viewed from the horizontal axis toward the wheel, the strut rotates in the clockwise direction.) The x axis of the coordinate system shown remains parallel to the horizontal axis and the y axis remains parallel to the strut as the wheel retracts. Let ω_W be the magnitude of the wheel's angular velocity when the airplane lifts off, and assume that it remains constant. Let ω_0 be the magnitude of the constant angular velocity of the strut about the horizontal axis as the landing gear is retracted. The magnitude of the angular velocity of the strut about its longitudinal axis also equals ω_0. The landing gear begins retracting at $t = 0$. Determine the wheel's angular velocity vector relative to the airplane as a function of time.

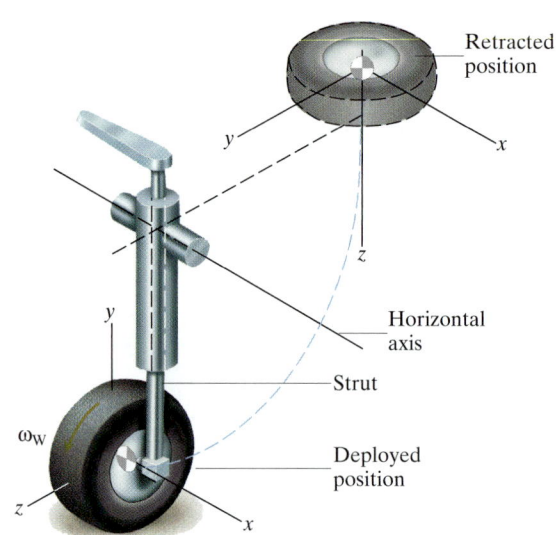

Problem 20.25

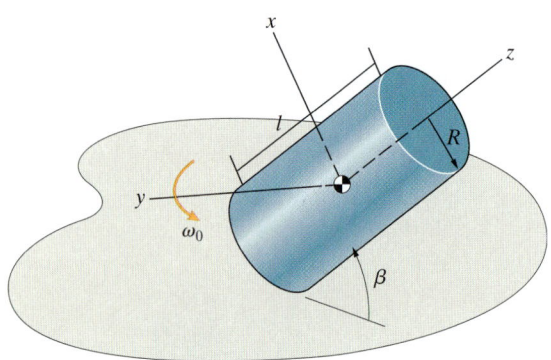

Problems 20.23/20.24

20.2 Euler's Equations

The three-dimensional equations of motion for a rigid body are called *Euler's equations*. They consist of Newton's second law,

$$\Sigma \mathbf{F} = m\mathbf{a}, \tag{20.6}$$

which states that the sum of the external forces on a rigid body equals the product of its mass and the acceleration of its center of mass, and equations of angular motion. In deriving the equations of angular motion, we consider first the special case of rotation of a rigid body about a fixed point and then general three-dimensional motion of a rigid body.

Rotation about a Fixed Point

Let m_i be the mass of the ith particle of a rigid body, and let $\mathbf{r}_i$ be its position relative to a point O that is fixed with respect to an inertial primary reference frame (Fig. 20.8). In Section 18.2, we showed that for an arbitrary system of particles, the sum of the moments about O equals the rate of change of the total angular momentum about O:

$$\Sigma \mathbf{M}_O = \frac{d\mathbf{H}_O}{dt}, \tag{20.7}$$

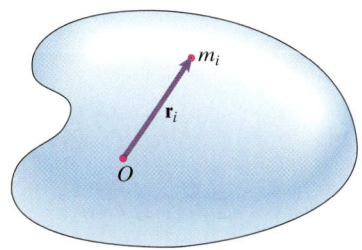

Figure 20.8
Mass and position of the ith particle of a rigid body.

where the total angular momentum is

$$\mathbf{H}_O = \sum_i \mathbf{r}_i \times m_i \frac{d\mathbf{r}_i}{dt}.$$

If the rigid body rotates about O with angular velocity $\boldsymbol{\omega}$, the velocity of the ith particle is $d\mathbf{r}_i/dt = \boldsymbol{\omega} \times \mathbf{r}_i$, and the angular momentum is

$$\mathbf{H}_O = \sum_i \mathbf{r}_i \times m_i(\boldsymbol{\omega} \times \mathbf{r}_i). \tag{20.8}$$

In Fig. 20.9, we introduce a secondary coordinate system with its origin at O. We express the vectors $\boldsymbol{\omega}$ and $\mathbf{r}_i$ in terms of their components in this coordinate system as

$$\boldsymbol{\omega} = \omega_x \mathbf{i} + \omega_y \mathbf{j} + \omega_z \mathbf{k}$$

and

$$\mathbf{r}_i = x_i \mathbf{i} + y_i \mathbf{j} + z_i \mathbf{k},$$

where (x_i, y_i, z_i) are the coordinates of the ith particle. Substituting these expressions into Eq. (20.8) and evaluating the cross products, we can write the resulting components of the angular momentum vector in the forms

$$H_{Ox} = I_{xx}\omega_x - I_{xy}\omega_y - I_{xz}\omega_z,$$

$$H_{Oy} = -I_{yx}\omega_x + I_{yy}\omega_y - I_{yz}\omega_z, \tag{20.9}$$

$$H_{Oz} = -I_{zx}\omega_x - I_{zy}\omega_y + I_{zz}\omega_z.$$

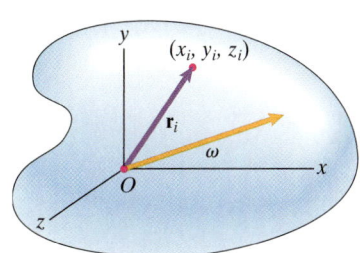

Figure 20.9
Secondary coordinate system with its origin at O.

The coefficients

$$I_{xx} = \sum_i m_i(y_i^2 + z_i^2),$$

$$I_{yy} = \sum_i m_i(x_i^2 + z_i^2), \tag{20.10}$$

$$I_{zz} = \sum_i m_i(x_i^2 + y_i^2)$$

are called the *moments of inertia* about the *x, y,* and *z* axes, and the coefficients

$$I_{xy} = I_{yx} = \sum_i m_i x_i y_i,$$

$$I_{yz} = I_{zy} = \sum_i m_i y_i z_i, \tag{20.11}$$

$$I_{xz} = I_{zx} = \sum_i m_i x_i z_i$$

are called the *products of inertia*. (Evaluation of the moments and products of inertia is discussed in the appendix to this chapter.)

To obtain the equations of angular motion, we must substitute the components of the angular momentum given by Eqs. (20.9) into Eq. (20.7). The secondary coordinate system in which these components are expressed is usually chosen to be body fixed, so it rotates relative to the primary reference frame with the angular velocity $\boldsymbol{\omega}$ of the rigid body. However, we have seen in the previous section that in some situations it is convenient to use a secondary coordinate system that rotates, but is not body fixed. We denote the secondary coordinate system's angular velocity vector by $\boldsymbol{\Omega}$, where $\boldsymbol{\Omega} = \boldsymbol{\omega}$ if the coordinate system is body fixed. Expressing the angular momentum vector in terms of its components as

$$\mathbf{H}_O = H_{Ox}\mathbf{i} + H_{Oy}\mathbf{j} + H_{Oz}\mathbf{k},$$

we obtain the derivative of $\mathbf{H}_O$ with respect to time:

$$\frac{d\mathbf{H}_O}{dt} = \frac{dH_{Ox}}{dt}\mathbf{i} + H_{Ox}\frac{d\mathbf{i}}{dt} + \frac{dH_{Oy}}{dt}\mathbf{j} + H_{Oy}\frac{d\mathbf{j}}{dt} + \frac{dH_{Oz}}{dt}\mathbf{k} + H_{Oz}\frac{d\mathbf{k}}{dt}.$$

Using this expression and writing the time derivatives of the unit vectors in terms of the angular velocity $\boldsymbol{\Omega}$ of the coordinate system,

$$\frac{d\mathbf{i}}{dt} = \boldsymbol{\Omega} \times \mathbf{i}, \qquad \frac{d\mathbf{j}}{dt} = \boldsymbol{\Omega} \times \mathbf{j}, \qquad \frac{d\mathbf{k}}{dt} = \boldsymbol{\Omega} \times \mathbf{k},$$

we can write Eq. (20.7) as

$$\Sigma\mathbf{M}_O = \frac{dH_{Ox}}{dt}\mathbf{i} + \frac{dH_{Oy}}{dt}\mathbf{j} + \frac{dH_{Oz}}{dt}\mathbf{k} + \mathbf{\Omega} \times \mathbf{H}_O.$$

Substituting the components of $\mathbf{H}_O$ from Eq. (20.9) into this equation, we obtain the equations of angular motion (see Problem 20.60):

$$\begin{aligned}
\Sigma M_{Ox} = {} & I_{xx}\frac{d\omega_x}{dt} - I_{xy}\frac{d\omega_y}{dt} - I_{xz}\frac{d\omega_z}{dt} \\
& - \Omega_z(-I_{yx}\omega_x + I_{yy}\omega_y - I_{yz}\omega_z) \\
& + \Omega_y(-I_{zx}\omega_x - I_{zy}\omega_y + I_{zz}\omega_z), \\[6pt]
\Sigma M_{Oy} = {} & -I_{yx}\frac{d\omega_x}{dt} + I_{yy}\frac{d\omega_y}{dt} - I_{yz}\frac{d\omega_z}{dt} \\
& + \Omega_z(I_{xx}\omega_x - I_{xy}\omega_y - I_{xz}\omega_z) \\
& - \Omega_x(-I_{zx}\omega_x - I_{zy}\omega_y + I_{zz}\omega_z), \\[6pt]
\Sigma M_{Oz} = {} & -I_{zx}\frac{d\omega_x}{dt} - I_{zy}\frac{d\omega_y}{dt} + I_{zz}\frac{d\omega_z}{dt} \\
& - \Omega_y(I_{xx}\omega_x - I_{xy}\omega_y - I_{xz}\omega_z) \\
& + \Omega_x(-I_{yx}\omega_x + I_{yy}\omega_y - I_{yz}\omega_z).
\end{aligned} \tag{20.12}$$

In carrying out this final step, *we have assumed that the moments and products of inertia are constants*. This is so when the secondary reference frame is body fixed, but must be confirmed when it is not. Equations (20.12) can be written as the matrix equation

$$\begin{bmatrix} \Sigma M_{Ox} \\ \Sigma M_{Oy} \\ \Sigma M_{Oz} \end{bmatrix} = \begin{bmatrix} I_{xx} & -I_{xy} & -I_{xz} \\ -I_{yx} & I_{yy} & -I_{yz} \\ -I_{zx} & -I_{zy} & I_{zz} \end{bmatrix}\begin{bmatrix} d\omega_x/dt \\ d\omega_y/dt \\ d\omega_z/dt \end{bmatrix}$$

$$+ \begin{bmatrix} 0 & -\Omega_z & \Omega_y \\ \Omega_z & 0 & -\Omega_x \\ -\Omega_y & \Omega_x & 0 \end{bmatrix}\begin{bmatrix} I_{xx} & -I_{xy} & -I_{xz} \\ -I_{yx} & I_{yy} & -I_{yz} \\ -I_{zx} & -I_{zy} & I_{zz} \end{bmatrix}\begin{bmatrix} \omega_x \\ \omega_y \\ \omega_z \end{bmatrix}, \tag{20.13}$$

where

$$\begin{bmatrix} I_{xx} & -I_{xy} & -I_{xz} \\ -I_{yx} & I_{yy} & -I_{yz} \\ -I_{zx} & -I_{zy} & I_{zz} \end{bmatrix} = [I] \tag{20.14}$$

is called the *inertia matrix* of the rigid body.

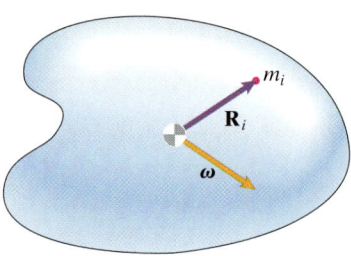

Figure 20.10
Position of the *i*th particle of a rigid body relative to the center of mass of the body.

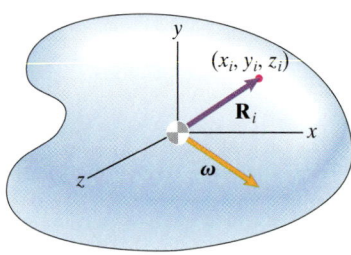

Figure 20.11
Coordinate system with its origin at the center of mass of the body.

General Three-Dimensional Motion

Let $\mathbf{R}_i$ be the position of the *i*th particle of a rigid body relative to the center of mass of the body (Fig. 20.10). In Section 18.2, we showed that the sum of the moments about the center of mass equals the rate of change of the total angular momentum of the body relative to its center of mass; that is,

$$\Sigma \mathbf{M} = \frac{d\mathbf{H}}{dt}, \qquad (20.15)$$

where the total angular momentum is

$$\mathbf{H} = \sum_i \mathbf{R}_i \times m_i \frac{d\mathbf{R}_i}{dt}.$$

In terms of the rigid body's angular velocity $\boldsymbol{\omega}$, the velocity of the *i*th particle is $d\mathbf{R}_i/dt = \boldsymbol{\omega} \times \mathbf{R}_i$, and the angular momentum is

$$\mathbf{H} = \sum_i \mathbf{R}_i \times m_i(\boldsymbol{\omega} \times \mathbf{R}_i). \qquad (20.16)$$

We introduce a secondary coordinate system with its origin at the center of mass (Fig. 20.11) and express the vectors $\boldsymbol{\omega}$ and $\mathbf{R}_i$ in terms of their components in this coordinate system as

$$\boldsymbol{\omega} = \omega_x \mathbf{i} + \omega_y \mathbf{j} + \omega_z \mathbf{k}$$

and

$$\mathbf{R}_i = x_i \mathbf{i} + y_i \mathbf{j} + z_i \mathbf{k}.$$

Substituting these expressions into Eq. (20.16) and evaluating the cross products, we obtain the components of the angular momentum vector in the forms

$$H_x = I_{xx}\omega_x - I_{xy}\omega_y - I_{xz}\omega_z,$$

$$H_y = -I_{yx}\omega_x + I_{yy}\omega_y - I_{yz}\omega_z, \qquad (20.17)$$

and

$$H_z = -I_{zx}\omega_x - I_{zy}\omega_y + I_{zz}\omega_z,$$

where the expressions for the moments and products of inertia are again given by Eqs. (20.10) and (20.11). Denoting the coordinate system's angular velocity vector by $\boldsymbol{\Omega}$, and following the same steps we used to obtain Eqs. (20.12), we obtain the equations of angular motion,

$$\Sigma M_x = I_{xx}\frac{d\omega_x}{dt} - I_{xy}\frac{d\omega_y}{dt} - I_{xz}\frac{d\omega_z}{dt}$$
$$- \Omega_z(-I_{yx}\omega_x + I_{yy}\omega_y - I_{yz}\omega_z)$$
$$+ \Omega_y(-I_{zx}\omega_x - I_{zy}\omega_y + I_{zz}\omega_z),$$

$$\Sigma M_y = -I_{yx}\frac{d\omega_x}{dt} + I_{yy}\frac{d\omega_y}{dt} - I_{yz}\frac{d\omega_z}{dt}$$
$$+ \Omega_z(I_{xx}\omega_x - I_{xy}\omega_y - I_{xz}\omega_z) \qquad (20.18)$$
$$- \Omega_x(-I_{zx}\omega_x - I_{zy}\omega_y + I_{zz}\omega_z),$$

$$\Sigma M_z = -I_{zx}\frac{d\omega_x}{dt} - I_{zy}\frac{d\omega_y}{dt} + I_{zz}\frac{d\omega_z}{dt}$$
$$- \Omega_y(I_{xx}\omega_x - I_{xy}\omega_y - I_{xz}\omega_z)$$
$$+ \Omega_x(-I_{yx}\omega_x + I_{yy}\omega_y - I_{yz}\omega_z),$$

which we can write as the matrix equation

$$\begin{bmatrix} \Sigma M_x \\ \Sigma M_y \\ \Sigma M_z \end{bmatrix} = \begin{bmatrix} I_{xx} & -I_{xy} & -I_{xz} \\ -I_{yx} & I_{yy} & -I_{yz} \\ -I_{zx} & -I_{zy} & I_{zz} \end{bmatrix} \begin{bmatrix} d\omega_x/dt \\ d\omega_y/dt \\ d\omega_z/dt \end{bmatrix}$$

$$+ \begin{bmatrix} 0 & -\Omega_z & \Omega_y \\ \Omega_z & 0 & -\Omega_x \\ -\Omega_y & \Omega_x & 0 \end{bmatrix} \begin{bmatrix} I_{xx} & -I_{xy} & -I_{xz} \\ -I_{yx} & I_{yy} & -I_{yz} \\ -I_{zx} & -I_{zy} & I_{zz} \end{bmatrix} \begin{bmatrix} \omega_x \\ \omega_y \\ \omega_z \end{bmatrix}. \qquad (20.19)$$

We have obtained equations that are identical in form to the equations of angular motion for rotation about a fixed point. Equations (20.12) and (20.13) are expressed in terms of the total moment about a fixed point O about which the rigid body rotates, and the moments and products of inertia and the components of the vectors are expressed in terms of a coordinate system with its origin at O. Equations (20.18) and (20.19) are expressed in terms of the total moment about the center of mass of the body, and the moments and products of inertia and the components of the vectors are expressed in terms of a coordinate system with its origin at the center of mass.

If the secondary coordinate system used to apply Eqs. (20.12), (20.13), (20.18), and (20.19) is body fixed, the terms $d\omega_x/dt$, $d\omega_y/dt$, and $d\omega_z/dt$ are the components of the rigid body's angular acceleration $\boldsymbol{\alpha}$. But this is not generally the case if the secondary coordinate system rotates but is not body fixed. [See Eq. (20.4).]

Using the Euler equations to analyze three-dimensional motions of rigid bodies typically involves three steps:

1. *Choose a coordinate system.* If an object rotates about a fixed point O, it is usually preferable to use a secondary coordinate system with its origin at O and express the equations of angular motion in the forms given by Eqs. (20.12) or (20.13). Otherwise, it is necessary to use a coordinate system with its origin at the center of mass of the body and express the equations of angular motion in the forms given by Eqs. (20.18) or (20.19). In either case, it is usually preferable to choose a coordinate system that simplifies the determination of the moments and products of inertia.

2. *Draw the free-body diagram.* Isolate the object and identify the external forces and couples acting on it.

3. *Apply the equations of motion.* Use Euler's equations to relate the forces and couples acting on the object to the acceleration of its center of mass and its angular acceleration vector.

Equations of Planar Motion

Here we demonstrate how the equations of angular motion for a rigid body in planar motion can be obtained from the three-dimensional equations. Consider a rigid body that rotates about a fixed axis L_O. We introduce a body-fixed secondary coordinate system with the z axis aligned with L_O, so that the rigid body's angular velocity vector is $\boldsymbol{\omega} = \omega_z\mathbf{k}$ (Fig. 20.12). Substituting $\Omega_x = \omega_x = 0$, $\Omega_y = \omega_y = 0$, and $\Omega_z = \omega_z$ into Eqs. (20.12), we find that the third equation reduces to $\Sigma M_{Oz} = I_{zz}(d\omega_z/dt)$. Introducing the simpler notation $\Sigma M_{Oz} = \Sigma M_O$, $I_{zz} = I_O$, and $\omega_z = \omega$, we obtain

$$\Sigma M_O = I_O \frac{d\omega}{dt}. \qquad (20.20)$$

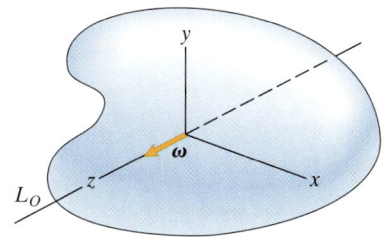

Figure 20.12

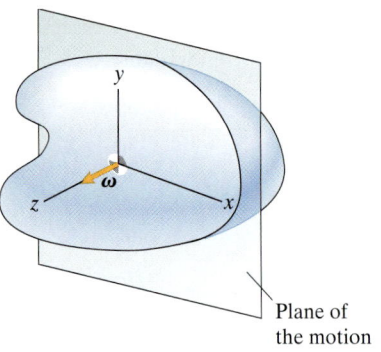

Figure 20.13

This is the equation we used in Chapter 18 to analyze the rotation of a rigid body about a fixed axis. [See Eq. (18.17).] The total moment about the fixed axis equals the product of the moment of inertia about the fixed axis and the angular acceleration.

For general planar motion, we introduce a body-fixed secondary coordinate system with its origin at the center of mass of the body and the z axis perpendicular to the plane of the motion (Fig. 20.13). The rigid body's angular velocity vector is $\boldsymbol{\omega} = \omega_z \mathbf{k}$. Substituting $\Omega_x = \omega_x = 0$, $\Omega_y = \omega_y = 0$, and $\Omega_z = \omega_z$ into Eqs. (20.18), the third equation reduces to $\Sigma M_z = I_{zz}(d\omega_z/dt)$. With the notation $\Sigma M_z = \Sigma M$, $I_{zz} = I$, and $\omega_z = \omega$, we obtain

$$\Sigma M = I \frac{d\omega}{dt}. \tag{20.21}$$

This is the equation of angular motion we used in Chapter 18 to analyze the general planar motion of a rigid body. [See Eq. (18.20).] The total moment about the center of mass of the body equals the product of the moment of inertia about the center of mass and the angular acceleration. (The term I is the moment of inertia about the axis through the center of mass that is perpendicular to the plane of the motion.)

Study Questions

1. What are Euler's equations?

2. How are the moments and products of inertia of a rigid body defined?

3. Suppose that a rigid body rotates about a fixed point O with a known angular velocity vector. If you know the moments and products of inertia of the body in terms of a body-fixed secondary reference frame with its origin at O, how can you determine the rigid body's angular momentum about O?

4. What are the differences between Eqs. (20.12) and Eqs. (20.18)?

Example 20.4   **Three-Dimensional Dynamics of a Plate**

During an assembly process, the 4-kg rectangular plate in Fig. 20.14 is held at O by a robotic manipulator. Point O is stationary. At the instant shown, the plate is horizontal, its angular velocity vector is $\boldsymbol{\omega} = 4\mathbf{i} - 2\mathbf{j}$ (rad/s), and its angular acceleration vector is $\boldsymbol{\alpha} = -10\mathbf{i} + 6\mathbf{j}$ (rad/s^2). Determine the couple exerted on the plate by the manipulator.

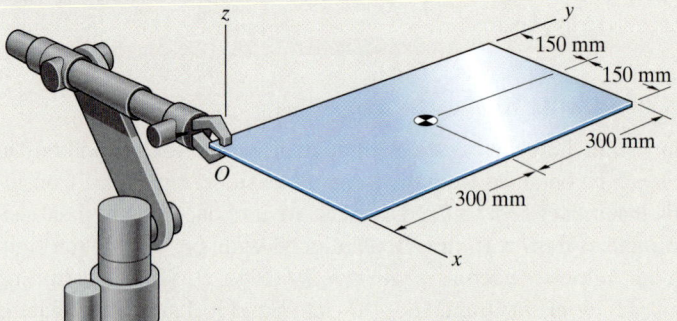

Figure 20.14

Strategy

The plate rotates about the fixed point O, so we can use Eq. (20.13) to determine the total moment exerted on the plate about O.

Solution

Draw the Free-Body Diagram   We denote the force and couple exerted on the plate by the manipulator by **F** and **C** (Fig. a).

Apply the Equations of Motion   The total moment about O is the sum of the couple exerted by the manipulator and the moment about O due to the plate's weight:

$$\Sigma \mathbf{M}_O = \mathbf{C} + (0.15\mathbf{i} + 0.30\mathbf{j}) \times [-(4)(9.81)\mathbf{k}]$$
$$= \mathbf{C} - 11.77\mathbf{i} + 5.89\mathbf{j} \ (\text{N-m}). \tag{1}$$

To obtain the unknown couple **C**, we can determine the total moment about O from Eq. (20.13).

We let the secondary coordinate system be body fixed, so its angular velocity $\boldsymbol{\Omega}$ equals the plate's angular velocity $\boldsymbol{\omega}$. We determine the plate's inertia matrix in Example 20.9, obtaining

$$[I] = \begin{bmatrix} 0.48 & -0.18 & 0 \\ -0.18 & 0.12 & 0 \\ 0 & 0 & 0.6 \end{bmatrix} \text{kg-m}^2.$$

Therefore, the total moment about O exerted on the plate is

$$\begin{bmatrix} \Sigma M_{Ox} \\ \Sigma M_{Oy} \\ \Sigma M_{Oz} \end{bmatrix} = \begin{bmatrix} I_{xx} & -I_{xy} & -I_{xz} \\ -I_{yx} & I_{yy} & -I_{yz} \\ -I_{zx} & -I_{zy} & I_{zz} \end{bmatrix} \begin{bmatrix} d\omega_x/dt \\ d\omega_y/dt \\ d\omega_z/dt \end{bmatrix}$$

$$+ \begin{bmatrix} 0 & -\omega_z & \omega_y \\ \omega_z & 0 & -\omega_x \\ -\omega_y & \omega_x & 0 \end{bmatrix} \begin{bmatrix} I_{xx} & -I_{xy} & -I_{xz} \\ -I_{yx} & I_{yy} & -I_{yz} \\ -I_{zx} & -I_{zy} & I_{zz} \end{bmatrix} \begin{bmatrix} \omega_x \\ \omega_y \\ \omega_z \end{bmatrix}$$

$$= \begin{bmatrix} 0.48 & -0.18 & 0 \\ -0.18 & 0.12 & 0 \\ 0 & 0 & 0.6 \end{bmatrix} \begin{bmatrix} -10 \\ 6 \\ 0 \end{bmatrix}$$

$$+ \begin{bmatrix} 0 & 0 & -2 \\ 0 & 0 & -4 \\ 2 & 4 & 0 \end{bmatrix} \begin{bmatrix} 0.48 & -0.18 & 0 \\ -0.18 & 0.12 & 0 \\ 0 & 0 & 0.6 \end{bmatrix} \begin{bmatrix} 4 \\ -2 \\ 0 \end{bmatrix}$$

$$= \begin{bmatrix} -5.88 \\ 2.52 \\ 0.72 \end{bmatrix} \text{N-m}.$$

We substitute this result into Eq. (1) to get

$$\Sigma \mathbf{M}_O = \mathbf{C} - 11.77\mathbf{i} + 5.89\mathbf{j} = -5.88\mathbf{i} + 2.52\mathbf{j} + 0.72\mathbf{k},$$

and solve for the couple **C**:

$$\mathbf{C} = 5.89\mathbf{i} - 3.37\mathbf{j} + 0.72\mathbf{k} \ (\text{N-m}).$$

Critical Thinking

Notice that we specified the angular velocity and angular acceleration of the plate and used the equations of angular motion to determine the couple exerted on the plate. In this chapter, we will discuss many examples in which an object's motion is specified and Euler's equations are used to determine the forces and couples acting on the object. The inverse problem, determining an object's three-dimensional motion when the forces and couples acting on it are known, is more difficult and must usually be solved by numerical methods.

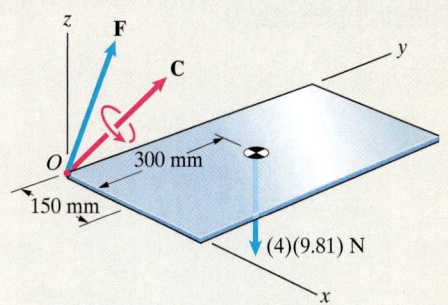

(a) Free-body diagram of the plate.

Example 20.5 | Three-Dimensional Dynamics of a Bar

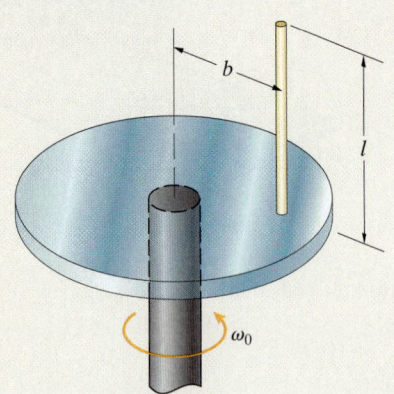

Figure 20.15

A slender vertical bar of mass m is rigidly attached to a horizontal disk rotating with constant angular velocity ω_0 (Fig. 20.15). What force and couple are exerted on the bar by the disk?

Strategy

The external forces and couples on the bar are its weight and the force and couple exerted on it by the disk. The angular velocity and angular acceleration of the bar are given, and we can determine the acceleration of its center of mass, so we can use the Euler equations to determine the total force and couple exerted on the bar.

Solution

Choose a Coordinate System In Fig. a, we place the origin of a body-fixed coordinate system at the center of mass of the bar with the y axis vertical and the x axis in the radial direction. With this orientation, we will obtain simple expressions for the bar's angular velocity vector and the acceleration of its center of mass.

Draw the Free-Body Diagram We draw the free-body diagram of the bar in Fig. a, showing the force $\mathbf{F}$ and couple $\mathbf{C}$ exerted by the disk.

Apply the Equations of Motion The acceleration of the center of mass of the bar due to its motion along its circular path is $\mathbf{a} = -\omega_0^2 b\,\mathbf{i}$. From Newton's second law,

$$\Sigma \mathbf{F} = \mathbf{F} - mg\mathbf{j} = m(-\omega_0^2 b\,\mathbf{i}),$$

we obtain the force exerted on the bar by the disk:

$$\mathbf{F} = -m\omega_0^2 b\,\mathbf{i} + mg\,\mathbf{j}.$$

The total moment about the center of mass is the sum of the couple $\mathbf{C}$ and the moment due to $\mathbf{F}$:

$$\Sigma \mathbf{M} = \mathbf{C} + \left(-\tfrac{1}{2}l\mathbf{j}\right) \times \left(-m\omega_0^2 b\mathbf{i} + mg\mathbf{j}\right)$$
$$= C_x\mathbf{i} + C_y\mathbf{j} + \left(C_z - \tfrac{1}{2}mlb\omega_0^2\right)\mathbf{k}.$$

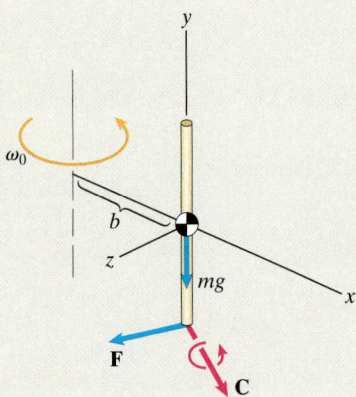

(a) Free-body diagram of the bar.

The bar's inertia matrix in terms of the coordinate system in Fig. a is

$$[I] = \begin{bmatrix} \tfrac{1}{12}ml^2 & 0 & 0 \\ 0 & 0 & 0 \\ 0 & 0 & \tfrac{1}{12}ml^2 \end{bmatrix},$$

and its angular velocity vector, $\boldsymbol{\omega} = \omega_0\mathbf{j}$, is constant. The equation of angular motion, from Eq. (20.19), is

$$\begin{bmatrix} C_x \\ C_y \\ C_z - \tfrac{1}{2}mlb\omega_0^2 \end{bmatrix} = \begin{bmatrix} 0 & 0 & \omega_0 \\ 0 & 0 & 0 \\ -\omega_0 & 0 & 0 \end{bmatrix} \begin{bmatrix} \tfrac{1}{12}ml^2 & 0 & 0 \\ 0 & 0 & 0 \\ 0 & 0 & \tfrac{1}{12}ml^2 \end{bmatrix} \begin{bmatrix} 0 \\ \omega_0 \\ 0 \end{bmatrix}.$$

The right side of this equation equals zero, so the components of the couple exerted on the bar by the disk are $C_x = 0$, $C_y = 0$, and $C_z = \tfrac{1}{2}mlb\omega_0^2$.

Alternative Solution   The bar rotates about a fixed axis, so we can also determine the couple **C** by using Eq. (20.13). Let the fixed point O be the center of the disk (Fig. b), and let the body-fixed coordinate system be oriented with the x axis through the bottom of the bar. Then the total moment about O is

$$\Sigma \mathbf{M}_O = \mathbf{C} + (b\mathbf{i}) \times (-m\omega_0^2 b\mathbf{i} + mg\mathbf{j}) + \left(b\mathbf{i} + \tfrac{1}{2}l\mathbf{j}\right) \times (-mg\mathbf{j})$$

$$= \mathbf{C}.$$

Thus, the only moment about O is the couple exerted by the disk. Applying the parallel-axis theorems (Eqs. 20.42), we get for the bar's moments and products of inertia (Fig. c),

$$I_{xx} = I_{x'x'} + (d_y^2 + d_z^2)\, m = \tfrac{1}{12}ml^2 + \left(\tfrac{1}{2}l\right)^2 m = \tfrac{1}{3}ml^2,$$

$$I_{yy} = I_{y'y'} + (d_x^2 + d_z^2)\, m = mb^2,$$

$$I_{zz} = I_{z'z'} + (d_x^2 + d_y^2)\, m = \tfrac{1}{12}ml^2 + \left[b^2 + \left(\tfrac{1}{2}l\right)^2\right] m = \tfrac{1}{3}ml^2 + mb^2,$$

$$I_{xy} = I_{x'y'} + d_x d_y m = 0 + (b)\left(\tfrac{1}{2}l\right) m = \tfrac{1}{2}mbl,$$

$$I_{yz} = I_{y'z'} + d_y d_z m = 0,$$

and

$$I_{zx} = I_{z'x'} + d_z d_x m = 0.$$

Substituting these results into Eq. (20.13), we obtain

$$\begin{bmatrix} C_x \\ C_y \\ C_z \end{bmatrix} = \begin{bmatrix} 0 & 0 & \omega_0 \\ 0 & 0 & 0 \\ -\omega_0 & 0 & 0 \end{bmatrix} \begin{bmatrix} \tfrac{1}{3}ml^2 & -\tfrac{1}{2}mbl & 0 \\ -\tfrac{1}{2}mbl & mb^2 & 0 \\ 0 & 0 & \tfrac{1}{3}ml^2 + mb^2 \end{bmatrix} \begin{bmatrix} 0 \\ \omega_0 \\ 0 \end{bmatrix}$$

$$= \begin{bmatrix} 0 \\ 0 \\ \tfrac{1}{2}mlb\omega_0^2 \end{bmatrix}.$$

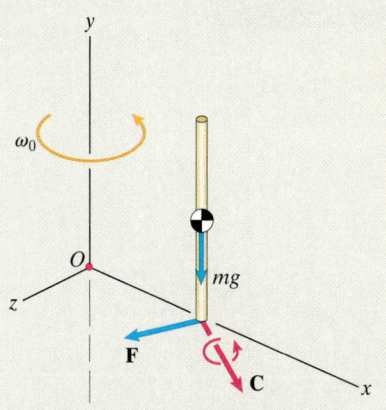

(b) Expressing the equation of angular motion in terms of the fixed point O

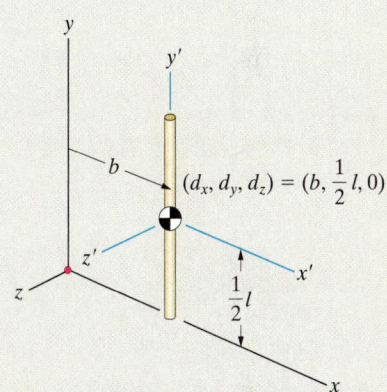

(c) Applying the parallel-axis theorem.

Critical Thinking

What is the physical interpretation of the force and couple exerted on the bar by the disk? The bar has no acceleration in the vertical direction, so the disk exerts a vertical force on the bar equal to the bar's weight. The bar has a normal component of acceleration due to its circular motion, which means that the disk must exert a force on it in the normal direction. Notice that if the bar were attached to the disk by a ball-and-socket support instead of a fixed support, it would rotate outward due to the disk's rotation. The couple exerted on the bar by the disk prevents it from rotating outward.

| **Example 20.6** | **Three-Dimensional Dynamics of a Tilted Cylinder** |

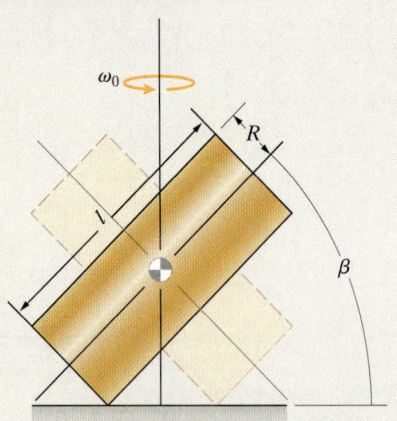

Figure 20.16

The tilted homogeneous cylinder in Fig. 20.16 undergoes a steady motion in which one end rolls on the floor while the center of mass of the cylinder remains stationary. The angle β between the cylinder axis and the horizontal remains constant, and the cylinder axis rotates about the vertical axis with constant angular velocity ω_0. The cylinder has mass m, radius R, and length l. What is ω_0?

Strategy

By expressing the equations of angular motion in terms of ω_0, we can determine the value of ω_0 necessary for the equations to be satisfied. Therefore, our first task is to determine the cylinder's angular velocity $\boldsymbol{\omega}$ in terms of ω_0. We can simplify this task by using a secondary coordinate system that is not body fixed.

Solution

Choose a Coordinate System We use a secondary coordinate system in which the z axis remains aligned with the cylinder axis and the y axis remains horizontal (Fig. a). The reason for this choice is that the angular velocity of the coordinate system is easy to describe—the coordinate system rotates about the vertical axis with the angular velocity ω_0—and the rotation of the cylinder relative to the coordinate system is also easy to describe. The angular velocity vector of the coordinate system is

$$\boldsymbol{\Omega} = \omega_0 \cos \beta \mathbf{i} + \omega_0 \sin \beta \mathbf{k}.$$

Relative to the coordinate system, the cylinder rotates about the z axis. Writing its angular velocity vector relative to the coordinate system as $\omega_{\text{rel}}\mathbf{k}$, we express the angular velocity vector of the cylinder as

$$\boldsymbol{\omega} = \boldsymbol{\Omega} + \omega_{\text{rel}}\mathbf{k} = \omega_0 \cos \beta \mathbf{i} + (\omega_0 \sin \beta + \omega_{\text{rel}})\mathbf{k}.$$

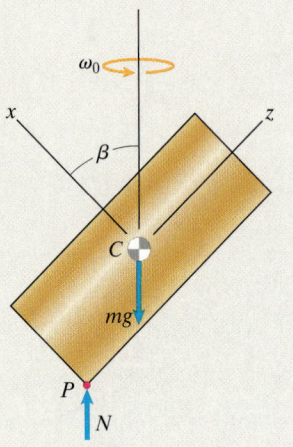

(a) Coordinate system with the z axis aligned with the cylinder axis and the y axis horizontal.

We can determine ω_{rel} from the condition that the velocity of the point P in contact with the floor is zero. Expressing the velocity of P in terms of the velocity of the center of mass C, we obtain

$$\mathbf{v}_P = \mathbf{v}_C + \boldsymbol{\omega} \times \mathbf{r}_{P/C}:$$

$$\mathbf{0} = \mathbf{0} + \left[\omega_0 \cos \beta \mathbf{i} + (\omega_0 \sin \beta + \omega_{\text{rel}})\mathbf{k}\right] \times \left[-R\mathbf{i} - \tfrac{1}{2}l\mathbf{k}\right]$$

$$= \left[\tfrac{1}{2}l\omega_0 \cos \beta - R(\omega_0 \sin \beta + \omega_{\text{rel}})\right]\mathbf{j}.$$

Solving for ω_{rel} yields

$$\omega_{\text{rel}} = \left[\frac{1}{2}\left(\frac{l}{R}\right) \cos \beta - \sin \beta\right]\omega_0.$$

Therefore, the cylinder's angular velocity vector is

$$\boldsymbol{\omega} = \omega_0 \cos \beta \mathbf{i} + \frac{1}{2}\left(\frac{l}{R}\right)\omega_0 \cos \beta \mathbf{k}.$$

Draw the Free-Body Diagram We draw the free-body diagram of the cylinder in Fig. a, showing the weight the of the cylinder and the normal force exerted by the floor. Because the cylinder's center of mass is stationary, the floor exerts no horizontal force on the cylinder, and the normal force is $N = mg$.

Apply the Equations of Motion The moment about the center of mass due to the normal force is

$$\Sigma\mathbf{M} = \left(mgR\sin\beta - \tfrac{1}{2}mgl\cos\beta\right)\mathbf{j}.$$

From Appendix C, the moments and products of inertia are

$$\begin{bmatrix} \tfrac{1}{4}mR^2 + \tfrac{1}{12}ml^2 & 0 & 0 \\ 0 & \tfrac{1}{4}mR^2 + \tfrac{1}{12}ml^2 & 0 \\ 0 & 0 & \tfrac{1}{2}mR^2 \end{bmatrix}.$$

Substituting our expressions for $\mathbf{\Omega}$, $\boldsymbol{\omega}$, $\Sigma\mathbf{M}$, and the moments and products of inertia into the equation of angular motion, Eq. (20.19), and evaluating the matrix products, we obtain the equation

$$mg\left(R\sin\beta - \tfrac{1}{2}l\cos\beta\right) = \left(\tfrac{1}{4}mR^2 + \tfrac{1}{12}ml^2\right)\omega_0^2\sin\beta\cos\beta$$

$$- \frac{1}{2}\left(\frac{1}{2}mR^2\right)\omega_0^2\left(\frac{l}{R}\right)\cos^2\beta.$$

We solve this equation for ω_0^2:

$$\omega_0^2 = \frac{g\left(R\sin\beta - \tfrac{1}{2}l\cos\beta\right)}{\left(\tfrac{1}{4}R^2 + \tfrac{1}{12}l^2\right)\sin\beta\cos\beta - \tfrac{1}{4}lR\cos^2\beta}. \qquad (1)$$

Critical Thinking

If our solution yields a negative value for ω_0^2 for a given value of β, the assumed steady motion of the cylinder is not possible. For example, if the cylinder's diameter is equal to its length ($2R = l$), we can write Eq. (1) as

$$\frac{R\omega_0^2}{g} = \frac{\sin\beta - \cos\beta}{\tfrac{7}{12}\sin\beta\cos\beta - \tfrac{1}{2}\cos^2\beta}.$$

In Fig. 20.17, we show the graph of this equation as a function of β. For values of β from approximately $40°$ to $45°$, there is no real solution for ω_0. Notice that at $\beta = 45°$, $\omega_0 = 0$, which means that the cylinder is stationary and balanced, with the center of mass directly above point P.

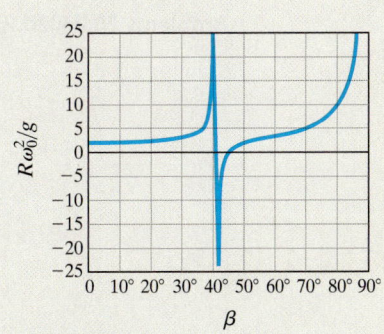

Figure 20.17
Graph of $R\omega_0^2/g$ as a function of β.

Problems

20.26 The airplane's inertia matrix in terms of the body-fixed coordinate system is shown. The airplane's angular velocity is zero when the pilot actuates the elevators and ailerons, subjecting the airplane to the couple $\Sigma \mathbf{M} = 6.4\mathbf{i} + 12.2\mathbf{j}$ (kN-m). What is the airplane's angular acceleration at that instant?

> *Strategy*: Use Eqs. (20.18) or Eq. (20.19). Because the coordinate system is body fixed, the terms $d\omega_x/dt$, $d\omega_y/dt$, and $d\omega_z/dt$ are the components of the angular acceleration.

20.27 The airplane's inertia matrix in terms of the body-fixed coordinate system is shown. At the present instant, the airplane's angular velocity and angular acceleration are $\boldsymbol{\omega} = 0.13\mathbf{i} + 0.24\mathbf{j} - 0.08\mathbf{k}$ (rad/s) and $\boldsymbol{\alpha} = 0.139\mathbf{i} + 0.108\mathbf{j}$ (rad/s^2). What is the total moment about the center of mass due to the forces and couples acting on the airplane?

20.28 A robotic manipulator moves a casting. The inertia matrix of the casting in terms of a body-fixed coordinate system with its origin at the center of mass is shown. At the present instant, the angular velocity and angular acceleration of the casting are $\boldsymbol{\omega} = 1.2\mathbf{i} + 0.8\mathbf{j} - 0.4\mathbf{k}$ (rad/s) and $\boldsymbol{\alpha} = 0.26\mathbf{i} - 0.07\mathbf{j} + 0.13\mathbf{k}$ (rad/s^2). What moment is exerted about the center of mass of the casting by the manipulator?

20.29 A robotic manipulator holds a casting. The inertia matrix of the casting in terms of a body-fixed coordinate system with its origin at the center of mass is shown. At the present instant, the casting is stationary. If the manipulator exerts a moment $\Sigma \mathbf{M} = 0.042\mathbf{i} + 0.036\mathbf{j} + 0.066\mathbf{k}$ (N-m) about the center of mass, what is the angular acceleration of the casting at that instant?

$$\begin{bmatrix} I_{xx} & -I_{xy} & -I_{xz} \\ -I_{yx} & I_{yy} & -I_{yz} \\ -I_{zx} & -I_{zy} & I_{zz} \end{bmatrix} = \begin{bmatrix} 46{,}000 & 0 & 0 \\ 0 & 113{,}000 & 0 \\ 0 & 0 & 134{,}000 \end{bmatrix} \text{kg-m}^2.$$

Problems 20.26/20.27

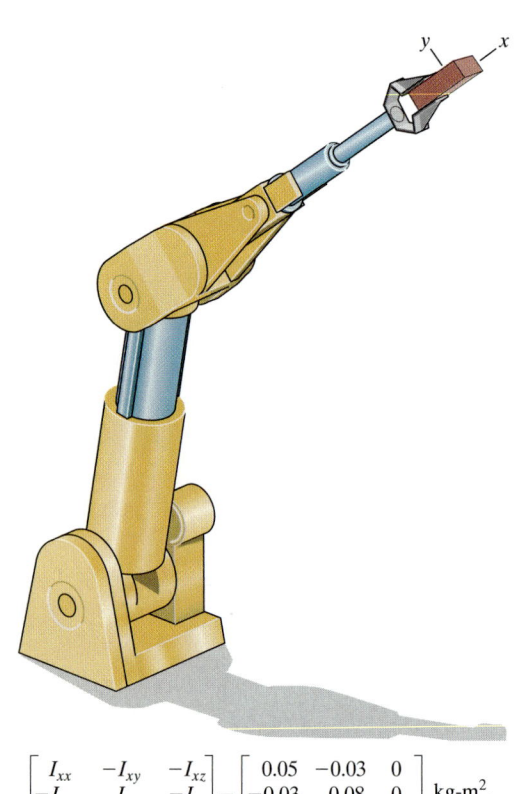

$$\begin{bmatrix} I_{xx} & -I_{xy} & -I_{xz} \\ -I_{yx} & I_{yy} & -I_{yz} \\ -I_{zx} & -I_{zy} & I_{zz} \end{bmatrix} = \begin{bmatrix} 0.05 & -0.03 & 0 \\ -0.03 & 0.08 & 0 \\ 0 & 0 & 0.04 \end{bmatrix} \text{kg-m}^2.$$

Problems 20.28/20.29

20.30 The rigid body rotates about the fixed point O. Its inertia matrix in terms of the body-fixed coordinate system is shown. At the present instant, the rigid body's angular velocity is $\boldsymbol{\omega} = 6\mathbf{i} + 6\mathbf{j} - 4\mathbf{k}$ (rad/s) and its angular acceleration is zero. What total moment about O is being exerted on the rigid body?

20.31 The rigid body rotates about the fixed point O. Its inertia matrix in terms of the body-fixed coordinate system is shown. At the present instant, the rigid body's angular velocity is $\boldsymbol{\omega} = 6\mathbf{i} + 6\mathbf{j} - 4\mathbf{k}$ (rad/s). The total moment about O due to the forces and couples acting on the rigid body is zero. What is its angular acceleration?

$$\begin{bmatrix} I_{xx} & -I_{xy} & -I_{xz} \\ -I_{yx} & I_{yy} & -I_{yz} \\ -I_{zx} & -I_{zy} & I_{zz} \end{bmatrix} = \begin{bmatrix} 4 & -2 & 0 \\ -2 & 3 & 1 \\ 0 & 1 & 5 \end{bmatrix} \text{ slug-ft}^2.$$

Problems 20.30/20.31

20.32 The dimensions of the 20-kg thin plate are $h = 0.4$ m and $b = 0.6$ m. The plate is stationary relative to an inertial reference frame when the force $F = 10$ N is applied in the direction perpendicular to the plate. No other forces or couples act on the plate. At the instant F is applied, what is the magnitude of the acceleration of point A relative to the inertial reference frame?

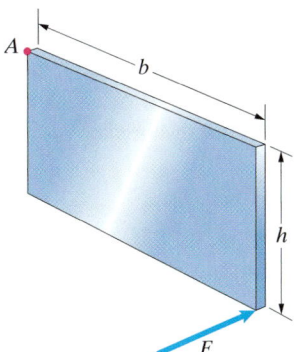

Problem 20.32

20.33 In terms of the coordinate system shown, the inertia matrix of the 6-kg slender bar is

$$\begin{bmatrix} I_{xx} & -I_{xy} & -I_{xz} \\ -I_{yx} & I_{yy} & -I_{yz} \\ -I_{zx} & -I_{zy} & I_{zz} \end{bmatrix} = \begin{bmatrix} 0.500 & 0.667 & 0 \\ 0.667 & 2.667 & 0 \\ 0 & 0 & 3.167 \end{bmatrix} \text{ kg-m}^2.$$

The bar is stationary relative to an inertial reference frame when the force $\mathbf{F} = 12\mathbf{k}$ (N) is applied at the right end of the bar. No other forces or couples act on the bar. Determine (a) the bar's angular acceleration relative to the inertial reference frame and (b) the acceleration of the right end of the bar relative to the inertial reference frame at the instant the force is applied.

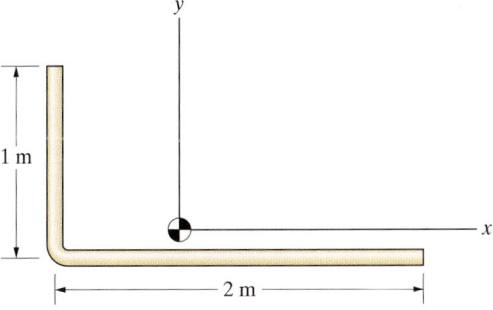

Problem 20.33

20.34 In terms of the coordinate system shown, the inertia matrix of the 12-kg slender bar is

$$\begin{bmatrix} I_{xx} & -I_{xy} & -I_{xz} \\ -I_{yx} & I_{yy} & -I_{yz} \\ -I_{zx} & -I_{zy} & I_{zz} \end{bmatrix} = \begin{bmatrix} 2 & -3 & 0 \\ -3 & 8 & 0 \\ 0 & 0 & 10 \end{bmatrix} \text{ kg-m}^2.$$

The bar is stationary relative to an inertial reference frame when a force $\mathbf{F} = 20\mathbf{i} + 40\mathbf{k}$ (N) is applied at the point $x = 1$ m, $y = 1$ m. No other forces or couples act on the bar. Determine (a) the bar's angular acceleration and (b) the acceleration of the point $x = -1$ m, $y = -1$ m, relative to the inertial reference frame, at the instant the force is applied.

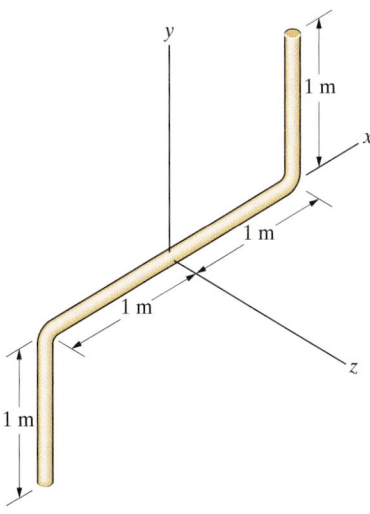

Problem 20.34

20.35 The inertia matrix of the 2.4-kg plate in terms of the given coordinate system is shown. The angular velocity vector of the plate is $\boldsymbol{\omega} = 6.4\mathbf{i} + 8.2\mathbf{j} + 14\mathbf{k}$ (rad/s), and its angular acceleration vector is $\boldsymbol{\alpha} = 60\mathbf{i} + 40\mathbf{j} - 120\mathbf{k}$ (rad/s^2). What are the components of the total moment exerted on the plate about its center of mass?

20.36 The inertia matrix of the 2.4-kg plate in terms of the given coordinate system is shown. At $t = 0$, the plate is stationary and is subjected to a force $\mathbf{F} = -10\mathbf{k}$ (N) at the point with coordinates (220, 0, 0) mm. No other forces or couples act on the plate. Determine (a) the acceleration of the plate's center of mass and (b) the plate's angular acceleration at the instant the force is applied.

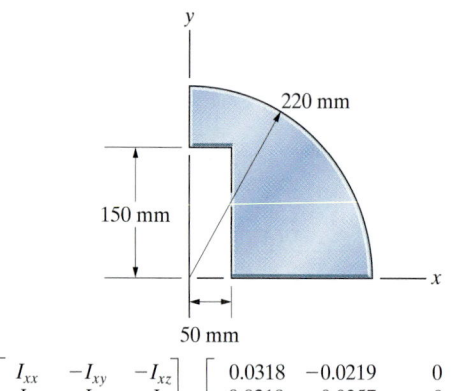

$$\begin{bmatrix} I_{xx} & -I_{xy} & -I_{xz} \\ -I_{yx} & I_{yy} & -I_{yz} \\ -I_{zx} & -I_{zy} & I_{zz} \end{bmatrix} = \begin{bmatrix} 0.0318 & -0.0219 & 0 \\ -0.0219 & 0.0357 & 0 \\ 0 & 0 & 0.0674 \end{bmatrix} \text{ kg-m}^2.$$

Problems 20.35/20.36

20.37 A 3-kg slender bar is rigidly attached to a 2-kg thin circular disk. In terms of the body-fixed coordinate system shown, the angular velocity vector of the composite object is $\boldsymbol{\omega} = 100\mathbf{i} - 4\mathbf{j} + 6\mathbf{k}$ (rad/s) and its angular acceleration is zero. What are the components of the total moment exerted on the object about its center of mass?

20.38 A 3-kg slender bar is rigidly attached to a 2-kg thin circular disk. At $t = 0$, the composite object is stationary and is subjected to the moment $\Sigma\mathbf{M} = -10\mathbf{i} + 10\mathbf{j}$ (N-m) about its center of mass. No other forces or couples act on the object. Determine the object's angular acceleration at $t = 0$.

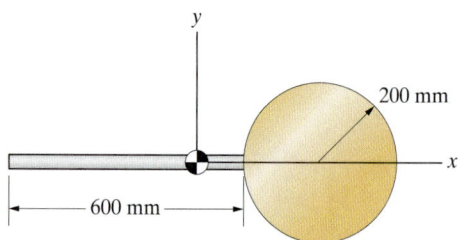

Problems 20.37/20.38

20.39 The vertical shaft supporting the dish antenna is rotating with a constant angular velocity of 1 rad/s. The angle $\theta = 30°$, $d\theta/dt = 20°/\text{s}$, and $d^2\theta/dt^2 = -40°/\text{s}^2$. The mass of the antenna is 280 kg, and its moments and products of inertia, in kg-m^2, are $I_{xx} = 140$, $I_{yy} = I_{zz} = 220$, $I_{xy} = I_{yz} = I_{zx} = 0$. Determine the couple exerted on the antenna by its support at A at the instant shown.

Problem 20.39

20.40 The 5-kg triangular plate is connected to a ball-and-socket support at O. If the plate is released from rest in the horizontal position, what are the components of its angular acceleration at that instant?

20.41 If the 5-kg plate is released from rest in the horizontal position, what force is exerted on it by the ball-and-socket support at that instant?

20.42 The 5-kg triangular plate is connected to a ball-and-socket support at O. If the plate is released in the horizontal position with angular velocity $\boldsymbol{\omega} = 4\mathbf{i}$ (rad/s), what are the components of its angular acceleration vector at that instant?

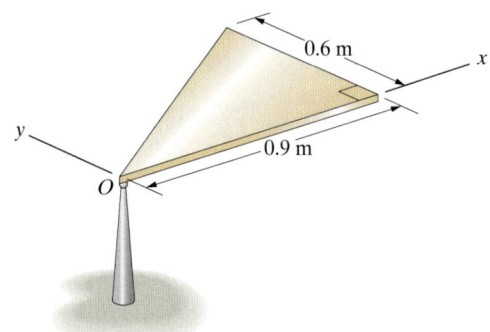

Problems 20.40–20.42

20.43 A subassembly of a space station can be modeled as two rigidly connected slender bars, each with a mass of 5000 kg. The subassembly is not rotating at $t = 0$, when a reaction control motor exerts a force $\mathbf{F} = 400\mathbf{k}$ (N) at B. What is the acceleration of point A relative to the center of mass of the subassembly at $t = 0$?

20.44 A subassembly of a space station can be modeled as two rigidly connected slender bars, each with a mass of 5000 kg. If the subassembly is rotating about the x axis at a constant rate of 1 revolution every 10 minutes, what is the magnitude of the couple its reaction control system is exerting on it?

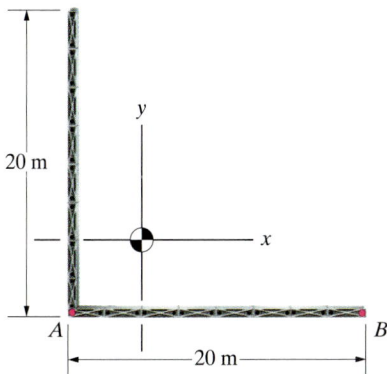

Problems 20.43/20.44

20.45 The thin circular disk of radius $R = 0.2$ m and mass $m = 4$ kg is rigidly attached to the vertical shaft. The plane of the disk is slanted at an angle $\beta = 30°$ relative to the horizontal. The shaft rotates with constant angular velocity $\omega_0 = 25$ rad/s. Determine the magnitude of the couple exerted on the disk by the shaft.

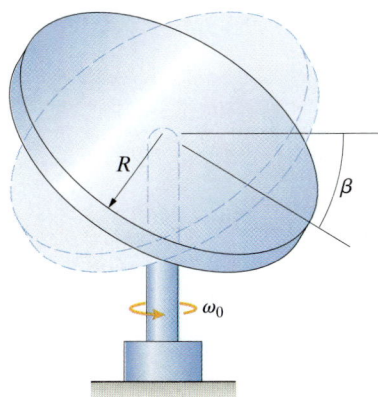

Problem 20.45

20.46 The slender bar of mass $m = 8$ kg and length $l = 1.2$ m is welded to a horizontal shaft that rotates with constant angular velocity $\omega_0 = 25$ rad/s. The angle $\beta = 30°$. Determine the magnitudes of the force $\mathbf{F}$ and couple $\mathbf{C}$ exerted on the bar by the shaft. (Write the equations of angular motion in terms of the body-fixed coordinate system shown.)

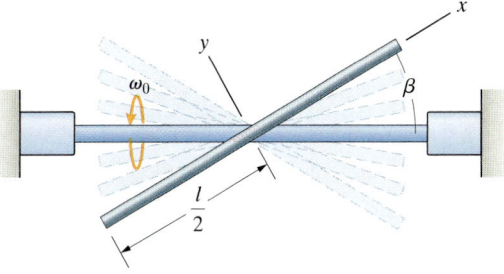

Problem 20.46

20.47 The slender bar of mass $m = 8$ kg and length $l = 1.2$ m is welded to a horizontal shaft that rotates with constant angular velocity $\omega_0 = 25$ rad/s. The angle $\beta = 30°$. Determine the magnitudes of the force $\mathbf{F}$ and couple $\mathbf{C}$ exerted on the bar by the shaft. (Write the equations of angular motion in terms of the body-fixed coordinate system shown. See Problem 20.98.)

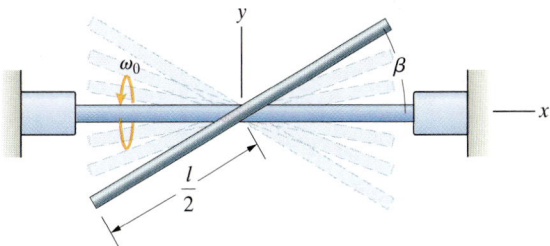

Problem 20.47

20.48 The slender bar of length l and mass m is pinned to the vertical shaft at O. The vertical shaft rotates with a constant angular velocity ω_0. Show that the value of ω_0 necessary for the bar to remain at a constant angle β relative to the vertical is

$$\omega_0 = \sqrt{3g/(2l \cos \beta)}.$$

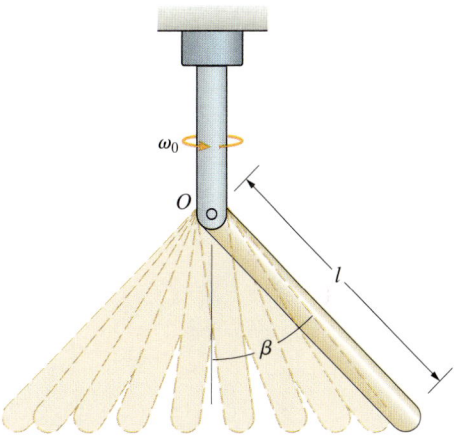

Problem 20.48

20.49 The vertical shaft rotates with constant angular velocity ω_0. The $35°$ angle between the edge of the 10-lb thin rectangular plate pinned to the shaft and the shaft remains constant. Determine ω_0.

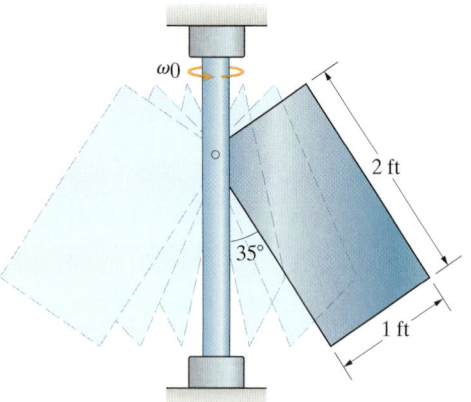

Problem 20.49

20.50 The radius of the 5-kg thin circular disk is $R = 0.2$ m. The disk is mounted on the horizontal shaft and rotates with constant angular velocity $\omega_d = 6$ rad/s relative to the shaft. The vertical shaft rotates with constant angular velocity $\omega_0 = 2$ rad/s. Determine the magnitude of the couple exerted on the disk by the horizontal shaft.

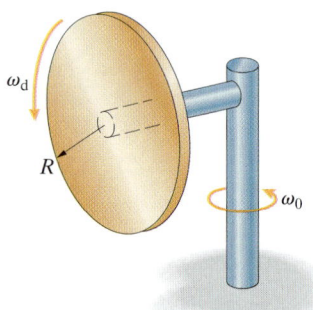

Problem 20.50

20.51 The thin triangular plate has ball-and-socket supports at A and B. The y axis is vertical. If the plate rotates with constant angular velocity ω_0, what are the horizontal components of the reactions on the plate at A and B?

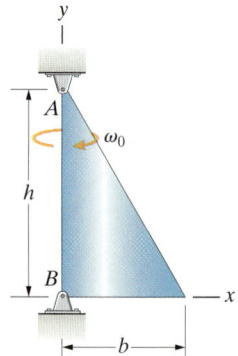

Problem 20.51

20.52 The 10-lb thin circular disk is rigidly attached to the 12-lb slender horizontal shaft. The disk and horizontal shaft rotate about the axis of the shaft with constant angular velocity $\omega_d = 20$ rad/s. The entire assembly rotates about the vertical axis with constant angular velocity $\omega_0 = 4$ rad/s. Determine the components of the force and couple exerted on the horizontal shaft by the disk.

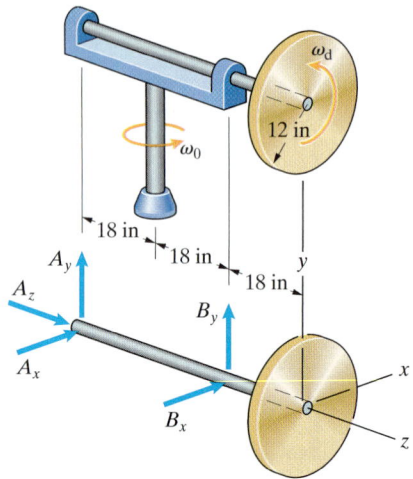

Problem 20.52

20.53 The Hubble telescope is rotating about its longitudinal axis with constant angular velocity ω_0. The coordinate system is fixed with respect to the solar panel. Relative to the telescope, the solar panel rotates about the x axis with constant angular velocity ω_x. Assume that the moments of inertia I_{xx}, I_{yy}, and I_{zz} are known, and $I_{xy} = I_{yz} = I_{zx} = 0$. Show that the moment about the x axis the servomechanisms must exert on the solar panel is

$$\Sigma M_x = (I_{zz} - I_{yy})\omega_0^2 \sin\theta \cos\theta.$$

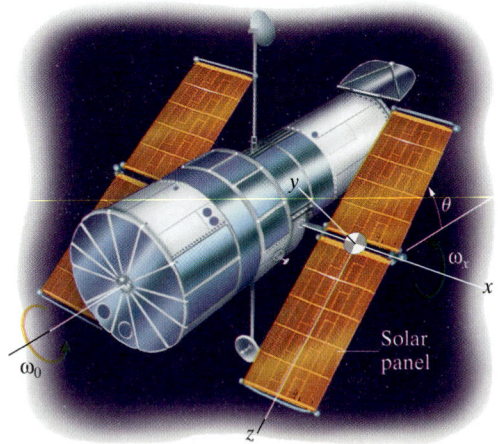

Problem 20.53

20.54 The thin rectangular plate is attached to the rectangular frame by pins. The frame rotates with constant angular velocity ω_0. Show that

$$\frac{d^2\beta}{dt^2} = -\omega_0^2 \sin\beta \cos\beta.$$

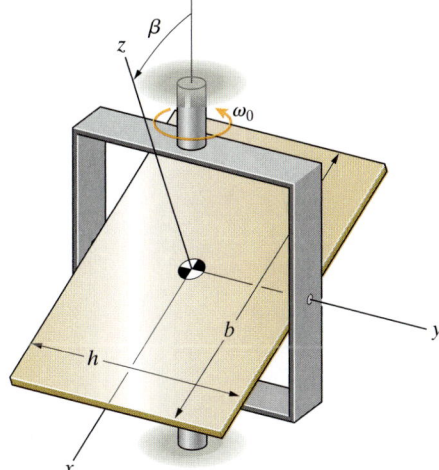

Problem 20.54

20.55* The axis of the right circular cone of mass m, height h, and radius R spins about the vertical axis with constant angular velocity ω_0. The center of mass of the cone is stationary, and its base rolls on the floor. Show that the angular velocity necessary for this motion is $\omega_0 = \sqrt{10g/3R}$.

Strategy: Let the z axis remain aligned with the axis of the cone and the x axis remain vertical.

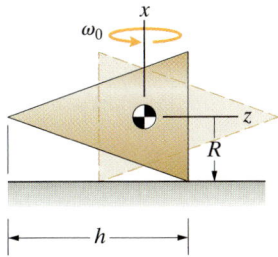

Problem 20.55

20.56* A thin circular disk of radius R and mass m rolls along a circular path of radius r. The magnitude v of the velocity of the center of the disk and the angle θ between the disk's axis and the vertical are constants. Show that v satisfies the equation

$$v^2 = \frac{\frac{2}{3}g \cot\theta(r - R\cos\theta)^2}{r - \frac{5}{6}R\cos\theta}.$$

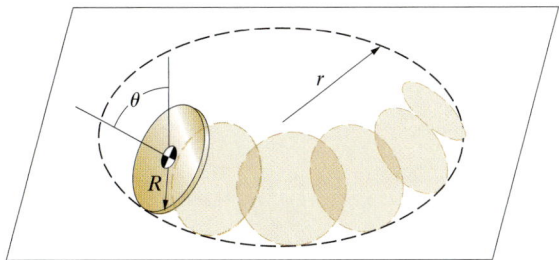

Problem 20.56

20.57* The two thin disks are rigidly connected by a slender bar. The radius of the large disk is 200 mm and its mass is 4 kg. The radius of the small disk is 100 mm and its mass is 1 kg. The bar is 400 mm in length and its mass is negligible. The composite object undergoes a steady motion in which it spins about the vertical y axis through its center of mass with angular velocity ω_0. The bar is horizontal during this motion and the large disk rolls on the floor. What is ω_0?

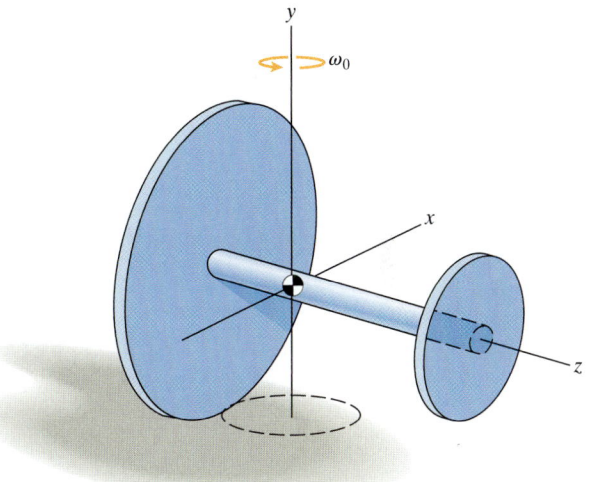

Problem 20.57

20.58 The view of an airplane's landing gear as seen looking from behind the airplane is shown in Fig. a. The radius of the wheel is 300 mm, and its moment of inertia is 2 kg-m². The airplane takes off at 30 m/s. After takeoff, the landing gear retracts by rotating toward the right side of the airplane, as shown in Fig. b. Determine the magnitude of the couple exerted by the wheel on its support. (Neglect the airplane's angular motion.)

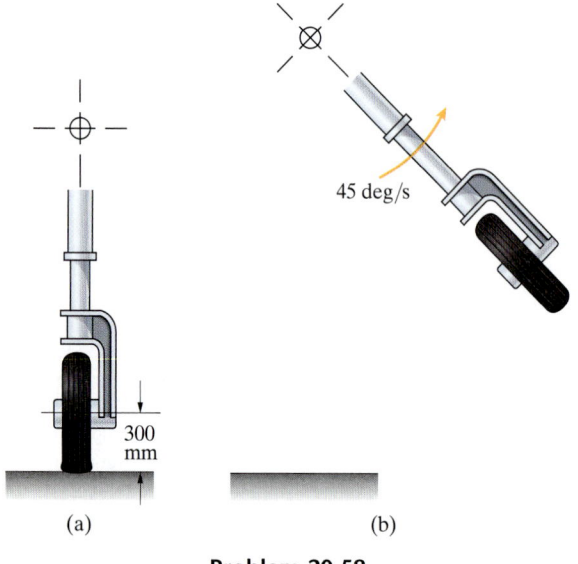

45 deg/s

(a)   (b)

Problem 20.58

20.59 If the rider turns to his left, will the couple exerted on the motorcycle by its wheels tend to cause the motorcycle to lean toward the rider's left side or his right side?

Problem 20.59

20.60* By substituting the components of $\mathbf{H}_O$ from Eqs. (20.9) into the equation

$$\Sigma \mathbf{M}_O = \frac{dH_{Ox}}{dt}\mathbf{i} + \frac{dH_{Oy}}{dt}\mathbf{j} + \frac{dH_{Oz}}{dt}\mathbf{k} + \mathbf{\Omega} \times \mathbf{H}_O,$$

derive Eqs. (20.12).

20.3 The Euler Angles

The equations of angular motion relate the total moment acting on a rigid body to its angular velocity and acceleration. If we know the total moment and the angular velocity, we can determine the angular acceleration. But how can we use the angular acceleration to determine the rigid body's angular position, or orientation, as a function of time? To explain how this is done, we must first show how to specify the orientation of a rigid body in three dimensions.

We have seen that describing the orientation of a rigid body in planar motion requires only the angle θ that specifies the body's rotation relative to some reference orientation. In three-dimensional motion, three angles are required. To understand why, consider a particular axis that is fixed relative to a rigid body. Two angles are necessary to specify the direction of the axis, and a third angle is needed to specify the rigid body's orientation about the axis. Although several systems of angles for describing the orientation of a rigid body are commonly used, the best-known system is the one called the Euler angles. In this section we define these angles and express the equations of angular motion in terms of them.

Objects with an Axis of Symmetry

We first explain how the Euler angles are used to describe the orientation of an object with an axis of rotational symmetry, because this case results in simpler equations of angular motion.

Definitions We assume that an object has an axis of rotational symmetry, and we introduce two reference frames: a secondary coordinate system xyz, with its z axis coincident with the object's axis of symmetry, and an inertial primary coordinate system XYZ. We begin with the object in a reference position in which xyz and XYZ are superimposed on each other (Fig. 20.18a).

Our first step is to rotate the object and the xyz system together through an angle ψ about the Z axis (Fig. 20.18b). In this intermediate orientation, we denote the secondary coordinate system by $x'y'z'$. Next, we rotate the object and the xyz system together through an angle θ about the x' axis (Fig. 20.18c). Finally, we rotate the object relative to the xyz system through an angle ϕ about the object's axis of symmetry (Fig. 20.18d). Notice that the x axis remains in the XY plane.

The angles ψ and θ specify the orientation of the secondary xyz system relative to the primary XYZ system. The angle ψ is called the *precession angle*, and θ is called the *nutation angle*. The angle ϕ specifying the rotation of the rigid body relative to the xyz system is called the *spin angle*. These three angles specify the orientation of the rigid body relative to the primary coordinate system and are called the *Euler angles*. We can obtain any orientation of the object relative to the primary coordinate system by appropriate choices of the Euler angles: We choose ψ and θ to obtain the desired direction of the axis of symmetry and then choose ϕ to obtain the desired rotational position of the object about its axis of symmetry.

Equations of Angular Motion To analyze an object's motion in terms of the Euler angles, we must express the equations of angular motion in terms of those angles. Figure 20.19a shows the rotation ψ from the reference orientation of the xyz system to its intermediate orientation $x'y'z'$. We represent the angular velocity of the coordinate system due to the rate of change of ψ by the angular velocity vector $\dot{\psi}$ pointing in the z' direction. (We use a dot to denote the derivative with respect to time.) Figure 20.19b shows

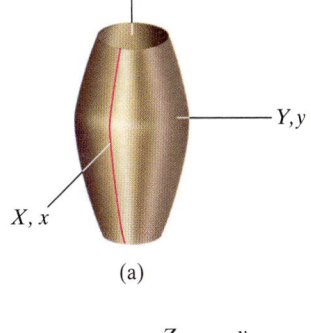

(a)

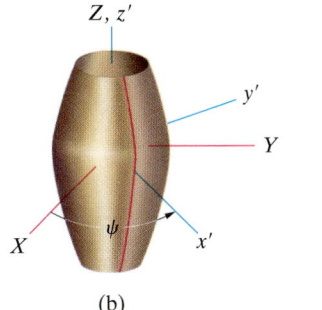

(b)

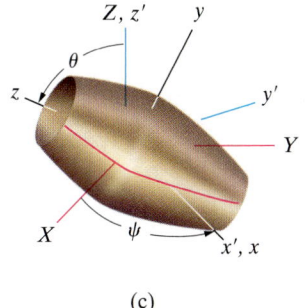

(c)

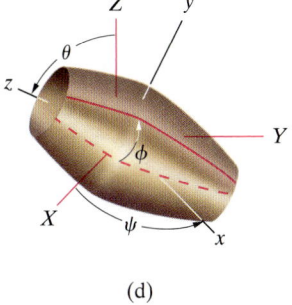

(d)

Figure 20.18
(a) The reference position.
(b) The rotation ψ about the Z axis.
(c) The rotation θ about the x' axis.
(d) The rotation ϕ of the object relative to the xyz system.

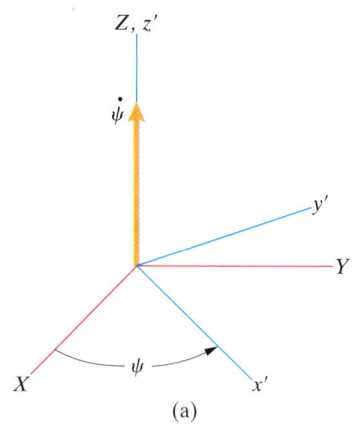

(a)

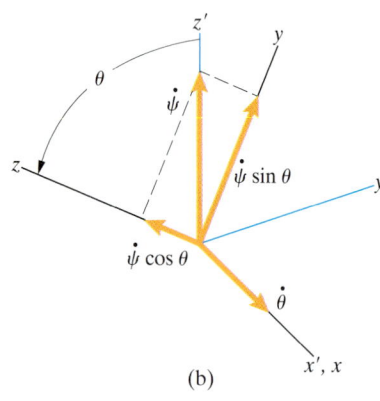

(b)

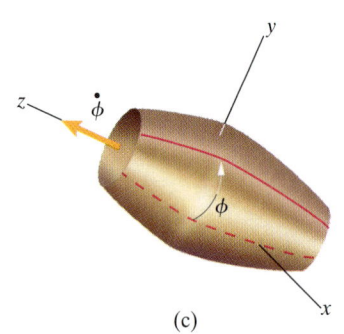

(c)

Figure 20.19
(a) The rotation ψ and the angular velocity $\dot\psi$.
(b) The rotation θ, the angular velocity $\dot\theta$, and the components of the angular velocity $\dot\psi$.
(c) The rotation ϕ and the angular velocity $\dot\phi$.

the second rotation θ. We represent the angular velocity due to the rate of change of θ by the vector $\dot\theta$ pointing in the x direction. We also resolve the angular velocity vector $\dot\psi$ into components in the y and z directions. The components of the angular velocity of the xyz system relative to the primary coordinate system are

$$\Omega_x = \dot\theta,$$

$$\Omega_y = \dot\psi \sin\theta,$$ (20.22)

and

$$\Omega_z = \dot\psi \cos\theta.$$

In Fig. 20.19c, we represent the angular velocity of the rigid body relative to the xyz system by the vector $\dot\phi$. Adding this angular velocity to the angular velocity of the xyz system, we obtain the components of the angular velocity of the rigid body relative to the XYZ system:

$$\omega_x = \dot\theta,$$

$$\omega_y = \dot\psi \sin\theta,$$ (20.23)

and

$$\omega_z = \dot\phi + \dot\psi \cos\theta.$$

Taking the derivatives of these equations with respect to time yields

$$\frac{d\omega_x}{dt} = \ddot\theta,$$

$$\frac{d\omega_y}{dt} = \ddot\psi \sin\theta + \dot\psi\dot\theta \cos\theta,$$ (20.24)

and

$$\frac{d\omega_z}{dt} = \ddot\phi + \ddot\psi \cos\theta - \dot\psi\dot\theta \sin\theta.$$

As a consequence of the object's rotational symmetry, the products of inertia I_{xy}, I_{xz}, and I_{yz} are zero and $I_{xx} = I_{yy}$. The inertia matrix is of the form

$$[I] = \begin{bmatrix} I_{xx} & 0 & 0 \\ 0 & I_{xx} & 0 \\ 0 & 0 & I_{zz} \end{bmatrix}.$$ (20.25)

Substituting Eqs. (20.22)–(20.25) into Eqs. (20.18), we obtain the equations of angular motion in terms of the Euler angles:

$$\Sigma M_x = I_{xx}\ddot\theta + (I_{zz} - I_{xx})\dot\psi^2 \sin\theta \cos\theta + I_{zz}\dot\phi\dot\psi \sin\theta,$$ (20.26)

$$\Sigma M_y = I_{xx}(\ddot\psi \sin\theta + 2\dot\psi\dot\theta \cos\theta) - I_{zz}(\dot\phi\dot\theta + \dot\psi\dot\theta \cos\theta),$$ (20.27)

$$\Sigma M_z = I_{zz}(\ddot\phi + \ddot\psi \cos\theta - \dot\psi\dot\theta \sin\theta).$$ (20.28)

To determine the Euler angles as functions of time when the total moment is known, these equations usually must be solved by numerical integration. However, we can obtain an important class of closed-form solutions by assuming a specific type of motion.

Steady Precession The motion called *steady precession* is commonly observed in tops and gyroscopes. The object's rate of spin $\dot{\phi}$ relative to the xyz coordinate system is assumed to be constant (Fig. 20.20). The nutation angle θ, the inclination of the *spin axis* z relative to the Z axis, is assumed to be constant, and the *precession rate* $\dot{\psi}$, the rate at which the xyz system rotates about the Z axis, is assumed to be constant. The last assumption explains the name given to this motion.

With these assumptions, Eqs. (20.26)–(20.28) reduce to

$$\Sigma M_x = (I_{zz} - I_{xx})\dot{\psi}^2 \sin\theta \cos\theta + I_{zz}\dot{\phi}\dot{\psi}\sin\theta, \qquad (20.29)$$

$$\Sigma M_y = 0, \qquad (20.30)$$

and

$$\Sigma M_z = 0. \qquad (20.31)$$

We discuss two examples: the steady precession of a spinning top and the steady precession of an axially symmetric object that is free of external moments.

Precession of a Top The peculiar behavior of a top (Fig. 20.21a) inspired some of the first studies of three-dimensional motions of rigid bodies. When a top is set into motion, its spin axis may initially remain vertical, a motion called *sleeping*. As friction reduces the spin rate, the spin axis begins to lean over and rotate about the vertical axis. This phase of the top's motion approximates steady precession. (The top's spin rate continuously decreases due to friction, whereas in steady precession we assume the spin rate to be constant.)

To analyze the motion, we place the primary coordinate system XYZ with its origin at the point of the top and the Z axis upward. Then we align the z axis of the xyz system with the spin axis (Fig. 20.21b). We assume that the top's point rests in a small depression so that it remains at a fixed point on the floor. The precession angle ψ and nutation angle θ specify the orientation of the spin axis, and the spin rate of the top relative to the xyz system is $\dot{\phi}$.

The top's weight exerts a moment $\Sigma M_x = mgh \sin\theta$ about the origin, and the moments $\Sigma M_y = 0$ and $\Sigma M_z = 0$. Substituting $\Sigma M_x = mgh \sin\theta$ into Eq. (20.29), we obtain

$$mgh = (I_{zz} - I_{xx})\dot{\psi}^2 \cos\theta + I_{zz}\dot{\phi}\dot{\psi}, \qquad (20.32)$$

and Eqs. (20.30) and (20.31) are identically satisfied. Equation (20.32) relates the spin rate, nutation angle, and rate of precession. For example, if we know the spin rate $\dot{\phi}$ and nutation angle θ, we can solve for the top's precession rate $\dot{\psi}$.

Moment-Free Steady Precession A spinning axisymmetric object that is free of external moments, such as an axisymmetric satellite in orbit, can exhibit a motion similar to the steady precessional motion of a top. This motion is observed when an American football is thrown in a "wobbly" spiral. To analyze it, we place the origin of the xyz system at the object's center of mass (Fig. 20.22a). Equation (20.29) then becomes

$$(I_{zz} - I_{xx})\dot{\psi} \cos\theta + I_{zz}\dot{\phi} = 0, \qquad (20.33)$$

and Eqs. (20.30) and (20.31) are identically satisfied. For a given value of the nutation angle, Eq. (20.33) relates the object's rates of precession and spin.

We can interpret Eq. (20.33) in a way that makes it possible to visualize the motion. We look for a point in the y–z plane at which the object's velocity relative to the center of mass is zero at the current instant. We want to find a point

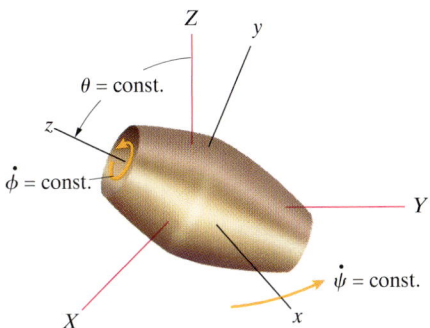

Figure 20.20
Steady procession.

(a)

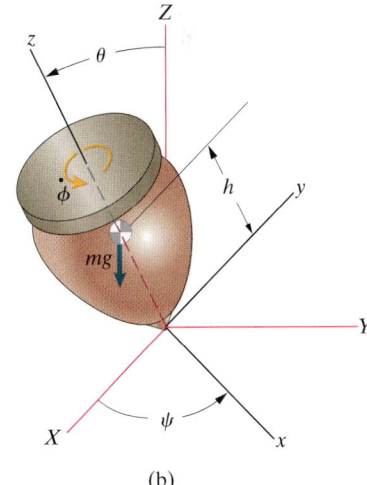

(b)

Figure 20.21
(a) A spinning top seems to defy gravity.
(b) The precession angle ψ and nutation angle θ specify the orientation of the spin axis.

with coordinates $(0, y, z)$ such that

$$\boldsymbol{\omega} \times (y\mathbf{j} + z\mathbf{k}) = [(\dot{\psi} \sin \theta)\mathbf{j} + (\dot{\phi} + \dot{\psi} \cos \theta)\mathbf{k}] \times [y\mathbf{j} + z\mathbf{k}]$$
$$= [z\dot{\psi} \sin \theta - y(\dot{\phi} + \dot{\psi} \cos \theta)]\mathbf{i} = 0.$$

This equation is satisfied at points in the y–z plane such that

$$\frac{y}{z} = \frac{\dot{\psi} \sin \theta}{\dot{\phi} + \dot{\psi} \cos \theta}.$$

The preceding relation is satisfied by points on the straight line at an angle β relative to the z axis in Fig. 20.22b, where

$$\tan \beta = \frac{y}{z} = \frac{\dot{\psi} \sin \theta}{\dot{\phi} + \dot{\psi} \cos \theta}.$$

Solving Eq. (20.33) for $\dot{\phi}$ and substituting the result into this equation, we obtain

$$\tan \beta = \left(\frac{I_{zz}}{I_{xx}}\right) \tan \theta.$$

If $I_{xx} > I_{zz}$, the angle $\beta < \theta$. In Fig. 20.22c, we show an imaginary cone of half-angle β, called the *body cone*, whose axis is coincident with the z axis. The body cone is in contact with a fixed cone, called the *space cone*, whose axis is coincident with the Z axis. If the body cone rolls on the curved surface of the space cone as the z axis precesses about the Z axis (Fig. 20.22d), the points of the body cone lying on the straight line in Fig. 20.22b have zero velocity relative to the XYZ system. That means that *the motion of the body cone*

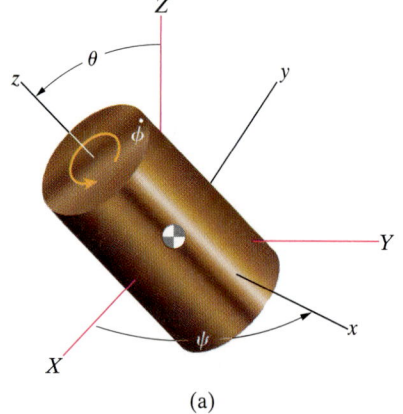

(a)

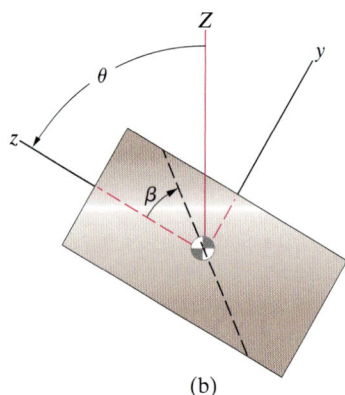

(b)

Figure 20.22
(a) An axisymmetric object.
(b) Points on the straight line at an angle β from the z axis are stationary relative to the XYZ coordinate system.
(c), (d) The body and space cones. The body cone rolls on the stationary space cone.
(e) When $\beta > \theta$, the interior surface of the body cone rolls on the stationary space cone.

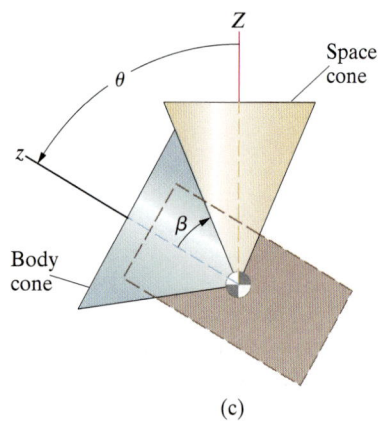

(c)

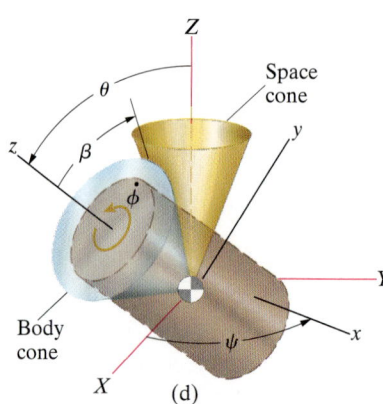

(d)

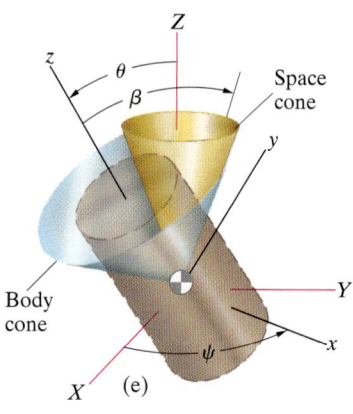

(e)

is identical to the motion of the object. The object's motion can be visualized by visualizing the motion of the body cone as it rolls around the outer surface of the space cone. This motion is called *direct precession.*

If $I_{xx} < I_{zz}$, the angle $\beta > \theta$. In this case, we must visualize the *interior* surface of the body cone rolling on the fixed space cone (Fig. 20.22e). This motion is called *retrograde precession.*

Arbitrary Objects

In our analysis of axially symmetric objects, we let the object move relative to the secondary xyz coordinate system, rotating about the z axis. As a consequence, only two angles—the precession angle ψ and nutation angle θ—are needed to specify the orientation of the xyz coordinate system, and this simplifies the equations of angular motion. The object must be axially symmetric about the z axis, so that the moments and products of inertia will not vary as it rotates. In the case of an arbitrary object, the moments and products of inertia will be constants only if the xyz coordinate system is body fixed. This means that three angles are needed to specify the orientation of the coordinate system, and the resulting equations of angular motion are more complicated.

Definitions   We begin with a reference position in which the body-fixed xyz and primary XYZ coordinate systems are superimposed on each other (Fig. 20.23a). First, we rotate the xyz system through the precession angle ψ about the Z axis (Fig. 20.23b) and denote it by $x'y'z'$ in this intermediate orientation. Then we rotate the xyz system through the nutation angle θ about the x' axis (Fig. 20.23c), denoting it now by $x''y''z''$. We obtain the final orientation of the xyz system by rotating it through the angle ϕ about the z'' axis (Fig. 20.23d). Notice that we use one more rotation of the xyz system than in the case of an axially symmetric object.

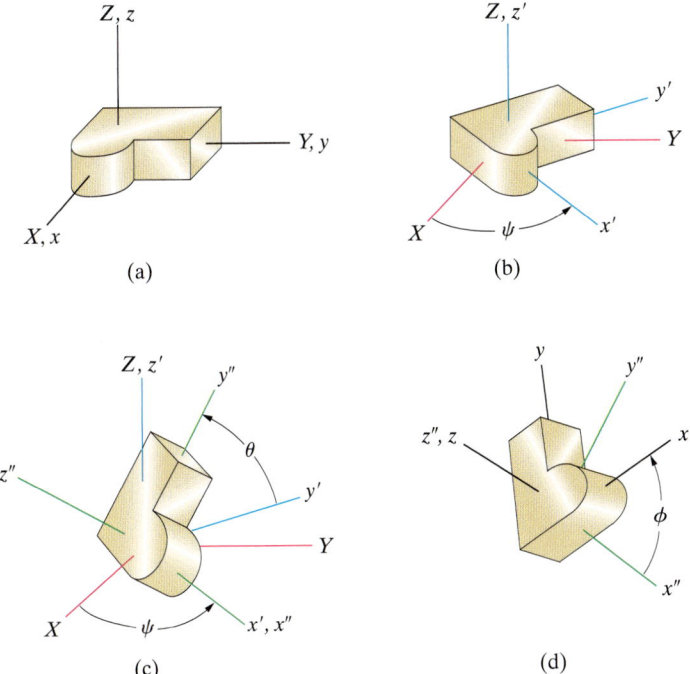

Figure 20.23
(a) The reference position.
(b) The rotation ψ about the Z axis.
(c) The rotation θ about the x' axis.
(d) The rotation ϕ about the z'' axis.

We can obtain any orientation of the body-fixed coordinate system relative to the reference coordinate system by these three rotations. We choose ψ and θ to obtain the desired direction of the z axis and then choose ϕ to obtain the desired orientation of the x and y axes.

Just as in the case of an object with rotational symmetry, we must express the components of the rigid body's angular velocity in terms of the Euler angles to obtain the equations of angular motion. Figure 20.24a shows the rotation ψ from the reference orientation of the xyz system to the intermediate orientation $x'y'z'$. We represent the angular velocity of the body-fixed coordinate system due to the rate of change of ψ by the vector $\dot{\psi}$ pointing in the z' direction. Figure 20.24b shows the next rotation θ that takes the body-fixed coordinate system to the intermediate orientation $x''y''z''$. We represent the angular velocity due to the rate of change of θ by the vector $\dot{\theta}$ pointing in the x'' direction. In this figure, we also show the components of the angular velocity vector $\dot{\psi}$ in the y'' and z'' directions. Figure 20.24c shows the third rotation ϕ that takes the body-fixed coordinate system to its final orientation defined by the three Euler angles. We represent the angular velocity due to the rate of change of ϕ by the vector $\dot{\phi}$ pointing in the z direction.

To determine ω_x, ω_y, and ω_z in terms of the Euler angles, we need to determine the components of the angular velocities shown in Fig. 20.24c in the x-, y-, and z-axis directions. The vectors $\dot{\phi}$ and $\dot{\psi}\cos\theta$ point in the z-axis direction. In Figs. 20.24d and e, which are drawn with the z axis pointing out of the page, we determine the components of the vectors $\dot{\psi}\sin\theta$ and $\dot{\theta}$ in the x- and y-axis directions.

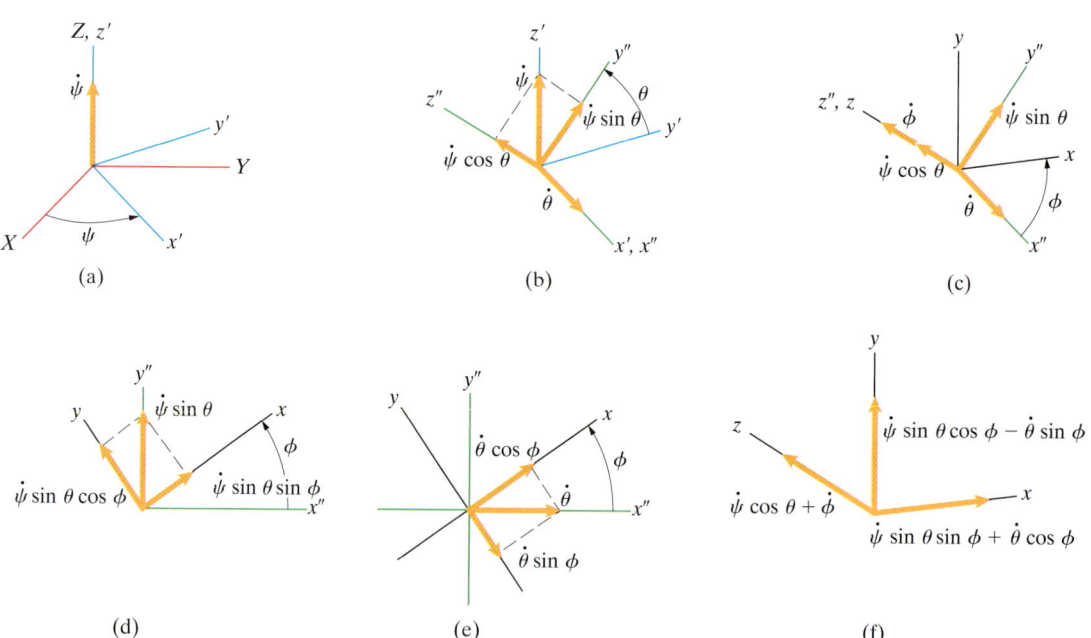

Figure 20.24
(a) The rotation ψ and the angular velocity $\dot{\psi}$.
(b) The rotation θ, the angular velocity $\dot{\theta}$, and the components of $\dot{\psi}$ in the $x''y''z''$ system.
(c) The rotation ϕ and the angular velocity $\dot{\phi}$.
(d), (e) The components of the angular velocities $\dot{\psi}\sin\theta$ and $\dot{\theta}$ in the xyz system.
(f) The angular velocities ω_x, ω_y, ω_z.

By summing the components of the angular velocities in the three coordinate directions (Fig. 20.24f), we obtain

$$\omega_x = \dot{\psi} \sin\theta \sin\phi + \dot{\theta} \cos\phi,$$

$$\omega_y = \dot{\psi} \sin\theta \cos\phi - \dot{\theta} \sin\phi, \qquad (20.34)$$

$$\omega_z = \dot{\psi} \cos\theta + \dot{\phi}.$$

The derivatives of these equations with respect to time are

$$\frac{d\omega_x}{dt} = \ddot{\psi} \sin\theta \sin\phi + \dot{\psi}\dot{\theta} \cos\theta \sin\phi + \dot{\psi}\dot{\phi} \sin\theta \cos\phi$$

$$+ \ddot{\theta} \cos\phi - \dot{\theta}\dot{\phi} \sin\phi,$$

$$\frac{d\omega_y}{dt} = \ddot{\psi} \sin\theta \cos\phi + \dot{\psi}\dot{\theta} \cos\theta \cos\phi - \dot{\psi}\dot{\phi} \sin\theta \sin\phi$$

$$- \ddot{\theta} \sin\phi - \dot{\theta}\dot{\phi} \cos\phi), \qquad (20.35)$$

$$\frac{d\omega_z}{dt} = \ddot{\psi} \cos\theta - \dot{\psi}\dot{\theta} \sin\theta + \ddot{\phi}.$$

Equations of Angular Motion With Eqs. (20.34) and (20.35), we can express the equations of angular motion in terms of the three Euler angles. To simplify the equations, *we assume that the body-fixed coordinate system xyz is a set of principal axes.* (See the appendix to this chapter.) Then the equations of angular motion, Eqs. (20.18), become

$$\Sigma M_x = I_{xx}\frac{d\omega_x}{dt} - (I_{yy} - I_{zz})\omega_y\omega_z,$$

$$\Sigma M_y = I_{yy}\frac{d\omega_y}{dt} - (I_{zz} - I_{xx})\omega_z\omega_x,$$

$$\Sigma M_z = I_{zz}\frac{d\omega_z}{dt} - (I_{xx} - I_{yy})\omega_x\omega_y.$$

Substituting Eqs. (20.34) and (20.35) into these relations, we obtain the equations of angular motion in terms of Euler angles:

$$\Sigma M_x = I_{xx}\ddot{\psi} \sin\theta \sin\phi + I_{xx}\ddot{\theta} \cos\phi$$

$$+ I_{xx}(\dot{\psi}\dot{\theta} \cos\theta \sin\phi + \dot{\psi}\dot{\phi} \sin\theta \cos\phi - \dot{\theta}\dot{\phi} \sin\phi)$$

$$- (I_{yy} - I_{zz})(\dot{\psi} \sin\theta \cos\phi - \dot{\theta} \sin\phi)(\dot{\psi} \cos\theta + \dot{\phi}),$$

$$\Sigma M_y = I_{yy}\ddot{\psi} \sin\theta \cos\phi - I_{yy}\ddot{\theta} \sin\phi \qquad (20.36)$$

$$+ I_{yy}(\dot{\psi}\dot{\theta} \cos\theta \cos\phi - \dot{\psi}\dot{\phi} \sin\theta \sin\phi - \dot{\theta}\dot{\phi} \cos\phi)$$

$$- (I_{zz} - I_{xx})(\dot{\psi} \cos\theta + \dot{\phi})(\dot{\psi} \sin\theta \sin\phi + \dot{\theta} \cos\phi),$$

$$\Sigma M_z = I_{zz}\ddot{\psi} \cos\theta + I_{zz}\ddot{\phi} - I_{zz}\dot{\psi}\dot{\theta} \sin\theta$$

$$- (I_{xx} - I_{yy})(\dot{\psi} \sin\theta \sin\phi + \dot{\theta} \cos\phi)(\dot{\psi} \sin\theta \cos\phi - \dot{\theta} \sin\phi).$$

If the Euler angles and their first and second derivatives with respect to time are known, Eqs. (20.36) can be solved for the components of the total moment. Or if the total moment, the Euler angles, and the first derivatives of the Euler angles are known, Eqs. (20.36) can be solved for the second derivatives of the Euler angles. In this way, the Euler angles can be determined as functions of time when the total moment is known as a function of time, but numerical integration is usually necessary.

Study Questions

1. How are the Euler angles defined?
2. What is steady precession?
3. How can the body and space cones be used to visualize moment-free steady precession?
4. If you know the Euler angles of a rigid body and their first derivatives with respect to time, how can you determine the rigid body's angular velocity vector?

Example 20.7 | Steady Precession of a Disk

The thin circular disk of radius R and mass m in Fig. 20.25 rolls along a horizontal circular path of radius r. The angle θ between the disk's axis and the vertical remains constant. Determine the magnitude v of the velocity of the center of the disk as a function of the angle θ.

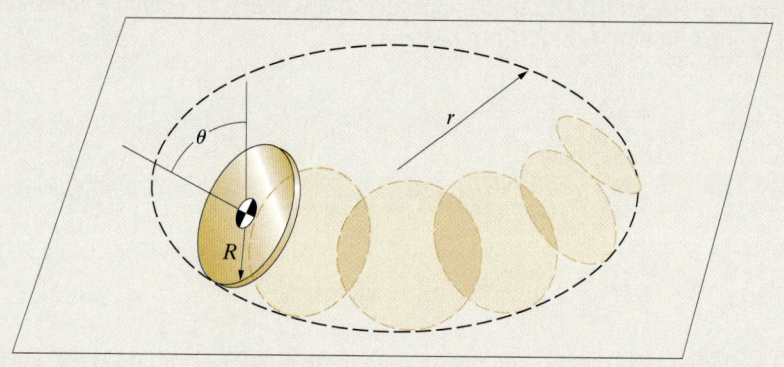

Figure 20.25

Strategy

We can obtain the velocity of the center of the disk by assuming that the disk is in steady precession and determining the conditions necessary for the equations of motion to be satisfied.

Solution

In Fig. a, we align the z axis of the secondary coordinate system with the disk's spin axis and assume that the x axis remains parallel to the surface on which the disk rolls. The angle θ is the nutation angle. The center of mass moves in a circular path of radius $r_G = r - R \cos \theta$. Therefore, the precession rate—the rate at which the x axis rotates in the horizontal plane—is

$$\dot{\psi} = \frac{v}{r_G}.$$

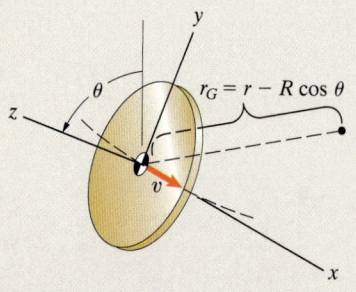

(a) Aligning the z axis with the spin axis. The x axis is horizontal.

From Eqs. (20.23), the components of the disk's angular velocity are

$$\omega_x = \dot{\theta} = 0,$$

$$\omega_y = \dot{\psi} \sin \theta = \frac{v}{r_G} \sin \theta,$$

and

$$\omega_z = \dot{\phi} + \dot{\psi} \cos \theta = \dot{\phi} + \frac{v}{r_G} \cos \theta,$$

where $\dot{\phi}$ is the spin rate. To determine $\dot{\phi}$, we use the condition that the velocity of the point of the disk that is in contact with the surface is zero. In terms of the velocity of the center of the disk, the velocity of the point of contact is

$$\mathbf{0} = v\mathbf{i} + \boldsymbol{\omega} \times (-R\mathbf{j}) = v\mathbf{i} + \begin{vmatrix} \mathbf{i} & \mathbf{j} & \mathbf{k} \\ 0 & \dfrac{v}{r_G}\sin\theta & \dot{\phi} + \dfrac{v}{r_G}\cos\theta \\ 0 & -R & 0 \end{vmatrix}.$$

Expanding the determinant and solving for $\dot{\phi}$, we obtain

$$\dot{\phi} = -\frac{v}{R} - \frac{v}{r_G}\cos\theta.$$

We draw the free-body diagram of the disk in Fig. b. Because the center of mass moves in the horizontal plane, its acceleration in the vertical direction is zero. Therefore, the normal force $N = mg$. The acceleration of the center of mass in the direction perpendicular to its circular path is $a_n = v^2/r_G$. Newton's second law in the direction perpendicular to the circular path is

$$T = m\frac{v^2}{r_G}.$$

Therefore, the components of the total moment about the center of mass are

$$\Sigma M_x = TR\sin\theta - NR\cos\theta = m\frac{v^2}{r_G}R\sin\theta - mgR\cos\theta,$$

$$\Sigma M_y = 0,$$

and

$$\Sigma M_z = 0.$$

Substituting our expressions for $\dot{\psi}$, $\dot{\phi}$, and ΣM_x into the equation of angular motion for steady precession [Eq. (20.29)] and solving for v, we obtain

$$v = \sqrt{\frac{\frac{2}{3}g\cot\theta(r - R\cos\theta)^2}{r - \frac{5}{6}R\cos\theta}}.$$

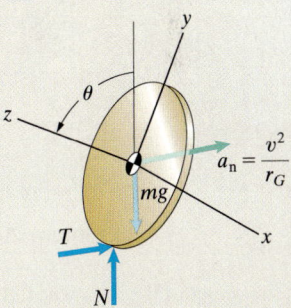

(b) Free-body diagram of the disk showing the normal acceleration of the center of mass.

Critical Thinking

Although the concept of steady precession was motivated by a spinning top, the resulting analysis can be adapted to a surprising variety of applications involving spinning symmetric objects. (See Problems 20.61–20.79.) While this example is rather artificial, it demonstrates the basic technique. You try to define a secondary coordinate system with its z axis coincident with the axis of symmetry of the spinning object such that the orientation and motion of the coordinate system correspond to those in our analysis of the steady precession of a top. Once that is achieved, you can analyze the motion with Eq. (20.29) or Eq. (20.33).

Problems

20.61 A ship has a turbine engine. The spin axis of the axisymmetric turbine is horizontal and aligned with the ship's longitudinal axis. The turbine rotates at 10,000 rpm. Its moment of inertia about its spin axis is 1000 kg-m². If the ship turns at a constant rate of 20 degrees per minute, what is the magnitude of the moment exerted on the ship by the turbine?

Strategy: Treat the turbine's motion as steady precession with nutation angle $\theta = 90°$.

Problem 20.61

20.62 The center of the car's wheel A travels in a circular path about O at 15 mi/h. The wheel's radius is 1 ft, and the moment of inertia of the wheel about its axis of rotation is 0.8 slug-ft². What is the magnitude of the total external moment about the wheel's center of mass?

Strategy: Treat the wheel's motion as steady precession with nutation angle $\theta = 90°$.

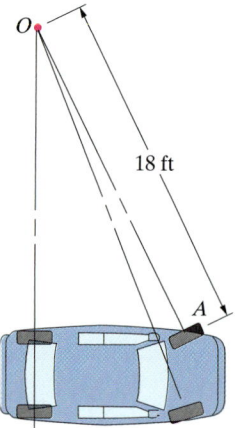

Problem 20.62

20.63 The radius of the 5-kg disk is $R = 0.2$ m. The disk is pinned to the horizontal shaft and rotates with constant angular velocity $\omega_d = 6$ rad/s relative to the shaft. The vertical shaft rotates with constant angular velocity $\omega_0 = 2$ rad/s. By treating the motion of the disk as steady precession, determine the magnitude of the couple exerted on the disk by the horizontal shaft.

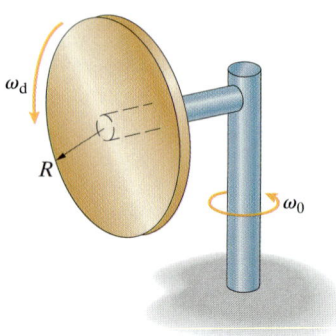

Problem 20.63

20.64 The helicopter is stationary. The z axis of the body-fixed coordinate system points downward and is coincident with the axis of the helicopter's rotor. The moment of inertia of the rotor about the z axis is 8600 kg-m². Its angular velocity is $-258\mathbf{k}$ (rpm). If the helicopter begins a pitch maneuver during which its angular velocity is $0.02\mathbf{j}$ (rad/s), what is the magnitude of the gyroscopic moment exerted on the helicopter by the rotor? Does the moment tend to cause the helicopter to roll about the x axis in the clockwise direction (as seen in the photograph) or the counterclockwise direction?

Problem 20.64

20.65 The bent bar is rigidly attached to the vertical shaft, which rotates with constant angular velocity ω_0. The disk of mass m and radius R is pinned to the bent bar and rotates with constant angular velocity ω_d relative to the bar. Determine the magnitudes of the force and couple exerted on the disk by the bar.

20.66 The bent bar is rigidly attached to the vertical shaft, which rotates with constant angular velocity ω_0. The disk of mass m and radius R is pinned to the bent bar and rotates with constant angular velocity ω_d relative to the bar. Determine the value of ω_d for which no couple is exerted on the disk by the bar.

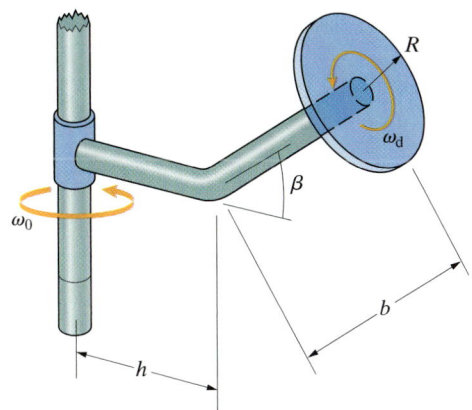

Problems 20.65/20.66

20.67 A thin circular disk undergoes moment-free steady precession. The z axis is perpendicular to the disk. Show that the disk's precession rate is $\dot{\psi} = -2\dot{\phi}/\cos\theta$. (Notice that when the nutation angle is small, the precession rate is approximately two times the spin rate.)

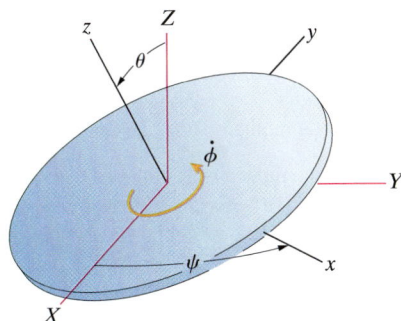

Problem 20.67

20.68 The rocket is in moment-free steady precession with nutation angle $\theta = 40°$ and spin rate $\dot{\phi} = 4$ revolutions per second. Its moments of inertia are $I_{xx} = 10,000$ kg-m^2 and $I_{zz} = 2000$ kg-m^2. What is the rocket's precession rate $\dot{\psi}$ in revolutions per second?

20.69 Sketch the body and space cones for the motion of the rocket in Problem 20.68.

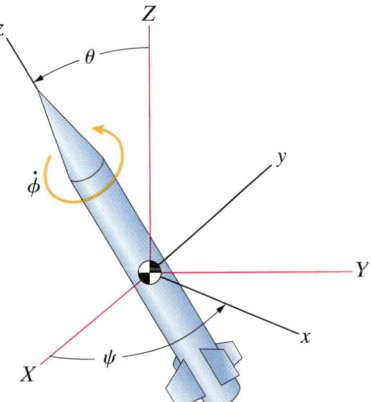

Problems 20.68/20.69

20.70 The top is in steady precession with nutation angle $\theta = 15°$ and precession rate $\dot{\psi} = 1$ revolution per second. The mass of the top is 8×10^{-4} slug, its center of mass is 1 in from the point, and its moments of inertia are $I_{xx} = 6 \times 10^{-6}$ slug-ft^2 and $I_{zz} = 2 \times 10^{-6}$ slug-ft^2. What is the spin rate $\dot{\phi}$ of the top in revolutions per second?

20.71 Suppose that top described in Problem 20.70 has a spin rate $\dot{\phi} = 15$ revolutions per second. Draw a graph of the precession rate (in revolutions per second) as a function of the nutation angle θ for values of θ from zero to 45°.

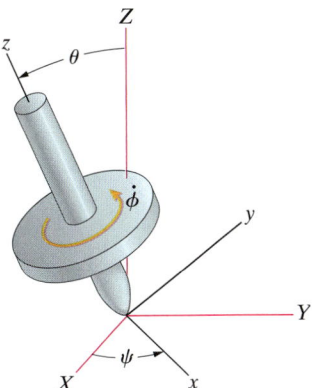

Problems 20.70/20.71

20.72 The rotor of a tumbling gyroscope can be modeled as being in moment-free steady precession. The moments of inertia of the gyroscope are $I_{xx} = I_{yy} = 0.04$ kg-m^2 and $I_{zz} = 0.18$ kg-m^2. The gyroscope's spin rate is $\dot{\phi} = 1500$ rpm and its nutation angle is $\theta = 20°$.
(a) What is the precession rate of the gyroscope in rpm?
(b) Sketch the body and space cones.

20.73 A satellite can be modeled as an 800-kg cylinder 4 m in length and 2 m in diameter. If the nutation angle is $\theta = 20°$ and the spin rate $\dot{\phi}$ is one revolution per second, what is the satellite's precession rate $\dot{\psi}$ in revolutions per second?

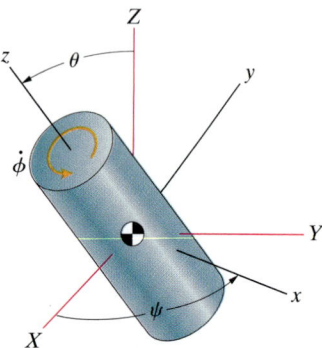

Problem 20.73

20.74* The top consists of a thin disk bonded to a slender bar. The radius of the disk is 30 mm and its mass is 0.008 kg. The length of the bar is 80 mm and its mass is negligible compared to the disk. When the top is in steady precession with a nutation angle of 10°, the precession rate is observed to be 2 revolutions per second in the same direction the top is spinning. What is the top's spin rate?

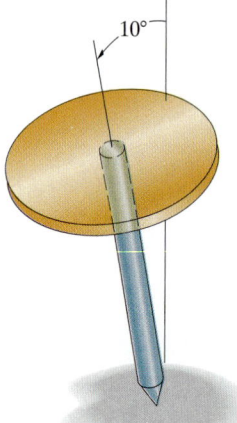

Problem 20.74

20.75 Solve Problem 20.58 by treating the motion as steady precession.

20.76* The two thin disks are rigidly connected by a slender bar. The radius of the large disk is 200 mm and its mass is 4 kg. The radius of the small disk is 100 mm and its mass is 1 kg. The bar is 400 mm in length and its mass is negligible. The composite object undergoes a steady motion in which it spins about the vertical y axis through its center of mass with angular velocity ω_0. The bar is horizontal during this motion and the large disk rolls on the floor. Determine ω_0 by treating the motion as steady precession.

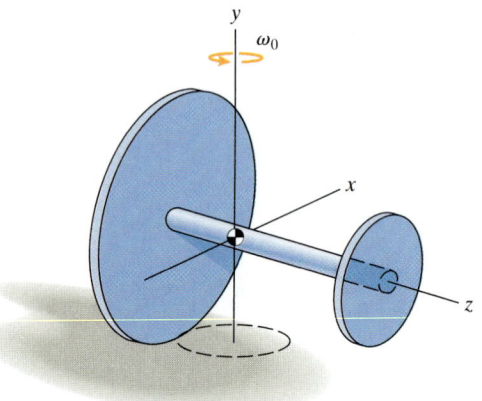

Problem 20.76

20.77* Suppose that you are testing a car and use accelerometers and gyroscopes to measure its Euler angles and their derivatives relative to a reference coordinate system. At a particular instant, $\psi = 15°$, $\theta = 4°$, $\phi = 15°$, the rates of change of the Euler angles are zero, and their second derivatives with respect to time are $\ddot{\psi} = 0$, $\ddot{\theta} = 1$ rad/s^2, and $\ddot{\phi} = -0.5$ rad/s^2. The car's principal moments of inertia, in kg-m^2, are $I_{xx} = 2200$, $I_{yy} = 480$, and $I_{zz} = 2600$. What are the components of the total moment about the car's center of mass?

20.78* If the Euler angles and their second derivatives for the car described in Problem 20.77 have the given values, but their rates of change are $\dot{\psi} = 0.2$ rad/s, $\dot{\theta} = -2$ rad/s, and $\dot{\phi} = 0$, what are the components of the total moment about the car's center of mass?

20.79* Suppose that the Euler angles of the car described in Problem 20.77 are $\psi = 40°$, $\theta = 20°$, and $\phi = 5°$, their rates of change are zero, and the components of the total moment about the car's center of mass are $\Sigma M_x = -400$ N-m, $\Sigma M_y = 200$ N-m, and $\Sigma M_z = 0$. What are the x, y, and z components of the car's angular acceleration?

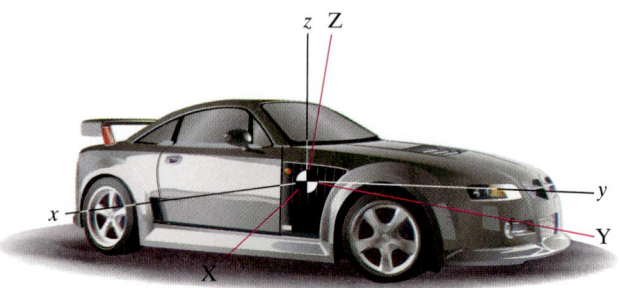

Problems 20.77–20.79

Appendix: Moments and Products of Inertia

To use the equations of motion to predict the behavior of a rigid body in three dimensions, the moments and products of inertia of the body, given by Eqs. (20.10) and (20.11), must be known. In this appendix, we demonstrate how the moments and products can be evaluated for simple objects such as slender bars and thin plates. We derive the parallel-axis theorems, which make it possible to determine the moments and products of inertia of composite objects. We also introduce the concept of principal axes, which simplifies the equations of angular motion.

Simple Objects

If we model a rigid body as a continuous distribution of mass, we can express its moments and products of inertia [Eqs. (20.10) and (20.11)] as

$$[I] = \begin{bmatrix} I_{xx} & -I_{xy} & -I_{xz} \\ -I_{yx} & I_{yy} & -I_{yz} \\ -I_{zx} & -I_{zy} & I_{zz} \end{bmatrix}$$

$$= \begin{bmatrix} \int_m (y^2 + z^2)\, dm & -\int_m xy\, dm & -\int_m xz\, dm \\ -\int_m yx\, dm & \int_m (x^2 + z^2)\, dm & -\int_m yz\, dm \\ -\int_m zx\, dm & -\int_m zy\, dm & \int_m (x^2 + y^2)\, dm \end{bmatrix}, \quad (20.37)$$

where x, y, and z are the coordinates of the differential element of mass dm (Fig. 20.26).

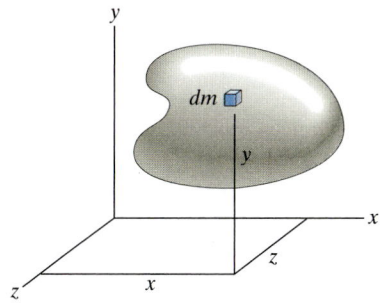

Figure 20.26
Determining the moments and products of inertia by modeling an object as a continuous distribution of mass.

Slender Bars   Let the origin of the coordinate system be at a slender bar's center of mass, with the x axis along the bar (Fig. 20.27a). The bar has length l, cross-sectional area A, and mass m. We assume that A is uniform along the length of the bar and that the material is homogeneous.

Consider a differential element of the bar of length dx at a distance x from the center of mass (Fig. 20.27b). The mass of the element is $dm = \rho A\, dx$, where ρ is the mass density. We neglect the lateral dimensions of the bar, assuming the coordinates of the differential element dm to be $(x, 0, 0)$. As a consequence of this approximation, the moment of inertia of the bar about the x axis is zero:

$$I_{xx} = \int_m (y^2 + z^2)\, dm = 0.$$

The moment of inertia about the y axis is

$$I_{yy} = \int_m (x^2 + z^2)\, dm = \int_{-1/2}^{1/2} \rho A x^2\, dx = \frac{1}{12} \rho A l^3.$$

Expressing this result in terms of the mass of the bar, $m = \rho A l$, we obtain

$$I_{yy} = \frac{1}{12} m l^2.$$

The moment of inertia about the z axis is equal to the moment of inertia about the y axis:

$$I_{zz} = \int_m (x^2 + y^2)\, dm = \frac{1}{12} m l^2.$$

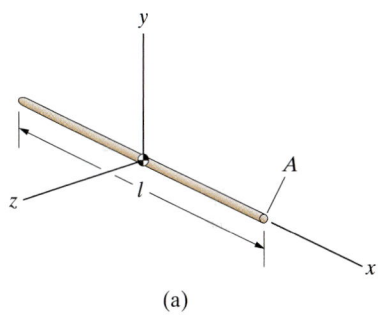

(a)

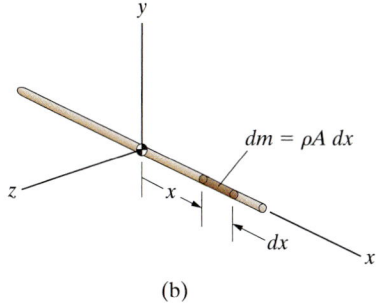

(b)

Figure 20.27
(a) A slender bar and a coordinate system with the x axis aligned with the bar.
(b) A differential element of mass of length dx.

Because the y and z coordinates of dm are zero, the products of inertia are zero, so the inertia matrix for the slender bar is

$$[I] = \begin{bmatrix} 0 & 0 & 0 \\ 0 & \frac{1}{12}ml^2 & 0 \\ 0 & 0 & \frac{1}{12}ml^2 \end{bmatrix}. \qquad (20.38)$$

It is important to remember that the moments and products of inertia depend on the orientation of the coordinate system relative to the object. In terms of the alternative coordinate system shown in Fig. 20.28, the bar's inertia matrix is

$$[I] = \begin{bmatrix} \frac{1}{12}ml^2 & 0 & 0 \\ 0 & 0 & 0 \\ 0 & 0 & \frac{1}{12}ml^2 \end{bmatrix}.$$

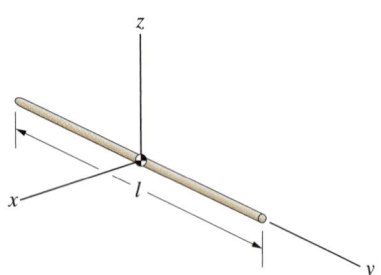

Figure 20.28
Aligning the y axis with the bar.

Thin Plates

Suppose that a homogeneous plate of uniform thickness T, area A, and unspecified shape lies in the x–y plane (Fig. 20.29a). We can express the moments of inertia of the plate in terms of the moments of inertia of its cross-sectional area.

By projecting an element of area dA through the thickness T of the plate (Fig. 20.29b), we obtain a differential element of mass $dm = \rho T\, dA$. We neglect the thickness of the plate in calculating the moments of inertia, so the coordinates of the element dm are $(x, y, 0)$. The plate's moment of inertia about the x axis is

$$I_{xx} = \int_m (y^2 + z^2)\, dm = \rho T \int_A y^2\, dA = \rho T I_x,$$

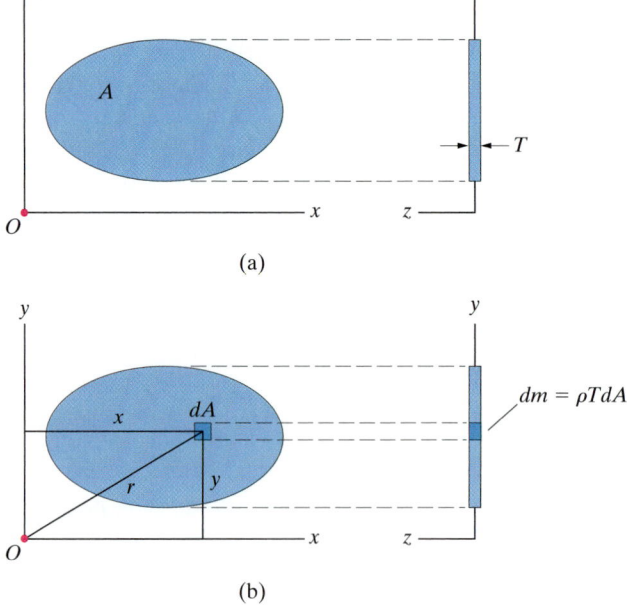

(a)

(b)

Figure 20.29
(a) A thin plate lying in the x–y plane.
(b) Obtaining a differential element of mass by projecting an element of area dA through the plate.

where I_x is the moment of inertia of the plate's cross-sectional area about the x axis. Since the mass of the plate is $m = \rho T A$, the product $\rho T = m/A$, and we obtain the moment of inertia in the form

$$I_{xx} = \frac{m}{A}I_x.$$

The moment of inertia about the y axis is

$$I_{yy} = \int_m (x^2 + z^2)\, dm = \rho T \int_A x^2\, dA = \frac{m}{A}I_y,$$

where I_y is the moment of inertia of the cross-sectional area about the y axis. The moment of inertia about the z axis is

$$I_{zz} = \int_m (x^2 + y^2)\, dm = \frac{m}{A}J_O,$$

where $J_O = I_x + I_y$ is the polar moment of inertia of the cross-sectional area. The product of inertia I_{xy} is

$$I_{xy} = \int_m xy\, dm = \frac{m}{A}I_{xy}^A,$$

where

$$I_{xy}^A = \int_A xy\, dA$$

is the product of inertia of the cross-sectional area. (We use a superscript A to distinguish the product of inertia of the plate's cross-sectional area from the product of inertia of its mass.) If the cross-sectional area A is symmetric about either the x axis or the y axis, $I_{xy}^A = 0$.

Because the z coordinate of dm is zero, the products of inertia I_{xz} and I_{yz} are zero. The inertia matrix for the thin plate is thus

$$[I] = \begin{bmatrix} \dfrac{m}{A}I_x & -\dfrac{m}{A}I_{xy}^A & 0 \\[2mm] -\dfrac{m}{A}I_{xy}^A & \dfrac{m}{A}I_y & 0 \\[2mm] 0 & 0 & \dfrac{m}{A}J_O \end{bmatrix}. \tag{20.39}$$

If the moments of inertia and product of inertia of the plate's cross-sectional area are known, (Eq. 20.39) can be used to obtain the moments and products of inertia of the plate.

Parallel-Axis Theorems

Suppose that we know an object's inertia matrix $[I']$ in terms of a coordinate system $x'y'z'$ with its origin at the center of mass of the object, and we want to determine the inertia matrix $[I]$ in terms of a parallel coordinate system xyz (Fig. 20.30). Let (d_x, d_y, d_z) be the coordinates of the center of mass in the xyz coordinate system. The coordinates of a differential element of mass dm in the xyz system are given in terms of its coordinates in the $x'y'z'$ system by

$$x = x' + d_x, \qquad y = y' + d_y, \qquad z = z' + d_z. \tag{20.40}$$

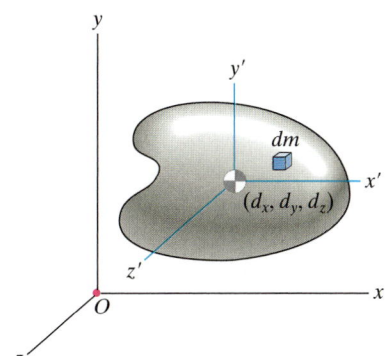

Figure 20.30
A coordinate system $x'y'z'$ with its origin at the center of mass and a parallel coordinate system xyz.

Substituting these expressions into the definition of I_{xx}, we obtain

$$
\begin{aligned}
I_{xx} = \int_m \left[(y')^2 + (z')^2 \right] dm + 2d_y \int_m y' \, dm \\
+ 2d_z \int_m z' \, dm + (d_y^2 + d_z^2) \int_m dm.
\end{aligned}
\tag{20.41}
$$

The first integral on the right is the object's moment of inertia about the x' axis. We can show that the second and third integrals are zero by using the definitions of the object's center of mass, expressed in terms of the $x'y'z'$ coordinate system:

$$
\overline{x}' = \frac{\displaystyle\int_m x' \, dm}{\displaystyle\int_m dm}, \qquad
\overline{y}' = \frac{\displaystyle\int_m y' \, dm}{\displaystyle\int_m dm}, \qquad
\overline{z}' = \frac{\displaystyle\int_m z' \, dm}{\displaystyle\int_m dm}.
$$

The object's center of mass is at the origin of the $x'y'z'$ system, so $\overline{x}' = \overline{y}' = \overline{z}' = 0$. Therefore, the second and third integrals on the right of Eq. (20.41) are zero, and we obtain

$$
I_{xx} = I_{x'x'} + (d_y^2 + d_z^2) \, m,
$$

where m is the mass of the object. Substituting Eqs. (20.40) into the definition of I_{xy}, we get

$$
\begin{aligned}
I_{xy} &= \int_m x'y' \, dm + d_x \int_m y' \, dm + d_y \int_m x' \, dm + d_x d_y \int_m dm \\
&= I_{x'y'} + d_x d_y m.
\end{aligned}
$$

Proceeding in this way for each of the moments and products of inertia, we obtain the *parallel-axis theorems*:

$$
\begin{aligned}
I_{xx} &= I_{x'x'} + (d_y^2 + d_z^2) \, m, \\
I_{yy} &= I_{y'y'} + (d_x^2 + d_z^2) \, m, \\
I_{zz} &= I_{z'z'} + (d_x^2 + d_y^2) \, m, \\
I_{xy} &= I_{x'y'} + d_x d_y m, \\
I_{yz} &= I_{y'z'} + d_y d_z m, \\
I_{zx} &= I_{z'x'} + d_z d_x m.
\end{aligned}
\tag{20.42}
$$

If an object's inertia matrix is known in terms of a particular coordinate system, these theorems can be used to determine its inertia matrix in terms of any parallel coordinate system. They can also be used to determine the inertia matrices of composite objects.

Moment of Inertia about an Arbitrary Axis

If we know a rigid body's inertia matrix in terms of a given coordinate system with origin O, we can determine its moment of inertia about an arbitrary axis through O. Suppose that the rigid body rotates with angular velocity $\boldsymbol{\omega}$ about an arbitrary fixed axis L_O through O, and let $\mathbf{e}$ be a unit vector with the same direction as $\boldsymbol{\omega}$ (Fig. 20.31). Then, in terms of the moment of inertia I_O about L_O, the rigid body's angular momentum about L_O is

$$H_O = I_O|\boldsymbol{\omega}|.$$

We can express the angular velocity vector as

$$\boldsymbol{\omega} = |\boldsymbol{\omega}|(e_x\mathbf{i} + e_y\mathbf{j} + e_z\mathbf{k}),$$

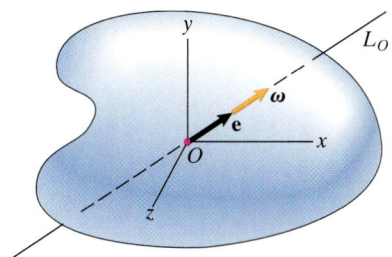

Figure 20.31
Rigid body rotating about L_O.

so that $\omega_x = |\boldsymbol{\omega}|e_x$, $\omega_y = |\boldsymbol{\omega}|e_y$, and $\omega_z = |\boldsymbol{\omega}|e_z$. Using these expressions and Eqs. (20.9), we obtain the angular momentum about L_O:

$$H_O = \mathbf{H}_O \cdot \mathbf{e} = (I_{xx}|\boldsymbol{\omega}|e_x - I_{xy}|\boldsymbol{\omega}|e_y - I_{xz}|\boldsymbol{\omega}|e_z)e_x$$

$$+ (-I_{yx}|\boldsymbol{\omega}|e_x + I_{yy}|\boldsymbol{\omega}|e_y - I_{yz}|\boldsymbol{\omega}|e_z)e_y$$

$$+ (-I_{zx}|\boldsymbol{\omega}|e_x - I_{zy}|\boldsymbol{\omega}|e_y + I_{zz}|\boldsymbol{\omega}|e_z)e_z.$$

Equating our two expressions for H_O yields

$$I_O = I_{xx}e_x^2 + I_{yy}e_y^2 + I_{zz}e_z^2 - 2I_{xy}e_xe_y - 2I_{yz}e_ye_z - 2I_{zx}e_ze_x. \quad (20.43)$$

Notice that the moment of inertia about an arbitrary axis depends on the products of inertia, in addition to the moments of inertia about the coordinate axes. If an object's inertia matrix is known, Eq. (20.43) can be used to determine the object's moment of inertia about an axis through O whose direction is specified by the unit vector $\mathbf{e}$.

Principal Axes

For *any* object and origin O, at least one coordinate system exists for which the products of inertia are zero:

$$[I] = \begin{bmatrix} I_{xx} & 0 & 0 \\ 0 & I_{yy} & 0 \\ 0 & 0 & I_{zz} \end{bmatrix}. \quad (20.44)$$

These coordinate axes are called *principal axes*, and the moments of inertia about them are called the *principal moments of inertia*.

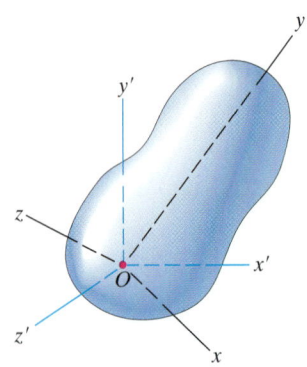

Figure 20.32
The $x'y'z'$ system with its origin at O and a set of principal axes xyz.

If the inertia matrix of a rigid body is known in terms of a coordinate system $x'y'z'$ and the products of inertia are zero, then $x'y'z'$ is a set of principal axes. Suppose that the products of inertia are not zero, and we want to find a set of principal axes xyz and the corresponding principal moments of inertia (Fig. 20.32). It can be shown that the principal moments of inertia are roots of the cubic equation

$$I^3 - (I_{x'x'} + I_{y'y'} + I_{z'z'})I^2$$

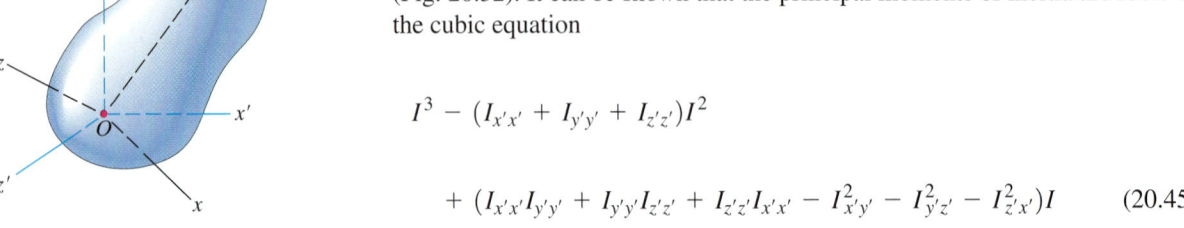

$$+ (I_{x'x'}I_{y'y'} + I_{y'y'}I_{z'z'} + I_{z'z'}I_{x'x'} - I_{x'y'}^2 - I_{y'z'}^2 - I_{z'x'}^2)I \qquad (20.45)$$

$$- (I_{x'x'}I_{y'y'}I_{z'z'} - I_{x'x'}I_{y'z'}^2 - I_{y'y'}I_{z'x'}^2 - I_{z'z'}I_{x'y'}^2 - 2I_{x'y'}I_{y'z'}I_{z'x'}) = 0.$$

For each principal moment of inertia I, the vector $\mathbf{V}$ with components

$$V_{x'} = (I_{y'y'} - I)(I_{z'z'} - I) - I_{y'z'}^2,$$

$$V_{y'} = I_{x'y'}(I_{z'z'} - I) + I_{x'z'}I_{y'z'}, \qquad (20.46)$$

$$V_{z'} = I_{x'z'}(I_{y'y'} - I) + I_{x'y'}I_{y'z'}$$

is parallel to the corresponding principal axis.

When the inertia matrix of an object is known in terms of a coordinate system with origin O, determining the associated principal moments of inertia and a set of principal axes involves two steps:

1. Determine the principal moments of inertia by obtaining the roots of Eq. (20.45).

2. If the three principal moments of inertia are distinct, substitute each one into Eqs. (20.46) to obtain the components of a vector parallel to the corresponding principal axis. The three principal axes can be denoted as x, y, and z arbitrarily, as long as the resulting coordinate system is right handed. If the three principal moments of inertia are equal, the moment of inertia about any axis through O has the same value, and any coordinate system with origin O is a set of principal axes. This is the case, for example, if the object is a homogeneous sphere with O at its center (Fig. 20.33a). If only two of the principal moments of inertia are equal, the third one can be substituted into Eqs. (20.46) to determine the associated principal axis. Then the moment of inertia about any axis through O that is perpendicular to the determined axis has the same value, so any coordinate system with origin O that has an axis coincident with the determined axis is a set of principal axes. This is the case when an object has an axis of rotational symmetry and O is on the axis (Fig. 20.33b).

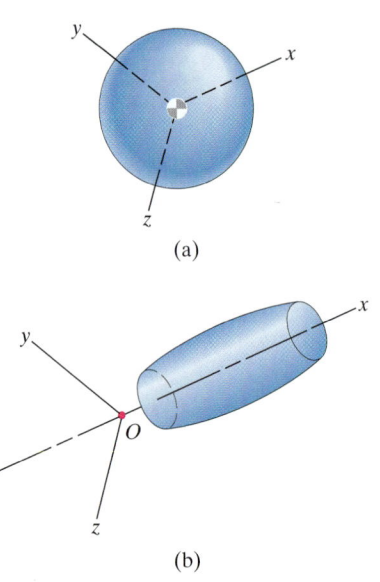

Figure 20.33
(a) A homogeneous sphere. Any coordinate system with its origin at the center is a set of principal axes.
(b) A rotationally symmetric object. The axis of symmetry is a principal axis, and any axis perpendicular to it is a principal axis.

Example 20.8 Parallel-Axis Theorems

The boom AB of the crane in Fig. 20.34 has a mass of 4800 kg, and the boom BC has a mass of 1600 kg and is perpendicular to AB. Modeling each boom as a slender bar and treating them as a single object, determine the moments and products of inertia of the object in terms of the coordinate system shown.

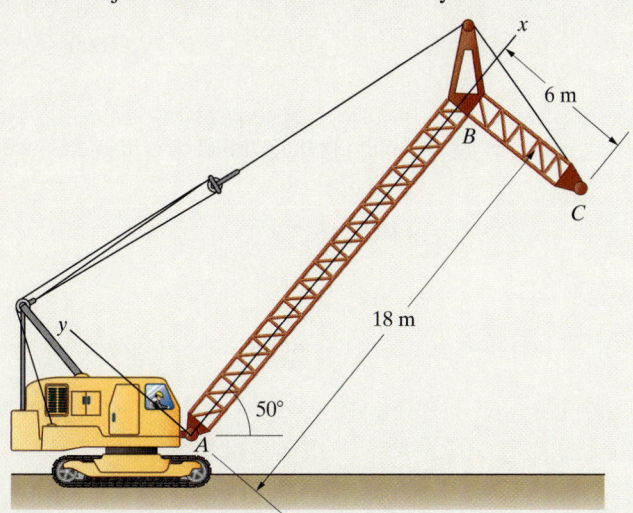

Figure 20.34

Strategy

We can apply the parallel-axis theorems to each boom to determine its moments and products of inertia in terms of the given coordinate system. The moments and products of inertia of the combined object are the sums of those for the two booms.

Solution

Boom AB In Fig. a, we introduce a parallel coordinate system $x'y'z'$ with its origin at the center of mass of boom AB. In terms of the $x'y'z'$ system, the inertia matrix of boom AB is

$$[I'] = \begin{bmatrix} 0 & 0 & 0 \\ 0 & \frac{1}{12}ml^2 & 0 \\ 0 & 0 & \frac{1}{12}ml^2 \end{bmatrix} = \begin{bmatrix} 0 & 0 & 0 \\ 0 & \frac{1}{12}(4800)(18)^2 & 0 \\ 0 & 0 & \frac{1}{12}(4800)(18)^2 \end{bmatrix} \text{ kg-m}^2.$$

The coordinates of the origin of the $x'y'z'$ system relative to the xyz system are $d_x = 9$ m, $d_y = 0$, $d_z = 0$. Applying the parallel-axis theorems, we obtain

$$I_{xx} = I_{x'x'} + (d_y^2 + d_z^2)m = 0,$$

$$I_{yy} = I_{y'y'} + (d_x^2 + d_z^2)m = \frac{1}{12}(4800)(18)^2 + (9)^2(4800)$$

$$= 518{,}400 \text{ kg-m}^2,$$

$$I_{zz} = I_{z'z'} + (d_x^2 + d_y^2)m = \frac{1}{12}(4800)(18)^2 + (9)^2(4800)$$

$$= 518{,}400 \text{ kg-m}^2,$$

$$I_{xy} = I_{x'y'} + d_x d_y m = 0,$$

$$I_{yz} = I_{y'z'} + d_y d_z m = 0,$$

and

$$I_{zx} = I_{z'x'} + d_z d_x m = 0.$$

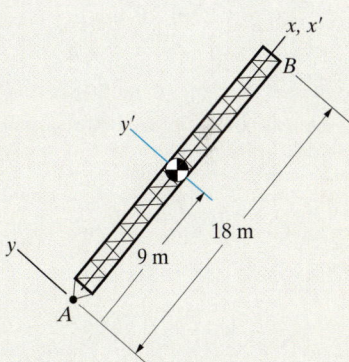

(a) Applying the parallel-axis theorems to boom AB.

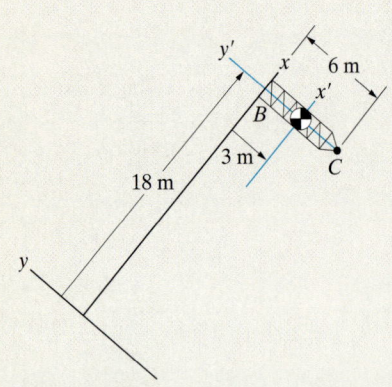

(b) Applying the parallel-axis theorems to boom BC.

Boom *BC* In Fig. b we introduce a parallel coordinate system $x'y'z'$ with its origin at the center of mass of boom *BC*. In terms of the $x'y'z'$ system, the inertia matrix of boom *BC* is

$$[I'] = \begin{bmatrix} \frac{1}{12}ml^2 & 0 & 0 \\ 0 & 0 & 0 \\ 0 & 0 & \frac{1}{12}ml^2 \end{bmatrix} = \begin{bmatrix} \frac{1}{12}(1600)(6)^2 & 0 & 0 \\ 0 & 0 & 0 \\ 0 & 0 & \frac{1}{12}(1600)(6)^2 \end{bmatrix} \text{kg-m}^2.$$

The coordinates of the origin of the $x'y'z'$ system relative to the xyz system are $d_x = 18$ m, $d_y = -3$ m, $d_z = 0$. Applying the parallel-axis theorems, we obtain

$$I_{xx} = I_{x'x'} + (d_y^2 + d_z^2)\, m = \frac{1}{12}(1600)(6)^2 + (-3)^2(1600)$$
$$= 19{,}200 \text{ kg-m}^2,$$

$$I_{yy} = I_{y'y'} + (d_x^2 + d_z^2)\, m = 0 + (18)^2(1600) = 518{,}400 \text{ kg-m}^2,$$

$$I_{zz} = I_{z'z'} + (d_x^2 + d_y^2)\, m = \frac{1}{12}(1600)(6)^2 + [(18)^2 + (-3)^2](1600)$$
$$= 537{,}600 \text{ kg-m}^2,$$

$$I_{xy} = I_{x'y'} + d_x d_y m = 0 + (18)(-3)(1600) = -86{,}400 \text{ kg-m}^2,$$

$$I_{yz} = I_{y'z'} + d_y d_z m = 0,$$

and

$$I_{zx} = I_{z'x'} + d_z d_x m = 0.$$

Summing the results for the two booms, we obtain the inertia matrix for the single object:

$$[I] = \begin{bmatrix} 19{,}200 & -(-86{,}400) & 0 \\ -(-86{,}400) & 518{,}400 + 518{,}400 & 0 \\ 0 & 0 & 518{,}400 + 537{,}600 \end{bmatrix}$$

$$= \begin{bmatrix} 19{,}200 & 86{,}400 & 0 \\ 86{,}400 & 1{,}036{,}800 & 0 \\ 0 & 0 & 1{,}056{,}000 \end{bmatrix} \text{kg-m}^2.$$

Critical Thinking

Most of the objects you will encounter in engineering will be assemblies of simpler parts, such as the crane's boom in this example. When the moments and products of inertia of the parts are known, the moments and products of inertia of the assembly can be determined using the procedure in this example. You apply the parallel-axis theorems to each part to determine its moments and products of inertia in terms of a given coordinate system, then sum the results for the parts to obtain the moments and products of inertia of the assembly in terms of that coordinate system. Notice that in order to apply the parallel-axis theorems, the moments and products of inertia of the parts must be expressed in terms of parallel coordinate systems.

Example 20.9 Inertia Matrix of a Plate

The 4-kg rectangular plate in Fig. 20.35 lies in the x–y plane of the body-fixed coordinate system.
(a) Determine the plate's moments and products of inertia.
(b) Determine the plate's moment of inertia about the diagonal axis L_O.
(c) If the plate is rotating about the fixed point O with angular velocity $\boldsymbol{\omega} = 4\mathbf{i} - 2\mathbf{j}$ (rad/s), what is the plate's angular momentum about O?

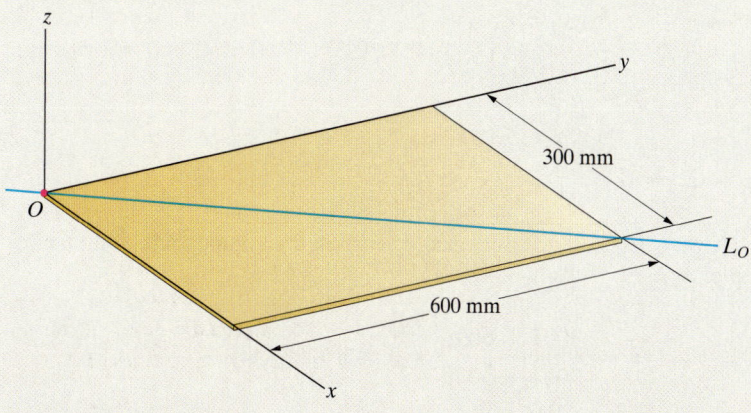

Figure 20.35

Strategy

(a) We can obtain the moments and products of inertia of the plate's rectangular area from Appendix B and use Eq. (20.39) to obtain the moments and products of inertia of the plate.
(b) Once we know the moments and products of inertia, we can use Eq. (20.43) to determine the moment of inertia about L_O.
(c) The angular momentum about O is given by Eqs. (20.9).

Solution

(a) From Appendix B, the moments of inertia of the plate's cross-sectional area are as follows (Fig. a):

$$I_x = \frac{1}{3}bh^3, \qquad I_y = \frac{1}{3}hb^3,$$

$$I_{xy}^A = \frac{1}{4}b^2h^2, \qquad J_O = \frac{1}{3}(bh^3 + hb^3).$$

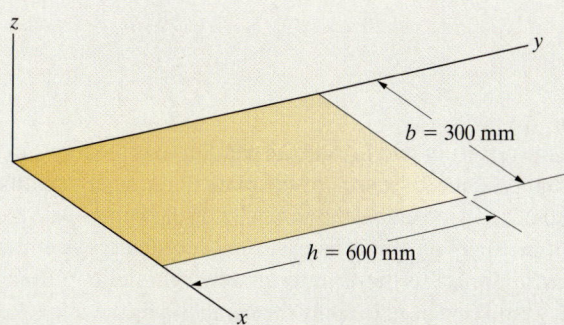

(a) Determining the moments of inertia of the plate's area.

Therefore, the moments and products of inertia of the plate are

$$I_{xx} = \frac{m}{A}I_x = \frac{(4)}{(0.3)(0.6)}\left(\frac{1}{3}\right)(0.3)(0.6)^3 = 0.48 \text{ kg-m}^2,$$

$$I_{yy} = \frac{m}{A}I_y = \frac{(4)}{(0.3)(0.6)}\left(\frac{1}{3}\right)(0.6)(0.3)^3 = 0.12 \text{ kg-m}^2,$$

$$I_{xy} = \frac{m}{A}I_{xy}^A = \frac{(4)}{(0.3)(0.6)}\left(\frac{1}{4}\right)(0.3)^2(0.6)^2 = 0.18 \text{ kg-m}^2,$$

$$I_{zz} = \frac{m}{A}J_O = \frac{(4)}{(0.3)(0.6)}\left(\frac{1}{3}\right)[(0.3)(0.6)^3 + (0.6)(0.3)^3] = 0.60 \text{ kg-m}^2,$$

and

$$I_{xz} = I_{yz} = 0.$$

(b) To apply Eq. (20.43), we must determine the components of a unit vector parallel to L_O:

$$\mathbf{e} = \frac{300\mathbf{i} + 600\mathbf{j}}{|300\mathbf{i} + 600\mathbf{j}|} = 0.447\mathbf{i} + 0.894\mathbf{j}.$$

The moment of inertia about L_O is

$$I_O = I_{xx}e_x^2 + I_{yy}e_y^2 + I_{zz}e_z^2 - 2I_{xy}e_xe_y - 2I_{yz}e_ye_z - 2I_{zx}e_ze_x$$

$$= (0.48)(0.447)^2 + (0.12)(0.894)^2 - 2(0.18)(0.447)(0.894)$$

$$= 0.048 \text{ kg-m}^2.$$

(c) The plate's angular momentum about O is

$$\begin{bmatrix} H_{Ox} \\ H_{Oy} \\ H_{Oz} \end{bmatrix} = \begin{bmatrix} I_{xx} & -I_{xy} & -I_{xz} \\ -I_{yx} & I_{yy} & -I_{yz} \\ -I_{zx} & -I_{zy} & I_{zz} \end{bmatrix} \begin{bmatrix} \omega_x \\ \omega_y \\ \omega_z \end{bmatrix}$$

$$= \begin{bmatrix} 0.48 & -0.18 & 0 \\ -0.18 & 0.12 & 0 \\ 0 & 0 & 0.6 \end{bmatrix} \begin{bmatrix} 4 \\ -2 \\ 0 \end{bmatrix}$$

$$= \begin{bmatrix} 2.28 \\ -0.96 \\ 0 \end{bmatrix} \text{ kg-m}^2/\text{s}.$$

Critical Thinking

Could you use the results of part (a) and the parallel-axis theorems to determine the moments and products of inertia of the plate in terms of any other parallel coordinate system? Remember that the parallel-axis theorems relate the moments and products of inertia of an object in terms of a coordinate system with its origin *at the center of mass* to those in terms of a parallel coordinate system. Therefore, you would first need to apply the parallel-axis theorems to the results of part (a) to determine the moments and products of inertia of the plate in terms of a parallel coordinate system with its origin at the center of mass.

Example 20.10 Principal Axes and Moments of Inertia

In terms of a coordinate system $x'y'z'$ with its origin at the center of mass, the inertia matrix of a rigid body is

$$[I'] = \begin{bmatrix} 4 & -2 & 1 \\ -2 & 2 & -1 \\ 1 & -1 & 3 \end{bmatrix} \text{kg-m}^2.$$

Determine the principal moments of inertia and the directions of a set of principal axes relative to the $x'y'z'$ system.

Strategy
The principal moments of inertia are the roots of Eq. (20.45). For each principal moment of inertia, Eqs. (20.46) give the components of a vector that is parallel to the corresponding principal axis.

Solution
Substituting the moments and products of inertia into Eq. (20.45), we obtain the equation

$$I^3 - 9I^2 + 20I - 10 = 0. \tag{1}$$

We show the value of the left side of this equation as a function of I in Fig. 20.36. The three roots, which are the values of the principal moments of inertia in kg-m², are $I_1 = 0.708$, $I_2 = 2.397$, and $I_3 = 5.895$.

Substituting the principal moment of inertia $I_1 = 0.708$ kg-m² into Eqs. (20.46) and dividing the resulting vector **V** by its magnitude, we obtain a unit vector parallel to the corresponding principal axis:

$$\mathbf{e}_1 = 0.473\mathbf{i} + 0.864\mathbf{j} + 0.171\mathbf{k}.$$

Substituting $I_2 = 2.397$ kg-m² into Eqs. (20.46), we obtain the unit vector

$$\mathbf{e}_2 = -0.458\mathbf{i} + 0.076\mathbf{j} + 0.886\mathbf{k},$$

and substituting $I_3 = 5.895$ kg-m² into Eqs. (20.46), we obtain the unit vector

$$\mathbf{e}_3 = 0.753\mathbf{i} - 0.497\mathbf{j} + 0.432\mathbf{k}.$$

We have determined the principal moments of inertia and the components of unit vectors parallel to the corresponding principal axes. In Fig. 20.37, we show the principal axes, arbitrarily designating them so that $I_{xx} = 5.895$ kg-m², $I_{yy} = 0.708$ kg-m², and $I_{zz} = 2.397$ kg-m².

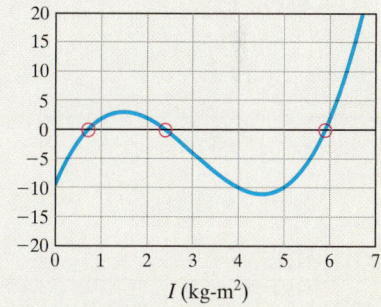

Figure 20.36
Graph of $I^3 - 9I^2 + 20I - 10$.

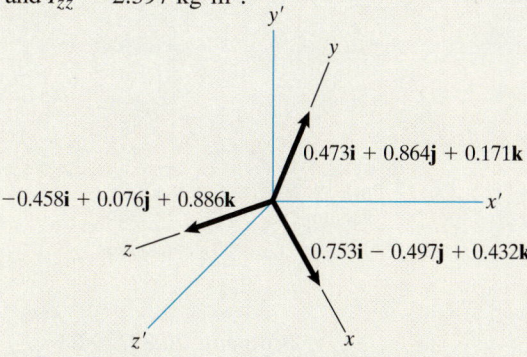

Figure 20.37
The principal axes. Our choice of which to call x, y, and z is arbitrary.

Critical Thinking

Many programmable calculators and computer programs are available that will determine roots of nonlinear algebraic equations. That is how we obtained the precise values of the roots of Eq. (1) in this example. We needed to determine all three roots of the cubic equation. Unless the program you use is designed to determine all of the roots of an Nth-order equation, you may find that the program continues to converge on a root that you have already determined instead of one you are still seeking. You can avoid this in a simple way. In this example, you would begin by asking the program to determine a root of Eq. (1). Suppose that the root obtained is 0.708. (This value is rounded off to three significant digits. In your computations, retain as much accuracy as your computer provides.) Then seek a root of the equation

$$\frac{I^3 - 9I^2 + 20I - 10}{I - 0.708} = 0.$$

In this way, you are "dividing out" the root you have determined. If the next root you obtain is 2.397, obtain the final root by seeking a root of the equation

$$\frac{I^3 - 9I^2 + 20I - 10}{(I - 0.708)(I - 2.397)} = 0.$$

Problems

20.80 The mass of the bar is 6 kg. Determine the moments and products of inertia of the bar in terms of the coordinate system shown.

20.81 The mass of the bar is 6 kg.
(a) Determine its moments and products of inertia in terms of a parallel coordinate system $x'y'z'$ with its origin at the bar's center of mass.
(b) If the bar is rotating with angular velocity $\boldsymbol{\omega} = 4\mathbf{i}$ (rad/s), what is its angular momentum about its center of mass?

20.82 The 4-kg thin rectangular plate lies in the x–y plane. Determine the moments and products of inertia of the plate in terms of the coordinate system shown.

20.83 If the 4-kg plate is rotating with angular velocity $\boldsymbol{\omega} = 6\mathbf{i} + 4\mathbf{j} - 2\mathbf{k}$ (rad/s), what is its angular momentum about its center of mass?

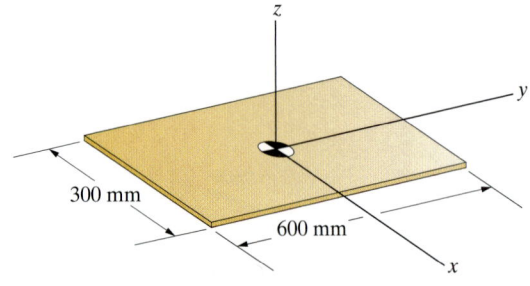

Problems 20.82/20.83

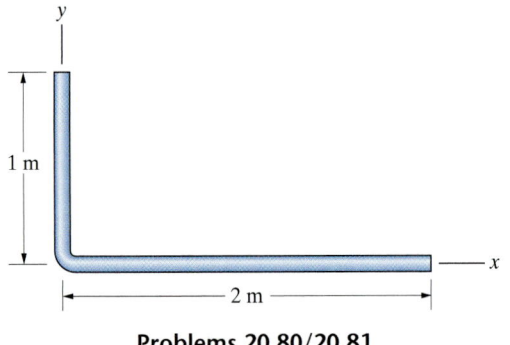

Problems 20.80/20.81

20.84 The 30-lb triangular plate lies in the *x–y* plane. Determine the moments and products of inertia of the plate in terms of the coordinate system shown.

20.85 The 30-lb triangular plate lies in the *x–y* plane.
(a) Determine its moments and products of inertia in terms of a parallel coordinate system $x'y'z'$ with its origin at the plate's center of mass.
(b) If the plate is rotating with angular velocity $\omega = 20\mathbf{i} - 12\mathbf{j} + 16\mathbf{k}$ (rad/s), what is its angular momentum about its center of mass?

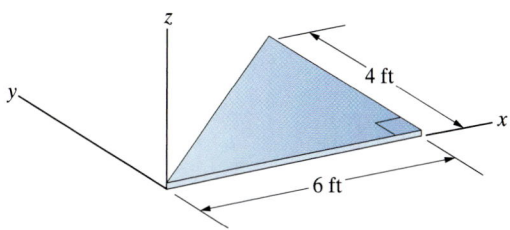

Problems 20.84/20.85

20.86 Determine the inertia matrix of the 2.4-kg steel plate in terms of the coordinate system shown.

20.87 The mass of the steel plate is 2.4 kg.
(a) Determine its moments and products of inertia in terms of a parallel coordinate system $x'y'z'$ with its origin at the plate's center of mass.
(b) If the plate is rotating with angular velocity $\omega = 20\mathbf{i} + 10\mathbf{j} - 10\mathbf{k}$ (rad/s), what is its angular momentum about its center of mass?

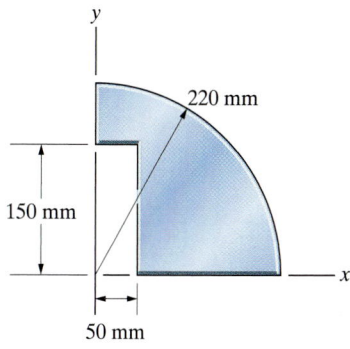

Problems 20.86/20.87

20.88 The slender bar of mass m rotates about the fixed point O with angular velocity $\omega = \omega_y\mathbf{j} + \omega_z\mathbf{k}$. Determine its angular momentum (a) about its center of mass and (b) about O.

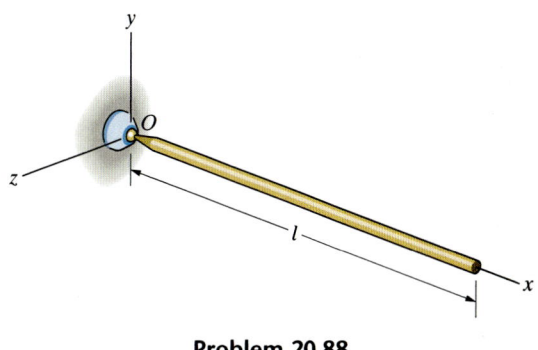

Problem 20.88

20.89 The slender bar of mass m is parallel to the x axis. If the coordinate system is body fixed and its angular velocity about the fixed point O is $\omega = \omega_y\mathbf{j}$, what is the bar's angular momentum about O?

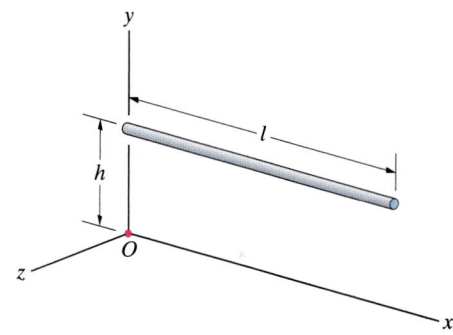

Problem 20.89

20.90 In Example 20.8, the moments and products of inertia of the object consisting of the booms AB and BC were determined in terms of the coordinate system shown in Fig. 20.34. Determine the moments and products of inertia of the object in terms of a parallel coordinate system $x'y'z'$ with its origin at the center of mass of the object.

20.91 Suppose that the crane described in Example 20.8 undergoes a rigid-body rotation about the vertical axis at 0.1 rad/s in the counterclockwise direction when viewed from above.

(a) What is the crane's angular velocity vector $\boldsymbol{\omega}$ in terms of the body-fixed coordinate system shown in Fig. 20.34?
(b) What is the angular momentum of the object consisting of the booms AB and BC about its center of mass?

20.92 A 3-kg slender bar is rigidly attached to a 2-kg thin circular disk. In terms of the body-fixed coordinate system shown, the angular velocity of the composite object is
$\boldsymbol{\omega} = 100\mathbf{i} - 4\mathbf{j} + 6\mathbf{k}$ (rad/s). What is the object's angular momentum about its center of mass?

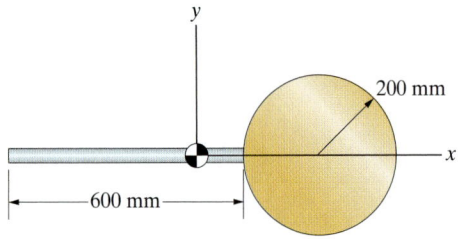

Problem 20.92

20.93* The mass of the homogeneous slender bar is m. If the bar rotates with angular velocity $\boldsymbol{\omega} = \omega_0(24\mathbf{i} + 12\mathbf{j} - 6\mathbf{k})$, what is its angular momentum about its center of mass?

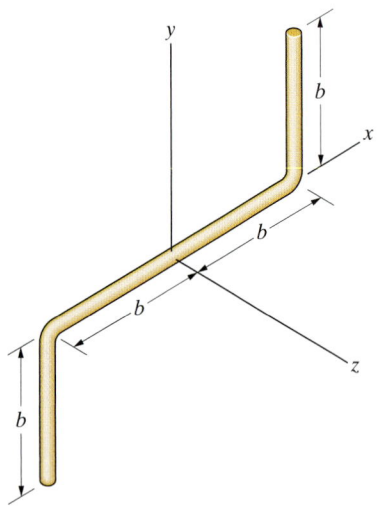

Problem 20.93

20.94* The 8-kg homogeneous slender bar has ball-and-socket supports at A and B.
(a) What is the bar's moment of inertia about the axis AB?
(b) If the bar rotates about the axis AB at 4 rad/s, what is the magnitude of its angular momentum about its axis of rotation?

20.95* The 8-kg homogeneous slender bar is released from rest in the position shown. (The x–z plane is horizontal.) What is the magnitude of the bar's angular acceleration about the axis AB at the instant of release?

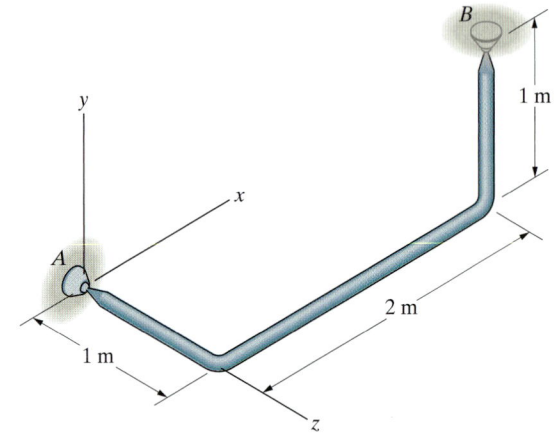

Problems 20.94/20.95

20.96 In terms of a coordinate system $x'y'z'$ with its origin at the center of mass, the inertia matrix of a rigid body is

$$[I'] = \begin{bmatrix} 20 & 10 & -10 \\ 10 & 60 & 0 \\ -10 & 0 & 80 \end{bmatrix} \text{kg-m}^2.$$

Determine the principal moments of inertia and unit vectors parallel to the corresponding principal axes.

20.97 For the steel plate and coordinate system in Problem 20.86, determine the principal moments of inertia and unit vectors parallel to the corresponding principal axes. Draw a sketch of the plate showing the principal axes.

20.98 The 1-kg, 1-m long slender bar lies in the $x'-y'$ plane. Its inertia matrix in kg-m² is

$$[I'] = \begin{bmatrix} \frac{1}{12}\sin^2\beta & -\frac{1}{12}\sin\beta\cos\beta & 0 \\ -\frac{1}{12}\sin\beta\cos\beta & \frac{1}{12}\cos^2\beta & 0 \\ 0 & 0 & \frac{1}{12} \end{bmatrix}.$$

Use Eqs. (20.45) and (20.46) to determine the principal moments of inertia and unit vectors parallel to the corresponding principal axes.

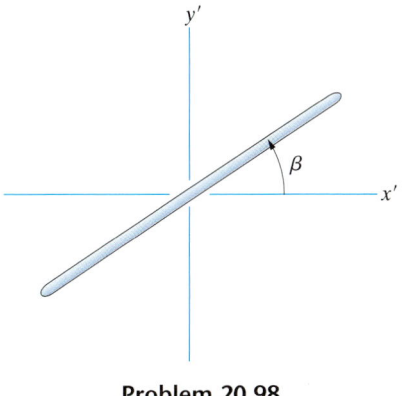

Problem 20.98

20.99* The mass of the homogeneous thin plate is 3 slugs. For the coordinate system shown, determine the plate's principal moments of inertia and unit vectors parallel to the corresponding principal axes.

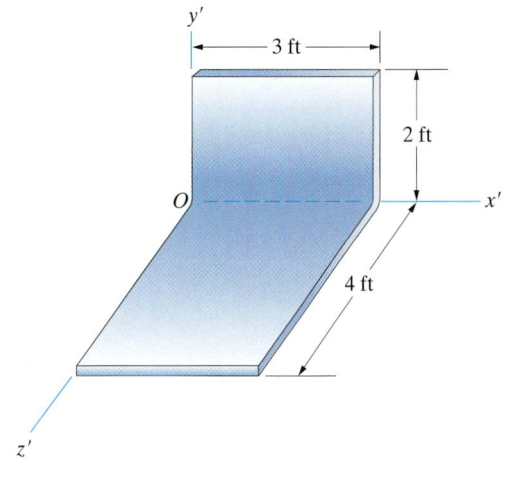

Problem 20.99

CHAPTER SUMMARY

Kinematics

The velocity of a point A of a rigid body relative to a given reference frame is given in terms of the velocity of a point B of the rigid body and the rigid body's angular velocity by (Fig. a)

$$\mathbf{v}_A = \mathbf{v}_B + \boldsymbol{\omega} \times \mathbf{r}_{A/B}. \qquad (20.1)$$

The acceleration of point A is given in terms of the acceleration of point B, the rigid body's angular acceleration, and its angular velocity by

$$\mathbf{a}_A = \mathbf{a}_B + \boldsymbol{\alpha} \times \mathbf{r}_{A/B} + \boldsymbol{\omega} \times (\boldsymbol{\omega} \times \mathbf{r}_{A/B}). \qquad (20.2)$$

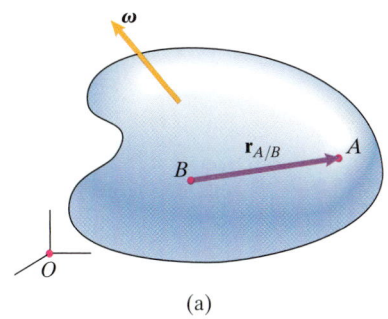

(a)

Consider a secondary reference frame xyz with angular velocity $\boldsymbol{\Omega}$ relative to a primary reference frame, and a rigid body with angular velocity $\boldsymbol{\omega}$ relative to the primary reference frame (Fig. b). If the secondary reference frame is body fixed, $\boldsymbol{\Omega} = \boldsymbol{\omega}$. The rigid body's angular acceleration relative to the primary reference frame is

$$\boldsymbol{\alpha} = \frac{d\omega_x}{dt}\mathbf{i} + \frac{d\omega_y}{dt}\mathbf{j} + \frac{d\omega_z}{dt}\mathbf{k} + \boldsymbol{\Omega} \times \boldsymbol{\omega}. \qquad (20.4)$$

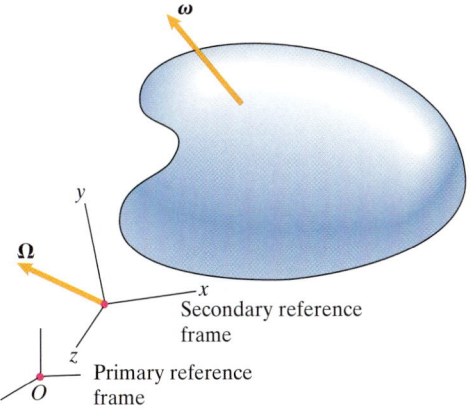

(b)

When the secondary reference frame is not body fixed, it is often convenient to express $\boldsymbol{\omega}$ as the sum of $\boldsymbol{\Omega}$ and the angular velocity $\boldsymbol{\omega}_{\mathrm{rel}}$ of the rigid body relative to the secondary reference frame:

$$\boldsymbol{\omega} = \boldsymbol{\Omega} + \boldsymbol{\omega}_{\mathrm{rel}}. \tag{20.5}$$

Euler's Equations

The three-dimensional equations of motion for a rigid body are called *Euler's equations*. They include Newton's second law and equations of angular motion. For a rigid body rotating about a fixed point O (Fig. c), the equations of angular motion are expressed in terms of the components of the total moment about O as

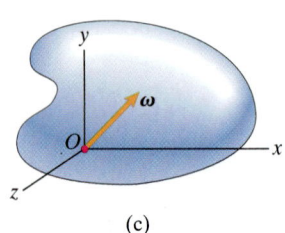

(c)

$$
\begin{aligned}
\Sigma M_{Ox} = \; & I_{xx}\frac{d\omega_x}{dt} - I_{xy}\frac{d\omega_y}{dt} - I_{xz}\frac{d\omega_z}{dt} \\
& - \Omega_z(-I_{yx}\omega_x + I_{yy}\omega_y - I_{yz}\omega_z) \\
& + \Omega_y(-I_{zx}\omega_x - I_{zy}\omega_y + I_{zz}\omega_z), \\[4pt]
\Sigma M_{Oy} = \; & -I_{yx}\frac{d\omega_x}{dt} + I_{yy}\frac{d\omega_y}{dt} - I_{yz}\frac{d\omega_z}{dt} \\
& + \Omega_z(I_{xx}\omega_x - I_{xy}\omega_y - I_{xz}\omega_z) \\
& - \Omega_x(-I_{zx}\omega_x - I_{zy}\omega_y + I_{zz}\omega_z), \\[4pt]
\Sigma M_{Oz} = \; & -I_{zx}\frac{d\omega_x}{dt} - I_{zy}\frac{d\omega_y}{dt} + I_{zz}\frac{d\omega_z}{dt} \\
& - \Omega_y(I_{xx}\omega_x - I_{xy}\omega_y - I_{xz}\omega_z) \\
& + \Omega_x(-I_{yx}\omega_x + I_{yy}\omega_y - I_{yz}\omega_z),
\end{aligned}
\tag{20.12}
$$

where $\boldsymbol{\omega}$ is the angular velocity of the rigid body and $\boldsymbol{\Omega}$ is the angular velocity of the chosen secondary reference frame with origin at O. If the secondary reference frame is body fixed, $\boldsymbol{\Omega} = \boldsymbol{\omega}$. In the case of general three-dimensional motion (Fig. d), the equations of angular motion [Eqs. (20.18)] are identical in form to those for motion about a fixed point, except that they are expressed in terms of the components of the total moment about the center of mass and the origin of the secondary reference frame is at the center of mass.

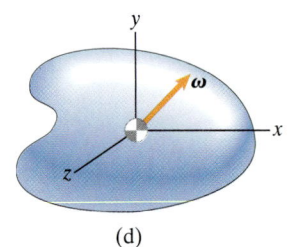

(d)

Euler Angles: Axisymmetric Objects

In the case of an object with an axis of rotational symmetry, the orientation of the xyz system relative to the reference XYZ system is specified by the *precession angle* ψ and the *nutation angle* θ (Fig. e). The rotation of the object relative to the xyz system is specified by the *spin angle* ϕ.

The components of the rigid body's angular velocity relative to the XYZ system are given by

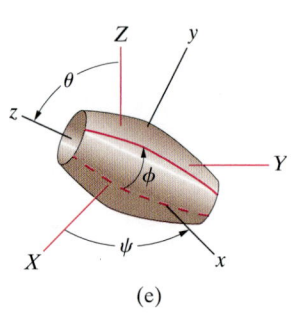

(e)

$$
\begin{aligned}
\omega_x &= \dot{\theta} \\
\omega_y &= \dot{\psi}\sin\theta, \\
\omega_z &= \dot{\phi} + \dot{\psi}\cos\theta.
\end{aligned}
\tag{20.23}
$$

The equations of angular motion expressed in terms of the Euler angles are

$$\Sigma M_x = I_{xx}\ddot{\theta} + (I_{zz} - I_{xx})\dot{\psi}^2 \sin\theta \cos\theta + I_{zz}\dot{\phi}\dot{\psi} \sin\theta, \quad (20.26)$$

$$\Sigma M_y = I_{xx}(\ddot{\psi} \sin\theta + 2\dot{\psi}\dot{\theta} \cos\theta) - I_{zz}(\dot{\phi}\dot{\theta} + \dot{\psi}\dot{\theta} \cos\theta), \quad (20.27)$$

$$\Sigma M_z = I_{zz}(\ddot{\phi} + \ddot{\psi} \cos\theta - \dot{\psi}\dot{\theta} \sin\theta). \quad (20.28)$$

In the *steady precession* of an axisymmetric spinning object, the spin rate $\dot{\phi}$, the nutation angle θ, and the precession rate $\dot{\psi}$ are assumed to be constant. With these assumptions, the equations of angular motion reduce to

$$\Sigma M_x = (I_{zz} - I_{xx})\dot{\psi}^2 \sin\theta \cos\theta + I_{zz}\dot{\phi}\dot{\psi} \sin\theta, \quad (20.29)$$

$$\Sigma M_y = 0, \quad (20.30)$$

$$\Sigma M_z = 0. \quad (20.31)$$

Moments and Products of Inertia

In terms of a given coordinate system xyz, the *inertia matrix* of an object is defined by

$$[I] = \begin{bmatrix} I_{xx} & -I_{xy} & -I_{xz} \\ -I_{yx} & I_{yy} & -I_{yz} \\ -I_{zx} & -I_{zy} & I_{zz} \end{bmatrix} \quad (20.37)$$

$$= \begin{bmatrix} \int_m (y^2 + z^2)\, dm & -\int_m xy\, dm & -\int_m xz\, dm \\ -\int_m yx\, dm & \int_m (x^2 + z^2)\, dm & -\int_m yz\, dm \\ -\int_m zx\, dm & -\int_m zy\, dm & \int_m (x^2 + y^2)\, dm \end{bmatrix},$$

where x, y, and z are the coordinates of the differential element of mass dm. The terms I_{xx}, I_{yy}, and I_{zz} are the *moments of inertia* about the x, y, and z axes, and I_{xy}, I_{yz}, and I_{zx} are the *products of inertia*.

If $x'y'z'$ is a coordinate system with its origin at the center of mass of an object and xyz is a parallel system (Fig. f), the *parallel-axis theorems* state that

$$I_{xx} = I_{x'x'} + (d_y^2 + d_z^2)\, m,$$

$$I_{yy} = I_{y'y'} + (d_x^2 + d_z^2)\, m,$$

$$I_{zz} = I_{z'z'} + (d_x^2 + d_y^2)\, m, \quad (20.42)$$

$$I_{xy} = I_{x'y'} + d_x d_y m,$$

$$I_{yz} = I_{y'z'} + d_y d_z m,$$

$$I_{zx} = I_{z'x'} + d_z d_x m.$$

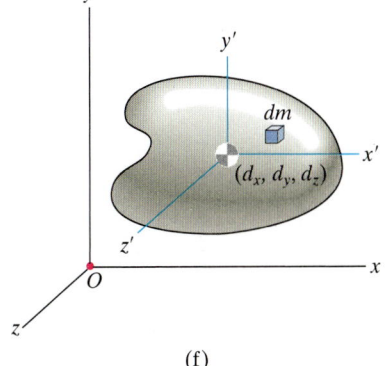

(f)

where (d_x, d_y, d_z) are the coordinates of the center of mass of the object in the xyz coordinate system.

The moment of inertia of the object about an axis through the origin O parallel to a unit vector $\mathbf{e}$ is given by

$$I_O = I_{xx}e_x^2 + I_{yy}e_y^2 + I_{zz}e_z^2 - 2I_{xy}e_x e_y - 2I_{yz}e_y e_z - 2I_{zx}e_z e_x. \quad (20.43)$$

Review Problems

20.100 The disk is pinned to the horizontal shaft and rotates relative to it with angular velocity ω_d. Relative to an earth-fixed reference frame, the vertical shaft rotates with angular velocity ω_0.
(a) Determine the disk's angular velocity vector ω relative to the earth-fixed reference frame.
(b) What is the velocity of point A of the disk relative to the earth-fixed reference frame?

20.101 The disk is pinned to the horizontal shaft and rotates relative to it with constant angular velocity ω_d. Relative to an earth-fixed reference frame, the vertical shaft rotates with constant angular velocity ω_0. What is the acceleration of point A of the disk relative to the earth-fixed reference frame?

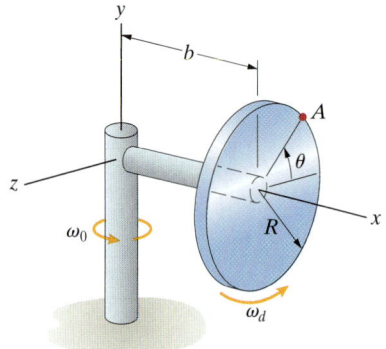

Problems 20.100/20.101

20.102 The cone is connected by a ball-and-socket joint at its vertex to a 100-mm post. The radius of its base is 100 mm, and the base rolls on the floor. The velocity of the center of the base is $\mathbf{v}_C = 2\mathbf{k}$ (m/s).
(a) What is the cone's angular velocity vector ω?
(b) What is the velocity of point A?

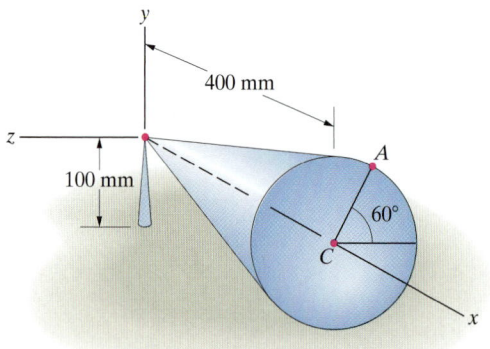

Problem 20.102

20.103 The mechanism shown is a type of universal joint called a yoke and spider. The axis L lies in the x–z plane. Determine the angular velocity ω_L and the angular velocity vector ω_S of the cross-shaped "spider" in terms of the angular velocity ω_R at the instant shown.

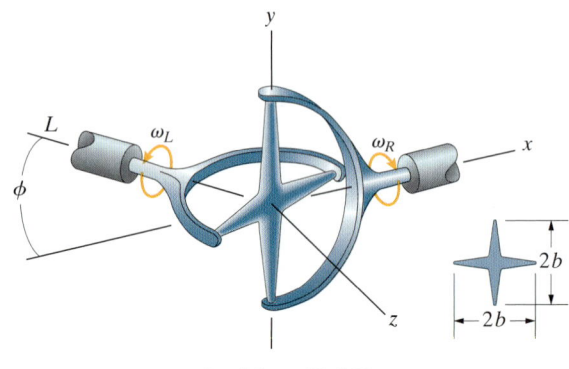

Problem 20.103

20.104 The inertia matrix of a rigid body in terms of a body-fixed coordinate system with its origin at the center of mass is

$$[I] = \begin{bmatrix} 4 & 1 & -1 \\ 1 & 2 & 0 \\ -1 & 0 & 6 \end{bmatrix} \text{kg-m}^2.$$

If the rigid body's angular velocity is $\omega = 10\mathbf{i} - 5\mathbf{j} + 10\mathbf{k}$ (rad/s), what is its angular momentum about its center of mass?

20.105 What is the moment of inertia of the rigid body described in Problem 20.104 about the axis that passes through the origin and the point $(4, -4, 7)$ m?

    Strategy: Determine the components of a unit vector parallel to the axis and use Eq. (20.43).

20.106 Determine the inertia matrix of the 0.6-slug thin plate in terms of the coordinate system shown.

20.107 At $t = 0$, the 0.6-slug thin plate has angular velocity $\boldsymbol{\omega} = 10\mathbf{i} + 10\mathbf{j}$ (rad/m) and is subjected to the force $\mathbf{F} = -10\mathbf{k}$ (lb) acting at the point $(0, 6, 0)$ in. No other forces or couples act on the plate. What are the components of its angular acceleration at that instant?

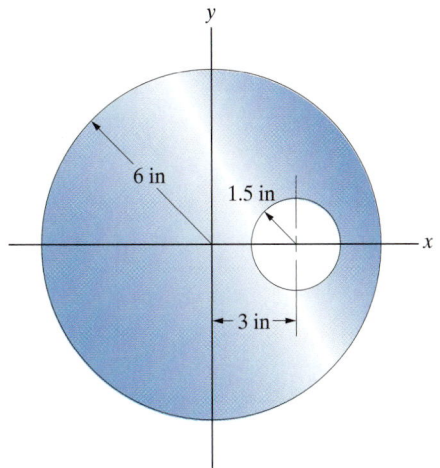

6 in

1.5 in

3 in

Problems 20.106/20.107

20.108 The inertia matrix of a rigid body in terms of a body-fixed coordinate system with its origin at the center of mass is

$$[I] = \begin{bmatrix} 4 & 1 & -1 \\ 1 & 2 & 0 \\ -1 & 0 & 6 \end{bmatrix} \text{kg-m}^2.$$

If the rigid body's angular velocity is $\boldsymbol{\omega} = 10\mathbf{i} - 5\mathbf{j} + 10\mathbf{k}$ (rad/s) and its angular acceleration is zero, what are the components of the total moment about its center of mass?

20.109 If the total moment about the center of mass of the rigid body described in Problem 20.108 is zero, what are the components of its angular acceleration?

20.110 The slender bar of length l and mass m is pinned to the L-shaped bar at O. The L-shaped bar rotates about the vertical axis with a constant angular velocity ω_0. Determine the value of ω_0 necessary for the bar to remain at a constant angle β relative to the vertical.

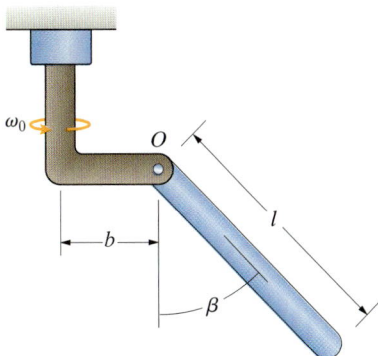

ω_0

O

b

l

β

Problem 20.110

20.111 A slender bar of length l and mass m is rigidly attached to the center of a thin circular disk of radius R and mass m. The composite object undergoes a motion in which the bar rotates in the horizontal plane with constant angular velocity ω_0 about the center of mass of the composite object and the disk rolls on the floor. Show that $\omega_0 = 2\sqrt{g/R}$.

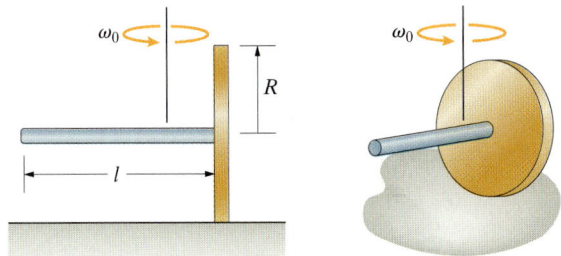

ω_0 ω_0

R

l

Problem 20.111

20.112* The thin plate of mass m spins about a vertical axis with the plane of the plate perpendicular to the floor. The corner of the plate at O rests in an indentation, so that it remains at the same point on the floor. The plate rotates with constant angular velocity ω_0 and the angle β is constant.
(a) Show that the angular velocity ω_0 is related to the angle β by

$$\frac{h\omega_0^2}{g} = \frac{2\cos\beta - \sin\beta}{\sin^2\beta - 2\sin\beta\cos\beta - \cos^2\beta}.$$

(b) The equation you obtained in (a) indicates that $\omega_0 = 0$ when $2\cos\beta - \sin\beta = 0$. What is the interpretation of this result?

20.113* In Problem 20.112, determine the range of values of the angle β for which the plate will remain in the steady motion described.

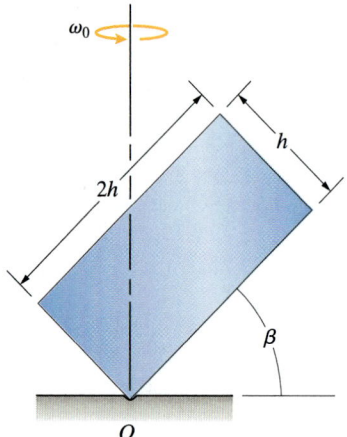

Problems 20.112/20.113

20.114 Arm BC has a mass of 12 kg, and its moments and products of inertia, in terms of the coordinate system shown, are $I_{xx} = 0.03$ kg-m², $I_{yy} = I_{zz} = 4$ kg-m², and $I_{xy} = I_{yz} = I_{xz} = 0$. At the instant shown, arm AB is rotating in the horizontal plane with a constant angular velocity of 1 rad/s in the counterclockwise direction viewed from above. Relative to arm AB, arm BC is rotating about the z axis with a constant angular velocity of 2 rad/s. Determine the force and couple exerted on arm BC at B.

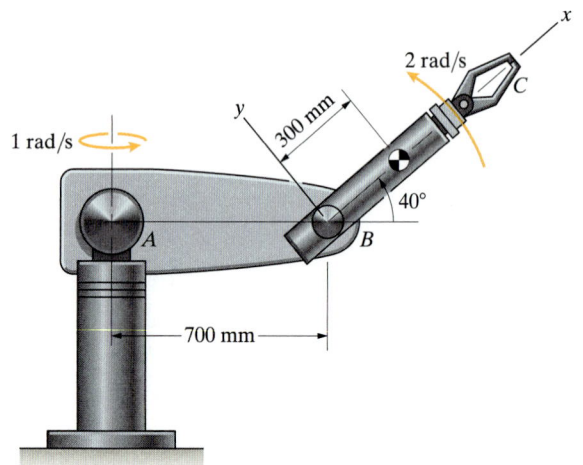

Problem 20.114

20.115 Suppose that you throw a football in a wobbly spiral with a nutation angle of 25°. The football's moments of inertia are $I_{xx} = I_{yy} = 0.003$ slug-ft² and $I_{zz} = 0.001$ slug-ft². If the spin rate is $\dot{\phi} = 4$ revolutions per second, what is the magnitude of the precession rate (the rate at which the football wobbles)?

20.116 Sketch the body and space cones for the motion of the football in Problem 20.115.

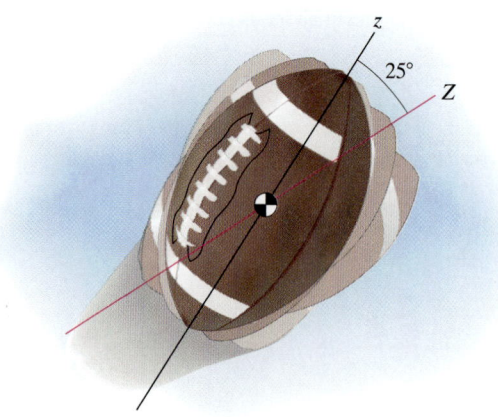

Problems 20.115/20.116

20.117 The mass of the homogeneous thin plate is 1 kg. For the coordinate system shown, determine the plate's principal moments of inertia and the directions of unit vectors parallel to the corresponding principal axes.

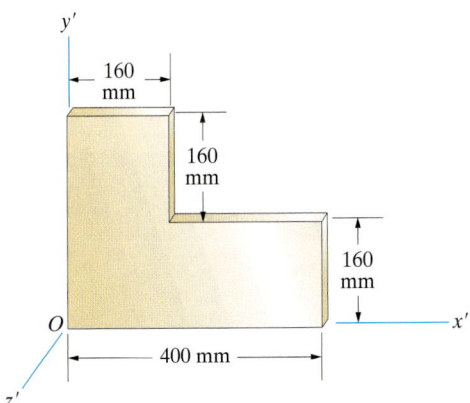

Problem 20.117

20.118 The airplane's principal moments of inertia, in slug-ft^2, are $I_{xx} = 8000$, $I_{yy} = 48,000$, and $I_{zz} = 50,000$.

(a) The airplane begins in the reference position shown and maneuvers into the orientation $\psi = \theta = \phi = 45°$. Draw a sketch showing the plane's orientation relative to the XYZ system.

(b) If the airplane is in the orientation described in (a), the rates of change of the Euler angles are $\dot{\psi} = 0$, $\dot{\theta} = 0.2$ rad/s, and $\dot{\phi} = 0.2$ rad/s, and the second derivatives of the angles with respect to time are zero, what are the components of the total moment about the airplane's center of mass?

20.119 What are the x, y, and z components of the angular acceleration of the airplane described in Problem 20.118?

20.120 If the orientation of the airplane in Problem 20.118 is $\psi = 45°$, $\theta = 60°$, and $\phi = 45°$, the rates of change of the Euler angles are $\dot{\psi} = 0$, $\dot{\theta} = 0.2$ rad/s, and $\dot{\phi} = 0.1$ rad/s, and the components of the total moment about the center of mass of the plane are $\Sigma M_x = 400$ ft-lb, $\Sigma M_y = 1200$ ft-lb, and $\Sigma M_z = 0$, what are the x, y, and z components of the airplane's angular acceleration?

Problems 20.118–20.120

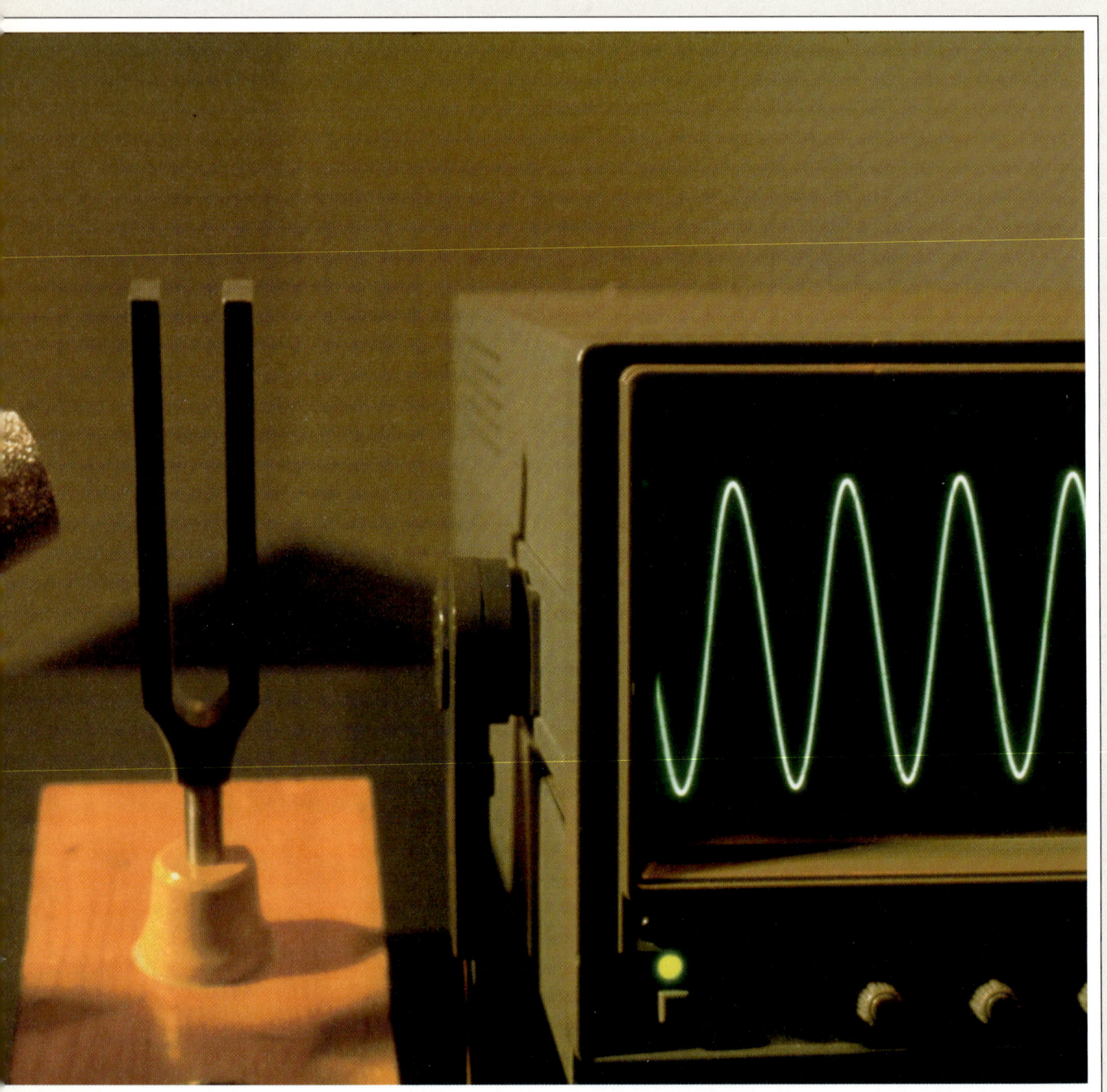

CHAPTER
21
Vibrations

Vibrations have been of concern in engineering since the beginning of the industrial revolution. Beginning with the development of electromechanical devices capable of creating and measuring mechanical vibrations, engineering applications of vibrations have included the various areas of acoustics, from architectural acoustics to earthquake detection and analysis. We consider vibrating systems that have one degree of freedom—that is, the position, or configuration, of a system can be specified by a single variable. The fundamental concepts we introduce, including amplitude, frequency, period, damping, and resonance, are also used in the analysis of systems with multiple degrees of freedom.

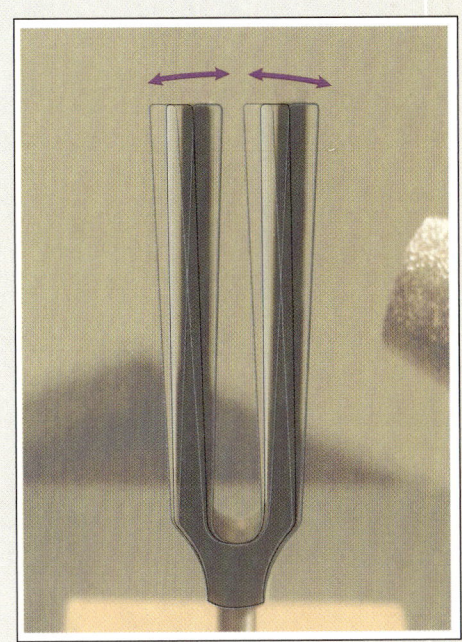

◀ Vibrations of the tuning fork's prongs create sound waves that register on the microphone and are displayed by the oscilloscope. In this chapter we analyze the vibrations of simple mechanical systems.

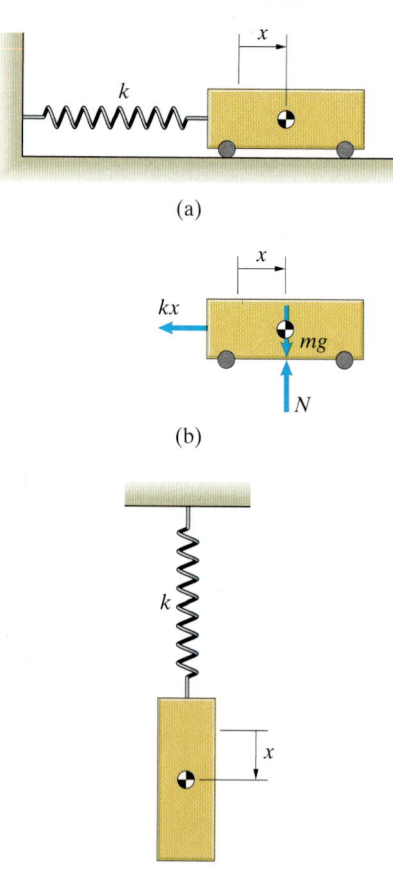

(a)

(b)

(c)

Figure 21.1
(a) The spring–mass oscillator has one degree of freedom.
(b) Free-body diagram of the mass.
(c) Suspending the mass.

21.1 Conservative Systems

We begin by presenting different examples of one-degree-of-freedom systems subjected to conservative forces, demonstrating that their motions are described by the same differential equation. We then examine solutions of this equation and use them to describe the vibrations of one-degree-of-freedom conservative systems.

Examples

The *spring–mass oscillator* (Fig. 21.1a) is the simplest example of a one-degree-of-freedom vibrating system. A single coordinate x measuring the displacement of the mass relative to a reference point is sufficient to specify the position of the system. We draw the free-body diagram of the mass in Fig. 21.1b, neglecting friction and assuming that the spring is unstretched when $x = 0$. Applying Newton's second law, we can write the equation describing the horizontal motion of the mass as

$$\frac{d^2x}{dt^2} + \frac{k}{m}x = 0. \tag{21.1}$$

We can obtain this equation by a different method that is very useful. The only force that does work on the mass, the force exerted by the spring, is conservative, which means that the sum of the kinetic and potential energies is constant:

$$\frac{1}{2}m\left(\frac{dx}{dt}\right)^2 + \frac{1}{2}kx^2 = \text{constant}.$$

Taking the derivative of this equation with respect to time, we can write the result as

$$\left(\frac{dx}{dt}\right)\left(\frac{d^2x}{dt^2} + \frac{k}{m}x\right) = 0,$$

again obtaining Eq. (21.1).

Suppose that the mass is suspended from the spring, as shown in Fig. 21.1c, and undergoes vertical motion. If the spring is unstretched when $x = 0$, it is easy to confirm that the equation of motion is

$$\frac{d^2x}{dt^2} + \frac{k}{m}x = g.$$

If the suspended mass is stationary, the magnitude of the force exerted by the spring must equal the weight ($kx = mg$), so the equilibrium position is $x = mg/k$. (Notice that we can also determine the equilibrium position by setting the acceleration equal to zero in the equation of motion.) Let us introduce a new variable $\tilde{x}$ that measures the position of the mass relative to its equilibrium position: $\tilde{x} = x - mg/k$. Writing the equation of motion in terms of this variable, we obtain

$$\frac{d^2\tilde{x}}{dt^2} + \frac{k}{m}\tilde{x} = 0, \tag{21.2}$$

which is identical to Eq. (21.1). The vertical motion of the mass in Fig. 21.1c relative to its equilibrium position is described by the same equation that describes the horizontal motion of the mass in Fig. 21.1a relative to its equilibrium position.

Now let's consider a different one-degree-of-freedom system. If we rotate the slender bar in Fig. 21.2a through some angle and release it, it will oscillate back and forth. (An object swinging from a fixed point is called a *pendulum*.) There is only one degree of freedom, since θ specifies the bar's position. Drawing the free-body diagram of the bar (Fig. 21.2b) and writing the equation of angular motion about A yields

$$\frac{d^2\theta}{dt^2} + \frac{3g}{2l}\sin\theta = 0. \tag{21.3}$$

We can also obtain this equation by using conservation of energy. The bar's kinetic energy is $T = \frac{1}{2}I_A(d\theta/dt)^2$. If we place the datum at the level of point A (Fig. 21.2b), the potential energy associated with the bar's weight is $V = -mg\left(\frac{1}{2}l\cos\theta\right)$, so

$$T + V = \frac{1}{2}\left(\frac{1}{3}ml^2\right)\left(\frac{d\theta}{dt}\right)^2 - \frac{1}{2}mgl\cos\theta = \text{constant}.$$

Taking the derivative of this equation with respect to time and writing the result in the form

$$\left(\frac{d\theta}{dt}\right)\left(\frac{d^2\theta}{dt^2} + \frac{3g}{2l}\sin\theta\right) = 0,$$

we obtain Eq. (21.3). Note that Eq. (21.3) does not have the same form as Eq. (21.1). However, if we express $\sin\theta$ in terms of its Taylor series,

$$\sin\theta = \theta - \frac{1}{6}\theta^3 + \frac{1}{120}\theta^5 + \cdots,$$

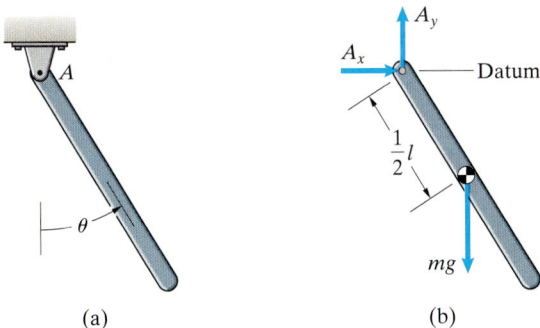

(a)   (b)

Figure 21.2
(a) A pendulum consisting of a slender bar.
(b) Free-body diagram of the bar.

and assume that θ remains small enough to approximate $\sin \theta$ by θ, then Eq. (21.3) becomes identical in form to Eq. (21.1):

$$\frac{d^2\theta}{dt^2} + \frac{3g}{2l}\theta = 0. \tag{21.4}$$

Our analyses of the spring–mass oscillator and the pendulum resulted in equations of motion that are identical in form. To accomplish this in the case of the suspended spring–mass oscillator, we had to express the equation of motion in terms of displacement relative to the equilibrium position. In the case of the pendulum, we needed to assume that the motions were small. But within those restrictions, the form of equation we obtained describes the motions of many one-degree-of-freedom conservative systems.

Solutions

Let us consider the differential equation

$$\frac{d^2x}{dt^2} + \omega^2 x = 0, \tag{21.5}$$

where ω is a constant. We have seen that with $\omega^2 = k/m$, this equation describes the motion of a spring–mass oscillator, and with $\omega^2 = 3g/2l$, it describes small motions of a suspended slender bar. Equation (21.5) is an *ordinary differential equation*, because it is expressed in terms of ordinary (not partial) derivatives of the dependent variable x with respect to the independent variable t. Also, Eq. (21.5) is *linear*, meaning that there are no nonlinear terms in x or its derivatives, and it is *homogeneous*, meaning that each term contains x or one of its derivatives. Finally, Eq. (21.5) has *constant coefficients*, meaning that the coefficients multiplying the dependent variable x or its derivative in each term do not depend on the independent variable t. The standard approach to solving a differential equation of this kind is to assume that the solution is of the form

$$x = Ce^{\lambda t}, \tag{21.6}$$

where C and λ are constants. Substituting this expression into Eq. (21.5) yields

$$Ce^{\lambda t}(\lambda^2 + \omega^2) = 0.$$

This equation is satisfied for any value of the constant C if $\lambda = i\omega$ or $\lambda = -i\omega$, where $i = \sqrt{-1}$, so there are two nontrivial solutions of the form of Eq. (21.6), which we write as

$$x = Ce^{i\omega t} + De^{-i\omega t}. \tag{21.7}$$

By using *Euler's identity* $e^{i\theta} = \cos\theta + i\sin\theta$, we can express Eq. (21.7) in the alternative form

$$x = A\sin\omega t + B\cos\omega t, \tag{21.8}$$

where A and B are arbitrary constants.

Although in practical applications Eq. (21.8) is usually the most convenient form of the solution of Eq. (21.5), we can describe the properties of the solution more easily by expressing it in the form

$$x = E \sin(\omega t - \phi), \tag{21.9}$$

where E and ϕ are constants. To show that this solution is equivalent to Eq. (21.8), we use the identity

$$E \sin(\omega t - \phi) = E(\sin \omega t \cos \phi - \cos \omega t \sin \phi)$$
$$= (E \cos \phi) \sin \omega t + (-E \sin \phi) \cos \omega t.$$

This expression is identical to Eq. (21.8) if the constants A and B are related to E and ϕ by

$$A = E \cos \phi \quad \text{and} \quad B = -E \sin \phi. \tag{21.10}$$

Equation (21.9) clearly demonstrates the oscillatory nature of the solution of Eq. (21.5). Called *simple harmonic motion*, it describes a sinusoidal function of ωt (Fig. 21.3). The positive constant E is called the *amplitude* of the vibration. By squaring Eqs. (21.10) and adding them, we obtain a relation between the amplitude and the constants A and B:

$$E = \sqrt{A^2 + B^2}. \tag{21.11}$$

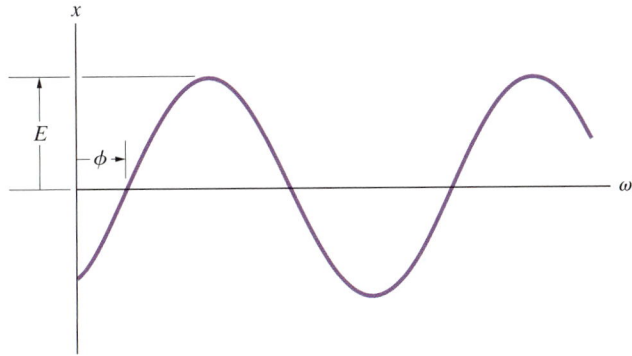

Figure 21.3
Graph of x as a function of ωt.

Equation (21.9) can be interpreted in terms of the uniform motion of a point along a circular path. We draw a circle whose radius equals the amplitude (Fig. 21.4) and assume that the line from O to P rotates in the counterclockwise direction with constant angular velocity ω. If we choose the position of P at $t = 0$ as shown, the projection of the line OP onto the vertical axis is $E \sin(\omega t - \phi)$. Thus, there is a one-to-one correspondence between the circular motion of P and Eq. (21.9). Point P makes one complete revolution, or *cycle*, during the time required for the angle ωt to increase by 2π radians. The

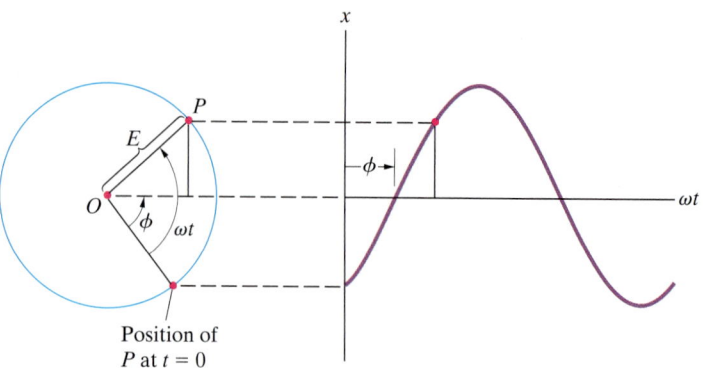

Figure 21.4
Correspondence of simple harmonic motion with circular motion
of a point.

time $\tau = 2\pi/\omega$ required for one cycle is called the *period* of the vibration.
Since τ is the time required for one cycle, its inverse $f = 1/\tau$ is the number
of cycles per unit time, or *frequency* of the vibration. The frequency is usually expressed in cycles per second, or *Hertz* (Hz). The effect of changing the period and frequency is illustrated in Fig. 21.5.

We see that the period and frequency are given by

$$\tau = \frac{2\pi}{\omega}, \tag{21.12}$$

$$f = \frac{\omega}{2\pi}. \tag{21.13}$$

A system's period and frequency are determined by its physical properties, and
do not depend on the functional form in which its motion is expressed. The frequency f is the number of revolutions the point P moves around the circular
path in Fig. 21.4 per unit time, so $\omega = 2\pi f$ is the number of radians per unit

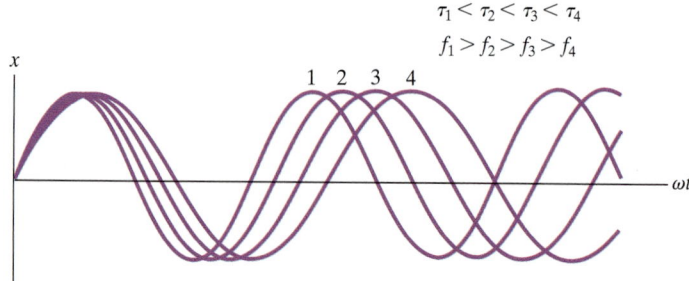

Figure 21.5
Effect of increasing the period (decreasing the frequency) of simple
harmonic motion.

time. Therefore, ω *is also a measure of the frequency* and is expressed in radians per second (rad/s).

Suppose that Eq. (21.9) describes the displacement of the spring–mass oscillator in Fig. 21.1a, so that $\omega^2 = k/m$. Then the kinetic energy of the mass is

$$T = \frac{1}{2}m\left(\frac{dx}{dt}\right)^2 = \frac{1}{2}mE^2\omega^2\cos^2(\omega t - \phi), \qquad (21.14)$$

and the potential energy of the spring is

$$V = \frac{1}{2}kx^2 = \frac{1}{2}mE^2\omega^2\sin^2(\omega t - \phi). \qquad (21.15)$$

The sum of the kinetic and potential energies, $T + V = \frac{1}{2}mE^2\omega^2$, is constant (Fig. 21.6). As the system vibrates, its total energy oscillates between kinetic and potential energy. Notice that the total energy is proportional to the square of the amplitude and the square of the natural frequency.

In summary, the motions of many one-degree-of-freedom conservative systems relative to an equilibrium position can be modeled by Eq. (21.5). To obtain the equation in that form, it may be necessary to linearize the system's equation of motion by assuming that the displacement from equilibrium is small, as we did in obtaining Eq. (21.4) from Eq. (21.3). Once the equation of motion has been expressed in the form of Eq. (21.5), the value of ω is known in terms of the physical parameters (spring constants and masses) of the system and can be used to determine the period and frequency of the system from Eqs. (21.12) and (21.13). The displacement of the system can be determined as a function of time from Eq. (21.8) or Eq. (21.9) if there is sufficient information to determine the arbitrary constants.

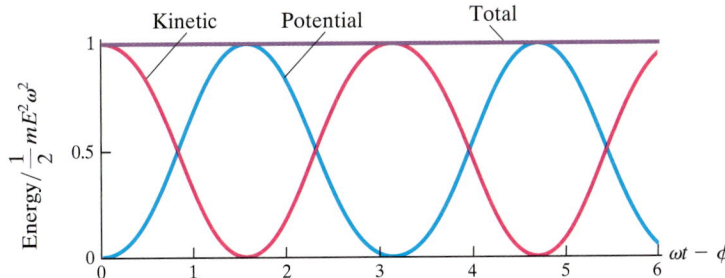

Figure 21.6
Kinetic, potential, and total energies of a spring–mass oscillator.

Study Questions

1. What is a one-degree-of-freedom system? Give examples.

2. How can you use conservation of energy to obtain the equation of motion for a conservative one-degree-of-freedom system?

3. If the motion of a one-degree-of-freedom system is described by Eq. (21.5), how can you determine the period and frequency of the system?

4. If you know the constants A and B in Eq. (21.8), how can you determine the amplitude of the vibrations the equation describes?

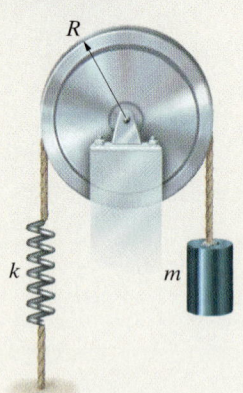

Figure 21.7

Example 21.1 **Vibration of a Conservative System**

The pulley in Fig. 21.7 has radius R and moment of inertia I, and the cable does not slip relative to the pulley. The mass m is displaced downward a distance h from its equilibrium position and released from rest at $t = 0$.
(a) What is the frequency of the resulting vibrations?
(b) Determine the position of the mass relative to its equilibrium position as a function of time.

Strategy

A single coordinate specifying the vertical displacement of the mass specifies the positions of both the mass and pulley, so there is one degree of freedom. We can obtain the equation of motion of the system either by writing the individual equations of motion of the mass and pulley or by using conservation of energy.

Solution

Let x be the downward displacement of the mass relative to its position when the spring is unstretched. We draw the free-body diagrams of the pulley and mass in Fig. a, where T_C is the tension in the cable and α is the angular acceleration of the pulley. Applying Newton's second law to the mass, we obtain

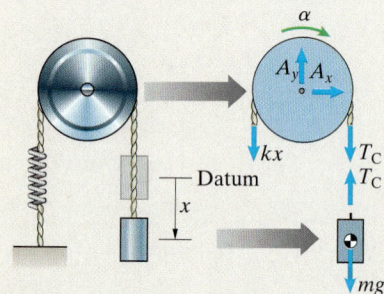

(a) Free-body diagrams of the pulley and mass.

$$mg - T_C = m\frac{d^2x}{dt^2}. \qquad (1)$$

The equation of angular motion for the pulley is

$$T_C R - (kx)R = I\alpha.$$

The angular acceleration of the pulley and the acceleration of the mass are related by $\alpha = (d^2x/dt^2)/R$, so we can write the equation of angular motion as

$$T_C - kx = \left(\frac{I}{R^2}\right)\frac{d^2x}{dt^2}.$$

Summing this equation and Eq. (1), we obtain the equation of motion:

$$\left(m + \frac{I}{R^2}\right)\frac{d^2x}{dt^2} + kx = mg. \qquad (2)$$

Alternative Method In terms of the velocity of the mass, the angular velocity of the pulley is $(dx/dt)/R$. Therefore, we can write the total kinetic energy of the mass and pulley as

$$T = \frac{1}{2}m\left(\frac{dx}{dt}\right)^2 + \frac{1}{2}I\left[\frac{1}{R}\left(\frac{dx}{dt}\right)\right]^2.$$

Placing the datum for the potential energy associated with the weight of the mass at $x = 0$, we obtain the total potential energy:

$$V = -mgx + \frac{1}{2}kx^2.$$

The sum of the kinetic and potential energies is constant:

$$T + V = \frac{1}{2}\left(m + \frac{I}{R^2}\right)\left(\frac{dx}{dt}\right)^2 - mgx + \frac{1}{2}kx^2 = \text{constant.}$$

Taking the derivative of this equation with respect to time, we again obtain Eq. (2).

By setting $d^2x/dt^2 = 0$ in Eq. (2), we see that the equilibrium position is $x = mg/k$. By expressing Eq. (2) in terms of a new variable $\tilde{x} = x - mg/k$ that measures the position of the mass relative to its equilibrium position, we obtain the equation of motion in the form of Eq. (21.5),

$$\frac{d^2\tilde{x}}{dt^2} + \omega^2\tilde{x} = 0,$$

where

$$\omega^2 = \frac{k}{m + I/R^2}.$$

(a) The frequency of vibration of the system is

$$f = \frac{\omega}{2\pi} = \frac{1}{2\pi}\sqrt{\frac{k}{m + I/R^2}}.$$

(b) From Eq. (21.8), we can write the general solution for $\tilde{x}$ in the form

$$\tilde{x} = A \sin \omega t + B \cos \omega t.$$

When $t = 0$, $\tilde{x} = h$ and $d\tilde{x}/dt = 0$. The derivative of the general solution is

$$\frac{d\tilde{x}}{dt} = A\omega \cos \omega t - B\omega \sin \omega t.$$

The initial conditions yield the equations

$$h = B$$

and

$$0 = A\omega,$$

so the position of the mass relative to its equilibrium position as a function of time is

$$\tilde{x} = h \cos \omega t.$$

Critical Thinking

By applying Newton's second law to the spring–mass oscillators in Fig. 21.1, we obtained equations of motion having the form of Eq. (21.5). Although the system in this example is more complex, it has only one degree of freedom, and the equation governing its motion is also identical in form to Eq. (21.5). Notice that in order to obtain it in that form, we needed to introduce a variable that measured the displacement of the system relative to its equilibrium position.

Example 21.2   Frequency of a System

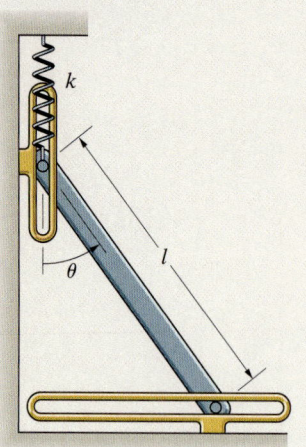

Figure 21.8

The spring attached to the slender bar of mass m in Fig. 21.8 is unstretched when $\theta = 0$. Neglecting friction, determine the frequency of small vibrations of the bar relative to its equilibrium position.

Strategy

The angle θ specifies the bar's position, so there is one degree of freedom. We can express the kinetic and potential energies in terms of θ and its time derivative and then take the derivative of the total energy with respect to time to obtain the equation of motion.

Solution

The kinetic energy of the bar is

$$T = \frac{1}{2}mv^2 + \frac{1}{2}I\left(\frac{d\theta}{dt}\right)^2,$$

where v is the velocity of the center of mass and $I = \frac{1}{12}ml^2$. The distance from the bar's instantaneous center to its center of mass is $\frac{1}{2}l$ (Fig. a), so $v = \left(\frac{1}{2}l\right)(d\theta/dt)$, and the kinetic energy is

$$T = \frac{1}{2}m\left[\frac{1}{2}l\left(\frac{d\theta}{dt}\right)\right]^2 + \frac{1}{2}\left(\frac{1}{12}ml^2\right)\left(\frac{d\theta}{dt}\right)^2 = \frac{1}{6}ml^2\left(\frac{d\theta}{dt}\right)^2.$$

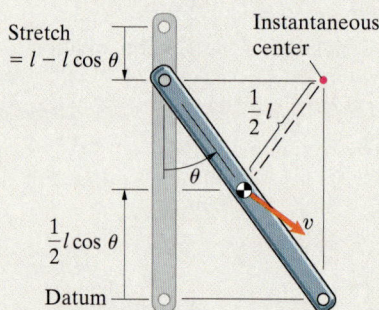

(a) Determining the velocity of the center of mass, the stretch of the spring, and the height of the center of mass above the datum.

In terms of θ, the stretch of the spring is $l - l\cos\theta$. We place the datum for the potential energy associated with the weight at the bottom of the bar (Fig. a), so the total potential energy is

$$V = mg\left(\frac{1}{2}l\cos\theta\right) + \frac{1}{2}k(l - l\cos\theta)^2.$$

The sum of the kinetic and potential energies is constant:

$$T + V = \frac{1}{6}ml^2\left(\frac{d\theta}{dt}\right)^2 + \frac{1}{2}mgl\cos\theta + \frac{1}{2}kl^2(1 - \cos\theta)^2 = \text{constant}.$$

Taking the derivative of this equation with respect to time, we obtain the equation of motion:

$$\frac{1}{3}ml^2\frac{d^2\theta}{dt^2} - \frac{1}{2}mgl\sin\theta + kl^2(1 - \cos\theta)\sin\theta = 0. \qquad (1)$$

To express this equation in the form of Eq. (21.5), we need to write it in terms of small vibrations relative to the equilibrium position. Let θ_e be the value of θ when the bar is in equilibrium. By setting $d^2\theta/dt^2 = 0$ in Eq. (1), we find that θ_e must satisfy the relation

$$\cos\theta_e = 1 - \frac{mg}{2kl}. \qquad (2)$$

We define $\widetilde{\theta} = \theta - \theta_e$, and expand $\sin\theta$ and $\cos\theta$ in Taylor series in terms of $\widetilde{\theta}$:

$$\sin\theta = \sin(\theta_e + \widetilde{\theta}) = \sin\theta_e + \cos\theta_e\widetilde{\theta} + \cdots,$$

$$\cos\theta = \cos(\theta_e + \widetilde{\theta}) = \cos\theta_e - \sin\theta_e\widetilde{\theta} + \cdots.$$

Substituting these expressions into Eq. (1), neglecting terms in $\widetilde{\theta}$ of second and higher orders, and using Eq. (2), we obtain

$$\frac{d^2\widetilde{\theta}}{dt^2} + \omega^2\widetilde{\theta} = 0,$$

where

$$\omega^2 = \frac{3g}{l}\left(1 - \frac{mg}{4kl}\right).$$

From Eq. (21.13), the frequency of small vibrations of the bar is

$$f = \frac{\omega}{2\pi} = \frac{1}{2\pi}\sqrt{\frac{3g}{l}\left(1 - \frac{mg}{4kl}\right)}.$$

Critical Thinking

This example demonstrates the advantage of using conservation of energy to obtain the equation of motion of a one-degree-of-freedom conservative system. You should confirm that you can also obtain the equation of motion by drawing the free-body diagram of the bar and applying Newton's second law and the equation of angular motion. But that procedure is considerably more involved, in part because it is necessary to consider the normal forces exerted on the ends of the bar. We were able to ignore them in applying conservation of energy because they do no work on the bar.

Problems

21.1 Confirm that $x = A \sin \omega t + B \cos \omega t$, where A and B are arbitrary constants, satisfies Eq. (21.5).

21.2 Confirm that $x = E \sin(\omega t - \phi)$, where E and ϕ are arbitrary constants, satisfies Eq. (21.5).

21.3 The mass $m = 4$ kg and the spring constant is $k = 64$ N/m. For vibration of the spring–mass oscillator relative to its equilibrium position, determine (a) the frequency in Hz and (b) the period.

21.4 The mass $m = 4$ kg and the spring constant is $k = 64$ N/m. The spring is unstretched when $x = 0$. At $t = 0$, $x = 0$ and the mass has a velocity of 2 m/s toward the right. What is the value of x at $t = 1$ s?

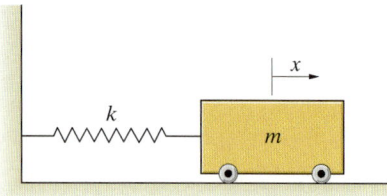

Problems 21.3/21.4

21.5 The mass $m = 4$ kg and the spring constant is $k = 64$ N/m. For vibration of the spring–mass oscillator relative to its equilibrium position, determine (a) the frequency in Hz and (b) the period.

21.6 The mass $m = 4$ kg and the spring constant is $k = 64$ N/m. The spring is unstretched when $x = 0$. At $t = 0$, $x = 0$ and the mass has a velocity of 2 m/s down the inclined surface. What is the value of x at $t = 0.8$ s?

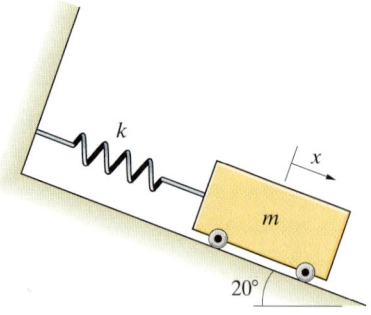

Problems 21.5/21.6

21.7 Suppose that in a mechanical design course you are asked to design a pendulum clock, and you begin with the pendulum. The mass of the disk is 2 kg. Determine the length L of the bar so that the period of small oscillations of the pendulum is 1 s. For this preliminary estimate, neglect the mass of the bar.

21.8 The mass of the disk is 2 kg. The length of the bar is $L = 60$ mm and its mass is negligible. At $t = 0$, the pendulum is vertical and it has a counterclockwise angular velocity of 35 degrees per second. What is the amplitude of the resulting vibrations in degrees?

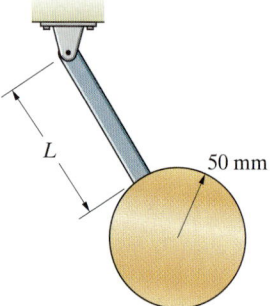

Problems 21.7/21.8

21.9 The suspended object A weighs 10 lb. The radius of the pulley is $R = 4$ in and its moment of inertia is 0.08 slug-ft². The spring constant is $k = 12$ lb/ft. For vibration of the system relative to its equilibrium position, determine (a) the frequency in Hz and (b) the period.

21.10 The suspended object A weighs 10 lb. The radius of the pulley is $R = 4$ in and its moment of inertia is 0.08 slug-ft². The spring constant is $k = 12$ lb/ft. The spring is unstretched when $x = 0$. At $t = 0$, the system is released from rest with $x = 0$. What is the value of x at $t = 3$ s?

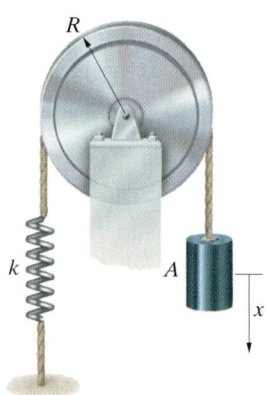

Problems 21.9/21.10

21.11 A "bungee jumper" who weighs 160 lb leaps from a bridge above a river. The bungee cord has an unstretched length of 60 ft, and it stretches an additional 40 ft before the jumper rebounds. Model the cord as a linear spring. When his motion has nearly stopped, what are the period and frequency of his vertical oscillations? (You can't model the cord as a linear spring during the early part of his motion. Why not?)

Problem 21.11

21.12 The spring constant is $k = 800$ N/m. The radius of the pulley is 120 mm and its moment of inertia is 0.03 kg-m^2. For vibration of the system relative to its equilibrium position, determine (a) the frequency in Hz and (b) the period.

21.13 The spring constant is $k = 800$ N/m. The radius of the pulley is 120 mm and its moment of inertia is 0.03 kg-m^2. The spring is unstretched when $x = 0$. At $t = 0$, $x = 0$ and the 20-kg mass is moving downward at 1 m/s. What is the value of x at $t = 2$ s?

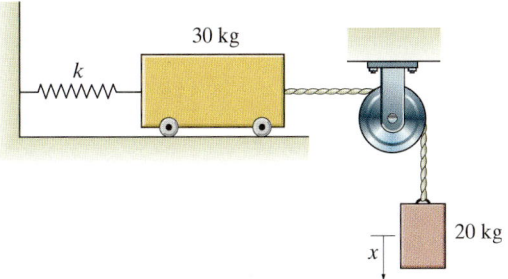

Problems 21.12/21.13

21.14 The 20-lb disk rolls on the horizontal surface. Its radius is $R = 6$ in. Determine the spring constant k so that the frequency of vibration of the system relative to its equilibrium position is $f = 1$ Hz.

21.15 The 20-lb disk rolls on the horizontal surface. Its radius is $R = 6$ in. The spring constant is $k = 15$ lb/ft. At $t = 0$, the spring is unstretched and the disk has a clockwise angular velocity of 2 rad/s. What is the amplitude of the resulting vibrations of the center of the disk?

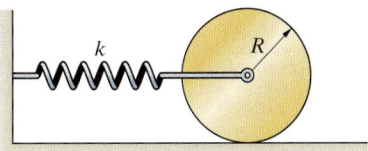

Problems 21.14/21.15

21.16 The 2-lb bar is pinned to the 5-lb disk. The disk rolls on the circular surface. What is the frequency of small vibrations of the system relative to its vertical equilibrium position?

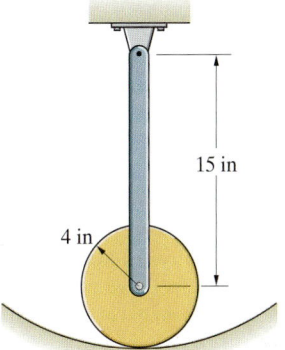

Problem 21.16

21.17 The mass of the suspended object A is 4 kg. The mass of the pulley is 2 kg and its moment of inertia is 0.018 N-m². The spring constant is $k = 150$ N/m. For vibration of the system relative to its equilibrium position, determine (a) the frequency in Hz and (b) the period.

21.18 The mass of the suspended object A is 4 kg. The mass of the pulley is 2 kg and its moment of inertia is 0.018 N-m². The spring constant is $k = 150$ N/m. The spring is unstretched when $x = 0$. At $t = 0$, the system is released from rest with $x = 0$. What is the velocity of the object A at $t = 1$ s?

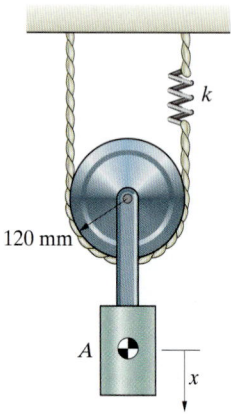

120 mm

A

x

Problems 21.17/21.18

21.19 The thin rectangular plate is attached to the rectangular frame by pins. The frame rotates with constant angular velocity $\omega_0 = 6$ rad/s. The angle β between the z axis of the body-fixed coordinate system and the vertical is governed by the equation

$$\frac{d^2\beta}{dt^2} = -\omega_0^2 \sin \beta \cos \beta.$$

Determine the frequency of small vibrations of the plate relative to its horizontal position. *Strategy:* By writing $\sin \beta$ and $\cos \beta$ in terms of their Taylor series and assuming that β is small, show that the equation governing β can be expressed in the form of Eq. (21.5).

21.20 Consider the system described in Problem 21.19. At $t = 0$, the angle $\beta = 0.01$ rad and $d\beta/dt = 0$. Determine β as a function of time.

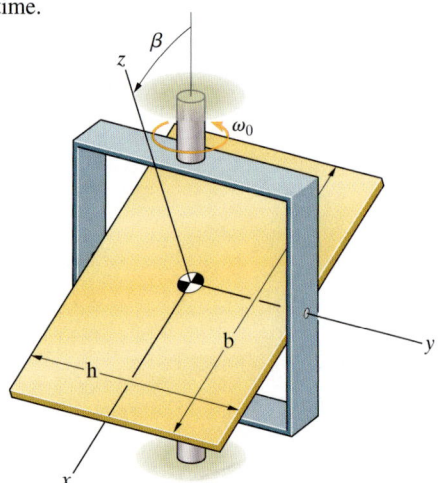

Problems 21.19/21.20

21.21 A slender bar of mass m and length l is pinned to a fixed support as shown. A torsional spring of constant k attached to the bar at the support is unstretched when the bar is vertical. Show that the equation governing small vibrations of the bar from its vertical equilibrium position is

$$\frac{d^2\theta}{dt^2} + \omega^2\theta = 0, \qquad \text{where } \omega^2 = \frac{\left(k - \frac{1}{2}mgl\right)}{\frac{1}{3}ml^2}.$$

21.22 The initial conditions of the slender bar in Problem 21.21 are

$$t = 0 \begin{cases} \theta = 0 \\ \dfrac{d\theta}{dt} = \dot{\theta}_0. \end{cases}$$

(a) If $k > \frac{1}{2}mgl$, show that θ is given as a function of time by

$$\theta = \frac{\dot{\theta}_0}{\omega} \sin \omega t, \quad \text{where } \omega^2 = \frac{\left(k - \frac{1}{2}mgl\right)}{\frac{1}{3}ml^2}.$$

(b) If $k < \frac{1}{2}mgl$, show that θ is given as a function of time by

$$\theta = \frac{\dot{\theta}_0}{2h}\left(e^{ht} - e^{-ht}\right), \quad \text{where } h^2 = \frac{\left(\frac{1}{2}mgl - k\right)}{\frac{1}{3}ml^2}.$$

Strategy: To do part (b), seek a solution of the equation of motion of the form $x = Ce^{\lambda t}$, where C and λ are constants.

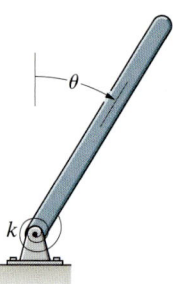

θ

k

Problems 21.21/21.22

21.23 Engineers use the device shown to measure an astronaut's moment of inertia. The horizontal board is pinned at O and supported by the linear spring with constant $k = 12$ kN/m. When the astronaut is not present, the frequency of small vibrations of the board about O is measured and determined to be 6.0 Hz. When the astronaut is lying on the board as shown, the frequency of small vibrations of the board about O is 2.8 Hz. What is the astronaut's moment of inertia about the z axis?

21.24 In Problem 21.23, the astronaut's center of mass is at $x = 1.01$ m, $y = 0.16$ m, and his mass is 81.6 kg. What is his moment of inertia about the z' axis through his center of mass?

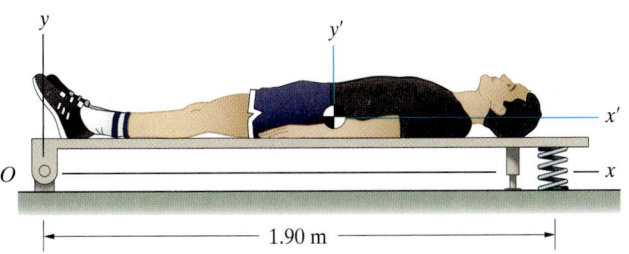

Problems 21.23/21.24

21.25* A floating sonobuoy (sound-measuring device) is in equilibrium in the vertical position shown. (Its center of mass is low enough that it is stable in this position.) The device is a 10-kg cylinder 1 m in length and 125 mm in diameter. The water density is 1025 kg/m³, and the buoyancy force supporting the buoy equals the weight of the water that would occupy the volume of the part of the cylinder below the surface. If you push the sonobuoy slightly deeper and release it, what is the frequency of the resulting vertical vibrations?

Problem 21.25

21.26* A disk rotates about a fixed *vertical* axis with constant angular velocity Ω. (The plane of the disk is horizontal.) A mass m slides in a smooth slot in the disk and is attached to a spring with constant k. The distance from the center of the disk to the mass when the spring is unstretched is r_0. Show that if $k/m > \Omega^2$, the frequency of vibration of the mass is $f = (1/2\pi)\sqrt{k/m - \Omega^2}$.

21.27* Suppose that at $t = 0$, the mass described in Problem 21.26 is located at $r = r_0$ and its radial velocity is $dr/dt = 0$. Determine the position r of the mass as a function of time.

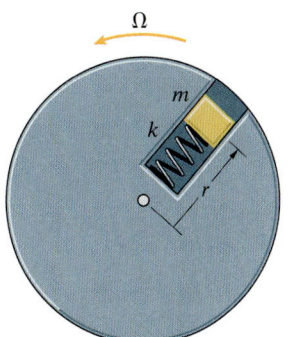

Problems 21.26/21.27

21.28 A homogeneous 100-lb disk with radius $R = 1$ ft is attached to two identical cylindrical steel bars of length $L = 1$ ft. The relation between the moment M exerted on the disk by one of the bars and the angle of rotation, θ, of the disk is

$$M = \frac{GJ}{L}\theta,$$

where J is the polar moment of inertia of the cross section of the bar and $G = 1.7 \times 10^9$ lb/ft² is the shear modulus of the steel. Determine the required radius of the bars if the frequency of rotational vibrations of the disk is to be 10 Hz.

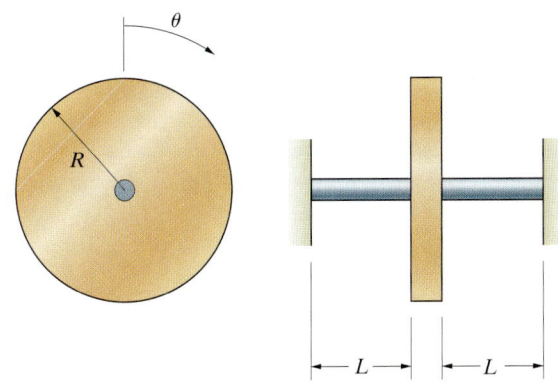

Problem 21.28

21.29 The moments of inertia of gears A and B are $I_A = 0.025$ kg-m^2 and $I_B = 0.100$ kg-m^2. Gear A is connected to a torsional spring with constant $k = 10$ N-m/rad. What is the frequency of small angular vibrations of the gears?

21.30 At $t = 0$, the torsional spring in Problem 21.29 is unstretched and gear B has a counterclockwise angular velocity of 2 rad/s. Determine the counterclockwise angular position of gear B relative to its equilibrium position as a function of time.

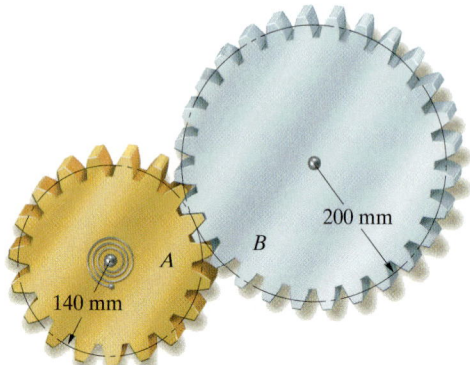

200 mm

B

A

140 mm

Problems 21.29/21.30

21.31 Each 2-kg slender bar is 1 m in length. What are the period and frequency of small vibrations of the system?

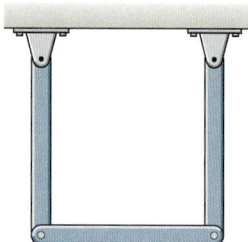

Problem 21.31

21.32* The masses of the slender bar and the homogeneous disk are m and m_d, respectively. The spring is unstretched when $\theta = 0$. Assume that the disk rolls on the horizontal surface.

(a) Show that the motion of the system is governed by the equation

$$\left(\frac{1}{3} + \frac{3m_d}{2m}\cos^2\theta\right)\frac{d^2\theta}{dt^2} - \frac{3m_d}{2m}\sin\theta\cos\theta\left(\frac{d\theta}{dt}\right)^2$$

$$-\frac{g}{2l}\sin\theta + \frac{k}{m}(1 - \cos\theta)\sin\theta = 0.$$

(b) If the system is in equilibrium at the angle $\theta = \theta_e$ and $\widetilde{\theta} = \theta - \theta_e$, show that the equation governing small vibrations relative to the equilibrium position is

$$\left(\frac{1}{3} + \frac{3m_d}{2m}\cos^2\theta_e\right)\frac{d^2\widetilde{\theta}}{dt^2}$$

$$+\left[\frac{k}{m}(\cos\theta_e - \cos^2\theta_e + \sin^2\theta_e) - \frac{g}{2l}\cos\theta_e\right]\widetilde{\theta} = 0.$$

21.33* The masses of the bar and disk in Problem 21.32 are $m = 2$ kg and $m_d = 4$ kg, respectively. The dimensions $l = 1$ m and $R = 0.28$ m, and the spring constant is $k = 70$ N/m.

(a) Determine the angle θ_e at which the system is in equilibrium.
(b) The system is at rest in the equilibrium position, and the disk is given a clockwise angular velocity of 0.1 rad/s. Determine θ as a function of time.

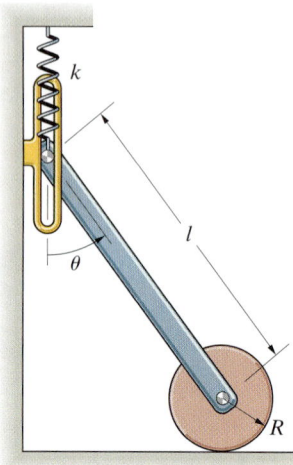

k

l

θ

R

Problems 21.32/21.33

21.34 The mass of each slender bar is 1 kg. If the frequency of small vibrations of the system is 0.935 Hz, what is the mass of the object A?

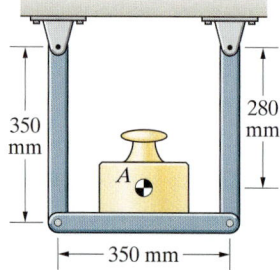

350 mm

280 mm

A

350 mm

Problem 21.34

21.35* The 4-kg slender bar is 2 m in length. It is held in equilibrium in the position $\theta_0 = 35°$ by a torsional spring with constant k. The spring is unstretched when the bar is vertical. Determine the period and frequency of small vibrations of the bar relative to the equilibrium position shown.

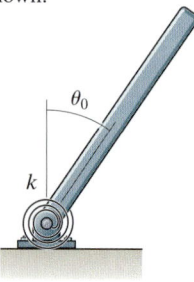

θ_0

k

Problem 21.35

21.2 Damped Vibrations

If the mass of a spring–mass oscillator is displaced and released, it will not vibrate indefinitely. It will slow down and eventually stop as a result of frictional forces, or *damping mechanisms*, acting on the system. Damping mechanisms damp out, or *attenuate*, the vibration. In some cases, engineers intentionally include damping mechanisms in vibrating systems. For example, the shock absorbers in a car are designed to damp out vibrations of the suspension relative to the frame. In the previous section we neglected damping, so the solutions we obtained describe only motions of systems over periods of time brief enough that the effects of damping can be neglected. We now discuss a simple method for modeling damping in vibrating systems.

The spring–mass oscillator in Fig. 21.9a has a *damping element*. The schematic diagram for the damping element represents a piston moving in a cylinder of viscous fluid. The force required to lengthen or shorten a damping element is defined to be the product of a constant c, the *damping constant*, and the rate of change of the length of the element (Fig. 21.9b). Therefore, the equation of motion of the mass is

$$-c\frac{dx}{dt} - kx = m\frac{d^2x}{dt^2}.$$

By defining $\omega = \sqrt{k/m}$ and $d = c/2m$, we can write this equation in the form

$$\frac{d^2x}{dt^2} + 2d\frac{dx}{dt} + \omega^2x = 0. \tag{21.16}$$

This equation describes the vibrations of many damped, one-degree-of-freedom systems. The form of its solution, and consequently the character of the predicted behavior of the system the equation describes, depends on whether the constant d is less than, equal to, or greater than ω. We discuss these cases in the sections that follow.

Subcritical Damping

If $d < \omega$, the system is said to be *subcritically damped*. Assuming a solution of the form

$$x = Ce^{\lambda t} \tag{21.17}$$

and substituting it into Eq. (21.16), we obtain

$$\lambda^2 + 2d\lambda + \omega^2 = 0.$$

This quadratic equation yields two roots for the constant λ that we can write as

$$\lambda = -d \pm i\omega_d,$$

where

$$\omega_d = \sqrt{\omega^2 - d^2}. \tag{21.18}$$

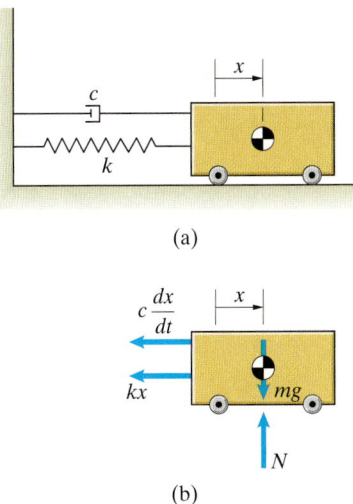

(a)

(b)

Figure 21.9
(a) Damped spring–mass oscillator.
(b) Free-body diagram of the mass.

Because we are assuming that $d < \omega$, the constant ω_d is a real number. The two roots for λ give us two solutions of the form of Eq. (21.17). The resulting general solution of Eq. (21.16) is

$$x = e^{-dt}(Ce^{i\omega_d t} + De^{-i\omega_d t}),$$

where C and D are constants. By using the Euler identity $e^{i\theta} = \cos \theta + i \sin \theta$, we can express this solution in the form

$$x = e^{-dt}(A \sin \omega_d t + B \cos \omega_d t), \tag{21.19}$$

where A and B are constants. Equation (21.19) is the product of an exponentially decaying function of time and an expression identical in form to the solution we obtained for an undamped system. The exponential function describes the expected effect of damping: The amplitude of the vibration attenuates with time. The coefficient d determines the rate at which the amplitude decreases.

Damping has an important effect in addition to causing attenuation. Because the oscillatory part of the solution is identical in form to Eq. (21.8), except that the term ω is replaced by ω_d, it follows from Eqs. (21.12) and (21.13) that the period and frequency of the damped system are

$$\tau_d = \frac{2\pi}{\omega_d}, \qquad f_d = \frac{\omega_d}{2\pi}. \tag{21.20}$$

From Eq. (21.18), we see that $\omega_d < \omega$, so *the period of the vibration is increased and its frequency is decreased as a result of subcritical damping.*

The rate of damping is often expressed in terms of the *logarithmic decrement* δ, which is the natural logarithm of the ratio of the amplitude at a time t to the amplitude at time $t + \tau_d$. Since the amplitude is proportional to e^{-dt}, we can obtain a simple relation between the logarithmic decrement, the coefficient d, and the period:

$$\delta = \ln\left[\frac{e^{-dt}}{e^{-d(t+\tau_d)}}\right] = d\tau_d.$$

Critical and Supercritical Damping

When $d \geq \omega$, the character of the solution of Eq. (21.16) is different from the case of subcritical damping. Suppose that $d > \omega$. When this is the case, the system is said to be *supercritically damped*. We again substitute a solution of the form

$$x = Ce^{\lambda t} \tag{21.21}$$

into Eq. (21.16), obtaining

$$\lambda^2 + 2d\lambda + \omega^2 = 0. \tag{21.22}$$

We can write the roots of this equation as

$$\lambda = -d \pm h,$$

where

$$h = \sqrt{d^2 - \omega^2}. \qquad (21.23)$$

The resulting general solution of Eq. (21.16) is

$$x = Ce^{-(d-h)t} + De^{-(d+h)t}, \qquad (21.24)$$

where C and D are constants.

When $d = \omega$, a system is said to be *critically damped*. Then the constant $h = 0$, so Eq. (21.22) has a repeated root $\lambda = -d$, and we obtain only one solution of the form (21.21). In this case, it can be shown that the general solution of Eq. (21.16) is

$$x = Ce^{-dt} + Dte^{-dt}, \qquad (21.25)$$

where C and D are constants.

Equations (21.24) and (21.25) indicate that the motion of a system is not oscillatory when $d \geq \omega$. These equations are expressed in terms of exponential functions, and do not contain sines and cosines. The condition $d = \omega$ defines the minimum amount of damping necessary to avoid oscillatory behavior, which is why it is referred to as the critically damped case. Figure 21.10 shows the effect of increasing amounts of damping on the behavior of a vibrating system.

The concept of critical damping has important implications in the design of many systems. For example, it is desirable to introduce enough damping into a car's suspension so that its motion is not oscillatory, but too much damping would cause the suspension to be too "stiff."

In summary, the motions of many damped one-degree-of-freedom systems can be modeled by Eq. (21.16). Once the equation of motion has been expressed in that form, the values of d and ω are known in terms of the physical parameters of the system. Those values determine whether the damping is subcritical, critical, or supercritical, which indicates the form of the solution of Eq. (21.16). If the system is subcritically damped, the period and frequency of its vibrations are given by Eqs. (21.20).

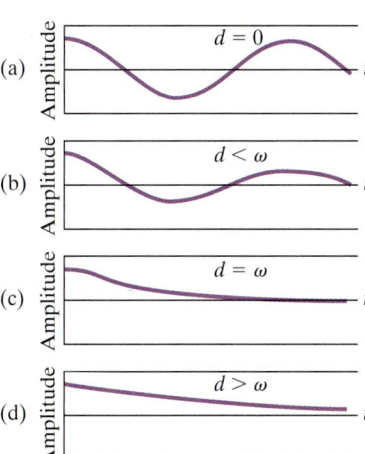

Figure 21.10
Amplitude history of a vibrating system that is (a) undamped; (b) subcritically damped; (c) critically damped; (d) supercritically damped.

Type of Damping	Solution	
$d < \omega$:	Subcritical	Eq. (21.19)
$d = \omega$:	Critical	Eq. (21.25)
$d > \omega$:	Supercritical	Eq. (21.24)

Study Questions

1. What is a damping element? How is the damping constant defined?

2. How do you determine whether the damping of a system whose motion is described by Eq. (21.16) is subcritical, critical, or supercritical?

3. Why can you calculate the period and frequency for a system that is subcritically damped, but not for one that is critically or supercritically damped?

4. What is the logarithmic decrement?

Example 21.3	Damped Spring–Mass Oscillator

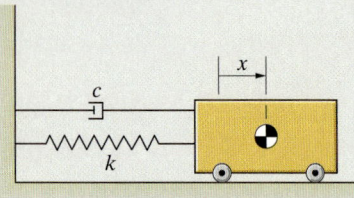

Figure 21.11

The damped spring–mass oscillator in Fig. 21.11 has mass $m = 2$ kg, spring constant $k = 8$ N/m, and damping constant $c = 1$ N-s/m. At $t = 0$, the mass is released from rest in the position $x = 0.1$ m. Determine its position as a function of time.

Strategy

The equation of motion for the damped spring-mass oscillator is given by Eq. (21.16). By calculating the values of ω and d, we will determine whether the damping is subcritical, critical, or supercritical and so choose the appropriate form of solution. Then we can use the given initial conditions to obtain the position as a function of time.

Solution

The constants $\omega = \sqrt{k/m} = 2$ rad/s and $d = c/2m = 0.25$ rad/s, so the damping is subcritical and the motion is described by Eq. (21.19). From Eq. (21.18),

$$\omega_d = \sqrt{\omega^2 - d^2} = 1.98 \text{ rad/s}.$$

From Eq. (21.19),

$$x = e^{-0.25t}(A \sin 1.98t + B \cos 1.98t),$$

and the velocity of the mass is

$$\frac{dx}{dt} = -0.25e^{-0.25t}(A \sin 1.98t + B \cos 1.98t)$$

$$+ e^{-0.25t}(1.98A \cos 1.98t - 1.98B \sin 1.98t).$$

From the conditions $x = 0.1$ m and $dx/dt = 0$ at $t = 0$, we obtain $A = 0.0126$ m and $B = 0.1$ m, so the position of the mass is

$$x = e^{-0.25t}(0.0126 \sin 1.98t + 0.1 \cos 1.98t) \text{ m}.$$

The graph of x for the first 10 s of motion in Fig. 21.12 clearly exhibits the attenuation of the amplitude.

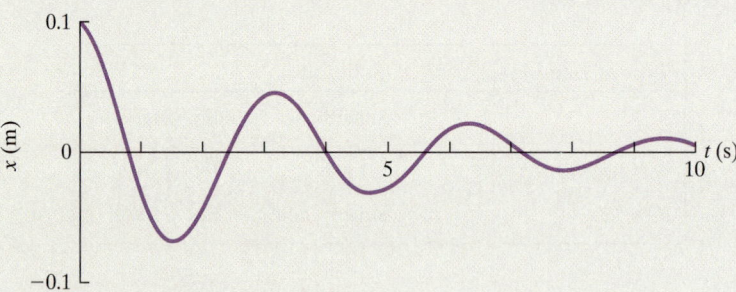

Figure 21.12
Position of the mass as a function of time.

Critical Thinking

Why should you want to analyze this very artificial system, which is the same one we used to obtain Eq. (21.16)? The reason is that the behaviors of many actual one-degree-of-freedom vibrating systems are described by equations identical in form to Eq. (21.16). By learning about the behavior of the simple system in Fig. 21.11, you are gaining insight into a large class of vibration problems of importance in engineering.

Example 21.4 Motion of a Damped System

The mass of the stepped disk in Fig. 21.13 is $m = 20$ kg, its radius is $R = 0.3$ m, and its moment of inertia is $I = \frac{1}{2}mR^2$. The spring constant is $k = 60$ N/m and the damping constant is $c = 24$ N-s/m. Determine the position of the center of the disk as a function of time if the disk is released from rest with the spring unstretched.

Strategy

We will obtain the equation of motion for the disk by drawing its free-body diagram and using Newton's second law and the equation of angular motion. If the resulting equation can be expressed in the form of Eq. (21.16), we can analyze its motion using the same approach we applied to the damped spring–mass oscillator in Example 21.3.

Solution

Let x be the downward displacement of the center of the disk relative to its position when the spring is unstretched. From the position of the disk's instantaneous center (Fig. a), we can see that the rate at which the spring is stretched is $2(dx/dt)$ and the rate at which the damping element is lengthened is $3(dx/dt)$. When the center of the disk is displaced a distance x, the stretch of the spring is $2x$.

We draw the free-body diagram of the disk in Fig. b, showing the forces exerted by the spring, the damping element, and the tension in the cable. Newton's second law is

$$mg - T - 2kx - 3c\frac{dx}{dt} = m\frac{d^2x}{dt^2},$$

and the equation of angular motion is

$$RT - R(2kx) - 2R\left(3c\frac{dx}{dt}\right) = \left(\frac{1}{2}mR^2\right)\alpha.$$

The angular acceleration is related to the acceleration of the center of the disk by $\alpha = (d^2x/dt^2)/R$. Eliminating T from Newton's second law and the equation of angular motion, we obtain the equation of motion:

$$\frac{3}{2}m\frac{d^2x}{dt^2} + 9c\frac{dx}{dt} + 4kx = mg.$$

By setting d^2x/dt^2 and dx/dt equal to zero in this equation, we find that the equilibrium position of the disk is $x = mg/4k$. Expressing the equation of motion in terms of the position of the center of the disk relative to its equilibrium position, $\tilde{x} = x - mg/4k$, we obtain

$$\frac{d^2\tilde{x}}{dt^2} + \left(\frac{6c}{m}\right)\frac{d\tilde{x}}{dt} + \left(\frac{8k}{3m}\right)\tilde{x} = 0.$$

This equation is identical in form to Eq. (21.16), where the constants are

$$d = \frac{6c}{2m} = \frac{(6)(24)}{(2)(20)} = 3.60 \text{ rad/s}$$

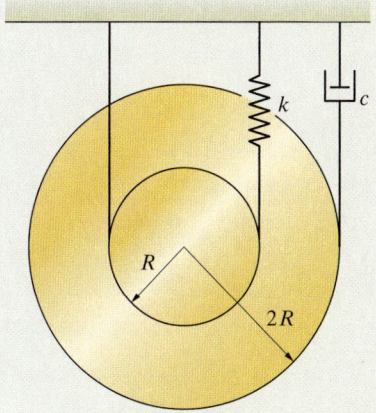

Figure 21.13

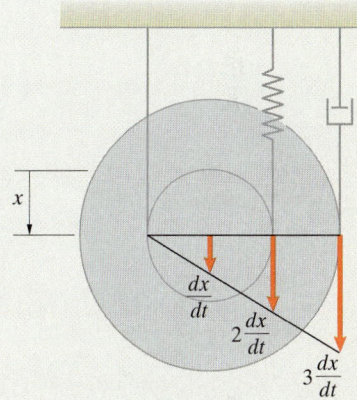

(a) Using the instantaneous center to determine the relationships between the velocities.

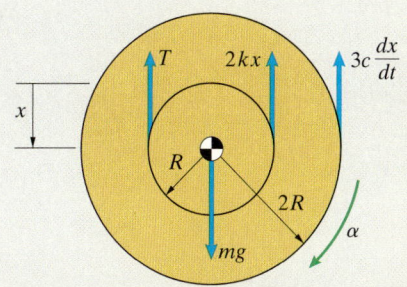

(b) Free-body diagram of the disk.

and

$$\omega = \sqrt{\frac{8k}{3m}} = \sqrt{\frac{(8)(60)}{(3)(20)}} = 2.83 \text{ rad/s}.$$

The damping is supercritical $(d > \omega)$ so the motion is described by Eq. (21.24) with $h = \sqrt{d^2 - \omega^2} = 2.23$ rad/s:

$$\widetilde{x} = Ce^{-(d-h)t} + De^{-(d+h)t} = Ce^{-1.37t} + De^{-5.83t}.$$

The velocity is

$$\frac{d\widetilde{x}}{dt} = -1.37Ce^{-1.37t} - 5.83De^{-5.83t}.$$

At $t = 0$, $\widetilde{x} = -mg/4k = -0.818$ m and $d\widetilde{x}/dt = 0$. From these conditions, we obtain $C = -1.069$ m and $D = 0.252$ m, so the position of the center of the disk relative to its equilibrium position is

$$\widetilde{x} = -1.069e^{-1.37t} + 0.252e^{-5.83t} \text{ m}.$$

The graph of the position for the first 4 s of motion is shown in Fig. 21.14.

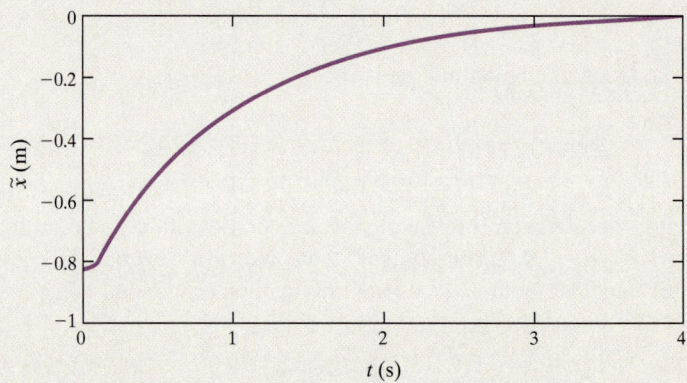

Figure 21.14
Position of the center of the disk as a function of time.

Critical Thinking

The disk in this example rotates and its center moves in the vertical direction. Why is this a one-degree-of-freedom system? As the center moves downward a distance x, the disk rotates through a clockwise angle $\theta = x/R$. Both the position of the center of the disk and the disk's angular position are specified when x is specified. The system has one degree of freedom.

Problems

21.36 The mass $m = 2$ kg and the spring constant is $k = 8$ N/m. Determine the frequency of vibration of the system relative to its equilibrium position if (a) $c = 0$ and (b) $c = 6$ N-s/m.

21.37 The mass $m = 2$ kg and the spring constant is $k = 8$ N/m. Determine the ranges of the damping constant c for which the damping of the system is subcritical, critical, and supercritical.

21.38 The mass $m = 2$ kg, the spring constant is $k = 8$ N/m, and the damping coefficient is $c = 4$ N-s/m. The spring is unstretched when $x = 0$. At $t = 0$, $x = 0$ and the mass has a velocity of 1 m/s toward the right. Determine x as a function of time.

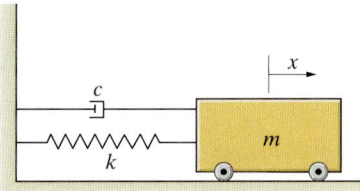

Problems 21.36–21.38

21.39 The mass $m = 2$ kg, the spring constant is $k = 8$ N/m, and the damping coefficient is $c = 12$ N-s/m. The spring is unstretched when $x = 0$. At $t = 0$, the mass is released from rest with $x = 0$. Determine the value of x at $t = 2$ s.

21.40 The mass $m = 0.15$ slugs, the spring constant is $k = 0.5$ lb/ft, and the damping coefficient is $c = 0.8$ lb-s/ft. The spring is unstretched when $x = 0$. At $t = 0$, the mass is released from rest with $x = 0$. Determine the value of x at $t = 2$ s.

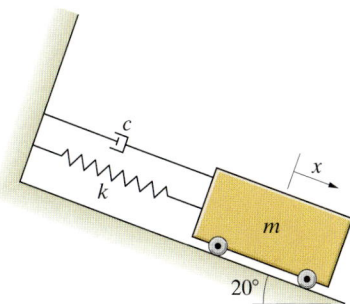

Problems 21.39/21.40

21.41 A 1270-kg test car moving at $v_0 = 2$ km/h collides with a rigid barrier at $t = 0$. As a result of the behavior of its energy-absorbing bumper, the response of the car to the collision can be simulated by the damped spring–mass oscillator shown with $k = 40$ kN/m and $c = 20$ kN-s/m. Assume that the 1270-kg mass is moving to the left with velocity $v_0 = 2$ km/h and the spring is unstretched at $t = 0$. Determine the car's deceleration (a) immediately after it contacts the barrier; (b) at $t = 0.1$ s; and (c) at $t = 0.2$ s.

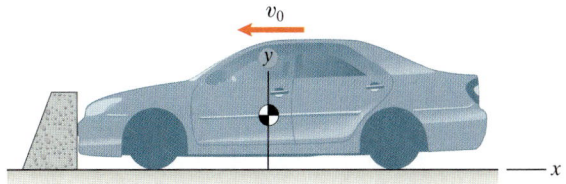

Car colliding with a rigid barrier

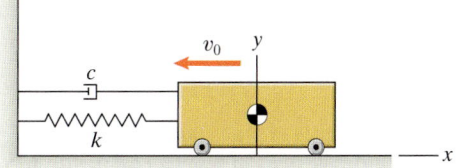

Simulation model

Problem 21.41

21.42 For small vertical displacements of the tire and wheel, the motion of the car's suspension can be modeled by the damped spring–mass oscillator with $m = 36$ kg and $k = 22$ kN/m. Determine the value of the damping constant c that must be provided by the suspension's shock absorber to achieve critical damping.

21.43 The motion of the car's suspension can be modeled by the damped spring–mass oscillator with $m = 36$ kg, $k = 22$ kN/m, and $c = 2.2$ kN-s/m. Assume that no external forces act on the tire and wheel. At $t = 0$, the spring is unstretched and the tire and wheel are given a velocity $dx/dt = 10$ m/s. Determine the position x as a function of time.

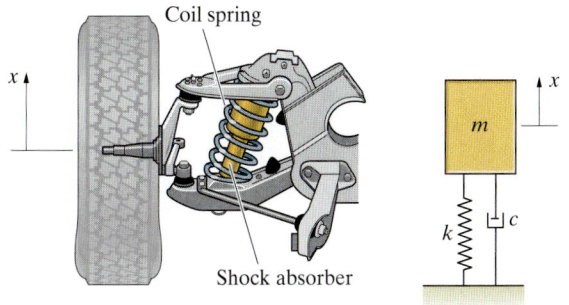

Problems 21.42/21.43

21.44 The 4-kg slender bar is 2 m in length. Aerodynamic drag on the bar and friction at the support exert a resisting moment about the pin support of magnitude $1.4(d\theta/dt)$ N-m, where $d\theta/dt$ is the angular velocity in rad/s.

(a) What are the period and frequency of small vibrations of the bar?
(b) How long does it take for the amplitude of vibration to decrease to one-half of its initial value?

21.45 The bar described in Problem 21.44 is given a displacement $\theta = 2°$ and released from rest at $t = 0$. What is the value of θ (in degrees) at $t = 2$ s?

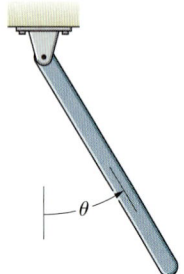

Problems 21.44/21.45

21.46 The radius of the pulley is $R = 100$ mm and its moment of inertia is $I = 0.1$ kg-m^2. The mass $m = 5$ kg, and the spring constant is $k = 135$ N/m. The cable does not slip relative to the pulley. The coordinate x measures the displacement of the mass relative to the position in which the spring is unstretched. Determine x as a function of time if $c = 60$ N-s/m and the system is released from rest with $x = 0$.

21.47 For the system described in Problem 21.46, determine x as a function of time if $c = 120$ N-s/m and the system is released from rest with $x = 0$.

21.48 For the system described in Problem 21.46, choose the value of c so that the system is critically damped, and determine x as a function of time if the system is released from rest with $x = 0$.

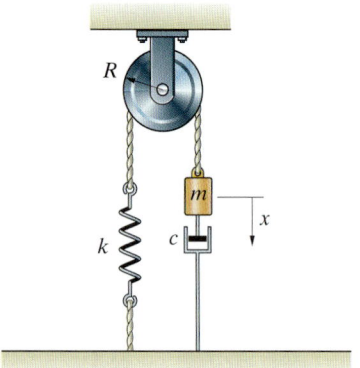

Problems 21.46–21.48

21.49 The spring constant is $k = 800$ N/m. The radius of the pulley is 120 mm and its moment of inertia is 0.03 kg-m^2. Determine the frequency of vibration of the system relative to its equilibrium position if (a) $c = 0$ and (b) $c = 40$ N-s/m.

21.50 The spring constant is $k = 800$ N/m and the damping coefficient is $c = 40$ N-s/m. The radius of the pulley is 120 mm and its moment of inertia is 0.03 kg-m^2. The spring is unstretched when $x = 0$. At $t = 0$, $x = 0$ and the 20-kg mass is moving downward at 1 m/s. What is the value of x at $t = 2$ s?

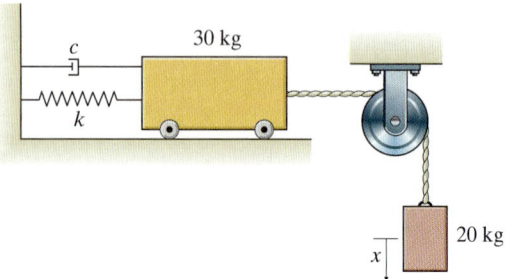

Problems 21.49/21.50

21.51 The homogeneous disk weighs 100 lb and its radius is $R = 1$ ft. It rolls on the plane surface. The spring constant is $k = 100$ lb/ft and the damping constant is $c = 3$ lb-s/ft. Determine the frequency of small vibrations of the disk relative to its equilibrium position.

21.52 In Problem 21.51, the spring is unstretched at $t = 0$ and the disk has a clockwise angular velocity of 2 rad/s. What is the angular velocity of the disk when $t = 3$ s?

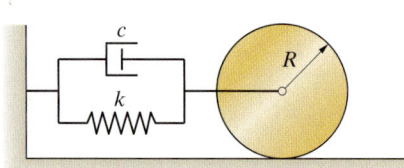

Problems 21.51/21.52

21.53 The moment of inertia of the stepped disk is I. Let θ be the angular displacement of the disk relative to its position when the spring is unstretched. Show that the equation governing θ is identical in form to Eq. (21.16), where

$$d = \frac{R^2 c}{2I} \quad \text{and} \quad \omega^2 = \frac{4R^2 k}{I}.$$

21.54 In Problem 21.53, the radius $R = 250$ mm, $k = 150$ N/m, and the moment of inertia of the disk is $I = 2$ kg-m^2.

(a) At what value of c will the system be critically damped?
(b) At $t = 0$, the spring is unstretched and the clockwise angular velocity of the disk is 10 rad/s. Determine θ as a function of time if the system is critically damped.
(c) Using the result of (b), determine the maximum resulting angular displacement of the disk and the time at which it occurs.

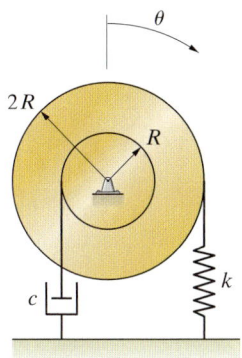

Problems 21.53/21.54

21.55 The moments of inertia of gears A and B are $I_A = 0.025$ kg-m^2 and $I_B = 0.100$ kg-m^2. Gear A is connected to a torsional spring with constant $k = 10$ N-m/rad. The bearing supporting gear B incorporates a damping element that exerts a resisting moment on gear B of magnitude $2(d\theta_B/dt)$ N-m, where $d\theta_B/dt$ is the angular velocity of gear B in rad/s. What is the frequency of small angular vibrations of the gears?

21.56 At $t = 0$, the torsional spring in Problem 21.55 is unstretched and gear B has a counterclockwise angular velocity of 2 rad/s. Determine the counterclockwise angular position of gear B relative to its equilibrium position as a function of time.

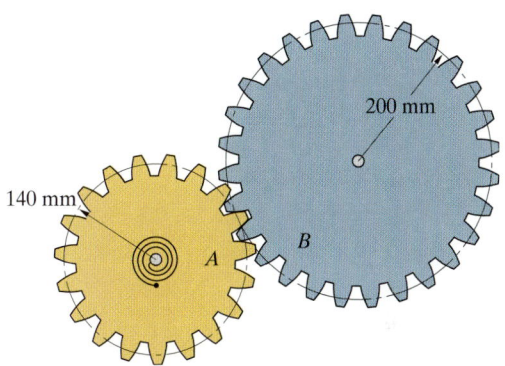

Problems 21.55/21.56

21.57 For the case of critically damped motion, confirm that the expression

$$x = Ce^{-dt} + Dte^{-dt}$$

is a solution of Eq. (21.16).

21.3 Forced Vibrations

The term *forced vibrations* means that external forces affect the vibrations of a system. Until now, we have discussed *free vibrations* of systems—vibrations unaffected by external forces. For example, during an earthquake, a building undergoes forced vibrations induced by oscillatory forces exerted on its foundations. After the earthquake subsides, the building vibrates freely until its motion damps out.

The damped spring–mass oscillator in Fig. 21.15a is subjected to a horizontal time-dependent force $F(t)$. From the free-body diagram of the mass (Fig. 21.15b), its equation of motion is

$$F(t) - kx - c\frac{dx}{dt} = m\frac{d^2x}{dt^2}.$$

Defining $d = c/2m$, $\omega^2 = k/m$, and $a(t) = F(t)/m$, we can write this equation in the form

$$\frac{d^2x}{dt^2} + 2d\frac{dx}{dt} + \omega^2 x = a(t). \tag{21.26}$$

We call $a(t)$ the *forcing function*. Equation (21.26) describes the forced vibrations of many damped one-degree-of-freedom systems. It is nonhomogeneous, because the forcing function does not contain x or one of its derivatives. Its general solution consists of two parts—the homogeneous and particular solutions:

$$x = x_h + x_p.$$

The *homogeneous solution* x_h is the general solution of Eq. (21.26) with the right side set equal to zero. Therefore, the homogeneous solution is the general solution for free vibrations, which we described in Section 21.2. The *particular solution* x_p is a solution that satisfies Eq. (21.26). In the sections that follow, we discuss the particular solutions for two types of forcing functions that occur frequently in applications.

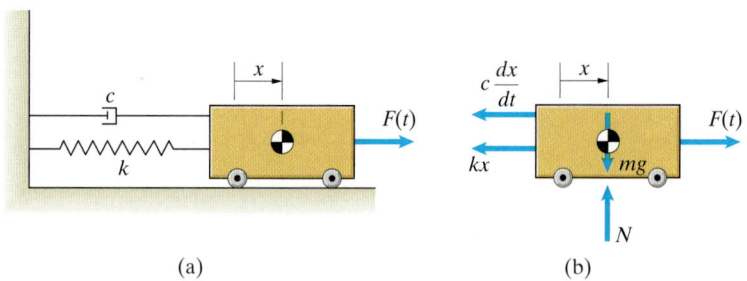

(a)   (b)

Figure 21.15
(a) A damped spring–mass oscillator subjected to a time-dependent force.
(b) Free-body diagram of the mass.

Oscillatory Forcing Function

Unbalanced wheels and shafts exert forces that oscillate at their frequency of rotation. When a car's wheels are out of balance, they exert oscillatory forces that cause vibrations passengers can feel. Engineers design electromechanical devices that transform oscillating currents into oscillating forces for use in testing vibrating systems. But the principal reason we are interested in this type of forcing function is that nearly any forcing function can be represented as a sum of oscillatory forcing functions with several different frequencies or with a continuous spectrum of frequencies.

By studying the motion of a vibrating system subjected to an oscillatory forcing function, we can determine the response of the system as a function of the frequency of the force. Suppose that the forcing function is an oscillatory function of the form

$$a(t) = a_0 \sin \omega_0 t + b_0 \cos \omega_0 t, \tag{21.27}$$

where a_0, b_0, and the frequency of the forcing function ω_0 are given constants. We can obtain the particular solution to Eq. (21.26) by seeking a solution of the form

$$x_p = A_p \sin \omega_0 t + B_p \cos \omega_0 t, \tag{21.28}$$

where A_p and B_p are constants we must determine. Substituting this expression and Eq. (21.27) into Eq. (21.26), we can write the resulting equation as

$$(-\omega_0^2 A_p - 2d\omega_0 B_p + \omega^2 A_p - a_0)\sin \omega_0 t$$

$$+ (-\omega_0^2 B_p + 2d\omega_0 A_p + \omega^2 B_p - b_0)\cos \omega_0 t = 0.$$

Equating the coefficients of $\sin \omega_0 t$ and $\cos \omega_0 t$ to zero and solving for A_p and B_p, we obtain

$$A_p = \frac{(\omega^2 - \omega_0^2)a_0 + 2d\omega_0 b_0}{(\omega^2 - \omega_0^2)^2 + 4d^2\omega_0^2}$$

and

$$B_p = \frac{-2d\omega_0 a_0 + (\omega^2 - \omega_0^2)b_0}{(\omega^2 - \omega_0^2)^2 + 4d^2\omega_0^2}. \tag{21.29}$$

Substituting these results into Eq. (21.28) yields the particular solution:

$$x_p = \left[\frac{(\omega^2 - \omega_0^2)a_0 + 2d\omega_0 b_0}{(\omega^2 - \omega_0^2)^2 + 4d^2\omega_0^2}\right]\sin \omega_0 t$$

$$+ \left[\frac{-2d\omega_0 a_0 + (\omega^2 - \omega_0^2)b_0}{(\omega^2 - \omega_0^2)^2 + 4d^2\omega_0^2}\right]\cos \omega_0 t. \tag{21.30}$$

The amplitude of the particular solution is

$$E_{\mathrm{p}} = \sqrt{A_{\mathrm{p}}^2 + B_{\mathrm{p}}^2} = \frac{\sqrt{a_0^2 + b_0^2}}{\sqrt{(\omega^2 - \omega_0^2)^2 + 4d^2\omega_0^2}}. \tag{21.31}$$

In Section 21.2, we showed that the solution of the equation describing free vibration of a damped system attenuates with time. For this reason, the particular solution for the motion of a damped vibrating system subjected to an oscillatory external force is also called the *steady-state solution*. The motion approaches the steady-state solution with increasing time. (See Example 21.5.)

To illustrate the effects of damping and the frequency of the forcing function on the amplitude of the particular solution, in Fig. 21.16 we plot the nondimensional expression $\omega^2 E_{\mathrm{p}}/\sqrt{a_0^2 + b_0^2}$ as a function of ω_0/ω for several values of the parameter d/ω. When there is no damping ($d = 0$), the amplitude of the particular solution approaches infinity as the frequency ω_0 of the forcing function approaches the frequency ω. When the damping is small, the amplitude of the particular solution approaches a finite maximum value at a value of ω_0 that is smaller than ω. The frequency at which the amplitude of the particular solution is a maximum is called the *resonant frequency*. (See Problem 21.67.)

The phenomenon of resonance is a familiar one in our everyday experience. For example, when a wheel of a car is out of balance, the resulting vibrations are noticed when the car is moving at a certain speed. At that speed, the wheel rotates at the resonant frequency of the car's suspension. Resonance is of practical importance in many applications, because relatively small oscillatory forces can result in large vibrational amplitudes that may cause damage or interfere with the functioning of a system. The classic example is soldiers marching across a bridge. If their steps in unison coincide with one of the bridge's resonant frequencies, they may damage the bridge even though it can safely support their weight.

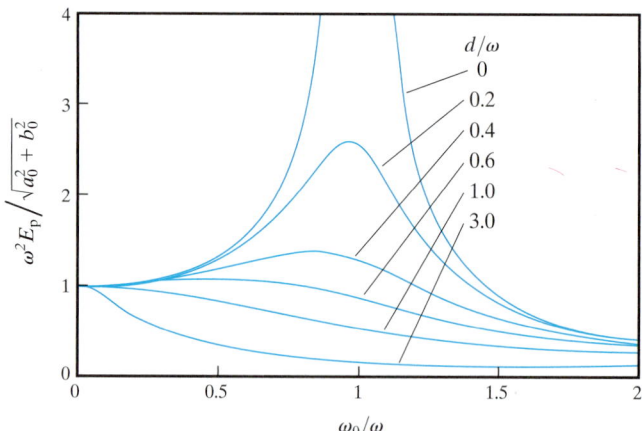

Figure 21.16
Amplitude of the particular (steady-state) solution as a function of the frequency of the forcing function.

Polynomial Forcing Function

Suppose that the forcing function $a(t)$ in Eq. (21.26) is a polynomial function of time; that is,

$$a(t) = a_0 + a_1 t + a_2 t^2 + \cdots + a_N t^N,$$

where $a_1, a_2, \ldots, a_N$ are given constants. This forcing function is important in applications because many smooth functions can be approximated by polynomials over a given interval of time. In this case we can obtain the particular solution of Eq. (21.26) by seeking a solution of the same form, namely,

$$x_p = A_0 + A_1 t + A_2 t^2 + \cdots + A_N t^N, \tag{21.32}$$

where $A_0, A_1, A_2, \ldots, A_N$ are constants to be determined.

For example, if $a(t) = a_0 + a_1 t$, Eq. (21.26) becomes

$$\frac{d^2 x}{dt^2} + 2d \frac{dx}{dt} + \omega^2 x = a_0 + a_1 t, \tag{21.33}$$

and we seek a particular solution of the form $x_p = A_0 + A_1 t$. Substituting this solution into Eq. (21.33), we can write the resulting equation as

$$(2d A_1 + \omega^2 A_0 - a_0) + (\omega^2 A_1 - a_1)t = 0.$$

This equation can be satisfied over an interval of time only if

$$2d A_1 + \omega^2 A_0 - a_0 = 0$$

and

$$\omega^2 A_1 - a_1 = 0.$$

Solving these two equations for A_0 and A_1, we obtain the particular solution:

$$x_p = \frac{a_0 - 2d a_1/\omega^2 + a_1 t}{\omega^2}.$$

It can be confirmed that this is a particular solution by substituting it into Eq. (21.33).

Study Questions

1. How do you determine the homogeneous solution of Eq. (21.26)?

2. Why is the particular solution for the motion of a damped vibrating system subjected to an oscillatory external force also called the steady-state solution?

3. What is the resonant frequency? Why is it significant?

4. If a vibrating system is subjected to a polynomial forcing function, how do you determine the particular solution of the equation describing the system's motion?

Example 21.5   Oscillatory Forcing Function

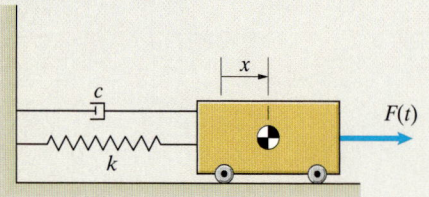

Figure 21.17

An engineer designing a vibration isolation system for an instrument console models the console and isolation system by the damped spring–mass oscillator in Fig. 21.17 with mass $m = 2$ kg, spring constant $k = 8$ N/m, and damping constant $c = 1$ N-s/m. To determine the system's response to external vibration, she assumes that the mass is initially stationary with the spring unstretched, and at $t = 0$ a force

$$F(t) = 20 \sin 4t \text{ N}$$

is applied to the mass.

(a) What is the amplitude of the particular (steady-state) solution?

(b) What is the position of the mass as a function of time?

Strategy

The forcing function is $a(t) = F(t)/m = 10 \sin 4t$ m/s^2, which is an oscillatory function of the form of Eq. (21.27) with $a_0 = 10$ m/s^2, $b_0 = 0$, and $\omega_0 = 4$ rad/s. The amplitude of the particular solution is given by Eq. (21.31), and the particular solution itself is given by Eq. (21.30). We must also determine whether the damping is subcritical, critical, or supercritical and choose the appropriate form of the homogeneous solution.

Solution

(a) The frequency of the undamped system is $\omega = \sqrt{k/m} = 2$ rad/s and the constant $d = c/2m = 0.25$ rad/s. Therefore, the amplitude of the particular solution is

$$E_p = \frac{a_0}{\sqrt{(\omega^2 - \omega_0^2)^2 + 4d^2\omega_0^2}} = \frac{10}{\sqrt{[(2)^2 - (4)^2]^2 + 4(0.25)^2(4)^2}}$$

$$= 0.822 \text{ m}.$$

(b) Since $d < \omega$, the system is subcritically damped and the homogeneous solution is given by Eq. (21.19). The frequency of the damped system is $\omega_d = \sqrt{\omega^2 - d^2} = 1.98$ rad/s, so the homogeneous solution is

$$x_h = e^{-0.25t}(A \sin 1.98t + B \cos 1.98t).$$

From Eq. (21.30), the particular solution is

$$x_p = -0.811 \sin 4t - 0.135 \cos 4t,$$

and the complete solution is

$$x = x_h + x_p$$

$$= e^{-0.25t}(A \sin 1.98t + B \cos 1.98t) - 0.811 \sin 4t - 0.135 \cos 4t.$$

At $t = 0$, $x = 0$ and $dx/dt = 0$. Using these conditions to determine the constants A and B, we obtain $A = 1.651$ m and $B = 0.135$ m. The position of the mass as a function of time is

$$x = e^{-0.25t}(1.651 \sin 1.98t + 0.135 \cos 1.98t)$$

$$- 0.811 \sin 4t - 0.135 \cos 4t \text{ m}.$$

Figure 21.18 shows the homogeneous, particular, and complete solutions for the first 25 s of motion of the system.

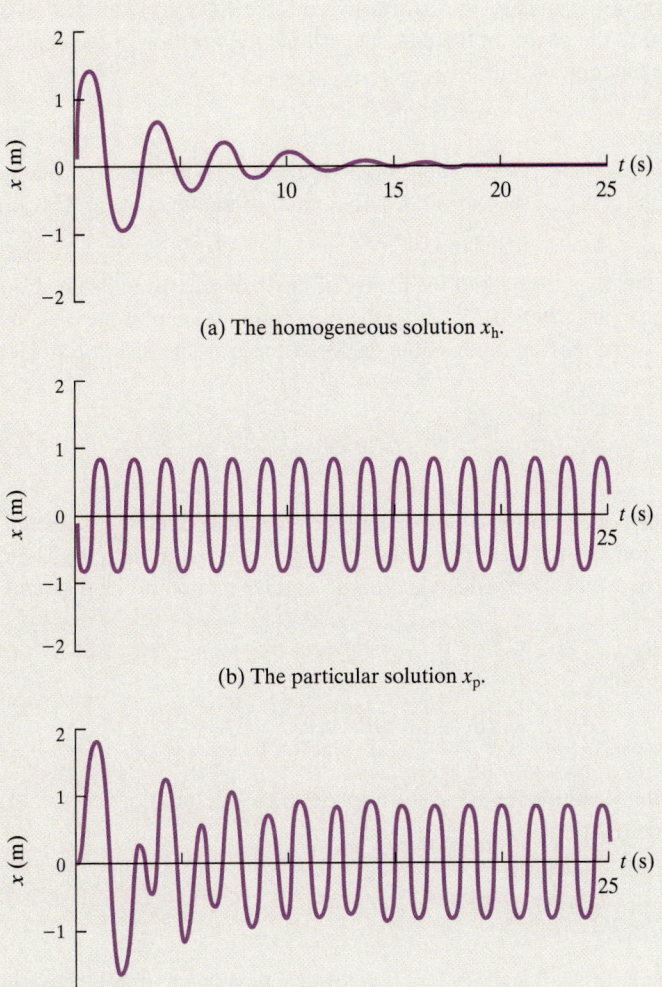

(a) The homogeneous solution x_h.

(b) The particular solution x_p.

(c) The total solution $x_h + x_p$.

Figure 21.18
The homogeneous, particular, and complete solutions.

Critical Thinking

This example clearly illustrates the response of a damped system to an oscillatory forcing function. Notice in Fig. 21.18c that the complete solution has an initial *transient* phase due to the homogeneous part of the solution. As time increases, the homogeneous solution attenuates and the complete solution approaches the particular solution. This is why the particular solution resulting from an oscillatory forcing function is also called the *steady-state solution*. The system achieves a steady state in which it oscillates with constant amplitude at the frequency of the forcing function.

Example 21.6   Polynomial Forcing Function

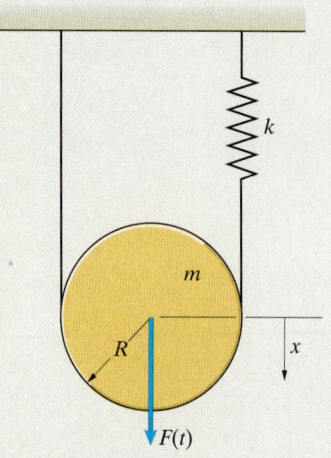

Figure 21.19

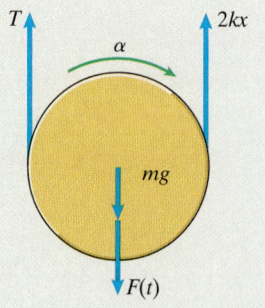

(a) Free-body diagram of the disk.

The homogeneous disk in Fig. 21.19 has radius $R = 2$ m and mass $m = 4$ kg. The spring constant is $k = 30$ N/m. The disk is initially stationary in its equilibrium position, and at $t = 0$ a downward force $F(t) = 12 + 12t - 0.6t^2$ N is applied to the center of the disk. Determine the position of the center of the disk as a function of time.

Strategy

The force $F(t)$ is a second-order polynomial, so we will seek a particular solution in the form of a second-order polynomial of the form of Eq. (21.32).

Solution

Let x be the displacement of the center of the disk relative to its position when the spring is unstretched. We draw the free-body diagram of the disk in Fig. a, where T is the tension in the cable on the left side of the disk. From Newton's second law,

$$F(t) + mg - 2kx - T = m\frac{d^2x}{dt^2}. \tag{1}$$

The angular acceleration of the disk in the clockwise direction is related to the acceleration of the center of the disk by $\alpha = (d^2x/dt^2)/R$. Using this expression, we can write the equation of angular motion of the disk as

$$\Sigma M = I\alpha:$$

$$TR - 2kxR = \left(\frac{1}{2}mR^2\right)\left(\frac{1}{R}\frac{d^2x}{dt^2}\right).$$

Solving this equation for T and substituting the result into Eq. (1), we obtain the equation of motion:

$$\frac{3}{2}m\frac{d^2x}{dt^2} + 4kx = F(t) + mg. \tag{2}$$

Setting $d^2x/dt^2 = 0$ and $F(t) = 0$ in this equation, we find that the equilibrium position of the disk is $x = mg/4k$. In terms of the position of the center of the disk relative to its equilibrium position $\tilde{x} = x - mg/4k$, Eq. (2) is

$$\frac{d^2\tilde{x}}{dt^2} + \frac{8k}{3m}\tilde{x} = \frac{2F(t)}{3m}.$$

This equation is identical in form to Eq. (21.26). Substituting the values of k and m and the polynomial function $F(t)$, we obtain

$$\frac{d^2\tilde{x}}{dt^2} + 20\tilde{x} = 2 + 2t - 0.1t^2. \tag{3}$$

Comparing this equation with Eq. (21.26), we see that $d = 0$ (there is no damping) and $\omega^2 = 20$ (rad/s)^2. From Eq. (21.19), the homogeneous solution is

$$\tilde{x}_h = A \sin 4.472t + B \cos 4.472t.$$

To obtain the particular solution, we seek a solution in the form of a polynomial of the same order as $F(t)$. That is,

$$\widetilde{x}_\mathrm{p} = A_0 + A_1 t + A_2 t^2,$$

where A_0, A_1, and A_2 are constants we must determine. We substitute this expression into Eq. (3) and collect terms of equal powers in t:

$$(2A_2 + 20A_0 - 2) + (20A_1 - 2)t + (20A_2 + 0.1)t^2 = 0.$$

This equation is satisfied if the coefficients multiplying each power of t equal zero, which yields

$$2A_2 + 20A_0 = 2,$$

$$20A_1 = 2,$$

and

$$20A_2 = -0.1.$$

Solving these three equations for A_0, A_1, and A_2, we obtain the particular solution:

$$\widetilde{x}_\mathrm{p} = 0.101 + 0.100t - 0.005t^2.$$

The complete solution is

$$\widetilde{x} = \widetilde{x}_\mathrm{h} + \widetilde{x}_\mathrm{p}$$

$$= A \sin 4.472t + B \cos 4.472t + 0.101 + 0.100t - 0.005t^2.$$

At $t = 0$, $\widetilde{x} = 0$ and $d\widetilde{x}/dt = 0$. Using these conditions to determine A and B, we obtain the position of the center of the disk (in meters) as a function of time:

$$\widetilde{x} = -0.022 \sin 4.472t - 0.101 \cos 4.472t + 0.101 + 0.100t - 0.005t^2.$$

The position is shown for the first 25 seconds of motion in Fig. 21.20. The undamped, oscillatory homogeneous solution is superimposed on the slowly varying particular solution.

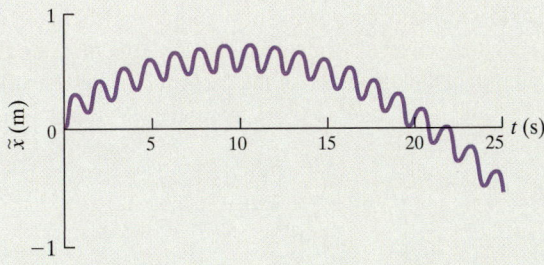

Figure 21.20
Position of the center of the disk as a function of time.

Critical Thinking

Forcing functions that arise in engineering applications can often be expressed as a Taylor series or approximated by a power series over some interval of time. When that is the case, you can analyze the response of the system during that interval of time by the approach used in this example.

Design Example 21.7 | Displacement Transducers

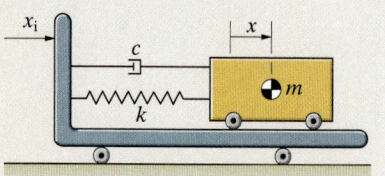

Figure 21.21

A damped spring–mass oscillator, or a device that can be modeled as a damped spring–mass oscillator, can be used to measure an object's displacement. Suppose that the base of the spring–mass oscillator in Fig. 21.21 is attached to an object and the coordinate x_i is a displacement to be measured relative to an inertial reference frame. The coordinate x measures the displacement of the mass relative to the base. When $x = 0$, the spring is unstretched. Suppose that the system is initially stationary and at $t = 0$ the base undergoes the oscillatory motion

$$x_i = a_i \sin \omega_i t + b_i \cos \omega_i t. \tag{1}$$

If $m = 2$ kg, $k = 8$ N/m, $c = 4$ N-s/m, $a_i = 0.1$ m, $b_i = 0.1$ m, and $\omega_i = 10$ rad/s, what is the resulting steady-state amplitude of the displacement of the mass relative to the base?

Strategy

We need to obtain the equation of motion for the mass, accounting for the effect of the oscillating base. To do so, we must write Newton's second law in terms of the acceleration of the mass *relative to the inertial reference frame.*

Solution

The acceleration of the mass relative to the base is d^2x/dt^2, so its acceleration relative to the inertial reference frame is $(d^2x/dt^2) + (d^2x_i/dt^2)$. Newton's second law for the mass is

$$-c\frac{dx}{dt} - kx = m\left(\frac{d^2x}{dt^2} + \frac{d^2x_i}{dt^2}\right).$$

We can write this equation as

$$\frac{d^2x}{dt^2} + 2d\frac{dx}{dt} + \omega^2 x = a(t),$$

where $d = c/2m = 1$ rad/s, $\omega = \sqrt{k/m} = 2$ rad/s, and the function

$$a(t) = -\frac{d^2x_i}{dt^2} = a_i\omega_i^2 \sin \omega_i t + b_i\omega_i^2 \cos \omega_i t. \tag{2}$$

Thus, we obtain an equation of motion identical in form to that for a spring–mass oscillator subjected to an oscillatory force. Comparing Eq. (2) with Eq. (21.27), we can obtain the amplitude of the particular (steady-state) solution from Eq. (21.31) by setting $a_0 = a_i\omega_i^2$, $b_0 = b_i\omega_i^2$, and $\omega_0 = \omega_i$:

$$E_p = \frac{\omega_i^2\sqrt{a_i^2 + b_i^2}}{\sqrt{(\omega^2 - \omega_i^2)^2 + 4d^2\omega_i^2}}. \tag{3}$$

Therefore, the steady-state amplitude of the displacement of the mass relative to its base is

$$E_p = \frac{(10)^2\sqrt{(0.1)^2 + (0.1)^2}}{\sqrt{[(2)^2 - (10)^2]^2 + 4(1)^2(10)^2}} = 0.144 \text{ m.}$$

Design Issues

A microphone transforms sound waves into a varying voltage that can be recorded or transformed back into sound waves by a speaker. A device that transforms a mechanical input into an electromagnetic output, or an electromagnetic input into a mechanical output, is called a transducer. Transducers can be used to measure displacements, velocities, and accelerations by transforming them into measurable voltages or currents.

In Fig. 21.21, the coordinate x_i is the displacement to be measured (the input). To use the spring–mass oscillator as a transducer, it would be designed to produce a voltage or current (the output) proportional to the displacement x. If the relationship between the input and output is known, the displacement x_i can be determined. Some seismographs (Fig. 21.22) measure motions of the earth in this way.

If the input is an oscillatory displacement given by Eq. (1), the amplitude of the output is given by Eq. (3). We can write the latter equation as

$$\frac{E_p}{E_i} = \frac{(\omega_i/\omega)^2}{\sqrt{[1 - (\omega_i/\omega)^2]^2 + 4(d/\omega)^2(\omega_i/\omega)^2}},$$

where $E_i = \sqrt{a_i^2 + b_i^2}$ is the amplitude of the input. In Fig. 21.23, we show the ratio E_p/E_i as a function of the ratio of the input frequency to the frequency of the undamped system (ω_i/ω) for several values of d/ω. If the parameters of the spring–mass oscillator are known, a graph of this type can be used to determine the amplitude of the input by measuring the amplitude of the output.

In practice the input displacement does not usually have a single frequency, but consists of a combination of different frequencies or even a continuous *spectrum* of frequencies. For example, the displacements resulting from earthquakes have a spectrum of frequencies. In that case, it is desirable for the ratio of the output amplitude to the input amplitude to be approximately constant over the range of the input frequencies. The response of the instrument is then said to be "flat." In Fig. 21.23, the response is approximately flat for frequencies ω_i greater than about 2ω if the damping of the system is chosen so that d/ω is in the range 0.6–0.7. Also, notice that making the frequency ω small increases the range of input frequencies over which the response of the instrument is flat. For that reason, seismographs are often designed with large masses and relatively weak springs.

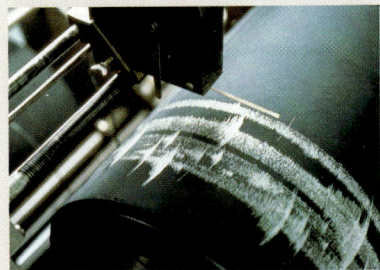

Figure 21.22
A seismograph that measures the local displacement of the earth.

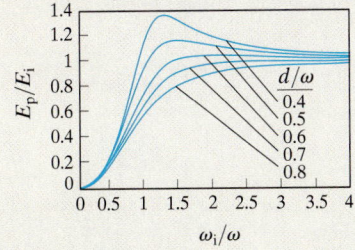

Figure 21.23
Ratio of the output amplitude to the input amplitude.

Problems

21.58 The mass $m = 1$ kg and the spring constant is $k = 100$ N/m. Let x be the position of the mass relative to its position when the spring is unstretched. The force $F(t) = 360 \sin 8t$ N. Determine the particular solution.

21.59 The mass $m = 1$ kg and the spring constant is $k = 100$ N/m. Let x be the position of the mass relative to its position when the spring is unstretched. The force $F(t) = 100 - 200t^2$ N. Determine the particular solution.

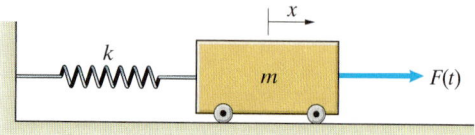

Problems 21.58/21.59

21.60 The damped spring–mass oscillator is initially stationary with the spring unstretched. At $t = 0$, a constant force $F(t) = 6$ N is applied to the mass.

(a) What is the steady-state (particular) solution?
(b) Determine the position of the mass as a function of time.

21.61 The damped spring–mass oscillator is initially stationary with the spring unstretched. At $t = 0$, a force $F(t) = 6 \cos 1.6t$ N is applied to the mass.

(a) What is the steady-state (particular) solution?
(b) Determine the position of the mass as a function of time.

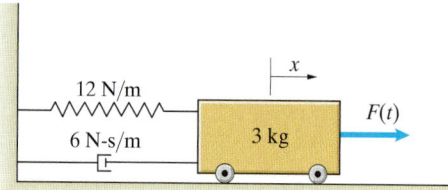

Problems 21.60/21.61

21.62 The disk with moment of inertia $I = 3$ kg-m^2 rotates about a fixed shaft and is attached to a torsional spring with constant $k = 20$ N-m/rad. At $t = 0$, the angle $\theta = 0$, the angular velocity is $d\theta/dt = 4$ rad/s, and the disk is subjected to a couple $M(t) = 10 \sin 2t$ N-m. Determine θ as a function of time.

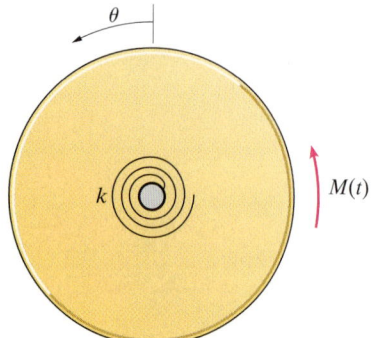

Problem 21.62

21.63 The stepped disk weighs 20 lb and its moment of inertia is $I = 0.6$ slug-ft^2. It rolls on the horizontal surface. The disk is initially stationary with the spring unstretched, and at $t = 0$ a constant force $F = 10$ lb is applied as shown. Determine the position of the center of the disk as a function of time.

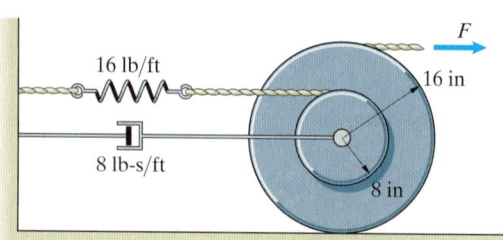

Problem 21.63

21.64* An electric motor is bolted to a metal table. When the motor is on, it causes the tabletop to vibrate horizontally. Assume that the legs of the table behave like linear springs, and neglect damping. The total weight of the motor and the tabletop is 150 lb. When the motor is not turned on, the frequency of horizontal vibration of the tabletop and motor is 5 Hz. When the motor is running at 600 rpm, the amplitude of the horizontal vibration is 0.01 in. What is the magnitude of the oscillatory force exerted on the table by the motor at its 600-rpm running speed?

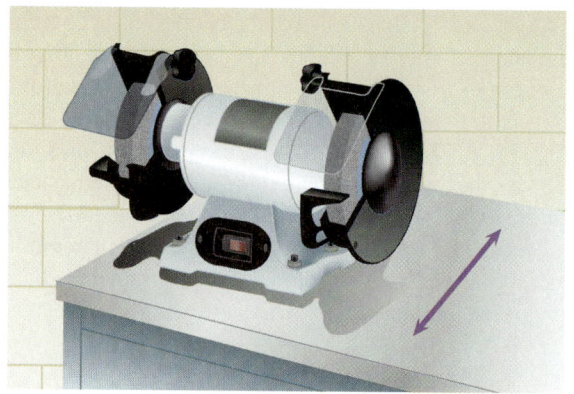

Problem 21.64

21.65 The moments of inertia of gears A and B are $I_A = 0.014$ slug-ft^2 and $I_B = 0.100$ slug-ft^2. Gear A is connected to a torsional spring with constant $k = 2$ ft-lb/rad. The system is in equilibrium at $t = 0$ when it is subjected to an oscillatory force $F(t) = 4 \sin 3t$ lb. What is the downward displacement of the 5-lb weight as a function of time?

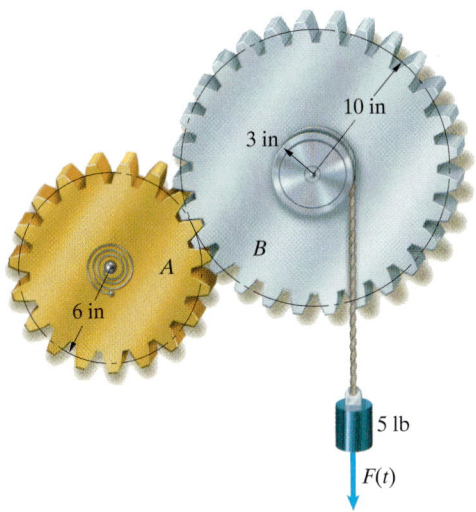

Problem 21.65

21.66* A 1.5-kg cylinder is mounted on a sting in a wind tunnel with the cylinder axis transverse to the direction of flow. When there is no flow, a 10-N vertical force applied to the cylinder causes it to deflect 0.15 mm. When air flows in the wind tunnel, vortices subject the cylinder to alternating lateral forces. The velocity of the air is 5 m/s, the distance between vortices is 80 mm, and the magnitude of the lateral forces is 1 N. If you model the lateral forces by the oscillatory function $F(t) = (1.0) \sin \omega_0 t$ N, what is the amplitude of the steady-state lateral motion of the sphere?

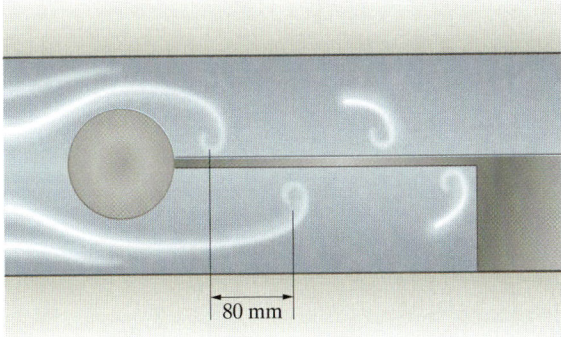

80 mm

Problem 21.66

21.67 Show that the amplitude of the particular solution given by Eq. (21.31) is a maximum when the frequency of the oscillatory forcing function is $\omega_0 = \sqrt{\omega^2 - 2d^2}$.

21.68* A sonobuoy (sound-measuring device) floats in a standing-wave tank. The device is a cylinder of mass m and cross-sectional area A. The water density is ρ, and the buoyancy force supporting the buoy equals the weight of the water that would occupy the volume of the part of the cylinder below the surface. When the water in the tank is stationary, the buoy is in equilibrium in the vertical position shown at the left. Waves are then generated in the tank, causing the depth of the water at the sonobuoy's position *relative to its original depth* to be $d = d_0 \sin \omega_0 t$. Let y be the sonobuoy's vertical position relative to its original position. Show that the sonobuoy's vertical position is governed by the equation

$$\frac{d^2 y}{dt^2} + \left(\frac{A\rho g}{m}\right)y = \left(\frac{A\rho g}{m}\right)d_0 \sin \omega_0 t.$$

21.69 Suppose that the mass of the sonobuoy in Problem 21.68 is $m = 10$ kg, its diameter is 125 mm, and the water density is $\rho = 1025$ kg/m^3. If $d = 0.1 \sin 2t$ m, what is the magnitude of the steady-state vertical vibrations of the sonobuoy?

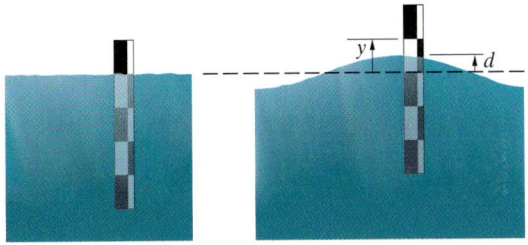

Problems 21.68/21.69

Design Experience

Problems 21.70–21.73 are related to Design Example 21.7.

21.70 The mass weighs 50 lb. The spring constant is $k = 200$ lb/ft, and $c = 10$ lb-s/ft. If the base is subjected to an oscillatory displacement x_i of amplitude 10 in and frequency $\omega_i = 15$ rad/s, what is the resulting steady-state amplitude of the displacement of the mass relative to the base?

21.71 The mass is 100 kg. The spring constant is $k = 4$ N/m, and $c = 24$ N-s/m. The base is subjected to an oscillatory displacement of frequency $\omega_i = 0.2$ rad/s. The steady-state amplitude of the displacement of the mass relative to the base is measured and determined to be 200 mm. What is the amplitude of the displacement of the base?

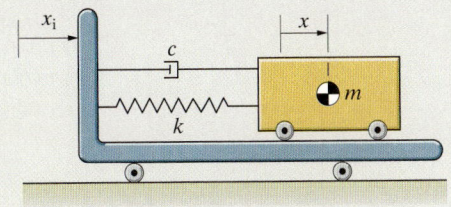

Problems 21.70/21.71

21.72 A team of engineering students builds the simple seismograph shown. The coordinate x_i measures the local horizontal ground motion. The coordinate x measures the position of the mass relative to the frame of the seismograph. The spring is unstretched when $x = 0$. The mass $m = 1$ kg, the spring constant $k = 10$ N/m, and $c = 2$ N-s/m. Suppose that the seismograph is initially stationary and that at $t = 0$ it is subjected to an oscillatory ground motion $x_i = 10 \sin 2t$ mm. What is the amplitude of the steady-state response of the mass?

21.73 In Problem 21.72, determine the position x of the mass relative to the base as a function of time.

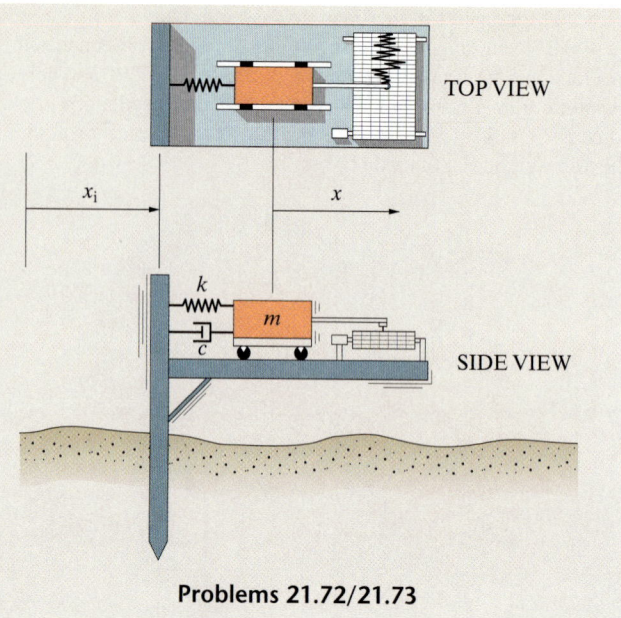

TOP VIEW

SIDE VIEW

Problems 21.72/21.73

CHAPTER SUMMARY

Conservative Systems

Small vibrations of many one-degree-of-freedom conservative systems relative to an equilibrium position are governed by the equation

$$\frac{d^2x}{dt^2} + \omega^2 x = 0, \tag{21.5}$$

where ω is a constant determined by the properties of the system. The general solution of this equation is

$$x = A \sin \omega t + B \cos \omega t \tag{21.8}$$

where A and B are constants. The general solution can also be expressed in the form

$$x = E \sin(\omega t - \phi), \tag{21.9}$$

where the constants E and ϕ are related to A and B by

$$A = E \cos \phi \quad \text{and} \quad B = -E \sin \phi. \tag{21.10}$$

The *amplitude* of the vibration is

$$E = \sqrt{A^2 + B^2}. \tag{21.11}$$

The *period* τ of the vibration is the time required for one complete oscillation, or *cycle*. The *frequency* f is the number of cycles per unit time. The period and frequency are related to ω by

$$\tau = \frac{2\pi}{\omega} \tag{21.12}$$

and

$$f = \frac{\omega}{2\pi}. \tag{21.13}$$

The term $\omega = 2\pi f$ is also a measure of the frequency and is expressed in rad/s.

Damped Vibrations

Small vibrations of many damped one-degree-of-freedom systems relative to an equilibrium position are governed by the equation

$$\frac{d^2x}{dt^2} + 2d\frac{dx}{dt} + \omega^2 x = 0. \tag{21.16}$$

Subcritical Damping If $d < \omega$, the system is said to be *subcritically damped*. In this case, the general solution of Eq. (21.16) is

$$x = e^{-dt}(A \sin \omega_d t + B \cos \omega_d t), \tag{21.19}$$

where A and B are constants and

$$\omega_d = \sqrt{\omega^2 - d^2}. \tag{21.18}$$

The period and frequency of the damped vibrations are

$$\tau_d = \frac{2\pi}{\omega_d} \quad \text{and} \quad f_d = \frac{\omega_d}{2\pi}. \tag{21.20}$$

Critical and Supercritical Damping If $d > \omega$, the system is said to be supercritically damped. The general solution is

$$x = Ce^{-(d-h)t} + De^{-(d+h)t}, \tag{21.24}$$

where C and D are constants and

$$h = \sqrt{d^2 - \omega^2}. \tag{21.23}$$

If $d = \omega$, the system is said to be *critically damped*. The general solution is

$$x = Ce^{-dt} + Dte^{-dt}, \tag{21.25}$$

where C and D are constants.

Forced Vibrations

The forced vibrations of many damped one-degree-of-freedom systems are governed by the equation

$$\frac{d^2x}{dt^2} + 2d\frac{dx}{dt} + \omega^2 x = a(t), \tag{21.26}$$

where $a(t)$ is the *forcing function*. The general solution of Eq. (21.26) is the sum of the homogeneous and particular solutions:

$$x = x_h + x_p.$$

The *homogeneous solution* x_h is the general solution of Eq. (21.26) with the right side set equal to zero, and the *particular solution* x_p is a solution that satisfies Eq. (21.26).

Oscillatory Forcing Function If $a(t)$ is an oscillatory function of the form

$$a(t) = a_0 \sin \omega_0 t + b_0 \cos \omega_0 t,$$

where a_0, b_0, and ω_0 are constants, the particular solution is

$$x_p = \left[\frac{(\omega^2 - \omega_0^2)a_0 + 2d\omega_0 b_0}{(\omega^2 - \omega_0^2)^2 + 4d^2\omega_0^2} \right] \sin \omega_0 t$$

$$+ \left[\frac{-2d\omega_0 a_0 + (\omega^2 - \omega_0^2)b_0}{(\omega^2 - \omega_0^2)^2 + 4d^2\omega_0^2} \right] \cos \omega_0 t, \qquad (21.30)$$

and its amplitude is

$$E_p = \frac{\sqrt{a_0^2 + b_0^2}}{\sqrt{(\omega^2 - \omega_0^2)^2 + 4d^2\omega_0^2}}. \qquad (21.31)$$

The particular solution for the motion of a damped vibrating system subjected to an oscillatory external force is also called the *steady-state solution*. The motion approaches the steady-state solution with increasing time.

Polynomial Forcing Function   If $a(t)$ is a polynomial of the form

$$a(t) = a_0 + a_1 t + a_2 t^2 + \cdots + a_N t^N,$$

where $a_1, a_2, \ldots, a_N$ are constants, the particular solution can be obtained by seeking a solution of the same form—that is,

$$x_p = A_0 + A_1 t + A_2 t^2 + \cdots + A_N t^N, \qquad (21.32)$$

where $A_0, A_1, A_2, \ldots, A_N$ are constants that must be determined.

Review Problems

21.74 The coordinate x measures the displacement of the mass relative to the position in which the spring is unstretched. The mass is given the initial conditions

$$t = 0 \begin{cases} x = 0.1 \text{ m}, \\ \dfrac{dx}{dt} = 0. \end{cases}$$

(a) Determine the position of the mass as a function of time.
(b) Draw graphs of the position and velocity of the mass as functions of time for the first 5 s of motion.

21.75 When $t = 0$, the mass is in the position in which the spring is unstretched and has a velocity of 0.3 m/s to the right. Determine the position of the mass as a function of time and the amplitude of the vibration
(a) by expressing the solution in the form given by Eq. (21.8) and
(b) by expressing the solution in the form given by Eq. (21.9).

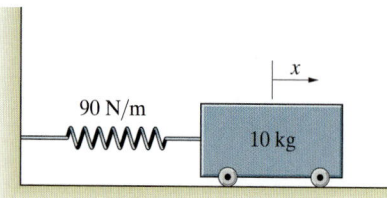

Problems 21.74/21.75

21.76 A homogeneous disk of mass m and radius R rotates about a fixed shaft and is attached to a torsional spring with constant k. (The torsional spring exerts a restoring moment of magnitude $k\theta$, where θ is the angle of rotation of the disk relative to its position in which the spring is unstretched.) Show that the period of rotational vibrations of the disk is $\tau = \pi R \sqrt{2m/k}$.

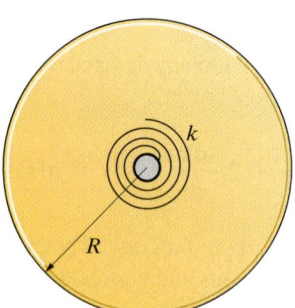

Problem 21.76

21.77 Assigned to determine the moments of inertia of astronaut candidates, an engineer attaches a horizontal platform to a vertical steel bar. The moment of inertia of the platform about L is 7.5 kg-m^2, and the frequency of torsional oscillations of the unloaded platform is 1 Hz. With an astronaut candidate in the position shown, the frequency of torsional oscillations is 0.520 Hz. What is the candidate's moment of inertia about L?

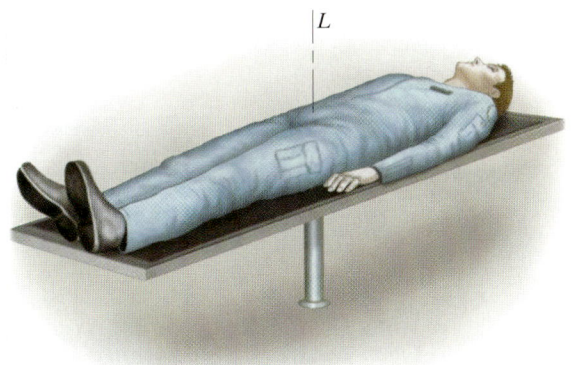

Problem 21.77

21.78 The 22-kg platen P rests on four roller bearings that can be modeled as 1-kg homogeneous cylinders with 30-mm radii. The spring constant is $k = 900$ N/m. What is the frequency of horizontal vibrations of the platen relative to its equilibrium position?

21.79 At $t = 0$, the platen described in Problem 21.78 is 0.1 m to the left of its equilibrium position and is moving to the right at 2 m/s. What are the platen's position and velocity at $t = 4$ s?

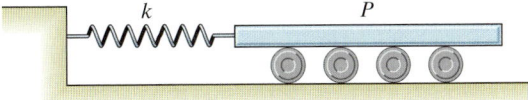

Problems 21.78/21.79

21.80 The moments of inertia of gears A and B are $I_A = 0.014$ slug-ft^2 and $I_B = 0.100$ slug-ft^2. Gear A is connected to a torsional spring with constant $k = 2$ ft-lb/rad. What is the frequency of angular vibrations of the gears relative to their equilibrium position?

21.81 The 5-lb weight in Problem 21.80 is raised 0.5 in from its equilibrium position and released from rest at $t = 0$. Determine the counterclockwise angular position of gear B relative to its equilibrium position as a function of time.

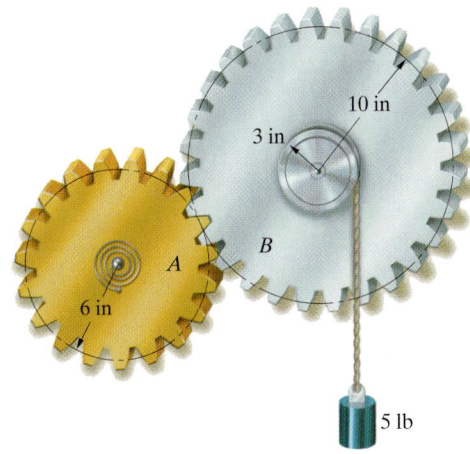

Problems 21.80/21.81

21.82 The mass of the slender bar is m. The spring is unstretched when the bar is vertical. The light collar C slides on the smooth vertical bar so that the spring remains horizontal. Determine the frequency of small vibrations of the bar.

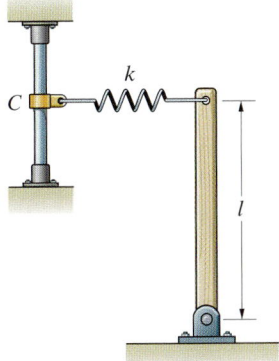

Problem 21.82

21.83 A homogeneous hemisphere of radius R and mass m rests on a level surface. If you rotate the hemisphere slightly from its equilibrium position and release it, what is the frequency of its vibrations?

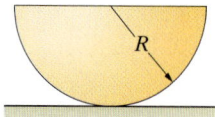

Problem 21.83

21.84 The frequency of the spring–mass oscillator is measured and determined to be 4.00 Hz. The oscillator is then placed in a barrel of oil, and its frequency is determined to be 3.80 Hz. What is the logarithmic decrement of vibrations of the mass when the oscillator is immersed in oil?

21.85 Consider the oscillator immersed in oil described in Problem 21.84. If the mass is displaced 0.1 m to the right of its equilibrium position and released from rest, what is its position relative to the equilibrium position as a function of time?

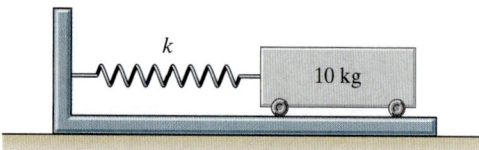

Problems 21.84/21.85

21.86 The stepped disk weighs 20 lb, and its moment of inertia is $I = 0.6$ slug-ft^2. It rolls on the horizontal surface. If $c = 8$ lb-s/ft, what is the frequency of vibration of the disk?

21.87 The stepped disk described in Problem 21.86 is initially in equilibrium, and at $t = 0$ it is given a clockwise angular velocity of 1 rad/s. Determine the position of the center of the disk relative to its equilibrium position as a function of time.

21.88 The stepped disk described in Problem 21.86 is initially in equilibrium, and at $t = 0$ it is given a clockwise angular velocity of 1 rad/s. Determine the position of the center of the disk relative to its equilibrium position as a function of time if $c = 16$ lb-s/ft.

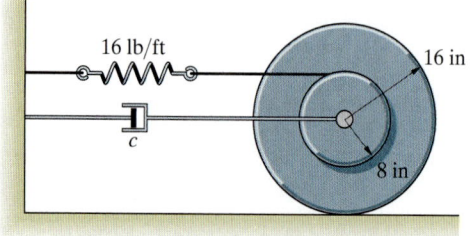

Problems 21.86–21.88

21.89 The 22-kg platen P rests on four roller bearings that can be modeled as 1-kg homogeneous cylinders with 30-mm radii. The spring constant is $k = 900$ N/m. The platen is subjected to a force $F(t) = 100 \sin 3t$ N. What is the magnitude of the platen's steady-state horizontal vibration?

21.90 At $t = 0$, the platen described in Problem 21.89 is 0.1 m to the right of its equilibrium position and is moving to the right at 2 m/s. Determine the platen's position relative to its equilibrium position as a function of time.

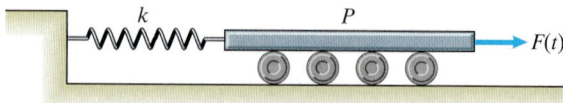

Problems 21.89/21.90

21.91 The moments of inertia of gears A and B are $I_A = 0.014$ slug-ft^2 and $I_B = 0.100$ slug-ft^2. Gear A is connected to a torsional spring with constant $k = 2$ ft-lb/rad. The bearing supporting gear B incorporates a damping element that exerts a resisting moment on gear B of magnitude $1.5(d\theta_B/dt)$ ft-lb, where $d\theta_B/dt$ is the angular velocity of gear B in rad/s. What is the frequency of angular vibration of the gears?

21.92 The 5-lb weight in Problem 21.91 is raised 0.5 in from its equilibrium position and released from rest at $t = 0$. Determine the counterclockwise angular position of gear B relative to its equilibrium position as a function of time.

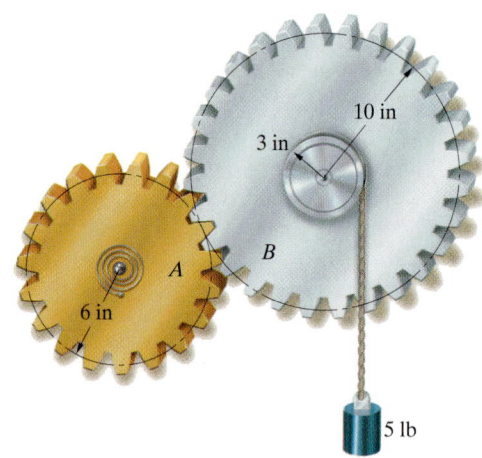

Problems 21.91/21.92

21.93 The base and mass m are initially stationary. The base is then subjected to a vertical displacement $h \sin \omega_i t$ relative to its original position. What is the magnitude of the resulting steady-state vibration of the mass m *relative to the base*?

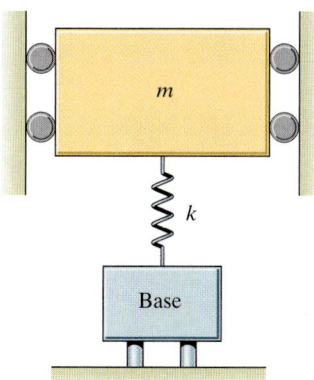

Problem 21.93

21.94* The mass of the trailer, not including its wheels and axle, is m, and the spring constant of its suspension is k. To analyze the suspension's behavior, an engineer assumes that the height of the road surface relative to its mean height is $h \sin(2\pi x/\lambda)$. Assume that the trailer's wheels remain on the road and its horizontal component of velocity is v. Neglect the damping due to the suspension's shock absorbers.

(a) Determine the magnitude of the trailer's vertical steady-state vibration *relative to the road surface*.
(b) At what velocity v does resonance occur?

21.95* The trailer in Problem 21.94, not including its wheels and axle, weighs 1000 lb. The spring constant of its suspension is $k = 2400$ lb/ft, and the damping coefficient due to its shock absorbers is $c = 200$ lb-s/ft. The road surface parameters are $h = 2$ in and $\lambda = 8$ ft. The trailer's horizontal velocity is $v = 6$ mi/h. Determine the magnitude of the trailer's vertical steady-state vibration relative to the road surface, (a) neglecting the damping due to the shock absorbers and (b) not neglecting the damping.

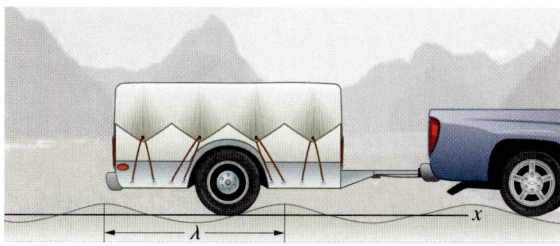

Problems 21.94/21.95

21.96* A disk with moment of inertia I rotates about a fixed shaft and is attached to a torsional spring with constant k. The angle θ measures the angular position of the disk relative to its position when the spring is unstretched. The disk is initially stationary with the spring unstretched. At $t = 0$, a time-dependent moment $M(t) = M_0(1 - e^{-t})$ is applied to the disk, where M_0 is a constant. Show that the angular position of the disk as a function of time is

$$\theta = \frac{M_0}{I} \left[-\frac{1}{\omega(1 + \omega^2)} \sin \omega t - \frac{1}{\omega^2(1 + \omega^2)} \cos \omega t \right.$$
$$\left. + \frac{1}{\omega^2} - \frac{1}{(1 + \omega^2)} e^{-t} \right].$$

Strategy: To determine the particular solution, seek a solution of the form

$$\theta_p = A_p + B_p e^{-t},$$

where A_p and B_p are constants that you must determine.

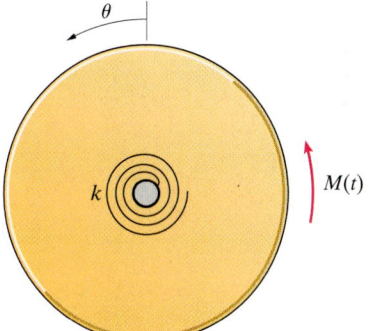

Problem 21.96

Review of Mathematics

A.1 Algebra

Quadratic Equations

The solutions of the quadratic equation

$$ax^2 + bx + c = 0$$

are

$$x = \frac{-b \pm \sqrt{b^2 - 4ac}}{2a}.$$

Natural Logarithms

The natural logarithm of a positive real number x is denoted by $\ln x$. It is defined to be the number such that

$$e^{\ln x} = x,$$

where $e = 2.7182\ldots$ is the base of natural logarithms.

Logarithms have the following properties:

$$\ln(xy) = \ln x + \ln y,$$

$$\ln(x/y) = \ln x - \ln y,$$

$$\ln y^x = x \ln y.$$

A.2 Trigonometry

The trigonometric functions for a right triangle are

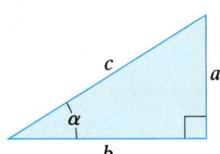

$$\sin \alpha = \frac{1}{\csc \alpha} = \frac{a}{c}, \qquad \cos \alpha = \frac{1}{\sec \alpha} = \frac{b}{c}, \qquad \tan \alpha = \frac{1}{\cot \alpha} = \frac{a}{b}.$$

The sine and cosine satisfy the relation

$$\sin^2 \alpha + \cos^2 \alpha = 1,$$

and the sine and cosine of the sum and difference of two angles satisfy

$$\sin(\alpha + \beta) = \sin \alpha \cos \beta + \cos \alpha \sin \beta,$$

$$\sin(\alpha - \beta) = \sin \alpha \cos \beta - \cos \alpha \sin \beta,$$

$$\cos(\alpha + \beta) = \cos \alpha \cos \beta - \sin \alpha \sin \beta,$$

$$\cos(\alpha - \beta) = \cos \alpha \cos \beta + \sin \alpha \sin \beta.$$

The **law of cosines** for an arbitrary triangle is

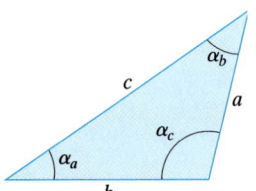

$$c^2 = a^2 + b^2 - 2ab \cos \alpha_c,$$

and the **law of sines** is

$$\frac{\sin \alpha_a}{a} = \frac{\sin \alpha_b}{b} = \frac{\sin \alpha_c}{c}.$$

A.3 Derivatives

$$\frac{d}{dx} x^n = nx^{n-1} \qquad \frac{d}{dx} \sin x = \cos x \qquad \frac{d}{dx} \sinh x = \cosh x$$

$$\frac{d}{dx} e^x = e^x \qquad \frac{d}{dx} \cos x = -\sin x \qquad \frac{d}{dx} \cosh x = \sinh x$$

$$\frac{d}{dx} \ln x = \frac{1}{x} \qquad \frac{d}{dx} \tan x = \frac{1}{\cos^2 x} \qquad \frac{d}{dx} \tanh x = \frac{1}{\cosh^2 x}$$

A.4 Integrals

$$\int x^n \, dx = \frac{x^{n+1}}{n+1} \qquad (n \neq -1)$$

$$\int x^{-1} \, dx = \ln x$$

$$\int \frac{dx}{a - bx^2} = \frac{1}{(ab)^{1/2}} \arctan \frac{(ab)^{1/2} x}{a}$$

$$\int \frac{dx}{a - bx^2} = \frac{1}{2(ab)^{1/2}} \ln \frac{a + x(ab)^{1/2}}{a - x(ab)^{1/2}}$$

$$\int \frac{x \, dx}{a - bx^2} = -\frac{1}{2b} \ln(a - bx^2)$$

$$\int (a + bx)^{1/2} \, dx = \frac{2}{3b}(a + bx)^{3/2}$$

$$\int x(a + bx)^{1/2} \, dx = -\frac{2(2a - 3bx)(a + bx)^{3/2}}{15b^2}$$

$$\int (1 + a^2x^2)^{1/2} \, dx = \frac{1}{2}\left\{ x(1 + a^2x^2)^{1/2} \right.$$

$$\left. + \frac{1}{a} \ln\left[x + \left(\frac{1}{a^2} + x^2 \right)^{1/2} \right] \right\}$$

$$\int x(1 + a^2x^2)^{1/2} \, dx = \frac{a}{3}\left(\frac{1}{a^2} + x^2 \right)^{3/2}$$

$$\int x^2(1 + a^2x^2)^{1/2} \, dx = \frac{1}{4}ax\left(\frac{1}{a^2} + x^2 \right)^{3/2}$$

$$- \frac{1}{8a^2} x(1 + a^2x^2)^{1/2} - \frac{1}{8a^3} \ln\left[x + \left(\frac{1}{a^2} + x^2 \right)^{1/2} \right]$$

$$\int (1 - a^2x^2)^{1/2} \, dx = \frac{1}{2}\left[x(1 - a^2x^2)^{1/2} + \frac{1}{a} \arcsin ax \right]$$

$$\int x(1 - a^2x^2)^{1/2} \, dx = -\frac{a}{3}\left(\frac{1}{a^2} - x^2 \right)^{3/2}$$

$$\int x^2(a^2 - x^2)^{1/2} \, dx = -\frac{1}{4}x(a^2 - x^2)^{3/2}$$

$$+ \frac{1}{8}a^2\left[x(a^2 - x^2)^{1/2} + a^2 \arcsin \frac{x}{a} \right]$$

$$\int \frac{dx}{(1 + a^2x^2)^{1/2}} = \frac{1}{a} \ln\left[x + \left(\frac{1}{a^2} + x^2 \right)^{1/2} \right]$$

$$\int \frac{dx}{(1 - a^2x^2)^{1/2}} = \frac{1}{a} \arcsin ax \quad \text{or} \quad -\frac{1}{a} \arccos ax$$

$$\int \sin x \, dx = -\cos x$$

$$\int \cos x \, dx = \sin x$$

$$\int \sin^2 x \, dx = -\frac{1}{2} \sin x \cos x + \frac{1}{2}x$$

$$\int \cos^2 x \, dx = \frac{1}{2} \sin x \cos x + \frac{1}{2}x$$

$$\int \sin^3 x \, dx = -\frac{1}{3} \cos x(\sin^2 x + 2)$$

$$\int \cos^3 x \, dx = \frac{1}{3} \sin x(\cos^2 x + 2)$$

$$\int \cos^4 x \, dx = \frac{3}{8}x + \frac{1}{4} \sin 2x + \frac{1}{32} \sin 4x$$

$$\int \sin^n x \cos x \, dx = \frac{(\sin x)^{n+1}}{n+1} \qquad (n \neq -1)$$

$$\int \sinh x \, dx = \cosh x$$

$$\int \cosh x \, dx = \sinh x$$

$$\int \tanh x \, dx = \ln \cosh x$$

$$\int e^{ax} \, dx = \frac{e^{ax}}{a}$$

$$\int xe^{ax} \, dx = \frac{e^{ax}}{a^2}(ax - 1)$$

A.5 Taylor Series

The Taylor series of a function $f(x)$ is

$$f(a + x) = f(a) + f'(a)x + \frac{1}{2!}f''(a)x^2 + \frac{1}{3!}f'''(a)x^3 + \cdots,$$

where the primes indicate derivatives.

Some useful Taylor series are

$$e^x = 1 + x + \frac{x^2}{2!} + \frac{x^3}{3!} + \cdots,$$

$$\sin(a + x) = \sin a + (\cos a)x - \frac{1}{2}(\sin a)x^2 - \frac{1}{6}(\cos a)x^3 + \cdots,$$

$$\cos(a + x) = \cos a - (\sin a)x - \frac{1}{2}(\cos a)x^2 + \frac{1}{6}(\sin a)x^3 + \cdots,$$

$$\tan(a + x) = \tan a + \left(\frac{1}{\cos^2 a}\right)x + \left(\frac{\sin a}{\cos^3 a}\right)x^2$$

$$+ \left(\frac{\sin^2 a}{\cos^4 a} + \frac{1}{3\cos^2 a}\right)x^3 + \cdots.$$

A.6 Vector Analysis

Cartesian Coordinates

The gradient of a scalar field ψ is

$$\nabla\psi = \frac{\partial\psi}{\partial x}\mathbf{i} + \frac{\partial\psi}{\partial y}\mathbf{j} + \frac{\partial\psi}{\partial z}\mathbf{k}.$$

The divergence and curl of a vector field $\mathbf{v} = v_x\mathbf{i} + v_y\mathbf{j} + v_z\mathbf{k}$ are

$$\nabla\cdot\mathbf{v} = \frac{\partial v_x}{\partial x} + \frac{\partial v_y}{\partial y} + \frac{\partial v_z}{\partial z},$$

$$\nabla\times\mathbf{v} = \begin{vmatrix} \mathbf{i} & \mathbf{j} & \mathbf{k} \\ \frac{\partial}{\partial x} & \frac{\partial}{\partial y} & \frac{\partial}{\partial z} \\ v_x & v_y & v_z \end{vmatrix}.$$

Cylindrical Coordinates

The gradient of a scalar field ψ is

$$\nabla\psi = \frac{\partial\psi}{\partial r}\mathbf{e}_r + \frac{1}{r}\frac{\partial\psi}{\partial\theta}\mathbf{e}_\theta + \frac{\partial\psi}{\partial z}\mathbf{e}_z.$$

The divergence and curl of a vector field $\mathbf{v} = v_r\mathbf{e}_r + v_\theta\mathbf{e}_\theta + v_z\mathbf{e}_z$ are

$$\nabla\cdot\mathbf{v} = \frac{\partial v_r}{\partial r} + \frac{v_r}{r} + \frac{1}{r}\frac{\partial v_\theta}{\partial\theta} + \frac{\partial v_z}{\partial z},$$

$$\nabla\times\mathbf{v} = \frac{1}{r}\begin{vmatrix} \mathbf{e}_r & r\mathbf{e}_\theta & \mathbf{e}_z \\ \frac{\partial}{\partial r} & \frac{\partial}{\partial\theta} & \frac{\partial}{\partial z} \\ v_r & rv_\theta & v_z \end{vmatrix}.$$

APPENDIX

B

Properties of Areas and Lines

B.1 Areas

The coordinates of the centroid of the area A are

$$\bar{x} = \frac{\displaystyle\int_A x \, dA}{\displaystyle\int_A dA}, \qquad \bar{y} = \frac{\displaystyle\int_A y \, dA}{\displaystyle\int_A dA}.$$

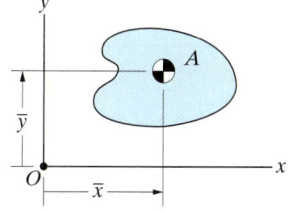

The moment of inertia about the x axis I_x, the moment of inertia about the y axis I_y, and the product of inertia I_{xy} are

$$I_x = \int_A y^2 \, dA, \qquad I_y = \int_A x^2 \, dA, \qquad I_{xy} = \int_A xy \, dA.$$

The polar moment of inertia about O is

$$J_O = \int_A r^2 \, dA = \int_A (x^2 + y^2) \, dA = I_x + I_y.$$

Area $= bh$

$$I_x = \frac{1}{3}bh^3, \qquad I_y = \frac{1}{3}hb^3, \qquad I_{xy} = \frac{1}{4}b^2h^2$$

$$I_{x'} = \frac{1}{12}bh^3, \qquad I_{y'} = \frac{1}{12}hb^3, \qquad I_{x'y'} = 0$$

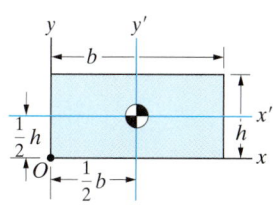

Rectangular area

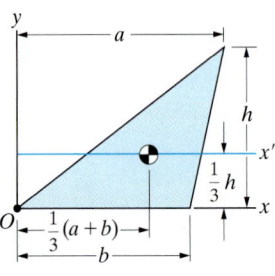

Triangular area

$$\text{Area} = \frac{1}{2}bh$$

$$I_x = \frac{1}{12}bh^3, \qquad I_y = \frac{1}{4}hb^3, \qquad I_{xy} = \frac{1}{8}b^2h^2$$

$$I_{x'} = \frac{1}{36}bh^3, \qquad I_{y'} = \frac{1}{36}hb^3, \qquad I_{x'y'} = \frac{1}{72}b^2h^2$$

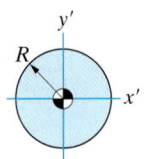

Triangular area

$$\text{Area} = \frac{1}{2}bh \qquad I_x = \frac{1}{12}bh^3, \qquad I_{x'} = \frac{1}{36}bh^3$$

Circular area

$$\text{Area} = \pi R^2 \qquad I_{x'} = I_{y'} = \frac{1}{4}\pi R^4, \qquad I_{x'y'} = 0$$

Semicircular area

$$\text{Area} = \frac{1}{2}\pi R^2 \qquad I_x = I_y = \frac{1}{8}\pi R^4, \qquad I_{xy} = 0$$

$$I_{x'} = \frac{1}{8}\pi R^4, \qquad I_{y'} = \left(\frac{\pi}{8} - \frac{8}{9\pi}\right)R^4, \qquad I_{x'y'} = 0$$

$$\text{Area} = \frac{1}{4}\pi R^2 \qquad I_x = I_y = \frac{1}{16}\pi R^4, \qquad I_{xy} = \frac{1}{8}R^4$$

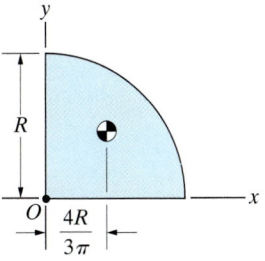

Quarter-circular area

$$\text{Area} = \alpha R^2$$

$$I_x = \frac{1}{4}R^4\left(\alpha - \frac{1}{2}\sin 2\alpha\right), \qquad I_y = \frac{1}{4}R^4\left(\alpha + \frac{1}{2}\sin 2\alpha\right),$$

$$I_{xy} = 0$$

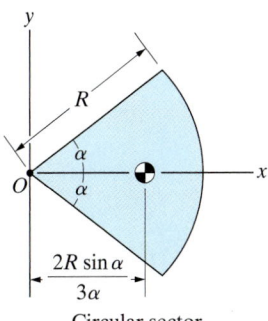

Circular sector

$$\text{Area} = \frac{1}{4}\pi ab$$

$$I_x = \frac{1}{16}\pi ab^3, \qquad I_y = \frac{1}{16}\pi a^3b, \qquad I_{xy} = \frac{1}{8}a^2b^2$$

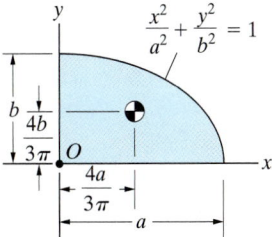

Quarter-elliptical area

$$\text{Area} = \frac{cb^{n+1}}{n+1}$$

$$I_x = \frac{c^3b^{3n+1}}{9n+3}, \qquad I_y = \frac{cb^{n+3}}{n+3}, \qquad I_{xy} = \frac{c^2b^{2n+2}}{4n+4}$$

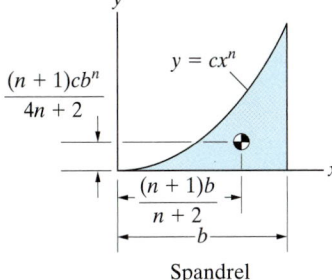

Spandrel

B.2 Lines

The coordinates of the centroid of the line L are

$$\bar{x} = \frac{\displaystyle\int_L x \, dL}{\displaystyle\int_L dL}, \qquad \bar{y} = \frac{\displaystyle\int_L y \, dL}{\displaystyle\int_L dL}, \qquad \bar{z} = \frac{\displaystyle\int_L z \, dL}{\displaystyle\int_L dL}.$$

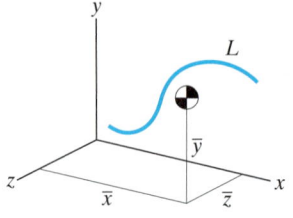

Length $= \pi R$

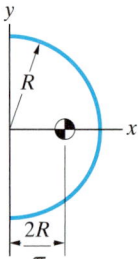

Semicircular arc

Length $= \frac{1}{2}\pi R$

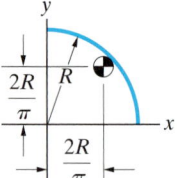

Quarter-circular arc

Length $= 2\alpha R$

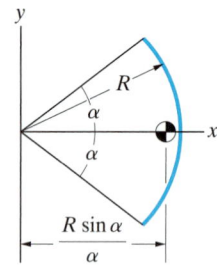

Circular arc

APPENDIX

C

Properties of Volumes and Homogeneous Objects

The moments and products of inertia of the object in terms of the xyz coordinate system are,

$$I_{x \text{ axis}} = I_{xx} = \int_m (y^2 + z^2) \, dm,$$

$$I_{y \text{ axis}} = I_{yy} = \int_m (x^2 + z^2) \, dm,$$

$$I_{z \text{ axis}} = I_{zz} = \int_m (x^2 + y^2) \, dm,$$

$$I_{xy} = \int_m xy \, dm, \quad I_{yz} = \int_m yz \, dm,$$

$$I_{zx} = \int_m zx \, dm.$$

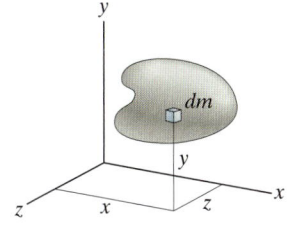

$$I_{x \text{ axis}} = 0, \qquad I_{y \text{ axis}} = I_{z \text{ axis}} = \frac{1}{3}ml^2,$$

$$I_{xy} = I_{yz} = I_{zx} = 0.$$

$$I_{x' \text{ axis}} = 0, \qquad I_{y' \text{ axis}} = I_{z' \text{ axis}} = \frac{1}{12}ml^2,$$

$$I_{x'y'} = I_{y'z'} = I_{z'x'} = 0.$$

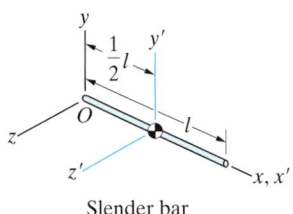

Slender bar

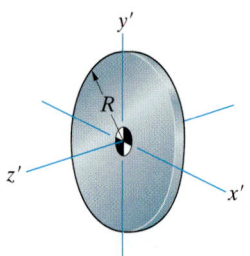

$$I_{x' \text{ axis}} = I_{y' \text{ axis}} = \frac{1}{4}mR^2, \qquad I_{z' \text{ axis}} = \frac{1}{2}mR^2,$$

$$I_{x'y'} = I_{y'z'} = I_{z'x'} = 0.$$

Thin circular plate

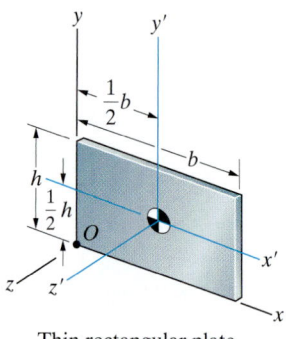

$$I_{x \text{ axis}} = \frac{1}{3}mh^2, \qquad I_{y \text{ axis}} = \frac{1}{3}mb^2, \qquad I_{z \text{ axis}} = \frac{1}{3}m(b^2 + h^2),$$

$$I_{xy} = \frac{1}{4}mbh, \qquad I_{yz} = I_{zx} = 0.$$

$$I_{x' \text{ axis}} = \frac{1}{12}mh^2, \qquad I_{y' \text{ axis}} = \frac{1}{12}mb^2, \qquad I_{z' \text{ axis}} = \frac{1}{12}m(b^2 + h^2),$$

$$I_{x'y'} = I_{y'z'} = I_{z'x'} = 0.$$

Thin rectangular plate

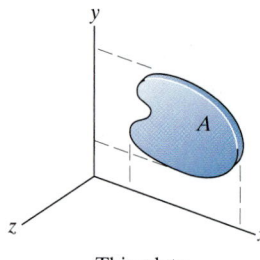

$$I_{x \text{ axis}} = \frac{m}{A}I_x, \qquad I_{y \text{ axis}} = \frac{m}{A}I_y, \qquad I_{z \text{ axis}} = I_{x \text{ axis}} + I_{y \text{ axis}},$$

$$I_{xy} = \frac{m}{A}I^A_{xy}, \qquad I_{yz} = I_{zx} = 0.$$

(The terms I_x, I_y, and I^A_{xy} are the moments and product of inertia of the plate's cross-sectional area A).

Thin plate

Volume $= abc$

$$I_{x'\text{ axis}} = \frac{1}{12}m(a^2 + b^2), \qquad I_{y'\text{ axis}} = \frac{1}{12}m(a^2 + c^2),$$

$$I_{z'\text{ axis}} = \frac{1}{12}m(b^2 + c^2), \qquad I_{x'y'} = I_{y'z'} = I_{z'x'} = 0.$$

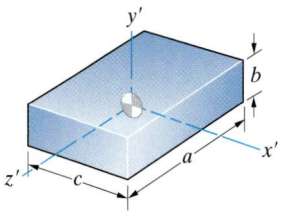

Rectangular prism

Volume $= \pi R^2 l$

$$I_{x\text{ axis}} = I_{y\text{ axis}} = m\left(\frac{1}{3}l^2 + \frac{1}{4}R^2\right), \qquad I_{z\text{ axis}} = \frac{1}{2}mR^2,$$

$$I_{xy} = I_{yz} = I_{zx} = 0.$$

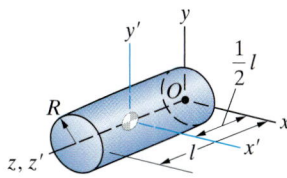

Circular cylinder

$$I_{x'\text{ axis}} = I_{y'\text{ axis}} = m\left(\frac{1}{12}l^2 + \frac{1}{4}R^2\right), \qquad I_{z'\text{ axis}} = \frac{1}{2}mR^2,$$

$$I_{x'y'} = I_{y'z'} = I_{z'x'} = 0.$$

Volume $= \frac{1}{3}\pi R^2 h$

$$I_{x\text{ axis}} = I_{y\text{ axis}} = m\left(\frac{3}{5}h^2 + \frac{3}{20}R^2\right), \qquad I_{z\text{ axis}} = \frac{3}{10}mR^2,$$

$$I_{xy} = I_{yz} = I_{zx} = 0.$$

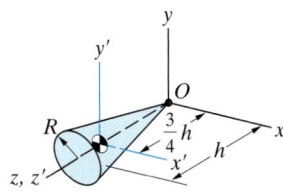

Circular cone

$$I_{x'\text{ axis}} = I_{y'\text{ axis}} = m\left(\frac{3}{80}h^2 + \frac{3}{20}R^2\right), \qquad I_{z'\text{ axis}} = \frac{3}{10}mR^2,$$

$$I_{x'y'} = I_{y'z'} = I_{z'x'} = 0.$$

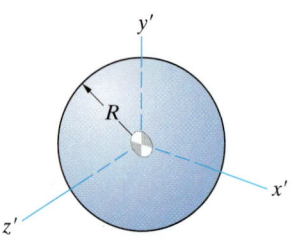

Sphere

$$\text{Volume} = \frac{4}{3}\pi R^3$$

$$I_{x'\text{ axis}} = I_{y'\text{ axis}} = I_{z'\text{ axis}} = \frac{2}{5}mR^2,$$

$$I_{x'y'} = I_{y'z'} = I_{z'x'} = 0.$$

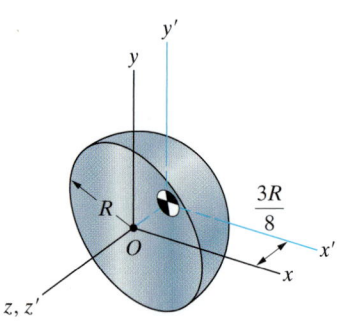

Hemisphere

$$\text{Volume} = \frac{2}{3}\pi R^3$$

$$I_{x\text{ axis}} = I_{y\text{ axis}} = I_{z\text{ axis}} = \frac{2}{5}mR^2$$

$$I_{x'\text{ axis}} = I_{y'\text{ axis}} = \frac{83}{320}mR^2, \qquad I_{z'\text{ axis}} = \frac{2}{5}mR^2$$

D

Spherical Coordinates

This appendix summarizes the equations of kinematics and vector calculus in spherical coordinates.

The position vector, velocity, and acceleration are

$$\mathbf{r} = r\mathbf{e}_r,$$

$$\mathbf{v} = \frac{dr}{dt}\mathbf{e}_r + r\frac{d\phi}{dt}\mathbf{e}_\phi + r\frac{d\theta}{dt}\sin\phi\,\mathbf{e}_\theta,$$

$$\mathbf{a} = \left[\frac{d^2r}{dt^2} - r\left(\frac{d\phi}{dt}\right)^2 - r\left(\frac{d\theta}{dt}\right)^2\sin^2\phi\right]\mathbf{e}_r$$

$$+ \left[r\frac{d^2\phi}{dt^2} + 2\frac{dr}{dt}\frac{d\phi}{dt} - r\left(\frac{d\theta}{dt}\right)^2\sin\phi\cos\phi\right]\mathbf{e}_\phi$$

$$+ \left[r\frac{d^2\theta}{dt^2}\sin\phi + 2\frac{dr}{dt}\frac{d\theta}{dt}\sin\phi + 2r\frac{d\phi}{dt}\frac{d\theta}{dt}\cos\phi\right]\mathbf{e}_\theta.$$

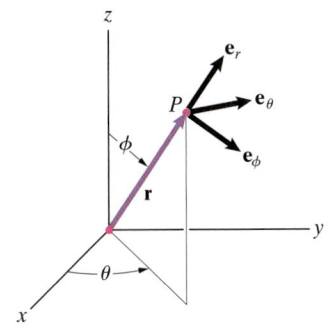

The gradient of a scalar field ψ is

$$\nabla\psi = \frac{\partial\psi}{\partial r}\mathbf{e}_r + \frac{1}{r}\frac{\partial\psi}{\partial\phi}\mathbf{e}_\phi + \frac{1}{r\sin\phi}\frac{\partial\psi}{\partial\theta}\mathbf{e}_\theta.$$

The divergence and curl of a vector field $\mathbf{v} = v_r\mathbf{e}_r + v_\theta\mathbf{e}_\theta + v_\phi\mathbf{e}_\phi$ are

$$\nabla\cdot\mathbf{v} = \frac{1}{r^2}\frac{\partial}{\partial r}(r^2v_r) + \frac{1}{r\sin\phi}\frac{\partial}{\partial\phi}(v_\phi\sin\phi) + \frac{1}{r\sin\phi}\frac{\partial v_\theta}{\partial\theta},$$

$$\nabla\times\mathbf{v} = \frac{1}{r^2\sin\phi}\begin{vmatrix} \mathbf{e}_r & r\mathbf{e}_\phi & r\sin\phi\,\mathbf{e}_\theta \\ \dfrac{\partial}{\partial r} & \dfrac{\partial}{\partial\phi} & \dfrac{\partial}{\partial\theta} \\ v_r & rv_\phi & r\sin\phi v_\theta \end{vmatrix}.$$

E

D'Alembert's Principle

This appendix describes an alternative approach for obtaining the equations of planar motion for a rigid body. By writing Newton's second law as

$$\Sigma \mathbf{F} + (-m\mathbf{a}) = \mathbf{0}, \tag{E.1}$$

we can regard it as an "equilibrium" equation stating that the sum of the forces, including an *inertial force* $-m\mathbf{a}$, equals zero (Fig. E.1). To state the equation of angular motion in an equivalent way, we use Eq. (18.19), which relates the total moment about a fixed point O to the acceleration of the center of mass and the angular acceleration in general planar motion:

$$\Sigma M_O = (\mathbf{r} \times m\mathbf{a}) \cdot \mathbf{k} + I\alpha.$$

We write this equation as

$$\Sigma M_O + [\mathbf{r} \times (-m\mathbf{a})] \cdot \mathbf{k} + (-I\alpha) = 0. \tag{E.2}$$

The term $[\mathbf{r} \times (-m\mathbf{a})] \cdot \mathbf{k}$ is the moment about O due to the inertial force $-m\mathbf{a}$. We can therefore regard this equation as an "equilibrium" equation stating that the sum of the moments about any fixed point, including the moment due to the inertial force $-m\mathbf{a}$ acting at the center of mass and an *inertial couple* $-I\alpha$, equals zero.

Stated in this way, the equations of motion for a rigid body are analogous to the equations for static equilibrium: The sum of the forces equals zero and the sum of the moments about any fixed point equals zero when we properly account for inertial forces and couples. This is called *D'Alembert's principle*.

If we define ΣM_O and α to be positive in the counterclockwise direction, the unit vector $\mathbf{k}$ in Eq. (E.2) points out of the page and the term $[\mathbf{r} \times (-m\mathbf{a})] \cdot \mathbf{k}$ is the counterclockwise moment due to the inertial force. This vector operation determines the moment, or we can evaluate it by using the fact that its magnitude is the product of the magnitude of the inertial force and the perpendicular distance from point O to the line of action of the force (Fig. E.2 a). The moment is positive if it is counterclockwise, as in Fig. E.2a, and negative if it is clockwise. Notice that the sense of the inertial couple is opposite to that of the angular acceleration (Fig. E.2b).

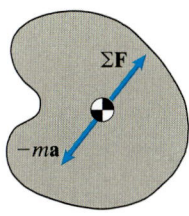

Figure E.1
The sum of the external forces and the inertial force is zero.

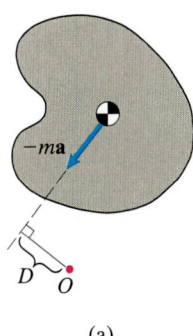

(a)

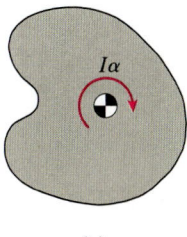

(b)

Figure E.2
(a) The magnitude of the moment due to the inertial force is $|-m\mathbf{a}|D$.
(b) A clockwise inertial couple results from a counterclockwise angular acceleration.

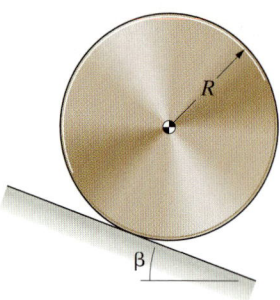

Figure E.3
Disk rolling on an inclined surface.

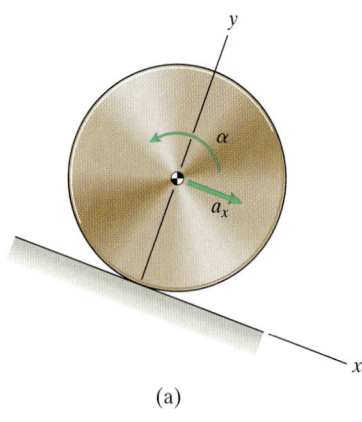

(a)

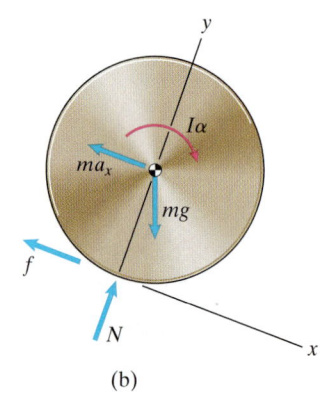

(b)

Figure E.4
(**a**) Acceleration of the center of the disk and its angular acceleration.
(**b**) Free-body diagram including the inertial force and couple.

As an example, consider a disk of mass m and moment of inertia I that is rolling on an inclined surface (Fig. E.3). We can use D'Alembert's principle to determine the disk's angular acceleration and the forces exerted on it by the surface. The angular acceleration of the disk and the acceleration of its center are shown in Fig. E.4a. In Figure E.4b, we draw the free-body diagram of the disk showing its weight, the normal and friction forces exerted by the surface, *and the inertial force and couple*. Equation (E.1) is

$$\Sigma \mathbf{F} + (-m\mathbf{a}) = 0:$$

$$(mg \sin \beta - f)\mathbf{i} + (N - mg \cos \beta)\mathbf{j} - ma_x\mathbf{i} = 0.$$

From this vector equation, we obtain the equations

$$mg \sin \beta - f - ma_x = 0,$$

$$N - mg \cos \beta = 0. \tag{E.3}$$

We now apply Eq. (E.2). By evaluating moments about the point where the disk is in contact with the surface, we can eliminate f and N from the resulting equation:

$$\Sigma M_O + [\mathbf{r} \times (-m\mathbf{a})] \cdot \mathbf{k} + (-I\alpha) = 0:$$

$$-R(mg \sin \beta) + R(ma_x) - I\alpha = 0. \tag{E.4}$$

The acceleration of the center of the rolling disk is related to the counterclockwise angular acceleration by $a_x = -R\alpha$. Substituting this relation into Eq. (E.4) and solving for the angular acceleration, we obtain

$$\alpha = -\frac{mgR \sin \beta}{mR^2 + I}.$$

From this result, we also know a_x and can therefore solve Eqs. (E.3) for the normal and friction forces, obtaining

$$N = mg \cos \beta, \qquad f = \frac{mgI \sin \beta}{mR^2 + I}.$$

In Eq. (E.4) we evaluated the moment due to the inertial force by simply multiplying the magnitude of the force and the perpendicular distance from O to its line of action, but we could have evaluated it with the vector expression:

$$[\mathbf{r} \times (-m\mathbf{a})] \cdot \mathbf{k} = [(R\mathbf{j}) \times (-ma_x\mathbf{i})] \cdot \mathbf{k} = R(ma_x).$$

Answers to Even-Numbered Problems

Chapter 12

12.2 (a) $e = 2.7183$.

(b) $e^2 = 7.3891$.

(c) $e^2 = 7.3892$.

12.4 7.32 m wide, 2.44 m high.

12.6 The 1-in. wrench fits the 25-mm nut.

12.8 343 mi/h.

12.10 (a) 5000 m/s; (b) 3.11 mi/s.

12.12 $g = 32.2$ ft/s^2.

12.14 0.310 m^2.

12.16 2.07×10^6 Pa.

12.18 $G = 3.44 \times 10^{-8}$ lb-ft^2/slug^2.

12.20 (a) The SI units of T are kg-m^2/s^2;

(b) $T = 73.8$ slug-ft^2/s^2.

12.22 (a) 4.448W.

(b) 4.448W / 9.81 = 0.453W.

12.24 (a) 491 N; (b) 81.0 N.

12.26 163 lb.

12.28 32.1 km.

12.30 345,000 km.

Chapter 13

13.2 (a) $v = \sin(5t)$ m/s. (b) $a = 2.04$ m/s^2.

13.4 (a) $v = 13.0$ m/s, $a = 1.28$ m/s^2.

(b) $v = 14.7$ m/s at $t = 6.67$ s.

13.6 (a) $v = 28$ m/s at $t = 4$ s. (b) $a = 0$.

13.8 (a) $v = -1.78$ m/s, $a = -11.2$ m/s^2.

(b) 2.51 m/s. (c) Zero.

13.10 (a) 0.628 m/s. (b) 3.95 m/s^2.

13.12 (a) $v = 126$ ft/s at $t = 11.8$ s.

(b) $a = 14.2$ ft/s^2 at $t = 4.2$ s.

13.14 $s = 100$ m, $v = 80$ m/s.

13.16 1.33 s.

13.18 (a) 0.0273 s. (b) 2.49 s, $v = 8020$ ft/s.

13.20 $a = -6.66$ m/s^2.

13.22 $s = 505$ ft.

13.24 $s = 1070$ m, $a = -24$ m/s^2.

13.26 6.69 s.

13.28 3630 ft.

13.30 Yes, the car travels 94.7 m in 5 s.

13.32 51.9 solar years.

13.34 (a) 45.5 s. (b) 3390 m.

13.36 10 s.

13.38 $v = 6.73$ ft/s, $a = -8.64$ ft/s^2.

13.40 $v = 0.602$ in/s.

13.42 $v = 3.33$ m/s.

13.44 $v = 2.80$ in/s.

13.46 11.2 km.

13.48 3240 ft.

13.50 2300 m.

13.52 42.5 m/s.

13.54 $v = 3$ ft/s.

13.56 $c = 38,900$.

13.58 $v = -2(1 - s^2)^{1/2}$ m/s.

13.60 (a) 1.29 ft. (b) 4.55 ft/s.

13.62 $v_0 = 35,900$ ft/s $= 24,500$ mi/hr.

13.66 1266 s, or 21.1 min.

13.68 $\mathbf{r} = 266.0\mathbf{i} + 75.3\mathbf{j} - 36.7\mathbf{k}$ (m).

13.70 (a) $R = 35.3$ m. (b) $R = 40.8$ m. (c) $R = 35.3$ m.

13.72 98.3 ft/s.

13.74 43.1 ft/s.

13.76 $31.2 < v_0 < 34.2$ ft/s.

13.78 (a) Yes. (b) No.

13.80 82.5 m.

13.82 $|\mathbf{v}| = 38.0$ ft/s, $t = 1.67$ s.

13.84 17.2 m.

13.86 $\mathbf{v} = 0.602\mathbf{i} - 4.66\mathbf{j}$ (m/s).

13.90 $\mathbf{a} = -0.099\mathbf{i} + 0.414\mathbf{j}$ (m/s^2).

13.94 (a) Positive.

    (b) Approximately 2π radians in 24 hours, or 7.27×10^{-5} rad/s.

13.96 $\omega = 3.51$ rad/s.

13.98 $\omega = 25.9$ rad/s.

13.100 (a) $\omega = 1.70$ rad/s. (b) $\theta = 1$ rad ($57.3°$).

13.102 $d\mathbf{e}/dt = -8.81\mathbf{i} + 13.4\mathbf{j}$.

13.104 (a) $|\mathbf{v}| = 3.35$ m/s. (b) $a_n = 140$ m/s^2, $a_t = 0$.

13.106 (a) 1730 rpm. (b) 3.01 rad/s^2.

13.108 (a) 106 rpm. (b) $\alpha = 0.0923$ rad/s^2.

13.110 (a) $\mathbf{v} = 16\mathbf{e}_t$ (m/s), $\mathbf{a} = 16\mathbf{e}_t + 64\mathbf{e}_n$ (m/s^2).

    (b) $s = 8$ m.

13.112 (a) $\mathbf{v}_B = 87.3\,\mathbf{e}_t$ (ft/s). (b) $\mathbf{v}_B = -61.7\mathbf{i} - 61.7\mathbf{j}$ (ft/s).

13.114 $|\mathbf{v}| = 401$ m/s, $|\mathbf{a}| = 0.0292$ m/s^2.

13.116 $\mathbf{v} = 18.3\mathbf{e}_t$ (m/s), $\mathbf{a} = 0.6\mathbf{e}_t + 6.68\mathbf{e}_n$ (m/s^2).

13.118 $\mathbf{v} = 12.05\mathbf{e}_t$ (m/s), $\mathbf{a} = 0.121\mathbf{e}_t + 2.905\mathbf{e}_n$ (m/s^2).

13.120 (a) $|\mathbf{a}| = 45.0$ ft/s^2. (b) $|\mathbf{a}| = 59.9$ ft/s^2.

13.122 (a) $\mathbf{a} = -4.75\mathbf{e}_t + 8.23\mathbf{e}_n$ (m/s^2).

13.124 (a) $\mathbf{v} = 15.2\mathbf{e}_t$ (m/s), $\mathbf{a} = -1.63\mathbf{e}_t + 9.67\mathbf{e}_n$ (m/s^2).

    (b) $\rho = 24.0$ m.

13.126 218 m.

13.128 $\rho = 81.0$ m.

13.130 $a_n = g/\sqrt{1 + (gt/v_0)^2}$.

13.132 $dy/dt = 0.260$ m/s, $d^2y/dt^2 = -0.150$ m/s^2.

13.134 (a) $\rho = 16.7$ ft. (b) $a_n = 77.3$ ft/s^2.

13.138 $\mathbf{a} = -4.4\mathbf{e}_r + 1.6\mathbf{e}_\theta$ (m/s^2).

13.140 $\mathbf{a}_A = 0.0167\mathbf{e}_r + 4.03\mathbf{e}_\theta$ (ft/s^2).

13.142 $\mathbf{v}_A = 2.22\mathbf{e}_r + 3.33\mathbf{e}_\theta$ (m/s).

13.144 (a) $\mathbf{v} = 0.32\mathbf{e}_r + 2.03\mathbf{e}_\theta$ (m/s).

    (b) $\mathbf{v} = 0.32\mathbf{i} + 2.03\mathbf{j}$ (m/s).

13.146 $\mathbf{a}_A = -58.0\mathbf{e}_r - 27.0\mathbf{e}_\theta$ (m/s^2).

13.148 $\mathbf{a}_A = -1.09\mathbf{e}_r$ (m/s^2).

13.152 $\mathbf{v} = \sqrt{v_0^2 + (\omega_0^2 - K)(r^2 - r_0^2)}\,\mathbf{e}_r + r\omega_0\mathbf{e}_\theta$.

13.154 $\mathbf{a} = 0$, $\alpha = -0.631$ rad/s^2.

13.156 (a) $\mathbf{a} = -3.52\mathbf{e}_r + 4.06\mathbf{e}_\theta$ (m/s^2).

    (b) $\mathbf{a} = -0.38\mathbf{i} - 5.36\mathbf{j}$ (m/s^2).

13.158 (a) $\mathbf{a} = -225\mathbf{e}_r - 173\mathbf{e}_\theta$ (ft/s^2).

    (b) $\mathbf{a} = -108\mathbf{i} - 263\mathbf{j}$ (ft/s^2).

13.160 $\mathbf{v} = 1047\mathbf{e}_\theta + 587\mathbf{e}_z$ (ft/s), $\mathbf{a} = -219{,}000\mathbf{e}_r$ (ft/s^2).

13.162 (a), (b) $\mathbf{v}_{A/B} = -3.66\mathbf{i} + 3.66\mathbf{j}$ (m/s).

13.164 $\mathbf{a}_{A/B} = 200\mathbf{i} + 200\mathbf{j}$ (ft/s^2).

13.166 $\mathbf{a}_{A/B} = 5\mathbf{i} - 2\mathbf{j}$ (ft/s^2).

13.168 $9.93°$ east of north, 42.0 min.

13.170 2.35 m/s, 342 s.

13.172 (a)

    (b) $v = 3.72$ m/s at $t = 2.31$ s.

13.174 $\theta_0 = 33.4°$.

13.176 $60.9°$ and $74.1°$.

13.178 $c = 1.31$ s^{-1}.

13.180 (b) $|\mathbf{a}|_{\min} = 2.07$ (m/s^2) at $t = 0.310$ s.

13.182 $|\mathbf{a}|_{\max} = 22.6$ m/s^2 at $r = 1.49$ m, $\theta = 20.5°$.

13.184 (a) 7.10 m. (b) 2.22 s. (c) 11.8 m/s.

13.186 $v = 42.3$ m/s.

13.188 13.1 s.

13.190 68.6 ft/s.

13.192 $|\mathbf{v}| = 2.19$ m/s, $|\mathbf{a}| = 5.58$ m/s^2.

13.194 $\mathbf{a} = -2.75\mathbf{e}_r - 4.86\mathbf{e}_\theta$ (m/s^2).

13.196 (a) $\mathbf{v} = -2.13\mathbf{e}_r + 6.64\mathbf{e}_\theta$ (m/s).

    (b) $\mathbf{v} = -5.90\mathbf{i} + 3.71\mathbf{j}$ (m/s).

Chapter 14

14.2 99.0 ft.

14.4 (a) 3.52 m/s. (b) 5.29 m.

14.6 (a) 4.03 m/s. (b) 6.04 m.

14.8 (a) $a = 3.43$ m/s^2. (b) $v = 14.8$ m/s.

14.10 $\mathbf{v} = 22\mathbf{i} + 2\mathbf{j} + 40\mathbf{k}$ (m/s).

14.12 $\Sigma\mathbf{F} = 2.40\mathbf{i} + 1.20\mathbf{j} + 2.08\mathbf{k}$ (kN).

14.14 18.8 ft/s^2.

14.16 (a) No. (b) Yes, $a_x = 0.5$ m/s^2. (c) $x = 2$ m.

14.18 $F_x = -198$ N, $F_y = 1800$ N.

14.20 $W = 88.3$ kN, $T = 61.7$ kN, $L = 66.9$ kN.

14.22 $L = 293.2$ kN, $D = 33.0$ kN.

14.24 $(66, 138, -58)$ m.

14.26 $F_x = 0.359$ N, $F_y = 19.888$ N.

14.28 1.30 m/s.

14.30 Velocity is 2.45 m/s, distance is 1.23 m.

14.32 $F = 331$ N, $a = 3.92$ m/s^2.

14.34 $\theta = 23.6°$.

14.36 4.43 ft up the surface.

14.38 (a) 1200 N. (b) 1.84 m/s.

14.40 (a) 2.1 ft. (b) 6.9 ft.

14.42 (a) $a = -5$ m/s^2. (b) $v = -2.24$ m/s.

14.44 (a) 773 ft/s^2 (24 g). (b) 28.0 ft/s.

14.46 (a) 11.9 ft. (b) 813 lb.

14.48 $y = -18.8$ mm.

14.50 (a) 50 s. (b) 40.8 N. (c) $4.8\mathbf{i} + \mathbf{j}$ (m/s).

14.52 2.06 m/s^2 up the bar.

14.54 $|\mathbf{a}_A| = 7.89$ ft/s^2.

14.56 $t = 0.600$ s.

14.58 $F_x = -0.544$ N.

14.60 $F_x = -73.4$ N, $F_y = 612$ N.

14.62 0.284 s.

14.64 $v_x = 0.486$ m/s, $v_y = -0.461$ m/s.

14.66 $\Sigma F_t = 0$, $\Sigma F_n = 373$ lb.

14.68 $a_t = 0.6$ m/s^2, $a_n = 6.68$ m/s^2.

14.70 Tension is 4.8 N, force is 0.6 N.

14.72 (a) $\Sigma F_t = 9740$ lb, $\Sigma F_n = 28{,}800$ lb.

(b) $d\theta/dt = 3.95°/$s.

14.76 $\theta = 49.9°$, $|\mathbf{v}| = 10.8$ ft/s.

14.78 $2.62 \le v \le 3.74$ m/s.

14.80 (a) 207,000 lb. (b) 41,700 ft.

14.82 (a) $v = 140$ m/s. (b) $\rho = 815$ m.

14.86 $\theta = L_0/R - \sqrt{(L_0/R)^2 - (2v_0/R)t}$.

14.88 $\rho = 697$ m, $\mathbf{e} = 0.916\mathbf{i} - 0.308\mathbf{j} + 0.256\mathbf{k}$.

14.90 $\beta = 68.2°$

14.92 11.4 m/s^2.

14.94 $\Sigma F_r = -16.8$ N, $\Sigma F_\theta = 20.7$ N.

14.96 $9.46\mathbf{e}_r + 3.44\mathbf{e}_\theta$ (N).

14.98 $v_r = 17.8$ m/s.

14.100 $|\mathbf{v}| = 2.89$ m/s, $T = 41.6$ N.

14.102 $N = 1.02$ lb.

14.104 $\mu_s = 0.406$. The mass slips toward O.

14.106 $-1.48\mathbf{e}_r - 0.20\mathbf{e}_\theta$ (lb).

14.108 $k = 2m\omega_0^2$.

14.110 $11.5\mathbf{e}_r + 44.2\mathbf{e}_\theta + 10\mathbf{e}_z$ (N).

14.112 $|\Sigma\mathbf{F}| = 8.36$ N at $t = 4.39$ s.

14.114 (a) $|\mathbf{v}| = 1020$ m/s. (b) $t = 2.36 \times 10^6$ s (27.3 days).

14.116

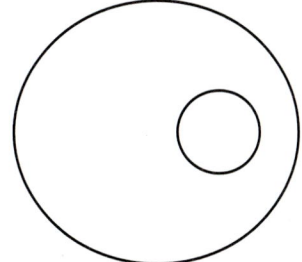

14.118 $v_0 = 35{,}500$ ft/s.

14.120

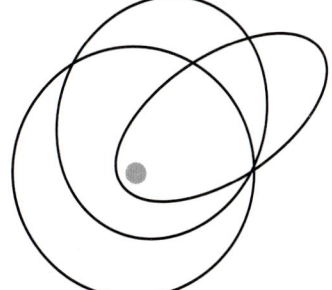

14.122

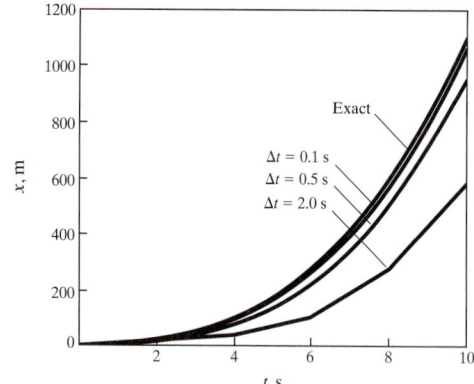

14.124

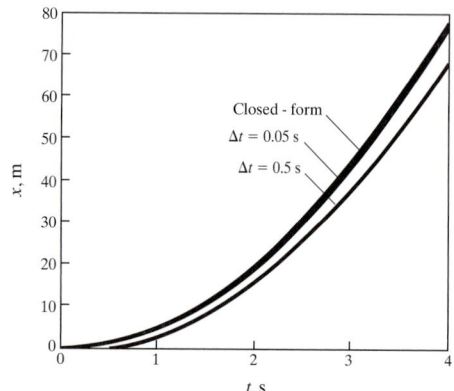

14.126

time, s	position, m	velocity, m/s
0.00	1.0000	0.0000
0.01	1.0000	−0.0100
0.02	0.9999	−0.0200
0.03	0.9997	−0.0300
0.04	0.9994	−0.0400
0.05	0.9990	−0.0500

14.128 $x = 22.5$ ft, $v_x = 23.1$ ft/s.

14.130

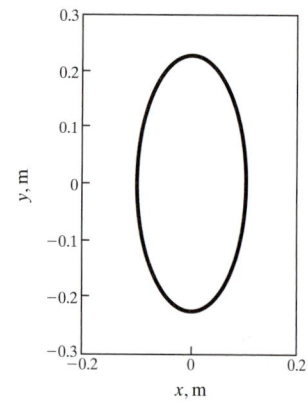

14.132 $C = 0.0217$.

14.134 $|\mathbf{a}| = 136.0$ ft/s^2 at $t = 0.084$ s.

14.136   (a) 34.6 ft/s^2, 3490 lb. (b) $\mu_s = 1.07$.

14.138   (a) $F_1 = 63$ kN, $F_2 = 126$ kN, $F_3 = 189$ kN.

(b) $F_1 = 75$ kN, $F_2 = 150$ kN, $F_3 = 225$ kN.

14.140   Deceleration is 1.49 m/s^2, compared with 8.83 m/s^2 on a level road.

14.142   14,300 ft (2.71 mi).

14.144   10.5 lb.

14.146 $a_A = 4.02$ ft/s^2, $T = 17.5$ lb.

14.148   29.7 ft.

14.150   9.30 N.

14.152 $\tan \alpha = v^2/\rho g$.

14.154 $\Sigma \mathbf{F} = -10.7\mathbf{e}_r + 2.55\mathbf{e}_\theta$ (N).

Chapter 15

15.2   5.82 m/s.

15.4   (a) 10.9 ft-lb. (b) 2.65 ft/s.

15.6 $P_{ave} = 1.62$ kW (kilowatts).

15.8   (a) $P_{max} = 4.53 \times 10^5$ ft-lb/s.

(b) $P_{ave} = 2.26 \times 10^5$ ft-lb/s.

15.10   (a) $P_{max} = 4.65$ MW (megawatts).

(b) $P_{ave} = 2.32$ MW (megawatts).

15.12   (a) −6.45 N-m. (b) 3.66 m/s.

15.14   6.00 m.

15.16   (a) 2160 ft-lb. (b) 7.39 lb.

15.18   (a) 196 N/m.

(b) 0.990 m/s at $s = 0.1$ m.

15.20   Work = 509 N-m, $v = 2.52$ m/s.

15.22 $v = 21.8$ m/s.

15.24   4.04 m/s.

15.26   2.58 ft/s.

15.28 $v = 1.72$ m/s.

15.30 $v = 1.14$ m/s.

15.32   (a), (b), (c), 117 ft/s.

15.34   (a) 5.98 m/s. (b) 5.56 m/s.

15.36 3.55 m/s.

15.38 12.7 ft/s.

15.40 $v_0 = 15.66$ m/s, $v_{top} = 7.00$ m/s.

15.42 11.2 ft/s.

15.44 39.3 ft/s, or 26.8 mi/hr.

15.46 -107 kN-m.

15.48 (a) $U_{12} = 0.210$ N-m.

(b) $v_2 = 1.45$ m/s.

15.50 621 lb/ft.

15.52 Distance is 40.3 mm, deceleration is 0.248 m/s^2.

15.54 2.18 m/s.

15.56 2.83 m/s.

15.58 (a) 5 ft. (b) 6.34 ft/s.

15.60 0.785 m.

15.62 0.385 m.

15.64 $v_2 = 7.03$ m/s.

15.66 5.77 m/s.

15.68 $k = 997$ lb/ft.

15.70 4.90 m/s.

15.72 193,000 lb.

15.74 1370 m/s.

15.76 2.25×10^{10} N-m.

15.78 (a) 50 ft-lb. (b) 50 ft-lb. (c) 13.9 ft/s.

15.80 (a), (b) 6.48 m.

15.82 3.95 m/s.

15.84 (a) $\alpha = 60°$.

(b) Before, 39.2 N; after, 58.9 N.

15.86 $\theta = 51.2°$.

15.88 (a) 687 N. (b) 883 N.

15.90 $W\left[1 + \sqrt{1 + 4C/W}\right]$.

15.92 1.20 m/s.

15.94 1.99 m/s.

15.96 $V = \frac{1}{2}kS^2 + \frac{1}{4}qS^4$.

15.98 $v = 2.30$ m/s.

15.100 $v = 8.45$ ft/s.

15.102 87.3 ft.

15.104 $v = 11.0$ km/s.

15.106 2880 m/s.

15.108 $v = 419$ m/s.

15.110 (a) $\mathbf{F} = -4x\mathbf{i} + \mathbf{j}$ (N).

(b) The work is -1 N-m for any path from position 1 to position 2.

15.112 $\mathbf{F} = -\left[k(r - r_0) + q(r - r_0)^3\right]\mathbf{e}_r$.

15.114 (a) $\mathbf{F} = (\sin\theta - 2r\cos^2\theta)\mathbf{e}_r$
$+ (\cos\theta + 2r - \sin\theta\cos\theta)\mathbf{e}_\theta$.

(b) The work is 2 ft-lb for any path from 1 to 2.

15.118 (a) and (c).

15.120 553 mm.

15.122 58.7 mm, 1280 N-m/s (watts).

15.124 2.18 m.

15.126 Maximum distance is 0.203 m.

15.128 $v = 6.32$ m/s.

15.130 $\theta_{max} = 51°$.

15.132 (a), (b) 193 ft.

15.134 He should choose (b). Impact velocity is 11.8 m/s, work is -251 kN-m. In (a), impact velocity is 13.9 m/s, work is -119 kN-m.

15.136 $v = 2.08$ m/s.

15.138 (a) 14.3×10^7 ft-lb.

(b) $v = 88\left[1 - e^{-(F_0/88m)t}\right]$.

15.140 $k = 163$ lb/ft.

15.142 (a) $k = 809$ N/m.

(b) $v_2 = 6.29$ m/s.

15.144 4.39 ft/s.

15.146 $h = 0.179$ m.

15.148 (a) $\theta = 27.9°$. (b) 159 lb. (c) 222 lb.

15.150 $v_1 = 4.73$ ft/s.

15.152 1.02 m.

15.154 2.00 m/s.

15.156 24.8 ft/s.

15.158 (a) $v = 0$.

(b) $v = \sqrt{gR_E/2} = 18,300$ ft/s, or 12,500 mi/hr.

15.160 (a) 11.3 MW (megawatts).

(b) 9.45 MW

Chapter 16

16.2 33.0 ft/s.

16.4 (a) 56,000 N-s. (b) 2.56 m/s (9.21 km/h).

16.6 6.07 m/s.

16.8 $6\mathbf{i} - 2\mathbf{j} + 56\mathbf{k}$ (m/s).

16.10 $\mathbf{v} = 4.29\mathbf{i}$ (ft/s).

16.12 (a) 335 kN-s. (b) 7.84 s.

16.14 4.15 m/s.

16.16 (a) 222 N-s. (b) 1.85 m/s.

16.18 0.199 s.

16.20 17.2 ft/s.

16.22 2.01 m/s.

16.24 (a) $\mathbf{v} = 10\mathbf{i} - 12.1\mathbf{j}$ (m/s).

    (b) $-589\mathbf{j}$ (N-s).

    (c) $\mathbf{v} = 10\mathbf{i} - 12.1\mathbf{j}$ (m/s).

16.26 (a) $F_x = -32\cos 4t$ N, $F_y = -32\sin 4t$ N.

    (b) $\mathbf{v} = 3.03\mathbf{i} - 2.61\mathbf{j}$ (m/s).

16.28 924 N.

16.30 37.3 m/s.

16.32 105 m/s.

16.34 (a) 8470 N. (b) 6.67 m/s^2.

16.36 75.1 lb (approximately 600 times the watch's weight).

16.38 Horizontal force is 0.234 N, vertical force is 0.364 N.

16.40 $-0.35\mathbf{i} + 122.96\mathbf{j}$ (N).

16.42 $|\mathbf{v}_A| = 2.07$ m/s, $|\mathbf{v}_B| = 2.76$ m/s.

16.44 (a), (b) 1.63 ft/s toward the right.

16.46 (a), (b) 1.64 m/s toward the right.

16.48 (a) $6.47 x 10^{-4}$ m/s toward the left.

    (b) Zero.

    (c) 4 m to his left.

16.50 2.41 m/s.

16.52 0.176 ft (2.11 in).

16.54 96.9 mm.

16.56 (a) $4\mathbf{i} + 30\mathbf{j}$ (mm/s). (b) $x = 6.2$ m, $y = 6$m.

16.58 $v_A = 6.08$ ft/s, $e = 0.197$.

16.60 $m_B = 6.4$ kg, $e = 0.8$.

16.62 0.0334 s.

16.64 $v_A\sqrt{mk/2}$.

16.66 $v_A = 7.70$ m/s.

16.68 3.00 m.

16.70 $e = 0.77$.

16.72 4.5 ft/s upward.

16.74 0.963 m/s toward the left.

16.76 $e = 0.673$.

16.78 $\mathbf{v}_A = -5.71\mathbf{i} + 4\mathbf{j} + 8\mathbf{k}$ (m/s),
$\mathbf{v}_B = 3.89\mathbf{i} - 6\mathbf{j} + 5\mathbf{k}$ (m/s).

16.80 $\mathbf{v}'_A = \mathbf{i} + \mathbf{j}$ (m/s), $\mathbf{v}'_B = -\mathbf{i} + \mathbf{j}$ (m/s).

16.84 $v_S = 35.3$ m/s.

16.86 (a) $\mathbf{v} = \dfrac{1}{2}t^2\mathbf{i} + 2t\mathbf{j}$ (m/s), $\mathbf{r} = \dfrac{1}{6}t^3\mathbf{i} + t^2\mathbf{j}$(m).

    (b), (c) $-432\mathbf{k}$ (kg-m^2/s).

16.88 $|v_r| = 2490$ m/s, $|v_\theta| = 4310$ m/s.

16.90 $\omega = 3.84$ rad/s, $v_r = \pm 0.262$ m/s.

16.92 2.31 m.

16.94 6 N-m.

16.96 (a) $-1440\mathbf{k}$ (kg-m^2/s). (b) 1.2 m/s.

16.100 103 lb.

16.102 750 N.

16.104 (a) $351\mathbf{i} - 849\mathbf{j}$ (N).

    (b) $1200\mathbf{i} - 1200\mathbf{j}$(N).

    (c) $2400\mathbf{i}$ (N).

16.106 $0.406\mathbf{k}$ (N-m).

16.108 $51.9\mathbf{i} - 282.2\mathbf{j}$ (N).

16.110 1530 m/s.

16.112 $v = v_f \ln\left(\dfrac{m_0}{m_0 - \dot{m}_f t}\right) - gt$.

16.114 (a) $F = 3s + 12/32.2$ lb.

    (b) 25.5 ft-lb.

16.116 $v = v_0/\left[1 + (\rho_L/2m)s\right]$.

16.118 6820 N.

16.120 18.5 kN.

16.122 (a) 5. (b) 43,500 lb. (c) 11.2 ft/s^2.

16.124 (a) $180\mathbf{i} + 360\mathbf{j}$ (lb-s).

    (b) $12\mathbf{i} + 44\mathbf{j}$ (ft/s).

16.126	877 kN.	**17.22**	$\mathbf{v}_A = -5\mathbf{i} + 8.66\mathbf{j}$ (m/s).

16.126 877 kN.

16.128 (a) 34.9 ft/s.

    (b) 15,700 ft-lb.

16.130 (a) 4.80 ft/s.

    (b) 415 ft.

16.132 (a) 3.45 m/s.

    (b) 18.7 kN.

16.134 (a) 8.23 kN.

    (b) 16.5 kN.

16.136 $|\mathbf{v}_A| = 2.46$ m/s, $|\mathbf{v}_B| = 2.90$ m/s.

16.138 1.57 ft.

16.140 $e = 0.304$.

16.142 2.30 m.

16.144 $x = -11.13$ in., $y = 6.42$ in.

16.146 $|v_r| = 11{,}550$ ft/s, $|v_\theta| = 9430$ ft/s.

16.148 867 lb (including the weight of the drum).

16.150 (a) 30.1 ft/s.

    (b) 46.8 ft/s.

Chapter 17

17.2 $\theta = 135°$, $|\mathbf{v}_A| = 2.33$ ft/s, $|\mathbf{a}_A| = 2.84$ ft/s^2.

17.4 (a) 1.2 rad/s clockwise, 0.4 rad/s^2 counterclockwise.

    (b) $|\mathbf{v}_A| = 2.4$ m/s, $|\mathbf{a}_A| = 2.99$ m/s^2.

17.6 (a) 2.67. (b) 2.67 rad/s.

17.8 $\mathbf{v}_A = -6.67\mathbf{i} - 6.67\mathbf{j}$ (ft/s), $\mathbf{a}_A = 66.7\mathbf{i} - 66.7\mathbf{j}$ (ft/s^2), $\mathbf{v}_B = 13.3\mathbf{j}$ (ft/s), $\mathbf{a}_B = -133\mathbf{i}$ (ft/s^2).

17.10 (a) 6.86 rad/s^2 clockwise.

    (b) $|a_t| = 5.71$ ft/s^2, $|a_n| = 5560$ ft/s^2.

17.12 $|\mathbf{v}_B| = 2.37$ m/s, $|\mathbf{a}_B| = 22.1$ m/s^2.

17.14 $\boldsymbol{\omega} = 30\mathbf{i}$ (rad/s).

17.16 $\boldsymbol{\omega}_{OQ} = -4\mathbf{k}$ (rad/s), $\boldsymbol{\omega}_{PQ} = 4\mathbf{k}$ (rad/s).

17.18 (a) $\mathbf{v}_{A/B} = 4.8\mathbf{j}$ (m/s).

    (b) $\boldsymbol{\omega} = 12\mathbf{k}$ (rad/s).

    (c) $\mathbf{v}_{A/B} = 4.8\mathbf{j}$ (m/s).

17.20 (a) $\mathbf{v}_{A/B} = -1.78\mathbf{i} - 5.21\mathbf{j}$ (m/s).

    (b) $\mathbf{v}_{B/A} = 1.78\mathbf{i} + 5.21\mathbf{j}$ (m/s).

17.22 $\mathbf{v}_A = -5\mathbf{i} + 8.66\mathbf{j}$ (m/s).

17.24 $\mathbf{v}_A = 5\mathbf{i}$ (m/s), $\mathbf{v}_B = -3.54\mathbf{i} - 3.54\mathbf{j}$ (m/s), $\mathbf{v}_C = -3.54\mathbf{i} + 3.54\mathbf{j}$ (m/s).

17.26 (a) $\boldsymbol{\omega} = 81.7\mathbf{k}$ (rad/s).

    (b) $-95.3\mathbf{i}$ (ft/s).

    (c) $-191\mathbf{i}$ (ft/s).

17.28 $\mathbf{v}_T = 12.4\mathbf{i} + 5.1\mathbf{j}$ (m/s).

17.30 $\mathbf{v}_G = 1.5\mathbf{i} - 0.546\mathbf{j}$ (m/s).

17.32 $\omega_{OQ} = 1.18$ rad/s clockwise, $\omega_{PQ} = 1.18$ rad/s counterclockwise.

17.34 $\omega_{BD} = 8$ rad/s clockwise, $\mathbf{v}_D = 6.40\mathbf{i} - 1.28\mathbf{j}$ (m/s).

17.36 $\omega_{CD} = 10$ rad/s counterclockwise.

17.38 $\omega_{BC} = 2.67$ rad/s counterclockwise, $\omega_{CD} = 2.67$ rad/s clockwise.

17.40 $\mathbf{v}_E = -12.3\mathbf{j}$ (m/s).

17.42 0.0857 rad/s counterclockwise.

17.44 $\omega_{AB} = 2.31$ rad/s clockwise, $v_B = 3.15$ m/s to the left.

17.46 $\mathbf{v}_C = 25.1\mathbf{i}$ in./s.

17.48 $\mathbf{v}_A = 1.2\mathbf{i} + 1.2\mathbf{j}$ (m/s).

17.50 $\omega_{BC} = 2.61$ rad/s, $\mathbf{v}_C = -9.1\mathbf{i}$ (m/s).

17.52 0.95 m/s.

17.54 $\mathbf{v}_C = 0.282\mathbf{i} + 0.202\mathbf{j}$ (m/s).

17.56 $\mathbf{v}_D = -0.557\mathbf{i} + 0.815\mathbf{j}$ (m/s).

17.58 $v_W = 0.2$ m/s, 4 rad/s counterclockwise.

17.60 35.5 rpm clockwise.

17.62 Angular velocity = 52.1 rad/s, or 497 rpm, $|\mathbf{v}_A| = 5.21$ m/s.

17.64 $x_C = 3$ m, $y_C = 0$, $v_B = 10$ m/s.

17.66 $x = 0.35$ ft, $y = -1.5$ ft.

17.68 $\mathbf{v}_B = 6\mathbf{i} + 3.46\mathbf{j}$ (ft/s).

17.70 (a) (3.46, 4) ft.

    (b) $\mathbf{v}_A = 2\mathbf{i}$ (ft/s).

17.72 6.57 in./s.

17.74 1.67 rad/s counterclockwise.

17.76 (a) $x = 225$ mm, $y = 225$ mm.

    (b) $\mathbf{v}_C = 13.5\mathbf{i}$ (m/s).

17.78 $\omega_{BC} = 5.33$ rad/s counterclockwise, $\omega_{CD} = 4.57$ rad/s clockwise.

17.80 (a) $x = -0.425$ m, $y = 0.737$ m.

   (b) $v_B = 2.31$ m/s, $\omega = 2.35$ rad/s counterclockwise.

17.82 (a), (b) $\mathbf{a}_A = -73.3\mathbf{i} + 27.0\mathbf{j}$ (m/s^2).

17.84 $\mathbf{a}_T = 2.02\mathbf{i} + 2.37\mathbf{j}$ (m/s^2).

17.86 $\mathbf{a}_A = -0.5\mathbf{j}$ (m/s^2), $\mathbf{a}_B = 0.3\mathbf{j}$ (m/s^2).

17.88 $\omega_{AC} = 0$, $\alpha_{AC} = 1.13$ rad/s^2 clockwise.

17.90 Acceleration $= 134$ in./s^2.

17.92 $\alpha_{OQ} = 1.39$ rad/s^2 clockwise,
   $\alpha_{PQ} = 1.39$ rad/s^2 counterclockwise.

17.94 1.77 rad/s^2 counterclockwise.

17.96 $\alpha_{BC} = 26.8$ rad/s^2 counterclockwise,
   $\alpha_{CD} = 12.3$ rad/s^2 counterclockwise.

17.98 $\mathbf{v}_C = 40.8\mathbf{i}$ (in./s), $\mathbf{a}_C = 6.98\mathbf{i}$ (in./s^2).

17.100 13.0 in/s^2 toward the left.

17.102 $\omega_{AB} = -0.879$ rad/s, $\alpha_{AB} = -1.06$ rad/s^2,
   $\omega_{BC} = -1.15$ rad/s, $\alpha_{BC} = -2.41$ rad/s^2.

17.104 $\mathbf{a}_E = -12.3\mathbf{j}$ (m/s^2).

17.106 $\mathbf{a}_C = -7.78\mathbf{i} - 33.54\mathbf{j}$ (m/s^2).

17.108 $\mathbf{a}_D = -0.135\mathbf{i} - 0.144\mathbf{j}$ (m/s^2).

17.110 $\omega_{AB} = 3.55$ rad/s clockwise, $\alpha_{AB} = 12.1$ rad/s^2 clockwise, $\omega_{BC} = 2.36$ rad/s counterclockwise, $\alpha_{BC} = 16.5$ rad/s^2 counterclockwise.

17.112 0.0571 rad/s^2 clockwise.

17.114 $\alpha_{\text{planet}} = 41.4$ rad/s^2 clockwise,
   $\alpha_{\text{sun}} = 82.9$ rad/s^2 counterclockwise.

17.116 $\alpha_{CD} = 26.5$ rad/s^2 clockwise,
   $\alpha_{DE} = 31.1$ rad/s^2 counterclockwise.

17.118 $\mathbf{a}_A = -200\mathbf{i} + 80\mathbf{j}$ (ft/s^2).

17.120 $\mathbf{a}_C = -8.80\mathbf{i} + 5.60\mathbf{j}$ (m/s^2).

17.122 $\omega_{AC} = 3$ rad/s counterclockwise,
   $\alpha_{AC} = 6$ rad/s counterclockwise.

17.124 0.549 m/s toward the left.

17.126 0.0972 rad/s counterclockwise.

17.128 $\omega_{BC} = 6.17°$ per second counterclockwise, rate of extension is 0.109 m/s.

17.130 $\omega_{AC} = 8.66$ rad/s counterclockwise, and bar AC slides through the sleeve at 5 m/s toward A.

17.132 $\omega_{AC} = 0.293$ rad/s clockwise,
   $\mathbf{v}_C = -0.738\mathbf{i} - 1.370\mathbf{j}$ (ft/s).

17.134 0.801 ft/s.

17.136 $\omega_{AB} = 2$ rad/s clockwise, $v_{B\,\text{rel}} = 2$ m/s toward C.

17.138 $\omega_{AB} = 5.18$ rad/s counterclockwise.

17.140 $\omega_{\text{plate}} = 2$ rad/s counterclockwise, and the velocity at which the pin slides relative to the slot is 0.2 m/s downward.

17.144 $\mathbf{a}_A = -0.00624\mathbf{i} + 0.0576\mathbf{j}$ (m/s^2).

17.146 $\mathbf{v}_A = -3.5\mathbf{i} + 0.5\mathbf{j} + 4\mathbf{k}$ (m/s),
   $\mathbf{a}_A = -10\mathbf{i} - 6.5\mathbf{j} - 19.25\mathbf{k}$ (m/s^2).

17.148 $-2\mathbf{i}$ (m/s).

17.150 $\mathbf{a}_{A\,\text{rel}} = -14\mathbf{i} - 2\mathbf{j}$ (ft/s^2).

17.152 (a) $\mathbf{v} = v\mathbf{j}$, $\mathbf{a} = -\left(v^2/R_E\right)\mathbf{i}$.

   (b) $\mathbf{v} = v\mathbf{j} - \omega_E R_E \cos L\mathbf{k}$,
   $\mathbf{a} = -\left(v^2/R_E + \omega_E^2 R_E \cos^2 L\right)\mathbf{i} + \omega_E^2 R_E \sin L \cos L\mathbf{j} + 2\omega_E v \sin L\mathbf{k}$.

17.154 (a) $-9.81\mathbf{k}$ (m/s^2). (b) $3.29\mathbf{i} - 9.81\mathbf{k}$ (m/s^2).

17.156 (a) $0.1\mathbf{i} + 0.1\mathbf{j}$ (m/s^2).

   (b) $0.125\mathbf{i} + 0.085\mathbf{j} + 0.106\mathbf{k}$ (m/s^2).

17.160 OQ: 9.29 rad/s counterclockwise;
   PQ: 2.92 rad/s clockwise.

17.162 $\mathbf{v}_C = -32.0\mathbf{j}$ (ft/s).

17.164 $\alpha_{AB} = 13.60 \times 10^3$ rad/s^2 clockwise,
   $\alpha_{BC} = 8.64 \times 10^3$ rad/s^2 counterclockwise.

17.166 $\mathbf{a}_D = -3490\mathbf{i}$ (in./s^2).

17.168 $\mathbf{a}_G = -8.18\mathbf{i} - 26.4\mathbf{j}$ (in./s^2).

17.170 $\mathbf{v}_C = -1.48\mathbf{i} + 0.79\mathbf{j}$ (m/s).

17.172 $\mathbf{a}_C = -2.99\mathbf{i} - 1.40\mathbf{j}$ (m/s^2).

17.174 Velocity is 55.9 in./s to the right; acceleration is 390 in./s^2 to the left.

17.176 $\omega_{BD} = 0.733$ rad/s counterclockwise.

17.178 $\omega_{AB} = 0.261$ rad/s counterclockwise,
   $\omega_{BC} = 2.80$ rad/s counterclockwise.

17.180 $\omega_{BC} = 1.22$ rad/s clockwise, 18.0 m/s from B toward C.

17.182 Velocity $= 6.89$ ft/s upward;
   acceleration $= 169$ ft/s^2 upward.

17.184 5.66 N.

17.186 0.103 rad/s counterclockwise.

Chapter 18

18.2 (a) 0.980 ft/s^2.

(b) $N_A = 57.7$ lb, $N_B = 172$ lb.

18.4 (a) 2.23 m/s^2.

(b) $N_A = 175$ kN, $N_B = 2770$ kN.

18.6 (a) 3.22 ft/s^2.

(b) $N_A = 182$ lb, $N_B = 58$ lb.

18.8 Time $= 0.923$ s, distance $= 2.31$ m.

18.10 (a) 11.4 rad/s.

(b) 4.52 revolutions.

18.12 (a) 9.52 rad/s.

(b) 8.21 revolutions.

18.14 (a) 565 rpm.

(b) 799 rpm.

18.16 $\omega_C = 57.5$ rad/s counterclockwise.

18.18 (a) 3.49 rad/s^2 counterclockwise.

(b) 4.20 rad/s^2 clockwise.

18.20 (a), (b) $\alpha = 11.4$ rad/s^2 counterclockwise,
$A_x = 0$, $A_y = 4.12$ lb.

18.22 $\alpha = 0$, $D_x = 0$, $D_y = 99.9$ N.

18.24 $B_x = 9.02$ kN, $B_y = 9.66$ kN.

18.26 $M_B = 27.1$ N-m counterclockwise,
$B_x = -11.0$ N, $B_y = 108.5$ N.

18.28 $\mathbf{a}_G = 0.1108\mathbf{i} - 0.0168\mathbf{j}$ (m/s^2),
$\alpha = -0.000427$ rad/s^2.

18.30 $\mathbf{F}_B = -19.1\mathbf{i} + 183.3\mathbf{j}$ (N), $\mathbf{M}_B = 62.6\mathbf{k}$ (N-m).

18.32 $\alpha = 0.200$ rad/s^2 clockwise.

18.34 $t_{\text{ring}}/t_{\text{disk}} = \sqrt{4/3}$.

18.36 1.01 m/s.

18.38 (a) 14.8 rad/s^2 clockwise.

(b) $\mu_s = 0.227$.

18.40 Velocity $= 3.81$ ft/s, time $= 1.97$ s.

18.42 (a) It doesn't slip, $\alpha = 22.2$ rad/s^2 clockwise.

(b) It does slip, $\alpha = 53.6$ rad/s^2 clockwise.

18.44 27.9 N.

18.46 2.35 rad/s^2 counterclockwise.

18.48 (a) 9.38 rad/s^2 clockwise.

(b) 9.57 rad/s^2 clockwise.

18.50 61.3 rad/s^2 clockwise.

18.52 3.16 ft/s^2.

18.54 $\mu_s = 0.108$.

18.56 2.26 m/s^2 to the left.

18.58 $B_x = -60.3$ lb, $B_y = -125.6$ lb,
$C_x = 11.4$ lb, $C_y = 135.6$ lb.

18.62 (a) 9.34 m/s^2. (b) 516 N-m.

18.64 $\alpha_{BC} = 17.0$ rad/s^2 counterclockwise.

18.66 $\alpha_{OQ} = 4.88$ rad/s^2 clockwise,
$\alpha_{PQ} = 4.88$ rad/s^2 counterclockwise.

18.70 $\alpha = 0.255$ rad/s^2 clockwise.

18.72 $|P| = 44.1$ kN.

18.74

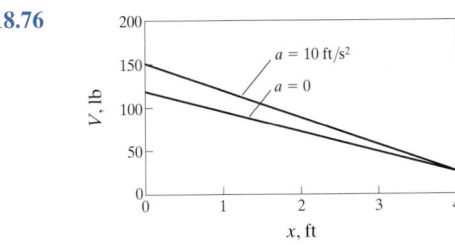

18.76

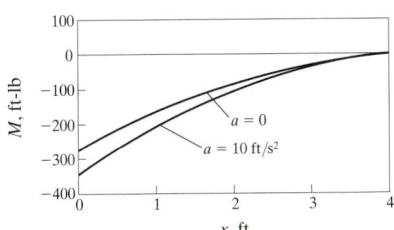

18.78

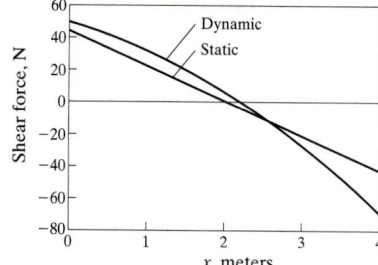

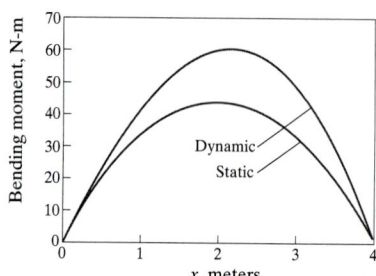

18.80

time, s	θ, rad	ω, rad/s	closed-form ω, rad/s
0.0	0.000	0.000	0.000
0.2	0.000	0.250	0.250
0.4	0.050	0.499	0.498
0.6	0.150	0.747	0.744
0.8	0.299	0.991	0.987
1.0	0.498	1.232	1.225

18.82

time, s	θ, rad	ω, rad/s
0.0	0.000	0.000
0.1	0.000	1.472
0.2	0.147	2.943
0.3	0.441	4.399
0.4	0.881	5.729
0.5	1.454	6.665

18.84

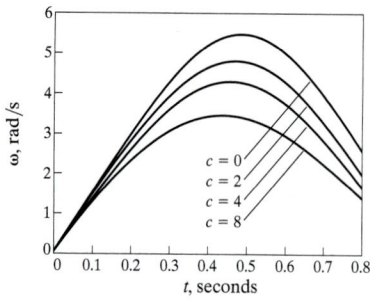

18.86

time, s	θ, rad	ω, rad/s
0.00	0.0000	0.0000
0.01	0.0000	0.1472
0.02	0.0015	0.2943
0.03	0.0044	0.4410
0.04	0.0088	0.5868
0.05	0.0147	0.7313

18.88 $I_0 = 14$ kg-m^2.

18.90 $I_{z\text{ axis}} = 15.1$ kg-m^2.

18.92 $I_{x\text{ axis}} = 2.667$ kg-m^2, $I_{y\text{ axis}} = 0.667$ kg-m^2, $I_{z\text{ axis}} = 3.333$ kg-m^2.

18.94 $I_{y\text{ axis}} = 1.99$ slug-ft^2.

18.96 20.8 kg-m^2.

18.98 $I_0 = \frac{17}{12} ml^2$.

18.100 $I_{z\text{ axis}} = 74.0$ kg-m^2.

18.102 $I_{z\text{ axis}} = 0.0803$ slug-ft^2.

18.104 3810 slug-ft^2.

18.106 $I_{z\text{ axis}} = 9.00$ kg-m^2.

18.108 $I_{y\text{ axis}} = 3.07$ kg-m^2.

18.110 $I_0 = 0.0188$ kg-m^2.

18.112 $I_{x\text{ axis}} = I_{y\text{ axis}} = m\left(\frac{3}{20} R^2 + \frac{3}{5} h^2\right)$.

18.114 $I_{x\text{ axis}} = 0.844$ kg-m^2.

18.116 $I_x = 0.0535$ kg-m^2.

18.118 $I_{x'} = 0.995$ kg-m^2.

18.120 $I_{z\text{ axis}} = 0.00911$ kg-m^2.

18.122 $I_0 = 0.00367$ kg-m^2.

18.124 $I_{z\text{ axis}} = 0.714$ slug-ft^2.

18.126 (a) 12.1 s.

    (b) 144 lb.

18.128 (a) 20 m/s^2.

    (b) $c \leq 49.1$ mm.

18.130 $I = 2.05$ kg-m^2.

18.132 40.2 kN.

18.134 $\alpha = -0.420$ rad/s^2, $F_x = 336$ N, $F_y = 1710$ N.

18.136 $\alpha = (g/l)[3(1 - \mu^2) \sin\theta - 6\mu \cos\theta]/(2 - \mu^2)$ counterclockwise.

18.138 $B_x = -1959$ N, $B_y = 1238$ N, $C_x = 2081$ N, $C_y = -922$ N.

18.140 $\alpha_{OA} = 0.425$ rad/s^2 counterclockwise, $\alpha_{AB} = 1.586$ rad/s^2 clockwise.

18.142 $\alpha_{HP} = 5.37$ rad/s^2 clockwise.

18.144 208 m/s^2 to the left.

Chapter 19

19.2 65.7 ft-lb.

19.4 (a) 1.53×10^{-4} rad/s.

(b) 0.381 W(watts, or N-m/s).

19.6 (a) 33,300 ft-lb.

(b) 12.9 rad/s.

19.8 (a) 3.84 rad/s.

(b) $A_x = 0$, $A_y = 196$ N

19.10 (a) 4.70 rad/s.

(b) $A_x = 25.0$ lb, $A_y = 28.3$ lb.

19.12 17.0 rad/s.

19.14 $A_x = 20.6$ N, $A_y = 61.4$ N.

19.16 0.731 rad = 41.9°.

19.18 $U_{12} = 5630$ N-m.

19.20 2.59 m/s to the right.

19.22 16.7 rad/s clockwise.

19.24 $v = 0.413$ m/s.

19.26 7.66 ft/s.

19.28 $s = 66.1$ m.

19.30 4.33 rad/s clockwise.

19.32 0.384 m/s.

19.34 (a) 0.522 rad/s. (b) 77.3°.

19.36 0.899 m/s.

19.38 2.57 rad/s counterclockwise.

19.40 4.52 rad/s counterclockwise.

19.42 2.80 rad/s counterclockwise.

19.44 2.34 rad/s.

19.46 $M = 28.2$ N-m.

19.48 (a) 250 N-m-s counterclockwise.

(b) 11.4 rad/s counterclockwise.

19.50 3 s.

19.52 (a) 126 rad/s. (b) 200 rad/s.

19.54 510 rad/s (4870 rpm).

19.56 $\omega_A = 0.0643$ rad/s clockwise, $\omega_B = 0.321$ rad/s counterclockwise.

19.58 0.115 rpm.

19.60 3.94 rad/s (37.6 rpm).

19.62 1.46 m/s.

19.64 14.5°.

19.66 $b = 0.183$ m.

19.68 0.00196 rad/s counterclockwise.

19.70 2.18 rad/s.

19.72 (a) $\omega = 16$ rad/s counterclockwise

(b) 100 N-m before and after

19.74 (a) $\omega_B = 3.81$ rad/s counterclockwise

(b) $\omega_C = 0.763$ rad/s clockwise.

19.78 Energy lost is $\frac{1}{6}mgl$.

19.80 5.96 ft/s.

19.82 9.6 rad/s counterclockwise

19.84 Velocity $= 5.77\mathbf{i} - 2.10\mathbf{j}$ (ft/s), angular velocity $= 3.15$ rad/s counterclockwise.

19.86 $\omega' = 0.375$ rad/s.

19.88 $\omega' = 0.3$ rad/s.

19.90 $\omega = 0.0997$ rad/s $= 5.71$ deg/s counterclockwise.

19.92 15.0 km/hr.

19.94 0.00336 rad/s clockwise.

19.96 $v = 2.05$ m/s.

19.98 $v = \sqrt{Fb/[\frac{1}{2}m_c + 2(m + I/R^2)]}$.

19.100 $\omega_S = 12.5$ rad/s.

19.102 (a) 1.13 ft. (b) 1.09 ft/s.

19.104 $P_{max} = 580$ ft-lb/s (1.05 hp), $P_{ave} = 290$ ft-lb/s (0.53 hp).

19.106 11.1 rad/s.

19.108 1.77 rad/s counterclockwise.

19.110 $N = (373/283)mg$.

19.112 $\omega' = 0.721$ rad/s.

19.114 $\omega' = \left(\frac{1}{3} + \frac{2}{3}\cos\beta\right)\omega$.

19.116 $b = 2$ ft.

19.118 $\omega = 8.51$ rad/s.

19.120 $\theta = 54.7°$, $\omega = 10.7$ rad/s counterclockwise.

19.122 $0.0822\mathbf{k}$ (rad/s).

19.124 0.641 rad/s clockwise, $\mathbf{v}' = -2.24\mathbf{i}$ (ft/s).

Chapter 20

20.2 $\mathbf{a}_A = -1.81\mathbf{i} - 0.317\mathbf{j} + 0.867\mathbf{k}$ (m/s²).

20.4 $\mathbf{a}_A = 40\mathbf{i} - 8\mathbf{j} - 56\mathbf{k}$ (m/s²).

20.6 $\mathbf{v}_C = 4\mathbf{i} + 4\mathbf{k}$(m/s), $\mathbf{v}_D = 4\mathbf{i} - 8\mathbf{j}$(m/s).

20.8 (a) $\boldsymbol{\omega} = 6\mathbf{i} + 4\mathbf{j}$ (rad/s).

  (b) $\mathbf{v}_A = 0.566\mathbf{i} - 0.849\mathbf{j} + 0.849\mathbf{k}$ (m/s).

20.10 (a) $\mathbf{v}_P = 48\mathbf{k}$ (ft/s).

  (b) $\boldsymbol{\alpha} = -288\mathbf{k}$ (rad/s²).

20.12 (a) $\boldsymbol{\omega}_{\text{rel}} = 6\mathbf{i}$ (rad /s).

  (b) $\mathbf{v}_P = -0.8\mathbf{i} + 1.2\mathbf{j} + 0.6\mathbf{k}$ (m/s).

20.14 (a) $\boldsymbol{\omega}_{\text{disk}} = \omega_{\text{d}}\cos\beta\mathbf{i} + \left(\omega_0 + \omega_d\sin\beta\right)\mathbf{j}$.

  (b) $\mathbf{v}_P = \left[\omega_{\text{d}}R - \omega_0(h + b\cos\beta - R\sin\beta)\right]\mathbf{k}$.

20.16 $\mathbf{v}_A = 80\mathbf{i}$ (mm/s),
  $\mathbf{v}_B = -102.6\mathbf{i} + 281.9\mathbf{j} - 56.4\mathbf{k}$ (mm/s).

20.18 (a) $\boldsymbol{\omega}_{BC} = 0.1\mathbf{j} + 0.4\mathbf{k}$ (rad/s).

  (b) $\mathbf{v}_C = -0.315\mathbf{i} + 0.085\mathbf{j} - 0.131\mathbf{k}$ (m/s).

20.20 (a) $\boldsymbol{\Omega} = 0.894\mathbf{i} + 1.789\mathbf{j}$ (rad/s).

  (b) $\boldsymbol{\omega}_{\text{rel}} = -4.47\mathbf{i}$ (rad/s).

20.22 $\mathbf{a} = -1.28\mathbf{i} + 0.64\mathbf{j} - 3.20\mathbf{k}$ (m/s²).

20.24 $\mathbf{v} = \mathbf{0}$.

20.26 $\boldsymbol{\alpha} = 0.139\mathbf{i} + 0.108\mathbf{j}$ (rad/s²).

20.28 $\Sigma\mathbf{M} = 0.0135\mathbf{i} - 0.0086\mathbf{j} + 0.01\mathbf{k}$ (N-m).

20.30 $\Sigma\mathbf{M} = -76\mathbf{i} + 36\mathbf{j} - 60\mathbf{k}$ (ft-lb).

20.32 $|\mathbf{a}_A| = 2.5$ m/s².

20.34 (a) $\boldsymbol{\alpha} = 28.57\mathbf{i} + 5.71\mathbf{j} - 2.00\mathbf{k}$ (rad/s²).

  (b) $\mathbf{a} = -0.33\mathbf{i} + 2.00\mathbf{j} - 19.52\mathbf{k}$ (m/s²).

20.36 (a) $\mathbf{a} = -4.17\mathbf{k}$ (m/s²).

  (b) $\boldsymbol{\alpha} = 49.5\mathbf{i} + 137.7\mathbf{j}$ (rad/s²).

20.38 $\boldsymbol{\alpha} = -500.0\mathbf{i} + 24.4\mathbf{j}$ (rad/s²).

20.40 $\boldsymbol{\alpha} = 14.5\mathbf{j}$ (rad/s²).

20.42 $\boldsymbol{\alpha} = 14.53\mathbf{j} + 4.65\mathbf{k}$ (rad/s²).

20.44 27.4 N-m.

20.46 $|\mathbf{F}| = 78.5$ N, $|\mathbf{M}| = 260$ N-m.

20.50 1.2 N-m.

20.52 $-10\mathbf{j} + 14.9\mathbf{k}$ (lb), $-12.4\mathbf{i}$ (ft-lb).

20.58 157 N-m.

20.62 21.5 ft-lb.

20.64 4650 N-m. Counterclockwise.

20.66 $\omega_{\text{d}} = -\frac{1}{2}\omega_0\sin\beta$.

20.68 $\dot{\psi} = 1.31$ rev/s.

20.70 29.1 rev/s.

20.72 (a) $\dot{\psi} = -2050$ rpm.

  (b)

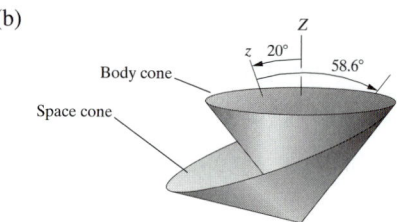

20.74 309 rad/s (49.1 revolutions per second).

20.76 $\omega_0 = 10.7$ rad/s.

20.78 $\Sigma\mathbf{M} = 2123\mathbf{i} - 155\mathbf{j} - 534\mathbf{k}$ (N-m).

20.80 $I_{xx} = 0.67$ kg-m², $I_{yy} = 5.33$ kg-m²,
  $I_{zz} = 6$ kg-m², $I_{xy} = I_{yz} = I_{zx} = 0$.

20.82 $I_{xx} = 0.12$ kg-m², $I_{yy} = 0.03$ kg-m²,
  $I_{zz} = 0.15$ kg-m², $I_{xy} = I_{yz} = I_{zx} = 0$.

20.84 $I_{xx} = 2.48$ slug-ft², $I_{yy} = 16.77$ slug-ft²,
  $I_{zz} = 19.25$ slug-ft², $I_{xy} = 5.59$ slug-ft², $I_{yz} = I_{zx} = 0$.

20.86 $I_{xx} = 0.0318$ kg-m^2, $I_{yy} = 0.0357$ kg-m^2,
$I_{zz} = 0.0674$ kg-m^2, $I_{xy} = 0.0219$ kg-m^2, $I_{yz} = I_{zx} = 0$.

20.88 (a) $\mathbf{H} = \frac{1}{12}ml^2(\omega_y\mathbf{j} + \omega_z\mathbf{k})$.

(b) $\mathbf{H}_O = \frac{1}{3}ml^2(\omega_y\mathbf{j} + \omega_z\mathbf{k})$.

20.90 $I_{x'x'} = 15,600$ kg-m^2, $I_{y'y'} = 226,800$ kg-m^2,
$I_{z'z'} = 242,400$ kg-m^2, $I_{x'y'} = -32,400$ kg-m^2,
$I_{y'z'} = I_{z'x'} = 0$.

20.92 $\mathbf{H} = 2.00\mathbf{i} - 1.64\mathbf{j} + 2.58\mathbf{k}$ (kg-m^2/s).

20.94 (a) $I = 3.56$ kg-m^2.

(b) 14.22 kg-m^2/s.

20.96 $I_1 = 16.15$, $I_2 = 62.10$, $I_3 = 81.75$ kg-m^2,
$\mathbf{e}_1 = 0.964\mathbf{i} - 0.220\mathbf{j} + 0.151\mathbf{k}$,
$\mathbf{e}_2 = -0.204\mathbf{i} - 0.972\mathbf{j} - 0.114\mathbf{k}$,
$\mathbf{e}_3 = 0.172\mathbf{i} + 0.079\mathbf{j} - 0.982\mathbf{k}$.

20.98 $I_1 = 0$, $I_2 = \frac{1}{12}$, $I_3 = \frac{1}{12}$ kg-m^2, $\mathbf{e}_1 = \cos\beta\mathbf{i} + \sin\beta\mathbf{j}$,
$\mathbf{e}_2 = -\sin\beta\mathbf{i} + \cos\beta\mathbf{j}$, $\mathbf{e}_3 = \mathbf{k}$.

20.100 (a) $\boldsymbol{\omega} = \omega_d\mathbf{i} + \omega_0\mathbf{j}$.

(b) $\mathbf{v}_A = -R\omega_0\cos\theta\mathbf{i} + R\omega_d\cos\theta\mathbf{j} + (R\omega_d\sin\theta - b\omega_0)\mathbf{k}$.

20.102 (a) $\boldsymbol{\omega} = 20\mathbf{i} - 5\mathbf{j}$ (rad/s).

(b) $\mathbf{v}_A = 0.25\mathbf{i} + 1.00\mathbf{j} + 3.73\mathbf{k}$ (m/s).

20.104 $\mathbf{a}_B = 166\mathbf{j}$ (ft/s^2),
$\boldsymbol{\alpha} = 7.31\mathbf{i} - 5.52\mathbf{j} + 22.89\mathbf{k}$ (rad/s^2).

20.106 $\mathbf{H} = 25\mathbf{i} + 50\mathbf{k}$ (kg-m^2/s).

20.108 $I_{xx} = 0.0398$ slug-ft^2, $I_{yy} = 0.0373$ slug-ft^2,
$I_{zz} = 0.0772$ slug-ft^2, $I_{xy} = I_{yz} = I_{zx} = 0$.

20.110 $\Sigma\mathbf{M} = -250\mathbf{i} - 250\mathbf{j} + 125\mathbf{k}$ (N-m).

20.112 $\omega_0 = \left[g\sin\beta/(\frac{2}{3}l\sin\beta\cos\beta + b\cos\beta)\right]^{1/2}$.

20.114 (b) If $\omega_0 = 0$, the plate is stationary. The solution of
the equation $2\cos\beta - \sin\beta = 0$ is the value of β for
which the center of mass of the plate is directly above
point O; the plate is balanced on one corner.

20.116 $\mathbf{F}_B = 52.72\mathbf{i} + 97.35\mathbf{j} + 9.26\mathbf{k}$ (N),
$\mathbf{M}_B = 0.05\mathbf{i} - 10.25\mathbf{j} + 30.63\mathbf{k}$ (N-m).

20.118

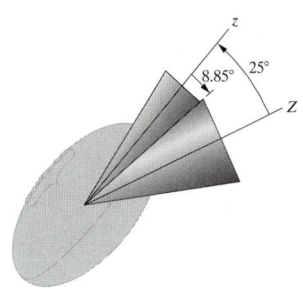

20.120 $\Sigma\mathbf{M} = -283\mathbf{i} - 2546\mathbf{j} - 800\mathbf{k}$ (ft-lb).

20.122 $\boldsymbol{\alpha} = 0.0535\mathbf{i} + 0.0374\mathbf{j} + 0.0160\mathbf{k}$ (rad/s^2).

Chapter 21

21.4 $x = -0.378$ m.

21.6 $x = 0.390$ m.

21.8 Amplitude $= 3.89°$.

21.10 $x = 1.41$ ft.

21.12 (a) $f = 0.624$ Hz.

(b) $\tau = 1.60$ s.

21.14 $k = 36.8$ lb/ft.

21.16 0.692 Hz.

21.18 0.287 m/s downward.

21.20 $\beta = 0.01\cos 6t$ rad.

21.24 $I_{(z' \text{ axis})} = 24.2$ kg-m^2.

21.28 0.39 in.

21.30 $\theta_B = 0.172\sin 11.6t$ rad.

21.32 $m_A = 4.38$ kg.

21.36 $f = 0.318$ Hz.
$f = 0.211$ Hz.

21.38 $x = 0.577e^{-t}\sin(1.73t)$ m.

21.40 $x = 2.38$ ft.

21.42 $c = 1780$ N-s/m.

21.44 (a) $\tau_d = 2.32$ s, $f_d = 0.431$ Hz.

(b) 5.28 s.

21.46 $x = e^{-2t}(-0.325 \sin 2.24t - 0.363 \cos 2.24t)$ $+ 0.363$ m.

21.48 $x = -(0.363 + 1.090t)e^{-3t} + 0.363$ m.

21.50 $x = 0.347$m.

21.52 0.153 rad/s clockwise.

21.54 (a) 277 N-s/m.

(b) $\theta = 10te^{-4.33t}$ rad.

(c) $\theta_{max} = 0.850$ rad at $t = 0.231$ s.

21.56 $\theta_B = 0.209e^{-6.62t} \sin 9.55t$ rad.

21.58 $x_p = 10 \sin 8t$ m.

21.60 (a) $x_p = 0.5$ m.

(b) $x = e^{-t}(-0.289 \sin 1.73t - 0.5 \cos 1.73t) + 0.5$ m.

21.62 $\theta = 0.581 \sin 2.58t + 1.250 \sin 2t$ rad.

21.64 11.5 lb.

21.66 0.113 mm.

21.70 16.5 in.

21.72 5.5 mm.

21.74 (a) $x = (0.1) \cos 3t$ m.

(b)

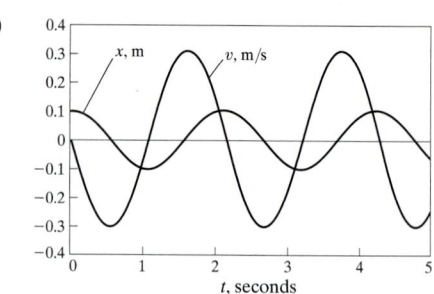

21.78 $f = 0.985$ Hz.

21.80 1.03 Hz.

21.82 $f = (1/2\pi)\sqrt{3[k/m) - (g/2l)]}$.

21.84 $\delta = 2.07$.

21.86 $f_d = 0.714$ Hz.

21.88 $x = 0.118e^{-2.68t} - 0.118e^{-14.01t}$ ft.

21.90 $\theta_B = e^{-5.05t}(0.244 \sin 3.45t + 0.167 \cos 3.45t)$ rad.

21.92 $x = 0.253 \sin 6.19t + 0.100 \cos 6.19t + 0.145 \sin 3.00t$ m.

21.94 (a) $E_p = (2\pi v/\lambda)^2 h/[(k/m) - (2\pi v/\lambda)^2]$.
(b) $v = \lambda\sqrt{k/m}/2\pi$.

Index

Properties of Volumes and Homogeneous Objects

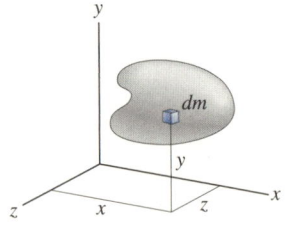

The moments and products of inertia of the object in terms of the xyz coordinate system are

$$I_{x\text{ axis}} = I_{xx} = \int_m \left(y^2 + z^2\right) dm,$$

$$I_{y\text{ axis}} = I_{yy} = \int_m \left(x^2 + z^2\right) dm,$$

$$I_{z\text{ axis}} = I_{zz} = \int_m \left(x^2 + y^2\right) dm,$$

$$I_{xy} = \int_m xy\, dm, \qquad I_{yz} = \int_m yz\, dm,$$

$$I_{zx} = \int_m zx\, dm.$$

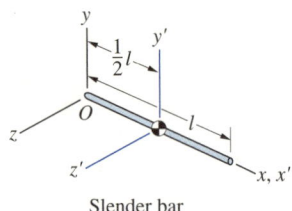

Slender bar

$$I_{x\text{ axis}} = 0, \qquad I_{y\text{ axis}} = I_{z\text{ axis}} = \frac{1}{3}ml^2,$$

$$I_{xy} = I_{yz} = I_{zx} = 0.$$

$$I_{x'\text{ axis}} = 0, \qquad I_{y'\text{ axis}} = I_{z'\text{ axis}} = \frac{1}{12}ml^2,$$

$$I_{x'y'} = I_{y'z'} = I_{z'x'} = 0.$$

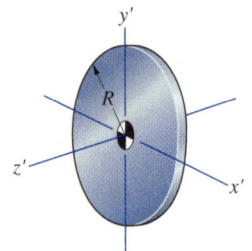

Thin circular plate

$$I_{x'\text{ axis}} = I_{y'\text{ axis}} = \frac{1}{4}mR^2, \qquad I_{z'\text{ axis}} = \frac{1}{2}mR^2,$$

$$I_{x'y'} = I_{y'z'} = I_{z'x'} = 0.$$

Thin rectangular plate

$$I_{x\text{ axis}} = \frac{1}{3}mh^2, \qquad I_{y\text{ axis}} = \frac{1}{3}mb^2, \qquad I_{z\text{ axis}} = \frac{1}{3}m(b^2 + h^2),$$

$$I_{xy} = \frac{1}{4}mbh, \qquad I_{yz} = I_{zx} = 0.$$

$$I_{x'\text{ axis}} = \frac{1}{12}mh^2, \qquad I_{y'\text{ axis}} = \frac{1}{12}mb^2, \qquad I_{z'\text{ axis}} = \frac{1}{12}m(b^2 + h^2),$$

$$I_{x'y'} = I_{y'z'} = I_{z'x'} = 0.$$